Manipulating the Mouse Embryo

A LABORATORY MANUAL   THIRD EDITION

Manipulating the Mouse Embryo

A LABORATORY MANUAL   THIRD EDITION

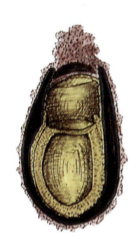

Andras Nagy
Samuel Lunenfeld Research Institute
Mount Sinai Hospital, Toronto

Marina Gertsenstein
Samuel Lunenfeld Research Institute
Mount Sinai Hospital, Toronto

Kristina Vintersten
European Molecular Biology Laboratory, Heidelberg

Richard Behringer
University of Texas, M.D. Anderson Cancer Center

COLD SPRING HARBOR LABORATORY PRESS
Cold Spring Harbor, New York

Manipulating the Mouse Embryo
A LABORATORY MANUAL
Third Edition

Publisher	John Inglis
Acquisition Editors	John Inglis and Judy Cuddihy
Developmental Editor	Judy Cuddihy
Project Coordinator	Mary Cozza
Production Editor	Patricia Barker
Desktop Editor	Daniel deBruin
Production Manager	Denise Weiss
Cover Design	Ed Atkeson

Front cover artwork (*paperback edition only*): 9.5-dpc embryo expressing GFP in the developing vascular system. (Courtesy of Andras Nagy.)

Title page illustration: Early postimplantation stages of mouse embryogenesis. Illustration by Rosa Beddington. (Printed, with permission, from John Skehel, Medical Research Council.)

Library of Congress Cataloging-in-Publication Data

Manipulating the mouse embryo : a laboratory manual / Andras Nagy ... [et al.].-- 3rd ed.
     p. cm.
   Includes bibliographical references and index.
   ISBN 0-87969-574-9 (cloth: alk.paper) -- ISBN 0-87969-591-9 (paperback : alk. paper)
     1. Mice--Embryos--Laboratory manuals. 2. Mice--Genetic engineering--Laboratory
manuals. 3. Transgenic mice--Laboratory manuals. 4. Mice as laboratory animals. I.
Nagy, Andras, 1951-

QL737.R6 M2468 2003
571.8'61935--dc21

2002034976

10 9 8 7 6 5 4

All Cold Spring Harbor Laboratory Press publications may be ordered directly from Cold Spring Harbor Laboratory Press, 500 Sunnyside Blvd., Woodbury, N.Y. 11797-2924. Phone: 1-800-843-4388 in Continental U.S. and Canada. All other locations: (516) 422-4100. FAX: (516) 422-4097. E-mail: cshpress@cshl.edu. For a complete catalog of all Cold Spring Harbor Laboratory Press publications, visit our World Wide Web Site http://www.cshlpress.com.

Contents

Preface

IT HAS BEEN 20 YEARS SINCE THE FIRST Cold Spring Harbor Laboratory Molecular Embryology of the Mouse course that led to the initial publication of this manual. In 1983, the generation of transgenic mice by pronuclear injection of zygotes had only recently been accomplished and was far from being routine. Embryonic stem cell lines had only recently been produced and their germ-line potential and usefulness had yet to be demonstrated. Homologous recombination was still a dream. Large-scale mutagenesis in the mouse existed only in the minds of a few pioneering individuals. In addition, the molecular embryology of the mouse was still in its infancy. As we reflect back in time, it is astonishing how much progress has been made in the fields of mouse developmental genetics and molecular embryology. We now have a publicly assembled and annotated mouse genome available to all researchers. Genomic and cDNA clones can be easily identified in the databases, ordered, and received within days. Any gene can be added to the mouse germ line or altered by design. Chromosomes can be engineered. Living embryonic tissues can fluoresce brightly with a myriad of designer colors. Embryos and gametes can be cryopreserved and archived. Mice can be cloned. There seem to be no limits for utilizing the mouse to address fundamental biological questions and provide novel biomedical insights for human biology and disease.

As the mouse developmental genetics and molecular embryology fields have progressed and evolved, so has the "Mouse Manual." The current edition is built upon the solid foundation of the previous editions and the efforts of the original editors, Brigid Hogan, Frank Costantini, Liz Lacy, and Rosa Beddington. This new edition has been significantly reorganized, incorporating many innovations since the publication of the second edition in 1994. New chapters and protocols have been added, including mouse cloning, intracytoplasmic sperm injection, artificial insemination, cryopreservation of embryos and gametes, and guidance for current vector designs. New techniques such as the introduction of foreign DNA into mouse embryos by electroporation have been added. In addition, chapters on generating and analyzing transgenic mice and chimeras have been considerably expanded. All of the surgical techniques have been moved to a single chapter. Methods to visualize living embryos and new reporter genes such as fluorescent proteins have also been added.

We are very grateful to the many people who generously helped us produce the present edition. They provided new and updated protocols, figures, and images, and served as an incredibly helpful source of expert information. We thank (in alphabetical order) Kathryn Anderson, Gusztav Belteki, Sally Camper, Chris Cretekos, S. K. Dey, Mary Dickinson, Hao Ding,

Scott Fraser, Yas Furuta, Joachim Gündel, Debrorah Guris, Kat Hadjantonakis, Jody Haigh, Britt Hansen, C.C. Hui, Akira Imamoto, Ian Jackson, Nancy Jenkins, Randy Johnson, Elizabeth Jones, Andrea Jurisicova, Monica Justice, Akio Kobayashi, Rashmi Kothary, Tilo Kunath, Kin Ming Kwan, Carlisle Landel, Carol Cutler Linder, Chengyu Liu, Tom Lufkin, William Mansfield, Jim Martin, Andy McMahon, Jennifer Merriam, Lluis Montoliu, Nagy lab (October 2002), Kazuhisa Nakashima, Atsuo Ogura, Noriko Osumi, Dmitry Ovchinnikov, Ginny Papaioannou, Anne Plueck-Becklas, Udo Ringeisen, Jaime Rivera, Liz Robertson, Merle Rosenzweig, Janet Rossant, Luis Gabriel Sanchez-Partida, Thom Saunders, Heike Schweizer, Jillian Shaw, Bill Shawlot, Michael Shen, Stanton Short, Davor Solter, Monika Szczygiel, Patrick Tam, Maki Wakamiya, Paul Wassarman, Michael Wilson, Werner Wittke, Chris Wylie, and Ryuzo Yanagimachi. We also thank the many individuals at the Cold Spring Harbor Laboratory Press who have helped to make this new edition a reality. We thank Mary Cozza, Pat Barker, Danny deBruin, Denise Weiss, Dave Crotty, Jan Argentine, and Executive Director John Inglis. We particularly thank our editor Judy Cuddihy for her wonderful enthusiasm, tremendous patience, friendly encouragement, and creative insights. Finally, it is our hope that the new edition of this manual will help train the future leaders and innovators of the mouse developmental genetics and molecular embryology fields.

A.N., M.G., K.V., R.B.

Manipulating the Mouse Embryo

A LABORATORY MANUAL   THIRD EDITION

Developmental Genetics and Embryology of the Mouse

Past, Present, and Future

THESE ARE PROBABLY THE BEST OF TIMES to be studying mouse genetics and embryology. Currently, the mouse germ line can be experimentally manipulated in almost every conceivable way either through direct injection of cloned DNA into zygotes or through the genetic modification of embryonic stem (ES) cells. Large-scale mutagenesis projects are yielding thousands of new mouse mutants. Fortunately, all of these mice do not have to be maintained "on the shelf" because mice can be archived by cryopreservation of embryos and gametes. Mice can now be routinely cloned by somatic nuclear transfer, creating new questions about genome programming. Finally, perhaps the most significant recent advance is the

Courtesy of Ian Jackson

availability of the first public annotated assembly of the mouse genome sequence (MGSC Version 3), which will greatly facilitate biomedical research using mice (Lindblad-Toh et al. 2001; http://www.ncbi.nlm.nih.gov).

CONTENTS

INTRODUCTION

There is a unique challenge to understanding how genes control the growth and differentiation of the mammalian embryo. To a large extent, this challenge is an intellectual one and derives from our curiosity to know how human form is generated and how it has evolved from that of simpler organisms. At a practical level, we also need to know how mutations and chemicals produce human malformations, congenital defects, and childhood cancers, and whether the productivity of agricultural animals can be improved. This knowledge, and the ability we now have to change the genetic program, must inevitably make a great impact on society and have far-reaching effects on the way in which we think about ourselves.

The roots of our knowledge about how genes control mammalian development can be traced back to experiments carried out in the early 1900s on the inheritance of coat colors in a variety of domestic animals. Since then, the mouse has become firmly established as the primary experimental mammal, and more information has accumulated on its genetics than on that of any other vertebrate, including humans. The mouse genome, which is contained on a haploid set of 20 chromosomes, has been assembled and predicts 46,370 gene models. A physical map composed of nearly 300 bacterial artificial chromosome (BAC) contigs with nearly 17,000 markers is also now available (Gregory et al. 2002). There is extensive linkage conservation or synteny between the mouse and human genomes, so that progress with the Human Genome Project has contributed to knowledge of the mouse genetic map and vice versa (Copeland et al. 1993; O'Brien et al. 1994, http://www.ncbi.nlm.nih.gov/Homology/).

The techniques of molecular biology, including whole-mount in situ hybridization, reverse transcriptase-polymerase chain reaction (RT-PCR), DNA microarrays and sophisticated imaging methods, are being used to reveal the temporal and spatial patterns of expression of specific genes at different stages of development. Novel cell-autonomous lineage markers have also been produced for following cell fate (see Table 2.5). However, the most compelling reason for excitement and optimism about studying developmental genetics in the mouse, instead of another vertebrate, is undoubtedly our ability to manipulate the genome of the mouse in a variety of different ways.

The first edition of *Manipulating the Mouse Embryo* (1986) emphasized the potential importance of introducing new genetic information into transgenic mice by microinjecting DNA into the pronucleus of the zygote or by infecting embryos with retroviral vectors. The targeting of mutations to specific genes by homologous recombination in pluripotential ES cells was still only a dream, tenaciously followed by a small group of scientists, who, like many others before them, persisted in the face of considerable skepticism from their contemporaries. Today, the technique has become routine, producing a wealth of often unexpected and therefore highly stimulating data about the in vivo function and interaction of genes in the context of the developing organism. The second edition of this manual (1994), like the first edition, provided a simple technical guide for scientists who wanted to learn some of the techniques for manipulating the mouse embryo and for introducing genes and mutations into mice. The current edition of this manual includes new and expanded chapters on ES cell genetic manipulations, mouse chimeras, mouse cloning, assisted reproduction strategies, and embryo and gamete cryopreservation. As before, we sincerely hope that making this information available to a wide audience will help to continue the spirit of international cooperation established by the first mouse geneticists.

MENDELIAN INHERITANCE AND LINKAGE: THE BEGINNINGS OF MOUSE GENETICS

Historians of science on both sides of the Atlantic acknowledge the American scientist William E. Castle as one of the founding fathers of mammalian genetics. As first director of the new Bussey Institute of Experimental Biology at Harvard, from 1909 to 1937, he encouraged work on the inheritance of variable characteristics in a wide range of organisms, including birds, cats, dogs, guinea pigs, rabbits, rats, and even mice (Russell 1954; Keeler 1978; Morse 1978, 1981). He was also responsible for introducing Thomas Hunt Morgan to *Drosophila* (Shine and Wrobel 1976). Castle had a profound influence on the course of mammalian genetics through the many scientists who came to visit or study at the Bussey Institute.

Of all the mammals studied by these early geneticists, the mouse became the mammal of choice because of its small size, resistance to infection, large litter size, and relatively rapid generation time (see Table 1.1). Mice were also favored because of the interesting pool of mutations affecting coat color and behavior that was readily available from breeders and collectors of pet mice, or mouse "fanciers." One of these mutants, *albino* (see Fig. 2.42), was used by Bateson in England, Cuenot in France, and Castle in the United States for the first breeding experiments demonstrating Mendelian inheritance in the mouse (for references, see Castle and Allen 1903). A few years later, *albino* and another old mutation of the mouse fanciers, *pink-eyed dilution* (see Fig. 2.42), were used by J.B.S. Haldane for the first demonstration of linkage in mice (Haldane et al. 1915). Sadly, this work was interrupted in 1914 when Haldane volunteered for service in the First World War, leaving his sister to continue their experiments for a while in the Department of Comparative Anatomy in Oxford (Clark 1984; N. Mitchison, pers. comm.). It was not until after the war that Haldane was able to turn his attention to the wider aspects of mammalian genetics and, along with others, begin developing mathematical models of inheritance and natural selection.

ORIGINS OF THE LABORATORY MOUSE

If Castle and Haldane are the founding fathers of mouse genetics, then the mother is undoubtedly Abbie E.C. Lathrop. A self-made woman, Lathrop established around 1900 a small mouse "farm" in Granby, Massachusetts, to breed mice as

TABLE 1.1. Some vital statistics of the European house mouse, *Mus musculus*, in the laboratory

Genome	
Number of chromosomes	40
Diploid DNA content	~6 pg (2.6 x 10⁹ bp)
Recombination units	1600 cM (2000 kb/cM)
Approximate number of genes[a]	46,370
Percent of genome as five families of highly repeated DNA sequences (B1, B2, R, MIF-1, and EC1)[b]	8–10%
Reproductive biology[c]	
Gestation time	19–20 days
Age at weaning	3 weeks
Age at sexual maturity	~6 weeks
Approximate weight	birth 1 g
	weaning 8–12 g
	adult 30–40 g (male >female)
Life span in laboratory	1.5–2.5 years
Average litter size[d]	~6–8
Total number of litters per breeding female	4–8

[a]Gene models, MGSC Version 3 (http://www.ncbi.nlm.nih.gov).

[b]Bennett et al. (1984).

[c]Parameters such as gestation time, weight, and life span vary between the different inbred strains. Details can be found in a number of books listed in Chapter 17; e.g., Altman and Katz (1979), Festing (1979), and Heiniger and Dorey (1980).

[d]Litter size depends on the number of oocytes liberated at ovulation and the rate of prenatal mortality, both of which vary with age of mother, parity, and environmental conditions (e.g., diet, stress, and presence of strange male) and with strain (reflecting genetic factors such as efficiency of placentation). Prenatal mortality in inbred strains can be ~10–20% (for references, see Boshier 1968).

pets. However, her mice were soon in demand as a source of experimental animals for the Bussey Institute and other American laboratories, and she gradually expanded her work to include quite sophisticated and well-documented breeding programs. For example, in collaboration with Leo Loeb, she carried out experiments to study the effects of genetic background, inbreeding, and pregnancy on the incidence of spontaneous tumors in her mice (Shimkin 1975; Morse 1978). As source material for the farm, Abbie Lathrop used wild mice trapped in Vermont and Michigan, fancy mice obtained from various European and North American sources, and imported Japanese "waltzing" mice. Waltzing mice had been bred as pets in China and Japan for many generations and were probably homozygous for a recessive mutation that causes a defect in the inner ear and thus nervous, circling, behavior. The Granby mouse farm was, to a large extent, the "melting pot" of the laboratory mouse, and, as shown in Figure 1.1, many of the old inbred strains can be traced back to the relatively small pool of founding mice that Lathrop maintained there. At present, more than 400 different inbred strains are available, and their origins and characteristics are listed in the *Mouse Genome Informatics* (MGI) database (http://www.informatics.jax.org/mgi-home/genealogy).

The formal systematics of the laboratory mouse is far from simple and reflects the existence of several subspecies of the European mouse species, *Mus musculus*, from which it was ultimately derived. The nature of this complexity has been revealed by the application of restriction-fragment-length polymorphism (RFLP) studies to mouse DNA. Analysis of the RFLP of mitochondrial

DNA (which is maternally inherited through the oocyte cytoplasm) has shown few differences among old established strains, compared with the wide variations seen among wild mice and newer strains derived from them. In fact, on the basis of mitochondrial DNA RFLPs, it has been argued that at least five of the primary strains (DBA, BALB/c, SWR, PL, and C57–C58) were derived originally from a single female of the subspecies *Mus musculus domesticus* (Ferris et al. 1982). This taxonomic group is found in western and southern Europe and is the source from which all wild mice in the northern parts of the United States were derived by migration with humans across the north Atlantic shipping lanes. A second taxonomic group or subspecies, *Mus musculus musculus*, is found in central and eastern Europe, Russia, and China, and only interbreeds with *domesticus* over a narrow band from north to south through central Europe (Fig. 1.2) (Bonhomme et al. 1984). In addition to having distinct mitochondrial DNA RFLPs, the two groups also show different patterns using DNA probes specific for the Y chromosome, which is inherited only through the male. Unexpectedly, in view of the mitochondrial RFLP data, many old inbred mouse strains, including A/J, BALB/c, C57BL/6, CBA/HeJ, C3H, DBA/2, 129/Sv, and 163/H, have Y chromosome RFLPs of the *musculus* type. The most likely explanation is that the Y chromosome came from Japanese pet mice; for example, those bred on the Granby mouse farm. A list of the origin of the Y chromosome of different inbred strains has been published (Nishioka 1987). In view of the mixed origin of the laboratory mouse, it has been agreed to refer to standard inbred strains as *Mus musculus* only (Auffray et al. 1990).

CREATION OF INBRED STRAINS AND OTHER RESOURCES OF MOUSE GENETICS

An inbred strain is defined as one that has been maintained for more than 20 generations of brother-to-sister mating and is essentially homozygous at all genetic loci, except for mutations arising spontaneously (Altman and Katz 1979; Morse 1981). The derivation of inbred strains represents one of the most important phases in the history of mouse genetics, and it revolutionized studies in cancer research, tissue transplantation, and immunology. One of the pioneers of the innovation was Clarence C. Little. He was originally a student of Castle at the Bussey Institute, where he studied the inheritance of mouse coat color, and he later went on to found the Roscoe B. Jackson Memorial Laboratory (usually known as The Jackson Laboratory) in Bar Harbor, Maine (Russell 1978; Morse 1981). Other pioneers were Lionelle Strong, Leo Loeb, and Jacob Furth (Morse 1978; Strong 1978). Among the first inbred strains were DBA, which was named after the coat color mutations it carried: *dilute* (*d*), *brown* (*b*), and *nonagouti* (*a*); and C57 and C58, which were derived from females 57 and 58 from the Granby mouse farm. While carrying out these early inbreeding experiments, both Little and Strong worked between 1918 and 1922 at the Carnegie Institution of Washington at Cold Spring Harbor, thus establishing the laboratory (then known as the Station for Experimental Evolution) as one of the birthplaces of mouse genetics (Keeler 1978; Strong 1978).

In deriving inbred strains, great tenacity was required to maintain the strict brother-to-sister matings through times when the breeding stocks reached a very low ebb due to disease or accidents, and accounts of these difficult times make fascinating reading (Morse 1978). It also required intellectual courage to challenge

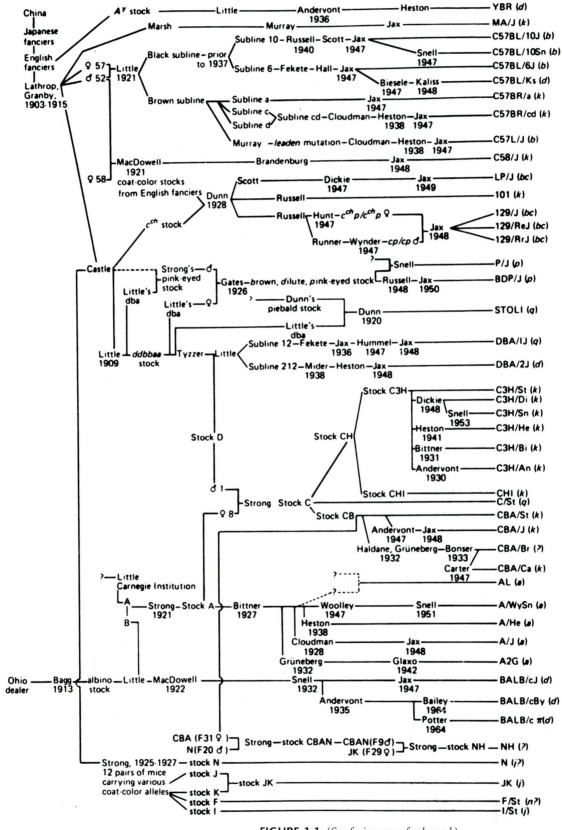

FIGURE 1.1. (*See facing page for legend.*)

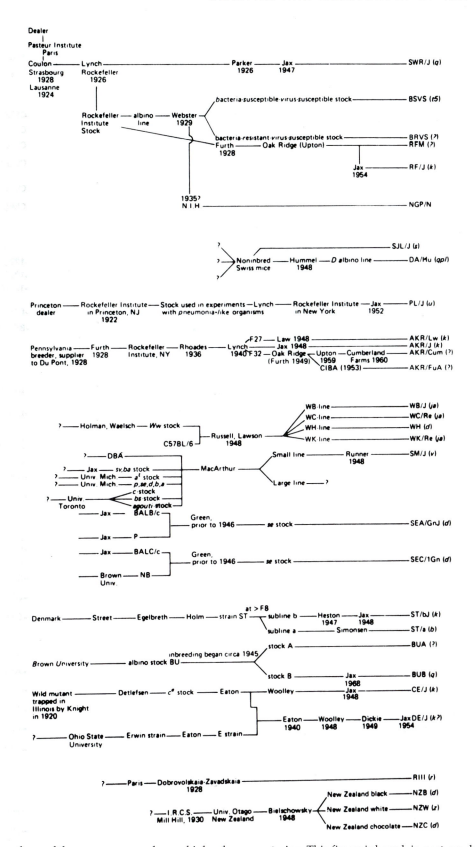

FIGURE 1.1. Genealogy of the more commonly used inbred mouse strains. This figure is based, in part, on data provided by Michael Potter and Rose Lieberman in 1967; it was extended by Jan Klein in 1975 and revised by Potter in 1978. H-2 haplotypes are shown in parentheses. (Reproduced, with permission, from Altman and Katz 1979.)

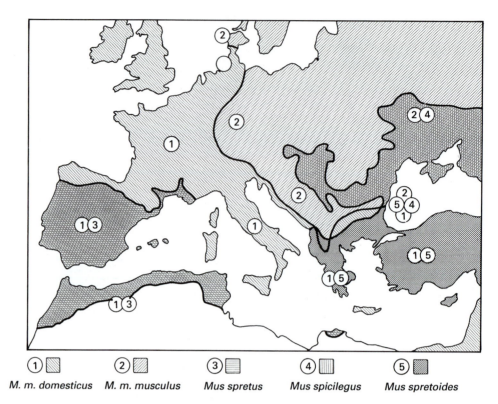

FIGURE 1.2. Geographical distribution of the five biochemical groups of the house mouse species complex in Europe. (Redrawn, with permission, from Bonhomme et al. 1984.)

the widely held belief that inbreeding to virtual homozygosity would be impossible due to recessive lethal mutations in the founding pairs. Each inbred strain has a standardized nomenclature, to indicate strain and substrain. Standard methods for maintaining breeding colonies and testing mice for genetic purity have been described previously (see, e.g., Nomura et al. 1985), and computerized databases for tracking breeding colonies are available (Silver 1993b). Unfortunately, newcomers to the field should be aware that examples of accidental cross-contamination of strains are by no means rare, even in the present day.

One of the driving forces behind the initial establishment of inbred strains was the need to rationalize studies on the genetics of cancer susceptibility. Inbred strains were also essential for solving the problem of why spontaneous tumors could be transplanted into some mice and not others. Although many groups studied this problem, a major contribution was made by Peter Gorer, working in Haldane's department at University College, London. Using A, C57BL, and DBA strains of mice and a transplantable A-strain tumor, he showed for the first time that mice resistant to tumor growth produced antibodies against antigens present not only on the tumor cells, but also on blood cells of strain-A mice. One particularly strong antigen was called Antigen II. In 1948, Gorer and the American geneticist George Snell together showed that the gene specifying Antigen II was closely linked to the *fused* (*Fu*) locus (now known to be on chromosome 17), and they called the gene *Histocompatibility-2*, or *H2* (Gorer et al. 1948). In a series of outstanding experiments, for which he was awarded the Nobel Prize in 1980, Snell went on to identify and map many of the minor histocompatibility loci as well. All of this work was carried out at The Jackson Laboratory and owes much to the unique environment built up there by C.C. Little and his colleagues. It was the first laboratory in which many inbred strains were maintained under conditions of strict breeding and health monitoring, and from the time of its foundation, a spirit of cooperation prevailed (Morse 1978; Russell 1978; Snell 1978).

To identify the histocompatibility genes, Snell developed the concept of congenic inbred strains, in which a short segment of the chromosome around a marker gene is transferred from one strain into an inbred genetic background by repeated backcrossing and selection. Like the inbred strains, congenic strains have a strict nomenclature (Snell 1978; Altman and Katz 1979; Morse 1981). For example, B6.C-$H1^b$ Tyr^c Hbb^d/By is a congenic strain in which the $H1^b$ allele derived from the BALB/cBy strain has been transferred onto the C57BL/6 inbred background. Amusingly, these mice are albino rather than black because the Tyr^c allele is tightly linked to the $H1^b$ allele derived from the albino BALB/cBy strain. Congenic strains carrying X-linked genes from wild mice have also been developed for studies on X-chromosome inactivation (Nielsen and Chapman 1977; Chapman et al. 1983). Many of the congenic strains originally developed by Snell and subsequently by others are widely available from commercial sources, including The Jackson Laboratory.

Another important innovation in mouse genetics was the development of recombinant inbred strains by Donald W. Bailey and Benjamin A. Taylor (Morse 1981). These strains were derived by crossing two different highly inbred progenitor strains and then inbreeding random pairs of the F_2 generation to produce a series of recombinant inbred or RI strains (Table 1.2). Their usefulness is in localizing within chromosomes any new locus that shows a polymorphism between the two progenitor strains. This is done by comparing the strain distribution pattern (SDP) of the new polymorphism with the many SDPs already established for enzyme, protein, or DNA RFLPs associated with known loci. One advantage of the system is that the data are cumulative; the patterns already published (Lyon and Searle 1989), or stored on computer at The Jackson Laboratory, provide a unique and expanding database for functional mapping of the mouse

TABLE 1.2. Schematized construction of eight RI strains

Progenitor inbred strains	AABBCC x aabbcc							
↓								
F_1	AaBbCc							
↓								
F_2	AaBbCc x AaBbCc							
↓								
Inbreeding for more than 20 generations								

RI Strain	1 AABBCC	2 AABBcc	3 AAbbCC	4 AAbbcc	5 aaBBCC	6 aaBBcc	7 aabbCC	8 aabbcc
A	A	A	A	A	a	a	a	a
B	B	B	b	b	B	B	b	b
C	C	c	C	c	C	c	C	c

Construction starts from two progenitor strains that have alternate alleles at three unlinked loci. The three alleles segregate and assort independently during the inbreeding process and eventually become genetically fixed. Each allele then has a unique strain distribution pattern (SDP).

genome. Another advantage is that because living animals are available, phenotypic differences in, for example, behavior or neurological responses can be studied (Takahashi et al. 1994). Thus, RI strains have been used to map loci affecting susceptibility of mouse strains to drug and alcohol addiction (Berrettini et al. 1994; Crabbe et al. 1994). One disadvantage of RI strains is that they are expensive to maintain; however, purified DNA is available from The Jackson Laboratory. Initially, it was often difficult to find polymorphisms among the progenitor strains. This is due in part to the rather restricted origin of laboratory mice, as discussed in the previous section. An alternative mapping technique was developed based on backcrossing F_1 hybrids between an inbred mouse strain and *M. spretus*, a wild mouse species found in Spain. Because *M. m. domesticus* and *M. spretus* are different species, the chances of finding an RFLP for any given DNA probe are much higher (Robert et al. 1985; Avner et al. 1988). A disadvantage of the system is that the F_1 males are sterile, so that once a cross has been made, the offspring cannot be bred to produce lines, and the amount of DNA is finite. Backcrosses between other inbred strains derived from wild mice (e.g., *M. castaneus*) and *M. domesticus* have also been established.

Wild mice have contributed to laboratory studies in other ways. For example, as shown originally by the German geneticist Alfred Gropp, they can be used to introduce cytogenetic variations into the karyotype of *M. m. domesticus*, which otherwise consists of 40 acrocentric chromosomes that are very difficult to distinguish. Gropp discovered in high Swiss valleys inbred groups of mice that have seven pairs of bi-armed (or Robertsonian fusion) chromosomes produced by the centric fusion of pairs of normal chromosomes (Gropp and Winking 1981). Individual Robertsonian chromosomes have been crossed into inbred laboratory strains where they can be used to generate embryos that are monosomic or trisomic for particular chromosomes (Epstein 1985) or have inherited two copies of a chromosome from one parent (Cattanach and Kirk 1985). They also provide markers for cytogenetic experiments. Since their discovery in mice of the Valle di Poschiavo, centric fusions have been found in mice in other localities and in laboratory strains. Like inbred strains, they have a strict nomenclature; for example, Rb (11.16)2H is a Robertsonian fusion involving chromosomes 11 and 16 and was the *second* of a series identified at the MRC Radiobiology Laboratory at Harwell (H) (Lyon and Searle 1989). For more information on the genetics and natural history of *M. m. domesticus* and its relatives, see the excellent symposium volume *Biology of the House Mouse* (Berry 1981) and the excellent book by Lee Silver (1994; publicly accessible through MGI).

ORIGINS OF DEVELOPMENTAL GENETICS OF THE MOUSE

Because of their availability from the mouse fanciers, many of the first mutants used in breeding experiments sported visible differences in coat color, hair morphology, and pigmentation patterns (see Fig. 2.42). In fact, these old mutations have proved to be an extremely valuable resource for studying a whole range of interesting biological problems, and many of the genes involved have now been cloned. For example, the *Dominant white spotting* (W) and *Steel* (Sl) pigmentation mutants have defects in the genes encoding, respectively, a transmembrane tyrosine kinase receptor and its ligand required for the growth and survival of melanocytes, primordial germ cells, and hematopoietic cells. The *waved-1* and *waved-2* hair mutations turned out to involve the genes encoding the transforming growth factor-α (TGF-α) and the epidermal growth factor (EGF) receptors, respectively (Luetteke et al. 1993, 1994; Mann et al. 1993). The *dilute* (*d*) mutation, which causes a dilution of pigmentation because the melanocytes cannot extend their dendritic cellular processes, is now known to have originated in a retroviral insertion into the *Myo5a* gene, encoding a novel myosin heavy chain expressed in the cytoplasm of melanocytes (Jenkins et al. 1981; Mercer et al. 1991). The *agouti* gene that is expressed in hair follicles and regulates pigment production by melanocytes (see Fig. 2.42) has also been cloned (Bultman et al. 1992).

Over the years, a large number of mouse mutants affecting other complex neurological, physiological, and morphogenetic processes have been identified. Some were uncovered during the early days of inbreeding as recessive mutations in wild or fancy mice. Others have arisen as spontaneous mutations in laboratory stocks of already inbred mice. Another important source has been the offspring of mice exposed to X-rays or chemical mutagens. Much of this work has been carried out in two laboratories established shortly after the Second World War in response to the need for research into the biological effects of radiation: the Oak Ridge National Laboratory in Tennessee, and the MRC Radiobiology Unit in Harwell, near Oxford. In addition to generating a whole range of important radiation-induced mutants and chromosomal rearrangements, these laboratories have done outstanding work on basic mouse genetics. For example, at Oak Ridge, Liane Russell mapped a series of overlapping deletions covering the *dilute-short ear* region on chromosome 9. This region encompasses several genes involved in pre- and postnatal development and is flanked by the *dilute* locus, which affects pigmentation, and the *short-ear* locus (see Fig. 2.43), which regulates the differentiation of the skeleton (Russell 1971; Rinchik et al. 1985). The retroviral insertion into the *dilute* gene described earlier provided the first molecular handle into the detailed analysis of the complex, and subsequent positional cloning led to the identification of the *short-ear* locus as the gene encoding the growth factor known as BMP5 (*bone morphogenetic protein-5*) (Kingsley et al. 1992). Work at Oak Ridge also generated an overlapping set of deletion mutants around the *albino* locus that has defined a set of loci affecting pre- and postnatal development (for review, see Holdener-Kenny et al. 1992). Recently, the embryonic ectoderm development (*eed*) locus has been cloned from this region and shown to encode a Polycomb group protein (Schumacher et al. 1996). On the other side of the Atlantic, Mary Lyon, working first at the Department of Genetics in Edinburgh and then at the MRC Radiobiology Laboratory at Harwell, was the first to describe the phenomenon of random X-chromosome inactivation in somatic tissues of female mice (Lyon 1961). At Harwell, she also generated many new, and for a long time unappreciated, ideas about the genetic organization of the *t* complex, which are described below (Lyon et al. 1979). By whatever route mouse mutants and chromosomal variants are derived, they are very expensive and time-consuming to isolate and maintain, and those that have been conserved and are cataloged are testament to an enormous amount of hard work, dedication, and foresight by many mouse geneticists.

Looking back, it is also easy to underestimate the painstaking work that went into describing the pathology and etiology of many of the early morphological mutants. It soon became apparent that to understand how a whole range of defects in the adult could be caused by mutation in a single gene, it was necessary to trace the mutant phenotype back to the early embryo. One geneticist who made a speciality of this approach was Hans Grüneberg, a refugee from Germany, who in 1938 was invited by Haldane to work at University College, London. Originally a physician, Grüneberg was motivated by a belief that mouse mutants could be used as models for understanding human congenital defects. For more than 40 years, he described a whole variety of mutants, in particular those with skeletal abnormalities. He traced many of them back to early postimplantation stages when they first showed signs of defects in the process of somite formation and differentiation. His books, *Animal Genetics and Medicine* (1947), *The Genetics of the Mouse* (1952), and *The Pathology of Development* (1963), are classics of their kind and were as influential as Ernst Hadorn's *Developmental Genetics and Lethal Factors*, which was written in 1955 and translated into English in 1961.

Foremost among the pioneers of mouse developmental genetics in the United States were L.C. Dunn, a contemporary of Thomas Hunt Morgan at Columbia University, and his colleagues Dorothea Bennett and Salome Glucksohn-Waelsch, originally a student of the renowned German embryologist, Hans Spemann. Dunn and his disciples can be credited with describing many homozygous lethal mouse mutants, but their most significant contribution has been to promote the genetic analysis of the *t* complex on chromosome 17. The first mutant forms of this complex were discovered by a Russian cancer research scientist, Nelly Dobrovolskaia-Zavadskaia, working at the Pasteur Institute in Paris on the effects of radiation. She found that one of the offspring of an X-irradiated mouse had a short tail, and she defined it as having a dominant mutation, *T* or *Brachyury* (see Fig. 2.43). Later, Dobrovolskaia-Zavadskaia crossed one of her *T* mice with a wild mouse she had caught and was surprised to find that the offspring had no tails at all. In fact, we now know that the wild mouse carried a variant form of the *t* complex which is called a *t* haplotype and which interacts with *T* to produce taillessness. In 1932, realizing the complexity of the system she had uncovered and the limitations of her own resources, Dobrovolskaia-Zavadskaia passed her mice on to Dunn.

Soon a number of different *t* haplotypes were discovered in wild mouse populations, and it was shown that embryos homozygous for different *t* haplotypes die at different stages of development (Bennett 1975). The recessive lethal mutations were maintained at high levels in the wild because males heterozygous for wild-type and *t*-haplotype forms of chromosome 17 transmit the latter to more than 90% of their offspring, a phenomenon known as transmission ratio distortion. Fifty years on, the *t* complex has been extensively analyzed by molecular techniques and has been shown to cover more than 16 cM of DNA, or about one third of chromosome 17 (equivalent to 1% of the entire genome), and to contain four large inversions, one of which includes the *H2* complex (Herrmann et al. 1986). It seems very likely that these inversions trap recessive embryonic lethal mutations in the many unrelated genes contained within the region by inhibiting recombination between wild-type and mutant chromosomes, and the complex is further held intact by transmission distortion (for reviews, see Silver 1988, 1993a; Schimenti 2000). Making use of a number of different *T* mutations and rearrangements within the *t* complex, Bernhard Herrmann, working in collaboration with Hans Lehrach and his colleagues at the EMBL in Germany, cloned the *T* gene in 1990 (Herrmann et al. 1990). This was the first example of the positional cloning of a mouse developmental mutation, and it has had a wide-ranging impact on the field. The *Brachyury* gene product turned out to be a DNA-binding protein and putative transcription factor expressed initially in the primitive streak of the embryo and then in the notochord and tailbud (Herrmann 1991; Kispert and Herrmann 1993). In homozygous mutant embryos, the absence of *T* leads to lethal developmental defects at ~10 dpc (Chesley 1935). Moreover, the *T* gene product and pattern of expression have been highly conserved during vertebrate evolution, and orthologs have been found in *Xenopus* (Smith et al. 1991), zebrafish (Schulte-Merker et al. 1992), and ascidians (Yasuo and Satoh 1993). Recent work has shown that a gene related to *Brachyury* is present in *Drosophila* and that *T* is a member of a large family of important developmental control genes encoding transcription factors with a so-called T-box DNA-binding domain (Bollag et al. 1994; for review, see Papaioannou 2001).

The nature of most of the *t*-associated genes, including those that regulate transmission distortion, has yet to be fully determined, but some progress has been made (Silver 1993a; Herrmann et al. 1999). Likewise, there are many other interesting "classical" mutations affecting embryonic pattern formation, organogenesis, and early neural development mapping elsewhere in the genome that must still be cloned. The public availability of the assembled mouse genome now provides the means to immediately identify candidate genes for mapped classical mutations. In addition, a cloned DNA fragment or a retrovirus used to generate transgenic mice may by chance insert into a gene controlling a specific morphogenetic process. The foreign DNA will then provide a unique handle for isolating the endogenous gene (for reviews, see Lock et al. 1991; Woychik and Alagramam 1998). One of the first examples of an embryonic lethal mutation caused by an experimentally induced retroviral integration was the *Mov13* strain in which the alpha 1(I) collagen gene had been disrupted (Jaenisch et al. 1983; Hartung et al. 1986). One of the initial examples of a transgene insertional mutation leading to identification of a developmental gene was the cloning of *limb deformity* (Woychik et al. 1985, 1990; Maas et al. 1990). Finally, studies in fields such as hematopoiesis, immunology, tumor biology, and developmental neurobiology are constantly leading to the identification of new growth factors, receptors, and cell–cell adhesion molecules. It is very likely that at least some of these proteins will be encoded by genes already identified by classical genetics. More systematic strategies for identifying new genes and cloning developmental mutants in the mouse are discussed at the end of this chapter.

A HERITAGE OF EXPERIMENTAL MOUSE EMBRYOLOGY

Mammalian genetics had a clear beginning with the rediscovery of Mendel's laws in 1900 and was initially championed by a relatively small number of enthusiasts, centered around leaders such as Castle, Little, and Haldane. In contrast, mammalian embryology is a much older science, and it would be beyond the scope of this chapter to trace the complex lineage of modern ideas back through many communities and continents to the classical experimental embryologists like Boveri, Roux, Spemann, Hadorn, Nieuwkoop, and Waddington. From the beginning, mammalian embryology was closely associated with studies into human and veterinary reproductive physiology, and it is through these links that social pressures have had their influence on academic research. For example, the accelerated pace of research into mammalian reproduction and embryology in the late 1950s and 1960s was due, in part, to the realization of the need for new methods of human population control and increased food production. Many laboratories in the United States and Europe were funded by the Population Council, Inc., and by the Ford Foundation. Other support came from bodies like the Agricultural Research Council in Great Britain that were anxious to see improvements in the fertility and yields of farm animals. For a history of mammalian embryology, see references in Needham (1959), Austin (1961), Oppenheimer (1967), and Mayr (1982).

Much of the early experimental work in mammalian embryology was done using rabbit embryos. This included accurate descriptions of preimplantation stages (Van Beneden 1875), oviduct transfer (Heape 1890), filming of morulae dividing in culture (Lewis and Gregory 1929), and other in vitro observations (for reviews, see Pincus 1936; Austin 1961). Rabbits were used initially because the oocytes are relatively large and easy to handle, being surrounded by a thick mucin coat, and the female ovulates only after mating, so that the age of the embryos can be timed quite precisely. However, these advantages were soon outweighed as more became known about the reproductive physiology and genetics of mice.

The first report of attempts to culture mouse embryos in vitro to the blastocyst stage was by John Hammond, Jr., son of his namesake, the great animal husbandry scientist who introduced the technique of artificial insemination for cattle. Working at the Strangeways Laboratory in Cambridge, England, Hammond Jr. succeeded in culturing eight-cell morulae and some four-cell-stage embryos to blastocysts, but embryos removed at the two-cell stage soon died (Hammond 1949). In 1956, important progress was made by an Australian veterinary scientist Wesley Whitten working at the Australian National University in Canberra after training in Oxford. The motivation behind these experiments was to obtain a defined medium in which the possible requirement of steroid hormones for embryo development could be tested. Using Krebs-Ringer's bicarbonate solution supplemented with glucose and bovine serum albumin, Whitten succeeded in culturing with high efficiency eight-cell mouse embryos to the blastocyst stage (Whitten 1956). Later, Whitten found that some two-cell-stage embryos developed into blastocysts with some modifications of his original medium (Whitten 1957). Whitten subsequently emigrated to the United States and continued his work at The Jackson Laboratory, but he also collaborated closely with John Biggers of the University of Pennsylvania in Philadelphia. It was in Biggers' laboratory that another veterinarian, Ralph Brinster, began his research career by defining the precise nutritional requirements of the preimplantation mouse embryo and, in the process, established the microdrop culture technique to routinely culture two-cell-stage mouse embryos to blastocysts (Brinster 1963, 1965, 1968).

These culture conditions, although in the end simple enough, opened up a whole new range of experiments. At the same time, the work of Anne McLaren in the United Kingdom on optimizing conditions for oviduct and uterine transfer (McLaren and Michie 1956) made it possible to overcome the final hurdle and routinely turn cultured embryos into live mice (McLaren and Biggers 1958). Together, these technical improvements meant that it was at last feasible to test the end result of experimental manipulations on large numbers of embryos. For example, Kristof Tarkowski in Warsaw was able to start analyzing the developmental potential of single mouse blastomeres, using the classical embryological approach of killing one blastomere and seeing how the other would develop (Tarkowski 1959). He was also able to make the first aggregation chimeras, an idea conceived and accomplished during a visit to the University of Bangor in north Wales (I. Wilson, pers. comm.). Tarkowski's original method involved breaking the zona pellucida mechanically and pushing the embryos together in a small drop of medium, which was technically extremely difficult (Tarkowski 1961). The whole process was made much easier by Beatrice Mintz in Philadelphia, who discovered that the zona could be gently digested by pronase (Mintz 1962). Subsequently, adult chimeras were generated and germ-line chimerism was documented (Mintz 1965).

Chimeras derived from embryos of two or more genotypes have been used to study such diverse topics as melanocyte migration and pigment patterns, sex determination, germ-cell differentiation, immunology, tumor clonality, size regulation, and cell lineage (for review, see McLaren 1976). The use of culture systems also led to the development of routine methods for both in vitro fertilization and the parthenogenetic activation of mouse oocytes. In addition, Ralph Brinster was able to carry out the first experiments on the injection of purified globin mRNA into mouse zygotes (Brinster et al. 1980) which, as described below, set the stage for the production of transgenic mice. With similar vision and persistence, Davor Solter, at the Wistar Institute in Philadelphia, succeeded in transferring nuclei between zygotes (McGrath and Solter 1983). This technology was crucial in revealing the importance of parental gene imprinting in mammalian development (Mann and Lovell-Badge 1984; McGrath and Solter 1984; Surani et al. 1984). Teruhiko Wakayama, working with Ryuzo Yanagimachi at the University of Hawaii in Honolulu, succeeded in cloning the first mouse, named Cumulina, and many others by directly injecting the nuclei of somatic cells into enucleated oocytes (Wakayama et al. 1998).

As far as studies on the postimplantation mouse embryo are concerned, there was considerable debate and confusion about the lineage of the different embryonic and extraembryonic tissues. The various conflicting theories have been summarized by Rossant and Papaioannou (1977). To resolve these problems, and to ask when early embryonic cells become committed to their developmental fate, Richard Gardner in Cambridge, England, developed the technique of generating chimeras by injecting isolated cells into host blastocysts (Gardner 1968). To test the developmental potential of different parts of the postimplantation embryo, several laboratories also developed methods for culturing isolated pieces of tissue in vitro and in ectopic sites. In this way, Nikola Skreb and his colleagues in Zagreb showed that the early embryonic ectoderm contains cells capable of contributing to all three germ layers of the fetus.

These studies on the pluripotentiality of cells from the normal embryo were complemented by the use of teratocarcinomas as a model system for studying early embryonic development, an approach pioneered by Leroy Stevens at The Jackson Laboratory and by Barry Pierce at the University of Colorado. Teratocarcinomas are gonadal tumors that contain a chaotic mixture of different tissue types, all derived from a population of undifferentiated stem cells known as embryonal carcinoma cells. Stevens first observed that male mice of the inbred

129 strain have a low incidence of testicular teratoma arising from primordial germ cells (Stevens and Little 1954). Stevens also identified modifier genes such as *ter* that increase the frequency of teratomas in the testis and eventually developed an inbred strain (129/Sv) in which the incidence is as high as 30%. Stevens also developed the LT strain in which about 50% of females develop ovarian teratocarcinomas.

The availability of transplantable teratocarcinomas inspired many new experiments, and it was not long before it was shown by Boris Ephrussi in France and Gordon Sato in the United States that cells from the tumors could be grown in vitro as cultures that consisted of both differentiated derivatives and undifferentiated embryonal carcinoma stem cells. The potential of this culture system for studying the biochemistry and molecular biology of early mammalian embryonic cells was also recognized by François Jacob, and because of his influence and the work of his research group in the Pasteur Institute, many cell biologists and biochemists were attracted to the teratocarcinoma system and the study of mouse developmental genetics (Jacob 1983).

Finally, the availability of teratocarcinomas led to the demonstration by Brinster (1974), Mintz and Illmensee (1975), and Papaioannou et al. (1975) that embryonal carcinoma (EC) stem cells could be injected into blastocysts to create adult chimeras that contained normal tissues derived from the EC cells. These studies opened up the exciting possibility of preselecting specific mutations in vitro in EC cells that could then be used to regenerate mice through chimeras. This idea was directly tested by Dewey et al. (1977), who chemically mutagenized EC cells in vitro and selected for HPRT-deficient variants. The HPRT-deficient EC cells successfully incorporated extensively into the somatic tissues of chimeras. However, EC cells rarely contributed to the germ line of chimeras (Stewart and Mintz 1982). Thus, it was not until the development of blastocyst-derived embryonic stem (ES) cell lines, independently by Martin Evans and Matt Kaufman in Cambridge, England (Evans and Kaufman 1981), and Gail Martin in San Francisco (Martin 1981), that integration of cultured cells into the germ line could be achieved with high efficiency and reproducibility. As described below, the availability of ES cells has made a significant impact on mammalian developmental genetics. ES cells have opened up exciting new approaches to studying gene function and interaction during embryonic development, and they are also widely used to engineer mouse models for human diseases and congenital abnormalities.

MANIPULATING THE MOUSE GENOME

The manipulation of the mouse genome required physical manipulations of the mouse embryo. In 1966, Teh Ping Lin at the University of California at San Francisco reported that macromolecules like bovine gamma globulin could be injected directly into the pronuclei of mouse zygotes and that the embryos survived this micromanipulation to yield viable mice at term, establishing the physical manipulations required to generate transgenic mice (Lin 1966). The first report of the direct introduction of new genetic material into the mouse embryo actually predates the widespread use of recombinant DNA techniques. In 1974, Rudolf Jaenisch and Beatrice Mintz reported that injection of purified SV40 DNA into the cavity of mouse blastocysts resulted in mice with viral DNA sequences detected in somatic tissues, suggesting that the exogenous DNA had integrated into the genome of embryonic cells. In addition, Jaenisch (1976) reported that the infection of zona pellucida-free four to eight cell stage embryos with the Moloney murine leukemia virus could lead to the stable incorporation of the retroviral genome into the mouse germ line. However, these approaches were not generally employed in subsequent attempts to introduce cloned genes into the somatic and germ line tissues of animals. In January, 1980, the first direct injections of nucleic acids and subsequent expression in fertilized mouse oocytes were reported by Ralph Brinster and colleagues (1980), presaging the injection of DNA into the pronuclei of zygotes. The first report of the successful introduction of a cloned gene into the somatic tissues of mice by direct injection into the pronuclei of zygotes was reported by Jon Gordon and Frank Ruddle and their colleagues at Yale in December, 1980 (Gordon et al. 1980). Shortly thereafter, several groups were also successful in introducing cloned genes into the somatic tissues of mice (Brinster et al., 1981b; Costantini and Lacy, 1981; E. Wagner et al., 1981; T. Wagner et al., 1981). More significantly, some of these introduced "transgenes" were found to express (Brinster et al., 1981b; E. Wagner et al., 1981; T. Wagner et al., 1981) and be transmitted to progeny, establishing that the animal germ line had been transformed with foreign DNA sequences (Brinster et al., 1981b; Costantini and Lacy, 1981; Gordon and Ruddle, 1981; T. Wagner et al., 1981). The structure, inheritance, and expression of foreign genes in transgenic mice and the applications this technique for the study of mouse development are discussed in Chapter 7. What is usually not appreciated is that the experimental modification of the animal germ line (transgenesis) was first established in mice. Some have even suggested that the experimental genetic manipulation of the animal germ line initially by pronuclear injection of cloned DNA into zygotes represents one of the most significant milestones in the history of human civilization.

The development of ES cells has significantly extended this primary achievement. ES cells were first derived directly from blastocysts in culture by Evans and Kaufman (1981) and Martin (1981). Shortly thereafter, they were shown to be capable of contributing to many different tissues in chimeras, including the germ line, when injected into host blastocysts and returned to a foster mother (Bradley et al. 1984). The first reports of genetic manipulation of ES cells were by Robertson et al. (1986), who demonstrated that cells containing integrated retroviruses could be transmitted through the germ line, and Gossler et al. (1986), who showed that the *neo*[r] gene introduced into ES cells by calcium phosphate precipitation could be similarly transmitted. Manipulation of a specific endogenous gene was shown by Hooper et al. (1987) and Kuehn et al. (1987), who like Dewey et al. (1977), preselected mutant ES cells defective in the X-linked gene, hypoxanthine phosphoribosyl transferase (*Hprt*), and used them to derive *Hprt*-deficient mice. Two methods used were selection of preexisting "spontaneous" mutant cells (Hooper et al. 1987) and disruption of a normal gene with a retrovirus (Kuehn et al. 1987). Interestingly, *Hprt*-deficient male mice did not show the symptoms of Lesch-Nyhan disease because of differences in the metabolism of purines by mouse and humans. Nevertheless, an important milestone had been achieved with the demonstration that it was possible to manipulate ES cells genetically in a controlled way in the culture dish and to introduce the mutation into the germ line. The stage was then set for experiments designed to target mutations to specific genes by homologous recombination in ES cells. This was initially achieved using *Hprt* (Doetschmann et al. 1987; Thomas and Capecchi 1987). Germ-line transmission of targeted mutations was first obtained with *Hprt* (Thompson et al. 1989), *c-abl* (Schwartzberg et al. 1989), and β_2-*microglobulin* (Zijlstra et al. 1989). To date, thousands of genes have been disrupted by homologous recombination and transmitted through the germ line, and a partial listing of the mouse lines with targeted mutations can be found in the *TBASE* database through The Jackson Laboratory (http://www.tbase.jax.org) or through the *Mouse Knockout and Mutation Database* by subscription through BioMedNet (http://www.bmn.com).

More recently, a myriad of sophisticated gene targeting strategies in ES cells have been developed to manipulate the mouse genome (Bradley and Liu 1996). Gene knock-in strategies have been devised to express heterologous genes in patterns of targeted endogenous loci (Hanks et al. 1995). In addition, conditional genetic manipulations (tissue-specific knockouts) in the mouse are now routine using site-specific DNA recombinases such as Cre and Flp to bypass early lethalities caused by constitutive mutations (for review, see Nagy 2000). Conditional reporter mouse lines (i.e., those that express cell-autonomous reporter genes like *lacZ*, human placental alkaline phosphatase, or fluorescent proteins depending on Cre or Flp expression) have also been generated (Lobe et al. 1999; Mao et al. 1999; Soriano 1999; Novak et al. 2000; Awatramani et al. 2001; Srinivas et al. 2001). Remarkably, large centimorgan genetic alterations in ES cells, including deletions, duplications, inversions, and translocations, are now possible using a method known as chromosome engineering (Ramirez-Solis et al. 1995; Smith et al. 1995; for review, see Yu and Bradley 2001) or radiation (You et al. 1997; Goodwin et al. 2001). It is likely that many more ways to manipulate the mouse genome will continue to be developed to reveal gene function during development.

THE SYSTEMATIC SEARCH FOR NEW GENES AND DEVELOPMENTAL MUTANTS IN THE MOUSE

A number of different strategies have been used to search systematically for novel genes regulating development in the mouse. One of the most fruitful strategies has been so-called "homology searching," which involves screening cDNA or genomic libraries or databases for mouse genes related to those that have been found to regulate growth and development in other organisms, in particular *Drosophila*, *C. elegans*, and *Xenopus*. This approach also includes searching for genes related to proto-oncogenes, growth factors, and receptors discovered in research with mammalian cells, and has led to the identification of highly conserved gene families encoding proteins with related functional motifs. One such example includes transcription factors with related DNA-binding domains such as the homeo, paired, zinc finger, winged-helix, and helix-loop-helix domains. Other examples are polypeptide-signaling molecules, including those related to growth factors and oncogene products (e.g., proteins related to FGFs, TGF-βs, EGFs, hedgehogs, and Wnts). To these can be added transmembrane receptors for different classes of signaling molecules, and proteins involved in axonal guidance, cell migration, and cell adhesion. The extraordinary success of this approach has led to the growing realization that not only have individual elements of developmental pathways such as polypeptide signaling molecules and transcription factors been conserved during evolution, but probably whole regulatory circuits have been conserved as well. This implies that some simple ancestral organism established basic mechanisms for determining anteroposterior, dorsoventral, and proximodistal axes and for determining the fate and proliferation of cells according to their position along these axes. Once these mechanisms were set up, it would be easier for future organisms to elaborate upon them by gene duplication and divergence to produce more complex morphologies than to completely re-invent alternative pathways. It remains to be seen to what extent this principle of conservation pervades embryonic development and whether screens based on alternative hypotheses lead to the discovery of whole new classes of developmental genes unique to vertebrates in general, and to mammals in particular. Meanwhile, it is obviously best to rely on several different experimental approaches to the problem, particularly since it is likely that some conserved genes will have diverged too far in nucleotide sequence to be detected by conventional methods for homology searching.

Another strategy to identify developmental genes is to carry out large-scale mutagenesis screens. Currently, one of the most popular protocols used by mouse geneticists is to treat male mice with the chemical ethylnitrosourea (ENU) (for review, see Justice et al. 1999) and to then screen offspring for dominant or recessive phenotypes. One of the first ENU-induced mutations in mice was reported by Popp et al. (1983), who identified a hemoglobin variant using an electrophoresis screen. Examples of interesting mutants generated by ENU mutagenesis are *Min* (*multiple intestinal neoplasia*), which has a mutation in the mouse ortholog of the human *APC* (*adenomatous polyposis colon*) gene (Su et al. 1992), and *Clock*, which has a defect in establishing circadian rhythm (Vitaterna et al. 1994). Recently, large-scale ENU screens for dominant mutations have been performed in the mouse (Hrabe de Angelis et al. 2000; Nolan et al. 2000) and some smaller-scale recessive screens have also been reported (Kasarskis et al. 1998; Herron et al. 2002). An important innovation for chemical mutagenesis screens has been the generation of mice with engineered coat-color-marked balancer chromosomes (Zheng et al. 1999). These balancer chromosomes hold great promise for efficiently carrying out chromosomal region-specific screens as well as facilitating the maintenance of lethal mutations (Justice et al. 1999). In the future, sensitized screens will also be performed to produce and screen for genetic modifiers (McDonald et al. 1990). Fortunately, the public availability of the assembled mouse genome is now making the molecular identification of ENU-induced mutations much easier and quicker. Gene-driven approaches have also been developed utilizing ENU-treated ES cells to identify mutations in specific genes (Chen et al. 2000; Munroe et al. 2000). In addition, cryopreserved sperm from ENU-mutagenized mice can be screened for specific gene mutations (Coghill et al. 2002). The mutant ES cell clones or sperm can then be used to regenerate mice.

The availability of ES cells opened up an entirely different and very powerful way of carrying out large-scale insertional mutagenesis screens. This potential was quickly recognized by Martin Evans and his group at Cambridge, who infected ES cells with a defective retrovirus carrying the *neo*[r] gene, which allowed cells with an insertion to be selected in culture (Robertson et al. 1986). These cells were then used to generate mouse lines from which homozygous mutants could be bred. One of these lines, 413.d, has been shown to have an insertion in *nodal*, which encodes a TGF-β-related gene expressed around the node during gastrulation (Conlon et al. 1991; Zhou et al. 1993).

Subsequently, insertional mutagenesis screens have evolved to use vectors that tag the insertion site with a reporter sequence such as *lacZ*, which both facilitates cloning and allows the expression pattern of the disrupted gene to be visualized. Different "trap" vectors have been designed for identifying coding sequences, enhancers, and promoters (Gossler et al. 1989; Friedrich and Soriano 1991; Skarnes et al. 1992; for review, see Stanford et al. 2001). Other innovations include the use of retroviral vectors designed to allow rapid sequencing of flanking DNA (von Melchner et al. 1992; Chen et al. 1994). This has facilitated the generation of large gene trap libraries of ES cell clones with sequence tags (Zambrowicz et al. 1998; Wiles et al. 2000). It is now worthwhile to first check the public ES cell gene trap consortium databases (ES cell gene trap databases include http://www.genetrap.de, http://www.genetrap.de/, http://baygenomics.ucsf.edu/, http://www.escells.ca/, http://cmhd.mshri.on.ca/) to determine whether your favorite gene has already been mutated to obtain the mutant clone to generate chimeras. Vectors have also been designed to select for insertion into genes encoding secreted or membrane-associated proteins (Skarnes et al. 1995). Screens can also be set up to select for insertions into genes that are specifically expressed when ES cells differentiate or that are activated or repressed in response to growth factors or other manipulations (Forrester et al. 1996; Medico et al. 2001; Vallis et al. 2002). Recently, the Sleeping Beauty transposon, originally derived from salmonid fish, has been modified to transpose in response to the Sleeping Beauty transposase in the mouse germ line (Dupuy et al. 2001; Fischer et al. 2001; Horie et al. 2001). Modifications to the Sleeping Beauty system hold great promise for development into a new powerful insertional mutagenesis tool in the mouse.

Future progress in the mouse genetics and embryology field appears to be limitless. Gene knockdowns using double-stranded RNA (RNA interference) have been performed in mice with initial success (McCaffrey et al. 2002), promising a new way to eliminate gene function. The introduction of nucleic acids (RNA and DNA) into pre- and postimplantation mouse embryos by electroporation will likely become more prevalent as a tool for modifying the mouse (Osumi and Inoue 2001; Grabarek et al. 2002). There will probably be more effort placed on improving techniques for temporally controlling gene expression in mice using drugs such as tetracycline, tamoxifen, and RU486 (Shin et al. 1999; Li et al. 2000; Cao et al. 2001). Unlike Cre and Flp recombinases that recognize homologous sites to mediate recombination, the integrase from the *Streptomyces* phage φC31 catalyzes recombination between heterologous sites, making possible precise unidirectional integration in mammalian cells (Groth et al. 2000). It appears that φC31 will be joining Cre and Flp as a standard tool for manipulating the mouse germ line (A. Nagy, pers. comm.). Tissue-specific expression of fluorescent proteins will facilitate the purification of cell types by flow sorting that can then be used for biochemical or molecular studies or transplantation back into the animal (Hadjantonakis and Nagy 2000). These are but a few of the new possibilities that await scientists wanting to understand the genetics and embryology of the mouse.

REFERENCES

Altman P.L. and Katz D.D., eds. 1979. *Inbred and genetically defined strains of laboratory animals.* Part 1: *Mouse and rat.* Federation of American Societies for Experimental Biology,

Bethesda, Maryland.

Anderson W.F., Killos L., Sanders-Haigh L., Kretschmer P.J., and Diacumakos E.G. 1980. Replication and expression of thymidine kinase and human globin genes microinjected into mouse fibroblasts. *Proc. Natl. Acad. Sci.* **77:** 5399–5403.

Auffray J.-C., Marshall J.T., Thaler L., and Bonhomme F. 1990. Focus on the nomenclature of European species of *Mus* mouse. *Mouse Genome* **88:** 7–8.

Austin C.R. 1961. *The mammalian egg.* Blackwell, London.

Avner P., Amar L., Dandolo L., and Guenet J.L. 1988. Genetic analysis of the mouse using interspecific crosses. *Trends Genet.* **4:** 18–23.

Awatramani R., Soriano P., Mai J.J., and Dymecki S. 2001. An Flp indicator mouse expressing alkaline phosphatase from the *ROSA26* locus. *Nat. Genet.* **29:** 257–259.

Bennett D. 1975. The T-locus of the mouse. *Cell* **6:** 441–454.

Bennett K.L., Hill R.E., Pietras D.F., Woodworth-Gutai M., Kane-Haas C., Houston J.M., Heath J.K., and Hastie N.D. 1984. Most highly repeated dispersed DNA families in the mouse genome. *Mol. Cell. Biol.* **4:** 1561–1571.

Berrettini W.H., Ferraro T.N., Alexander R.C., Buchberg A.M., and Vogel W.H. 1994. Quantitative trait loci mapping of three loci controlling morphine preference using inbred mouse strains. *Nat. Genet.* **7:** 54–58.

Berry R.J. 1981. *Biology of the house mouse.* The Zoological Society of London, Academic Press.

Bollag R.J., Siegfried Z., Cebra-Thomas J.A., Garvey N., Davison E.M., and Silver L.M. 1994. An ancient family of embryonically expressed mouse genes sharing a conserved protein motif with the T locus. *Nat. Genet.* **7:** 383–389.

Bonhomme F., Catalan U., Britton-Davidian J., Chapman V.M., Moriwaki K., Nevo E., and Thaler L. 1984. Biochemical diversity and evolution in the genus *Mus. Biochem. Genet.* **22:** 275–303.

Boshier D.P. 1968. The relationship between genotype and reproductive performance before parturition in mice. *J. Reprod. Fertil.* **15:** 427–435.

Bradley A. and Liu P. 1996. Target practice in transgenics. *Nat. Genet.* **14:** 121–123.

Bradley A., Evans M., Kaufman M.H., and Robertson E. 1984. Formation of germ-line chimaeras from embryo-derived teratocarcinoma cell lines. *Nature* **309:** 255–256.

Brinster R.L. 1963. A method for in vitro cultivation of mouse ova from two-cell to blastocyst. *Exp. Cell Res.* **32:** 205–208.

———. 1965. Studies on the development of mouse embryos in vitro. II. The effect of energy source. *J. Exp. Zool.* **158:** 59–68.

———. 1968. In vitro culture of mammalian embryos. *J. Anim. Sci.* **27:** 1–14.

———. 1974. The effect of cells transferred into the mouse blastocyst on subsequent development. *J. Exp. Med.* **140:** 1049–1056.

Brinster R.L., Chen H.Y., Trumbauer M.E., and Avarbock M.R. 1980. Translation of globin messenger RNA by the mouse ovum. *Nature* **283:** 499–501.

Brinster R.L., Chen H.Y., Trumbauer M., Senear A.W., Warren R., and Palmiter R.D. 1981. Somatic expression of herpes thymidine kinase in mice following injection of a fusion gene into eggs. *Cell* **27:** 223–231.

Bultman S.J., Michaud E.J., and Woychik R.P. 1992. Molecular characterization of the mouse agouti locus. *Cell* **71:** 1195–1204.

Cao T., Longley M.A., Wang X.J., and Roop D.R. 2001. An inducible mouse model for epidermolysis bullosa simplex: Implications for gene therapy. *J. Cell Biol.* **152:** 651–656.

Capecchi M.R. 1980. High efficiency transformation by direct microinjection of DNA into cultured mammalian cells. *Cell* **22:** 479–488.

Castle W.E. and Allen G.M. 1903. The heredity of albinism. *Proc. Am. Acad. Arts Sci.* **38:** 603–621.

Cattanach B.M. and Kirk M. 1985. Differential activity of maternally and paternally derived chromosome regions in mice. *Nature* **315:** 496–498.

Chapman V.M., Kratzer P.G., and Quarantillo B.A. 1983. Electrophoretic variation for X chromosome-linked hypoxanthine phosphoribosyl transferase (HPRT) in wild-derived mice. *Genetics* **103:** 785–795.

Chen J., DeGregori J., Hicks G., Roshon M., Shcerer C., Shi E., and Ruley H.E. 1994. Gene trap retroviruses. In *Methods in molecular genetics* (ed. Adolph K.W.), pp. 123–140. Academic Press, New York.

Chen Y., Yee D., Dains K., Chatterjee A., Cavalcoli J., Schneider E., Om J., Woychik R.P., and Magnuson T. 2000. Genotype-based screen for ENU-induced mutations in mouse embryonic stem cells. *Nat. Genet.* **24:** 314–317.

Chesley P. 1935. Development of the short-tailed mutant in the house mouse. *J. Exp. Zool.* **70:** 429–459.

Clark R. 1984. *The life and work of J.B.S. Haldane.* Oxford University Press, England.

Coghill E.L., Hugill A., Parkinson N., Davison C., Glenister P., Clements S., Hunter J., Cox R.D., and Brown S.D. 2002. A gene-driven approach to the identification of ENU mutants in the mouse. *Nat. Genet.* **30:** 255–256.

Conlon F.L., Barth K.S., and Robertson E.J. 1991. A novel retrovirally induced embryonic lethal mutation in the mouse: Assessment of the developmental fate of embryonic stem cells homozygous for the 413.d proviral integration. *Development* **111:** 969–981.

Copeland N.G., Jenkins N.A., Gilbert D.J., Eppig J.T., Maltais L.J., Miller J.C., Dietrich W.F., Weaver A., Lincoln S.E., Steen R.G. et al. 1993. A genetic linkage map of the mouse: Current applications and future prospects. *Science* **262:** 57–66.

Costantini F. and Lacy E. 1981. Introduction of a rabbit beta-globin gene into the mouse germ line. *Nature* **294:** 92–94.

Crabbe J.C., Belknap J.K., and Buck K.J. 1994. Genetic animal models of alcohol and drug abuse. *Science* **264:** 1715–1723.

Dewey M.J., Martin, Jr. D.W., Martin G.R., and Mintz B. 1977. Mosaic mice with terato-carcinoma-derived mutant cells deficient in hypoxanthine phosphoribosyltransferase. *Proc. Natl. Acad. Sci.* **74:** 5564–5568.

Doetschman T., Gregg R.G., Maeda N., Hooper M.L., Melton D.W., Thompson S., and Smithies O. 1987. Targetted correction of a mutant HPRT gene in mouse embryonic stem cells. *Nature* **330:** 576–578.

Dupuy A.J., Fritz S., and Largaespada D.A. 2001. Transposition and gene disruption in the male germline of the mouse. *Genesis* **30:** 82–88.

Epstein C.P. 1985. Mouse monosomies and trisomies as experimental systems for study-ing mammalian aneuploidy. *Trends Genet.* **1:** 129–134.

Evans M.J. and Kaufman M.H. 1981. Establishment in culture of pluripotential cells from mouse embryos. *Nature* **292:** 154–156.

Ferris S.D., Sage R.D., and Wilson A.C. 1982. Evidence from mtDNA sequences that com-mon laboratory strains of inbred mice are descended from a single female. *Nature* **295:** 163–165.

Festing M.F.W. 1979. *Inbred strains in biomedical research.* Oxford University Press, England.

Fischer S.E., Wienholds E., and Plasterk R.H. 2001. Regulated transposition of a fish trans-poson in the mouse germ line. *Proc. Natl. Acad. Sci.* **98:** 6759–6764.

Forrester L.M., Nagy A., Sam M., Watt A., Stevenson L., Bernstein A., Joyner A.L., and Wurst W. 1996. An induction gene trap screen in embryonic stem cells: Identification of genes that respond to retinoic acid in vitro. *Proc. Natl. Acad. Sci.* **93:** 1677–1682.

Friedrich G. and Soriano P. 1991. Promoter traps in embryonic stem cells: A genetic screen to identify and mutate developmental genes in mice. *Genes Dev.* **5:** 1513–1523.

Gardner R.L. 1968. Mouse chimeras obtained by the injection of cells into the blastocyst. *Nature* **220:** 596–597.

Goodwin N.C., Ishida Y., Hartford S., Wnek C., Bergstrom R.A., Leder P., and Schimenti J.C. 2001. DelBank: A mouse ES-cell resource for generating deletions. *Nat. Genet.* **28:** 310–311.

Gordon J.W. and Ruddle F.H. 1981. Integration and stable germ line transmission of genes injected into mouse pronuclei. *Science* **214:** 1244–1246.

Gordon J.W., Scangos G.A., Plotkin D.J., Barbosa J.A., and Ruddle F.H. 1980. Genetic transformation of mouse embryos by microinjection of purified DNA. *Proc. Natl. Acad. Sci.* **77:** 7380–7384.

Gorer P.A., Lyman S., and Snell G.D. 1948. Studies on the genetic and antigenic basis of tumour transplantation. Linkage between a histocompatibility gene and "fused" in mice. *Proc. R. Soc. Lond. B.* **135:** 499–505.

Gossler A., Joyner A.L., Rossant J., and Skarnes W.C. 1989. Mouse embryonic stem cells and reporter constructs to detect developmentally regulated genes. *Science* **244:** 463–465.

Gossler A., Doetschman T., Korn R., Serfling E., and Kemler R. 1986. Transgenesis by means of blastocyst-derived embryonic stem cell lines. *Proc. Natl. Acad. Sci.* **83:** 9065–9069.

Grabarek J.B., Plusa B., Glover D.M., and Zernicka-Goetz M. 2002. Efficient delivery of dsRNA into zona-enclosed mouse oocytes and preimplantation embryos by electroporation. *Genesis* **32:** 269–276.

Gregory S.G., Sekhon M., Schein J., Zhao S., Osoegawa K., Scott C.E., Evans R.S., Burridge P.W., Cox T.V., Fox C.A., Hutton R.D., Mullenger I.R., Phillips K.J., Smith J., Stalker J., Threadgold G.J., Birney E., Wylie K., Chinwalla A., Wallis J., Hillier L., Carter J., Gaige T., Jaeger S., Kremitzki C., Layman D., Maas J., McGrane R., Mead K., Walker R., Jones S., Smith M., Asano J., Bosdet I., Chan S., Chittaranjan S., Chiu R., Fjell C., Fuhrmann D., Girn N., Gray C., Guin R., Hsiao L., Krzywinski M., Kutsche R., Lee S.S., Mathewson C., McLeavy C., Messervier S., Ness S., Pandoh P., Prabhu A.L., Saeedi P., Smailus D., Spence L., Stott J., Taylor S., Terpstra W., Tsai M., Vardy J., Wye N., Yang G., Shatsman S., Ayodeji B., Geer K., Tsegaye G., Shvartsbeyn A., Gebregeorgis E., Krol M., Russell D., Overton L., Malek J.A., Holmes M., Heaney M., Shetty J., Feldblyum T., Nierman W.C., Catanese J.J., Hubbard T., Waterston R.H., Rogers J., De Jong P.J., Fraser C.M., Marra M., McPherson J.D., and Bentley D.R. 2002. A physical map of the mouse genome. *Nature* **418:** 743–750.

Gropp A. and Winking H. 1981. Robertsonian translocations: Cytology, meiosis, segregation patterns and biological consequences of heterozygosity. *Symp. Zool. Soc. Lond.* **47:** 141–181.

Groth A.C., Olivares E.C., Thyagarajan B., and Calos M.P. 2000. A phage integrase directs efficient site-specific integration in human cells. *Proc. Natl. Acad. Sci.* **97:** 5995–6000.

Grüneberg H. 1947. *Animal genetics and medicine.* Hamish Hamilton, London.

———. 1952. *The genetics of the mouse,* 2nd edition. Martinus Nijhoff, The Hague.

———. 1963. *The pathology of development: A study of inherited skeletal disorders in mammals.* Blackwell, Oxford.

Hadjantonakis A.K. and Nagy A. 2000. FACS for the isolation of individual cells from transgenic mice harboring a fluorescent protein reporter. *Genesis* **27:** 95–98.

Hadorn E. 1961. *Developmental genetics and lethal factors.* Methuen, London.

Haldane J.B.S., Sprunt A.D., and Haldane N.M. 1915. Reduplication in mice. *J. Genet.* **5:** 133–135.

Hammond J. 1949. Recovery and culture of tubal mouse ova. *Nature* **163:** 28–29.

Hanks M., Wurst W., Anson-Cartwright L., Auerbach A.B., and Joyner A.L. 1995. Rescue of the *En-1* mutant phenotype by replacement of *En-1* with *En-2. Science* **269:** 679–682.

Hartung S., Jaenisch R., and Breindl M. 1986. Retrovirus insertion inactivates mouse alpha 1(I) collagen gene by blocking initiation of transcription. *Nature* **320:** 365–367.

Heape W. 1890. Preliminary note on the trans-plantation and growth mammalian ova within a uterine foster mother. *Proc. R. Soc. Lond. B.* **48:** 457.

Heiniger H.-J. and Dorey J.J. 1980. *Handbook on genetically standard Jax mice.* The Jackson Laboratory, Bar Harbor, Maine.

Herrmann B.G. 1991. Expression pattern of the Brachyury gene in whole-mount T^Wis/T^Wis mutant embryos. *Development* **113:** 913–917.

Herrmann B.G., Koschorz B., Wertz K., McLaughlin K.J., and Kispert A. 1999. A protein kinase encoded by the *t* complex responder gene causes non-mendelian inheritance. *Nature* **402:** 141–146.

Herrmann B.G., Labeit S., Poustka A., King T.R., and Lehrach H. 1990. Cloning of the T gene required in mesoderm formation in the mouse. *Nature* **343:** 617–622.

Herrmann B., Bucan M., Mains P.E., Frischauf A.M., Silver L.M., and Lehrach H. 1986. Genetic analysis of the proximal portion of the mouse *t* complex: Evidence for a second inversion within *t* haplotypes. *Cell* **44:** 469–476.

Herron B.J., Lu W., Rao C., Liu S., Peters H., Bronson R.T., Justice M.J., McDonald J.D., and Beier D.R. 2002. Efficient generation and mapping of recessive developmental mutations using ENU mutagenesis. *Nat. Genet.* **30:** 185–189.

Hogan B., Costantini F., and Lacy E. 1986. *Manipulating the mouse embryo,* 1st Edition. Cold Spring Harbor Laboratory, Cold Spring Harbor, New York.

Hogan B., Beddington R., Costantini F., and Lacy E. 1994. *Manipulating the mouse embryo,* 2nd Edition. Cold Spring Harbor Laboratory Press, Cold Spring Harbor, New York.

Holdener-Kenny B., Sharan S.K., and Magnuson T. 1992. Mouse albino-deletions: From genetics to genes in development. *BioEssays* **14:** 831–839.

Hooper M., Hardy K., Handyside A., Hunter S., and Monk M. 1987. HPRT-deficient (Lesch-Nyhan) mouse embryos derived from germline colonization by cultured cells. *Nature* **326:** 292–295.

Horie K., Kuroiwa A., Ikawa M., Okabe M., Kondoh G., Matsuda Y., and Takeda J. 2001. Efficient chromosomal transposition of a Tc1/mariner-like transposon Sleeping Beauty in mice. *Proc. Natl. Acad. Sci.* **98:** 9191–9196.

Hrabe de Angelis M.H., Flaswinkel H., Fuchs H., Rathkolb B., Soewarto D., Marschall S., Heffner S., Pargent W., Wuensch K., Jung M., Reis A., Richter T., Alessandrini F., Jakob T., Fuchs E., Kolb H., Kremmer E., Schaeble K., Rollinski B., Roscher A., Peters C., Meitinger T., Strom T., Steckler T., Holsboer F., Klopstock T., Gekeler F., Schindewolf C., Jung T., Avraham K., Behrendt H., Ring J., Zimmer A., Schughart K., Pfeffer K., Wolf E., and Balling R. 2000. Genome-wide, large-scale production of mutant mice by ENU mutagenesis. *Nat. Genet.* **25:** 444–447.

Jacob F. 1983. Concluding remarks. *Cold Spring Harbor Conf. Cell Proliferation* **10:** 683–687.

Jaenisch R. 1976. Germ line integration and Mendelian transmission of the exogenous Moloney leukemia virus. *Proc. Natl. Acad. Sci.* **73:** 1260–1264.

Jaenisch R., Harbers K., Schnieke A., Lohler J., Chumakov I., Jahner D., Grotkopp D., and Hoffmann E. 1983. Germline integration of Moloney murine leukemia virus at the *Mov13* locus leads to recessive lethal mutation and early embryonic death. *Cell* **32:** 209–216.

Jenkins N.A., Copeland N.G., Taylor B.A., and Lee B.K. 1981. Dilute (*d*) coat colour mutation of DBA/2J mice is associated with the site of integration of an ecotropic MuLV genome. *Nature* **293:** 370–374.

Justice M.J., Noveroske J.K., Weber J.S., Zheng B., and Bradley A. 1999. Mouse ENU mutagenesis. *Hum. Mol. Genet.* **8:** 1955–1963.

Kasarskis A., Manova K., and Anderson K.V. 1998. A phenotype-based screen for embryonic lethal mutations in the mouse. *Proc. Natl. Acad. Sci.* **95:** 7485–7490.

Keeler C. 1978. How it began. In *Origins of inbred mice* (ed. Morse H.C.) pp. 179–192. Academic Press, New York.

Kingsley D.M., Bland A.E., Grubber J.M., Marker P.C., Russell L.B., Copeland N.G., and Jenkins N.A. 1992. The mouse short ear skeletal morphogenesis locus is associated with defects in a bone morphogenetic member of the TGF beta superfamily. *Cell* **71:** 399–410.

Kispert A. and Herrmann B.G. 1993. The Brachyury gene encodes a novel DNA binding protein. *EMBO J.* **12:** 3211–3220.

Kuehn M.R., Bradley A., Robertson E.J., and Evans M.J. 1987. A potential animal model for Lesch-Nyhan syndrome through introduction of HPRT mutations into mice. *Nature* **326:** 295–298.

Lewis W.H. and Gregory P.W. 1929. Cine-matographs of living developing rabbit eggs. *Science* **69:** 226–229.

Li M., Indra A.K., Warot X., Brocard J., Messaddeq N., Kato S., Metzger D., and Chambon P. 2000. Skin abnormalities generated by temporally controlled RXRalpha mutations in mouse epidermis. *Nature* **407:** 633–636.

Lin T.P. 1966. Microinjection of mouse eggs. *Science* **151:** 333–337.

Lindblad-Toh K., Lander E.S., McPherson J.D., Waterston R.H., Rodgers J., and Birney E. 2001. Progress in sequencing the mouse genome. *Genesis* **31:** 137–141.

Lobe C.G., Koop K.E., Kreppner W., Lomeli H., Gertsenstein M., and Nagy A. 1999. Z/AP, a double reporter for cre-mediated recombination. *Dev. Biol.* **208:** 281–292.

Lock L.F., Jenkins N.A., and Copeland N.G. 1991. Mutagenesis of the mouse germline using retroviruses. *Curr. Top. Microbiol. Immunol.* **171:** 27–41.

Luetteke N.C., Qiu T.H., Peiffer R.L., Oliver P., Smithies O., and Lee D.C. 1993. TGFα deficiency results in hair follicle and eye abnormalities in targeted and waved-1 mice. *Cell* **73:** 263–278.

Luetteke N.C., Phillips H.K., Qiu T.H., Copeland N.G., Earp H.S., Jenkins N.A., and Lee D.C. 1994. The mouse waved-2 phenotype results from a point mutation in the EGF receptor tyrosine kinase. Genes Dev **8:** 399–413.

Lyon M.F. 1961. Gene action in the X-chromosome of the mouse (*Mus musuls L.*). *Nature* **190:** 372–373.

Lyon, M.F. and A.G. Searle, eds. 1989. *Genetic variants and strains of the laboratory mouse*, 2nd Edition. Oxford University Press, England.

Lyon M.F., Evans E.P., Jarvis S.E., and Sayers I. 1979. *t*-Haplotypes of the mouse may involve a change in intercalary DNA. *Nature* **279:** 38–42.

Maas R.L., Zeller R., Woychik R.P., Vogt T.F., and Leder P. 1990. Disruption of formin-encoding transcripts in two mutant limb deformity alleles. *Nature* **346:** 853–855.

Mann G.B., Fowler K.J., Gabriel A., Nice E.C., Williams R.L., and Dunn A.R. 1993. Mice with a null mutation of the TGFα gene have abnormal skin architecture, wavy hair, and curly whiskers and often develop corneal inflammation. *Cell* **73:** 249–261.

Mann J.R. and Lovell-Badge R.H. 1984. Inviability of parthenogenones is determined by pronuclei, not egg cytoplasm. *Nature* **310:** 66–67.

Mao X., Fujiwara Y., and Orkin S.H. 1999. Improved reporter strain for monitoring Cre recombinase-mediated DNA excisions in mice. *Proc. Natl. Acad. Sci.* **96:** 5037–5042.

Martin G.R. 1981. Isolation of a pluripotent cell line from early mouse embryos cultured in medium conditioned by teratocarcinoma stem cells. *Proc. Natl. Acad. Sci.* **78:** 7634–7638.

Mayr E. 1982. *The growth of biological thought, diversity, evolution and inheritance.* Belknap Press of Harvard University Press, Cambridge.

McCaffrey A.P., Meuse L., Pham T.T., Conklin D.S., Hannon G.J., and Kay M.A. 2002. RNA interference in adult mice. *Nature* **418:** 38–39.

McDonald J.D., Bode V.C., Dove W.F., and Shedlovsky A. 1990. *Pahhph-5:* A mouse mutant deficient in phenylalanine hydroxylase. *Proc. Natl. Acad. Sci.* **87:** 1965–1967.

McGrath J. and Solter D. 1983. Nuclear transplantation in the mouse embryo by microsurgery and cell fusion. *Science* **220:** 1300–1302.

———. 1984. Completion of mouse embryogenesis requires both the maternal and paternal genomes. *Cell* **37:** 179–183.

McLaren A. 1976. *Mammalian chimeras.* Cambridge University Press, England.

McLaren A. and Biggers J.D. 1958. Successfull development and birth of mice cultivated in vitro as early embryos. *Nature* **182:** 877–878.

McLaren A. and Michie D. 1956. Studies on the transfer of fertilized mouse eggs to uterine foster-mothers. I. Factors affecting the implantation survival of native and transferred eggs. *J. Exp. Biol.* **33:** 394–416.

Medico E., Gambarotta G., Gentile A., Comoglio P.M., and Soriano P. 2001. A gene trap vector system for identifying transcriptionally responsive genes. *Nat. Biotechnol.* **19:** 579–582.

Mercer J.A., Seperack P.K., Strobel M.C., Copeland N.G., and Jenkins N.A. 1991. Novel myosin heavy chain encoded by murine dilute coat colour locus. *Nature* **349:** 709–713.

Mintz B. 1962. Formation of genotypically mosaic mouse embryos. *Am. Zool.* **2:** 432.

———. 1965. Genetic mosaicism in adult mice of quadriparental lineage. *Science* **148:** 1232–1233.

Mintz B. and Illmensee K. 1975. Normal genetically mosaic mice produced from malignant teratocarcinoma cells. *Proc. Natl. Acad. Sci.* **72:** 3585–3589.

Morse H.C. 1978. *Origins of inbred mice.* Academic Press, New York.

———. 1981. The laboratory mouse—A historical perspective. In *The mouse in the biomedical research. History, genetics and wild mice* (ed. Foster H.L. et al.), vol. 1, pp. 1–16.

Academic Press, New York.

Munroe R.J., Bergstrom R.A., Zheng Q.Y., Libby B., Smith R., John S.W., Schimenti K.J., Browning V.L., and Schimenti J.C. 2000. Mouse mutants from chemically mutagenized embryonic stem cells. *Nat. Genet.* **24:** 318–321.

Nagy A. 2000. Cre recombinase: The universal reagent for genome tailoring. *Genesis* **26:** 99–109.

Needham J. 1959. *A history of embryology.* Cambridge University Press, England.

Nielsen J.T., and Chapman V.M. 1977. Electrophoretic variation for x-chromosome-linked phosphoglycerate kinase (PGK-1) in the mouse. *Genetics* **87:** 319–325.

Nishioka Y. 1987. Y-chromosomal DNA polymorphism in mouse inbred strains. *Genet. Res.* **50:** 69–72.

Nolan P.M., Peters J., Strivens M., Rogers D., Hagan J., Spurr N., Gray I.C., Vizor L., Brooker D., Whitehill E., Washbourne R., Hough T., Greenaway S., Hewitt M., Liu X., McCormack S., Pickford K., Selley R., Wells C., Tymowska-Lalanne Z., Roby P., Glenister P., Thornton C., Thaung C., Stevenson J.A., Arkell R., Mburu P., Hardisty R., Kiernan A., Erven A., Steel K.P., Voegeling S., Guenet J.L., Nickols C., Sadri R., Nasse M., Isaacs A., Davies K., Browne M., Fisher E.M., Martin J., Rastan S., Brown S.D., and Hunter J. 2000. A systematic, genome-wide, phenotype-driven mutagenesis programme for gene function studies in the mouse. *Nat. Genet.* **25:** 440–443.

Nomura T., Esaki K., and Tomita T. 1985. *ICLAS manual for genetic monitoring of inbred mice.* University of Tokyo Press, Japan.

Novak A., Guo C., Yang W., Nagy A., and Lobe C.G. 2000. Z/EG, a double reporter mouse line that expresses enhanced green fluorescent protein upon Cre-mediated excision. *Genesis* **28:** 147–155.

O'Brien, S.J., J.E. Womack, L.A. Lyons, K.J. Moore, and N.A. Jenkins. 1994. Anchored reference loci for comparative genome mapping in mammals. *Nature Genet.* **3:** 103–112.

Oppenheimer J.M. 1967. *Essays on the history of embryology and biology.* The MIT Press, Cambridge.

Osumi N. and Inoue T. 2001. Gene transfer into cultured mammalian embryos by electroporation. *Methods* **24:** 35–42.

Papaioannou V.E. 2001. T-box genes in development: From hydra to humans. *Int. Rev. Cytol.* **207:** 1–70.

Papaioannou V.E., McBurney M.W., Gardner R.L., and Evans M.J. 1975. Fate of teratocarcinoma cells injected into early mouse embryos. *Nature* **258:** 70–73.

Pincus G. 1936. *The eggs of mammals.* Macmillan, New York.

Popp R.A., Bailiff E.G., Skow L.C., Johnson F.M., and Lewis S.E. 1983. Analysis of a mouse alpha-globin gene mutation induced by ethylnitrosourea. *Genetics* **105:** 157–167.

Ramirez-Solis R., Liu P., and Bradley A. 1995. Chromosome engineering in mice. *Nature* **378:** 720–724.

Rinchik E.M., Russell L.B., Copeland N.G., and Jenkins N.A. 1985. The dilute short ear (*dse*) complex of the mouse: Lessons from a fancy mutation. *Trends Genet.* **1:** 170–176.

Robert B., Barton P., Minty A., Daubas P., Weydert A., Bonhomme F., Catalan J., Chazottes D., Guenet J.L., and Buckingham M. 1985. Investigation of genetic linkage between myosin and actin genes using an interspecific mouse back-cross. *Nature* **314:** 181–183.

Robertson E., Bradley A., Kuehn M., and Evans M. 1986. Germ-line transmission of genes introduced into cultured pluripotential cells by retroviral vector. *Nature* **323:** 445–448.

Rossant J. and Papaioannou V.E. 1977. The biology of embryogenesis. In *Concepts in mammalian embryogenesis* (ed. Sherman M.I.), pp. 1–36. The MIT Press, Cambridge.

Russell E.S. 1954. One man's influence: A tribute to William Ernest Castle. *J. Hered.* **45:** 210–213.

———. 1978. Origins and history of mouse inbred strains: Contribution of Clarence Cook Little. In *Origins of inbred mice* (ed. Morse H.C.), pp. 33–43. Academic Press, New York.

Russell L.B. 1971. Definition of functional units in a small chromosomal segment of the mouse and its use in interpreting the nature of radiation-induced mutations. *Mutat. Res.* **11:** 107–123.

Schimenti J. 2000. Segregation distortion of mouse *t* haplotypes: The molecular basis emerges. *Trends Genet.* **16:** 240–243.

Schumacher A., Faust C., and Magnuson T. 1996. Positional cloning of a global regulator of anterior-posterior patterning in mice. *Nature* **384:** 648.

Schulte-Merker S., Ho R.K., Herrmann B.G., and Nüsslein-Volhard C. 1992. The protein product of the zebrafish homologue of the mouse T gene is expressed in nuclei of the germ ring and the notochord of the early embryo. *Development* **116:** 1021–1032.

Schwartzberg P.L., Goff S.P., and Robertson E.J. 1989. Germ-line transmission of a c-*abl* mutation produced by targeted gene disruption in ES cells. *Science* **246:** 799–803.

Shimkin M.B. 1975. A. E. C. Lathrop (1868–1918): Mouse woman of Granby. *Cancer Res.* **35:** 1597–1598.

Shin M.K., Levorse J.M., Ingram R.S., and Tilghman S.M. 1999. The temporal requirement for endothelin receptor-B signalling during neural crest development. *Nature* **402:** 496–501.

Shine I. and Wrobel S. 1976. *Thomas Hunt Morgan, pioneer of genetics.* University of Kentucky Press, Lexington.

Silver L.M. 1988. Mouse *t* haplotypes: A tale of tails and a misunderstood selfish chromosome. *Curr. Top. Microbiol. Immunol.* **137:** 64–69.

———. 1993a. The peculiar journey of a selfish chromosome: Mouse t haplotypes and meiotic drive. *Trends Genet.* **9:** 250–254.

———. 1993b. Record keeping and database analysis of breeding colonies. In *Guide to techniques in mouse development methods in enzymology* (ed. Wassarman P.M. and DePamphilis M.L.), vol. 225, pp. 3–5. Academic Press, San Diego.

———. 1994. *Mouse genetics: Concepts and applications.* Oxford University Press, England.

Skarnes W.C., Auerbach B.A., and Joyner A.L. 1992. A gene trap approach in mouse embryonic stem cells: The lacZ reported is activated by splicing, reflects endogenous gene expression, and is mutagenic in mice. *Genes Dev.* **6:** 903–918.

Skarnes W.C., Moss J.E., Hurtley S.M., and Beddington R.S. 1995. Capturing genes encoding membrane and secreted proteins important for mouse development. *Proc. Natl. Acad. Sci.* **92:** 6592–6596.

Smith A.J., De Sousa M.A., Kwabi-Addo B., Heppell-Parton A., Impey H., and Rabbitts P. 1995. A site-directed chromosomal translocation induced in embryonic stem cells by Cre-*loxP* recombination. *Nat. Genet.* **9:** 376–385.

Smith J.C., Price B.M., Green J.B., Weigel D., and Herrmann B.G. 1991. Expression of a *Xenopus* homolog of Brachyury (T) is an immediate-early response to mesoderm induction. *Cell* **67:** 79–87.

Snell G.D. 1978. Congenic resistant strains of mice. In *Origins of inbred mice* (ed. Morse H.C.), pp. 119–155. Academic Press, New York.

Soriano P. 1999. Generalized *lacZ* expression with the ROSA26 Cre reporter strain. *Nat. Genet.* **21:** 70–71.

Srinivas S., Watanabe T., Lin C.S., William C.M., Tanabe Y., Jessell T.M., and Costantini F. 2001. Cre reporter strains produced by targeted insertion of EYFP and ECFP into the ROSA26 locus. *BMC Dev. Biol.* **1:** 4.

Stanford W.L., Cohn J.B., and Cordes S.P. 2001. Gene-trap mutagenesis: Past, present and beyond. *Nat. Rev. Genet.* **2:** 756–768.

Stevens L.C. and Little C.C. 1954. Spontaneous testicular teratomas in an inbred strain of mice. *Proc. Natl. Acad. Sci.* **40:** 1080–1087.

Stewart T.A. and Mintz B. 1982. Recurrent germ-line transmission of the teratocarcinoma genome from the METT-1 culture line to progeny in vivo. *J. Exp. Zool.* **224:** 465–469.

Strong L.C. 1978. Inbred mice in science. In *Origins of inbred mice* (ed. Morse H.C.), pp. 45–66. Academic Press, New York.

Su L.K., Kinzler K.W., Vogelstein B., Preisinger A.C., Moser A.R., Luongo C., Gould K.A., and Dove W.F. 1992. Multiple intestinal neoplasia caused by a mutation in the murine homolog of the APC gene. *Science* **256:** 668–670.

Surani M.A., Barton S.C., and Norris M.L. 1984. Development of reconstituted mouse eggs suggests imprinting of the genome during gametogenesis. *Nature* **308:** 548–550.

Takahashi J.S., Pinto L.H., and Vitaterna M.H. 1994. Forward and reverse genetic approaches to behavior in the mouse. *Science* **264:** 1724–1733.

Tarkowski A.K. 1959. Experimental studies on regulation in the development of isolated blastomeres of noues eggs. *Acta Theriol.* **3:** 191–267.

———. 1961. Mouse chimeras developed from fused eggs. *Nature* **184:** 1286–1287.

Thomas K.R. and Capecchi M.R. 1987. Site-directed mutagenesis by gene targeting in mouse embryo-derived stem cells. *Cell* **51:** 503–512.

Thompson S., Clarke A.R., Pow A.M., Hooper M.L., and Melton D.W. 1989. Germ line transmission and expression of a corrected HPRT gene produced by gene targeting in embryonic stem cells. *Cell* **56:** 313–321.

Vallis K.A., Chen Z., Stanford W.L., Yu M., Hill R.P., and Bernstein A. 2002. Identification of radiation-responsive genes in vitro using a gene trap strategy predicts for modulation of expression by radiation in vivo. *Radiat. Res.* **157:** 8–18.

Van Beneden M.E. 1875. La maturation de l'oeuf, la fecondation, et les premieres phases du developpment embryonnaire de mammifere d'apres de recherches faites chez le lapin. *Bull. Acad. R. Belg. Cl. Sci.* **40:** 686–736.

Vitaterna M.H., King D.P., Chang A.M., Kornhauser J.M., Lowrey P.L., McDonald J.D., Dove W.F., Pinto L.H., Turek F.W., and Takahashi J.S. 1994. Mutagenesis and mapping of a mouse gene, *Clock*, essential for circadian behavior. *Science* **264:** 719–725.

von Melchner H., DeGregori J.V., Rayburn H., Reddy S., Friedel C., and Ruley H.E. 1992. Selective disruption of genes expressed in totipotent embryonal stem cells. *Genes Dev.* **6:** 919–927.

Wagner E.F., Stewart T.A., and Mintz B. 1981. The human beta-globin gene and a functional viral thymidine kinase gene in developing mice. *Proc. Natl. Acad. Sci.* **78:** 5016–5020.

Wagner T.E., Hoppe P.C., Jollick J.D., Scholl D.R., Hodinka R.L., and Gault J.B. 1981. Microinjection of a rabbit beta-globin gene into zygotes and its subsequent expression in adult mice and their offspring. *Proc. Natl. Acad. Sci.* **78:** 6376–6380.

Wakayama T., Perry A.C., Zuccotti M., Johnson K.R., and Yanagimachi R. 1998. Full-term development of mice from enucleated oocytes injected with cumulus cell nuclei. *Nature* **394:** 369–374.

Whitten W.K. 1956. Culture of tubal mouse ova. *Nature* **177:** 96.

———. 1957. Culture of tubal ova. *Nature* **179:** 1081–1082.

Wiles M.V., Vauti F., Otte J., Fuchtbauer E.M., Ruiz P., Fuchtbauer A., Arnold H.H., Lehrach H., Metz T., von Melchner H., and Wurst W. 2000. Establishment of a gene-trap sequence tag library to generate mutant mice from embryonic stem cells. *Nat. Genet.* **24:** 13–14.

Woychik R.P. and Alagramam K. 1998. Insertional mutagenesis in transgenic mice generated by the pronuclear microinjection procedure. *Int. J. Dev. Biol.* **42:** 1009–1017.

Woychik R.P., Maas R.L., Zeller R., Vogt T.F., and Leder P. 1990. "Formins": Proteins deduced from the alternative transcripts of the limb deformity gene. *Nature* **346:** 850–853.

Woychik R.P., Stewart T.A., Davis L.G., D'Eustachio P., and Leder P. 1985. An inherited limb deformity created by insertional mutagenesis in a transgenic mouse. *Nature* **318:** 36–40.

Yasuo H. and Satoh N. 1993. Function of vertebrate T gene. *Nature* **364:** 582–583.

You Y., Bergstrom R., Klemm M., Lederman B., Nelson H., Ticknor C., Jaenisch R., and Schimenti J. 1997. Chromosomal deletion complexes in mice by radiation of embryonic stem cells. *Nat. Genet.* **15:** 285–288.

Yu Y. and Bradley A. 2001. Engineering chromosomal rearrangements in mice. *Nat. Rev. Genet.* **2:** 780–790.

Zambrowicz B.P., Friedrich G.A., Buxton E.C., Lilleberg S.L., Person C., and Sands A.T. 1998. Disruption and sequence identification of 2,000 genes in mouse embryonic stem cells. *Nature* **392:** 608–611.

Zheng B., Sage M., Cai W.W., Thompson D.M., Tavsanli B.C., Cheah Y.C., and Bradley A. 1999. Engineering a mouse balancer chromosome. *Nat. Genet.* **22:** 375–378.

Zhou X., Sasaki H., Lowe L., Hogan B.L., and Kuehn M.R. 1993. *Nodal* is a novel TGF-β-like gene expressed in the mouse node during gastrulation. *Nature* **361:** 543–547.

Zijlstra M., Li E., Sajjadi F., Subramani S., and Jaenisch R. 1989. Germ-line transmission of a disrupted β2-microglobulin gene produced by homologous recombination in embryonic stem cells. *Nature* **342:** 435–438.

Summary of Mouse Development

THIS CHAPTER PROVIDES A GENERAL overview of mouse development for those researchers who are completely new to the field of mouse embryology. Like other specialized fields, the field of mouse embryology has a distinct language comprising many unique terms that must be learned to understand the specific details of how a mouse embryo forms. Furthermore, understanding how a mouse embryo develops normally is essential for determining how defects arise in a mutant. This knowledge can be used to help distinguish primary defects from secondary abnormalities. For detailed accounts of mouse development, some excellent textbooks are listed in Chapter 17. We also provide a very short section on the genetics of coat color for researchers who become interested in this topic after handling mice for the first time. Finally, a few illustrations of normal and mutant mice are shown to provide at least some idea of the genetic variation that exists in the laboratory mouse.

CONTENTS

MOUSE DEVELOPMENT

Embryonic development of the mouse begins with fertilization of the oocyte by the sperm. One important feature of mouse embryogenesis is that early development is much slower than in such model organisms as the sea urchin, *Drosophila*, and *Xenopus*. By 24 hours after fertilization, embryos of these species are well on their way to becoming free-living, feeding larvae, and they contain more than 60,000 cells, organized into many different tissue layers. In contrast, the mouse embryo is still at the two-cell stage and will continue to divide slowly without any increase in mass as it moves along the oviduct into the uterus for implantation 4.5 days after fertilization. This slow development allows the uterine tissue time to prepare for receiving the embryo. The embryo, in its turn, generates the first two tissue lineages (the trophectoderm and the primitive endoderm) that make major contributions to the placenta and the extraembryonic yolk sacs required for successful interaction with the mother. Once implantation has been achieved, a dramatic increase in the growth rate of the embryo occurs, particularly in the small group of pluripotential cells known as the epiblast, or primitive ectoderm, from which the fetus will develop. The epiblast is in many ways the equivalent of the cellular blastoderm of *Drosophila*, or the blastodisc of the chick. Between the fifth and tenth day after fertilization, the three primary germ layers—the ectoderm, mesoderm, and definitive endoderm—are formed as a result of gastrulation, and the basic body plan and organ primordia of the future mouse are established. To summarize briefly, the notochord is laid down along the midline of the anteroposterior axis, and the paraxial mesoderm on either side of the notochord is divided into reiterated pairs of somite blocks, generating an obvious segmented pattern, while the intermediate and lateral plate mesoderm remain unsegmented. The neural plate is induced and folds up into the neural tube, which is subdivided into forebrain, midbrain, hindbrain, and spinal cord. The placodes of the nose, ear, and lens are formed from the surface ectoderm. The neural crest cells start their migration, and the heart and circulatory system and limb buds are established. It is during this early postimplantation period that many of the genes controlling the differentiation and morphogenesis of the adult organs are gradually brought into play.

The gestation period for the mouse embryo is 19–20 days, depending on the strain. Figure 2.1 and Table 2.1 show the timing of the different stages, based on the development of F_1 hybrids between C57BL/6 females and CBA males; for some inbred strains, such as C3H, the process is somewhat slower. Figure 2.2 shows a series of whole-mount embryos at representative stages of mouse development. Figure 2.3 is a summary of our current knowledge of tissue lineages in the mouse embryo.

Generally, mouse development is staged by the number of days postcoitum (dpc). In this manual we use the following convention for timing pregnancy and the age of embryos. Assuming that fertilization takes place around midnight during a 7 p.m. to 5 a.m. dark cycle, then at noon on the following day (i.e., the day on which the vaginal plug is found), the embryos are aged "half-day postcoitum" or "0.5 dpc." According to this convention, the day on which the plug is found is day 1 of pregnancy. At noon on the next day, the embryos are 1.5 dpc, and so on. An 11.0-dpc embryo would have been dissected around midnight of the 11th day of pregnancy. Embryos within a litter can vary with regard to stage of development. Therefore, morphological criteria are used when more precise embryo staging is required. A morphological staging system of mouse embryonic development was developed by Theiler (see Table 2.1) (Theiler 1972, 1983). A detailed morphological staging system of gastrulation-stage mouse embryos was developed by Downs and Davies (1993). Embryos at somite stages can be staged by the number of somite pairs. At later somite stages, it is possible to stage the

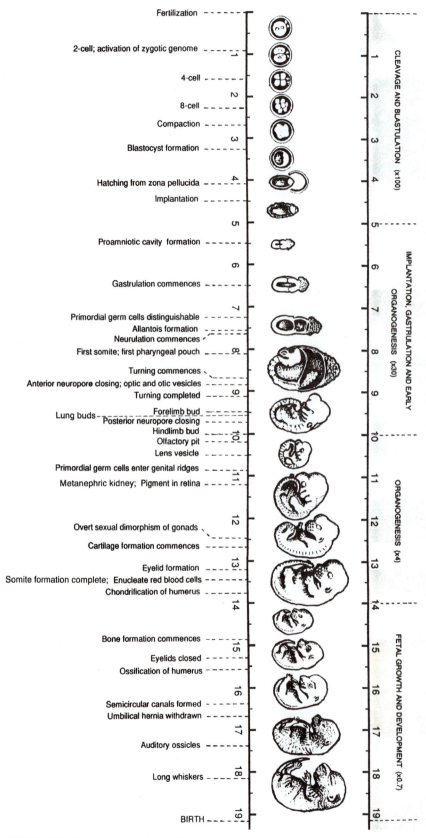

FIGURE 2.1. Time course of development. (*0–5 days*) Cleavage and blastulation; (*5–10 days*) implantation, gastrulation, and early organogenesis; (*10–14 days*) organogenesis; (*14–19 days*) fetal growth and development. For more details, see Table 2.1.

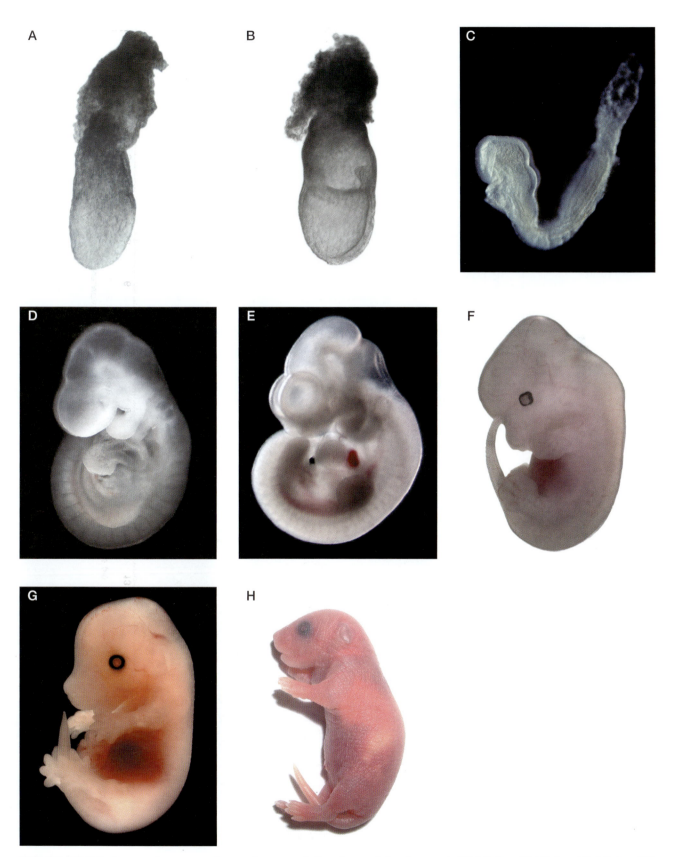

FIGURE 2.2. Whole-mount mouse embryos at representative stages of development. (*A*) 6.5 dpc. (*B*) 7.5 dpc; (*C*) 8.5 dpc; (*D*) 9.5 dpc; (*E*) 10.5 dpc; (*F*) 12.5 dpc; (*G*) 14.5 dpc; (*H*) newborn. Embryos shown in *C-G* have been dissected from their extraembryonic membranes. See Figure 2.17 for anatomical details.

TABLE 2.1. Development of the mouse embryo

Stage[a]	Age (dpc)	Features	System				
			extraembryonic	circulation	intestinal tract	nervous/sensory	urogenital
1	0–1	one-cell zygote					
2	1	two-cell embryo					
3	2	morula, 4–16 cells					
4	3	morula-blastocyst	trophectoderm formed				
5	4	free blastocyst without zona					
6	4.5	implanting blastocyst	primitive endoderm				
7	5	prestreak					
8	6	proamniotic cavity in primitive ectoderm	Reichert's membrane forming anterior visceral endoderm	ectoplacental cone fills with maternal blood			
9	6.5	embryonic axes morphologically obvious					
10	7	early–mid-primitive streak	amnion forming allantois appearing				
11	7.5	late primitve streak	allantois contracts chorion			neural plate	
12	8	1–7 somites		bloodislands in visceral yolk sac first aortic arch	foregut pocket		germ cells near base of allantois pronephros
13	8.5	8–12 somites; turning of embryo		paired heart primordia fusing anteriorly	hindgut pocket thyroid rudiment, second pharyngeal pouch, hepatic diverticulum	neural folds, otic placode neural folds close at level of somites 4–5	
14	9	13–20 somites	blood circulates in visceral yolk sac	heart begins to beat, three paired aortic arches	oral plate ruptures	anterior neuropore closes, olfactory placode	pronephric duct still solid

15	9.5	21-29 somites; forelimb bud at level of somites 8-12	common ventricle and atrium, dorsal aortae fused		posterior neuropore closes, otic vesicle	
16	10	30-34 somites; hindlimb bud at level of somites 23-28	primary bronchi	lung primorida pancreas evagination, vitelline duct closed	lens placode	Wolffian ducts contact cloaca in older specimens
17	10.5	35-39 somites; tail rudiment	sixth aortic arch	umbilical loop, cloacal membrane	deep lens pit	mesonephric tubules
18	11	40-44 somites	spleen primordium		lens vesicle closing, rims of olfactory placode fusing	distinct genital ridge
19	11.5	6-7 mm, forefoot plate	partitioned atrium, unpaired ventricle	bucconasal membrane	lens vesicle detached	ureteric buds Müllerian ducts
20	12	7-9 mm, hindfoot plate	partition of arterial trunk begins	tongue, thymus and parathyroid primordium	pineal body evaginates	sexual differentiation gonads in older specimens
21	13	9-10 mm, whisker rudiments	aortic and pulmonary trunks separated	palatine processes vertical, dental laminae	lens solid	cloaca subdivided
22	14	11-12 mm	interventricular septum closed		ganglionic cells of retina	separate opening of ureter into urogenital sinus
23	15	12-14 mm	palatine processes fused	coronary vessels		
24	16	14-17	Reichert's membrane breaks down	reposition of umbilical hernia	eyelids fusing	
25	17	17-20 mm	alveolar ducts of lung	ciliary body delineated		large central glomeruli in kidney
26	18	19.5-22.5	pancreatic islands of Langerhans	iris and ciliary body	solid cord of prostate cells	
27	19	23-27 mm	birth		testis cords still solid	

[a]Adapted from Thieler (1972, 1983) based on the development of C57BL/6 × CBA F_1 hybrids. Embryos of some inbred lines of mice may develop more slowly.

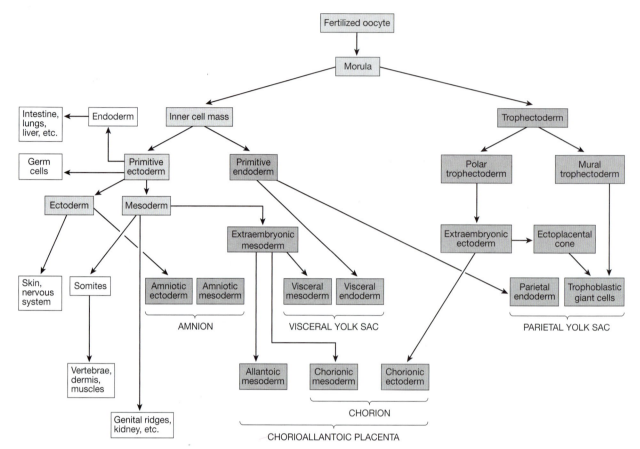

FIGURE 2.3. Summary of the lineages of tissues constituting the mouse embryo. (*Light gray areas*) All tissues that will give rise to the embryo proper and extraembryonic cells; (*dark gray areas*) extraembryonic tissues; (*open areas*) tissues of the embryo proper. (Adapted from Gardner 1983.)

embryos by the number of tail somite pairs (i.e., the number of somite pairs caudal to the posterior limit of the hindlimb bud). The day of birth is considered 1 day postpartum (dpp).

Although embryonic development starts with fertilization, both the oocyte and the sperm are themselves products of complex maturational processes initiated when the primordial germ cells enter the genital ridges. This summary of mouse development thus begins with a description of the origin and growth of the germ cells, which are among the most fascinating cells in the entire organism.

Origin of the Germ Line

The primordial germ cells (PGCs) are derived from the epiblast (Falconer and Avery 1978; Gardner et al. 1985; Lawson and Hage 1994). There is no evidence for a segregated germ line in the inner cell mass of blastocysts up to 4.5 dpc because single epiblast cells at this stage can give rise to both somatic tissue and gametes (Gardner et al. 1985). Cell fate mapping studies have shown that the proximal epiblast adjacent to the extraembryonic ectoderm is the region of the prestreak-stage embryo in which PGCs are derived (Lawson and Hage 1994). However, at this stage, the PGC lineage is still not yet distinct because single cell-labeling experiments show that the cellular progeny contain both PGCs and somatic cell types (Lawson and Hage 1994). PGCs are first detected during gastrulation at 7.0 dpc as a distinct cell population located in the extraembryonic mesoderm of the posterior amniotic fold (Fig. 2.4B) (Ginsburg et al. 1990). Subsequently, they come to underlie the posterior part of the primitive streak (Ozdenski 1967) and become incorporated into the base of the allantois, forming a coherent cluster of ~75 cells at the headfold (presomite) stage (Fig. 2.4A). PGCs are distinguished by their large round shape and high expression levels of tissue-nonspecific alkaline phos-

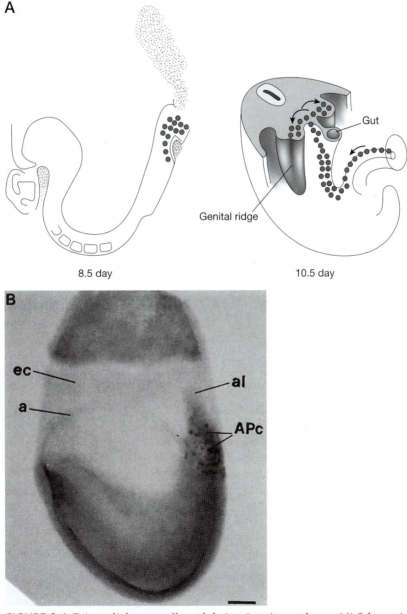

FIGURE 2.4. Primordial germ cells and their migration pathway. (*A*) Schematic representation of the localization of PGCs at the base of the allantois around the hindgut pocket in an 8.5-dpc embryo, and the migration of PGCs along the hindgut and into the genital ridges in a 10.5-dpc embryo. (*B*) 7.5-dpc embryo (neural-fold stage) stained in whole mount for alkaline phosphatase activity. Note the cluster of PGCs around the base of the allantois. (Reprinted, with permission, from Ginsburg et al. 1990.)

phatase (TNAP) (Hahnel et al. 1990) and high levels of *Pou5f1* (*Oct3/4*) mRNA (Rosner et al. 1990; Scholer et al. 1990a,b). Evidence that germ cells may have segregated as a separate lineage by mid-gastrulation comes from experiments in which small fractions of the 7.0- and 7.5-dpc embryonic region of the gastrula, composed of all three germ layers, were isolated in culture. The ability of these fractions to generate large, alkaline-phosphatase-positive cells was assessed. These studies showed that the potential to give rise to PGCs is restricted to the posterior aspect of the embryo (Snow 1981).

Alkaline phosphatase staining has also been used to follow the complex migration of PGCs from the base of the allantois to the genital ridges (Figs. 2.4 and 2.5) (Clark and Eddy 1975; Eddy et al. 1981; Eddy and Hahnel 1983). More recently, a green fluorescent protein (GFP) reporter has been used to follow PGC cell movements in vivo (Fig. 2.5) (Molyneaux et al. 2001). At 8.0 dpc, the small nest of PGCs begins to disperse, and by the early somite stage, the majority of alkaline-phosphatase-positive cells (~200 in all) are found intercalated into the epithelium of the hindgut (Ginsburg et al. 1990). The PGCs leave the gut endoderm and traverse the dorsal mesentery toward the coelomic angles, the first cells reaching the genital ridges by 10.5–11.5 dpc (Figs. 2.4A and 2.5). By 13 dpc, ~25,000 PGCs have colonized each gonad primordium. The genital ridges arise from intermediate mesoderm adjacent to the mesonephros and are first visible as a distinct urogenital ridge at ~10.0 dpc (see Lateral Plate and Intermediate Mesoderm and Their Derivatives, p. 90). However, the first signs of overt sexual dimorphism in the gonadal primordia are not apparent until 12.5 dpc. During their migration from the allantois, PGCs divide approximately once every 16 hours (Tam and Snow 1981). Their movement from the gut to the genital ridges involves active migrato-

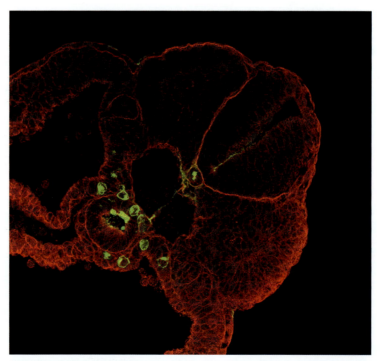

FIGURE 2.5. Primordial germ cell migration. Cross section of a 9.5-dpc mouse embryo through the hindgut showing PGCs immunostained for SSEA1 (*green*). The PGCs will subsequently migrate into the developing gonads.

ry movements, probably dependent on a suitable substratum and possibly involving chemotactic signals to ensure that the majority of cells converge on the genital ridge (Fig. 2.6). Female germ cells that migrate into prospective ovaries cease proliferation and enter meiosis, becoming arrested at prophase of the first meiotic division. In contrast, male germ cells entering prospective testes continue to proliferate until mitotic arrest occurs at about 14.0 dpc. It is not uncommon for some PGCs to lodge in mesonephric tissue or the adrenals adjacent to the genital ridges and thus fail to enter the developing gonads (McLaren 1984). Such ectopic PGCs, regardless of their genetic sex, tend to enter meiotic arrest, characteristic of the female developmental pathway (McLaren 1984). As discussed below, this situation highlights the importance of the gonadal somatic tissue in influencing the meiotic or mitotic behavior of PGCs.

Several mutations are known to affect germ cell proliferation and cause sterility in homozygotes by severely reducing the number of germ cells reaching the genital ridges. The molecular nature of two of these mutations has been defined. *White spotting* (W) and *viable white spotting* (Wv) reduce both the proliferation and migration of primordial germ cells (Mintz and Russell 1957). In contrast, in mice homozygous for mutations in *Steel* (*Sl*), the germ cells migrate toward the genital ridges, but once there, this already depleted population fails to proliferate and all PGCs degenerate (McCoshen and McCallion 1975). Both mutations also affect neural crest cells and progenitors of the hematopoietic system, two other populations of cells that must move as individual cells over long distances in the embryo. The pigment cell descendants of the neural crest from *Sl/Sl*, but not *W/W*, embryos can be rescued by providing them with a wild-type environment (Mayer 1973), suggesting that the Sl product acts non-cell-autonomously. Cloning of the *Sl* gene has confirmed this by showing that the gene product is a peptide-signaling molecule, Steel factor (SF) (now known as kit-ligand [KL], see Witte 1990). The various *W* mutations have been shown to be disruptions of the proto-oncogene c-*kit* (Chabot et al. 1988; Nocka et al. 1989), which is the transmembrane tyrosine kinase receptor for SF/KL and is expressed in PGCs.

Certainly, SF/KL is required for both survival and proliferation of PGCs in vitro, and it has been shown that primordial germ cells can give rise to perma-

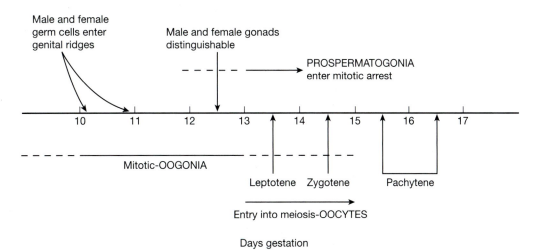

FIGURE 2.6. Timing of female and male germ cell development in the genital ridge and early gonad of the mouse. (Adapted from Monk and McLaren 1981.)

nent pluripotent cell lines in vitro if cultured in the presence of a variety of peptide growth factors, including SF/KL (Matsui et al. 1992; Resnick et al. 1992). These cells, known as embryonic germ (EG) cells (see Fig. 2.33, below), appear to be equivalent to embryonic stem (ES) cells, at least with respect to their developmental potential, capacity to form chimeras, and ability to give rise to functional sperm (Labosky et al. 1994; Stewart et al. 1994).

Sex Determination and Differentiation

The local gonadal environment has a profound influence on germ cell differentiation because, in general, the genetic sex of the gonad determines the type of gamete produced. In XX↔XY chimeras, formed by aggregating morulae (see Chapter 11) or by injecting cells into the blastocyst (see Chapter 11), XX germ cells in a predominantly XX gonadal environment will enter meiotic arrest, typical of the female pathway, whereas in XY gonadal soma, they become mitotically arrested, the characteristic prelude to spermatogenesis (see McLaren 1984). Such experiments, in combination with the study of *W* mutants where the number of primordial germ cells is severely reduced in the genital ridge, have identified the somatic tissue of the gonad as the primary determinant of sexual differentiation. Three principal cell types make up the somatic element of the male gonad: Sertoli cells, Leydig cells, and connective tissue (for review, see Swain and Lovell-Badge 1999). Sertoli cells have been favored as the responsible cell type because they are the first to differentiate. Moreover, in XX↔XY chimeras that are phenotypically male, the vast majority of Sertoli cells are XY, whereas Leydig cells and connective tissue demonstrate more variation in genotypic sex (Burgoyne et al. 1988).

It therefore would seem that a testis-determining gene carried on the Y chromosome (*Tdy*) acts earlier in development than any ovarian determining counterpart to direct the indifferent gonad into testis differentiation. Thus, testis determination is essentially synonymous with sex determination, and *Tdy* has been rechristened *Sry* (sex-determining region Y gene). This gene encodes a DNA-binding protein with an HMG motif (Gubbay et al. 1990; Sinclair et al. 1990). It is expressed at 10.5 dpc, when the urogenital ridge first appears, and during the critical phase (11.5–12.5 dpc) when the indifferent gonad acquires sexually dimorphic characteristics. In addition, its expression is limited to somatic cells, since normal levels of *Sry* are found in *W/W* embryos (Koopman et al. 1990). Furthermore, in situ hybridization has shown that transcription at 11.5 dpc is restricted to the genital ridge and does not extend into the adjacent mesonephros. This candidate transcription factor is thus likely to initiate testis differentiation, thereby inhibiting ovary formation. Support for this idea came from the observation that XX transgenic mice that express *Sry* develop a testis and male phenotype (Koopman et al. 1991) and that mutation of *Sry* causes male to female sex reversal (Lovell-Badge and Robertson 1990).

Normal testis differentiation, and in particular the development of testis cords, requires a contribution of cells derived from the adjacent mesonephric region. Cells from this region migrate into developing testes and contribute to peritubular myoid cells and other interstitial populations (Buehr et al. 1993; Tilmann and Capel 2002). If these cells are prevented from entering the gonad, the testis fails to develop normal cords.

Prior to sex determination of the gonads, both primordia for the male and female reproductive tracts (the Wolffian or mesonephric ducts and Müllerian or paramesonephric ducts) are present in both XX and XY fetuses. In the female, the Wolffian ducts will degenerate, whereas in the male, the Müllerian ducts regress. In the absence of testis formation, the Wolffian duct does not survive, but this degeneration is not dependent on the presence of an ovary, indicating that regression is the default pathway (Austin and Edwards 1981). In the male, testosterone produced by the Leydig cells of the testis ensures survival and differentiation of the Wolffian duct into the vas deferens, seminal vesicles, and epididymis. In addition, anti-Müllerian hormone (AMH, also known as Müllerian inhibiting substance, MIS), a very early product of the Sertoli cells and a member of the transforming growth factor-β (TGF-β) superfamily, causes the regression of the Müllerian ducts (Josso et al. 1998). In female transgenic mice overexpressing AMH, the Müllerian ducts are eliminated, and in male *Amh* knockout mice, the Müllerian ducts differentiate into oviducts and a uterus (Behringer et al. 1990; Behringer et al. 1994).

Spermatogenesis

Spermatogenesis is one of the most exquisite examples of a continuously synchronized and spatially organized sequence of differentiation (for review, see Hecht 1986). Since the time taken for spermatozoa to differentiate from a stem cell is more or less constant (~5 weeks in the mouse), any transverse section of a seminiferous tubule will contain a stereotyped array of cells from the basement lamina to the luminal surface, reflecting succeeding waves of spermatogenesis passing along the tubule. Unlike oogenesis, spermatogenesis relies on a population of true stem cells, which are capable of self-renewal as well as producing progeny for differentiation into spermatozoa (Fig. 2.7) (for review, see Brinster 2002). These stem cells, which are the direct descendants of primordial germ cells, are large cells known as type-A spermatogonia. They first appear about 3–7 days after birth and lie on the basement membrane surrounding the seminiferous tubules. Following division, some daughters of type-A spermatogonia differentiate into intermediate spermatogonia, and these in turn develop into type-B spermatogonia. Type-B spermatogonia, which are also located adjacent to the basement membrane, are smaller cells and behave like a transition population in that they can divide to produce more type-B spermatogonia. Consequently, type-B spermatogonia are more numerous than type A. Type-B spermatogonia enlarge and move away from the basement membrane, toward the lumen of the seminiferous tubule, thereby transforming into primary spermatocytes. It is at this stage that meiosis commences. During the first meiotic prophase, homologous chromosomes, including the X and Y chromosomes, pair, and crossing-over takes place. The products of the first meiotic division are known as secondary spermatocytes, with each nucleus containing 20 chromosomes composed of two sister chromatids. The second meiotic division, in which the sister chromatids separate, produces spermatids containing a haploid genome. No further division occurs, and the subsequent differentiation into mature spermatozoa, which occurs at the luminal surface, involves extrusion of cytoplasm together with extensive differentiation. Finally, mature spermatozoa are released into the lumen, leaving superfluous cytoplasm, known as residual bodies, on the luminal surface.

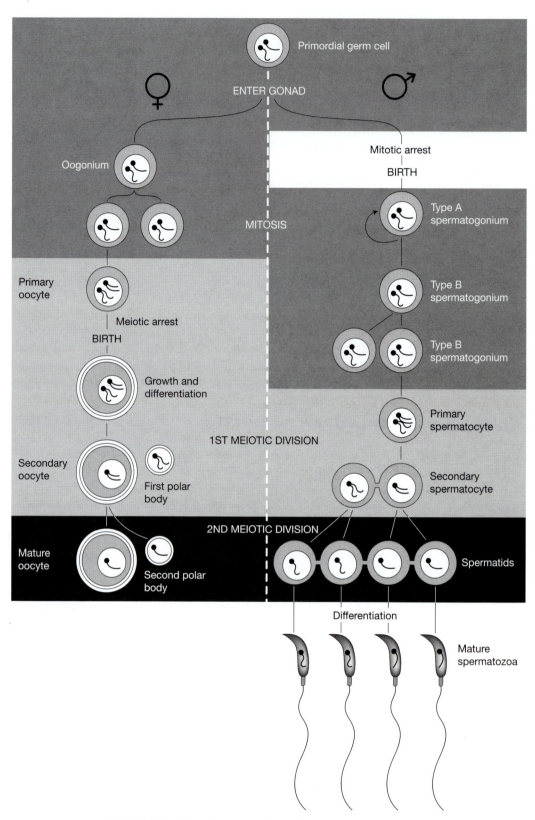

FIGURE 2.7. (*See facing page for legend.*)

The synchrony of spermatogenesis is due in part to incomplete cytokinesis so that all descendants of a type-B spermatogonium remain attached by cytoplasmic bridges, which are only lost when mature spermatozoa are released into the lumen of the seminiferous tubule. This means that although meiosis has already occurred, the complex differentiation required for spermatozoan development occurs within a shared cytoplasm and in the presence of most products of a diploid genome (Braun et al. 1989).

The testis comprises an array of seminiferous tubules that originate from the testis cords apparent in developing male gonads at 12.5 dpc. Germ cells that do not have a Y chromosome are at a selective disadvantage during spermatogenesis. They can, albeit rarely, enter meiosis, but they degenerate before the first meiotic metaphase. Therefore, the Y chromosome, in addition to carrying the sex-determining gene, is thought also to contain a gene or genes essential for normal spermatogenesis (Burgoyne 1987; Delbridge and Graves 1999).

Oogenesis

By 5 days after birth, all oocytes are in the diplotene stage of the prophase of the first meiotic division. They are therefore diploid but contain four times the haploid amount of DNA (4C). During the prolonged resting or dictyate stage, the paired homologous chromosomes are fully extended, and transcription of oocyte (maternal) mRNA takes place. Studies on X chromosome activity have shown that only one X is active in XX primordial germ cells and 11.5-dpc oogonia, but that by 12.5 dpc, both X chromosomes become active (Monk and McLaren 1981; McLaren 1983). In addition, it has been shown that the oocyte genome is globally undermethylated as compared with the sperm genome (Monk et al. 1987; Kafri et al. 1992).

Each oocyte is contained within a follicle composed of multiple layers of follicle or granulosa cells, which are of the same embryonic origin as Sertoli cells of the testis and have various roles in oocyte growth and differentiation (Figs. 2.8 and 2.9) (Richards et al. 1987; Matzuk et al. 2002). The follicle cells immediately surrounding the oocyte have numerous projections that form specialized junctions with the oocyte (Fig. 2.9). These junctional complexes involve gap junctions and allow metabolite transfer. They are maintained even when the follicle cells and oocyte are gradually separated by the deposition of the zona pellucida, a layer of extracellular material synthesized and deposited by the growing oocyte (Bleil and Wassarman 1980a,b; Greve and Wassarman 1985). The zona is com-

FIGURE 2.7. Schematic representation of spermatogenesis and oogenesis in the mouse showing the difference in phases of mitosis and meiosis in males and females. Primordial germ cells (PGCs) first reach the gonads at ~11.0 dpc. In the male, germ cells enter mitotic arrest at 14.0 dpc, whereas female germ cells at this time become arrested in the first meiotic prophase. In the adult female, ovulation and resumption of meiosis are initiated by hormonal stimuli. In the male, type-A spermatogonia are first identified at 3–7 dpp, and thereafter, spermatogenesis is maintained by stem-cell renewal and differentiation. Two homologous chromosomes (one from the father and one from the mother) are shown in the nucleus of the PGC. Sometime between the origin of the PGCs and its maturation in the gonad, parental imprinting of the chromosomes is erased and a new imprint is imposed.

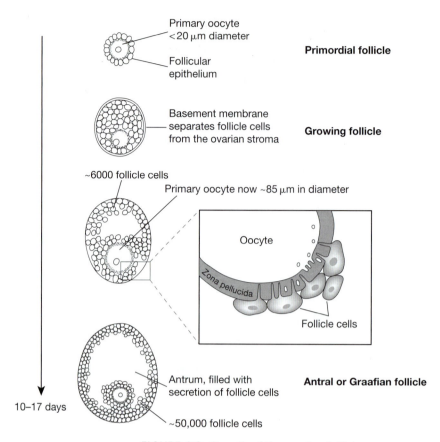

Primary oocyte
<20 μm diameter

Follicular epithelium

Primordial follicle

Basement membrane separates follicle cells from the ovarian stroma

Growing follicle

~6000 follicle cells

Primary oocyte now ~85 μm in diameter

Oocyte

Zona pellucida

Follicle cells

Antrum, filled with secretion of follicle cells

Antral or Graafian follicle

10–17 days

~50,000 follicle cells

FIGURE 2.8. Growth of the ovarian follicle.

posed of three major acidic, sulfated glycoproteins (ZP1, M_r 200,000; ZP2, M_r 120,000; ZP3, M_r 83,000) and reaches a thickness of about 7 μm. It has been shown that ZP3 functions as a sperm receptor and initiates the acrosome reaction, which must occur if a sperm is to fertilize the oocyte. Molecular analysis has shown that the mouse *ZP3* gene encodes a polypeptide of 402 amino acids, which is extensively glycosylated with both N-linked and O-linked complex oligosaccharides (for review, see Wassarman 1990). Sperm binding is localized not to the polypeptide backbone of ZP3, but to oligosaccharides associated with a glycopeptide fragment from the carboxy-terminal region of the intact protein (Rosiere and Wassarman 1992).

Apart from studies on the synthesis and processing of the zona glycoproteins, which together constitute about 10% of the total protein synthesis, relatively little is known about the gene activity of growing oocytes. Several groups have carried out two-dimensional gel electrophoretic analysis of total [^{35}S]methionine-labeled proteins synthesized by maturing oocytes and unfertilized oocytes (Van Blerkom 1981; Howlett and Bolton 1985). In addition, the synthesis of a number of specific proteins has been reported (for review, see Schultz 1986). For example, about 1.3% of the total protein synthesis of oocytes is devoted to tubulin (Schultz et al. 1979) and about 0.9% is devoted to actin in the midgrowth phase (Bachvarova et al. 1989). Transcripts for a variety of genes have been localized in mouse oocytes by in situ hybridization or reverse transcriptase polymerase chain reaction (RT-PCR) (e.g., genes encoding the TGF-β-related pro-

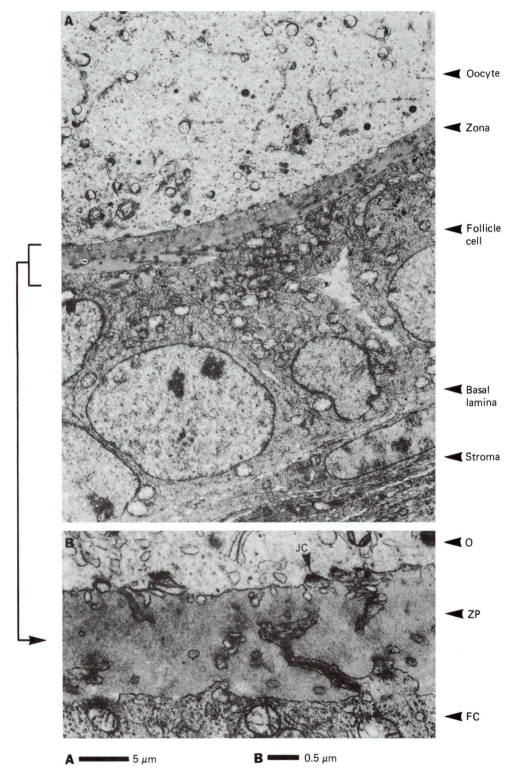

FIGURE 2.9. Relationship between oocyte and follicle cells. (O) Oocyte; (JC) junctional complex; (ZP) zona pellucida; (FC) follicle cell. Bars: (*A*) 5 μm; (*B*) 0.5 μm.

teins, bone morphogenetic protein-6 [BMP6] and activin, and the transcription factor, Pou5f1 [Oct3/4]), but the biological significance of these RNAs, and whether they are translated into functional protein, is not always clear (Lyons et al. 1989; Scholer et al. 1990a,b; Albano et al. 1993). In some cases, it is known that RNAs accumulated in the growing oocyte are not polyadenylated or translated until meiotic maturation (e.g., transcripts for c-*mos* and tissue plasminogen activator; see below). In other cases, strong evidence exists that genes expressed in oocytes have a role in their growth and development. For example, it has been shown that growing oocytes (as well as PGCs) express the transmembrane tyrosine kinase receptor, c-*kit*, the product of the *W* locus. In contrast, somatic cells of the ovary, including the follicle cells, express SF, the product of the *Steel* (*Sl*) locus and the ligand for c-*kit*. Mutations in the *Sl* gene that decrease the expression of SF in the ovary, for example, *Steel panda* (*Sl*pan), inhibit the growth of oocytes and result in a reduced number of follicles and their arrest at the one-layered cuboidal stage (Manova et al. 1990, 1993; Manova and Bachvarova 1991; Huang et al. 1993).

Surprisingly, more than half of the primordial follicles present in the mouse ovary at birth degenerate before 3–5 weeks of age, but little is known about the hormonal and local factors controlling this loss (Faddy et al. 1983). The female mouse reaches sexual maturity at ~6 weeks of age, depending on the strain and environmental conditions. By this time, each ovary contains ~10^4 oocytes at different stages of maturity. Techniques have been established for isolating and culturing both immature oocytes from preantral follicles and mature oocytes from antral follicles (see Chapter 14; Eppig and Telfer 1993; O'Brien et al. 1993). Maturation of oocytes from antral follicles occurs spontaneously under these conditions; they can be fertilized in vitro and develop normally thereafter. Oocytes from preantral follicles, on the other hand, must be cultured for several days with their surrounding follicle cells before they can be fertilized in vitro.

Ovulation

As the oocyte increases in size, it gradually acquires the competence to enter the final stages of meiosis in response to either the correct hormonal stimulus (in vivo) or release from the follicle (in vitro). Ovulation requires the coordinated response of both the follicle cells and the oocyte, and under optimal conditions, it occurs spontaneously once every 4 days. However, cycle length can be influenced by many environmental factors and can be induced artificially by hormone injection (see Chapter 3, Inducing Superovulation). In any one natural cycle, only a few follicles respond to an increase in the level of follicle-stimulating hormone (FSH), which is produced by the pituitary. The stimulated follicle cells break contact with the oocyte and increase their synthesis and secretion of high-molecular-weight proteoglycans and tissue plasminogen activator. At the same time, the follicle accumulates fluid, swells, and moves toward the periphery of the ovary, ready for the final maturation and release of the oocyte. The mature, fluid-filled follicle units are known as antral or graafian follicles, after the scientist Regnier de Graaf who first described them in 1672. For an extensive review of the biosynthetic activity of follicle cells in vivo and in culture, see Hsueh et al. (1984) and Richards et al. (1987).

Ovulation occurs in response to a surge in the level of luteinizing hormone (LH), also produced by the pituitary. After LH stimulation, the oocyte undergoes nuclear maturation (Fig. 2.10). The nucleus (which is also known as the germinal vesicle) loses its membrane (a process known as germinal vesicle breakdown), and the chromosomes assemble on the spindle and move toward the periphery

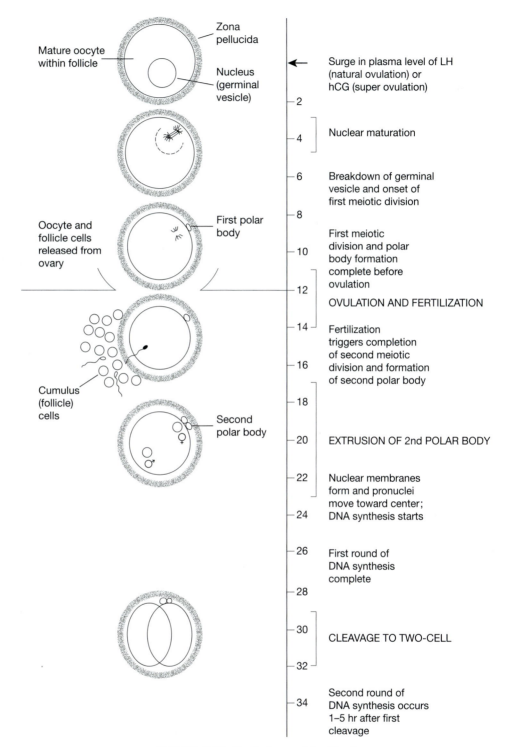

FIGURE 2.10. Ovulation and fertilization.

of the cell where the first meiotic division takes place. One set of homologous chromosomes, surrounded by a small amount of cytoplasm, is extruded as the first polar body, whereas the other set remains in metaphase II. It is in this state of arrest after the first meiotic division that the oocyte is finally released from the follicle, and meiosis does not resume until after fertilization.

Studies in a number of laboratories have implicated the mouse proto-oncogene, c-*mos*, encoding a cytoplasmic serine/threonine protein kinase, in the process of meiosis. Transcripts for c-*mos* accumulate in the growing oocyte but are not polyadenylated or translated until resumption of meiosis and are then degraded by the two-cell stage. Injection of antisense oligonucleotides to c-*mos* into an oocyte that has initiated meiosis results in failure to proceed beyond metaphase II. It has been suggested that the absence of c-Mos protein results in a destabilization of a protein required to maintain meiosis (Mutter et al. 1988; O'Keefe et al. 1989, 1991; Paules et al. 1989). However, in mice lacking functionally active c-*mos*, oocytes undergo parthenogenetic activation, indicating that c-*mos* is required for metaphase arrest at meiosis II (Colledge et al. 1994; Hashimoto et al. 1994).

Meiotic maturation of the oocytes also triggers the synthesis and secretion of the protease, tissue-type plasminogen activator (tPA). The protein is synthesized on mRNA already present in the oocyte but is not translated until meiotic maturation. Activation of tPA mRNA involves addition of a poly(A) tail, and this is controlled by an AU-rich cytoplasmic polyadenylation element (CPE) in the 3′-noncoding region. This CPE has been identified in several maternal mRNAs, in addition to that for tPA (Huarte et al. 1988a,b; Strickland et al. 1988; Sallés et al. 1992).

Each ovulated oocyte is surrounded by its zona and a mass of follicle cells (cumulus cells) with their associated proteoglycan. The oocytes are swept into the open end, or infundibulum, of the oviduct by the action of the numerous cilia on the surface of the oviduct epithelium. Other cells in the epithelium are secretory, and, at the time of ovulation, the section of the tube adjacent to the infundibulum becomes engorged and enlarged to form an amulla where fertilization takes place. In a natural ovulation, 8–12 oocytes are released (depending on the mouse strain), but the process is not synchronous and occurs over a period of 2–3 hours. After ovulation, the follicle cells remaining in the ovary differentiate into steroid-secreting cells (luteinized granulosa cells), which help to maintain pregnancy. Counting the number of bright yellow corpora lutea near the surface of the ovary is a way of determining how many oocytes were, in fact, released. Some properties of the ovulated unfertilized mouse oocyte are given in Table 2.2.

Parthenogenesis

In the mouse, oocytes can be induced to initiate embryonic development in the absence of sperm. Parthenogenetic activation of unfertilized oocytes can be elicited by exposing them to a variety of agents, including alcohol (see Chapter 13), hyaluronidase, the Ca^{++} ionophore A23187, Ca^{++}/Mg^{++}-free medium, heat/cold shock, and anesthetics. In addition, about 10% of the oocytes of the LT/Sv strain of mice undergo spontaneous activation with high frequency, either in the oviduct or in the ovary. Those LT parthenogenetic embryos that implant develop to the gastrula stage (7 dpc) and then become disorganized and die, whereas those remaining in the ovary give rise to teratomas. A similar phenomenon of high-frequency parthenogenetic activation of oocytes resulting in the formation of ovarian teratomas has been reported in female mice deficient in c-*mos* (Colledge et al. 1994; Hashimoto et al. 1994).

TABLE 2.2. Some properties of the ovulated, unfertilized oocyte

Diameter	85 μm
Volume	279 pl (volume of pronucleus 1 pl)
Protein	23 ng
Total DNA	8 pg
Mitochondrial DNA	2–3 pg (note that much of the DNA of the unfertilized oocyte is mitochondrial)
Number of mitochondria	10^5
Genomic DNA (haploid number of chromosomes but diploid [2C] amount of DNA)	6 pg
Ribosomal RNA	0.2–0.4 ng
Poly(A)	0.7 pg (120–200 nucleotides long)
Poly(A)$^+$RNA	exact amount not determined
Transfer RNA	0.14 ng

The genotype of the parthenogenetic embryo (also known as a parthenogenone, parthenogenote, or parthenote) can vary depending on the experimental conditions and, in particular, on the postovulatory age of the activated oocyte. The most important factor may be the state and orientation of the cytoskeletal elements in the oocyte when activation takes place. The following are among the possible genotypes arising from the parthenogenetic activation of an oocyte from an F_1 (heterozygous) female (Fig. 2.11):

1. Uniform haploid (second polar body successfully extruded).

2. Mosaic haploid (second polar body behaves as a normal blastomere).

3. Heterozygous diploid (results from suppression of second polar body formation or from fusion of the pronucleus and the second polar body). The heterozygosity in these parthenogenotes is the result of recombination during meiosis.

4. Homozygous diploid (results from diploidization of the haploid female genome).

Note also that up to 20% of oocytes activated by alcohol may be aneuploid as a result of nondisjunction (Kaufman 1982, 1983b). Most parthenogenotes, and particularly the uniform haploids, die before the blastocyst stage. A minority continue to develop past implantation, for example, up to the gastrula stage (LT/Sv and mosaic haploid parthenogenotes) and the early limb bud stage (heterozygous diploids; Kaufman et al. 1977). However, normal development to term has not been obtained from any class. The reason for this lies in the inactivation of specific genes by imprinting during oogenesis (see Imprinting, p. 104). ES cells can be derived from diploid parthenogenetic blastocysts, and these will differentiate into a wide range of cell types in culture and contribute to the formation of many tissues in chimeras (see Chapter 11).

Fertilization

Approximately 58×10^6 sperm are released into the female reproductive tract per male ejaculation. Some sperm reach the ampulla of the oviduct within 5 minutes, but they are not competent for fertilization for about 1 hour. This process of maturation is known as capacitation, but its mechanism is unknown. To reach the

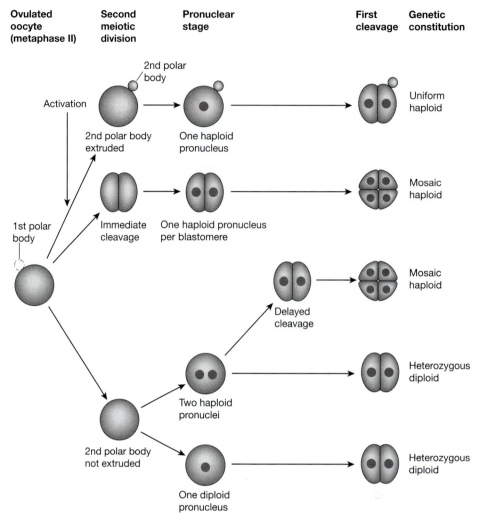

FIGURE 2.11. Possible products of parthenogenetic activation of mouse oocytes. Embryos of different genotype are produced depending on whether or not a second polar body is extruded and on the timing and nature of the first cell division.

surface of the oocyte, the sperm must penetrate first the cumulus mass and then the zona pellucida. As described earlier, the glycoprotein ZP3 has been identified as the sperm-binding protein in the zona. In many mammals, this binding is highly (but not absolutely) species-specific, and this prevents penetration by sperm from other species. ZP3 also triggers the acrosomal reaction, a process in which the acrosome (a secretory vacuole-like structure in the sperm head) fuses with the plasma membrane of the sperm head, releasing various hydrolytic enzymes. Unless the acrosomal reaction takes place, the sperm cannot fertilize the oocyte.

Fusion of the posterior part of the sperm head with the oocyte membrane triggers the cascade of reactions known as fertilization. One very early event is a change in the oocyte surface, inhibiting the fusion of additional sperm. Another event is the Ca^{++}-dependent release (exocytosis) of the cortical granules positioned beneath the plasma membrane. This event initiates the "zona reaction," which involves both cross-linking of the glycoproteins of the zona and modification of the ZP3 glycoprotein so that it no longer binds sperm or elicits the acrosomal reaction. These events also help to prevent polyspermy. During fertilization, the head, midpiece, and a large part of the tail of the sperm are all incorporated into the oocyte cytoplasm. The midpiece of the sperm contributes paternal centrioles and mitochondria to the zygote, but the latter are enormously diluted out by the mitochondria of the oocyte.

Fertilization triggers the second meiotic division and extrusion of the second polar body (see Table 2.3). Nuclear membranes, including nuclear lamin proteins, then form around the maternal and paternal chromosomes, forming separate haploid male and female pronuclei that move toward the center of the oocyte. DNA replication takes place during this migration. The pronuclei do not fuse, but the membranes break down and the chromosomes assemble on the spindle; the first cleavage occurs soon after. Because of asynchrony in ovulation and fertilization, the first cleavage occurs over a number of hours in a population of naturally fertilized zygotes. More synchronous development can be obtained by in vitro fertilization (see Chapter 14). Unfertilized oocytes remain viable for about 12 hours, and sperm for about 6 hours. The sequence of events from onset of nuclear maturation of the unfertilized oocyte to blastocyst formation is illustrated in Figs. 2.10, 2.12, and 2.13.

Recently, evidence has been provided which suggests that the position of second polar body formation and the sperm entry point (SEP) into the fertilized oocyte predetermine two axes of asymmetry in the resulting blastocyst (Piotrowska and Zernicka-Goetz 2001; for review, see Tam et al. 2001). These studies suggest that the embryonic axes (i.e., the body plan) may be determined at fertilization.

Oocyte and Zygote Cytoskeletal Organization

Throughout the cytoplasm of the oocyte, there is a complex matrix of cytoskeletal elements, including actin, tubulin, and certain cytokeratins (Fig. 2.14) (Lehtonen et al. 1983a; Maro et al. 1984; Schatten et al. 1985). The different systems presumably help to coordinate events at the cell surface with changes in the pronuclei as they migrate toward the center of the zygote. This migration is inhibited by both cytochalasin B (which inhibits actin polymerization) and colcemid (which inhibits tubulin polymerization), and both inhibitors are required together in nuclear transfer experiments to enable the nuclei to be withdrawn

TABLE 2.3. Timing of events during the first cell cycle of embryos derived from (C57BL x CBA F_1 x CFLP (outbred) following in vitro fertilization

Event[a]	Hours postinsemination
Extrusion of second polar body	2–5
Formation of pronucleus	4–7
Formation of pronucleus	6–9
DNA replication	11–18
Cleavage	17–20

Information supplied by S. Howlet, Department of Anatomy, University of Cambridge, U.K.
[a]For morphology, see Figure 2.12.

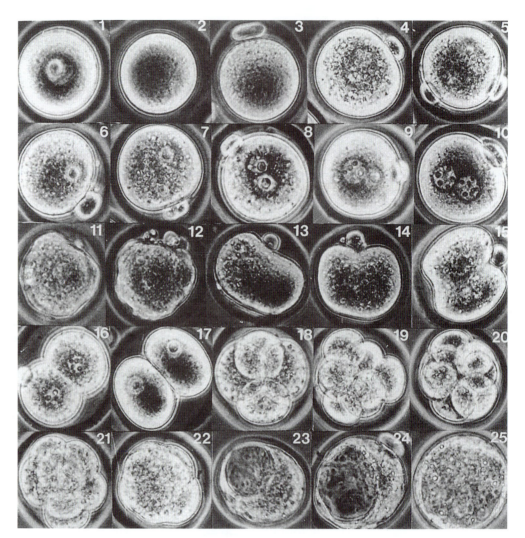

FIGURE 2.12. Morphology of preimplantation mouse development. (*1*) Preovulatory oocyte with germinal vesicle intact. (*2*) Preovulatory oocyte showing breakdown of germinal vesicle (GVBD) (2.5–4.5 hr post-hCG). (*3* and *4*) Extrusion of first polar body (~10 hr post-hCG) followed by ovulation (~11–13.5 hr post-hCG) and fertilization. (*5*) Resumption of meiosis by female set of chromosomes and extrusion of second polar body (occurs over the period 17–33 hr post-hCG). (*5* and *6*) Decondensation of sperm nucleus and formation of male pronucleus. (*7*) Formation of nuclear membrane around haploid set of female chromosomes to form female pronucleus which is subcortical, near the second polar body, and smaller than the male pronucleus (process complete in the majority [75%] of the embryos by ~26 hr post-hCG). (*8*) Migration of pronuclei to center of zygote. (*9* and *10*) Formation of visible nucleoli within both pronuclei. DNA replication (complete in the majority of embryos by ~28 hr post-hCG). (*11* and *12*) Breakdown of pronuclear membranes and disappearances of visible nucleoli. Ruffling of embryo surface indicating reorganization of the cytoskeleton preparatory to cleavage (obscured from ~27 hr post-hCG until cleavage is completed). (*13*) Elongation of embryo. (*14* and *15*) Formation of "waist." (*16*) Newly formed two-cell embryos with visible nucleoli (the majority of embryos have cleaved by ~32 hr post-hCG). (*17*) Later-stage two-cell embryo with nuclei visible. (*18–25*) Later stages of preimplantation development. (*18*) Four-cell embryo. (*19–20*) Six- to eight-cell embryo. (*21*) Compacting eight-cell embryo. (*22*) Compacted 8- to 16-cell embryo. (*23* and *24*) Early blastocysts. (*25*) Fully expanded blastocyst. (All timings are given in hours after injection of hCG. Information and figure provided by Dr. H. Pratt, Department of Anatomy, University of Cambridge.)

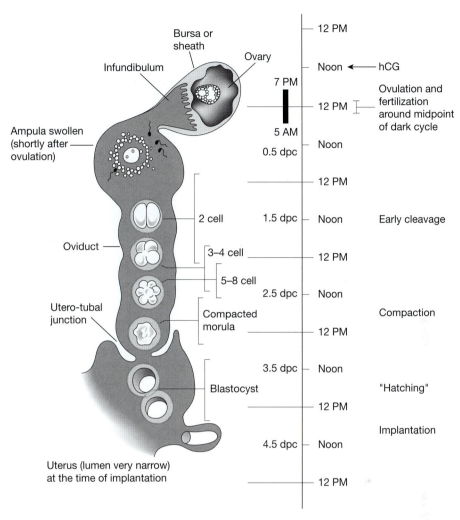

FIGURE 2.13. Summary of preimplantation development.

into a karyoplast (see Chapter 13). The earliest developmental changes in actin organization in the oocyte are seen at fertilization (Maro et al. 1984). In the ovulated oocyte, the plasma membrane above the meiotic spindle is devoid of concanavalin A (Con A)-binding sites and microvilli and is underlaid by an actin-rich subcortical layer. Fertilization results in the formation of a second ConA-free zone, the fertilization cone, which is around the site of sperm entry. The plasma membrane in this region is also underlaid by an actin-rich layer. As the pronuclei move toward the center of the zygote, the distribution of actin filaments becomes more uniform and the ConA-free regions disappear.

Early Cleavage: One-cell Embryo to Eight-cell Uncompacted Morula

Despite the small size of cleavage-stage mouse embryos, a considerable amount of information is available about changes in the pattern of RNA and protein synthesis during preimplantation development (for reviews, see Schultz 1986;

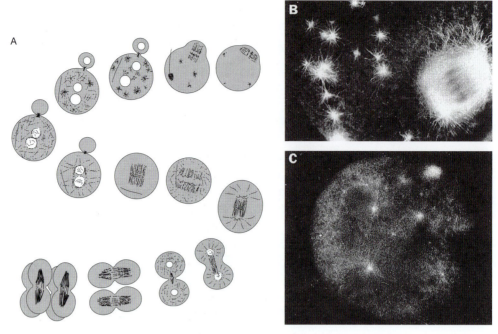

FIGURE 2.14. Distribution of tubulin in early mouse embryos. (*A*) Schematic representation of tubulin distribution from unfertilized oocyte through to eight-cell stage. (*B*) Unfertilized oocyte showing cytoplasmic asters. (*C*) Pronucleus-stage fertilized oocyte. (Photograph provided by Dr. Gerald Schatten, University of Pittsburgh School of Medicine.)

Kidder 1992). In summary, up to the mid two-cell stage (27 hours postfertilization), the embryo appears to rely largely on protein and RNA synthesized during oogenesis. By the mid two-cell stage, many embryonic genes are switched on. Coincidentally, much of the maternally inherited mRNA appears to be degraded rapidly, but maternally coded proteins can persist beyond this time. Three principal methods have been used to study protein synthesis and gene expression in the preimplantation embryo: (1) two-dimensional SDS-polyacrylamide gel electrophoresis of polypeptides produced in vivo or in vitro from embryonic RNA added to the reticulocyte lysate protein synthesis system, (2) amplification of specific embryonic RNAs using techniques based on RT-PCR, and (3) construction and screening of cDNA libraries.

Two-dimensional gel electrophoresis has revealed a number of changes in the pattern of [^{35}S]methionine-labeled proteins synthesized by early-cleavage embryos (see, e.g., Latham et al. 1991). It has been difficult to time these changes precisely, particularly in relation to other cellular events (e.g., DNA synthesis and cell division) because of asynchrony, both within a population of normally fertilized embryos and between blastomeres of individual embryos. This asynchrony can be reduced by in vitro fertilization, by picking out embryos after cleavage to two cells, and by dissociating and recombining groups of blastomeres at specific stages of the cell cycle before metabolic labeling with radioactive precursors.

Several processes may be responsible for the following changes in the pattern of protein synthesis seen after fertilization.

1. Increased turnover rates of some proteins made on stable maternal mRNAs. Evidence for such mechanisms has been found in mammalian oocytes (Howlett and Bolton 1985).

2. Posttranslational modification of proteins synthesized on either maternal or embryonic RNA. Some evidence exists for modification of proteins by phosphorylation, glycosylation, or proteolytic cleavage (Van Blerkom 1981; Cascio and Wassarman 1982; Pratt et al. 1983).

3. Selective use of subspecies of maternal mRNAs. There is clear evidence from the study of carefully timed embryos and from comparison of in vitro and in vivo translation products that some mRNA species are used or suppressed selectively.

4. Specific degradation of maternal mRNAs carried over from the oocyte. At the two-cell stage, there is a sharp fall in the level of total and poly(A)$^+$ RNAs (Clegg and Piko 1983), in the translation of globin mRNA injected into the zygote (Brinster et al. 1980), and in the translation of proteins thought to be encoded by maternal RNA (for review, see Johnson 1981; Pratt et al. 1983). In a particularly clear series of experiments using northern hybridization to embryonic RNA, Giebelhaus et al. (1983) showed a marked loss at the two-cell stage of the pools of actin and histone H3 mRNAs present in the oocyte. Subsequently, the level of these mRNAs in the embryo began to rise after the four-cell stage, when transcription from the embryonic genome is under way.

5. Synthesis of proteins on mRNAs transcribed de novo from the embryonic genome. Experiments in which oocytes are incubated in the presence of the RNA polymerase inhibitor α-amanitin have shown that new RNA synthesis is required for development beyond the two-cell stage and for the synthesis of many new proteins (Flach et al. 1982). Transcription of both ribosomal and poly(A)$^+$ RNAs, which is apparently down-regulated at the time of germinal vesicle breakdown, is thought to resume at a low level around the mid two-cell stage. Timing of the onset of synthesis of paternally coded proteins has come from studies on the expression of genetic enzyme variants or antigens, and the synthesis of a few specific proteins has been followed by metabolic labeling and immunoprecipitation.

ANALYSIS OF EMBRYONIC GENE EXPRESSION USING RT-PCR TECHNIQUES AND cDNA LIBRARIES

The RT-PCR technique has opened up new horizons for the analysis of gene expression in preimplantation mouse embryos. Total RNA isolated from small numbers of oocytes or cleavage-stage embryos, or from single blastocysts, is first used as a template for reverse transcription primed by oligo(dT). Specific sequences in the transcribed cDNA are then amplified by PCR. One of the first applications of this technique to preimplantation mouse embryos was the identification of RNAs for a number of growth factors (Rappolee et al. 1988). Maternal transcripts for platelet-derived growth factor A (PDGF-A) and transforming growth factor-α (TGF-α) were shown to be present in unfertilized, ovulated oocytes and to decline during the two-cell stage, only to reappear as zygotic transcripts in late cleavage–early blastocyst-stage embryos. TGF-β1 transcripts were not present in the oocyte but increased in amount during cleavage. Other studies have used RT-PCR to identify transcripts for insulin-like growth factor II (IGFII), IGFII receptor, IGFI receptor, insulin receptor, and epidermal growth factor (EGF) receptor during preimplantation development (Kidder 1992; Rappolee et al. 1992; Wiley et al. 1992). In many cases, specific antisera have revealed the presence of protein corresponding to the mRNAs found by RT-PCR or by in situ hybridization (Rappolee et al. 1988; Palmieri et al. 1992; Paria et al. 1992; Wiley et al. 1992; Albano et al. 1993). These findings have raised the possibility that growth factors produced by the preimplantation embryo may act in an autocrine loop to promote cell proliferation and/or survival. Although fertilized oocytes can develop into blastocysts in defined medium lacking serum or growth factors, it is now known that certain purified factors added to the medium (e.g., IGFII, TGF-α, and insulin) can significantly enhance the growth rate and final cell number of preimplantation embryos (Harvey and Kaye 1990, 1992; Rappolee et al. 1992).

Another approach to the identification of genes expressed in preimplantation mouse embryos has been the construction and screening of large and representative cDNA libraries from different stages (Weng et al. 1989; Rothstein et al. 1992). Libraries have been screened with probes for known mouse genes or evolutionarily conserved genes from other species, or used for subtractive hybridization to yield cDNAs for genes differentially expressed at two different stages. Recent advances in DNA microarray technologies using probes generated from limiting amounts of tissue hold new promise for identifying differentially expressed genes in early mouse embryos.

Developmental Potential of Embryonic Nuclei

Nuclear transplantation experiments in *Xenopus* have shown that blastula-stage nuclei can support the development of an enucleated oocyte to the tadpole stage. In contrast, nuclear transplantation experiments in the mouse by McGrath and Solter (1984a,b,c) present a different picture. These authors found that even nuclei transplanted from cleavage-stage embryos into enucleated zygotes are unable to support development beyond the blastocyst stage. Subsequently, however, Kanagawa and colleagues found that the nuclei from two-, four-, and eight-cell-stage embryos in the early stage of each cell cycle can be reprogrammed when transplanted into enucleated mature oocytes to generate mice (Cheong et al. 1993). In addition, both inner cell mass (ICM) and trophoblast nuclei from blastocysts can be reprogrammed by serial nuclear transfer to generate live mice (Tsunoda and Kato 1998). Moreover, after transfer into enucleated oocytes, adult somatic nuclei can be used to clone mice (Wakayama et al. 1998). To account for these results, one has to assume that oocyte cytoplasm is able to reprogram somatic nuclei and that this capacity is lost soon after fertilization. In addition, a close synchrony between nuclear events, the cell cycle stage of the donor nucleus, and many other factors determine the success of nuclear transfer (Solter 2000). Therefore, under the appropriate conditions, both embryonic and adult somatic cells retain totipotency.

Compaction and the Formation of the Blastocyst: The First Differentiation Events

Details of the timing of early events in preimplantation mouse development and the experimental evidence for changes in cell potency and fate have been reviewed by Pederson (1986). Up to the early eight-cell stage, there is good evidence that the blastomeres of the mouse embryo are equipotent. Single blastomeres from two-cell and four-cell morulae can each give rise to a mouse. Early eight-cell-stage blastomeres cannot generate a mouse by themselves, but when recombined with genetically marked morulae, they can give rise to a wide range of different tissues in chimeric offspring (Kelly 1977). As cleavage proceeds to the 16-cell stage, however, there is a gradual restriction in the developmental potency of the cells, eventually resulting in the generation of two distinct lineages: the trophectoderm (TE) and the ICM. This differentiation process starts with com-

TABLE 2.4. Changes occurring in blastomeres during compaction

1. Increased Ca⁺⁺-dependent adhesiveness, both to each other and to lectin-coated beads.
2. Increased spreading on adhesive surfaces, using lamellipodia-like cell processes.
3. Ability to express contact-induced cell polarization as shown by regionalization of membrane and cytoplasmic domains (microvilli, lectin-binding sites, and intracellular organelles).
4. Establishment of gap-junction-mediated intercellular communication (ionic coupling and dye transfer) between all cells of the morula unit.
5. Gradual development of apical, zonular tight junctions between outside cells, generating an impermeable outer epithelial layer.

Agents that inhibit features of compaction include cytochalasin B, tunicamycin, low Ca⁺⁺, rabbit anti-F9 embryonal carcinoma serum or Fab fragments, and some monoclonal antibodies against uvomorulin all tend to prevent or reverse the cell-spreading effects but not polarization.

paction, when the blastomeres flatten and increase their contact with each other and develop distinct apical and basal membrane and cytoplasmic domains (polarization). Cells that end up on the inside of the compacted embryo give rise to the ICM, whereas the outer cells give rise to the trophectoderm. However, even at late morula stages, normal blastocysts can be formed from exclusively inside or exclusively outside cells. Eventually, however, the process becomes irreversible, with the formation of a fully expanded blastocyst consisting of a hollow vesicle of trophectoderm surrounding a fluid-filled cavity and a small group of ICM cells. The trophectoderm has all the features of a true epithelium, with apical junctional complexes forming a complete permeability seal against the outside environment. The trophectoderm overlying the ICM is known as the polar TE, and the cells surrounding the blastocyst cavity constitute the mural TE.

Some of the cellular changes associated with compaction are listed in Table 2.4. It should be stressed that these changes do not occur synchronously within all the cells of one embryo. Likewise, the cell cycles are not synchronized (Graham and Deussen 1978). The changes associated with compaction clearly point to alterations in both the surface properties of the cells and the organization of the cytoskeleton. The molecular basis of these changes, the signal(s) eliciting them, and their relationship to each other and to the cell cycle are areas of active research.

Changes in Cell Adhesiveness with Compaction

The process of compaction is regulated by changes in cell–cell adhesion and perhaps extracellular matrix proteins. Compaction is completely inhibited by polyvalent rabbit antibodies (either whole serum or immunoglobulin G [IgG] Fab fragments) against a surface glycoprotein originally known as uvomorulin (Hyafil et al. 1980, 1981; Peyrieras et al. 1983). It is known that uvomorulin is a member of a diverse family of Ca⁺⁺-dependent transmembrane cell adhesion molecules known as cadherins and is identical to E-cadherin (for review, see Yagi and Takeichi 2000). Synthesis of uvomorulin/E-cadherin is therefore not unique to compacting morulae, but antibody-blocking experiments point to its fundamental importance at this stage of development. Uvomorulin/E-cadherin (*Cdh1*) RNA and protein synthesis is initiated in the morula around the four-cell stage prior to compaction, and a key event appears to be its trafficking to the basolateral domains of the cell surface, a process that may be associated with posttranslational modifications of the protein (for review, see Kidder 1992). E-cadherin is a transmembrane protein, and its cytoplasmic region is specifically complexed with a number of molecules, including β-catenin, γ-catenin or plakoglobin, and α-catenin. These may mediate interaction of E-cadherin both with the cytoskeletal system and with other proteins of adherens junctions and transduce signals generated by cell–cell interactions from the cell surface to the nucleus (Ozawa et al. 1990; for review, see Yagi and Takeichi 2000). E-cadherin knockout embryos die around the time of implantation, but the mutant morulae compact like wild-type embryos because of maternal E-cadherin (Larue et al. 1994; Riethmacher et al. 1995). At the blastocyst stage, mutant embryos fail to form trophectoderm or a blastocyst cavity.

Cell Polarization with Compaction

One essential feature of compaction is the polarization of the blastomeres, so that they show distinct apical and basolateral membrane domains reminiscent of many epithelial tissues. These domains are clearly seen by scanning electron microscopy of compacted embryos that have been dissociated by incubation in the absence of calcium (Reeve and Ziomek 1981); the outer poles of the cells have numerous microvilli, whereas the inner surfaces are smooth. Transmembrane receptors such as the EGF receptor may become preferentially localized to one domain, for example, the basolateral surface of the TE cells (Dardik et al. 1992). Conversely, the Na$^+$ glucose transport protein has been localized to the apical domain of polarized blastomeres (Wiley et al. 1991). Cytoplasm organelles also appear to be polarized after compaction, with nuclei taking up a basal position. The onset of polarization can be followed in vitro by incubating pairs of isolated precompaction blastomeres. During culture, the microvillous surfaces and ConA-binding sites always develop at the poles opposite the points of cell–cell contact (Ziomek and Johnson 1980; Johnson and Ziomek 1981). An important question under investigation is whether this redistribution of plasma membrane domains precedes, or results from, a reorganization of cytoskeletal elements. The rules that emerge about polarization of mouse embryos at the morula stage may apply more generally to the differentiation of epithelial tissues from nonpolarized precursor cells at later stages of development.

Segregation of the Trophectoderm and Inner Cell Mass Cell Lineages

As outlined above and in Table 2.4, compaction is associated with cellular polarization. This property forms the basis of a polarization hypothesis to account for the differentiation of the two distinct cell lineages of the blastocyst—the trophectoderm and the ICM. Cleavage planes through compacted morula cells horizontal to the polarized axis will generate basal or inside cells, and apical or outside cells, each inheriting different membrane and cytoplasmic molecules (e.g., plasma membrane glycoprotein receptors and cytoskeletal organizing centers). These inherited molecules are thought to be responsible for initiating differences in the developmental potentials of the inner and outer cells. According to this hypothesis, differentiation is the result of cellular polarization elicited early in compaction. According to an alternative inside/outside microenvironment hypothesis, differentiation does not occur until after a network of tight junctions between the outer cells has formed. This generates distinct inside and outside microenvironments to which the cells respond; the inside cells become ICM and the outside cells become trophectoderm. Recent molecular studies suggest that specific levels of the transcription factor Pou5f1 (Oct 3/4) are important for maintaining pluripotency, and that reduction in these levels leads to differentiation into trophectoderm (Niwa et al. 2000). For a full description of compaction and a discussion of the polarization and microenvironment theories, see Johnson and Ziomek (1981), Pratt et al. (1981), Gardner (1983), Johnson et al. (1986), and Pratt (1989).

Implantation in the Uterus

During the fifth day of development, the blastocyst hatches from the zona pellucida and is ready for implantation. Hatching may be affected by a trypsin-like enzyme that digests the glycoprotein matrix of the zona pellucida and is synthesized by cells in the mural trophoblast (Wassarman et al. 1984), but uterine enzymes probably have a major role in vivo. Escape from the zona may also be facilitated by rhythmic expansion and contraction of the blastocyst. Hatching is independent of the uterine environment and will occur in vitro. At the time of implantation, the walls of the uterus become tightly apposed so that the uterine lumen is more or less occluded, and changes in the surface of the uterine epithelium make it a conducive surface for blastocyst attachment. The mouse blastocyst first adheres by its abembryonic pole (the mural trophectoderm farthest from the ICM) to the anti-mesometrial uterine wall (Fig. 2.15). No preformed attachment sites exist within the uterus, and the more-or-less even spacing of implantation sites is thought to result from the peristaltic movements of the uterus. Blastocyst attachment induces the formation of a uterine crypt (Fig. 2.15) and also stimulates the uterine stroma to form a spongy mass of cells known as decidual tissue. The process is known as the decidual reaction, and the mass of decidual cells around a single embryo can be referred to as the "deciduum," meaning "the thing that falls off." Alternatively, many texts refer to "the decidua," although this is, strictly speaking, the plural of deciduum. The decidual reaction only occurs in a uterus appropriately primed by progesterone and estrogen, but it will occur in response to stimuli other than the embryo, such as mechanical trauma or oil droplets. Thus, the decidual reaction is dependent on high levels of estrogen during estrus, followed by a few days when progesterone predominates, and finally, a small surge in estrogen on the fourth day of gestation. This final surge of estrogen coincides with a surge in uterine expression of leukemia-inhibiting factor (DIA-LIF) (Bhatt et al. 1992; Smith et al. 1992), and mutant female mice with nonfunctional DIA-LIF cannot support implantation (Stewart et al. 1992). Therefore, the production of this cytokine by the uterine endometrial glands seems to be an essential ingredient in the maternal initiation of implantation.

The decidual reaction involves a rapid increase in the permeability of local capillaries, causing the uterine stroma to become swollen and edematous. The stromal cells in the decidual tissue proliferate, increase in size, and establish numerous tight junctional complexes with their neighbors (for review, see Finn 1971). In due course, the epithelium separating the blastocyst from the stroma is eroded, and again this is not embryo-dependent, because it also occurs in artificially induced deciduoma. However, degeneration of the epithelium allows the trophoblast cells, which phagocytose the moribund epithelial cells, to invade the deciduum. The invasive nature of trophoblast cells may be associated with their synthesis of the protease urokinase-type plasminogen activator, as well as various metalloproteinases and their inhibitors (Strickland and Richards 1992). It is not known what limits trophoblast invasion to the uterus, but it is thought that the deciduum must have a role in restricting the distribution of these cells, whose behavior otherwise much resembles that of metastatic tumor cells. Blastocysts in vitro will simulate implantation, in that they will attach to plastic or extracellular matrix substrata and spread out over the dish (Enders et al. 1981).

If estrogen is absent on the fourth day of gestation, which is the case in lactating females or in females that have been ovariectomized after fertilization (see Chapter 6), the blastocysts will not implant but instead enter a quiescent phase known as "delay" (diapause) (Mantalenakis and Ketchel 1966; Yoshinaga and Adams 1966). This state can be maintained for up to 10 days, but it can be reversed at any time by the removal of suckling young or the administration of estrogen. Cell proliferation and DNA synthesis cease within a few days of the onset of delay (McLaren 1968), but the differentiation of primitive endoderm occurs at the same time that it would in normal blastocysts (Gardner et al. 1988). The mechanism by which small quantities of estrogen influence the progression of development is not known, but it is possible that the production and availability of DIA-LIF might mediate the effect, since this cytokine supports the continued proliferation of ES cells and prevents their differentiation in vitro (Smith et al. 1992).

The orientation of implantation has been studied to investigate whether there is any relationship between asymmetries in implantation and the future definitive axes of the embryo (Smith 1985; N.A. Brown et al. 1992). Histological sections of peri-implantation embryos have shown that the three primary axes of the embryo (anteroposterior, dorsoventral, right–left) correlate with the three axes of the uterine horn (oviduct–cervix [long] axis, mesometrial–antimesometrial axis, and right–left axis; Fig. 2.15). The anteroposterior axis of the streak-stage embryo is perpendicular to the long axis of the uterus, and the primitive streak lies toward either the right or left side of the uterine horn. The dorsal side of the embryo at the gastrula stage lies toward the mesometrium, but once the embryo has turned (see Embryonic Turning, p. 83), it is the right–left axis that is parallel to the mesometrial–antimesometrial axis of the uterus, the right side of the embryo facing the placenta. The orientation of C-shaped 9.5-dpc. embryos in the uterus is always that either the head is toward the oviduct and the dorsal surface is toward the left wall, or the head is toward the cervix and the dorsal surface is toward the right wall. The dorsoventral relationship derives from the asymmetric implantation of embryos in mice, whereby the abembryonic pole of the blastocyst (that farthest away from the ICM) always attaches to the antimesometrial wall of the uterus. The anteroposterior relationship has been ascribed to early asymmetry in the blastocyst, which is preserved during implantation such that the longest side of the blastocyst comes to abut either the left wall or right wall of the uterus (Fig. 2.15), with no intermediate positions observed. It has yet to be proved experimentally whether this asymmetry in the blastocyst has any causal role in determining the anteroposterior axis. Examination of the correlation between asymmetry in the ectoplacental cone, visualized as a distinct tilt presumed to derive from the original asymmetry of the blastocyst, and the location

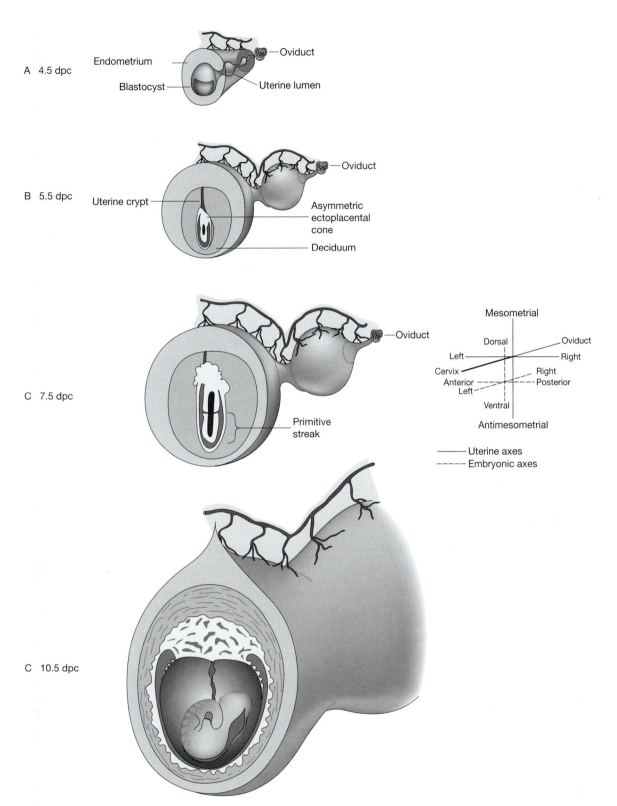

FIGURE 2.15. Schematic representation of implantation and the orientation of the embryo with respect to the uterus in vivo. (*A*) The blastocyst implants such that the ICM is located toward the mesometrial aspect of the uterus. The longest side of the asymmetrical blastocyst attaches either to the right or left wall of the uterus. (*B*) This asymmetry is evident as a slightly tilted ectoplacental cone at 5.5 dpc. (*C*) The primitive streak (defining the posterior aspect of the embryo) can form at either pole of the plane of tilt, such that the posterior axis of the embryo coincides with either the right wall or the left wall of the uterus.

of the primitive streak indicates that the anteroposterior axis of the embryo is not random with respect to this ectoplacental cone asymmetry but that its polarity is, in that the primitive streak may form either on the same side as the tilt or on the opposite side (Gardner et al. 1992).

Trophectoderm and Its Derivatives

As described above, an essential feature of the differentiation of the trophectoderm of the early blastocyst is the organization of the cells into a typical epithelium. The cells have apical junctional complexes and distinct apical and basal membrane domains. The junctional complexes involve extensive desmosomes (Fig. 2.16), and associated with these are large numbers of intermediate filament bundles (Jackson et al. 1980).

During postimplantation development, the trophectoderm does not remain as a simple epithelium but becomes regionally specialized with respect to morphology and growth potential (for review, see Gardner 1983). One subpopulation, the mural trophectoderm, is derived from the cells that surround the blastocyst cavity but are not in contact with the ICM. These cells cease division and become large and can contain up to 1000 times the haploid amount of DNA due to chromosomes becoming polytene (Varmuza et al. 1988). These are the so-called primary trophoblastic giant cells. In contrast, the trophectoderm cells in close proximity to the ICM and its derivatives remain diploid and continue to proliferate rapidly.

After implantation, this population of so-called polar trophectoderm spreads in several directions. First, some cells migrate around the embryo, replacing the primary mural trophoblastic giant cells and themselves becoming polyploid. Second, a finger-like projection of polar trophectoderm penetrates down into the blastocyst cavity, forming the extraembryonic ectoderm of the pregastrula and pushing the ICM derivatives ahead of itself. This projection develops a central cavity and becomes epithelial. After the formation of the extraembryonic mesoderm, the extraembryonic ectoderm retracts toward the placenta, where it forms the chorion. Finally, some trophectoderm cells continue to penetrate into the endometrium, forming the bulk of the placenta. Some of these cells, and cells of the chorion, also become polyploid (secondary giant cells).

Proliferation of the early trophectoderm appears to be controlled by its proximity to ICM derivatives; in the absence of ICM derivatives, TE cells do not proliferate but instead become giant, which has obvious advantages in preventing continued growth of trophectoderm if the embryo dies in utero. It appears that FGF4 may be the factor produced by the ICM and epiblast that acts through the receptor FGFR2 to maintain a proliferating population of trophoblast cells. This conclusion is based on the expression pattern of *Fgf4* and its requirement for the establishment of trophoblast stem (TS) cells (Tanaka et al. 1998).

Studies have been initiated to characterize genes specifically expressed in trophoblast cells. For example, differential screening of a cDNA library made from 13.5-dpc mouse placenta yielded a TE-specific cDNA encoding a novel, secreted protein expressed in differentiated cells (Lescisin et al. 1988). In situ hybridization studies have shown that the proto-oncogene, c-*fms*, which encodes the receptor for the cytokine, CSF-1, is expressed at high levels in the trophoblast from about 9.5 dpc (Regenstreif and Rossant 1989). At the same time, the ligand, CSF-1, is expressed in the uterus, specifically in the epithelium. This finding raised the

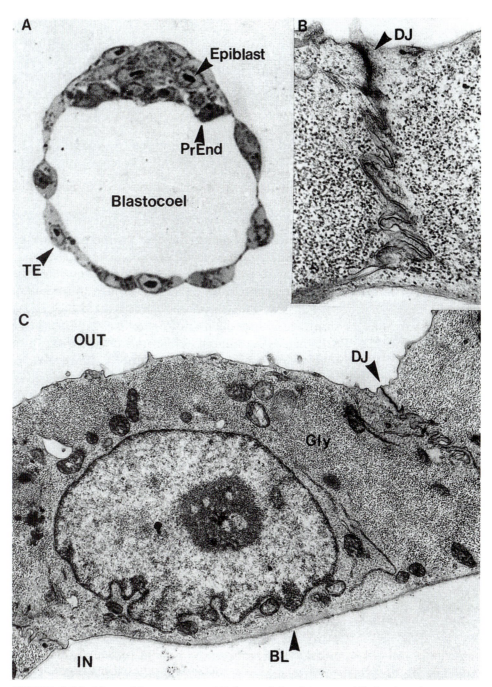

FIGURE 2.16. Mouse blastocyst at ~4.5 days of development. (*A*) Section showing outer epithelial vesicle of trophectoderm (TE) surrounding the blastocoel cavity and the epiblast and primitive endoderm (PrEnd). (*B*) Junction between two trophectoderm cells showing desmosomal junction (DJ) and interdigitation of the plasma membranes. (*C*) Trophectodermal cells showing abundant cytoplasmic glycogen, desmosomal junction, and thin basal lamina (BL) on the internal surface.

possibility that the c-*fms*/CSF-1 system has a paracrine role in promoting the growth of the embryo and its interaction with the uterus. Subsequent studies have shown that the CSF-1 gene is inactive in osteopetrotic (*op/op*) mutant mice (Yoshida et al. 1990). Homozygous *op/op* females have a markedly reduced fertility when mated with $op/+$ or $+/+$ males, but placental and fetal weights of the implanted embryos are normal. These and other results suggest that maternal CSF-1 is not absolutely required for embryo development but nevertheless may contribute to reproductive success at several different stages, for example, during ovulation, implantation, and colonization of the uterus with macrophages (Pollard et al. 1991). Experiments have shown that the *Mash2* gene, which encodes a basic helix–loop–helix transcription factor of the *achaete-scute* family, is specifically required for normal development of the trophoblast lineage. Embryos lacking *Mash2* die from placental failure around 10 dpc (Guillemot et al. 1994).

Formation of the Primitive Endoderm and Ectoderm: The Second Round of Differentiation

Like the formation of the trophectoderm lineage, the second differentiation event in mouse embryogenesis is also characterized by the appearance of an epithelial layer—in this case, the primitive endoderm—on the free surface of a group of nonpolarized cells, the ICM (see Fig. 2.17, below). The remaining core of ICM cells then becomes organized into a layer known as the primitive ectoderm. (The primitive endoderm is also known as the hypoblast, and the primitive ectoderm is known as the epiblast or embryonic ectoderm.) The differentiation of the primitive endoderm begins around 4.0 dpc, shortly before implantation, when there are only 20–40 cells in the ICM. Because of the small number of cells involved, it has so far been very difficult to make precise statements about the sequence of cellular and molecular changes involved in this differentiation and how they are related to the cell cycle and to intercellular communication and organization (for reviews of primitive endoderm and ectoderm differentiation, see Gardner 1983; Hogan et al. 1983).

Results from injection chimera experiments indicate that primitive endoderm cells do not colonize the endodermal tissues of the fetus, but only the extraembryonic parietal and visceral endoderm in the yolk sacs surrounding the developing embryo (Gardner 1982, 1983). Similar experiments have shown that the primitive ectoderm lineage gives rise to the ectodermal, mesodermal, and endodermal tissues of the fetus; to the germ cells; and to the mesodermal components of the extraembryonic membranes and placenta (Gardner and Rossant 1979). These lineages are summarized in Figure 2.3.

Recent studies have revealed that the position of ICM cells that generate primitive endoderm relative to the polar body in the blastocyst biases their directional movements during pre- and early-streak-stage development (Weber et al. 1999). Primitive endoderm cells located centrally tend to contribute to distal visceral endoderm, whereas primitive endoderm cells located near or away from the polar body tend to contribute to posterior visceral endoderm. These studies suggest that the directional movements of the visceral endoderm may influence the specification of the anterior–posterior axis that this ultimately derives from the geometry of the oocyte, sperm entry point, and blastocyst (for review, see Tam et al. 2001).

LINEAGE MARKERS USED WITH MOUSE EMBRYOS

Some of the initial genetically determined lineage markers available for use with mouse embryos were the GPI allozymes GPI-1AA and GPI-1BB. These differ in electrophoretic mobility and can be assayed in tissue homogenates (see, e.g., Gardner and Rossant 1979). Their sensitivity and precision were therefore limited. Fortunately, a variety of markers are now available that can be used at the cellular level on fixed or even viable cells (see Table 2.5).

The Primitive Ectoderm Lineage

At the time of implantation (~4.5 dpc), the blastocyst is composed of three distinct tissue lineages: trophectoderm, primitive endoderm, and epiblast (Fig. 2.17). The epiblast comprises the smallest population, consisting of only 20–25 cells, situated between the polar trophectoderm and the primitive endoderm. The cells are interconnected by gap junctions, but they are apolar. Shortly after implantation, the epiblast cells organize into a simple epithelium surrounding a small central cavity—the proamniotic cavity (Fig. 2.17). This process is associated with visible cell death and probably a high degree of cell intermixing. Like all epithelia, the epiblast cells are now polarized, attached at their apices by junctional complexes, and they contain a subapical concentration of cytokeratin polypeptides (Jackson et al. 1981). Their basal surface lies on a continuous basal lamina, which separates them from the primitive endoderm (Leivo et al. 1980). At the interface with the extraembryonic ectoderm, it appears that epithelial continuity is lost, although the two populations are tightly apposed, and they form discrete populations with respect to gap junctional communication (Lo and Gilula 1979).

During the early postimplantation period, epiblast cells stain positively for alkaline phosphatase, produced from the embryonic and tissue-nonspecific alkaline phosphatase genes (Hahnel et al. 1990). They also react with anti-SSEA-1 and anti-uvomorulin antibodies and express high levels of *Pou5f1* (*Oct3/4*) (see Beddington and Lawson 1990). These markers distinguish epiblast during the early stages of gastrulation but disappear as the embryo differentiates into the definitive germ layers. However, some of them persist as markers of primordial germ cells. From implantation to the onset of gastrulation, the epiblast population undergoes subtle changes. The cells lose their capacity to colonize the blastocyst, surface polysaccharides are modified, changes in polypeptide synthesis can be detected, X-inactivation occurs (see below), and the length of the cell cycle decreases (see Gardner and Beddington 1988). In addition, during this time, the overall methylation of the genome increases. Studies have shown that CpG islands of specific genes undergo programmed methylation changes that are stage-specific. In the early embryo, a general wave of demethylation occurs, so that by the 16-cell and blastocyst stages most CpG sites are demethylated. The DNA gradually becomes remethylated, and by 6.5 dpc, almost all genes studied show the methylation pattern characteristic of adult somatic tissues (Monk et al. 1987; Kafri et al. 1992).

Primitive Ectoderm Cells Divide Rapidly

Unlike most lower vertebrates, gastrulation in the mouse is associated with rapid cell proliferation. As shown in Tables 2.6–2.8, analysis of cell number and mitotic index in an outbred stock of mouse has demonstrated that the cells of the primitive ectoderm (epiblast) divide rapidly between 5.5 and 7.5 dpc (Snow 1977). At 5.5 dpc, the epiblast consists of about 120 cells, and this increases to 660 cells by

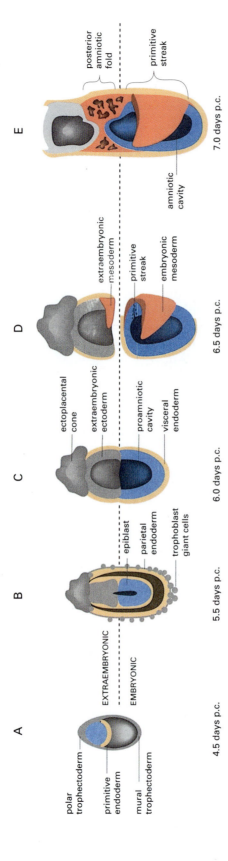

FIGURE 2.17. Schematic representation of mouse development from implantation up to the neurula stage (5 somites). (*A*) The embryo is composed of three distinct tissue lineages at the time of implantation: trophectoderm (*gray*), primitive endoderm (*beige*), and epiblast (*blue*). The primitive endoderm subsequently differentiates into parietal (*B*) and visceral (*C*) endoderm. The epiblast becomes organized into an epithelium surrounding the proamniotic cavity (*C*). The primitive streak forms at ~6.5 dpc (*D*) and mesoderm (*red*) emerges as two wings of tissue underlying the epiblast and extending into the extraembryonic region. Lacunae appear in the extraembryoinc mesoderm (*E*) and coalesce to form the cavity of the visceral yolk sac (exocoelom) bounded by the chorion mesometrially and the amnion antimesometrially (*F*).

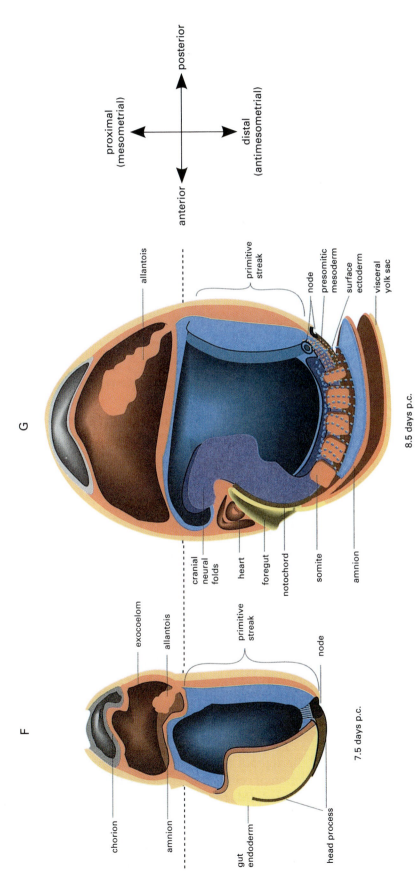

proximal
(mesometrial)

distal
(antimesometrial)

posterior

anterior

allantois

primitive
streak

node

presomitic
mesoderm

surface
ectoderm

visceral
yolk sac

cranial
neural
folds

heart

foregut

notochord

somite

amnion

G

8.5 days p.c.

exocoelom

allantois

primitive
streak

node

chorion

amnion

gut
endoderm

head process

F

7.5 days p.c.

FIGURE 2.17. (*continued*) The streak extends toward the distal tip of the gastrula (*E*), and by 7.5 dpc (*F*), the node is clearly visible at the anterior end of the primitive streak. The head process and subsequently the notochord (*brown*) arise from the node and underlie the midline of the neurectoderm (*G*). Definitive gut endoderm (*yellow*) is also laid down from the anterior primitive streak and is initially contiguous with the notochord. At the posterior end of the streak, the allantois forms (*F*). By 8.5 dpc (*G*), the neurectoderm (*purple*) has become organized into distinctive neural folds. The heart develops rapidly and the more anterior paraxial mesoderm is arranged into paired somite blocks. Caudally, the primitive streak continues to supply additional mesoderm for future trunk structures.

TABLE 2.5 Cell-autonomous markers used to follow cell lineages during mouse development

Marker	Reference
A. *Dyes, enzymes, or nucleic acids for microinjection*	
1. DiI, DiO	Serbedzija et al. (1989); Beddington (1994)
2. Lysinated dextran rhodamine and rhodamine-conjugated dextran	Gimlich and Braun (1985); Lawson et al. (1991)
3. Horseradish peroxidase	Lawson et al. (1991)
4. *EGFP* mRNA	Ciemerych et al. (2000)
B. *Transgenic mouse lines*	
1. Line carrying about 1000 copies of the β-globin gene; ES cell lines are available; for staining technique, see Chapter 16, DNA-DNA In Situ Hybridization	Lo (1986); Lo et al. (1987)
2. *LacZ* driven by a constitutive promoter	Beddington et al. (1989); Tan et al. (1993); Kisseberth et al. (1999)
3. Gene trap lines (e.g., *Gtrosa26*)	Friedrich and Soriano (1991)
4. Human placental alkaline phosphatase	DePrimo et al. (1996)
5. Fluorescent proteins (FP)	Okabe et al. (1997); Hadjantonakis et al. (1998); Hadjantonakis et al. (2002)
6. Homologous recombination of inactive duplication in *lacZ* driven by *hprt*	Bonnerot and Nicolas (1993)
7. Conditionally expressed *lacZ*, AP, and FP reporters	Lobe et al. (1999); Mao et al. (1999); Soriano (1999); Awatramani et al. (2001); Novak et al. (2001); Srinivas et al. (2001)
8. Replication-defective retrovirus containing *lacZ* reporter	Sanes et al. (1986); Price et al. (1987); DeGregori et al. (1994)
C. *Genetic differences between inbred mouse strains*	
1. Monoclonal antibodies specific for H-2^b and H-2^k	Ponder et al. (1983)
2. Satellite DNA sequence distribution between *M. musculus* and *M. caroli*	Rossant (1985)
3. Null mutation in cytoplasmic malic enzyme (*Mod-1*^{a} vs. *Mod-1*^{b})	Gardner (1984)
4. Carbohydrate polymorphism recognized by *Dolichos biflorus* agglutinin	Schmidt et al. (1985)
5. Y-specific probe in XX-XY chimeras	Jones and Singh (1981); Bishop et al. (1985)
6. Monoclonal antibody (OX7) specific for the Thy-1.1 allele of Thy-1, a surface glycoprotein on T lymphocytes and certain other tissues, including fibroblasts and embryonic brain cells; has been used to follow grafts of embryonic tissue into adult brain; congenic pairs of Thy-1.1 and Thy-1.2 mice are available	John et al. (1972); Morris and Barber (1983)
7. High- and low-activity alleles of β-glucuronidase; use limited to central nervous system	Mullen (1977)

6.5 dpc when gastrulation begins. To account for the number of ectoderm and mesoderm cells present at 7.0 dpc, given that mesoderm cells are derived from the epiblast following ingression through the primitive streak, the cell cycle of the epiblast at 6.5 dpc is estimated to be as short as 4.4 hours. Since the epiblast also gives rise to definitive endoderm during this period, this may be an underestimate. Studies using a variety of techniques have indicated that the division rate is not uniform throughout the gastrulation-stage embryo but is significantly faster in a subpopulation of cells. In rat embryos, this population has been localized to the primitive streak (MacAuley et al. 1993), whereas in the mouse, using different techniques, a more limited distribution at the anterior of the primitive streak has been proposed by Snow (1977). It is unclear whether a real species difference exists or whether the apparent difference is due to the techniques used. The mesoderm emerging from the primitive streak proliferates more slowly, with a cell cycle time of ~8–10 hours.

SUMMARY OF MOUSE DEVELOPMENT ■ 71

The Epiblast Is a Pluripotent Tissue

Several lines of evidence show that epiblast cells, or at least some of them, are pluripotent up until the later stages of gastrulation. It is these experiments that have identified the epiblast as the sole founder tissue of the fetus, including both the somatic tissues and the germ line.

1. When *single* epiblast cells from 4.5-dpc blastocysts are injected into 3.5-dpc blastocysts of a different genotype, their cellular progeny in chimeras populate all somatic tissues of the fetus or liveborn mouse as well as germ cells (Gardner and Rossant 1979; Gardner et al. 1985). They also contribute to extraembryonic mesodermal components of the chorioallantoic placenta (chorionic mesoderm and allantois) and the amnion. This ability to colonize the blastocyst is lost by 5.5 dpc.

2. Blastocysts can give rise to cell lines in vitro (ES cells) that are totipotent (Evans and Kaufman 1981; Martin and Lock 1983). ES cells inoculated into adult syngeneic mice give rise to teratocarcinomas (see Teratocarcinoma Cells and Embryonic Stem Cells, p. 101, and Fig. 2.33, p. 103), which are tumors composed of a variety of both differentiated cell types and undifferentiated stem cells known as embryonal carcinoma (EC) cells; EC cells are responsible for the progressive growth and transplantability of the tumor. Furthermore, ES cells injected into 3.5-dpc blastocysts contribute to the soma and germ line of adult chimeras (Bradley et al. 1984). ES cells can also be derived from blastocysts arrested in delay (see Chapter 8), when the epiblast lineage has probably already segregated (Gardner et al. 1988), further emphasizing the pluripotency of epiblast tissue.

3. Isolated epiblast tissue from 6.5- or 7.5-dpc embryos of all mouse strains tested can give rise to teratocarcinomas, containing both pluripotent EC cells and a large repertoire of differentiated tissues, at a very high frequency when transplanted into an ectopic site in adult syngeneic mice (see Fig. 2.33) (Diwan and Stevens 1976). Very small pieces of epiblast (~100 cells) from the anterior region of the 7.5-dpc embryo also generate teratocarcinomas (Beddington 1983). Because primordial germ cells are located at the posterior aspect of the embryo at this stage (see Origin of the Germ Line, p. 38), the potential to give rise to pluripotent EC cells must be a property of epiblast tissue and not just of a subpopulation of PGCs within it. This potential to generate teratocarcinomas is lost by 8.5 dpc (Damjanov et al. 1971).

TABLE 2.6. Total cell numbers in the embryonic germ layers

Age (dpc)	Number of embryos (no. of litters)	Endoderm[a]	Mesoderm[a]	Epiblast/Ectoderm[a]
5.5	14(5)	95	–	120
6	6(3)	130	–	250
6.5	13(5)	250	–	660
7	7(2)	430	1220	3290
7.5	16(5)	680	6230	8060

Reprinted, with permission, from Snow (1977).
[a]Number of cells.

TABLE 2.7. Mean cell cycle times required to account for the growth of the epiblast

Age (dpc)	5.5		6		6.5		7		7.5
Number of cells	120		250		660		4510		14,290
Number of divisions		1.04		1.32		2.71		1.58	
Mean cell cycle		11.5		9.1		4.4		6.7	

Reprinted, with permission, from Snow (1977).

4. When preimplantation embryos are transferred to an ectopic site, they develop more or less normally to the pregastrula stage (equivalent to 6.0 dpc of normal development) before becoming disorganized and forming either a teratoma (differentiated tissues but no EC cell population) or a teratocarcinoma. Similary, in the LT strain of mice, which shows a high incidence of ovarian teratomas, spontaneous parthenogenetic activation of oocytes is followed initially by apparently normal development to the pregastrula and even early gastrulation stages, before embryogenesis degenerates into teratocarcinogenesis (for review, see Martin 1980; Stevens 1983).

5. Single 7.5-dpc epiblast cells injected with a lineage tracer into the intact embryo (see Table 2.5) can give rise to derivatives in a variety of embryonic tissues (Lawson et al. 1991).

Gastrulation: Formation of Mesoderm and Definitive Endoderm

Gastrulation is the process by which the bilaminar 6.0-dpc embryo is transformed into a multilayered, three-chambered conceptus and the embryo itself acquires the full array of fetal primordia arranged according to the characteristic vertebrate body plan. All of these definitive tissues of the embryo and the extraembryonic mesoderm components are derived from the simple epithelium of the epiblast, which contains about 800 cells at the onset of gastrulation (Snow 1977). Not surprisingly, such reorganization and production of new tissues require an extremely complex and coordinated combination of morphogenesis, proliferation, cytodifferentiation, and pattern formation. Conceptualizing the morphogenetic movements associated with gastrulation is particularly difficult with respect to the mouse embryo, because the epiblast is shaped like a cup. However, the movements are thought to be essentially the same as those seen in the chick embryo, where the blastoderm is a planar sheet (Fig. 2.18).

Gastrulation begins at about 6.5 dpc when the primitive streak forms in a localized region of the epiblast adjacent to the embryonic/extraembryonic junction (Fig. 2.17). Although the signals that lead to the initiation of gastrulation at a particular point along the embryonic/extraembryonic junction are still largely

TABLE 2.8. Estimated cell cycle times (hours) for various regions of the embryo

Age (dpc)	6.4	7	7.5
Epiblast/ectoderm	4.8	7.2	8.1
Mesoderm		22.2	13.9
Proliferative zone	2.2	3.2	3.6
Other epiblast	5.1	7.5	8.5

Reprinted, with permission, from Snow (1977).

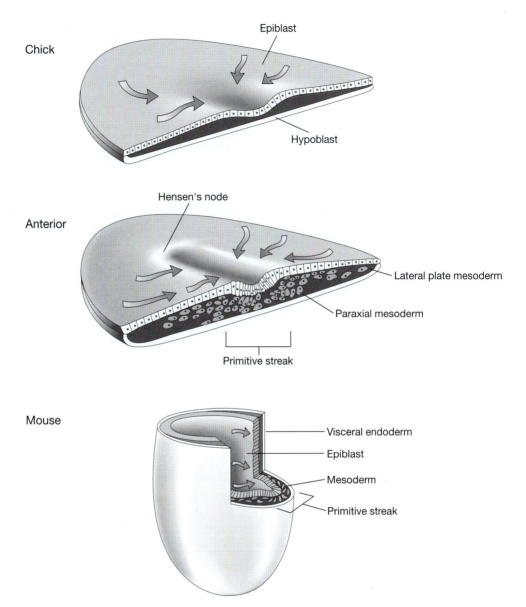

Chick

Epiblast

Hypoblast

Anterior

Hensen's node

Lateral plate mesoderm

Paraxial mesoderm

Primitive streak

Mouse

Visceral endoderm

Epiblast

Mesoderm

Primitive streak

FIGURE 2.18. Schematic representation of gastrulation in the chick and mouse embryo. (*A*) Chick blastoderm before and after formation of the primitive streak and Hensen's node. Arrows mark the direction of migration of cells within the epithelial sheet of the epiblast. Mesoderm cells are delaminating from the epiblast and accumulating between the upper and lower epithelial sheets. (*B*) Primitive-streak-stage mouse embryo in section. If flattened out, it would resemble the chick blastoderm above.

unknown, recent studies indicate essential roles for the secreted factors Nodal and Wnt3 in this process (Conlon et al. 1994; Liu et al. 1999). The primitive streak marks the future posterior of the embryo. The epithelial continuity of the epiblast is lost in the streak region (~10–15 cell diameters in width), and cells move or delaminate through the streak to emerge between the epiblast and visceral endoderm as a new intermediate layer of mesoderm (Fig. 2.19). Some cells also intercalate into the outer visceral endoderm layer to provide the first cohort of defin-

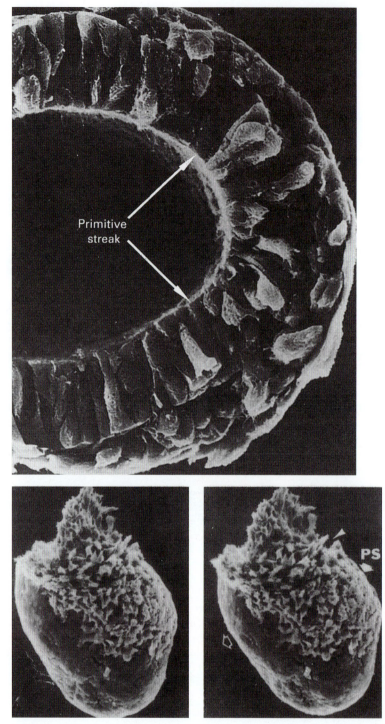

FIGURE 2.19. (*Top*) Transverse cross-fracture through the primitive streak of an early primitive-streak-stage embryo showing the ingression of epiblast cells and the lateral spread of mesoderm cells from the midline. (*Bottom*) Stereo pair showing the cellular organization of newly formed mesoderm. Arrow marks anterior. (PS) Primitive streak. (Photograph provided by Dr. Patrick Tam, Children's Medical Research Institute, Sydney.)

itive or gut endoderm. The origin of the streak marks the posterior aspect of the embryo and provides the first unequivocal morphological definition of the anteroposterior axis. However, it is clear that once the streak is fully extended,

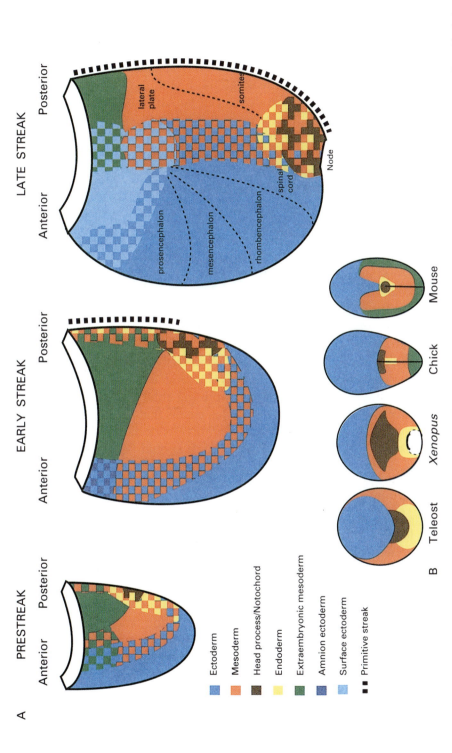

FIGURE 2.20. Fate maps of epiblast before and during gastrulation. (*A*) The epiblast origin of prospective tissues is plotted onto the left half of the embryo. Considerable overlap occurs between the boundaries of different prospective tissues. (*B*) The overall topography of the mouse fate map is comparable to that of other vertebrates during gastrulation. The embryos are depicted flattened out and viewed from the dorsal aspect.

the epiblast in its immediate vicinity shows differential transcriptional activity according to its anteroposterior position relative to the streak.

As gastrulation progresses, the streak elongates from its origin at the embryonic/extraembryonic junction toward the distal tip of the embryo, eventually extending down its entire side. In reality, this elongation is probably due to extension of the streak proximally by the addition of new delaminating tissue behind the anterior aspect of the early streak. The location of epiblast cells that will pass through the streak (all those with the exception of prospective neurectoderm and surface ectoderm) can be surmised from the fate maps shown in Figure 2.20. In addition to providing an avenue through which epiblast cells can pass to form the new mesoderm and endoderm layers, the primitive streak is an actively proliferating cell population (Hashimoto and Nakatsuji 1989; MacAuley et al. 1993). However, it is not clear whether the streak also contains a resident population of proliferating stem cells in addition to the transitory population of epiblast cells. Clonal fate map studies (see below) indicate that only in the anterior third of the streak is there a population of cells that has some of the properties of a stem cell pool: Such cells give rise both to progeny in the new germ layers and to descendants that remain within the streak. More caudally, the streak seems only to be a thoroughfare for delaminating epiblast traffic (Lawson and Pedersen 1992).

During the first 24 hours of gastrulation, nascent mesoderm moves in two directions. The most posterior mesoderm pushes its way proximally into the extraembryonic region, displacing the extraembryonic ectoderm toward the ectoplacental cone (Fig. 2.17E). It also moves laterally around the circumference of the embryo in both the embryonic and extraembryonic regions (Fig. 2.17E). Definitive endoderm emerges from the anterior aspect of the streak and moves predominantly anteriorward. This means that the posterior regions of the streak may remain associated with primitive endoderm for most of the gastrulation phase (Tam and Beddington 1987). It should be emphasized that much of this apparent rostral or lateral movement of new tissue is actually due to rapid expansion of the embryo by proliferation of tissues anterior and lateral to the primitive streak, rather than to extensive migratory movements.

In the posterior extraembryonic region, and to a lesser extent anteriorly and laterally in the extraembryonic region, mesoderm accumulates and increases in volume due to the acquisition of intercellular lacunae (Fig. 2.17E). These lacunae will eventually coalesce to form a new, mesoderm-lined cavity, the exocoelom (Fig. 2.17F). The expanding extraembryonic mesoderm population pushes the proximal rim of epiblast and the distal rim of extraembryonic ectoderm toward the center of the proamniotic cavity; this results in the formation of distinct bulges, most pronounced posteriorly, which are known as the amniotic folds (Fig. 2.17E). These folds meet and fuse so that a new chamber is formed in the embryo, separated from the embryonic region by the amnion and from the ectoplacental cone by the chorion (Fig. 2.17F). The mesodermal and visceral endoderm walls of this chamber will expand to form the visceral yolk sac. Following the formation of the exocoelom, some, if not all, of the nascent extraembryonic mesoderm still emerging from the posterior streak gives rise to a distinct structure, the allantois (Fig. 2.17F), which grows mesometrially across the exocoelom to fuse with the chorion. This will form a major component of the chorioallantoic placenta and provide a direct link for nutrient and waste exchange between fetus and placenta.

Fate Maps of Gastrulation

Two different methods have been used to construct fate maps of the gastrulating epiblast, and both have generated comparable results. Because labeling of specific cells requires direct access to the embryo, these studies have used whole-embryo culture (see Chapter 5) to follow development during gastrulation and early organogenesis. The first method employed was to graft orthotopically 7.5-dpc [³H]thymidine-labeled epiblast tissue into embryos of the same stage (Beddington 1981, 1982). The second method is more precise in that it involves less perturbation to the embryo and follows the development of descendant clones derived from a single epiblast cell injected iontophoretically with HRP or HRP and rhodamine dextran (Lawson et al. 1991; Lawson and Pederson 1992). The fate maps generated for pre- and early-streak embryos by such a clonal analysis, together with maps derived from orthotopic grafting in the late-streak embryo (Beddington 1981, 1982; Tam 1989), are illustrated in Figure 2.20. Three important points emerge from this work. The fate map of the mouse epiblast during gastrulation is clearly similar to those charted in the chick, *Xenopus*, and teleosts at a similar stage; i.e., the topography of prospective tissues appears to be well conserved during vertebrate evolution (Beddington and Smith 1993). The borders between different prospective tissues are not absolute but show a considerable degree of overlap, and a single epiblast cell can give rise to derivatives in all three germ layers. Therefore, the fate maps do not reflect a predetermined mosaic of cells already committed to forming specific tissues (Lawson et al. 1991). This interpretation is supported by the observation that heterotopically grafted epiblast tends to give rise to tissues characteristic of its new location, rather than adhering to its original developmental fate (Beddington 1982).

Generation of Regional Diversity in the Mesoderm

Little is understood about diversification within the mesoderm population. Indeed, it is not clear that "mesoderm," as a homogeneous progenitor tissue for the various different mesoderm derivatives, ever exists. Rather, it is possible that cells emerging from the primitive streak are already committed to specific mesodermal fates and express different genes. Clearly, the axial mesoderm of the head process and notochord shows a unique pattern of gene expression (see p. 82, Fig. 2.22). Similarly, other genes such as *Msx1*, *Msx2*, and *Lhx1* (*Lim1*) are expressed in lateral mesoderm but not more medially (Davidson and Hill 1991; Barnes et al. 1994), whereas genes such as the *forkhead* domain gene, *MF-1*, follistatin, *Sek*, and *Mox1* and *Mox2* (Candia et al. 1992; Nieto et al. 1992; Sasaki and Hogan 1993; Albano et al. 1994) appear to demarcate presomitic or paraxial mesoderm. Certainly, fate-mapping studies have demonstrated that different mesodermal derivatives arise from different rostro-caudal regions of the primitive streak (Tam and Beddington 1987, 1992; Tam et al. 1993; Lawson and Pedersen 1992), and there is evidence that different genes are expressed within different subpopulations of the primitive streak (Fig. 2.22). Recently, functional studies have shown that Amnionless, a product of the visceral endoderm that encodes a putative BMP regulator, is required for the development of the middle streak region to specify trunk mesoderm (Kalantry et al. 2001). Thus, the visceral endoderm appears to have a role in the regional specification of the primitive streak.

By 8.5 dpc, the diversity of mesodermal derivatives is already apparent. In the extraembryonic region, mesoderm has formed part of the chorion, part of the amnion, the mesothelium lining the exocoelom, capillary endothelium, fibroblasts, hematopoietic precursors (the blood islands), and the allantois. Within the embryo, at least seven different categories of mesoderm can be distinguished on the basis of morphology and anatomy: cranial mesoderm, cardiac mesoderm, somites and presomitic mesoderm, intermediate mesoderm, lateral plate mesoderm (splanchnopleure and somatopleure), blood vessels, and notochord. Undoubtedly, future molecular descriptions will demonstrate an even greater variety of mesodermal components.

The Anterior Visceral Endoderm

Gene expression studies have revealed that the visceral endoderm adjacent to the epiblast of pre-streak- and early-streak-stage embryos is molecularly heterogeneous (Hermesz et al. 1996; Thomas and Beddington 1996; Thomas et al. 1998). It was found that the anterior region of the epiblast-associated visceral endoderm (anterior visceral endoderm, AVE) expressed a specific set of genes (for reviews, see Tam and Behringer 1997; Beddington and Robertson 1999). Dye-labeling studies revealed that visceral endoderm cells located at the distal tip of the prestreak embryo move to the anterior region at the junction of the epiblast and extraembryonic ectoderm prior to gastrulation. Removal of the AVE from streak-stage embryos that were subsequently cultured in vitro resulted in anterior defects (Thomas and Beddington 1996). Futhermore, chimera studies have revealed that *Nodal*, *Lhx1* (*Lim1*), and *Otx2* are required in the visceral endoderm (probably the AVE) for the induction of anterior fates in the overlying epiblast (Varlet et al. 1997; Rhinn et al. 1998; Shawlot et al. 1999). These findings have established the importance of the extraembryonic tissues, specifically the visceral endoderm, for initiating anterior–posterior patterning in the early mouse embryo.

The Node

At the most anterior aspect of the streak, there is a specialized structure about 20 cells in diameter, known as Hensen's node or simply the node (Figs. 2.17F and 2.21), which is equivalent to Hensen's node in the chick and the dorsal blastopore lip of *Xenopus*; this structure has a crucial role in organizing and patterning the midline axis of the embryo. Figure 2.22 illustrates some examples of differential gene expression along the length of the streak and within and surrounding the node. The morphology and developmental fate of the mouse node have been reviewed previously (Sulik et al. 1994; Tam and Behringer 1997; Beddington and Robertson 1999). It is first clearly recognizable at the late primitive-streak stage when the streak extends to the distal tip of the gastrula (for conventions used in staging gastrulating mouse embryos, see Downs and Davies 1993). However, fate maps (Fig. 2.20) indicate that a region with the developmental fate characteristics of the later node exists at mid-streak stages, although at this stage, it does not appear as a discrete morphological structure. At later stages, the node region is recognized by a slight indentation at the distal tip of the gastrula, and there is no visceral endoderm overlying this region. Consequently, the node is a bilaminar structure, with dorsal and ventral layers, in contrast to the trilaminar composition of the rest of the embryonic region of the gastrula. The two layers of the node may be more intimately associated than the definitive germ layers (which are separated from each other by basal lamina) since the node cannot be separated into constituent tissue layers by conventional enzyme digestion (see Chapter 5).

The ventral layer of the node is derived from the overlying epiblast and is initially intercalated laterally with the epithelial layer of definitive endoderm, which is also derived from the epiblast. At this stage, the ventral cells of the node are known as the notochordal plate; each cell in this layer is distinguished by the presence of a single, motile, central cilium (Fig. 2.21C,D). Later, the notochord separates as a rod of cells from the endoderm, which becomes a continuous epithelial layer. By analogy with the chick embryo, the notochord precursors probably undergo extensive cellular rearrangements during gastrulation and become extended and intercalated along the anteroposterior axis (Fig. 2.21C) (Jurand 1974; Poelman 1981; Sausedo and Schoenwolf 1994; Sulik et al. 1994).

DiI (1-1-dioctadecyl-3,3,3′,-3′-tetramethylindocarbocyanine perchlorate) labeling experiments (Beddington 1994) and injection of horseradish peroxidase (HRP) into individual cells in the node region (Lawson and Pederson 1992) indicate that the node contains a population of resident cells responsible for providing most of the axial mesoderm. Just anterior of the notochordal plate is a population of compact midline mesoderm cells known as the head process, first distinguishable at the late primitive-streak stage (~7.5 days; Jurand 1962; Poelman 1981; Tam et al. 1983). The ultimate fate of these cells is presently unknown. Unlike the cells of the notochordal plate, head-process cells are not intercalated with the endoderm but emerge between the ectoderm and primitive endoderm. They underlie the future forebrain and possibly also the midbrain neurectoderm and are thought to be equivalent to the prechordal plate of the avian embryo. Indeed, given the confusion over the term "head process," prechordal plate is probably a more precise term (Sulik et al. 1994). The generally accepted view is that in the mouse, the head process or prechordal plate mesoderm originates from the epiblast at the node and migrates anteriorly. However, some of the axial mesoderm underlying the future fore/midbrain may derive from the leading edges of the lateral wings of mesoderm, which emerge from the early primitive streak and meet in the anterior midline (Fig. 2.17E; Tam et al. 1993 and pers. comm.). The most anterior mesoderm derived from the primitive streak gives rise to cardiogenic mesoderm.

Definitive endoderm also emerges from the node region and flanks the developing notochord (Figs. 2.17F and 2.21B) (Beddington 1981; Lawson et al. 1991; Tam and Beddington 1992; Tam et al. 1993). There is also evidence from chick and mouse that floor plate cells of the neural tube are derived from cells situated in the node region (Selleck and Stern 1991; Sulik et al. 1994).

In recent years, evidence has emerged that the mouse node has "organizing" properties analogous to those of the amphibian dorsal blastopore lip and Hensen's node of the chick. Heterotopic grafting of the node to a posterolateral position in a late-streak-stage mouse embryo results in the induction of a second neural axis (Beddington 1994). Similarly, grafts of the anterior primitive streak region from early streak-stage embryos (called the early gastrula organizer, EGO) prior to the formation of the definitive node also induce secondary axes (Tam et al. 1997). However, in both situations, the induced secondary axes lacked anterior neural character, suggesting that the node lacks anterior organizing activity. Recently, Tam and colleagues showed that the anterior region of the primitive streak at the midstreak stage is capable of inducing a secondary axis with anteri-

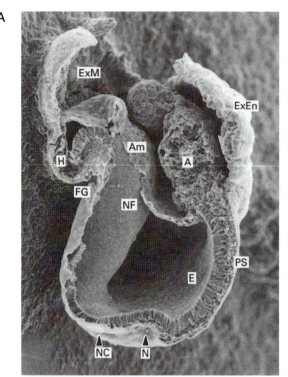

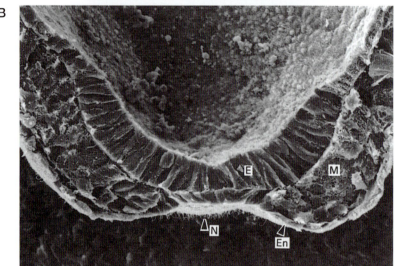

FIGURE 2.21. Scanning electron micrograph of neural-fold-stage embryo showing the position of the node and notochord. (*A*) Midsagittal section showing the allantois (A), amnion (Am), embryonic ectoderm (E), extraembryonic endoderm (ExEn), extraembryonic mesoderm of the yolk sac (ExM), foregut diverticulum (FG), presumptive heart (H), node (N), notochord (NC), and neural folds (NF). (*B*) Cross section through a similar stage embryo immediately anterior of the node. Note that in the midline, there are only two cell layers; the ciliated notochord cells (NC) tightly apposed to the embryonic ectoderm (E) which will form the neural plate. The paraxial mesoderm (M) lies on either side of the notochord, between the ectoderm and the definitive endoderm (En), which is still adjacent to or continuous with the notochord. (Photographs provided by Dr. R.E. Poelman, Rijksuniversiteit, Leiden, The Netherlands.)

or neural fates (Kinder et al. 2001). These studies suggest that the organizer of the mouse gastrula changes during development in its ability to specify the anterior–posterior axis.

C

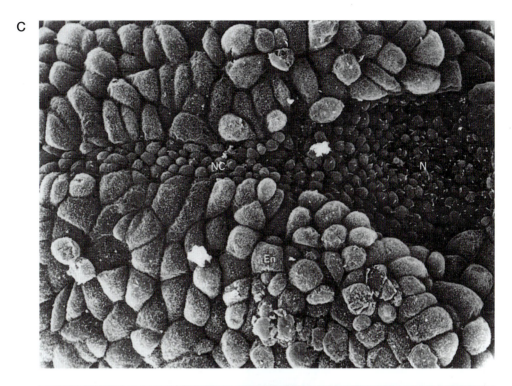

D

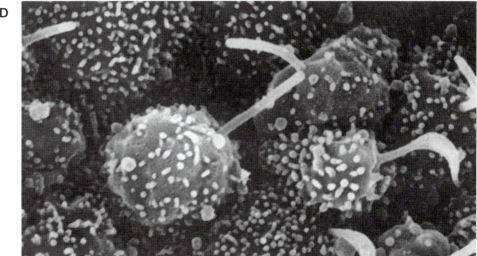

FIGURE 2.21. (*continued*) (*C,D*) Scanning electron micrograph of the ventral surface of 7.5-dpc embryo showing the morphology of the node and notochord cells. (C) Cells in the node (N) and notochord (NC) have microvilli and a large single cilium as well. This is shown at higher magnification in *D*. In contrast, adjacent endoderm cells (En) have only a large number of microvilli. (Photographs provided by Dr. K. Sulik, University of North Carolina, Chapel Hill.)

The Tail Bud

The tail bud replaces the primitive streak as the source of new caudal tissue on the 10th day of gestation. It is thought that this transition from a streak to solely a tail-bud supply of tissue coincides with closure of the posterior neuropore and corresponds to the lumbosacral region of the anteroposterior axis. The remnant of the primitive streak, which is located immediately behind the posterior neuropore, becomes, or is replaced by, a population of mesodermal cells at the distal tip of the growing tail. Tail-bud tissue divides rapidly (Tam and Beddington

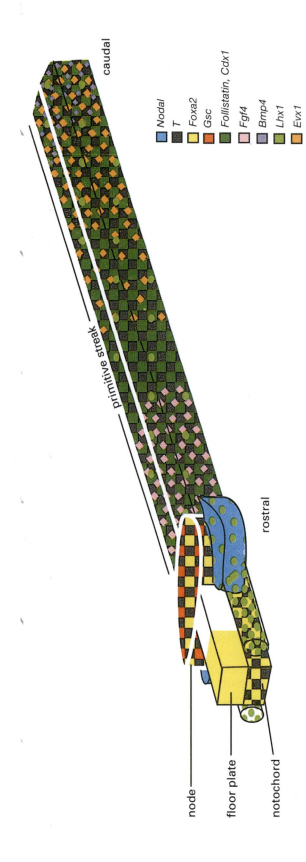

FIGURE 2.22. Some examples of localized gene expression in the primitive streak, node, and emerging notochord, adjacent mesoderm, and floor plate. Many more than the ten genes depicted show regional expression in this part of the embryo, but even this small subset of genes illustrates that the origin of different tissues from different levels of the streak and node results from differential gene expression.

1986) and has the potential to give rise to somitic and lateral mesoderm, neural tube, notochord, and gut endoderm; however, if it is removed, both somitogenesis and axial elongation are arrested (Tam 1984; Tam and Tan 1992). The tail bud thus appears to behave like a pluripotent progenitor population for tail development. The mechanisms by which this tail-bud population is maintained and the cues that control the final size of the body axis have yet to be elucidated. However, a recent study reinvestigating a tail-bud structure called the ventral ectodermal ridge (VER) suggests that the tail-bud VER produces signals that are required for somitogenesis (Goldman et al. 2000).

Embryonic Turning

At the early somite stage, the mouse embryo is U-shaped, and, relative to the amniotic cavity, the germ layers are initially inverted in the mouse conceptus, the ectoderm (neural tube and surface ectoderm) lies on the inside of the conceptus, and the endoderm (gut) lies on the outside in the yolk sac cavity. Turning effectively reverses this topography to generate the typical C-shaped conformation of a vertebrate embryo (for a detailed description, see Kaufman 1992). Figure 2.23 illustrates the basic movements effected by turning, which involve both the caudal and rostral sections of the embryo rotating in opposite directions while the whole conceptus rotates 180° anticlockwise about the midpoint of the U. These movements are also responsible for the complete envelopment of the embryo by its embryonic and extraembryonic membranes.

Somites and Their Derivatives

The somites and their derivatives, in particular the vertebrae and ribs, represent the most obvious example of segmental pattern in vertebrate embryos. Their genesis and fate have been most extensively studied in the chick embryo. Somites arise from strips of mesoderm, the presomitic mesoderm, or segmental plate, which lie on either side of the neural tube immediately anterior to the primitive streak or, at later stages, the tail bud (Figs. 2.24 and 2.25). The presomitic mesoderm and its derivatives are known collectively as the paraxial mesoderm. In the mouse, somitogenesis commences at about 7.75 dpc and continues up to the 14th day of gestation (Tam and Tan 1992), with ~65 pairs of somites being generated altogether, in a rostral-to-caudal gradient of maturation. Discrete somites are not evident in the cranial region anterior to the otic vesicle, although some paraxial mesoderm is present, arranged into seven pairs of metameric condensations, known as somitomeres and recognized by scanning electron microscopy (Fig. 2.24) (Meier and Tam 1982). Most of the skull is derived from the neural crest (see below), and the principal derivatives of cranial paraxial mesoderm are the musculature of the eyes and the branchial arches. Somitomeres have also been observed in the presomitic mesoderm prior to overt somite segmentation. However, lineage studies make it unlikely that a presomitic somitomere constitutes the direct precursor of a somite (Tam and Beddington 1986). Cell mixing within each strip of presomitic mesoderm indicates that cells may be moving between somitomeres, and therefore, they do not represent a constant population of cells. Once somites have formed, little or no cell mixing occurs between them.

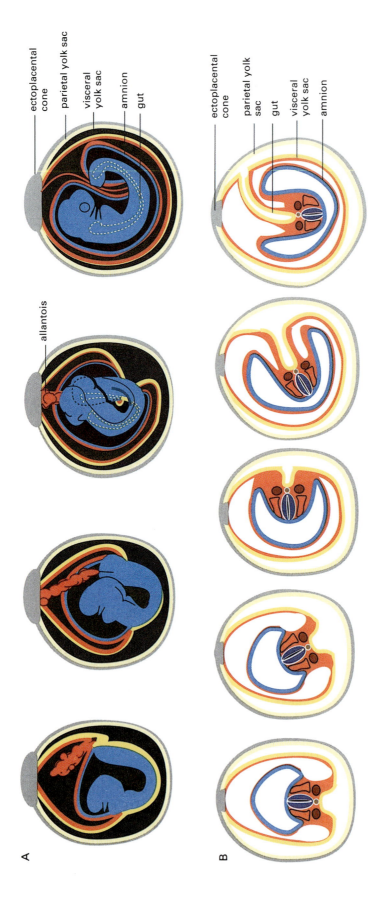

FIGURE 2.23. Schematic representation of turning of the mouse embryo (after Kaufman 1992). (*A*) Changes in the conformation of the embryo and arrangement of embryonic and extraembryonic membranes from 8.5 dpc to 9.5 dpc. (*B*) Vertical sections through the mid-trunk region of the embryo during turning to illustrate how the amnion comes to envelop the embryo while contained within the visceral yolk sac.

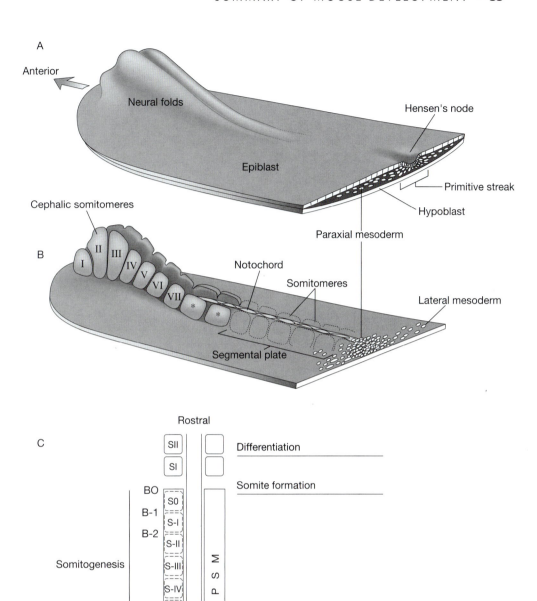

FIGURE 2.24. Schematic representation of a neural-fold-stage chick or mouse embryo. (*A*) Viewed with ectoderm in position. (*B*) The ectoderm and neural tube have been removed to reveal the underlying mesoderm. (*) The first somite blocks to condense; (I–VII) represent the cranial somitomeres detected only by stereo scanning electron microscopy. The fate of somitomeres and their contribution to craniofacial development are discussed in Trainer et al. (1994). In vivo, there are up to six somitomeres in the segmental plate of the mouse embryo. (*C*) Somite nomenclature system. Prospective somites in the presomitic mesoderm are numbered beginning with Somite S0, i.e., the next one to be segmented, in negative Roman numerals (-I, -II, -III, etc.) rostrally to caudally. Each prospective somite is approximately the size of the newest formed somite (SI). The borders between prospective somites are numbered beginning with border B0 between S-I and S0, B1 between S0 and S-I, etc. Somites that have formed (SI, SII, etc.) are numbered according to Ordahl (1993).(Redrawn from Pourquie and Tam 2001.)

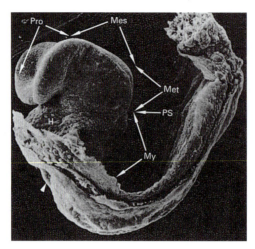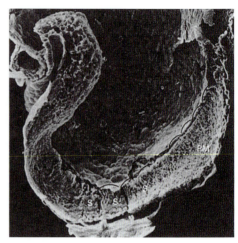

FIGURE 2.25. Scanning electron micrograph of the neural-fold-stage mouse embryo (8.5 dpc). (*Left*) External morphology showing headfold with preotic sulcus (PS), and prosencephalon (Pro), mesencephalon (Mes), metencephalon (Met), and myelencephalon (My) regions of the prospective brain. The metencephalon/myelencephalon boundary is approximately between rhombomeres R1/R2 or R2/R3, just rostral to Krox20 expression in R3. White arrowhead is foregut and H is heart. (*Right*) Right half of the embryo from which the neuroepithelium has been removed. There are three somites (S) and seven somitomeres in the cranial region and six in the presomite mesoderm. (Photos courtesy of Patrick Tam, Children's Medical Research Institute, Sydney.)

Periodically (approximately every hour for the initial 6–10 somites, 1.5–2.0 hours for trunk somites, and 2.0–3.0 hours for tail somites) and apparently synchronously, mesoderm at the rostral aspect of each strip of presomitic mesoderm becomes organized into an epithelial sphere and a new pair of somites is formed. Thus, the file of somites is extended by the addition of new somites caudally. The establishment of the borders between somites and their anterior–posterior polarity involves the Notch signaling pathway. Mutations in genes in the Notch pathway, including *Notch1*, *Delta1*, *Delta3*, *lunatic fringe*, and a *suppressor of hairless* ortholog (*Rbpsuh*), all cause defects in somitogenesis (Swiatek et al. 1994; Conlon et al. 1995; Oka et al. 1995; Hrabé de Angelis et al. 1997; Evrard et al. 1998; Kusumi et al. 1998; Zhang and Gridley 1998). Recent findings suggest that somite formation is regulated by very dynamic cycling waves of gene expression emanating from the posterior presomitic mesoderm that may also require Notch signaling (for review, see Pourquié 2001). These studies are currently being interpreted relative to a proposed "clock-and-wavefront" model (Cooke and Zeeman 1976). Recently, a new nomenclature for prospective somites and cyclic phases of gene expression in the presomitic mesoderm has been proposed (Fig. 2.24C) (Pourquié and Tam 2001).

Once formed, each somite goes through a similar maturation sequence. The epithelial structure is maintained for about 10 hours, after which cells on the ventral margin of the block disperse and move toward the notochord; these cells constitute the sclerotome, the precursor of the vertebrae and ribs (Fig. 2.26). Dorsally, the epithelial organization of the somite is retained in the dermomyotome population. The medial myotome component gives rise to the muscles of the vertebrae and back, whereas the lateral myotome cells generate the muscles of the body wall and limbs. The dermatome provides the dermis of the skin of the trunk and

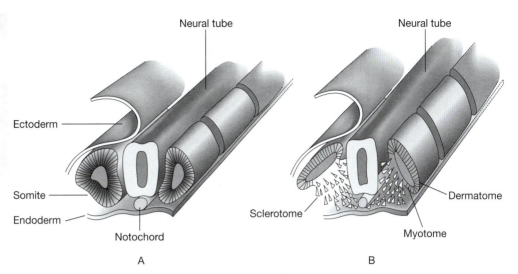

FIGURE 2.26. Schematic representation of the differentiation of the somites in the trunk region of the mouse embryo. The ectoderm has been pulled back on one side to reveal the somites adjacent to the neural tube. (*A*) The somite blocks have an epithelial organization. (*B*) Some cells migrate around the notochord to form the sclerotome cells of the vertebrae. The remainder differentiate into dermatome and myotome.

tail. A family of basic helix–loop–helix proteins, which include Myf5, MyoD, myogenin, and Myf6, has been implicated in myogenesis in vitro and in vivo, and the genes are expressed in a defined sequence in the myotome region of developing somites from 8.0 dpc: *Myf5* is expressed first, then *Myog*, then *Myf6*, and finally *Myod1* (Sassoon et al. 1989; Bober et al. 1991; Ott et al. 1991). Myogenin knockout mice have a severe muscle deficiency (Hasty et al. 1993; Nabeshima et al. 1993). In contrast, *Myod1*, *My5*, and *Myf6* single-knockout mice form muscle (Rudnicki et al. 1992; Braun and Arnold 1995; Patapoutian et al. 1995; Zhang et al. 1995). However, mice mutant for both *Myf5* and *Myod1* do lack muscle (Rudnicki et al. 1993). These and other findings form the basis for a transcriptionally regulated model of myogenesis.

Experiments in the chick and mouse suggest that patterning of the somites extends beyond their visible separation into distinct epithelial blocks and that early in development, they acquire anteroposterior and dorsoventral positional cues. For example, the rostral and caudal halves of each somite appear to have different properties, and this is reflected in the fact that both neural crest cells and motor axons are excluded from the caudal half of somites (Serbedzija et al. 1990). Furthermore, experiments in avian embryos indicate that each vertebra is derived from the caudal sclerotome of one somite and the rostral sclerotome of the adjacent somite (Goldstein and Kalcheim 1992). Somites are not equivalent along the anteroposterior axis. In the chick, transplantation of somites has demonstrated that they develop autonomously with respect to sclerotomal derivatives. Thus, a somite from the future thoracic region (Fig. 2.27) will give rise to an ectopic rib if transplanted to the prospective cervical region. However, in such experiments, the dermomyotome exhibits greater developmental lability and will give rise to skeletal muscle derivatives appropriate to its new location. The expression of certain genes (e.g., members of the HOX family; Fig. 2.27) provides visible evidence for anteroposterior patterning within the paraxial meso-

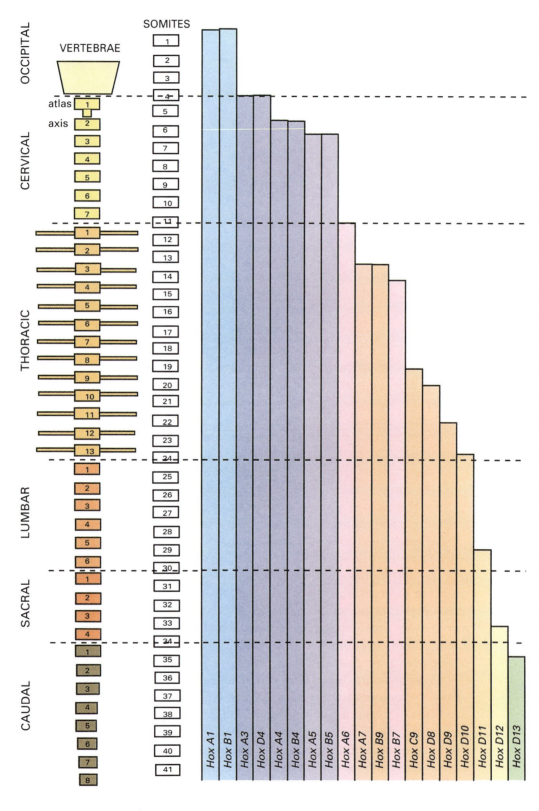

FIGURE 2.27. Diagram showing the relationship between specific somites and their vertebral derivatives and the anterior boundary of expression of different *Hox* genes in the paraxial mesoderm. Due to natural degeneration of one of the occipital somites, the somite number ascribed to each vertebra varies according to the method of counting used. Therefore, correspondence of boundaries with respect to a specific somite number can only be considered accurate to within ±1 somite. Vertebrae serve as more reliable and standardized landmarks.

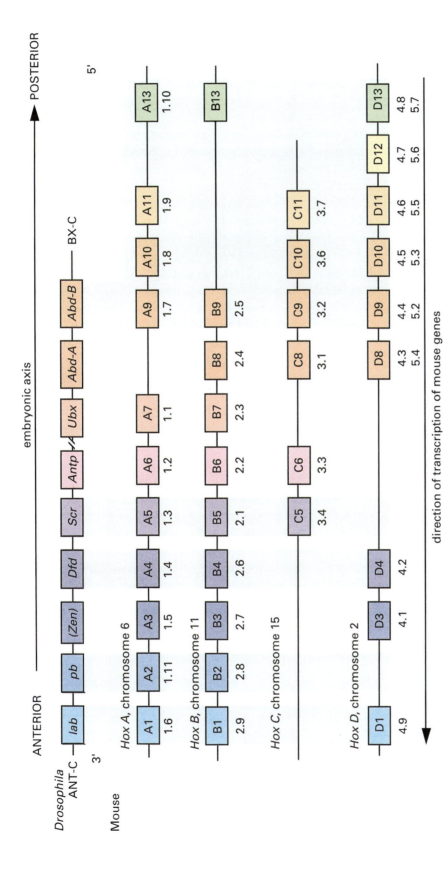

FIGURE 2.28. Alignment of the vertebrate *Hox* complexes into 13 paralogous groups and comparison with *Drosophila* HOM-C (after McGinnis and Krumlauf 1992). The current nomenclature for each gene is given within the box, and the old name is given below the box. The gray areas indicate homology with specific genes of *Drosophila* HOM-C. (*lab*) *labial;* (*pb*) *proboscipedia;* (*Dfd*) *deformed;* (*Scr*) *Sex Combs Reduced;* (*Antp*) *antennapedia;* (*Ubx*) *Ultrabithorax;* (*Abd*) *Abdominal.*

derm, and the combination of *Hox* genes expressed may define the axial character of the vertebrae. For example, a null mutation in *Hoxc8* results in a homeotic transformation such that the vertebrae derived from somites expressing the highest levels of *Hoxc8* acquire a more anterior character (Le Mouellic et al. 1992). This is consistent with the notion that as more 5′ *Hox* genes are activated in more caudal somites (see Fig. 2.28), a more posterior identity to the vertebral derivatives is ascribed (Kessel and Gruss 1991).

There is good evidence from experiments in the chick and mouse that the dorsoventral patterning of the somites, i.e., whether the cells will give rise to muscle (dorsal) or cartilaginous sclerotome (ventral), is controlled by a diffusible factor(s) from the notochord and neural tube. These signals act early in the development of a somite and are of at least two kinds. In one case, the notochord and the floor plate produce ventralizing signals that induce the adjacent somite to differentiate into sclerotome. If a supernumerary notochord or floor plate is grafted in a chick embryo dorsally between the somite and the neural tube, the nearby cells will change their fate and develop not into muscle but into cartilage (Pourquié et al. 1993). Sonic hedgehog appears to be an important factor for maintaining the ventralization of somites. On the other hand, it has been shown that local factors produced by the adjacent neural tube are required for the dorsalization of the somite and for promoting the differentiation of muscle precursors. If the neural tube is removed, axial muscles (but not muscles derived from the lateral part of the somite) fail to develop (Christ et al. 1992; Rong et al. 1992). Therefore, it seems that the dorsoventral patterning of the somite is regulated by different factors produced in different regions of the neural tube. These dorsalizing signals include BMP4, noggin, Wnt1, Wnt3a, Wnt4, and neurotrophin 3. Regionalization of the neural tube is in turn influenced by the floor plate and, ultimately, by the notochord (see Generation of Regional Diversity in the Neural Tube, p. 93). This hierarchy of patterning controls appears to underlie the abnormal differentiation of the somites in mouse mutants in which the growth and differentiation of the notochord are defective, e.g., *Brachyury* (Dietrich et al. 1994a), *Danforth's short tail* (Koseki et al. 1993), *pintail*, and *truncate* (Lyon and Searle 1989). Moreover, the ventralizing effect of the notochord on the somite is mediated, at least in part, by the expression of the paired-domain transcription factor, Pax1, and mutations in *Pax1* affect sclerotome differentiation, as first reported in different mutations of the undulated (*un*) locus (Balling et al. 1988, 1992; Koseki et al. 1993).

Lateral Plate and Intermediate Mesoderm and Their Derivatives

The mesoderm lateral to the somites is subdivided into intermediate mesoderm (adjacent to the somites) and lateral mesoderm (on the lateral margins of the embryo). Lateral mesoderm is split in two to form the coelom. Dorsally, the somatopleure is associated with overlying ectoderm and continuous with the mesoderm of the amnion; ventrally, the splanchnopleure is associated with endoderm and continuous with the visceral yolk sac mesoderm (Fig. 2.17G). Lateral mesoderm provides much of the mesenchyme involved in development of the viscera and connective tissue of the limbs, the mesothelial lining of the body cavity, and the mesenteries.

In the caudal part of the trunk region and proximal tail of 9.0- to 9.5-dpc embryos, the intermediate mesoderm differentiates into the pronephric duct and nephric primordium. Soon afterward (9.5 dpc), the urogenital ridge can be discerned extending from the mid-trunk to the mid-tail region. The lateral aspect of the urogenital ridge forms the mesonephros (visible at 10.0 dpc), and the medial portion forms the gonads. Although vesicles and ducts differentiate within the mesonephros, and show some segmental organization, they probably never function as an excretory system, and the mesonephros is to all intents and purposes a vestigial structure, although it can influence gonad differentiation. In addition, caudally, the mesonephric duct gives rise to an important offshoot, the ureteric bud, which is required to induce the formation of the definitive metanephric kidney. The mesoderm induced to form the kidney primordium (metanephric blastema) is caudal intermediate mesoderm.

Kidney differentiation and morphogenesis are probably some of the best-studied organogenetic processes in mammals (Saxen 1987; for review, see Vainio and Lin 2002) and serve as an excellent example of the complexities involved in establishing causal mechanisms responsible for tissue interactions. Kidney development has been shown to depend on reciprocal interactions between the epithelium of the ureter and the metanephric mesenchyme: The ureter induces the mesenchyme to become epithelial and form tubules, whereas the mesenchyme induces the ureter to branch. Although the metanephric mesenchyme can be induced by tissues other than the ureteric bud (e.g., embryonic spinal cord), no other mesenchyme in the embryo will form tubules in response to the inducing signal. This suggests that the caudal intermediate mesenchyme has already acquired special but, as yet, undefined properties predisposing it to nephric development. The initial induction requires direct cell contact between inducer and mesenchyme, and this appears to trigger a complex cascade of interdependent events. The fact that kidney development can be analyzed in vitro, by employing organ culture methods, bodes well for establishing the biological role of these molecules: More sophisticated, controlled, and reliable tissue recombination experiments and more precise application of specific stimulatory or inhibitory molecules can be carried out in vitro than in vivo. These kinds of studies provide an important complementary approach to molecular genetic analysis.

Limb Formation

Limb development has been used as an experimental system for studying vertebrate pattern formation for many years. These classical embryological experiments were based on tissue ablation and grafting, and they defined three important regions of the limb bud responsible for establishing the characteristic pattern of the mature limb: the apical ectodermal ridge, the underlying progress zone, and the zone of polarizing activity. The theoretical models of pattern formation derived from these experiments have provided the foundation for interpreting the complex patterns of gene expression revealed by more molecular techniques (for reviews, see Johnson and Tabin 1997; Capdevila and Izpisúa Belmonte 2001).

The limb buds develop as outgrowths of lateral mesoderm enveloped in surface ectoderm. In the mouse, the forelimb buds first appear at ~9.0–9.5 dpc adjacent to somites 7–12 (for a useful staging regime, see Wanek et al. 1989). By 9.5–10 dpc, the hindlimb buds are distinguishable at the level of somites 23–28. Lineage studies in chick/quail chimeras have shown that the bone and cartilaginous elements (including muscle tendons) are derived from the lateral mesoderm of the initial outgrowth, whereas the limb musculature originates separately from myotome cells migrating from the somite into the limb bud at a later stage. As the limb buds grow, the surface ectoderm overlying their distal tip thickens into a distinct structure known as the apical ectodermal ridge (AER). Grafting experiments have identified the cells of the AER as essential for maintaining the proliferation and survival of the underlying progress zone (PZ). Removal of the AER or PZ during limb bud outgrowth results in a truncated limb lacking distal structures. Conversely, replacing the PZ of an older limb bud with that of a younger limb bud results in an elongated limb with duplicated proximodistal elements. For example, the usual forelimb proximodistal pattern of humerus/radius + ulnar/digits may be transformed into humerus/radius + ulnar/a second radius + ulnar/digits. This indicates that the PZ contains information for patterning the proximodistal axis of the limb, and it appears to be the age of the PZ rather than the age of the AER that is important. The expression of various polypeptide signaling molecules, e.g., members of the Wnt family (Gavin et al. 1990), BMP family (Lyons et al. 1992), and FGF4/FGF8 (Niswander and Martin 1992; Crossley and Martin 1995), has been detected in the AER; these factors may be responsible for inducing and maintaining the PZ. Recently, however, the progress zone model has been questioned. New data from chick embryo manipulations and mouse *Fgf4/Fgf8* conditional knockouts in the developing limb suggest that the early limb bud mesenchyme is prepatterned and that FGFs from the AER act as mitogens and survival factors for the subjacent mesenchyme cells (Dudley et al. 2002; Sun et al. 2002).

The anteroposterior patterning of the limb requires a region of the limb bud mesenchyme situated at its posterior margin immediately proximal to the PZ. These cells, which are morphologically indistinguishable from the rest of the limb mesoderm, constitute the zone of polarizing activity (ZPA). If an additional ZPA is grafted to the anterior margin of a limb bud, mirror-image duplications of the AP axis result, the elements forming nearest to both the host and grafted ZPA always being posterior in character. These duplications occur quantally and stepwise, in that extra *whole* digits, as opposed to partial elements, are always formed and their anteroposterior character follows a predictable sequence. This all-or-none kind of differentiation indicates that the pattern is established by threshold responses within a continuous gradient. In both the chick and mouse, the node at the anterior aspect of the primitive streak can reproduce the effect of a transplanted ZPA when grafted to the anterior margin of a developing chick limb bud (Hogan et al. 1992). Identical duplications also result from anterior implantation of a bead soaked in retinoic acid (RA; Tickle et al. 1982). Studies have shown that in the mouse and chick, the ZPA (as well as the node) expresses the gene *sonic hedgehog*, which encodes an extracellular signaling protein (Echelard et al. 1993; Krauss et al. 1993; Riddle et al. 1993). In chick embryos, ectopic anterior expression of *sonic hedgehog* in the limb bud produces anteroposterior duplications. Application of RA in an anterior bead induces the expression of *sonic hedgehog*, but it is not yet clear whether RA is involved in regulation of *sonic hedgehog* in vivo. What is clear is that RA signaling is required for limb bud outgrowth (Niederreither et al. 1999). Members of the *HoxA* and *HoxD* families are expressed in a region-specific manner within the limb, consistent with a role in determining the final anteroposterior pattern. For example, posterior cells express *Hoxd5* to *Hoxd9*, whereas anterior cells express only *Hoxd13*. If anterior mesoderm from the mouse limb bud is transplanted into the posterior region of a chick limb bud, *Hoxd11* expression is induced in the mouse cells. Thus, whatever the morphogen responsible for initiating the anteroposterior polarity of the limb, members of the *HoxA* and *D* families are likely to be involved in translating this information into the final pattern of cartilaginous elements.

Neurulation: Formation of the Nervous System

The brain and spinal cord (central nervous system [CNS]) are derived from a thickened, medial strip of ectoderm anterior (or rostral) to the primitive streak. This is the neural plate, which is first distinguishable soon after 7.5 dpc and elongates by the addition of newly differentiated neurectoderm at its posterior (or caudal) aspect. As with lower vertebrates, the medial epiblast is probably induced to form neurectoderm by vertical signals from the underlying mesoderm (prechordal and notochordal) emerging from the front of the primitive streak, and patterning is also influenced by planar signals propagated within the ectoderm itself. After its initial formation, the neural plate changes shape dramatically, and its lateral edges elevate to form the neural folds. Consequently, in cross section, the neurectoderm comes to resemble a "V," with the "hinge point" being the midline overlying the notochord. The cells of this medial hinge point will give rise to the floor plate and probably originate in the node (Lawson and Pedersen 1992; Schoenwolf et al. 1992; Sulik et al. 1994). The lateral edges of the neural folds become apposed to each other and then fuse to form the neural tube.

Neural tube closure initiates at three locations, between the cervical and hindbrain boundary, the midbrain and forebrain boundary, and the anterior extremity of the forebrain, at three different times during development (Copp et al. 1990). The initial site of neural tube closure is at the level of the fourth and fifth somites, at about 8.0–8.5 dpc. Subsequently, fusion proceeds both rostrally and caudally in a zipper-like fashion from the initial sites of fusion. Closure of the anterior neuropore is complete by the 15- to 20-somite stage (9.0 dpc), whereas the posterior neuropore remains open until the 32-somite stage (10.0 to 10.5 dpc). The most posterior regions of the mouse spinal cord form by a distinct process called secondary neurulation in which the primordium of the neural tube, called the medullary cord, forms by canalization (Schoenwolf 1984). Neurulation is one of the most widely studied morphogenetic processes in vertebrate development, and a variety of motive forces are implicated in folding the simple sheet of neurepithelium and causing it to roll up into a tube. These forces include cell autonomous factors influencing cell shape (such as microfilaments, microtubules, and cortical cytoplasmic flow), cell division, convergent extension cell movements, and expansion of the extracellular matrix underlying the neural plate, as well as influences from the underlying mesenchyme (for reviews, see Copp et al. 1990; Schoenwolf and Smith 1990).

Generation of Regional Diversity in the Neural Tube

It is becoming increasingly evident that the developing nervous system is a highly patterned tissue from a very early stage. At a gross level, there is obvious regional differentiation along the anteroposterior axis, generating morphologically distinct forebrain (prosencephalon), midbrain (mesencephalon), hindbrain (rhombencephalon), and spinal cord domains. Like the paraxial mesoderm, this pattern is reflected in the restricted anteroposterior expression domains of cer-

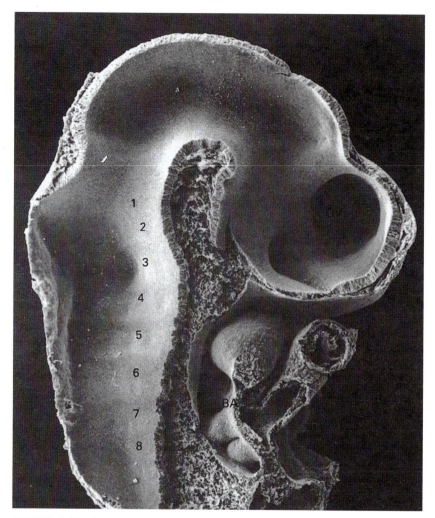

FIGURE 2.29. Scanning electron micrograph of a midsagittal cut through a 10-dpc mouse embryo illustrating the position of the rhombomeres (hindbrain neuromeres) (1–8). (OV) Optic vesicle; (H) heart; (BA) branchial arches. (Photograph provided by Dr. K. Sulik, University of North Carolina, Chapel Hill.)

tain genes. The various aspects of anteroposterior patterning and segmentation in the early CNS have been studied most extensively in the developing hindbrain. More than 175 years ago, von Baer (1828) described periodic swellings (neuromeres) along the neural tube reminiscent of some segmental organization (Fig. 2.29). More recently, the eight particularly pronounced neuromeres in the hindbrain region (called rhombomeres) have been reexamined in the chick in the context of their relationship with segmentally distributed nerves, such as the motor neurons or sensory ganglia, emanating from this region of the brain (Lumsden 1991). These studies have revealed a convincing correspondence between individual rhombomeres and specific neuronal pathways and have indicated that individual rhombomeres come to comprise segregated tissue lineages (i.e., cells from one rhombomere can no longer populate adjacent rhombomeres). Consequently, a rather precise relationship exists between particular rhombomeres, or pairs of rhombomeres, and cranial motor nerves, sensory ganglia, and neural crest migration into adjacent branchial arches (Fig. 2.30A). However, these lineage relationships in the chick may not be a hard-and-fast pro-

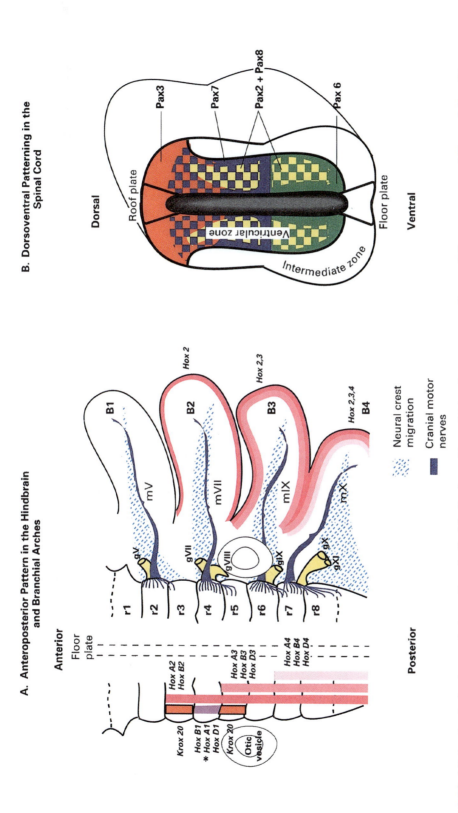

FIGURE 2.30. (A) Anatomical anteroposterior pattern in the hindbrain and branchial arches corresponding to different patterns of *Hox* and *Krox 20* gene expression. Asterisk indicates that *Hox A1* subsequently regresses caudally from this anterior boundary, whereas *Hox B1* persists as a single stripe of expression in r4. (r1–8) Rhombomeres; (gV–XI) cranial ganglia; (B1-4) branchial arches. The anatomical relationships between specific rhombomeres and the origin of cranial nerves and neural crest are based on studies in the chick and should not be taken as a definite map of the mouse hindbrain. (B) Transverse section through the spinal cord. Dorsoventral and mediolateral patternings are evident from the different domains of expression of members of the *Pax* gene family.

totype for mice, and the correspondence between specific rhombomeres and the origin of specific cranial nerves may not follow the same precise periodicity. This anatomical and cell lineage work has been augmented by descriptions of gene expression patterns in the hindbrain region, which have revealed an intriguing correlation between the boundaries of gene expression and rhombomere boundaries (Fig. 2.30A). One of the earliest genes to be expressed in a restricted pattern encodes a zinc finger protein, called Krox20. In situ hybridization shows two early and distinct stripes of *Krox20* expression corresponding to the position of rhombomeres 3 and 5 (Wilkinson et al. 1989). Subsequently, the mature anterior boundaries of the more 3′ members of *Hox* gene clusters have been shown to coincide with the anterior boundaries of odd-numbered rhombomeres, thus revealing a more striking two-rhombomere periodicity. Studies have shown that Krox20 can directly *trans*-activate *Hoxb2*, whose expression is highest in rhombomeres 3–5 (Sham et al. 1993). Furthermore, mutations in *Hox* genes with different anterior boundaries in the hindbrain (*Hoxa3*, anterior boundary r5; *Hoxa1*, anterior boundary r4) result in complementary phenotypes, indicating that *Hox* genes may have a determinative role in specifying derivatives of the hindbrain only in their most rostral domains of expression and that different hindbrain derivatives may be specified by different *Hox* genes (Chisaka and Capecchi 1991; Lufkin et al. 1991).

The segmental organization of the hindbrain is now well established, but less is known about the patterning of the midbrain and forebrain. A neuromeric model of forebrain organization has been proposed, based on the temporal and spatial expression of a number of genes (Puelles and Rubenstein 1993; for review, see Rubenstein et al. 1998). These include the mouse orthologs of the *Drosophila orthodenticle* and *empty spiracle* genes (*Otx1, Otx2, Emx1,* and *Emx2*) (Holland et al. 1992; Simeone et al. 1992) and members of the Wnt, Pax, Pou, homeodomain, helix–loop–helix, and other gene families. A somewhat different model for the segmental organization of the chick embryonic diencephalon has been put forward by Figdor and Stern (1993).

At the cellular level, the initially columnar epithelium of the neural plate acquires a more complex stratified arrangement, and once the neural tube has formed, proliferating cells become restricted to the ventricular layer, adjacent to its lumen. In due course, the differentiation of specific neurons follows a stereotyped dorsoventral pattern: Motor neurons are located ventrally in the spinal cord, whereas commissural neurons and the neural crest (see below) arise in dorsal positions. This dorsoventral pattern is also presaged by the differential expression of genes, such as the *Pax* and *Lhx* (*Lim*) genes, in restricted domains (Fig. 2.30B) (Tsuchida et al. 1994; for review, see Deutsch and Gruss 1991). It is thought that the dorsoventral polarity of the neural plate is established early in development before closure of the neural tube, at least in part by interaction with the notochord, which underlies the ventral midline. The neurectoderm cells lying immediately above the notochord are induced to differentiate into the floor plate, a small group of wedge-shaped cells extending along the ventral midline. In avian embryos, the addition of an ectopic notochord or floor plate next to the neural tube, or their deletion, results in marked but predictable changes in the dorsoventral pattern of the neural tube. In particular, the immediate proximity of a notochord induces a new floor plate and ventralizes adjacent neural tissue (Yamada et al. 1991, 1993; Goulding et al. 1993; Placzek et al. 1993). A model has been developed in which the *forkhead* domain gene, *Foxa2* (*Hnf3b*), is an important regulator of floor plate development in the mouse. It is hypothesized that *Foxa2* expression in the notochord induces the expression of *sonic hedgehog*, which encodes an extracellular signaling molecule (Echelard et al. 1993; Sasaki and Hogan 1994). This, in turn, induces the expression of *Foxa2* in the ventral midline cells of the neural tube, and this activates the transcription of other floor plate marker genes. Thus, misexpression of *Foxa2* in the dorsal midbrain leads to the ectopic expression of floor plate genes (Sasaki and Hogan 1994).

In addition to ventral signals, there may also be dorsal influences emanating from the roof plate (Basler et al. 1993) and/or the dorsal ectoderm, and the correct dorsoventral patterning of the neural tube probably requires an interplay between such dorsal and ventral signaling systems (for review, see Jessell 2000). A late function of the floor plate is guidance of axonal projections by contact-mediated and diffusible signals that promote directed outgrowth of commissural axons (for review, see Dodd and Jessell 1993).

Neural Crest

The neural crest is a transient population of cells that originates in the dorsal part of the neural tube at the junction between neurepithelium and surface ectoderm. Crest cells migrate extensively as single cells and move away from the neural tube to ventral and dorsolateral locations where they differentiate into a wide variety of cell types. Their derivatives include adrenomedullary cells, bone and cartilage, melanocytes, glial and Schwann cells, and several different kinds of neurons (sensory cranial nerves; parasympathetic, sympathetic, and sensory ganglia). Their developmental potential is influenced by their axial location, because cranial neural crest can give rise to bone, cartilage, and odontoblasts (one of the principal precursors in tooth formation), whereas the trunk crest cannot differentiate into these different cell types. In the head, the neural crest gives rise to most of the skull bones, with only minor components of the posterior skull being derived from cephalic mesoderm or from the first few somites (Couly et al. 1993).

Neural crest cells in the mouse first appear on the dorsal surface of the neural tube at a level corresponding to the third or fourth somite rostral to the most recently formed somite (Erickson and Weston 1983). Thus, they both emerge and migrate in a rostral-to-caudal sequence. Migration of neural crest cells in the mouse has been studied most thoroughly by labeling the plasma membrane of cells with the lipid-soluble, hydrophobic fluorescent dye DiI (Serbedzija et al. 1990, 1991, 1992; Osumi-Yamashita et al. 1994). Dye injected into the neural tube of embryos subsequently cultured in vitro shows that trunk neural crest cells emerge between 8.5 and 10.5 dpc (Serbedzija et al. 1990). Two pathways of migration are evident: (1) a ventral pathway through the rostral half of somites and (2) a dorsolateral pathway between the dermamyotome and the epidermis (Fig. 2.31). The ventral migration is composed of two phases, an early phase between 8.5 and 9.5 dpc, in which neural crest cells reach more ventral destinations such as the sympathetic ganglia and dorsal aorta, and a later phase between 9.5 and 10.5 dpc, which contributes to the dorsal root ganglia and Schwann cells of the motor axons. In the sacral region, neural crest cells emigrating from the neural tube also populate the enteric nervous system of the postumbilical gut (Serbedzija et al. 1991). In contrast to the situation for the trunk, cranial neural

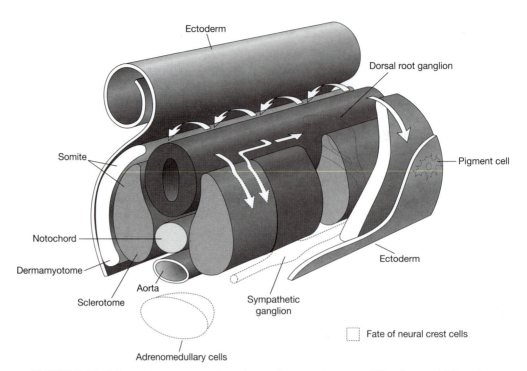

Ectoderm

Dorsal root ganglion

Somite

Pigment cell

Notochord

Ectoderm

Dermamyotome

Aorta

Sclerotome

Sympathetic
ganglion

☐ Fate of neural crest cells

Adrenomedullary cells

FIGURE 2.31. Schematic representation of neural crest migration. (*Closed arrows*) Neural crest cells arising from the dorsal neural tube migrate either ventrally through the rostral half of the somites or dorsolaterally between the dermamyotome and the epidermis. (Redrawn from Wolpert.)

crest migration follows a caudal-to-rostral sequence, beginning first in the midbrain region at about the 5-somite stage and in the forebrain region at about the 10-somite stage. At all levels, cell emigration lasts ~9–12 hours. Neural crest cells emerging from the hindbrain region follow segmental pathways in that substantial streams of crest are only observed lateral to r2, r4, and r6 and populate the adjacent first, second, and third branchial arches, respectively. However, DiI labeling of r3 and r5 shows that at least some neural crest differentiates in these rhombomeres and joins the segmental pathways from adjacent rhombomeres. This argues that the segmental distribution of crest in the hindbrain region is not due to any segment-specific failure to form neural crest cells (Serbedzija et al. 1992), although the relatively small contribution from r3 and r5 may result from rhombomere-specific cell death of neural crest cells (Graham et al. 1993).

It remains somewhat controversial whether the crest migration pathways are truly specific, paved, as it were, with particular extracellular molecules that direct crest cells to their final destination. Certainly, a variety of extracellular matrix molecules, such as hyaluronic acid, collagen types I and III, laminin, and fibronectin, have been described along the neural crest migration routes. In the chick, antibodies against the integrin fibronectin receptor perturb cranial neural crest migration (Bronner-Fraser 1985). However, it is possible that the so-called pathways are simply the only navigable routes available to migrating cells, other regions being impenetrable due to preexisting embryonic structures or an extracellular matrix incompatible with cell movement. As yet, it is not understood what prompts the neural crest cells to delaminate in the first place or what makes them stop and differentiate at one of their potential destinations. Certain genes, such as *W* (*Dominant white spotting*) and *Sl* (*Steel*), affect neural crest migration and differentiation. The demonstration that *Sl/Sl* neural crest melanoblasts could be rescued by provision of wild-type skin, whereas *W/W* melanoblasts could not, was the first indication that *W* and *Sl* encoded interacting receptor and ligand molecules (Mayer 1973).

Formation of the Branchial Arches and the Pharyngeal Region

The six pairs of branchial (or visceral) arches, which arise in the pharyngeal region of the embryo in cranial caudal sequence starting at 8.0 dpc, constitute major building blocks of the head and neck (with the exception of the fifth arch, which is rudimentary). Each arch, which grows from its distal tip, is an ectoderm-covered bar of mesenchyme (much of it neural crest in origin) that curves laterally and ventrally around the future oral cavity (Sulik and Schoenwolf 1985). The most rostral branchial arch (BA1) forms first, and bifurcates distally so that the maxillary prominence grows out from its dorsal surface, while the remainder of the arch (the mandibular prominence) continues to grow more ventrally. The maxillary and mandibular prominences become more overtly separated at 11 dpc and contribute to the upper and lower jaw, respectively. The second arch makes a major contribution to the neck region, effectively overgrowing the smaller third and fourth arches. The relationship between the more rostral branchial arches (BA1–BA4) and the hindbrain, with respect to *Hox* gene expression, neural crest migration, and innervation, is shown in Figure 2.30A.

The branchial arches are separated from each other externally by deep ectodermal grooves (visceral grooves) and internally by similar endodermal invaginations (pharyngeal pouches; Fig. 2.32), and each arch receives its own blood supply from a specific offshoot of the aorta (aortic arches). The first branchial or pharyngeal pouch, possibly in association with the second pouch, gives rise to the eustachian tube and contributes to the middle ear cavity, whereas the third pouch gives rise to the thymus and parathyroid glands, and the fourth to ultimobranchial bodies. The ultimobranchial bodies fuse with the thyroid and are thought to provide its complement of parafollicular cells. The thyroid itself differentiates from an endodermal thickening in the floor of the pharynx, which enlarges to form a ventral diverticulum.

The roof of the primitive mouth cavity, which is lined with epithelium continuous with surface ectoderm, gives rise to Rathke's pouch. This enlarges dorsally and abuts the infundibulum, a ventral outgrowth from the floor of the diencephalon. Together, the two form the pituitary gland, located immediately ventroposterior to the optic chiasma. Rathke's pouch gives rise to the anterior lobe (or adenohypophysis) of the pituitary, and the infundibulum gives rise to the posterior part (or neurohypophysis). The outgrowth of Rathke's pouch marks the anterior limit of the notochord. The parotid salivary glands also originate from the ectodermal lining of the roof of the stomodeum, whereas the submandibular and sublingual salivary glands arise from endoderm in the floor of the oral cavity.

Gut Development

The early development of the gut is poorly understood. Cell lineage studies in the blastocyst and during gastrulation (see above) have demonstrated that the

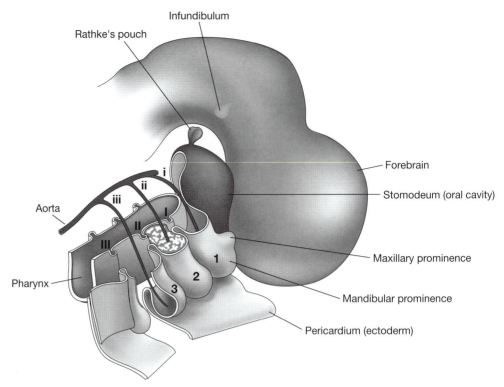

FIGURE 2.32. Schematic representation of the relationship between the branchial arches (1, 2, 3), pharyngeal pouches (I, II, III), and aortic arches (i, ii, iii).

epiblast and not the primitive endoderm is the precursor of the fetal gut. However, it is not clear exactly how definitive endoderm intercalates into the preexisting visceral embryonic endoderm layer nor how long residual primitive endoderm descendants may persist in the developing gut. The majority of epiblast cells are recruited into the gut primordium through the anterior part of the primitive streak; because they emerge on the ventral surface of the gastrula, the future roof of the gut is located medially, whereas the prospective ventral gut resides more laterally (Beddington 1981; Poelmann 1981; Lawson et al. 1987, 1991). Before the gut becomes a separate tube, the lateral margins of the definitive endoderm are continuous with the visceral yolk sac endoderm, and medially, the notochord is intercalated into the roof of the gut. The notochord detaches itself from the endoderm to form a discrete midline rod. Rostrally, the blind ending of the gut tube in the pharynx abuts the ectodermal lining of the primitive oral cavity (the stomodeum), but this ectoendodermal membrane (the buccopharyngeal membrane) ruptures at about 9.0 dpc due to a combination of cell death and cell rearrangements, so that the gut opens into the amniotic cavity. Later, a similar process occurs in the hindgut to generate the cloacal membrane (10–10.5 dpc), which ruptures to provide the anal opening. The gut endoderm gives rise to a number of derivatives, such as those originating from the pharyngeal pouches (see section above), most of which are derived from its ventral aspect (e.g., thyroid, ventral pharyngeal endoderm [first evident at 8- to 10-somite stage]; lung buds, ventral endoderm of the caudal pharyngeal region [first evident at 22- to

28-somite stage]; liver, ventral endoderm at the foregut–midgut junction [first evident at 8- to 10-somite stage]; and pancreas, rostral midgut [first evident at 25- to 30-somite stage]). The signals prompting the outgrowth of endoderm epithelium at the appropriate position for the various organ primordia are not well understood, although it is believed that the endoderm acquires its own rostrocaudal polarity early during development. Studies have shown that genes encoding DNA-binding transcription factors of the forkhead domain family are expressed in an anteroposterior sequence along the definitive gut from the gastrulation stage (Monaghan et al. 1993; Sasaki and Hogan 1993). Functional studies have shown that the transcription factors Sox17 and Mixl1 are essential for gut endoderm development (Hart et al. 2002; Kanai-Azuma et al. 2002). The subsequent differentiation and morphogenesis of each gut derivative invariably require intimate cooperation and interaction with adjacent lateral plate mesenchyme (for review, see Saxen et al. 1976).

Teratocarcinoma Cells and Embryonic Stem Cells

Transplantation of epiblast dissected from postimplantation embryos up to midgastrulation to a well-vascularized site in an immunologically compatible host results in the formation of tumors. These tumors usually contain a chaotic array of differentiated tissues, such as skin, bone, striated muscle, and nervous tissue, and represent a caricature of normal development (for reviews, see Stevens 1967; Damjanov and Solter 1974). They contain many of the normal derivatives of epiblast (the fetal tissues) and even structures (such as hair follicles) requiring local tissue interactions for their formation. However, highly sophisticated organotypic differentiation dependent on complex tissue interactions, such as liver or kidney formation, does not occur, nor is there any indication of the organized pattern of tissues characteristic of the embryo. In addition to these benign, differentiated tissues, a high percentage of these tumors contain a population of undifferentiated proliferating cells known as EC cells, which closely resemble ICM or epiblast cells in their morphology and biochemical characteristics (Diwan and Stevens 1976). EC cells behave as true malignant stem cells in that even a single EC cell is capable of giving rise to a new teratocarcinoma, complete with its own repertoire of mature tissues and dividing stem-cell population (Kleinsmith and Pierce 1964). Therefore, it is the EC cells that are responsible for generating not only more EC cells, but also the plethora of differentiated tissues. Early embryos from other mammalian species transplanted to an ectopic site can generate benign teratomas with a characteristic array of chaotic differentiated derivatives. However, the production of EC cells in such transplants appears to be a peculiarity of the mouse, although EC cells are found in human tumors, also called teratocarcinomas, derived from germ cells in the gonads. Postimplantation embryos from any strain of mouse can be used to generate teratocarcinomas containing EC cells, but strain differences, presumably reflecting differences in the immune system, have been observed in the ability of host animals to support progressive growth of teratocarcinomas (Solter et al. 1975). For example, C57BL/6 and AKR strains are nonpermissive for EC cell growth, although embryos of these strains give rise to a normal percentage of teratocarcinomas if transplanted to F_1 hosts.

As with most malignant tumor stem cells, EC cells can be isolated in culture and propagated as permanent cell lines. Under appropriate conditions, these cells can be induced to differentiate into a number of different cell types in vitro, thereby providing a convenient differentiation system in culture (for reviews, see Graham 1977; Martin 1980). One of the most remarkable features of EC cells is that when injected back into a normal blastocyst, they are able to resume embryonic development and participate in the formation of a normal chimeric mouse, derived partly from the host blastocyst and partly from the tumor stem cells (Brinster 1974; Mintz and Illmensee 1975; Papaioannou et al. 1975). EC cells also occasionally give rise to normal gametes in these chimeras and therefore could transmit their genotype to subsequent generations (Stewart and Mintz 1982). This astonishing triangular relationship among embryo, tumor, and tissue culture cell line promised a completely unexpected route for undertaking sophisticated molecular genetics in the mouse. Here was a population of "embryonic cells" that could be grown in culture and subjected to molecular genetic manipulation and selection and then be returned to an embryo where they could form a chimera and transmit their modified genome to future generations of mice (Dewey et al. 1977). However, the realization of such an experimental genetic program required that EC cells colonize the germ line at a high frequency, and this was not the case. Their checkered history, including proliferation as a malignant stem cell in vivo, may select for many genomic changes incompatible with gametogenesis (for review, see Papaioannou and Rossant 1983).

If cell lines could be isolated directly from the embryo, without resorting to a tumor phase, colonization of the germ line might be improved. This has been achieved (Evans and Kaufman 1981; Martin 1981), and ES cells can be recovered routinely from both normal and delayed blastocysts of at least two inbred mouse strains (129 and C57BL/6). If due care is taken in their maintenance, they can be grown for many generations in culture and will still contribute to the germ line of chimeras at a high frequency (Bradley et al. 1984). Consequently, ES cell lines provide one of the most powerful tools in mammalian genetics. Not only can they serve as a substrate for random mutagenesis, but they can also be used to select for extremely rare events such as homologous recombination (see Chapter 9). This allows the production of specific mutations in known genes and, following the establishment of new lines of mice carrying the mutation, a detailed examination of the mutant phenotype (for reviews, see Bradley and Liu 1996; Yu and Bradley 2001).

In addition to their use as a genetic tool, ES cells are at present the only normal stem-cell population that can be maintained indefinitely in vitro without losing stem-cell properties. Thus, they provide an invaluable resource for investigating the parameters of stem-cell growth and maintenance and identifying factors involved in sustaining their proliferation at the expense of differentiation (e.g., DIA-LIF; Smith and Rathjen 1991). Furthermore, progress has been made toward developing culture conditions that select for the differentiation of ES cells into populations of tissue-specific stem cells, such as hematopoietic and neuronal stem cells that can be isolated in vitro (see, e.g., Lindenbaum and Grosveld 1991; Wiles and Keller 1991; Kyba et al. 2002; Wichterle et al. 2002). Like EC cells, ES cells are tumorigenic if introduced into adult mice, where they form teratocarcinomas. They therefore also provide an interesting population for studying factors involved in the suppression of malignancy. The interrelationship of tumor cells, embryonic stem cells, and the normal embryo is schematically represented in Figure 2.33.

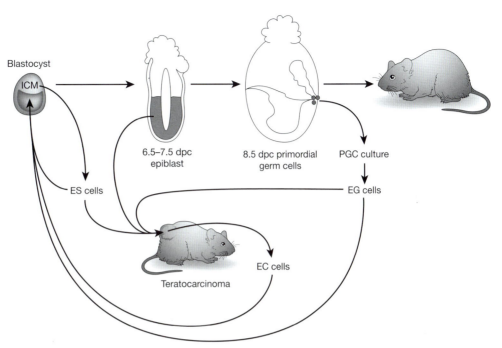

FIGURE 2.33. Interrelationship between the early embryo, embryonic stem (ES) cells, and embryonic carcinoma (EC) cells. ES cells can be derived in vitro from the ICM of blastocysts. Like the epiblast or pregastrula- and gastrula-stage embryos, ES cells form transplantable teratocarcinomas if inoculated into syngeneic adult hosts. The stem cells of these tumors (EC cells), like ES cells, can participate in normal development if introduced into a blastocyst. Both can contribute to the somatic tissues of liveborn chimeras, but, unlike EC cells, ES cells colonize the germ line at a high frequency, allowing their genotype to be transmitted to subsequent generations. EG cell lines, showing all the characteristics of ES cells, can be isolated directly in vitro from the posterior portion of 8.5-dpc embryos, the region that contains the primordial germ-cell population.

Size Regulation

The mouse embryo has a remarkable capacity to compensate for either a substantial increase or decrease in cell number (for review, see Snow et al. 1981). If two (Tarkowski 1961), or even nine (Petters and Markert 1980), morulae are aggregated together (Chapter 11, Making Aggregation Chimeras), and the resulting giant blastocysts are transferred to pseudopregnant recipients, the ensuing offspring are normal in size. In quadruple-size embryos, some regulation of cell number occurs before implantation, although the ICM-to-trophectoderm ratio remains disproportionately high (Rands 1987). However, the most dramatic phase of regulation occurs between 5.5 and 6.5 dpc and appears to involve an increase in cell cycle time simultaneously in all tissues of the conceptus without a significant increase in cell death (Lewis and Rossant 1982). By the onset of gastrulation, cell numbers are more or less normal. Furthermore, the timetable of development is not severely perturbed in giant embryos, although there is some evidence that proamnion formation may be advanced, indicating that this process is dependent on total cell number, rather than the number of cell divisions that have occurred during preceding development (Lewis and Rossant 1982).

Likewise, if cell number in the preimplantation embryo is reduced by either destroying cells or separating blastomeres, the final size of liveborn offspring is not affected. However, compensatory growth occurs quite late in development. Half-sized embryos, produced from development of a single blastomere isolated at the two-cell stage, appear morphologically normal, but they are half the size of control embryos up to 10 dpc (Tarkowski 1959). By 11.5 dpc, they have regained normal size; although it is not known how this is achieved, it may be significant that it coincides with the maturation of a fully functional placenta. A similar timetable of "catch up" growth is seen in XO embryos that are small and retarded in growth up to 10.5 dpc but close to normal at 12.5 dpc. However, XO embryos are invariably underweight at birth (Burgoyne et al. 1983). A substantial decrease in cell number can also be tolerated by gastrulating embryos. Administration of mitomycin C, which randomly kills cells, to 6.5- to 7.5-dpc embryos in utero can rapidly reduce total cell number to 15% of that found in untreated embryos. However, these diminutive embryos complete gastrulation on schedule and regain normal fetal size and weight by 13.5 dpc. Elevated proliferation in all tissues is first evident at the neurula stage. Remarkably, only subtle developmental abnormalities can be detected in these conceptuses, and the defects are more consistent with some degree of asynchronous development between different tissues during organogenesis than with destruction of specific progenitor populations. This implies that the epiblast must remain a developmentally labile tissue during gastrulation if the embryo can recover from such a drastic teratogenic insult. However, the primordial germ cell population, although showing some compensatory expansion between 9.5 and 10.5 dpc, remains about half the size of the same population in normal embryos (Snow et al. 1981), and defects are found in specific vertebrae (Gregg and Snow 1983).

Imprinting

Imprinting is the phenomenon whereby the activity of some genes is influenced by their parental origin. Its occurrence in mammals has only relatively recently been recognized and was deduced from a number of different lines of research, including classical genetic studies and studies on the pattern of X-inactivation and the development of diploid parthenogenetic, gynogenetic, and androgenetic embryos (for reviews, see Solter 1988; Cattanach and Beechey 1990; Surani et al. 1990). If mice heterozygous for reciprocal or Robertsonian translocations are crossed, some offspring will inherit regions of a particular autosomal chromosome only from one parent. Despite diploidy being maintained, it emerged that anomalous phenotypes are seen if certain chromosomes or regions of chromosomes are inherited from only one parent.

Parthenogenetic embryos derived from activated oocytes (see Parthenogenesis, Chapter 13) can develop apparently normally up to the blastocyst stage. A minority of these embryos form pregastrulas following implantation, and occasional development to early limb bud (25 somites) stage has been observed. However, none of the embryos progresses to term, and all have very poorly developed extraembryonic tissues. Nuclear transplantation (Chapter 13) has been used to reconstitute fertilized oocytes with pronuclei either only of paternal origin (androgenotes) or only of maternal origin (gynogenotes). In both cases, diploidy and even heterozygosity are maintained, but neither class of embryo develops normally. Gynogenotes can develop up to about 10 dpc (similar to parthenogenotes) but subsequently die with extremely poor development of extraembryonic tissues (Surani and Barton 1983). Androgenotes fare even worse, exhibiting poor preimplantation development and seldom forming embryos with recognizable somites, although the extraembryonic tissues in this case are relatively well developed. Control embryos, on the other hand, subjected to the same manipulations but reconstituted with a pronucleus from each parent develop normally. When chimeras are made between parthenogenotes and normal embryos, development can continue to term, although the chimeric offspring tend to be small, and parthenogenetic cells are selected against in the latter part of gestation and are particularly underrepresented in skeletal muscle and liver. In addition, the parthenogenetic cells contribute poorly, if at all, to the trophoblast, probably because polar trophectoderm lacking a paternal genome does not proliferate in response to a signal from the underlying ICM. In contrast, similar chimeras made with androgenotes tend to be larger than control embryos and show a disproportionate androgenetic contribution to mesodermal derivatives compared with ectodermal tissues. Androgenetic ES cells, derived from blastocysts developed from reconstituted zygotes, retain their imprint (at least in part) despite several passages in vitro. Chimeras made between these cells and wild-type embryos do not survive if the ES cell contribution is high, and even when the contribution is low, the offspring consistently show dramatic skeletal defects and die at an early age (Mann et al. 1990).

Evidence that development to term requires the presence of both a maternal and a paternal genome has been augmented by the identification of specific genes whose expression is influenced by their parental origin. The first example was identified when examining the phenotype of offspring derived from mice containing a targeted null mutation in the insulin-like growth factor II (*Igf2*) gene (DeChiara et al. 1991). When the *Igf2* mutation was inherited from the father, heterozygous offspring were ~60% the size of normal littermates. In contrast, heterozygotes inheriting the mutation from the mother were normal in size. This led to the conclusion that in normal individuals, only the paternal *Igf2* allele is transcriptionally active, and this was subsequently confirmed by molecular analysis. Interestingly, this imprinting is tissue-specific in that expression of maternal *Igf2* is seen in the choroid plexus late in gestation. It has now been demonstrated that the IGFII receptor (thought to be important in reducing levels of circulating IGFII) is imprinted in a reciprocal manner, with only the maternal allele being active (Barlow et al. 1991).

Currently, it is known that more than half of the mouse chromosomes have imprinted loci (chromosomes 2, 6, 7, 9–12, 14, 15, and 17–19), and nearly 60 imprinted genes have been identified (see http://www.mgu.har.mrc.-ac.uk/imprinting). The molecular mechanisms that lead to gene imprinting are still not clear (for review, see Tilghman 1999). Methylation has been invoked as one means of achieving a heritable, epigenetic, and reversible (in the germ line) state of gene activity, but it may be more important in maintaining the imprinted state rather than in setting it up. Other proposed mechanisms include enhancer competition, chromatin insulators, transcriptional antisense interference, posttranscriptional antisense interference, and chromatin propagation. At present, it is not known when the imprint in PGCs is changed, such that the parental imprint apparent in somatic cells of the embryo is erased and the new

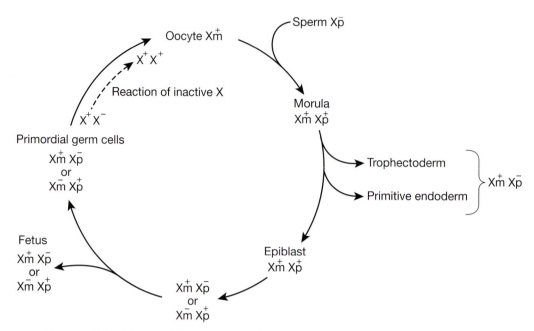

FIGURE 2.34. Timing of X-inactivation during development. Shown are the changes in X chromosome activity in the female mouse embryo. (m) Maternal; (p) paternal; (+) active; (–) inactive. (After Monk and Grant 1990.)

imprint that will be inherited by the next generation is established. Erasure and setting up of a new imprint could be a simultaneous or a multistep process.

X-Inactivation

In female eutherian (placental) mammals, one of the two X chromosomes is inactivated in all somatic cells, which results in dosage compensation for X-linked genes (Lyon 1961). This inactivation is random with respect to which X chromosome is silenced, except in the trophoblast and primitive endoderm lineages, where the paternally derived X chromosome is preferentially inactivated (Takagi and Sasaki 1975; West et al. 1977). The timing of X-inactivation during development is illustrated in Figure 2.34 (for review, see Chapman 1986). Cytogenetic and biochemical studies have indicated that inactivation occurs first in the trophoblast and primitive endoderm of the blastocyst and subsequently in the early postimplantation epiblast. X-inactivation also occurs in female primordial germ cells, probably at the same time inactivation occurs in the epiblast, although both X chromosomes appear to be active during oogenesis. The activity of X chromosomes in EC and ES cell cultures in vitro has also been studied, and there is some evidence from certain cell lines that both X chromosomes are active before differentiation (implying that the cell lines were derived from embryonic cells prior to X-inactivation). However, once differentiation is induced, X-inactivation occurs. In ES cells known to contain two active X chromosomes, complete loss or partial deletion of one X chromosome occurs, which suggests that inadequate dosage compensation is deleterious to stem-cell growth (Rastan and Robertson 1985).

In general, X-inactivation, once initiated, is stably inherited by all clonal descendants. Unlike the situation in humans, where some X-linked genes coding for proteins escape inactivation, all such X-linked genes appear to be transcriptionally silent on the inactivated mouse X chromosome (Ashworth et al. 1990). Classic genetic experiments identified a *cis*-acting locus called the X-inactivation-center (*Xic*) that regulates X chromosome inactivation (for review, see Brockdorff 2002). The X-linked *Xist* gene has been described in humans and the mouse that maps to the *Xic* locus, and, more importantly, it has been shown to be actively transcribed only from the inactive X chromosome (Borsani et al. 1991; C.J. Brown et al. 1991). The *Xist* RNA is noncoding and coats the inactive X chromosome. It has been proposed that the accumulation of *Xist* RNA along the X chromosome recruits gene silencing complexes to establish a heritable repressed state.

Extraembryonic Tissues

The extraembryonic tissues form an integral part of the life support system essential for the maintenance, nourishment, and protection of the fetus within the uterus (for review, see Rossant and Cross 2002). These tissues constitute the placenta, the parietal yolk sac (parietal endoderm and trophoblast), the visceral yolk sac (visceral endoderm and mesoderm), and the amnion (mesoderm and ectoderm) (Fig. 2.35). Studies on gene expression in extraembryonic tissues have focused on proteins synthesized at high levels to fulfill certain specialized functions. Examples cited below are the synthesis of fetal serum proteins such as α-fetoprotein (AFP), transferrin, and apolipoproteins by the visceral endoderm and the production of extracellular matrix glycoproteins such as laminin and type IV collagen, which form part of the basement membrane laid down by the parietal endoderm. Other studies have revealed more esoteric properties of these tissues. For example, in the trophoblast and the parietal and visceral endoderm (but not in the visceral mesoderm or amnion), the paternal X chromosome is specifically

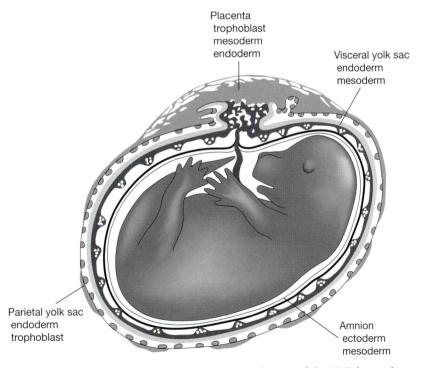

FIGURE 2.35. Placenta and extraembryonic membranes of the 13.5-dpc embryo.

inactivated (Takagi and Sasaki 1975; Kratzer et al. 1983; Lyon and Rastan 1984), and both repetitive and single-copy DNA sequences are undermethylated (Chapman et al. 1984). In addition, in all extraembryonic tissues, and in particular in the amnion, there is a high level of expression of the proto-oncogene, c-*fos* (Muller et al. 1983; Curran et al. 1984). The physiological significance of many of these observations is not yet known.

Extraembryonic Endoderm: The Primitive Endoderm Gives Rise Only to Visceral and Parietal Endoderm

The endoderm cells of the yolk sacs surrounding the mouse embryo are derived from a precursor pool of about 20 bipotential primitive endoderm (PrE) cells (Fig. 2.17A,B). Lineage studies with Gpi allozyme-marked cells have shown that the primitive endoderm does not contribute to the definitive endoderm of the adult mouse (Gardner and Rossant 1979; Gardner 1982, 1983) but only to visceral and parietal extraembryonic endoderm. This lineage study has been confirmed using the $Mod1^+/Mod1^n$ marker system (Gardner 1984).

PrE cells first differentiate on the blastocyst cavity surface of the ICM at 4–4.5 dpc and can be distinguished from the 20 or so primitive ectoderm cells by a number of morphological features (Nadijcka and Hillman 1974), in particular, a more extensive endoplasmic reticulum swollen with secretory material (presumably including type IV collagen, laminin, and fibronectin; see below). The PrE cells do not, however, form a well-defined polarized epithelium at this stage (see, e.g., Enders et al. 1978). As the primitive and extraembryonic ectoderm layers grow and elongate to form the core of the pregastrula (see Fig. 2.17), the outer endoderm cells differentiate into two morphologically and biochemically distinguishable subpopulations, the visceral endoderm (VE) and the parietal endoderm (PE). Endoderm cells that remain in contact with the embryonic and extraembryonic ectoderm become organized into a distinct epithelium of VE cells, with apical desmosomal junctions and microvilli, and many small and larger vacuoles and organelles are distributed in the cytoplasm in a polarized manner. VE cells surrounding the primitive ectoderm are known as visceral embryonic endoderm and tend to be flatter or squamous, whereas those surrounding the extraembryonic ectoderm (visceral extraembryonic endoderm) are columnar and have a very vacuolated cytoplasm and numerous microvilli (Fig. 2.36) (Hogan and Tilly 1981). A completely different morphology is seen in the parietal PE cells (Figs. 2.36 and 2.37). These first appear at the time of implantation (Enders et al. 1978) when primitive endoderm cells migrate onto the inner surface of the trophectoderm, which is covered by a thin basal lamina containing fibronectin and laminin (Wartiovaara et al. 1979; Leivo et al. 1980). In contrast to the VE cells, PE cells are individual and migratory, do not form specialized intracellular junctions, do not have an obvious polarity in the distribution of their intracellular organelles, and coexpress vimentin and cytokeratins (Lane et al. 1983; Lehtonen et al. 1983b). The most significant feature of PE cells is their enormously enlarged endoplasmic reticulum filled with secretory material, including components of the thick basement membrane (Reichert's membrane), which is laid down by the PE cells (see below). In vivo, the apical surfaces of PE and VE cells are closely apposed (Fig. 2.35), and the limited space between the two is filled with secretions of both the PE and VE cells and substances from the maternal circulation that have been filtered through Reichert's membrane.

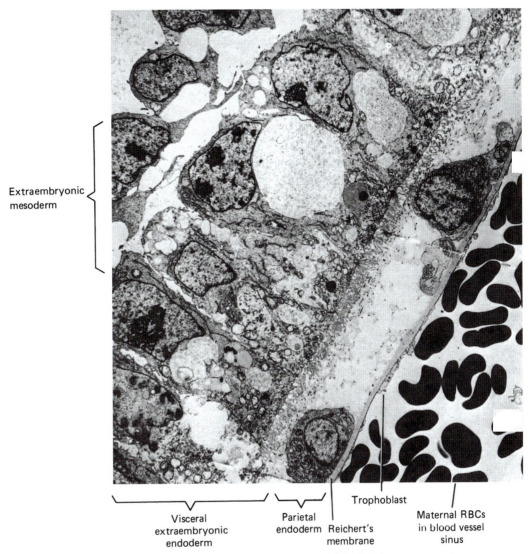

Extraembryonic mesoderm

Visceral extraembryonic endoderm

Parietal endoderm

Reichert's membrane

Trophoblast

Maternal RBCs in blood vessel sinus

FIGURE 2.36. Visceral and parietal endoderm of the 7.5-dpc mouse embryo.

Gene Expression in Visceral Endoderm

The most important functions of the VE are (1) absorption, i.e., uptake of substances from the maternal circulation that have filtered through Reichert's membrane into the cavity of the parietal yolk sac, and (2) secretion, i.e., production and secretion of serum components and other proteins such as AFP, transferrin, high- and low-density apolipoproteins, and α_1-antitrypsin. The VE therefore performs some of the same functions as both the fetal large intestine and liver (Meehan et al. 1984), even though these tissues are derived from a completely different lineage (see Fig. 2.3).

The morphology of visceral endoderm cells is highly specialized for absorption because the cells have numerous apically located microvilli and coated pits, as well as lysosomes, etc. (Figs. 2.36 and 2.38). Because the cells are polarized, it is likely that the absorptive functions are located at the apical surface, whereas secretion of AFP, transferrin, and other serum components may take place via the

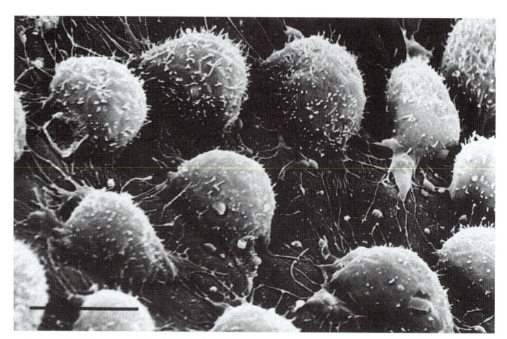

FIGURE 2.37. Scanning electron micrograph of parietal endoderm cells of 10.5-dpc rat embryo attached to Reichert's membrane. Bar, 10 μm. (Photograph provided by Dr. Stephanie Ellington, Department of Physiology, University of Cambridge.)

basal surface, but this question has not been resolved. In the visceral yolk sac, the basal surface is adjacent to the mesoderm layer containing fetal blood vessels. The direction of secretion of other VE products such as plasminogen activator is a matter of speculation.

One of the primary "markers" of the VE is AFP, a glycoprotein (M_r 68,000) made only by the VE, the fetal or regenerating liver, and localized regions of the embryonic intestine. AFP represents ~25% of the total protein synthesis of the visceral endoderm in 15.5-dpc mouse embryos, and its mRNA represents 15% of total poly(A)$^+$ RNA (Andrews et al. 1982a,b; Janzen et al. 1982). The precise function of AFP is unknown, but because it is the major γ-globulin in fetal blood, it may fulfill the same role as serum albumin in adult blood. The AFP gene is closely related to the albumin gene, probably by duplication and divergence of a common ancestral sequence (Gorin and Tilghman 1980; Eiferman et al. 1981).

Immunoperoxidase staining of sections of mouse embryos of different stages demonstrated that AFP is first detected in the visceral embryonic endoderm at 7 dpc (Dziadek and Adamson 1978; Dziadek and Andrews 1983). It is absent from VE around the extraembryonic ectoderm as a result of the inhibitory influence of this tissue. If visceral extraembryonic endoderm is separated from the underlying extraembryonic ectoderm, the cells will start to synthesize AFP within 12 hours (Dziadek 1978). In situ hybridization studies have shown that AFP mRNA is present in all VE cells of the 14-dpc visceral yolk sac but is absent from the mesodermal cells (Dziadek and Andrews 1983).

Gene Expression in Parietal Endoderm

The most obvious feature of PE cells is the fact that they are specialized for synthesizing and secreting a thick basement membrane known as Reichert's mem-

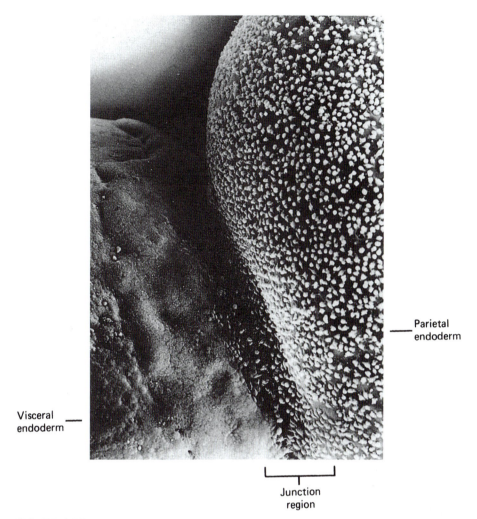

Parietal
endoderm

Visceral
endoderm

Junction
region

FIGURE 2.38. Junction region between parietal and visceral endoderm in the 7.5-dpc mouse embryo. The parietal endoderm on Reichert's membrane has been folded back to reveal the underlying visceral endoderm layer around the embryo proper. Magnification, 440x.

brane between themselves and the trophectoderm (Fig. 2.36). Until about 16 days of gestation (when it breaks down), Reichert's membrane is one of the major barriers between the maternal and fetal environments in the mouse. This is because in the rodent, the endothelial cells of the maternal blood vessels break down to produce large sinuses, and the trophectoderm cells lying between these blood sinuses and the embryo do not remain as a continuous shell below the placenta but gradually die away. Although it is assumed that Reichert's membrane acts as a passive filter, keeping out maternal cells and large molecules, there is in fact little hard information about its permeability properties, and even less is known about the relationship between the function of Reichert's membrane and its structure and composition. PE cells synthesize large amounts of basement membrane components, including laminin, entactin, type IV procollagen, and heparan sulfate proteoglycan. The mature cells do not synthesize fibronectin (Hogan 1980; Smith and Strickland 1981; Amenta et al. 1983), and this extracellular protein is not considered to be a structural component of Reichert's membrane (Semoff et al. 1982). However, it is likely that primitive endoderm cells do make fibronectin and possibly use it in migration on the trophectoderm (Hogan et al. 1983). RNA from PE cells has been used to isolate cDNA clones for both type IV procollagen and laminin (Kurkinen et al. 1983a,b; Barlow et al. 1984), and Reichert's membrane is an excellent model system for studying basement membrane synthesis, assembly, and remodeling (Hogan et al. 1984). PE cells also synthesize large amounts of tissue plasminogen activator, which has a molecular weight, antigenicity, and inhibitor sensitivity different from those of the urokinase-type plasminogen activator made by VE (Marotti et al. 1982).

Differentiation of the Extraembryonic Mesoderm

Some of the mesoderm cells generated by the early primitive streak contribute to the different extraembryonic tissues listed below (see Fig. 2.17 and Gardner 1983).

1. *Amnion*. The amnion is generated from both embryonic ectoderm and mesoderm. It appears first as a fold and then as a continuous roof over the top of the cup-shaped primitive ectoderm (Fig. 2.17). It then expands rapidly, and with the turning of the embryo, it forms a thin membrane surrounding the fetus. The ectodermal and mesodermal cells of the amnion have very different morphologies and are separated by a basement membrane (Fig. 2.39).

2. *Allantois*. The allantois starts as a finger-like projection of mesodermal cells from the posterior margin of the embryonic ectoderm where the primitive streak first arises. It expands upward, fuses with the chorion, is a major component of the labyrinth of the placenta, and gives rise to the blood vessels of the umbilical cord.

3. *Mesoderm of the visceral yolk sac*. Mesodermal cells generated from the posterior primitive streak migrate onto the inner surface of the visceral endoderm and give rise to the first hematopoietic tissue of the fetus in the form of blood islands in the visceral yolk sac. Mesodermal cells that cover the chorion also contribute to the placenta.

The Structure and Function of the Placenta

By midgestation, the placenta has become a very complex organ consisting of both fetal and maternal tissues and blood cells. The development of the placenta is described in detail by Theiler (1972), and only a schematic representation of two stages is given in Figure 2.40. An important feature distinguishing the mouse placenta from that of humans is that the maternal blood vessels in the mouse break down, so that the blood cells come into direct contact with the fetal trophoblast. Listed below are the major fetal tissues of the mouse placenta:

1. *Trophoblast*. In the outer spongiotrophoblast layer closest to the uterine decidual tissue, most of the trophoblast cells are polyploid and giant, whereas in the inner labyrinth layer, many of the cells are diploid. In both layers are maternal blood sinuses containing maternal blood cells, and the outer layer, even after dissection from the uterus, is contaminated with maternal decidual cells (Rossant and Croy 1985).

2. *Mesoderm.* After the allantois fuses with the chorion, it gives rise to fetal capillaries and blood vessels. These mingle with the trophoblast cells and with maternal blood sinuses in the labyrinth.

3. *Visceral and parietal endoderm.* The crypts of Duval contain both visceral and parietal endoderm. The visceral cells synthesize AFP, and the parietal cells make basement membrane.

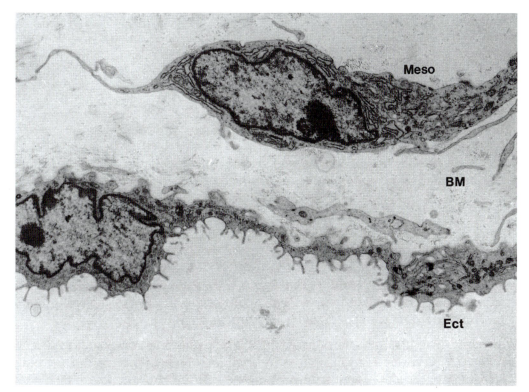

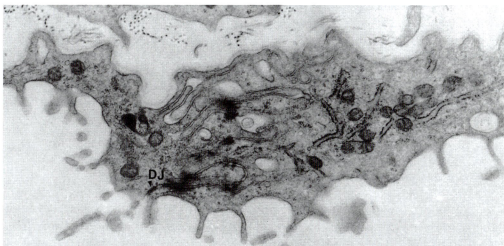

FIGURE 2.39. (*Top*) Section through the amnion of a 10.5-dpc embryo. (Meso) Cell derived from mesoderm; (Ect) cell derived from ectoderm; (BM) basement membrane; (DJ) desmosomal junction. The mesodermal cells are not joined by specialized junctions. (*Bottom*) Higher magnification of an ectodermal cell (Ect) in top panel.

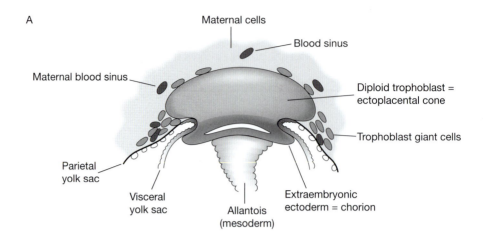

FIGURE 2.40. Schematic representation of the mouse placenta at 8.5 dpc (*A*) and 14.5 dpc (*B*). Note that in the older embryo, both the outer and inner placenta have a significant contribution of maternal cells. The best source of trophoblast tissue free of maternal contamination is therefore the 7.5- and 8.5-dpc ectoplacental cone (see Rossant and Croy 1985).

Apart from having an important function in the transfer of nutrients and metabolites into and out of the fetal circulation, the placenta synthesizes many steroid, polypeptide, and prostaglandin-type hormones that are required for the coordination of maternal and embryo physiology during pregnancy (see, e.g., Soares et al. 1985).

THE ADULT MOUSE

Mouse Coat Color and Its Genetics

As discussed above, the genes affecting coat color were among the first to be tested for Mendelian inheritance in mice. Since these early studies, more than 50

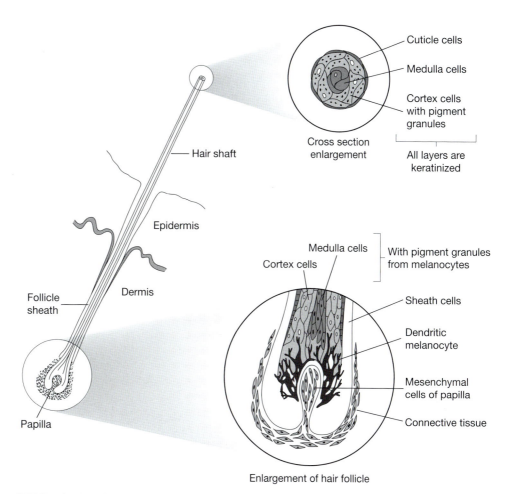

FIGURE 2.41. Schematic representation of a hair follicle to show the relative arrangement of cells derived from the epidermis (sheath, cortex, medulla), the dermis (mesenchymal cells of papilla and connective tissue), and the neural crest (melanocyte).

genes have been identified that affect hair growth and pigmentation. Here, only the briefest outline of the subject is given to stimulate the curiosity of researchers who will be handling mice for the first time. For a full and scholarly account of the genetics of coat color, see Silvers (1979) and Grüneberg (1952). Various aspects of hair follicle development and coat color genetics have been reviewed by Hardy (1992), Jackson (1993), and Barsh (1996).

During embryogenesis, each hair follicle develops from an epidermal thickening or placode that penetrates into the dermis and surrounds a condensation of mesodermal cells known as the dermal papilla (Fig. 2.41). In the mouse, there are ~50 hair follicles/mm^2, or ~500,000 over the entire skin (Potten 1985), and they are formed between 14 dpc and 3 days after birth. Melanocyte precursor cells (melanoblasts) migrate from the neural crest into the hair follicles and take up a position above the dermal papilla (Fig. 2.41). There are about 20–30 dendritic melanocytes in each follicle, and evidence suggests that they are derived by division from one founder melanoblast (Potten 1985). Each melanocyte has the capacity to synthesize two kinds of pigments, pheomelanin (yellow) and eumelanin (black or brown). Both pigments are derived from tyrosine and involve an initial conversion with the copper-containing enzyme tyrosinase. After this initial step, different enzyme systems generate the alternate chromatophores, which are then linked to proteins and incorporated into pigment granules of different size and shape. During active phases of the hair growth cycle, the melanocytes secrete these granules, which are taken up by the cortical and medullary epidermal cells of the hair shaft. Several different kinds of hair are found on the body, besides the whiskers (vibrissae); the three kinds of larger overhairs are the monotrich, awl, and auchene; and the more abundant, smaller underhairs are known as zigzags. Their morphology is determined by the mesodermal component of the follicle.

Since the genes affecting hair color were among the first to be studied, the loci were named alphabetically. Here, A (agouti), b (brown), c (albino), d (dilute), and p (pink-eyed dilution) are discussed. The wild-type alleles in the European house mouse are probably A^w, B^+, C^+, D^+, P^+ (Grüneberg 1952; Bultman et al. 1994). All of these genes have been cloned, and their analysis has shed light on a number of important problems in cell biology and development. This rich harvest is likely to continue as more genes are characterized. The genotypes of some common inbred mouse strains are provided in Table 2.9.

A (agouti) Chromosome 2

Hair from an agouti mouse is black with a subapical band of yellow (Fig. 2.42A). This pattern is generated by the influence of the hair follicle which transiently inhibits the production of black pigment by the melanocytes during an early phase of the growth cycle. Nonagouti (a/a) mice are therefore almost completely plain black (e.g., C57BL) or brown (e.g., C57BR), apart from a few yellow hairs on the ears and around the genitals (Fig. 2.42C, D). In newborn mice, the switch from eumelanin to pheomelanin production occurs between 3 and 6 days after birth so that chimeric pups made up of cells derived from agouti and nonagouti strains of mice cannot be distinguished before this time.

The *agouti* gene was isolated by a positional cloning strategy based on a radiation-induced chromosomal rearrangement with a breakpoint in the A gene (Bultman et al. 1992). It encodes a small, secreted protein and is expressed in the

TABLE 2.9. Coat color genotypes of common inbred mouse strains

Inbred strain (abbreviation)	Coat color phenotype	Allele at coat color loci[a]				
		a	*Tyrp1[b]*	*Tyr[c]*	*Myo5a[d]*	*p*
129S6/SvEvTac	agouti	A^w	+	+	+	+
129X1/SvJ	light chinchilla or albino	A^w	+	Tyr^{c-ch}/Tyr^c	+	*p*
BALB/c (C)	albino	+	$Tyrp1^b$	Tyr^c	+	+
C3H (C3)	agouti	+	+,	+	+	+
C57BL/6 (B6)	nonagouti (black)	*a*	+	+	+	+
CBA (CB)	agouti	+	+	+	+	+
DBA/2 (D2)	dilute brown nonagouti	*a*	$Tyrp1^b$	+	$Myo5a^d$	+
FVB	albino	+	+	Tyr^c	+	+
SJL (S or J)	albino	+	+	Tyr^c	+	*p*

[a]All loci are homozygous except for strain 129X1/SvJ in which two coat-color phenotypes occur because the *Tyr* locus is forced to remain heterozygous.

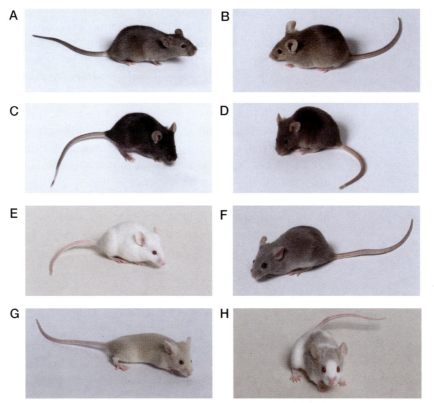

FIGURE 2.42. Mouse coat colors. (*A*) Agouti (*A/A, Tyrp1+/Tyrp1+*). An agouti mouse is generally described as "brownish-gray," and this is the basic color of wild *M. musculus domesticus* mice. (*B*) Brown agouti or cinnamon (*A/A, Tyrp1^b/Tyrp1^b*). A yellow band of pheomelanin on a brown hair results in this golden-colored mouse. (*C*) Black nonagouti (*a/a; Tyrp1+/Tyrp1+*). This is the color of C57BL/6 mice. (*D*) Brown nonagouti (*a/a; Tyrp1^b/Tyrp1^b*). This chocolate-colored mouse can be found among the F_1 intercross progeny of B6D2F1 mice that are used to generate transgenic mice. (*E*) Albino (*Tyr^c/Tyr^c*). Albino mice were recorded in Greek and Roman times, and the *c* locus was the first to be studied for Mendelian inheritance in mice. (*F*) Dilute brown nonagouti (*a/a; Tyrp1^b/Tyrp1^b; Myo5a^d/Myo5a^d*). DBA/2 mice are dilute brown nonagouti. (*G*) Pink-eyed dilution (*p/p*). The mouse shown is (*A/A; Tyrp1+/Tyrp1+; Tyr+/Tyr+; Myo5a+/Myo5a+; p/p*). Pink-eyed dilution is one of the old mutations of the mouse fancy. (*H*) Piebald (*s/s*) mice have large irregular white spots devoid of melanocytes. This very old mutation of the mouse fancy is caused by a mutation in the endothelin receptor type B gene (*Ednrb*).

hair follicles and not in the melanocytes (Miller et al. 1993). As a result of studies on another coat-color gene, known as the *extension* locus, which was shown to code for the melanocyte-stimulating hormone (MSH) receptor (Robbins et al. 1993), a model was proposed for the mechanism of action of the agouti protein (for review, see Jackson 1993). MSH normally stimulates the melanocyte to produce black rather than yellow pigment. However, the function of the agouti protein is to antagonize the action of MSH with its receptor so that the melanocyte transiently produces yellow pigment during the period when A is expressed. This explains why the overall yellow coat color resulting from the constitutive expression of agouti in *A^y* mutant mice (see below) can be overcome by injection of MSH, whereas similar treatment cannot reverse the effect of point mutations in the receptor encoded by the *extension* gene (for references, see Silvers 1979).

Molecular studies strongly suggest that the recessive allele *a* (nonagouti) is generated by insertion of a VL30 retrovirus-like transposable element in the first intron of the gene (Bultman et al. 1994). Reversion of *a* to *A* and *A*ʷ (white-bellied agouti, in which the ventral part of the coat is cream and the dorsal part is agouti) occurs at a high frequency due to recombination involving the VL30 sequences. It is likely that the true wild-type allele of the gene, present in most wild mice, is *A*ʷ. The mutation lethal yellow (*A*ʸ) has been particularly well studied because it is a homozygous lethal, with *A*ʸ/*A*ʸ embryos dying around implantation. Heterozygous *A*/*A*ʸ mice are completely yellow and show a variety of abnormalities, including increased susceptibility to spontaneous and induced tumors, obesity, and insulin-resistant diabetes. Studies have shown that the *A*ʸ mutation results from a deletion that brings the *A* gene under the control of an adjacent gene, *Raly*, which is constitutively expressed in all tissues (Michaud et al. 1993, 1994). The recessive lethality is unrelated to the *A* gene but is the result of the disruption in *Raly*, a ubiquitously expressed gene that encodes an RNA-binding protein implicated in premessenger RNA processing. The dominant pleiotypic effects of *A*ʸ, on the other hand, appear to be the result of ectopic expression of *A*.

b *(brown) Chromosome 4*

The wild-type allele at this locus, *B*, produces black eumelanin, whereas the more recessive allele, *b*, produces brown, an old mutation of the mouse fancy (Fig. 2.42D). *A*/*A*, *b*/*b* mice are known as cinnamon, a color produced by their yellow-banded brown hairs (Fig. 2.42B). The brown locus has been cloned (*Tyrp1*) and encodes a tyrosinase-related protein that catalyzes the oxidation of dihydroxyindole carboxylic acid, a more downstream step in the melanin biosynthesis (Jackson 1988).

c *(albino) Chromosome 7*

The wild-type (*C*) allele is dominant over all mutations at this locus and encodes the tyrosinase enzyme that is essential for pigment biosynthesis (Kwon et al. 1987). These mutants result in a deletion or alteration in the structure of tyrosinase and do not affect the number or distribution of the melanocytes. Albino mice (*Tyr*ᶜ/*Tyr*ᶜ) have no pigment at all either in the coat or in the eyes (Fig. 2.42E), whereas other mutants at the *c* locus have altered pigmentation, e.g., chinchilla *Tyr*ᶜ⁻ᶜʰ. It is important to realize that not all "albino" mice have the same coat-color genes; in the presence of *Tyr*ᶜ/*Tyr*ᶜ, the effects of changes at the *A*, *Tyrp1*, and *d* loci are masked.

d *(dilute) Chromosome 9*

The dilute locus is one of a class of genes affecting coat color through an alteration in the morphology of the melanocytes. This is an old mutation of the mouse fancy originally called Maltese dilution. In *d*/*d* mice, the pigment granules are clumped, and the melanocytes are less dendritic than those in wild-type mice. This causes a lightening (i.e., dilution) of the coat color (Fig. 2.42). Other *d* alleles also cause severe neurological defects. It is known that the *d* mutation in DBA/2J inbred mice is caused by integration of an ecotropic murine leukemia virus (MLV) retrovirus genome into the *Myo5a* gene, which encodes an unconventional myosin (Mercer et al. 1991). The gene is normally expressed in melanocytes and in neurons of the central and peripheral nervous systems.

p (pink-eyed dilution) Chromosome 7

The pink-eyed dilution locus affects both eye and coat pigmentation. Homozygous mutants have pink eyes with minimal pigmentation in the retina and choroids. Black pigmentation is greatly reduced, but yellow pigmentation is only minimally affected. This is also a very old mutation of the mouse fancy. Black agouti pink-eyed dilution (A/A, $Tyrp1^+/ Tyrp1^+$, p/p) mice are "fawn" colored (Fig. 2.42G). Black nonagouti pink-eyed dilution (a/a, $Tyrp1^+/ Tyrp1$, p/p) mice are "silver" colored. The gene encoded by the pink-eyed dilution locus has been cloned (Gardner et al. 1992). The *p* gene product is an integral membrane protein of eumelanosomes.

Spotting Mutations

A particularly interesting class of pigmentation mutants are those known as "spotting" (e.g., *Dominant white spotting* [*W*], *piebald* [*s*], *splotch* [*Sp*], and *belted* [*bt*]). An example of *piebald* is shown in Figure 2.42H. Among other factors, spotting genes affect the migration, viability, or differentiation of melanoblasts, the precursors of the melanocytes that arise from neural crest cells. Between 8.5 and 9 days of gestation, neural crest cells are still located near the neural tube; by 11 dpc, they have reached the skin of the trunk, and by 12 days, they have reached the limbs. Genes affecting melanocyte migration and/or survival are therefore more likely to result in the lack of pigment in the hair follicles on the ventral surface, forehead, and extremities, since the cells must migrate farther to reach these locations.

The *W* gene encodes the transmembrane tyrosine kinase, c-*kit*, which is expressed in melanoblasts, as well as germ cells and hematopoietic cells (Keshet et al. 1991) (see Origin of the Germ Line, p. 38). Homozygous null mutations of c-*kit* result in the complete absence of melanocytes in the skin. In heterozygous animals, some mutations cause a less severe "piebald" pigmentation trait both in mice and in humans (Fleischman et al. 1991; Giebel and Spritz 1991). Other mutations of *W* result in characteristic patterns of white (e.g., *W*sash; Duttlinger et al. 1993).

Morphological and Behavioral Mutants

The majority of mice present in most laboratory colonies appear overtly normal, only varying in coat color. However, morphological and behavioral mutants that are viable are readily identified as being variant. Some of the more commonly known morphological mutants are shown in Figure 2.43. These include *Brachyury* (*T*), *Small eye* (*Pax6*), *short ear* (*Bmp5*), *dwarf* (*Pit1*), *nude* (*Foxn1*), and *Extra toes* (*Gli3*).

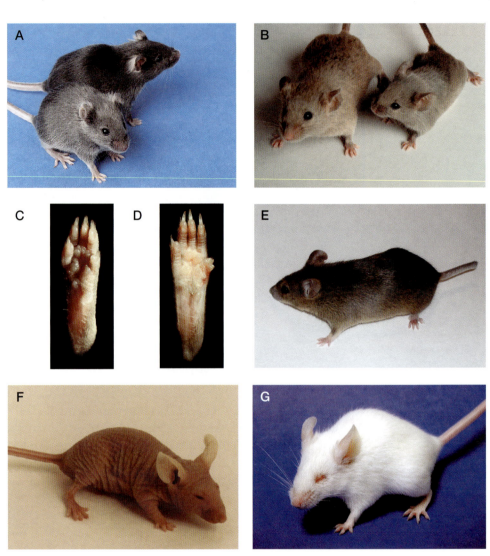

FIGURE 2.43. Morphological mutations in the mouse. (*A*) *short ear (se)* is a recessive viable and fertile mutation that causes defects in the ear cartilage and skeleton. *short ear* is caused by a mutation in the *Bmp5* locus. (*B*) *dwarf (dw)* is a recessive mutation that causes dwarfism due to growth hormone deficiency. *dwarf* is caused by a mutation in the *Pit1* locus that encodes a POU-domain transcription factor. (*C* and *D*) *Extra toes (Xt)* is a semidominant mutation that results in preaxial polydactyly of the limbs. *Extra toes* homozygous mutants are embryonic lethal with multiple abnormalities, including neural tube and limb defects. *Extra toes* is caused by a mutation in the *Gli3* gene that encodes a zinc finger transcription factor. (*E*) *Brachyury (T)* is a semidominant mutation that causes shortened tails with variable length, depending on genetic background. *Brachyury* homozygous mutants are embryonic lethal with no notochord and severe posterior abnormalities. *Brachyury* encodes a T-box transcription factor. (*F*) *nude (nu)* is a recessive mutation that causes hair defects and an absence of the thymus and therefore of T cells, resulting in immunodeficiency. *nude* is caused by a mutation in the *Foxn1* gene that encodes a forkhead transcription factor. Nude mice are widely used in immunological research and for the establishment of transplantable human tumors (Nomura et al. 1977). (*G*) *Small eye (Sey)* is a semidominant mutation that causes micropthalmia. *Small eye* homozygous mutants do not survive birth and lack eyes and nasal tissue. *Small eye* is caused by a mutation in the *Pax6* locus that encodes a paired homeodomain transcription factor.

REFERENCES

Albano R.M., Groome N., and Smith J.C. 1993. Activins are expressed in preimplantation mouse embryos and in ES and EC cells and are regulated on their differentiation. *Development* **117:** 711–723.

Albano R.M., Arkell R., Beddington R.S.P., and Smith J.C. 1994. Inhibin subunits and follistatin during postimplantation mouse development; decidual expression of activin and expression of follistatin in primitive streak, somites, and hindbrain. *Development* **120:** 803–813.

Amenta P.S., Clark C.C., and Martinez-Hernandez A. 1983. Deposition of fibronectin and laminin in the basement membrane of the rat parietal yolk sac: Immunohistochemical and biosynthetic studies. *J. Cell Biol.* **96:** 104–111.

Andrews G.K., Dziadek M., and Tamaoki T. 1982a. Expression and methylation of the mouse α-fetoprotein gene in embryonic, adult, and neoplastic tissues. *J. Biol. Chem.* **257:** 5148–5153.

Andrews G.K., Janzen R.G., and Tamaoki T. 1982b. Stability of α-fetoprotein messenger RNA in mouse yolk sac. *Dev. Biol.* **89:** 111–116.

Ashworth A., Rastan S., Lovell-Badge R., and Kay G.F. 1990. X inactivation may explain the difference in viability of XO humans and mice. *Nature* **351:** 406–408.

Austin C.R. and Edwards R.G. 1981. *Mechanisms of sex differentiation in animals and man.* Academic Press, London.

Awatramani R., Soriano P., Mai J.J., and Dymecki S. 2001. An Flp indicator mouse expressing alkaline phosphatase from the *ROSA26* locus. *Nat. Genet.* **29:** 257–259.

Bachvarova R., Cohen E.M., De Leon V., Tokunaga K., Sakiyama S., and Paynton B.V. 1989. Amounts and modulation of actin mRNAs in mouse oocytes and embryos. *Development* **106:** 561–565.

Balling R., Deutsch U., and Gruss P. 1988. Undulated, a mutation affecting the development of the mouse skeleton, has a point mutation in the paired box of Pax1. *Cell* **55:** 531–535.

Balling R., Lau C.F., Dietrich S., Wallin J., and Gruss P. 1992. Development of the skeletal system. *CIBA Found. Symp.* **165:** 132–143.

Barlow D.P., Green N.R., Kurkinen M., and Hogan B.L.M. 1984. Sequencing of laminin B chain cDNAs reveal C-terminal regions of coiled-coil alpha helix. *EMBO J.* **3:** 2355–2362.

Barlow D.P., Stroger R., Herrmann B.G., Saito K., and Scweifer N. 1991. The mouse insulin-like growth factor type II receptor is imprinted and closely linked to the Tme locus. *Nature* **349:** 84–87.

Barnes J.D., Crosby J.L., Jones C.M., Wright C.V.E., and Hogan B.L.M. 1994. Embryonic expression of Lim-1, the mouse homolog of *Xenopus* neurogenesis. *Dev. Biol.* **161:** 168–178.

Barsh G.S. 1996. The genetics of pigmentation: From fancy genes to complex traits. *Trends Genet.* **12:** 299–305.

Basler K., Edlund T., Jessell T.M., and Yamada T. 1993. Control of cell pattern in the neural tube: Regulation of cell differentiation by dorsalin-1, a novel TGFβ family member. *Cell* **73:** 687–702.

Beddington R.S.P. 1981. An autoradiographic analysis of the potency of embryonic ectoderm in the 8th day postimplantation mouse embryo. *J. Embryol. Exp. Morphol.* **64:** 87–104.

———. 1982. An autoradiographic analysis of tissue potency in different regions of the embryonic ectoderm during gastrulation in the mouse. *J. Embryol. Exp. Morphol.* **69:** 265–285.

———. 1983. Histogenic and neoplastic potential of different regions of the mouse embryonic egg cylinder. *J. Embryol. Exp. Morphol.* **75:** 189–204.

———. 1994. Induction of a second neural axis by the mouse node. *Development* **120:** 613–620.

Beddington R.S. and Lawson K.A. 1990. Clonal analysis of cell lineages. In

Postimplantation mammalian embryos: A practical approach (ed. A.J. Copp and D.L. Cockroft), pp. 267–292. IRL Press at Oxford University Press, England.

Beddington R.S. and Robertson E.J. 1999. Axis development and early asymmetry in mammals. *Cell* **96:** 195–209.

Beddington R.S.P. and Smith J.C. 1993. The control of vertebrate gastrulation: Inducing signals and responding genes. *Curr. Opin. Genet. Dev.* **3:** 655–661.

Beddington R.S.P., Morgernstern J., Land H., and Hogan A. 1989. An in situ transgenic enzyme marker for the midgestation mouse embryo and the visualization of inner cell mass clones during early organogenesis. *Development* **106:** 37–46.

Behringer R.R., Cate R.L., Froelick G.J., Palmiter R.D., and Brinster R.L. 1990. Abnormal sexual development in transgenic mice chronically expressing Mullerian inhibiting substance. *Nature* **345:** 167–170.

Behringer R.R., Finegold M.J., and Cate R.L. 1994. Mullerian-inhibiting substance function during mammalian sexual development. *Cell* **79:** 415–425.

Bhatt H., Brunet L.J., and Stewart C.L. 1992. Uterine expression of leukemia inhibitory factor coincides with the onset of blastocyst implantation. *Proc. Natl. Acad. Sci.* **88:** 11408–11412.

Bishop C.E., Boursot P., Baron B., Bonhomme F., and Hatat D. 1985. Most classical *Mus musculus domesticus* laboratory mouse strains carry a *Mus musculus musculus* Y chromosome. *Nature* **315:** 70–72.

Bleil J.D. and Wassarman P.M. 1980a. Structure and function of the zona pellucida: Identification and characterisation of the proteins of the mouse oocyte zona pellucida. *Dev. Biol.* **76:** 185–202.

———. 1980b. Synthesis of zona pellucida proteins by denuded and follicle-enclosed mouse oocytes during culture in vitro. *Proc. Natl. Acad. Sci.* **77:** 1029–1033.

Bober E., Lyons G.E., Braun T., Cossu G., Buckingham M.J., and Arnold H. 1991. The muscle regulatory gene, myf-6, has a biphasic pattern of expression during early mouse development. *J. Cell Biol.* **113:** 1255–1265.

Bonnerot C. and Nicolas J.-F. 1993. Clonal analysis in the intact mouse embryos by intragenic homologous recombination. *C.R. Acad. Sci. Paris. Ser. III* **316:** 1207–1217.

Borsani G., Tonlorenzi R., Simmler M.C., Dandolo L., Arnaud D., Capra V., Grompe M., Pizzuti A., Muzny D., Lawrence C., Willard H., Avner P., and Ballabio A. 1991. Characterization of a murine gene expressed from the inactive X chromosome. *Nature* **351:** 325–329.

Bradley A. and Liu P. 1996. Target practice in transgenics. *Nat. Genet.* **14:** 121–123.

Bradley A., Evans M., Kaufman M.H., and Robertson E. 1984. Formation of germ-line chimaeras from embryo-derived teratocarcinoma cell lines. *Nature* **309:** 255–256.

Braun R.E., Behringer R.R., Peschon J.J., Brinster R.L., and Palmiter R.D. 1989. Genetically haploid spermatids are phenotypically diploid. *Nature* **337:** 373–376.

Braun T. and Arnold H.H. 1995. Inactivation of *Myf-6* and *Myf-5* genes in mice leads to alterations in skeletal muscle development. *EMBO J.* **14:** 1176–1186.

Brinster R.L. 1974. The effect of cells transferred into mouse blastocyst on subsequent development. *J. Exp. Med.* **140:** 1049–1056.

———. 2002. Germline stem cell transplantation and transgenesis. *Science* **296:** 2174–2176.

Brinster R.L., Chen H.Y., Trumbauer M.E., and Avarbock M.R. 1980. Translation of globin mRNA by the mouse ovum. *Nature* **283:** 499–501.

Brockdorff N. 2002. X-chromosome inactivation: Closing in on proteins that bind *Xist* RNA. *Trends Genet.* **18:** 352–358.

Bronner-Fraser M. 1985. Alteration in neural crest cell migration by a monoclonal antibody that affects cell adhesion. *J. Cell Biol.* **101:** 610–617.

Brown C.J., Ballabio A., Rupert J.L., Lafreniere R.G., Grompe M., Tonlorenzi R., and Willard H. 1991. A gene from the region of the human X inactivation center is expressed exclusively from the inactive X chromosome. *Nature* **349:** 38–44.

Brown N.A., McCarthy A., and Seo J. 1992. Development of the left-right axes. *CIBA Found. Symp.* **165:** 144–154.

Buehr M., Gu S., and McLaren A. 1993. Mesonephric contribution to testis differentiation in the fetal mouse. *Development* **117:** 273–281.

Bultman S.J., Michaud E.J., and Woychik R.P. 1992. Molecular characterization of the mouse agouti locus. *Cell* **71:** 1195–1204.

Bultman S.J., Klebig M.L., Michaud E.J., Sweet H.O., Davisson M.T., and Woychik R.P. 1994. Molecular analysis of reverse mutations from nonagouti (a) to black-and-tan (at) and white-bellied agouti (Aw) reveals alternative forms of agouti transcripts. *Genes Dev.* **8:** 481–490.

Burgoyne P. 1987. The role of the mammalian Y chromosome in spermatogenesis. *Development* (suppl.) **101:** 133–141.

Burgoyne P.S., Tam P.P.L., and Evans E.P. 1983. Retarded development of XO conceptuses during early pregnancy in the mouse. *J. Reprod. Fertil.* **68:** 387–393.

Burgoyne P.S., Buehr M., Koopman P.R.J., and McLaren A. 1988. Cell autonomous action of the testis-determining gene: Sertoli cells are exclusively XY in XX-XY mouse testes. *Development* **102:** 443–450.

Candia A.F., Hu J., Crosby J., Lalley P.A., Noden D., Nadeau J.H., and Wright C.V.E. 1992. Mox-1 and Mox-2 define a novel homeobox gene subfamily and are differentially expressed during mesodermal patterning in mouse embryos. *Development* **116:** 1123–1136.

Capdevila J. and Izpisúa Belmonte J.C. 2001. Patterning mechanisms controlling vertebrate limb development. *Annu. Rev. Cell Dev. Biol.* **17:** 87–132.

Cascio S.M. and Wassarman P.M. 1982. Program of early development in the mammal: Post-transcriptional control of a class of proteins synthesised by mouse oocytes and early embryos. *Dev. Biol.* **89:** 397–408.

Cattanach B. and Beechey C.V. 1990. Autosomal and X-chromosome imprinting. *Development* (suppl.) **90:** 63–72.

Chabot B., Stephenson D.A., Chapman V.M., and Besmer P.B. 1988. The proto-oncogene c-kit encoding a transmembrane tyrosine kinase receptor maps to the mouse W locus. *Nature* **325:** 88–89.

Chapman V.M. 1986. X chromosome regulation in oogenesis and early mammalian development. In *Experimental approaches to mammalian embryonic development* (ed. J. Rossant and R.A. Pedersen), pp. 365–398. Cambridge University Press, England.

Chapman V., Forrester L., Sanford J., Hastie N., and Rossant J. 1984. Cell lineage specific undermethylation of mouse repetitive DNA. *Nature* **307:** 284–286.

Cheong H.T., Takahashi Y., and Kanagawa H. 1993. Birth of mice after transplantation of early cell-cycle-stage embryonic nuclei into enucleated oocytes. *Biol. Reprod.* **48:** 958–963.

Chisaka O. and Capecchi M.R. 1991. Regionally restricted developmental defects resulting from targeted disruption of the mouse homeobox gene Hox-1.5. *Nature* **350:** 473–479.

Christ B., Brand-Saberi B., Grim M., and Wilting J. 1992. Local signalling in dermomyotomal cell type specification. *Anat. Embryol.* **186:** 505–510.

Ciemerych M.A., Mesnard D., and Zernicka-Goetz M. 2000. Animal and vegetal poles of the mouse egg predict the polarity of the embryonic axis, yet are nonessential for development. *Development* **127:** 3467–3474.

Clark J.M. and Eddy E.M. 1975. Fine structural observations on the origin and association of primordial germ cells in the mouse. *Dev. Biol.* **47:** 136–155.

Clegg K.B. and Piko L. 1983. Poly(A) length, cytoplasmic adenylation and synthesis of poly(A)+ RNA in early mouse embryos. *Dev. Biol.* **95:** 331–341.

Colledge W.H., Carlton M.B.L., Udy G.B., and Evans M.J. 1994. Disruption of c-mos causes parthenogenetic development of unfertilized mouse eggs. *Nature* **370:** 65–68.

Conlon F.L., Lyons K.M., Takaesu N., Barth K.S., Kispert A., Herrmann B., and Robertson E.J. 1994. A primary requirement for *nodal* in the formation and maintenance of the primitive streak in the mouse. *Development* **120:** 1919–1928.

Conlon R.A., Reaume A.G., and Rossant J. 1995. *Notch1* is required for the coordinate segmentation of somites. *Development* **121:** 1533–1545.

Cooke J. and Zeeman E.C. 1976. A clock and wavefront model for control of the number of repeated structures during animal morphogenesis. *J. Theor. Biol.* **58:** 455–476.

Copp A.J., Brook F.A., Estibeiro J.P., Shum A.S., and Cockroft D.L. 1990. The embryonic development of mammalian neural tube defects. *Prog. Neurobiol.* **35:** 363–403.

Couly G.F., Coltey P.M., and LeDouarin N.M. 1993. The triple origin of skull in higher vertebrates: A study in quail-chick chimeras. *Development* **117:** 409–429.

Crossley P.H. and Martin G.R. 1995. The mouse *Fgf8* gene encodes a family of polypeptides and is expressed in regions that direct outgrowth and patterning in the developing embryo. *Development* **121:** 439–451.

Curran T., Miller A.D., Zokas L., and Verma I.M. 1984. Viral and cellular fos proteins: A comparative analysis. *Cell* **36:** 259–268.

Damjanov I. and Solter D. 1974. Experimental teratoma. *Curr. Top. Pathol.* **59:** 69–130.

Damjanov I., Solter D., and Skreb N. 1971. Teratocarcinogenesis as related to the age of embryos grafted under the kidney capsule. *Wilhelm Roux' Arch. Entwicklungsmech. Org.* **173:** 282–284.

Dardik A., Smith R.M., and Schultz R.M. 1992. Colocalization of transforming growth factor-a and a functional epidermal growth factor receptor (EGFR) to the inner cell mass and preferential localization of the EGFR on the basolateral surface of the trophectoderm in the mouse blastocyst. *Dev. Biol.* **154:** 396–409.

Davidson D.R. and Hill R.E. 1991. Msh-like genes: A family of homeobox genes with wide ranging expression during vertebrate development. *Semin. Dev.* **2:** 405–414.

DeChiara T.M., Robertson E.J., and Efstratiadis A. 1991. Parental imprinting of the mouse insulin-like growth factor II gene. *Cell* **64:** 849–859.

DeGregori J., Russ A., von Melchner H., Rayburn H., Priyaranjan P., Jenkins N.A., Copeland N.G., and Ruley H.E. 1994. A murine homolog of the yeast RNA1 gene is required for postimplantation development. *Genes Dev.* **8:** 265–276.

Delbridge M.L. and Graves J.A. 1999. Mammalian Y chromosome evolution and the male-specific functions of Y chromosome-borne genes. *Rev. Reprod.* **4:** 101–109.

DePrimo S.E., Stambrook P.J., and Stringer J.R. 1996. Human placental alkaline phosphatase as a histochemical marker of gene expression in transgenic mice. *Transgenic Res.* **5:** 459–466.

Deutsch U. and Gruss P. 1991. Murine paired domain proteins as regulatory factors of embryonic development. *Semin. Dev. Biol.* **2:** 415–424.

Dewey M.J., Martin D.W. Jr., Martin G.R., and Mintz B. 1977. Mosaic mice with teratocarcinoma-derived mutant cells deficient in hypoxanthine phosphoribosyltransferase. *Proc. Natl. Acad. Sci.* **74:** 5564–5568.

Dietrich S., Schubert F.R., and Gruss P. 1994. Altered Pax gene expression in murine notochord mutants: The notochord is required to initiate and maintain ventral identity in the somite. *Mech. Dev.* **44:** 189–207.

Diwan S.B. and Stevens L.C. 1976. Development of teratomas from ectoderm of mouse egg cylinders. *J. Natl. Cancer Inst.* **57:** 937–942.

Dodd J. and Jessell T.M. 1993. Axon guidance in the mammalian spinal cord. In *Cell-cell signaling in vertebrate development* (ed. E.J. Robertson et al.), pp. 81–95. Academic Press, New York.

Downs K.M. and Davies T. 1993. Staging of gastrulating mouse embryos by morphological landmarks in the dissecting microscope. *Development* **118:** 1255–1266.

Dudley A.T., Ros M.A., and Tabin C.J. 2002. A re-examination of proximodistal patterning during vertebrate limb development. *Nature* **418:** 539–544.

Duttlinger R., Manova K., Chu T.Y., Gyssler C., Zelenetz A.D., Bachvarova R.F., and Besmer P. 1993. W-sash affects positive and negative elements controlling c-kit expression: Ectopic c-kit expression at sites of kit-ligand expression affects melanogenesis. *Development* **118:** 705–717.

Dziadek M. 1978. Modulation of alpha-foetoprotein synthesis in the early postimplantation mouse embryo. *J. Embryol. Exp. Morphol.* **46:** 135–146.

Dziadek M. and Adamson E. 1978. Localisation and synthesis of alpha-foetoprotein in post-implantation mouse embryos. *J. Embryol. Exp. Morphol.* **43:** 289–313.

Dziadek M.A. and Andrews G.K. 1983. Tissue specificity of alpha-fetoprotein messenger

RNA expression during mouse embryo-genesis. *EMBO J.* **2:** 549–554.

Echelard Y., Epstein D.J., St-Jacques B., Shen L., Mohler J., McMahon J.A., and McMahon A.P. 1993. Sonic hedgehog, a member of a family of putative signaling molecules, is implicated in the regulation of CNS polarity. *Cell* **75:** 1417–1430.

Eddy E.M. and Hahnel A.C. 1983. Establishment of the germ cell line in mammals. In *Current problems in germ cell differentiation* (ed. A. McLaren and C.C. Wylie), pp. 41–70. Cambridge University Press, England.

Eddy E.M., Clark J.M., Gong D., and Fenderson B.A. 1981. Origin and migration of primordial germ cells in mammals. *Gamete Res.* **4:** 333–362.

Eiferman F.A., Young P.R., Scott R.W., and Tilghman S.M. 1981. Intragenic amplification and divergence in the mouse gene. *Nature* **294:** 713–718.

Enders A.C., Chavez P.J., and Schlafke S. 1981. Comparison of implantation in utero and in vitro. In *Cellular and molecular aspects of implantation* (ed. S.R. Glasser and D.W. Bullock), pp. 365–382. Plenum Press, New York.

Enders A.C., Given R.L., and Schlafke S. 1978. Differentiation and migration of endoderm in the rat and mouse at implantation. *Anat. Rec.* **190:** 65–78.

Eppig J.J. and Telfer E.E. 1993. Isolation and culture of oocytes. *Methods Enzymol.* **225:** 77–84.

Erickson C.A. and Weston J.A. 1983. An SEM analysis of neural crest cell migration in the mouse. *J. Embryol. Exp. Morphol.* **74:** 97–118.

Evans M.J. and Kaufman M.H. 1981. Establishment in culture of pluripotential cells from mouse embryos. *Nature* **292:** 154–156.

Evrard Y.A., Lun Y., Aulehla A., Gan L., and Johnson R.L. 1998. lunatic fringe is an essential mediator of somite segmentation and patterning. *Nature* **394:** 377–381.

Faddy M.J., Gosden R.G., and Edwards R.G. 1983. Ovarian follicle dynamics in mice: A comparative study of three inbred strains and a F1 hybrid. *J. Endocrinol.* **96:** 23–24.

Falconer D.S. and Avery P.J. 1978. Variability of chimeras and mosaics. *J. Embryol. Exp. Morphol.* **43:** 195–219.

Figdor M.C. and Stern C.D. 1993. Segmental organization of embryonic diencephalon. *Nature* **363:** 630–634.

Finn C.A. 1971. The biology of decidual cells. *Adv. Reprod. Physiol.* **5:** 1–26.

Flach G., Johnson M.H., Braude P.R., Taylor R.A.S., and Bolton V.N. 1982. The transition from maternal to embryonic control in the 2-cell mouse embryo. *EMBO J.* **1:** 681–686.

Fleischman R.A., Saltman D.L., Stastny V., and Zneimer S. 1991. Deletion of the c-kit proto-oncogene in the human developmental defect piebald trait. *Proc. Natl. Acad. Sci.* **88:** 10885–10889.

Friedrich G. and Soriano P. 1991. Promoter traps in embryonic stem cells: A genetic screen to identify and mutate developmental genes in mice. *Genes Dev.* **5:** 1513–1523.

Gardner J.M., Nakatsu Y., Gondo Y., Lee S., Lyon M.F., King R.A., and Brilliant M.H. 1992. The mouse pink-eyed dilution gene: Association with human Prader-Willi and Angelman syndromes. *Science* **257:** 1121–1124.

Gardner R.L. 1982. Investigation of cell lineage and differentiation in the extraembryonic endoderm of the mouse embryo. *J. Embryol. Exp. Morphol.* **68:** 175–198.

———. 1983. Origin and differentiation of extra-embryonic tissues in the mouse. *Int. Rev. Exp. Pathol.* **24:** 63–133.

———. 1984. An in situ cell marker for clonal analysis of development of the extraembryonic endoderm in the mouse. *J. Embryol. Exp. Morphol.* **80:** 251–288.

Gardner R.L. and R.S.P. Beddington. 1988. Multi-lineage "stem cells" in the mammalian embryo. *J. Cell Sci.* (suppl.) **10:** 11–27.

Gardner R.L. and Rossant J. 1979. Investigation of the fate of 4.5 d post coitum mouse ICM cells by blastocyst injection. *J. Embryol. Exp. Morphol.* **52:** 141–152.

Gardner R.L., Davies T.J., and Carey M.S. 1988. Effect of delayed implantation on differentiation of the extra-embryonic endoderm in the mouse blastocyst. *Placenta* **9:** 343–359.

Gardner R.L., Meredith M.R., and Altman D.G. 1992. Is the anterior-posterior axis of the fetus specified before implantation in the mouse? *J. Exp. Zool.* **264:** 437–443.

Gardner R.L., Lyon M.F., Evans E.P., and Burtenshaw M.D. 1985. Clonal analysis of X-chromosome inactivation and the origin of the germ line in the mouse embryo. *J. Embryol. Exp. Morphol.* **88:** 349–363.

Gavin B.J., McMahon J.A., and McMahon A.P. 1990. Expression of multiple novel Wnt-1/int-1-related genes during fetal and adult mouse development. *Genes Dev.* **4:** 2319–2332.

Giebel L.B. and Spritz R.A. 1991. Mutation of the KIT (mat/stem cell growth factor receptor) proto-oncogene in human piebaldism. *Proc. Natl. Acad. Sci.* **88:** 8696–8699.

Gielbelhaus D.H., Heikkila J.J., and Schultz G.A. 1983. Changes in the quantity of histone and actin mRNA during the development of pre-implantation mouse embryos. *Dev. Biol.* **98:** 148–154.

Gimlich R.L. and Braun J. 1985. Improved fluorescent compounds for tracing cell lineage. *Dev. Biol.* **109:** 509–514.

Ginsburg M., Snow M.H.L., and McLaren A. 1990. Primordial germ cells in the mouse embryo during gastrulation. *Development* **110:** 521–528.

Goldman D.C., Martin G.R., and Tam P.P. 2000. Fate and function of the ventral ectodermal ridge during mouse tail development. *Development* **127:** 2113–2123.

Goldstein R.S. and Kalcheim C. 1992. Determination of epithelial half-somites in skeletal morphogenesis. *Development* **116:** 441–445.

Gorin M.B. and Tilghman S.M. 1980. Structure of the gene in the mouse. *Proc. Natl. Acad. Sci.* **77:** 1351–1355.

Goulding M.D., Lumsden A., and Gruss P. 1993. Signals from the notochord and floor plate regulate the region-specific expression of two Pax genes in the developing spinal cord. *Development* **117:** 1001–1016.

Graham A., Heyman I., and Lumsden A. 1993. Even-numbered rhombomeres control the apoptotic elimination of neural crest cells from odd-numbered rhombomeres in the chick hindbrain. *Development* **119:** 233–245.

Graham C.F. 1977. Teratocarcinoma cells and normal mouse embryogenesis. In *Concepts in mammalian embryogenesis* (ed. M.I. Sherman.), pp. 315–394. MIT Press, Cambridge, Massachusetts.

Graham C.F. and Deussen Z.A. 1978. Features of cell lineage in preimplantation mouse embryos. *J. Embryol. Exp. Morphol.* **48:** 53–72.

Gregg B.C. and Snow M.H.L. 1983. Axial abnormalities following disturbed growth in mitomycin C-treated mouse embryos. *J. Embryol. Exp. Morphol.* **73:** 135–149.

Greve J.M. and Wassarman P.M. 1985. Mouse egg extracellular coat is a matrix of interconnected filaments possessing a structural repeat. *J. Mol. Biol.* **181:** 253–264.

Grüenberg H. 1952. *The genetics of the mouse*, 2nd edition. Martinus Nijhoff, The Hague.

Gubbay J., Collignon J., Koopman P., Capel B., Economou A., Munsterberg A., Vivian N., Goodfellow P., and Lovell-Badge R. 1990. A gene mapping to the sex-determining region of the mouse Y chromosome is a member of a novel family of embryonically expressed genes. *Nature* **346:** 245–250.

Guillemot F., Nagy A., Auerbach A., Rossant J., and Joyner A.L. 1994. Essential role of Mash-2 in extraembryonic development. *Nature* **371:** 333–336.

Hadjantonakis A.K., Macmaster S., and Nagy A. 2002. Embryonic stem cells and mice expressing different GFP variants for multiple non-invasive reporter usage within a single animal. *BMC Biotechnol.* **2:** 11.

Hadjantonakis A.K., Gertsenstein M., Ikawa M., Okabe M., and Nagy A. 1998. Generating green fluorescent mice by germline transmission of green fluorescent ES cells. *Mech. Dev.* **76:** 79–90.

Hahnel A.C., Rappolee D.A., Millan J.L., Manes T., Ziomek C.A., Theodosiou N.G., Werb Z., Pederson R.A., and Schultz G.A. 1990. Two alkaline phosphatase genes are expressed during early development in the mouse embryo. *Development* **110:** 555–564.

Hart A.H., Hartley L., Sourris K., Stadler E.S., Li R., Stanley E.G., Tam P.P., Elefanty A.G., and Robb L. 2002. *Mixl1* is required for axial mesendoderm morphogenesis and patterning in the murine embryo. *Development* **129:** 3597–3608.

Hardy M.H. 1992. The secret life of the hair follicle. *Trends Genet.* **8:** 55–61.

Hashimoto K. and Nakatsuji N. 1989. Formation of the primitive streak and mesoderm cells in mouse embryos—Detailed scanning electron microscopical study. *Dev. Growth Differ.* **31:** 209–218.

Hashimoto N., Watanabe N., Furuta Y., Tamemoto H., Sagata N., Yokoyama M., Okazaki K., Nagayoshi M., Takeda N., Ikawa Y., and Aizawa S. 1994. Parthenogenetic activation of oocytes in c-mos-deficient mice. *Nature* **370:** 68–71.

Harvey M.B. and Kaye P.L. 1990. Insulin increases the cell number of the inner cell mass and stimulates morphological development of mouse blastocysts in vitro. *Development* **110:** 963–967.

———. 1992. IGF-2 stimulates growth and metabolism of early mouse embryos. *Mech. Dev.* **38:** 169–174.

Hasty P., Bradley A., Morris J.H., Edmondson D.G., Venuti J.M., Olson E.N., and Klein W.H. 1993. Muscle deficiency and neonatal death in mice with a targeted mutation in the myogenin gene. *Nature* **364:** 501–506.

Hecht N.B. 1986. Regulation of gene expression during mammalian spermatogenesis. In *Experimental approaches to mammalian embryonic development* (ed. J. Rossant and R.A. Pedersen), pp. 151–193. Cambridge University Press, New York.

Hermesz E., Mackem S., and Mahon K.A. 1996. *Rpx*: A novel anterior-restricted homeobox gene progressively activated in the prechordal plate, anterior neural plate and Rathke's pouch of the mouse embryo. *Development* **122:** 41–52.

Hogan B.L.M. 1980. High molecular weight extracellular proteins synthesized by endoderm cells derived from mouse teratocarcinoma cells and normal extra-embryonic membranes. *Dev. Biol.* **76:** 275–285.

Hogan B.L.M. and Tilly R. 1981. Cell interactions and endoderm differentiation in cultured mouse embryos. *J. Embryol. Exp. Morphol.* **62:** 379–394.

Hogan B.L.M., Barlow D.P., and Kurkinen M. 1984. Reichert's membrane as a model system for biosynthesis of basement membrane components. *CIBA Found. Symp.* **108:** 60–69.

Hogan B.L.M., Barlow D.P., and Tilly R. 1983. F9 teratocarcinoma cells as a model for the differentiation of parietal and visceral endoderm in the mouse embryo. *Cancer Surv.* **2:** 115–140.

Hogan B.L.M., Thaller C., and Eichele G. 1992. Evidence that Hensen's node is a site of retinoic acid synthesis. *Nature* **359:** 237–241.

Holland P., Ingham P., and Krauss S. 1992. Mice and flies head to head. *Nature* **358:** 627–628.

Howlett S.K. and Bolton V.N. 1985. Sequence and regulation of morphological and molecular events during the first cell cycle of mouse embryogenesis. *J. Embryol. Exp. Morphol.* **87:** 175–206.

Hrabe de Angelis M., McIntyre J. 2nd, and Gossler A. 1997. Maintenance of somite borders in mice requires the Delta homologue *DlI1*. *Nature* **386:** 717–721.

Hsueh A.J.W., Adashi E.Y., Jones P.B.C., and Welsh T.H. 1984. Hormonal regulation of the differentiation of cultured ovarian granulosa cells. *Endocrine Rev.* **5:** 76–127.

Huang E.J., Manova K., Packer A.I., Sanchez S., Bachvarova R.F., and Besmer P. 1993. The murine steel panda mutation affects kit ligand expression and growth of early ovarian follicles. *Dev. Biol.* **157:** 100–109.

Huarte J., Belin D., and Vassalli J.-D. 1988a. Plasminogen activator in mouse and rat oocytes: Induction during meiotic maturation. *Cell* **43:** 551–558.

Huarte J., Belin D., Vassalli A., Strickland S., and Vassalli J.-D. 1988b. Meiotic maturation of mouse oocytes triggers the translation and polyadenylation of dormant tissue-type plasminogen activator mRNA. *Genes Dev.* **1:** 1201– 1211.

Hyafil F., Babinet C., and Jacob F. 1981. Cell-cell interactions in early embryogenesis: A molecular approach to the role of calcium. *Cell* **26:** 447–454.

Hyafil F., Morello D., Babinet C., and Jacob F. 1980. A cell surface glycoprotein involved in the compaction of embryonal carcinoma cells and cleavage stage embryos. *Cell* **21:** 927–934.

Jackson B.W., Grund C., Winter S., Franke W.W., and lllmensee K. 1981. Formation of cytoskeletal elements during mouse embryogenesis. II. Epithelial differentiation and intermediate-sized filaments in early postimplantation embryos. *Differentiation* **20:** 203–216.

Jackson B.W., Grund C., Schmid E., Burki K., Franke W.W., and Illmensee K. 1980. Formation of cytoskeletal elements during mouse embryogenesis. I. Intermediate filaments of the cytokeratin type and desmosomes in preimplantation embryos. *Differentiation* **17:** 161–179.

Jackson I. 1993. Color-coded switches. *Nature* **362:** 587–588.

Jackson I.J. 1988. A cDNA encoding tyrosinase-related protein maps to the brown locus in mouse. *Proc. Natl. Acad. Sci.* **85:** 4392–4396.

Janzen R.G., Andrews G.L., and Tamaoki T. 1982. Synthesis of secretory proteins in developing mouse yolk sac. *Dev. Biol.* **90:** 18–23.

Jessell T.M. 2002. Neuronal specification in the spinal cord: Inductive signals and transcriptional codes. *Nat. Rev. Genet.* **1:** 20–29.

John M., Carswell E., Boyse E.A., and Alexander G. 1972. Production of θ antibody by mice that fail to reject θ incompatible skin grafts. *Nature New Biol.* **238:** 57–58.

Johnson M.H. 1981. The molecular and cellular basis of preimplantation mouse development. *Biol. Rev.* **56:** 463–498.

Johnson M.H. and Ziomek C.A. 1981. The foundation of two distinct cell lineages within the mouse morula. *Cell* **24:** 71–80.

Johnson M.H., Chisholm J.S., Fleming T.P., and Houliston E. 1986. A role for cytoplasmic determinants in the development of the mouse early embryo? *J. Embryol. Exp. Morphol.* (suppl.) **97:** 97–121.

Johnson R.L. and Tabin C.J. 1997. Molecular models for vertebrate limb development. *Cell* **90:** 979–990.

Jones K.W. and Singh L. 1981. Conserved repeated DNA sequences in vertebrate sex chromosomes. *Hum. Genet.* **58:** 46–53.

Josso N., Racine C., di Clemente N., Rey R., and Xavier F. 1998. The role of anti-Mullerian hormone in gonadal development. *Mol. Cell. Endocrinol.* **145:** 3–7.

Jurand, A. 1962. The development of the notochord in chick embryos. *J. Embryol. Exp. Morphol.* **10:** 602–621.

———. 1974. Some aspects of the development of the notochord in mouse embryos. *J. Embryol. Exp. Morphol.* **32:** 1–33.

Kafri T., Ariel M., Brandeis M., Shemer R., Urven L., McCarrey J., Cedar H., and Razin A. 1992. Developmental pattern of gene-specific DNA methylation in the mouse embryo and germ line. *Genes Dev.* **6:** 705–714.

Kalantry S., Manning S., Haub O., Tomihara-Newberger C., Lee H.G., Fangman J., Disteche C.M., Manova K., and Lacy E. 2001. The *amnionless* gene, essential for mouse gastrulation, encodes a visceral-endoderm-specific protein with an extracellular cysteine-rich domain. *Nat. Genet.* **27:** 412–416.

Kanai-Azuma M., Kanai Y., Gad J.M., Tajima Y., Taya C., Kurohmaru M., Sanai Y., Yonekawa H., Yazaki K., Tam P.P., and Hayashi Y. 2002. Depletion of definitive gut endoderm in *Sox17*-null mutant mice. *Development* **129:** 2367–2379.

Kaufman M.H. 1982. The chromosome complement of single-pronuclear haploid mouse embryos following activation by ethanol treatment. *J. Embryol. Exp. Morphol.* **71:** 139–154.

———. 1983b. Ethanol-induced chromosomal abnormalities at conception. *Nature* **302:** 258–260.

———. 1992. *The atlas of mouse development.* Academic Press, London.

Kaufman M.H., Barton S.C., and Surani M.A.H. 1977. Normal post-implantation development of mouse parthenogenetic embryos to the forelimb bud stage. *Nature* **265:** 53–55.

Kelly S.J. 1977. Studies on the development potential of 4- and 8-cell stage mouse blastomeres. *J. Exp. Zool.* **200:** 365–376.

Keshet E., Lyman S.D., Williams D.E., Anderson D.M., Jenkins N.A., Copeland N.G., and Parada L.F. 1991. Embryonic RNA expression patterns of the c-kit receptor and its cognate ligand suggest multiple functional roles in mouse development. *EMBO J.* **9:** 2425–2435.

Kessel M. and Gruss P. 1991. Homeotic transformations of murine vertebrae and concomitant alteration of Hox codes induced by retinoic acid. *Cell* **67:** 89–104.

Kidder G.M. 1992. The genetic program for preimplantation development. *Dev. Genet.* **13:** 319–325.

Kinder S.J., Tsang T.E., Wakamiya M., Sasaki H., Behringer R.R., Nagy A., and Tam P.P. 2001. The organizer of the mouse gastrula is composed of a dynamic population of progenitor cells for the axial mesoderm. *Development* **128:** 3623–3634.

Kisseberth W.C., Brettingen N.T., Lohse J.K., Sandgren E.P. 1999. Ubiquitous expression of marker transgenes in mice and rats. *Dev. Biol.* **214:** 128–138.

Kleinsmith, L.J. and Pierce G.B. 1964. Multipotentiality of single embryonal carcinoma cells. *Cancer Res.* **24:** 1544–1552.

Koopman P., Gubbay J., Vivian N., Goodfellow P., and Lovell-Badge R. 1991. Male development of chromosomally female mice transgenic for Sry. *Nature* **351:** 117–121.

Koopman P., Munsterberg A., Capel B., Vivian N., and Lovell-Badge R. 1990. Expression of a candidate sex-determining gene during mouse testis differentiation. *Nature* **348:** 450–452.

Koseki H., Wallin J., Wilting J., Mizutani Y., Kispert A., Ebensperger C., Herrmannn B.G., Christ B., and Balling R. 1993. A role for Pax-1 as a mediator of notochordal signals during the dorsoventral specification of vertebrae. *Development* **119:** 649–660.

Krauss S., Concordet J.-P., and Ingham P.W. 1993. A functionally conserved homolog of the *Drosophila* segment polarity gene hh is expressed in tissues with polarizing activity in zebrafish embryos. *Cell* **75:** 1431–1444.

Kratzer P.G., Chapman V.M., Lambert H., Evans R.E., and Liskay R.M.. 1983. Differences in the DNA of the inactive X chromosomes of fetal and extraembryonic tissues of mice. *Cell* **33:** 37–42.

Kurkinen M., Barlow D.P., Jenkins J.R., and Hogan B.L.M. 1983a. In vitro synthesis of laminin and entactin polypeptides. *J. Biol. Chem.* **258:** 6543–6548.

Kurkinen M., Barlow D.P., Helfman D., William J.G., and Hogan B.L.M. 1983b. cDNAs for basement membrane components. Type IV collagen. *Nucleic Acid Res.* **11:** 6199–6209.

Kusumi K., Sun E.S., Kerrebrock A.W., Bronson R.T., Chi D.C., Bulotsky M.S., Spencer J.B., Birren B.W., Frankel W.N., and Lander E.S. 1998. The mouse *pudgy* mutation disrupts Delta homologue *Dll3* and initiation of early somite boundaries. *Nat. Genet.* **19:** 274–278.

Kwon B.S., Haq A.K., Pomerantz S.H., and Halaban R. 1987. Isolation and sequence of a cDNA clone for human tyrosinase that maps at the mouse albino locus. *Proc. Natl. Acad. Sci.* **84:** 7473–7477.

Kyba M., Perlingeiro R.C., and Daley G.Q. 2002. HoxB4 confers definitive lymphoid-myeloid engraftment potential on embryonic stem cell and yolk sac hematopoietic progenitors. *Cell* **109:** 29–37.

Labosky P.A., Barlow D.P., and Hogan B.L.M. 1994. Embryonic germ cell lines and their derivation from mouse primordial germ cells. *CIBA Found. Symp.* **182:** 157–168.

Lane E.B., Hogan B.L.M., Kurkinen M., and Garrels J.I. 1983. Coexpression of vimentin and cytokeratins in parietal endoderm cells of the early mouse embryo. *Nature* **303:** 701–704.

Larue L., Ohsugi M., Hirchenhain J., and Kemler R. 1994. E-cadherin null mutant embryos fail to form a trophectoderm epithelium. *Proc. Natl. Acad. Sci.* **91:** 8263–8267.

Latham K.E., Garrels J.I., Chang C., and Solter D. 1991. Quantitative analysis of protein synthesis in mouse embryos. 1. Extensive reprogramming at the one- and two-cell stage. *Development* **112:** 921–932.

Lawson K.A. and Hage W.J. 1994. Clonal analysis of the origin of primordial germ cells in the mouse. *Ciba Found. Symp.* **182:** 68–84.

Lawson, K.A. and R.A. Pedersen. 1992. Clonal analysis of cell fate during gastrulation and early neurulation in the mouse. *Ciba Found. Symp.* **165:** 3–26.

Lawson K.A. Meneses J.J., and Pedersen R.A. 1991. Clonal analysis of epiblast fate during germ layer formation in the mouse embryo. *Development* **113:** 891–911.

Lawson K.A., Pedersen R.A., and van der Geer S. 1987. Cell fate, morphogenetic movement and population kinetics of embryonic endoderm at the time of germ layer formation in the mouse. *Development* **101:** 627–652.

Lehtonen E., Lehto V.-P., Paasivuo R., and Virtanen I. 1983a. Parietal and visceral endoderm differ in their expression of intermediate filaments. *EMBO J.* **2:** 1023–1028.

Lehtonen E., Lehto V.P., Vartio T., Badley R.A., and Virtanen I. 1983b. Expression of cytokeratin polypeptides in mouse oocytes and preimplantation embryos. *Dev. Biol.* **100:** 158– 165.

Leivo I., Vaheri A., Timpl R., and Wartiovaara J.L. 1980. Appearance and distribution of collagens and laminin in the early mouse embryo. *Dev. Biol.* **76:** 100–114.

Le Mouellic H., Lallemand Y., and Brulet P. 1992. Homeosis in the mouse induced by a null mutation in the Hox-3.1 gene. *Cell* **69:** 251–264.

Lescisin K.R., Varmuza S., and Rossant J. 1988. Isolation and characterization of a novel trophoblast-specific cDNA in the mouse. *Genes Dev.* **2:** 1639–1646.

Lewis N.E. and Rossant J. 1982. Mechanism of size regulation in mouse embryo aggregates. *J. Embryol. Exp. Morphol.* **72:** 169–181.

Lindenbaum M.H. and Grosveld F. 1991. An in vitro globin switching model based on differentiated embryonic stem cells. *Genes Dev.* **4:** 2075–2085.

Liu P., Wakamiya M., Shea M.J., Albrecht U., Behringer R.R., and Bradley A. 1999. Requirement for *Wnt3* in vertebrate axis formation. *Nat. Genet.* **22:** 361–365.

Lo C.W. 1986. Localization of low abundance DNA sequences in tissue sections by in situ hybridization. *J. Cell Sci.* **81:** 143–162.

Lo C.W. and Gilula N. 1979. Gap junctional communication in the postimplantation mouse embryo. *Cell* **18:** 411–422.

Lo C.W., Coulling M., and Kirby C. 1987. Tracking of mouse cell lineage using microinjected DNA sequences: Analyses using genomic Southern blotting and tissue section in situ hybridizations. *Differentiation* **35:** 37–44.

Lobe C.G., Koop K.E., Kreppner W., Lomeli H., Gertsenstein M., and Nagy A. 1999. Z/AP, a double reporter for cre-mediated recombination. *Dev. Biol.* **208:** 281–292.

Loutit J.F. and Cattanach B.M. 1983. Haematopoietic role for Patch (Ph) revealed by new W mutant (wct) in mice. *Genet. Res.* **42:** 23–39.

Lovell-Badge R. and Robertson E. 1990. XY female mice resulting from a heritable mutation in the primary testis-determining gene, *Tdy*. *Development* **109:** 635–646.

Lufkin T., Dierich A., LeMeur M., Mark M., and Chambon P. 1991. Disruption of the Hox-1.6 homeobox gene results in defects in a region corresponding to its rostral domain of expression. *Cell* **66:** 1105–1119.

Lumsden A.G.S. 1991. The development and significance of hindbrain segmentation. *Semin. Dev. Biol.* **1:** 117–126.

Lyon M.F. 1961. Gene action in the X chromosome of the mouse (*Mus musculus* L.). *Nature* **190:** 372–373.

Lyon M.F. and Rastan S. 1984. Parental source of chromosome implantation and its relevance for X-chromosome inactivation. *Differentiation* **26:** 63–67.

Lyon M.F. and Searle A.G., eds. 1989. *Genetic variants and strains of the laboratory mouse,* 2nd edition. Oxford University Press, England.

Lyons K.M., Jones C.M., and Hogan B.L.M. 1992. The TGF-β-related DVR gene family in mammalian development. *Ciba Found. Symp.* **165:** 219–230.

Lyons K.M., Pelton R.W., and Hogan B.L.M. 1989. Patterns of expression of murine Vgr-1 and BMP-2a RNA suggest that transforming growth factor-β-like genes co-ordinately regulate aspects of embryonic development. *Genes Dev.* **3:** 1657–1668.

MacAuley A., Werb Z., and Mirkes P.E. 1993. Characterization of the unusually rapid cell cycles during rat gastrulation. *Development* **117:** 873–883.

Mann J.R., Gadi I., Harbison M.L., Abbondanzo S.J., and Stewart C.L. 1990. Androgenetic mouse embryonic stem cells are pluripotent and cause skeletal defects in chimeras: Implications for genetic imprinting. *Cell* **62:** 251–260.

Manova K. and Bachvarova R.F. 1991. Expression of c-kit encoded at the W locus of mice in developing embryonic germ cells and presumptive melanoblasts. *Dev. Biol.* **146:** 312–324.

Manova K., Nocka K., Besmer P., and Bachvarova R.F. 1990. Gonadal expression of c-kit encoded at the W locus of the mouse. *Development* **110:** 1057–1069.

Manova K., Huang E.J., Angeles M., De Leon V., Sanchez S., Pronovost S.M., Besmer P., and Bachvarova R.F. 1993. The expression pattern of the c-kit ligand in gonads of mice supports a role for the c-kit receptor in oocyte growth and in proliferation of spermatogonia. *Dev. Biol.* **157:** 85–99.

Mantalenakis S.J. and Ketchel M.M. 1966. Frequency and extent of delayed implantation in lactating rats and mice. *J. Reprod. Fertil.* **12:** 391–394.

Mao X., Fujiwara Y., and Orkin S.H. 1999. Improved reporter strain for monitoring Cre recombinase-mediated DNA excisions in mice. *Proc. Natl. Acad. Sci.* **96:** 5037–5042.

Maro B., Johnson M.H., Pickering S.J., and Flach G. 1984. Changes in actin distribution during fertilization of the mouse egg. *J. Embryol. Exp. Morphol.* **81:** 211–237.

Marotti K.R., Belin D., and Strickland S. 1982. The production of distinct forms of plasminogen activator by mouse embryonic cells. *Dev. Biol.* **90:** 154–159.

Martin G.R. 1980. Teratocarcinoma and mammalian embryogenesis. *Science* **209:** 768–776.

———. 1981. Isolation of a pluripotent cell line from early mouse embryos cultured in medium conditioned by teratocarcinoma stem cells. *Proc. Natl. Acad. Sci.* **78:** 7634–7636.

Martin G.R. and Lock L.F. 1983. Pluripotent cell lines derived from early mouse embryos cultured in medium conditioned by teratocarcinoma stem cells. *Cold Spring Harbor Conf. Cell Proliferation* **10:** 635–646.

Matsui Y., Zsebo K., and Hogan B.L.M. 1992. Derivation of pluripotent embryonic stem cells from murine primordial germ cells in culture. *Cell* **70:** 841–847.

Matzuk M.M., Burns K.H., Viveiros M.M., and Eppig J.J. 2002. Intercellular communication in the mammalian ovary: Oocytes carry the conversation. *Science* **296:** 2178–2180.

Mayer T.C. 1973. Site of gene action in Steel mice: Analysis of the pigment defect by mesodermal-ectodermal recombinations. *J. Exp. Zool.* **184:** 345–352.

McCoshen J.A. and McCallion D.J. 1975. A study of the primordial germ cells during their migratory phase in Steel mutant mice. *Experientia* **31:** 589–590.

McGinnis W. and Krumlauf R. 1992. Homeobox genes and axial patterning. *Cell* **68:** 283–302.

McGrath J. and Solter D. 1984a. Maternal T-hp lethality in the mouse is a nuclear, not cytoplasmic defect. *Nature* **308:** 550–551.

———. 1984b. Completion of mouse embryogenesis requires both the maternal and paternal genomes. *Cell* **37:** 179–183.

———. 1984c. Inability of mouse blastomere nuclei transferred to enucleated zygotes to support development in vitro. *Science* **226:** 1317–1319.

McLaren A. 1968. A study of blastocysts during delay and subsequent implantation in lactating mice. *J. Endocrinol.* **42:** 453–463.

McLaren A. 1983. Does the chromosomal sex of a mouse germ cell affect its development. In *7th Symposium of British Society for Developmental Biology* (ed. A. McLaren and C.C. Wylie), pp. 225–240. Cambridge University Press, England.

———. 1984. Germ cell lineages. In *Chimeras in developmental biology* (ed. N.L. Douarin and A. McLaren), pp. 111–129. Academic Press, London.

Meehan R.R., Barlow D.P., Hill R.E., Hogan B.L.M., and Hastie N.D. 1984. Pattern of serum protein gene expression in mouse visceral yolk sac and fetal liver. *EMBO J.* **3:** 1881–1885.

Meier S. and Tam P.P.L. 1982. Metameric pattern development in the embryonic axis of the mouse. I. Differentiation of the cranial segments. *Differentiation* **21:** 95–108.

Mercer J.A., Seperack P.K., Strobel M.C., Copeland N.G., and Jenkins N.A. 1991. Novel myosin heavy chain encoded by murine dilute coat color locus. *Nature* **349:** 709–713.

Michaud E.J., Bultman S.J., Stubbs L.J., and Woychik R.P. 1993. The embryonic lethality of homozygous lethal yellow mice (Ay/Ay) is associated with the disruption of a novel RNA-binding protein. *Genes Dev.* **7:** 1203–1213.

Michaud E.J., Bultman S.J., Klebig M.L., van Vugt M.J., Stubbs L.J., Russell L.B., and Woychik R.P. 1994. A molecular model for the genetic and phenotypic characteristics of the mouse lethal yellow (Ay) mutation. *Proc. Natl. Acad. Sci.* **91:** 2562–2566.

Miller M.W., Duhl D.M.J., Vrieling H., Cordes S.P., Ollman M.M., Winkes B.M., and Barsh G.S. 1993. Cloning of the mouse agouti gene predicts a secreted protein ubiquitously expressed in mice carrying the lethal yellow mutation. *Genes Dev.* **7:** 454–467.

Mintz B. and Illmensee K. 1975. Normal genetically mosaic mice produced from malignant teratocarcinoma cells. *Proc. Natl. Acad. Sci.* **72:** 3585-3589.

Mintz B. and Russell E.S. 1957. Gene induced embryological modifications of primordial germ cells. *J. Exp. Zool.* **134:** 207–239.

Molyneaux K.A., Stallock J., Schaible K., and Wylie C. 2001. Time-lapse analysis of living mouse germ cell migration. *Dev. Biol.* **240:** 488–498.

Monaghan A.P., Kaestner K.H., Grau E., and Schutz G. 1993. Postimplantation expression patterns indicate a role for the mouse forkhead/HNF-3α, β and γ genes in determination of the definitive endoderm, chordamesoderm and neuroectoderm. *Development* **119:** 567–578.

Monk M. and McLaren A. 1981. X-chromosome activity in foetal germ cells of the mouse. *J. Embryol. Exp. Morphol.* **63:** 75–84.

Monk M., Boubelik M., and Lehnert S. 1987. Temporal and regional changes in DNA methylation in the embryonic, extraembryonic and germ cell lineages during mouse embryo development. *Development* **99:** 371–382.

Morris R.J. and Barber P.C. 1983. Fixation of Thy-1 in nervous tissue for immunochemistry. *J. Histol. Cytochem.* **31:** 263–274.

Mullen R.J. 1977. Site of Ped gene action and Purkinje cell mosaicism in cerebella of chimeric mice. *Nature* **270:** 245–247.

Muller R., Verma I.M., and Adamson E.D. 1983. Expression of c-onc genes: c-fos transcripts accumulate to high levels during development of mouse placenta, yolk sac and amnion. *EMBO J.* **2:** 679–684.

Mutter G.L., Grills G.S., and Wolgemuth D.J. 1988. Evidence for the involvement of the proto-oncogene c-mos in mammalian meiotic maturation and possibly very early embryogenesis. *EMBO J.* **7:** 683–689.

Nabeshima Y., Hanaoka K., Hayasaka M., Esumi E., Li S., Nonaka I., and Nabeshima Y. 1993. Myogenin gene disruption results in perinatal lethality because of severe muscle defect. *Nature* **364:** 532–535.

Nadijcka M. and Hillman N. 1974. Ultrastructural studies of the mouse blastocyst substages. *J. Embryol. Exp. Morphol.* **32:** 675–695.

Niederreither K., Subbarayan V., Dolle P., and Chambon P. 1999. Embryonic retinoic acid synthesis is essential for early mouse post-implantation development. *Nat. Genet.* **21:** 444–448.

Nieto M.A., Gilardi-Hebenstreit P., Charnay P., and Wilkinson D.G. 1992. A receptor protein tyrosine kinase implicated in the segmental patterning of the hindbrain and mesoderm. *Development* **116:** 1137–1150.

Niswander L. and Martin G.R. 1992. Fgf-4 expression during gastrulation, myogenesis, limb and tooth development in the mouse. *Development* **115:** 755–768.

Niwa H., Miyazaki J., and Smith A.G. 2000. Quantitative expression of *Oct-3/4* defines differentiation, dedifferentiation or self-renewal of ES cells. *Nat. Genet.* **24:** 372–376.

Nocka K., Majumder S., Chabot B., Ray P., Cervone M., Bernstein A., and Besmer P. 1989. Expression of c-kit gene products in known cellular targets of W mutations in normal and W mutant mice—Evidence for an impaired c-kit kinase in mutant mice. *Genes Dev.* **3:** 816–826.

Nomura T., Ohsawa N., Tamaoki N., and Fujiura K., eds. 1977. *Proceedings of the 2nd International Workshop on Nude Mice.* University of Tokyo Press, Japan.

Novak A., Guo C., Yang W., Nagy A., and Lobe C.G. 2000. Z/EG, a double reporter mouse line that expresses enhanced green fluorescent protein upon Cre-mediated excision. *Genesis* **28:** 147–155.

O'Brien M.J., Wigglesworth K., and Eppig J.J. 1993. Mouse oocyte and embryo culture.

Methods Reprod. Toxicol. **3:** 128–141.

Oka C., Nakano T., Wakeham A., de la Pompa J.L., Mori C., Sakai T., Okazaki S., Kawaichi M., Shiota K., Mak T.W., and Honjo T. 1995. Disruption of the mouse RBP-J kappa gene results in early embryonic death. *Development* **121:** 3291–3301.

Okabe M., Ikawa M., Kominami K., Nakanishi T., and Nishimune Y. 1997. "Green mice" as a source of ubiquitous green cells. *FEBS Lett.* **407:** 313–319.

O'Keefe S.J., Kiessling A.A., and Cooper G.M. 1991. The c-mos gene product is required for cyclin B accumulation during meiosis of mouse eggs. *Proc. Natl. Acad. Sci.* **88:** 7869–7872.

O'Keefe S.J., Wolfes H., Kiessling A.A., and Cooper G.M. 1989. Microinjection of antisense c-mos oligonucleotides prevents meiosis II in the maturing mouse egg. *Proc. Natl. Acad. Sci.* **86:** 7038–7042.

Ordahl C.P. 1993. Myogenic lineages within the developing somite. In Bernfield M. (Ed.) *Molecular basis of morphogenesis.* John Wiley and Sons, New York.

Osumi-Yamashita N., Ninomiya Y., Doi H., and Eto K. 1994. The contribution of both forebrain and midbrain crest cells to the mesenchyme in the frontonasal mass of mouse embryos. *Dev. Biol.* **164:** 409–419.

Ott M.-O., Bober E., Lyons G., Arnold H., and Buckingham M. 1991. Early expression of the myogenic regulatory gene myf-5 in precursor cells of skeletal muscle in the mouse embryo. *Development* **111:** 1097–1107.

Ozawa M., Ringwald M., and Kemler R. 1990. Uvomorulin-catenin complex formation is regulated by a specific domain in the cytoplasmic region of the cell adhesion molecule. *Proc. Natl. Acad. Sci.* **87:** 4246–4250.

Ozdenski, W. 1967. Observations on the origin of the primordial germ cells in the mouse. *Zool. Pol.* **17:** 65–78.

Palmieri S.L., Payne J., Stiles C.D., Bigger J.D., and Mercola M. 1992. Expression of mouse PDGF-A and PDGF α-receptor genes during pre- and post-implantation development: Evidence for a developmental shift from an autocrine to a paracrine mode of action. *Mech. Dev.* **39:** 181–191.

Papaioannou V.E. and Rossant J. 1983. Effects of the embryonic environment on proliferation and differentiation of embryonal carcinoma cells. *Cancer Surv.* **2:** 165–183.

Papaioannou V.E., McBurney M.W., Gardner R.L., and Evans M.J. 1975. Fate of teratocarcinoma cells injected into early mouse embryos. *Nature* **258:** 70–73.

Paria B.C., Jones K.L., Flanders K.C., and Dey S.K. 1992. Localization and binding of transforming growth factor-β isoforms in mouse preimplantation embryos and in delayed and activated blastocysts. *Dev. Biol.* **151:** 91–104.

Patapoutian A., Yoon J.K., Miner J.H., Wang S., Stark K., and Wold B. 1995. Disruption of the mouse *MRF4* gene identifies multiple waves of myogenesis in the mouse. *Development* **121:** 3347–3358.

Paules R.S., Buccione R., Moschel R.C., Vande Woude G.F., and Eppig J.J. 1989. Mouse Mos proto-oncogene product is present and functions during oogenesis. *Proc. Natl. Acad. Sci.* **86:** 5395–5399.

Pedersen R.A. 1986. Potency, lineage and allocation in preimplantation mouse embryos. In *Experimental approaches to mammalian embryonic development* (ed. J. Rossant and R.A. Pedersen), pp. 3–33. Cambridge University Press, New York.

Petters R.M. and Markert C.L. 1980. Production and reproductive performance of hexaparental and octoparental mice. *J. Hered.* **71:** 70–74.

Peyrieras N., Hyafil F., Louvard D., Ploegh H.L., and Jacob F. 1983. Uvomorulin, a nonintegral membrane protein of early mouse embryo. *Proc. Natl. Acad. Sci.* **80:** 6274–6277.

Piotrowska K. and Zernicka-Goetz M. 2001. Role for sperm in spatial patterning of the early mouse embryo. *Nature* **409:** 517–521.

Placzek M., Jessell T.M., and Dodd J. 1993. Induction of floor plate differentiation by contact-dependent, homeogenetic signals. *Development* **117:** 205–218.

Poelmann R.E. 1981. The head process and the formation of the definitive endoderm in the mouse embryo. *Anat. Embryol.* **162:** 41–49.

Pollard J.W., Hunt J.S., Wiktor-Jedrzejczak W., and Stanley E.R. 1991. A pregnancy defect in the osteopetrotic (op/op) mouse demonstrates the requirement for CSF-1 in female fertility. *Dev. Biol.* **148:** 273–283.

Ponder B.A.T., Wilkinson M.M., and Wood M. 1983. H2 antigens as a marker of cellular genotype in chimeric mice. *J. Embryol. Exp. Morphol.* **76:** 83–93.

Potten C.S. 1985. *Radiation and skin.* Taylor and Francis, London.

Pourquié O. 2001. Vertebrate somitogenesis. *Annu. Rev. Cell Dev. Biol.* **17:** 311–350.

Pourquié O. and Tam P.P. 2001. A nomenclature for prospective somites and phases of cyclic gene expression in the presomitic mesoderm. *Dev. Cell* **1:** 619–620.

Pourquié O., Coltey M., Teillet M.A., Ordahl C., Le Douarin N.M. 1993. Control of dorsoventral patterning of somitic derivatives by notochord and floor plate. *Proc. Natl. Acad. Sci.* **90:** 5242–5246.

Pratt H. 1989. Marking time and making space: Chronology and topography in the early mouse embryo. *Int. Rev. Cytol.* **117:** 99–130.

Pratt H.P.M., Bolton V.N., and Gudgeon K.A. 1983. The legacy from the oocyte and its role in controlling early development of the mouse embryo. *CIBA Found. Symp.* **98:** 197–227.

Price J., Turner D., and Cepko C. 1987. Lineage analysis in the vertebrate nervous system by retrovirus mediated gene transfer. *Proc. Natl. Acad. Sci.* **84:** 158–160.

Puelles L. and Rubenstein J.L.R. 1993. Expression patterns of homeobox and other putative regulatory genes in the embryonic mouse forebrain suggest a neuromeric organization. *Trends Neurosci.* **16:** 472–479.

Rands G. 1987. Size regulation in the mouse embryo. *J. Embryol. Exp. Morphol.* **94:** 139–148.

Rappolee D.A., Brenner C.A., Schultz R., Mark D., and Werb Z. 1988. Developmental expression of PDGF, TGF-α, and TGF-β genes in preimplantation mouse embryos. *Science* **241:** 1823–1825.

Rappolee D.A., Sturm K.S., Behrendtsen O., Schultz G.A., Pederson R.A., and Werb Z. 1992. Insulin-like growth factor II acts through an endogenous growth pathway regulated by imprinting in early mouse embryos. *Genes Dev.* **6:** 939–952.

Rastan S. and Robertson E.J. 1985. X-chromosome deletions in embryo-derived (EK) cell lines associated with lack of X-chromosome inactivation. *J. Embryol. Exp. Morphol.* **90:** 379–388.

Reeve W.J.D. and Ziomek C.A. 1981. Distribution of microvilli on dissociated blastomeres from mouse embryos: Evidence for surface polarization at compaction. *J. Embryol. Exp. Morphol.* **62:** 339–350.

Regenstreif L.J. and Rossant J. 1989. Expression of the c-fms proto-oncogene and of the cytokine, CSF-1, during mouse embryogenesis. *Dev. Biol.* **133:** 284–294.

Resnick J.L., Bixler L.S., Cheng L., and Donovan P.J. 1992. Long-term proliferation of mouse primordial germ cells in culture. *Nature* **359:** 550–551.

Rhinn M., Dierich A., Shawlot W., Behringer R.R., Le Meur M., and Ang S.L. 1998. Sequential roles for Otx2 in visceral endoderm and neuroectoderm for forebrain and midbrain induction and specification. *Development* **125:** 845–856.

Richards J.S., Jahnsen T., Hedin L., Lifka J., Ratoosh S., Durica J.M., and Goldring N.B. 1987. Ovarian follicular development: From physiology to molecular biology. *Recent Prog. Hormone Res.* **43:** 231–270.

Riddle R.D., Johnson R.L., Laufer E., and Tabin C. 1993. Sonic hedgehog mediates the polarizing activity of the ZPA. *Cell* **75:** 1401–1416.

Riethmacher D., Brinkmann V., and Birchmeier C. 1995. A targeted mutation in the mouse E-cadherin gene results in defective preimplantation development. *Proc. Natl. Acad. Sci.* **92:** 855–859.

Robbins L.S., Nadeau J.H., Johnson K.R., Kelly M.A., Roselli-Rehfuss L., Baack E., Mountjoy K.G., and Cone R.D. 1993. Pigmentation phenotypes of variant extension locus alleles result from point mutations that alter MSH receptor function. *Cell* **72:** 827–834.

Rong P.M., Teillet M.-A., Ziller C., and Le Douarin N.M. 1992. The neural tube/notochord is necessary for vertebral but not limb and body wall striated muscle differentiation. *Development* **115:** 657–672.

Rosiere T.K. and Wasserman P.M. 1992. Identification of a region of mouse zona pelluci-da glycoprotein mZP3 that possesses sperm receptor activity. *Dev. Biol.* **154:** 309–317.

Rosner M.H., Vigano M.A., Ozato K., Timmons P.M., Poirier F., Rigby P.W.J., and Staudt L.M. 1990. A POU-domain transcription factor in early stem cells and germ cells of the mammalian embryo. *Nature* **345:** 686–692.

Rossant J. 1985. Interspecific cell markers and lineage in mammals. *Philos. Trans. R. Soc. Lond. B.* **312:** 91–100.

Rossant J. and Cross J.A. 2002. Extraembryonic lineages. In *Mouse development: Patterning, morphogenesis, organogenesis* (ed. J. Rossant and P.P.L. Tam), pp. 155–180. Academic Press, San Diego.

Rossant J. and Croy B.A. 1985. Genetic identification of tissue of origin of cellular popu-lations within the mouse placenta. *J. Embryol. Exp. Morphol.* **86:** 177–189.

Rothstein J.L., Johnson D., DeLoia J.A., Skowronski J., Solter D., and Knowles B. 1992. Gene expression during preimplantation mouse development. *Genes Dev.* **6:** 1190–1201.

Rubenstein J.L., Shimamura K., Martinez S., and Puelles L. 1998. Regionalization of the prosencephalic neural plate. *Annu. Rev. Neurosci.* **21:** 445–477.

Rudnicki M.A., Braun T., Hinuma S., and Jaenisch R. 1992. Inactivation of MyoD in mice leads to up-regulation of the myogenic HLH gene Myf-5 and results in apparently normal muscle development. *Cell* **71:** 383–390.

Rudnicki M.A., Schnegelsberg P.N., Stead R.H., Braun T., Arnold H.H., and Jaenisch R. 1993. MyoD or Myf-5 is required for the formation of skeletal muscle. *Cell* **75:** 1351–1359.

Sallés F.J., Darrow A.L., O'Connell M.L., and Strickland S. 1992. Isolation of novel murine maternal mRNAs regulated by cytoplasmic polyadenylation. *Genes Dev.* **6:** 1202–1212.

Sanes J.R., Rubenstein J.L.R., and Nicolas J.-F. 1986. Use of a recombinant retrovirus to study post-implantation cell lineage in mouse embryos. *EMBO J.* **5:** 3133–3142.

Sasaki H. and Hogan B.L.M. 1993. Differential expression of multiple fork head related genes during gastrulation and axial pattern formation in the mouse embryo. *Development* **118:** 47–59.

Sasaki H. and Hogan B.L.M.. 1994. HNF-3β as a regulator of floorplate development. *Cell* **76:** 103–115.

Sassoon D., Lyons G., Wright W., Lin V., Lasar A., Weintraub H., and Buckingham M. 1989. Expression of two myogenic regulatory factors myogenin and MyoD1 during mouse embryogenesis. *Nature* **341:** 303–307.

Sausedo R.A. and Schoenwolf G.C. 1994. Quantitative analyses of cell behaviours under-lying notochord formation and extension in mouse embryos. *Anat. Rec.* **239:** 103–112.

Saxen, L. 1987. *Organogenesis of the kidney.* Cambridge University Press, England.

Saxen L., Karkinen-Jaaskelainen M., Lehtonen E., Nordling S., and Wartiovaara J. 1976. Inductive tissue interactions. In *Cell surface interactions in embryogenesis* (ed. G. Poste and G.L. Nicholson), pp. 331–407. North-Holland, Amsterdam.

Schatten G., Simmerly C., and Schatten H. 1985. Microtubule configuration during fertil-ization, mitosis, and early development in the mouse and the requirement for egg microtubule mediated motility during mammalian fertilization. *Proc. Natl. Acad. Sci.* **82:** 4152– 4156.

Schmidt G.H., Wilkinson M.M., and Ponder B.A. 1985. Cell migration pathway in the intestional epithelium: An in situ marker system using mouse aggregation chimeras. *Cell* **40:** 425–429.

Schoenwolf G.C. 1984. Histological and ultrastructural studies of secondary neurulation in mouse embryos. *Am. J. Anat.* **169:** 361–376.

Schoenwolf G. and Smith J.L. 1990. Mechanisms of neurulation: Traditional viewpoint and recent advances. *Development* **109:** 243–270.

Schoenwolf G.C., Garcia-Martinez V., and Dias M.-S. 1992. Mesoderm movement and fate during avian gastrulation and neurulation. *Dev. Dynam.* **193:** 235–248.

Scholer H.R., Dressler G.R., Balling R., Rohdewohld H., and Gruss P. 1990a. Oct-4: A germline-specific transcription factor mapping to the mouse t-complex. *EMBO J.* **9:**

2185–2195.

Scholer H., Ruppert S., Suzuki N., Chowdhury K., and Gruss P. 1990b. New type of POU domain in germ line-specific protein Oct-4. *Nature* **344:** 435–439.

Schultz R.M. 1986. Molecular aspects of mammalian oocyte growth and maturation. In *Experimental approaches to mammalian embryonic development* (ed. J. Rossant and R.A. Pederson), pp. 195–237. Cambridge University Press, England.

Schultz R.M., Letourneau G.E., and Wassarman P.M. 1979. Program of early development in the mammal. Changes in the patterns and absolute rates of tubulin and total protein synthesis during oocyte growth in the mouse. *Dev. Biol.* **73:** 120–133.

Selleck M.A. and Stern C.D. 1991. Fate mapping and cell lineage analysis of Hensen's node in the chick embryo. *Development* **112:** 615–626.

Semoff S., Hogan B.L.M., and Hopkins C.R. 1982. Localisation of fibronectin, laminin and entactin in Reichert's membrane by immunoelectron microscopy. *EMBO J.* **1:** 1171–1175.

Serbedzija G.N., Bronner-Fraser M., and Fraser S.E. 1989. A vital dye analysis of the timing and pathways of avian trunk neural crest cell migration. *Development* **106:** 809–816.

———. 1992. Vital dye analysis of cranial neural crest cell migration in the mouse embryo. *Development* **116:** 297–307.

———. 1990. Pathways of trunk neural crest cell migration in the mouse embryo as revealed by vital dye labelling. *Development* **108:** 605–612.

Serbedzija G.N., Burgan S., Fraser S.E., and Bronner-Fraser M. 1991. Vital dye labelling demonstrates a sacral neural crest cell contribution to the enteric nervous system of chick and mouse embryos. *Development* **111:** 857–866.

Sham M.H., Vesque C., Nonchev S., Marshall H., Frain M., Das Gupta R., Whiting G.J., Wilkinson D., Charnay P., and Krumlauf R. 1993. The zinc finger gene Krox 20 regulates Hox B2 (Hox 2.8) during hindbrain segmentation. *Cell* **72:** 183–196.

Shawlot W., Wakamiya M., Kwan K.M., Kania A., Jessell T.M., Behringer R.R. 1999. Lim1 is required in both primitive streak-derived tissues and visceral endoderm for head formation in the mouse. *Development* **126:** 4925–4932.

Silvers W.K. 1979. *The coat colors of mice: A model for mammalian gene action and interaction.* Springer-Verlag, New York.

Simeone A., Acampora D., Gulisano M., Stornaiuolo A., and Boncinelli E. 1992. Nested expression domains of four homeobox genes in developing rostral brain. *Nature* **358:** 687–690.

Sinclair A.H., Berta P., Palmer M.S., Hawkins J.R., Griffiths B.L., Smith M.J., Foster J.W., Frischauf A.M., Lovell-Badge R., and Goodfellow P.N. 1990. A gene from the human sex-determining region encodes a protein with homology to a conserved DNA-binding motif. *Nature* **346:** 240–244.

Smith A.G. and Rathjen P.D. 1991. Embryonic stem cells, differentiation inhibiting activity, and the mouse embryo. *Semin. Dev. Biol.* **2:** 317–328.

Smith A.G., Nichols J., Robertson M., and Rathjen P. 1992. Differentiation inhibiting activity (DIA-LIF) and mouse development. *Dev. Biol.* **151:** 339–351.

Smith K.K. and Strickland S. 1981. Structural components and characteristics of Reichert's membrane, an extraembryonic basement membrane. *J. Biol. Chem.* **256:** 4654–4661.

Smith L.J. 1985. Embryonic axis orientation in the mouse and its correlation with blastocyst relationships to the uterus. II. Relationships from 4.5 to 9.5 days. *J. Embryol. Exp. Morphol.* **89:** 15–35.

Snow M.H.L. 1977. Gastrulation in the mouse: Growth and regionalization of the epiblast. *J. Embryol. Exp. Morphol.* **42:** 293–303.

———. 1981. Autonomous development of parts isolated from primitive-streak-stage mouse embryos. Is development clonal? *J. Embryol. Exp. Morphol.* (suppl.) **65:** 269–287.

Snow M.H.L., Tam P.P.L., and McLaren A. 1981. On the control and regulation of size and morphogenesis in mammalian embryos. In *Levels of genetic control in development* (ed. S. Subtleny and U.K. Abbott), pp. 201–217. Alan Liss, New York.

Soares M.J., Julian J.A., and Glasser S.R. 1985. Trophoblast giant cell release of placental lactogens: Temporal and regional characteristics. *Dev. Biol.* **107:** 520–526.

Solter D. 1988. Differential imprinting and expression of maternal and paternal genomes. *Annu. Rev. Genet.* **22:** 127–146.

———. 2000. Mammalian cloning: Advances and limitations. *Nat. Rev. Genet.* **1:** 199–207.

Solter D., Adams N., Damjanov I., and Koprowski H. 1975. Control of teratocarcino-genesis. In *Teratomas and differentiation* (ed. M.I. Sherman and D. Solter), pp. 139–159. Academic Press, New York.

Soriano P. 1999. Generalized *lacZ* expression with the ROSA26 Cre reporter strain. *Nat. Genet.* **21:** 70–71.

Stevens L.C. 1967. The biology of teratomas. *Adv. Morphog.* **6:** 1–31.

———. 1983. Testicular, ovarian and embryo-derived teratomas. *Cancer Surv.* **2:** 75–91.

Stewart C.L., Gadi I., and Bhatt H. 1994. Stem cells from primordial germ cells can reenter the germ line. *Dev. Biol.* **161:** 626–628.

Stewart C.L., Kaspar P., Brunet L.J., Bhatt H., Gadi I., Kontgen F., and Abbondanzo S.J. 1992. Blastocyst implantation depends on maternal expression of leukemia inhibitory factor. *Nature* **359:** 76–79.

Stewart T.A. and Mintz B. 1982. Recurrent germ line transmission of the teratocarcinoma genome from the METT-1 culture line to progeny in vivo. *J. Exp. Zool.* **224:** 465–471.

Strickland S. and Richards W.G. 1992. Invasion of the trophoblasts. *Cell* **71:** 355–357.

Strickland S., Huarte J., Belin D., Vassalli A., Rickles R.J., and Vassalli J.-D. 1988. Antisense RNA directed against the 3′ noncoding region prevents dormant mRNA activation in mouse oocytes. *Science* **241:** 680–684.

Srinivas S., Watanabe T., Lin C.S., William C.M., Tanabe Y., Jessell T.M., and Costantini F. 2001. Cre reporter strains produced by targeted insertion of EYFP and ECFP into the ROSA26 locus. *BMC Dev. Biol.* **1:** 4.

Sulik K.K. and Schoenwolf G.C. 1985. Highlight of craniofacial morphogenesis in mammalian embryos, as revealed by scanning electron microscopy. *Scanning Electron Microsc.* **4:** 1735–1752.

Sulik K., Dehart D.B., Inagaki T., Carson J.L., Vrablic T., Gesteland K., and Schoenwolf G.C. 1994. Morphogenesis of the murine node and notochordal plate. *Dev. Dynamics* **201:** 260–278.

Sun X., Mariani F.V., and Martin G.R. 2002. Functions of FGF signalling from the apical ectodermal ridge in limb development. *Nature* **418:** 501–508.

Surani, M.A.H. and S.C. Barton. 1983. Development of gynogenetic eggs in the mouse: Implications for parthenogenetic embryos. *Science* **222:** 1034–1036.

Surani M.A., Kothary R., Allen N.D., Singh P.B., Fundele R., Ferguson-Smith A.C., and Barton S.C. 1990. Genome imprinting and development in the mouse. *Development* (suppl.) **90:** 89–98.

Swain A. and Lovell-Badge R. 1999. Mammalian sex determination: A molecular drama. *Genes Dev.* **13:** 755–767.

Swiatek P.J., Lindsell C.E., del Amo F.F., Weinmaster G., and Gridley T. 1994. *Notch1* is essential for postimplantation development in mice. *Genes Dev.* **8:** 707–719.

Takagi N. and Sasaki M. 1975. Preferential inactivation of the paternally derived X chromosome in the extraembryonic membranes of the mouse. *Nature* **256:** 640–642.

Tam P.P.L. 1984. The histogenetic capacity of tissues in the caudal end of the embryonic axis of the mouse. *J. Embryol. Exp. Morphol.* **82:** 253–266.

———. 1989. Regionalisation of the mouse embryonic ectoderm: Allocation of prospective ectodermal tissues during gastrulation. *Development* **107:** 55–67.

Tam P.P.L. and Beddington R.S.P. 1986. The metameric organisation of the presomitic mesoderm and somite specification in the mouse embryo. In *Somites in developing embryos* (ed. R. Bellairs et al.), pp. 17–36. Plenum Press, New York.

———. 1987. The formation of mesodermal tissues in the mouse embryo during gastrulation and early organogenesis. *Development* **99:** 109–126.

———. 1992. Establishment and organisation of germ layers in the gastrulating mouse embryo. *CIBA Found. Symp.* **165:** 27–49.

Tam P.P. and Behringer R.R. 1997. Mouse gastrulation: The formation of a mammalian body plan. *Mech. Dev.* **68:** 3–25.

Tam P.P.L. and Snow M.H.L. 1981. Proliferation and migration of primordial germ cells during compensatory growth in mouse embryos. *J. Embryol. Exp. Morphol.* **64:** 133–147.

Tam P.P.L. and Tan S.-S. 1992. The somitogenetic potential of cells in the primitive streak and the tail bud of the organogenesis-stage mouse embryo. *Development* **115:** 703– 715.

Tam P.P.L., Williams E.A., and Chan W.Y. 1993. Gastrulation in the mouse: Ultrastructural and molecular aspects of germ layer morphogenesis. *Microscop. Res. Tech.* **26:** 301–328.

Tam P.P., Steiner K.A., Zhou S.X., and Quinlan G.A. 1997. Lineage and functional analyses of the mouse organizer. *Cold Spring Harb. Symp. Quant. Biol.* **62:** 135–144.

Tam P.P., Gad J.M., Kinder S.J., Tsang T.E., Behringer R.R. 2001. Morphogenetic tissue movement and the establishment of body plan during development from blastocyst to gastrula in the mouse. *BioEssays* **23:** 508–517.

Tan S.S., Williams E.A., and Tam P.P.L. 1993. X-chromosome inactivation occurs at different times in different tissues of the post-implantation mouse embryo. *Nature Genet.* **3:** 170–174.

Tanaka S., Kunath T., Hadjantonakis A.K., Nagy A., and Rossant J. 1998. Promotion of trophoblast stem cell proliferation by FGF4. *Science* **282:** 2072–2075.

Tarkowski A.K. 1959. Experimental studies on regulation in the development of isolated blastomeres of mouse eggs. *Acta Theriol.* **3:** 191–267.

———. 1961. Mouse chimeras developed from fused eggs. *Nature* **184:** 1286–1287.

Theiler K. 1972. *The house mouse.* Springer-Verlag, New York.

———. 1983. *Embryology in the mouse in biomedical research,* vol. 3. Academic Press, New York.

Thomas P. and Beddington R. 1996. Anterior primitive endoderm may be responsible for patterning the anterior neural plate in the mouse embryo. *Curr. Biol.* **6:** 1487–1496.

Thomas P.Q., Brown A., and Beddington R.S. 1998. *Hex:* A homeobox gene revealing periimplantation asymmetry in the mouse embryo and an early transient marker of endothelial cell precursors. *Development* **125:** 85–94.

Tickle C., Alberts B., Wolpert L., and Lee J. 1982. Local application of retinoic acid to the limb bud mimics the action of the polarizing region. *Nature* **296:** 564–566.

Tilghman S.M. 1999. The sins of the fathers and mothers: Genomic imprinting in mammalian development. *Cell* **96:** 185–193.

Tilmann C. and Capel B. 2002. Cellular and molecular pathways regulating mammalian sex determination. *Recent Prog. Horm. Res.* **57:** 1–18.

Trainor P.A., Tan S.-S., and Tam P.P.L. 1994. Cranial paraxial mesoderm: Regionalization of cell fate and impact on craniofacial development in mouse embryos. *Development* **120:** 2397–2408.

Tsuchida T., Ensini M., Morton S.B., Baldassare M., Edlund T., Jessell T.M., and Pfaff S.L. 1994. Topographic organization of embryonic motor neurons defined by expression of LIM homeobox genes. *Cell* **79:** 957–970.

Tsunoda Y. and Kato Y. 1998. Not only inner cell mass cell nuclei but also trophectoderm nuclei of mouse blastocysts have a developmental totipotency. *J. Reprod. Fertil.* **113:** 181–184.

Vainio S. and Lin Y. 2002. Coordinating early kidney development: lessons from gene targeting. *Nat. Rev. Genet.* **3:** 533–543.

Van Blerkom J. 1981. Structural relationship and posttranslational modification of stage-specific proteins synthesised during early preimplantation development in the mouse. *Proc. Natl. Acad. Sci.* **78:** 7629–7633.

Varlet I., Collignon J., and Robertson E.J. 1997. *nodal* expression in the primitive endoderm is required for specification of the anterior axis during mouse gastrulation. *Development* **124:** 1033–1044.

Varmuza S., Prideaux V., Kothary R., and Rossant J. 1988. Polytene chromosomes in mouse trophoblast giant cells. *Development* **102:** 127–134.

von Baer K.E. 1828. *Uber die Entwicklungsgeschichte der Thiere.* Konigsberg, E. Prussia.

Wakayama T., Perry A.C., Zuccotti M., Johnson K.R., and Yanagimachi R. 1998. Full-term development of mice from enucleated oocytes injected with cumulus cell nuclei. *Nature* **394:** 369–374.

Wanek N., Muneoka K., Holler-Dinsmore G., Burton R., and Bryant S.V. 1989. A staging system for mouse limb development. *J. Exp. Zool.* **249:** 41–49.

Wartiovaara J., Leivo I., and Vaheri A. 1979. Expression of the cell surface-associated glycoprotein, fibronectin, in the early mouse embryo. *Dev. Biol.* **69:** 247–257.

Wassarman P.M. 1990. Profile of a mammalian sperm receptor. *Development* **108:** 1–17.

Wassarman P.M., Greve J.M., Perona R.M., Roller R.J., and Salzmann G.S. 1984. How mouse cells put on and take off their extracellular coat. In *Molecular biology of development* (ed. E. Davidson and R. Firtel), pp. 213–225. Alan Liss, New York.

Weber R.J., Pedersen R.A., Wianny F., Evans M.J., and Zernicka-Goetz M. 1999. Polarity of the mouse embryo is anticipated before implantation. *Development* **126:** 5591–5598.

Weng D.E., Morgan R.A., and Gearhart J.D. 1989. Estimates of mRNA abundance in the mouse blastocyst based on cDNA library analysis. *Mol. Reprod. Dev.* **1:** 233–241.

West J.D., Frels I., Chapman V.M., and Papaioannou V.E. 1977. Preferential expression of the maternally derived X chromosome in the mouse yolk sac. *Cell* **12:** 873–882.

Wichterle H., Lieberam I., Porter J.A., and Jessell T.M. 2002. Directed differentiation of embryonic stem cells into motor neurons. *Cell* **110:** 385–397.

Wiles M.V. and Keller G. 1991. Multiple haemopoietic lineages develop from embryonic stem (ES) cells in culture. *Development* **111:** 259–267.

Wiley L.M., Lever J.E., Pape C., and Kidder G.M. 1991. Antibodies to a renal Na+/glucose cotransport system localize to the apical plasma membrane domain of polar mouse embryo blastomeres. *Dev. Biol.* **143:** 149–161.

Wiley L.M., Wu J.-X., Harari I., and Adamson E.D. 1992. Epidermal growth factor receptor mRNA and protein increase after the four-cell preimplantation stage in murine development. *Dev. Biol.* **149:** 247–260.

Wilkinson D.G., Bhatt S., Chavier P., Bravo R., and Charnay P. 1989. Segment-specific expression of a zinc-finger gene in the developing nervous system of the mouse. *Nature* **337:** 461–464.

Witte O. 1990. Steel locus defines new multipotent growth factor. *Cell* **63:** 5–6.

Wolpert L., Beddington R., Lawrence P., and Jessell T.M. 2002. *Principles of development*, 2nd ed. Oxford University Press, United Kingdom.

Yagi T. and Takeichi M. 2000. Cadherin superfamily genes: Functions, genomic organization, and neurologic diversity. *Genes Dev.* **14:** 1169–1180.

Yamada, T., S.L. Pfaff, T. Edlund, and T.M. Jessell. 1993. Control of cell pattern in the neural tube: Motor neuron induction by diffusible factors from notochord and floor plate. *Cell* **73:** 673–686.

Yamada T., Placzek M., Tanaka H., Dodd J., and Jessell T.M. 1991. Control of cell pattern in the developing nervous system: Polarizing activity of the floor plate and notochord. *Cell* **64:** 635–647.

Yoshida H., Hayashi S., Kunisada T., Ogawa M., Nishikawa S., Okamura H., Sudo T., Shultz L.D., and Nishikawa S. 1990. The murine mutation osteopetrosis is in the coding region of the macrophage colony stimulating factor gene. *Nature* **345:** 442–444.

Yoshinaga K. and Adams C.E. 1966. Delayed implantation in the spayed, progesterone treated adult mouse. *J. Reprod. Fertil.* **12:** 593–595.

Yu Y. and Bradley A. 2001. Engineering chromosomal rearrangements in mice. *Nat. Rev. Genet.* **2:** 780–790.

Zhang N. and Gridley T. 1998. Defects in somite formation in *lunatic fringe*-deficient mice. *Nature* **394:** 374–377.

Zhang W., Behringer R.R., and Olson E.N. 1995. Inactivation of the myogenic bHLH gene *MRF4* results in up-regulation of myogenin and rib anomalies. *Genes Dev.* **9:** 1388–1399.

Ziomek C.A. and Johnson M.H. 1980. Cell surface interaction induces polarization of mouse 8-cell blastomeres at compaction. *Cell* **21:** 935–942.

CHAPTER 3

Production of Transgenic and Chimeric Mice
General Issues

To GENERATE GENETICALLY ENGINEERED MICE successfully and efficiently, the investigator must be able to routinely obtain both a large number of preimplantation-stage embryos and an ample supply of pseudopregnant recipients. In addition, the health of the animals must be maintained and good breeding practices followed. This chapter describes a mouse colony suitable for transgenic and gene-targeting experiments. It is written with the assumption that an animal facility is available which is equipped with proper ventilation, temperature, humidity, and light controls and which has provisions for cage washing, food storage, an adequate water supply, and veterinary care. The strains, ages,

and numbers of mice that we describe will, of course, need to be modified according to each investigator's experimental requirements. Although this section is aimed primarily at researchers interested in the genetic modification of the mouse, it will also be useful for those scientists who want to perform other types of studies using mouse embryos. In this chapter, we do not cover breeding strategies that are required for more sophisticated genetic experiments, such as gene mapping or the derivation of congenic strains; for this information, see Lyon and Searle (1989), Foster et al. (1981, 1982, 1983), and http://www.informatics.jax.org/silver/. In addition, we do not give a detailed description of the reproductive physiology of mice; for this information, see Whittingham and Wood (1983).

CONTENTS

Animals maintained in a colony for the production of transgenic and knockout mice can be divided into the following categories, which are discussed separately below:

- Female mice (donors) for matings to produce zygotes for DNA injection

- Female mice (donors) for matings to produce blastocysts or cleavage-stage embryos for ES cell chimeras produced by injection or aggregation

- Fertile stud male mice

- Sterile stud male mice to induce pseudopregnancy in females

- Female mice to serve as pseudopregnant recipients

PATHOGEN CONTROL IN EXPERIMENTAL MOUSE COLONIES

Ideally, experimental mice should be maintained in an environment free of pathogens (specific-pathogen-free [SPF]). Although an increasing number of animal facilities are set up to prevent infection with common murine pathogens, varying levels of containment have demonstrated varying degrees of success in preventing infection of the colony. Often, the operation of animal facilities and the procedures for animal husbandry are determined by institutional authorities and are therefore beyond the control of the individual investigator. In some cases, investigators may have a say in the design of mouse facilities or the establishment of husbandry protocols. Therefore, although a full discussion of animal husbandry and animal facility design is beyond the scope of this chapter, it is important to consider the advantages and disadvantages of various containment procedures from the point of view of the investigator who uses the mouse for experimental genetic manipulations.

Usually, in "conventional" colonies, animals from suppliers whose stocks are believed to be free of pathogens are imported without quarantine. Although various precautions may be taken to reduce the likelihood of the introduction or spread of pathogens, these animals may be taken out of the facility for experimental procedures and then returned to the facility without quarantine. However, such colonies usually become infected with common pathogens, particularly mouse hepatitis virus (MHV).

Recently, more and more animal facilities have been constructed as SPF facilities and are behind some kind of physical barrier. In the strictest barrier facilities, every person must shower and change into sterile outfits (hats, gowns, shoe covers, masks, and gloves) before entering the facility, and all incoming materials, including food, must be autoclaved or otherwise sterilized (facility level containment). Often, notebooks and other personal items cannot be brought into the facility. An additional level of containment is usually provided at the room and cage level using microisolator cages or racks with individually ventilated cages (IVC). All new animals entering such a facility (including presumably pathogen-free animals purchased from suppliers) must be rederived (see Chapter 15) or at least quarantined and tested for pathogens. In less stringent "modified barrier" facilities, some of these requirements (e.g., showering, changing into sterile clothes) may not be required or may be replaced by other procedures. In all barrier facilities, animals that have been removed from the facility cannot be returned without going through quarantine, in case they have been exposed to pathogens. It is also critical that personnel who have been in recent contact with any rodents (e.g., mice from other facilities, pets at home) do not enter the facility for 1–5 days, depending on the facility's regulations.

Generally, the health status and the design of the mouse colony will determine the operating procedures in the SPF facility. In the strictest SPF facilities, all animals used in transgenic production are bred within the barrier and all experimental manipulations, such as microinjection of cells or DNA into embryos, embryo transfers, and tissue biopsies, must be performed inside the facility. In addition, any cell lines to be injected into embryos or mice (e.g., embryonic stem [ES] cell lines) must first be screened for murine pathogens by mouse antibody production (MAP) testing (for further information, see Chapter 8). Ideally, all genetic alteration experiments should be performed in a designated tissue culture area used only for MAP-tested ES cells and feeders. If this is not done, recipients should be kept in isolation until their screening is performed. After such screening, the weaned offspring can be introduced into the barrier. In some facilities, this screening procedure is done routinely, even when MAP-tested ES cells are used. It may be practical to have a tissue culture area close to the microinjection laboratory, thus reducing the need to transport ES cells over long distances.

As embryos are manipulated in vitro, and later reintroduced into SPF foster mothers, the location and cleanliness of the lab where manipulations take place are of major importance. Three approaches can be taken: (1) The lab is located within the barrier animal facility, and DNA, ES cells, equipment, supplies, and personnel are introduced behind the barrier through decontamination. Embryo donors and recipients are bred within the barrier. (2) The embryo donors are located outside the barrier to allow for their import, and recipients are bred within the barrier. Harvested and washed embryos enter the barrier for further manipulations. (3) The lab is located outside the barrier, and the embryos are introduced behind the barrier after manipulations by embryo transfer.

These systems have advantages and disadvantages. As discussed above, it is generally more difficult to work inside the barrier. Sterile clothing, masks, and gloves must be worn all the time. One major concern is that powdered gloves can cause clogged needles used for pronuclear microinjections (see Chapter 7). Every item introduced behind the barrier is a potential risk and needs to be decontaminated. An incubator necessary for embryo culture before, during, and after manipulations may present a culture environment for pathogens that are not welcome inside the barrier.

On the other hand, when the lab is located outside the barrier, the problem of transporting manipulated embryos into the barrier has to be solved (see Chapter 15). Even though such a lab is located outside the barrier, it should be regarded as a clean area similar to tissue culture rooms. HEPA-filtered incoming air and medium with antibiotics should be used, and personnel and tissues from non-SPF facilities should not enter the lab, in order to minimize the risk of mouse pathogen contamination.

There is a strong and growing trend in new SPF facilities to house embryo donors outside the barrier. In this case, animal donors may be purchased from approved commercial suppliers, hormone-primed, and mated outside the barrier, thus making their quarantine unnecessary. Then, all manipulations are done outside the barrier as described above, and manipulated embryos are brought inside the barrier via embryo transfers (see Chapter 15). The isolation and screening of foster mothers after weaning may be required in such a case. However, this method may significantly reduce the time and cost usually associated with the breeding of the embryo donors in SPF facilities.

In SPF facilities, embryo transfers are often performed in a laminar flow hood or in a designated HEPA-filtered operation room. Similar procedures are used for the rederivation of imported or contaminated mouse strains, as this way is often used to import mouse strains into the barrier (see Chapter 15). Preimplantation-stage embryos with intact zonae pellucidae are harvested from the contaminated donors, washed extensively using antibiotic-containing medium, and cultured for 24 hours (an optional but suggested step; see Hill and Stalley 1991) before being transferred into SPF recipients (Chapter 6). Contaminated donors are dissected in the quarantine (containment) area, and embryos are collected in a laminar flow hood preferably located separately from the lab used for manipulating embryos from SPF donors. It is important to avoid exposure of lab personnel to contaminated donors. This can be achieved either by designating personnel only for the recovery of contaminated embryos or by having preimplantation-stage embryos shipped from the original source for the immediate embryo transfer (for more details on rederivation, see Chapter 15). In either case, foster mothers should be screened after weaning to ensure the clean health status of rederived animals.

Thus, the important considerations regarding SPF and non-SPF colonies are:

1. **Breeding and survival.** The animals are healthier, live longer, and are not lost as a result of infections in SPF colonies. In contrast, in colonies infected by MHV, although most mice appear to be unaffected, young mice are sometimes lost at 2–3 weeks of age. Other viruses, such as Sendai virus, can cause severe respiratory problems in some strains.

2. **Experimental results.** For certain experiments, particularly those involving the immune system, viral or parasitic infections can affect results, even when the mice are apparently healthy.

3. **Convenience.** Access to animals is more difficult and more time-consuming in SPF colonies. Some experimental equipment must be set up within the animal facility, and investigators must spend more time in the animal facility and plan their schedules more carefully.

4. **Cost.** Maintaining mice under barrier conditions requires more elaborate husbandry procedures, and therefore greater labor costs, as well as additional equipment (e.g., laminar flow hoods for cage changing, microisolator cages, and IVC racks). Additional experimental equipment (e.g., microscopes) must be purchased specifically for use in the animal facility if micromanipulation or embryo transfers are to be performed.

5. **Transportation of strains**. Strains of mice obtained from other investigators may carry pathogens and thus must be rederived into the SPF facility by embryo transfer or Caesarean section. On the other hand, transgenic or mutant strains from SPF colonies can in some cases be transported to other facilities without the need for rederivation by embryo procedures. However, in many strict colonies, no animals can be imported without rederivation regardless of their origin. Even if the option of an animal's import is possible, animals must be quarantined, and thus 3- to 4-week-old females cannot be available for superovulation. In cases where the manipulations are done within the barrier, embryo donors must be bred within the facility. This can raise costs and space requirements considerably but may provide more consistent results.

Each investigator should consider the above issues when deciding what is best for one's particular situation and needs.

BREEDING COLONY TO SUPPLY DONOR AND/OR RECIPIENT MICE

To produce a supply of mice, a number of breeding cages should be established, each containing 1 female and 1 male. The progeny should be weaned and separated by sex when they reach 3 weeks of age. For maximal production, one litter should be weaned before the next litter is born to prevent the older pups from trampling the younger ones and not allowing them to feed. A breeding colony of 40–50 mating pairs should yield 20–30 female progeny per week, depending on the strain (and an equivalent number of males, most of which are culled). To increase the yield of female progeny per cage, 2 females may be placed with each male. The male progeny are culled at birth (males can be identified during the first few days after birth by the presence of a pigmented spot in the scrotal area). Note that it takes 3–4 weeks from the time a breeding colony is set up to the time the first progeny are born, and another 4–8 weeks before these progeny reach reproductive age. Because it takes several months to establish a new colony, such a breeding colony must be maintained even when the progeny are not required.

Weaning must be done before the female progeny become sexually mature (4–8 weeks depending on the strain) to avoid their mating with their fathers. Males of most strains that are weaned together before maturity (from the same or separate cages) may be caged together indefinitely as long as they are not exposed to females. Note that males of a few strains, such as BALB/c, cannot be caged together after 6–8 weeks of age because they are very aggressive. Mature males that have been kept in separate cages or have been exposed to females must not be placed in the same cage or they will fight. Females generally do not fight and thus may be caged with other females at any time. If fighting is observed in a cage with males, the animals should be separated immediately. Male mice tend to bite each other in the scrotal area, which may lead to breeding difficulties in the future.

To maintain a breeding colony at maximum efficiency, follow these few simple procedures:

1. Leave the male in the cage with the pregnant female. This is because immediately after giving birth, a female mouse goes into postpartum estrus and will mate with the male. In this way, a good breeding pair will produce a litter about every 3 weeks.

2. After the litter is weaned and separated by sex, record their dates of birth on new cage cards. Mice of significantly different ages (>1 week) should not be mixed, because only females in a certain age range can be used for efficient superovulation.

3. Keep a record of the number of progeny weaned from each mating cage on the cage card. The breeding pair should be replaced if

    a. they fail to produce progeny within about 2 months of the initial pairing

    b. they stop producing progeny for more than 2 months

    c. they begin to produce litters of significantly smaller size at weaning (i.e., 2–4 pups vs. 8–10 pups)

    d. they are older than ~9 months of age, or they have produced more than six litters

In setting up a breeding colony of an inbred strain, it is best to use commercially obtained inbred mice as breeders and to use their offspring for experiments, rather than to perpetuate the colony. Otherwise, the characteristics of the inbred strain might gradually change as the breeding colony accumulates spontaneous mutations. For a SPF barrier facility, it may be difficult to purchase a new breeding colony every 6 months. A good solution is to purchase the original breeding colony from a reputable vendor and to use these animals for in-house breeding for four generations, after which the breeding colony is replaced. The next-generation breeders should be chosen from the highest performing breeding pairs, and care should be taken to choose them from the parents' "latest" litters.

SETTING UP NATURAL MATINGS

One option for obtaining preimplantation-stage embryos for microinjection or other experiments is to set up natural matings; i.e., matings in which the timings of ovulation and fertilization are controlled by environmental conditions. Natural mating is used routinely to produce pseudopregnant females to serve as recipients and to obtain postimplantation embryos for various experiments. Females maintained on a constant light–dark cycle tend to ovulate once every 4–5 days, 3–5 hours after the onset of the dark period. The female mouse estrous cycle is divided into four phases: proestrus (development of ovarian follicles), estrus (ovulation), metestrus (formation of corpora lutea), and diestrus (beginning of follicle development for next ovulation, elimination of previous oocytes). Males maintained under the same conditions will copulate with females in estrus (i.e., ovulating females) at about the midpoint of the dark period. This means that fertilization takes place 1–2 hours after ovulation. For microinjection experiments, a convenient lighting cycle is 7 p.m. to 5 a.m. dark, 5 a.m. to 7 p.m. light (midpoint when fertilization takes place is 12 midnight). Interruptions of the light cycle may have an effect on breeding performance and should be strictly avoided.

Although induction of estrus by male pheromones (the Whitten effect; see p. 154) is often used to increase the number of plugged females, arguably the more efficient way of obtaining naturally mated females is to select their estrous stage by the appearance of the external genitalia (usually easier to see in strains with no or light skin pigmentation). It is possible to identify females in estrus by examining the color, moistness, and degree of swelling of the vagina. Table 3.1, from Champlin et al. (1973), describes the appearance of the vagina at different phases of the estrous cycle. To set up matings, females (6 weeks to 4 months of age) are examined in the afternoon, and those in estrus are placed with males (one or two females in each cage with one male). The morning after mating, the females are checked for the presence of a copulation plug in the vagina (vaginal plug) and are then removed from the male's cage. The plug consists of coagulated proteins from the male seminal fluid and can easily be seen in most strains. In some cases, however, the plug is small and lies deep in the vagina and can only be seen by using a blunt probe (e.g., Fisher 08-995 or Fine Science Tools 10088-15: http://www.finescience.com). Usually, 50% or more of the selected females will mate, and each will contain between 5 and 15 fertilized oocytes, depending on the strain. The presence of a vaginal plug indicates that the mating occurred, but it does not mean that a pregnancy will result. It is important to check vaginal plugs early in the morning because they fall out or are no longer detectable ~12 hours after mating, or sometimes earlier. Pseudopregnant females used for oviduct transfers can be checked for additional signs of ovulation during the surgery to assess their pregnancy (see Chapter 6).

TABLE 3.1. Estrous cycle indicators

Stage of estrous cycle	Appearance of the vagina
Diestrus	vagina has a small opening; tissues are blue and very moist
Proestrus	vagina is gaping; tissues are reddish-pink and moist; numerous longitudinal folds or striations are visible on both the dorsal and ventral lips
Estrus	vaginal signs are similar to proestrus, but the tissues are lighter pink and less moist, and the striations are more pronounced
Metestrus 1	vaginal tissues are pale and dry; dorsal lip is not as edematous as in estrus
Metestrus 2	similar to metestrus 1, but the lip is less edematous and has receded; whitish cellular debris may line the inner walls or partially fill the vagina

Data from Champlin et al. (1973).

For additional information on the mating behavior of mice, see Whitten and Champlin (1978).

INDUCING SUPEROVULATION

For experiments that require large numbers of preimplantation embryos, such as microinjection of zygotes, gonadotropins are often administered to females prior to mating to increase the number of oocytes that are ovulated; i.e., to induce superovulation. Pregnant mare's serum gonadotropin (PMSG) is used to mimic the oocyte maturation effect of the endogenous follicle-stimulating hormone (FSH), and human chorionic gonadotropin (hCG) is used to mimic the ovulation induction effect of luteinizing hormone (LH). Efficient induction of superovulation in mice depends on several variables: the age and weight of the females, the dose of the gonadotropins and their time of administration, and the mouse strain used. In addition, the number of superovulated oocytes that actually become fertilized depends on the reproductive performance of the stud males. These important features of the superovulation procedure are discussed below.

Influence of Age and Weight

The sexual maturity of the female is a major factor affecting the number of oocytes that are superovulated. The best age for superovulation varies from strain to strain, but usually lies between 3 and 6 weeks of age, during the prepubescent stage of development. For example, the optimal age for superovulation of BALB/cGa mice is 21 days (Gates 1971). By this stage of development, a wave of follicle maturation has taken place that increases the number of follicles capable of responding maximally to FSH. However, age is not always a reliable indicator of the sexual maturity of a female mouse; the nutritional status and health of a female can also affect follicular maturation. Underweight and/or sick animals tend to be retarded in development, and thus they yield a reduced number of oocytes after superovulation.

Mice at the precise age and weight required may not always be available from commercial suppliers. Commercial animals are usually weaned in a batch at the end or beginning of a week, and therefore, the actual age of individuals within a consignment will vary. In addition, the breeding conditions for commercial mice are not designed to produce animals of maximum weight for a given age. The transportation, environment, food, and light-cycle changes add stress to animals from vendors, and these animals need time to adjust to new conditions before being used. Thus, to obtain females at the optimal age and weight for superovulation, it is often preferable to establish one's own breeding colony. Procedures for setting up a breeding colony are described above (p. 145). In addition to those procedures, it is also advisable to maintain the breeding pairs on a high-fat diet (Purina Mouse Chow 5020) and to cull the males in the litters within 7 days of birth to ensure an ample supply of milk to the females. Such steps should enable the female offspring to obtain an optimal weight by 3 weeks of age.

Although female mice obtained from a commercial supplier will usually not yield the maximum number of superovulated oocytes obtainable in a given strain, such mice, superovulated between 3 and 6 weeks of age, will produce a significantly larger number of oocytes than will naturally ovulating females at any age. Oocytes produced from commercially supplied superovulated 3- to 6-week-old females are suitable for many types of experiments, not just microinjection. Six- to 8- or even 10-week-old females can produce good yields of embryos for some strains such as FVB and BALB/c.

Dose of Gonadotropins

Although the recommended dose of PMSG for most strains is 5 IU injected intraperitoneally, animals of some strains might respond better to 2.5 or up to 10 IU. PMSG is generally supplied as a lyophilized powder (e.g., PMSG from NIH National Hormone & Peptide Program, http://www.humc.edu/hormones; Folligon from Intervet, http://www.intervet.com; Sigma G4527, http://www.sigmaaldrich.com; Calbiochem 367222, http://www.calbiochem.com/). For administration, the PMSG is resuspended at 50 IU/ml in sterile PBS or distilled water and then divided into convenient aliquots such that a dose of 5 IU in 0.1 ml is injected into each animal. It can be stored in this form at –20°C for at least 1 month, but it does not last indefinitely. Alternatively, aliquots of a 500 IU/ml stock are stored for a month or two frozen at –20°C; each aliquot is diluted directly before use to 50 IU/ml. In both cases, the thawed and diluted aliquot is used immediately and not refrozen. Care should be taken to dispense hormones in a sterile manner because they are easily degraded by bacterial contamination.

hCG is the second gonadotropin that is administered to induce the rupture of the matured follicles. Generally, injections of 5 IU are administered, although a dose of 2.5 IU may be sufficient to assure ovulation in some strains. hCG is also supplied commercially as a lyophilized powder (e.g., Chorulon from Intervet, http://www.intervet.com; A.P.L. from Wyeth-Ayerst Laboratories, http://www.wyeth.com; Sigma C8554, http://www.sigmaaldrich.com). hCG is resuspended at 500 IU/ml in sterile PBS or distilled water, divided into convenient aliquots, and then stored, protected from light, at –20°C. To administer the hormone, an aliquot of hCG is diluted with sterile PBS or distilled water to give a final concentration of 50 IU/ml; 0.1 ml is then injected into each animal if a 5-IU dose is used.

Time of Administration of the Gonadotropins

The times that the PMSG and hCG are administered relative to each other and to the light–dark cycle of the mouse room will affect both the developmental uniformity and the number of oocytes that are recovered from superovulated female mice. For most strains, a 42- to 48-hour interval between the PMSG injection and the hCG injection has been found to be optimal in terms of oocyte yield. Generally, ovulation takes place between 10 and 13 hours after injection of hCG, but to control the time of ovulation precisely, it is important to administer the hCG prior to the release of endogenous LH. The time at which endogenous LH is released in response to PMSG is regulated by the light–dark cycle. Thus, animals purchased for superovulation from a commercial supplier should be given a few days to adjust to the light–dark cycle of the mouse room before the administration of the PMSG. The time of endogenous LH release will vary depending on the strain, but a reasonable estimate for most strains is between 15 and 20 hours after the midpoint of the second dark period following injection of PMSG (Gates 1971). For example, on a 5 a.m. to 7 p.m. light cycle, PMSG is administered between 1 p.m. and 2 p.m., and the hCG is administered 46–48 hours later, usually between 12 noon and 1 p.m., and thus at least 2–3 hours before endogenous LH release. In some strains of mice, the amount of endogenous LH released in response to PMSG in a prepubescent mouse may not be sufficient to induce ovulation (see Gates 1971); in these cases, the timing of the hCG injection is not so critical. After the administration of the hCG, one female is placed in a cage with one stud male, and the female is checked for a copulation plug the next morning. If space allows, it is possible to reuse unplugged superovulated females for natural mating 12–14 days later when necessary. Superovulating these females for a second time is possible, although not as efficient as the first superovulation.

Strain of Mouse for Superovulation

Inbred and F_1 hybrid strains vary with respect to the numbers of oocytes that are produced after superovulation. The same strain purchased from different vendors or bred in-house may differ in its response to superovulation. In general, the conditions of superovulation should be optimized empirically in each laboratory. All the parameters discussed above, such as the age and weight of animals, the dose of gonadotropins, time of injections, and light cycle in the animal room, should be considered.

Reproductive Performance of the Stud Males

To maximize the number of fertilized oocytes recovered from a superovulated female, it is critical to use stud males with a good plugging performance and a high sperm count. Procedures for maintaining such stud males are discussed below (p. 153).

PRODUCTION OF EMBRYO DONORS AND RECIPIENTS

Female Mice for Matings to Produce Zygotes for DNA Injection

The first decision to be made when setting up a colony for the production of transgenic mice is which mouse strain should be used to generate zygotes for microinjection. In general, because of the relatively poor reproductive performance of inbred mice, the production of zygotes, the generation of transgenic mice, and their subsequent breeding are more efficient when F_2 zygotes are used for microinjection (Brinster et al. 1985). Therefore, if an inbred genetic background is not critical to the experiment, injections most often are performed with F_2 hybrid zygotes generated from matings between F_1 hybrid male and female mice (e.g., [C57BL/6 x CBA]F_1 female x [C57BL/6 x CBA]F_1 male). F_2 hybrid zygotes derived from several different F_1 hybrids have been used successfully to produce transgenic mice. These F_1 hybrids include C57BL/6 x SJL, C57BL/6 x CBA, C3H x C57BL/6, C3H x DBA/2, and C57BL/6 x DBA/2.

In-house breeding of F_1 embryo donors is costly and requires a lot of space. Essentially, three breeding colonies must be maintained: one for each inbred strain, and a third for the F_1 production. In many cases, when a C57BL/6 colony is also maintained for blastocyst production, and outbred CD1 (ICR) are maintained for fosters, (C57BL/6 x CD1)F$_1$ embryos can be used successfully for pronuclear injection (K.Vintersten, pers. comm.). Outbred strains like Swiss Webster (SW) and CD1 (ICR) can also be used as donors for pronuclear injections, although zygotes from these strains lyse more easily than hybrid zygotes and have higher rates of one-cell arrested embryos after overnight culture. However, their use might be justified by their relatively low cost when embryo donors are purchased from commercial suppliers.

For many experiments, it is important to introduce the transgene into a defined genetic background, and, in these cases, the genetic advantages of inbred mice outweigh their embryological and economic disadvantages. Examples include the transfer of one allele of a mouse gene into a strain carrying a different allele (see, e.g., Grosschedl et al. 1984; Tronik et al. 1987) and experiments where the transgene is anticipated to affect a phenotype that has been characterized in a specific inbred background or is expected to be susceptible to genetic background effects. To date, the inbred strains most widely used include FVB/N, C57BL/6, BALB/c, and C3H. FVB/N is an inbred strain popular in transgenic experiments because of the large size of the pronuclei in the zygotes of this strain and their resistance to lysis, as well as the relatively good reproductive characteristics (Taketo et al. 1991). However, the FVB/N strain has not been as extensively characterized genetically as some of the older inbred strains.

Superovulated females are used almost exclusively over naturally ovulated females for the production of zygotes (see p. 146 and 148 for details). On average, 20–50 oocytes can be recovered from a 3- to 6-week-old superovulated C57BL/6 (abbreviated B6) or (B6 x CBA)F$_1$ female mouse, whereas only 6–8 and 8–10 oocytes, respectively, can be recovered from a naturally ovulated B6 or (B6 x CBA)F$_1$ female mouse. Superovulation of female mice minimizes the labor and expense involved in obtaining sufficient numbers of zygotes for microinjection. For example, to recover ~200 zygotes, one would need to dissect oviducts from only 7–10 superovulated females compared to 20–30 naturally ovulated females. Consequently, the number of female mice maintained for the isolation of zygotes is minimized by the use of superovulation.

Table 3.2 lists the number of female mice used and the number of zygotes recovered in a typical microinjection experiment using superovulated (B6 x SJL)F$_1$ hybrid female mice. Ten 3-week-old (B6 x SJL)F$_1$ females are injected with PMSG between 1 and 2 p.m. (assuming a light period of 5 a.m. to 7 p.m.) and with hCG 46–48 hours later. After the administration of hCG, 1 female is placed in a cage with 1 stud (B6 x SJL)F$_1$ male. Typically, 7–10 of the females will be plugged, yielding a total of 200–300 zygotes. Approximately 90% of the oocytes will usually be fertilized, as evidenced by the presence of two pronuclei. Two polar bodies are also usually visible. The remaining 10% will have no pronuclei; these oocytes have not been fertilized (one polar body) or have already undergone pronuclear breakdown in preparation for the first cleavage (two polar bodies). Some zygotes will have three pronuclei, most likely due to polyspermy (two sperm fertilized the same oocyte). These zygotes will not develop and can be discarded (see Fig. 7.1, p. 311).

Zygotes with two pronuclei are microinjected with a solution of cloned DNA (see Chapter 7). If not performed correctly, the microinjection procedure can lead to the immediate lysis of the injected zygote. The goal is to achieve a high percentage of the injected zygotes surviving the injection procedure. Although a

TABLE 3.2. A typical microinjection experiment using (B6 x SJL)F$_1$ superovulated females

Number of females set up to mate	10
Number of plugs obtained	7–10
Total number of zygotes isolated	200–300
Number of injectable zygotes	180–270
Number of surviving embryos	160–240
Number of transfers	6–10
Number of pups born	36–60
Number of transgenic mice	7–12

50% survival rate is workable, with practice it is possible to achieve >90% survival of the injected zygotes from hybrid strains. The zygotes that survive the microinjection are then transferred into the oviducts of pseudopregnant recipient females (see Chapter 6), usually 20–30 embryos per mouse. In a typical day, two to six oviduct transfers are performed. However, it should be noted that very skilled microinjectionists are capable of injecting sufficient numbers of zygotes with high efficiency to yield up to ten transfers per day.

If the strain to be used as the zygote donor is available commercially and can be imported into the colony, it is generally more convenient (and usually less expensive) to purchase females of breeding age regularly, rather than to raise them in the colony. Many inbred strains and F$_1$ hybrids are available from commercial suppliers (see Appendix 3), who will ship mice of a specified age weekly to the laboratory. In addition, when microinjection experiments are performed only sporadically during the year, it is definitely more economical to purchase females when they are needed, rather than to maintain the breeding colony. However, it may be necessary to place an order weeks or months in advance, because large numbers of females of certain popular strains may not be available from suppliers on short notice.

In many barrier facilities, mice cannot be imported without a quarantine period of 2–3 weeks, making it impossible to import young females. Therefore, either a breeding colony to supply embryo donors should be set up, or purchased donors should be housed outside the barrier as described above.

Female Mice for Matings to Supply Blastocysts or Cleavage-stage Embryos for ES Cell Chimeras Produced by Injection or Aggregation

For experiments involving injection of ES cells into blastocysts or aggregations with cleavage-stage embryos (see Chapter 11), the mouse strain used for embryo donors (the "host" strain) will often be different from the strains used for DNA microinjection into zygotes. Therefore, if both DNA injection and ES cell chimera experiments are to be performed, it will be necessary to breed or purchase additional strains of mice. The choice of strains will depend to a large extent on the genetic origin of the ES cell line to be used. It is important to choose the right coat-color combination determined by the genotypes of ES cells and host embryos for easy detection of chimerism. The most commonly used strains of mice used as a host for 129-derived ES cell chimeras are C57BL/6 for blastocyst injection and the outbred albino stock CD1 (ICR) for aggregations (for more details on background combinations of ES cells and host embryos, see Chapter

11).

The following details are given for 129-derived ES cells. Both natural matings and superovulation can be used for C57BL/6 blastocyst production. Females must be sexually mature when natural matings are used. Best results are obtained when females are at least 8–10 weeks of age. Females that fail to mate should be placed back in a stock cage of females; they can be used again when they enter estrus. Unfortunately, C57BL/6 mice do not produce good yields of blastocysts from natural matings. Furthermore, the embryos are not very well synchronized, which makes it necessary to culture less developed ones for several hours or even overnight before injection. Therefore, it is necessary to maintain a relatively large stock of breeding females to mate with stud males. Females in estrus can be selected before mating, or unselected females can be left in the cages with males (up to 3 females per male), and plugged females can be identified daily.

Superovulation of C57BL/6 females should at least double the number of embryos that can be collected from each donor female when the conditions such as the age of donors, hormone dose, light cycle, and time of injection are optimized. Most commonly, prepubescent, 12.5-g C57BL/6 females at 3–4 weeks of age are used; superovulation of older females is also possible, although there may be a reduction in the number of embryos.

Superovulation often lowers the quality of the blastocysts. Many blastocysts obtained from superovulated females are of a quality that makes them difficult to inject (for more details, see Chapter 11). Therefore, the yield of normal, usable blastocysts is not always increased sufficiently to justify superovulation. One approach that improves embryo quality after superovulation is collecting morula-stage embryos from the oviducts at 2.5 dpc and culturing them overnight (see Chapter 4).

For the production of aggregation chimeras with 129-derived ES cells, superovulated albino outbred CD1 (ICR) females produce good yields of 8-cell to morula-stage embryos (see Chapter 11).

Fertile Stud Male Mice

A transgenic mouse facility that uses 150–200 zygotes each day will require ~30–50 fertile stud males of the appropriate strain (or more, if multiple strains are used). A similar number of fertile stud males are required to obtain 50–100 blastocysts per day for ES cell injection.

Male mice reach sexual maturity at ~6–8 weeks of age. Males to be used as studs should be placed in individual cages 1 week before they are presented with a female, because the dominant male may suppress testosterone synthesis, and consequently sperm production, in his littermates. New stud males are often set up with females for about a week's practice before they are used in embryo production. Stud males must be kept in separate cages at all times to avoid fighting and injury.

Most F_1 hybrid males can be used as studs for ~1 year, whereas inbred males should be replaced at 6–8 months of age because their reproductive performance tends to decrease after this time. For zygote production, 1 superovulated female is placed in a cage with 1 stud male, usually in the afternoon, and is checked the next morning for a copulation plug. The mating or "plugging" performance of the male is recorded on each male's cage card by noting the date a female was presented and the presence (+) or absence (–) of a plug. A normal male will plug a superovulated female nearly every time one is presented; if a male fails to plug a superovulated female several times in a row, or if his "plugging average" is less than 60–80%, he should be replaced.

The male's sperm count will be depressed for 1–3 days after mating. This period, however, depends on the genetic background of the males. To ensure the fertilization of a maximum number of embryos, a stud male should not be used for several days after plugging a female. However, if the supply of stud males is limited, a rest of 1 day between matings is usually sufficient for most strains. With these requirements, more stud males must be maintained than would be used on any one night. For example, to obtain 150–200 zygotes on a given day, ~10 superovulated females would be mated with 10 stud males. If 150–200 zygotes are required every day, it will be necessary to maintain ~30–50 stud males, so that each set of 10 males can be rested for 1–3 days between matings. If space allows, it is suggested that C57BL/6 males be given 7–10 days' rest between matings.

For natural matings, 1 or 2 females selected for being in estrus are placed with 1 male; usually, one-third to two-thirds of the males will mate with 1 of the females, each yielding ~6–8 blastocysts. Often only half of these blastocysts will be of the right developmental stage for injection. If a male does not plug either of the 2 females (which should be noted on the cage card), he can be used again for mating the next night; males that do plug should be rested for 1–3 days. Female mice must be exposed to male pheromones to establish a regular estrous cycle; females enter estrus within a couple of days after being exposed to a male, and most of them plug on the third night (the so-called "Whitten effect"). On the basis of this effect, and if space allows, the females not selected for estrus are left in a cage with a male for 3 days, and plugged females are removed every day, with the highest plug rate reached by the third day.

Sterile Male Mice to Induce Pseudopregnancy in Females

Genetically sterile or vasectomized males are required for matings to generate pseudopregnant recipients. For vasectomy (see Chapter 6), males of at least 2 months of age from any strain with a good breeding performance (such as an outbred stock or F_1 hybrid strain) are suitable. Before using a vasectomized male in an experiment, it may be tested for sterility by mating at least 1 week after surgery. If the vasectomy is successful, none of the plugged females will become pregnant. Vasectomized males may also be purchased from some suppliers (see Appendix 3). A colony of 20 sterile males should be sufficient to produce 4–8 pseudopregnant females 5 days a week. Vasectomized males may be mated every night, but preferably every other night. In our experience, these mice tend to maintain a high plugging rate for at least 1 year. A plugging record (as described above for the fertile males) should be kept for each sterile male. Any vasectomized mouse that fails to produce a copulation plug in a female four to six times in a row should be replaced.

The same sterile males are used to produce pseudopregnant females for embryo transfers to the oviduct or to the uterus (see Chapter 6). However, when both types of experiments are performed simultaneously, additional sterile stud males will be required.

Female Mice to Serve as Pseudopregnant Recipients

Pseudopregnant mice for embryo transfer recipients are prepared by mating females in natural estrus with vasectomized or genetically sterile males. Pseudopregnant females at 0.5 dpc are used for oviduct transfers and at 2.5 dpc for uterine transfers. The same pool of pseudopregnant females can be used for both kinds of transfers. The females should be at least 6–8 weeks of age and ideally weigh between 25 and 35 grams. Overweight mice are difficult to use for oviduct transfer because of the large fat pads surrounding the ovaries, and underweight mice show reduced pregnancy rates. Underweight mice should be kept until they gain weight, and overweight females should be periodically culled from the stock.

Either outbred or F_1 hybrid mice make suitable recipients. Females of certain outbred stocks (e.g., CD1 mice from Charles River Laboratories, or ICR mice from Taconic or Harlan Sprague Dawley) have very large infundibula, which make oviduct transfer easier for less experienced investigators, and they are generally good mothers. Some investigators prefer to use F_1 hybrid females (e.g., [B6 x CBA]F_1), which, although their infundibula are smaller, make exceptionally good mothers, rearing litters as small as 2 pups.

In some cases, it may be desirable to use coat-color differences between the strain of the embryo donor and the strain of the pseudopregnant recipient to ensure that any mice born actually derive from the donor embryos. However, if the vasectomy was performed correctly and/or tested, coat-color markers are not necessary.

If females in estrus are selected properly, the plug rate should be around 50% or higher. Placing 2 females in a cage with 1 male increases the probability that each male will produce a plug, because not all females judged to be in estrus will actually be in estrus. If the number of vasectomized males is limiting, this will result in the greatest number of plugs per male. On the other hand, if the number of available females in estrus is limiting (e.g., if only 10 females in estrus are available on a particular day), 1 female should be placed in a cage with 1 male to maximize the number of plugs produced. With properly selected females, the necessary number of pseudopregnant plugs should be produced by either of the above strategies. Plan the matings so that an excess of pseudopregnant plugs is obtained. Sometimes an attempt to transfer embryos to a particular pseudopregnant female will fail, or the female used for the surgery will not show proper signs of ovulation, and it will then be necessary to use an additional pseudopregnant recipient.

Save the extra pseudopregnant females if space allows. They can be reused in matings once they begin to cycle again (10–14 days after plugging). Females that fail to mate should be placed in a stock cage to be used again when they enter estrus.

In general, females enter estrus and ovulate every 4–5 days, depending on the strain. Thus, in a colony of randomly cycling females, 20–25% of the mice should be in estrus at any one time. Approximately 50–75 mice (and sometimes more) must be examined to obtain 10–15 females in estrus. Therefore, to obtain 30 pseudopregnant females per week, it is necessary to maintain a steady-state stock of at least 100 females between the ages of 2 and 5 months. The rate at which new females will have to be produced (or purchased) will depend on the number of pseudopregnant females actually used per week, since unused pseudopregnant females can be recycled.

Females that are housed in large groups and isolated from males tend to synchronize their estrous cycle. Thus, if potential foster females are housed in groups and if each group of females is always kept together, their cycles will synchronize within the cage but will be different between other cages. Each day different groups of females can be used for mating. Alternatively, the "Whitten effect" (p. 154) may be used, or the whole colony of females can be searched for estrus signs and the nonplugged females mixed between cages.

Occasionally, only very few embryos develop to term in a pseudopregnant recipient. In these cases, the fetuses tend to grow very large and the mother may have difficulty giving birth (dystocia). To save the litter, it may be necessary to sacrifice the recipient to deliver the pups by Caesarean section and then foster them to a new mother. To ensure that a suitable mother is available for fostering, mate several females with fertile males 1–2 days earlier than the matings to produce pseudopregnant recipients. Most mice give birth on day 19.5 dpc, although this may vary plus or minus 1 day among strains.

If a pseudopregnant recipient has not given birth by the afternoon of the expected delivery day, the litter should be delivered immediately by Caesarean section. The mother to be used for fostering should have given birth by that time; ideally, the foster mother should be a female that has successfully raised a litter before. Most often the recipients from the same experiment that have given birth may be used for fostering, and additional mating is not necessary. When fostering, it is helpful to use coat-color differences to distinguish the potential transgenic pups from those belonging to the foster mother. If this is not possible, pups from the original litter can be marked by clipping the end of the tail (for more details, see Chapter 6 and Protocol 6.5).

Occasionally, 10–17 pups may be born to 1 mother if for some reason large numbers of embryos had been transferred. Care should be taken to avoid "runted" pups, because they are more likely to be sterile as adults. It is advisable to foster a few pups to another lactating female.

Although most mouse colonies are maintained on a low-fat diet (4.5–6% fat) to avoid obesity-related breeding problems, recipients after surgery can benefit from a high-fat (9–11% fat) or mixed high-/low-fat diet, which may reduce cannibalism of pups. If possible, a single pregnant female should be housed in a cage (or at maximum 2 pregnant females per cage) and not disturbed within 2–3 days of delivery (i.e., change cage in advance and provide nesting material). Nesting material (Kimwipes, cotton nestlets, shredded paper towels) may help in decreasing cannibalism and in transfer of young pups during cage changes. Noise, vibration, and smells may cause stress in mice, which decreases reproductive performance and can lead to abortion and cannibalism of neonates.

NOMENCLATURE FOR GENETICALLY ENGINEERED MICE

The Rules and Guidelines for Gene, Allele, and Mutation Nomenclature were completely rewritten in 2000 by the International Committee on Standardized Genetic Nomenclature for Mice (Chairperson: Ian Jackson; E-mail: Ian.Jackson@hgu.mrc.ac.uk) to explain the existing rules more clearly. Some existing gene names and symbols may retain names derived from earlier conventions that no longer apply. See http://www.informatics.jax.org/mgi-home/nomen/gene.shtml.

See also http://www.informatics.jax.org/mgihome/nomen/table.shtml for a more detailed discussion of nomenclature.

The Mouse Genome Database (MGD) serves as a central repository of gene names and symbols to avoid use of the same name for different genes or use of multiple names for the same gene. The MGD Nomenclature Committee (nomen@informatics.jax.org) provides advice and assistance in assigning new names and symbols as unique identifiers. A Web tool for proposing a new locus symbol is located at the MGD site (http://www.informatics.jax.org).

A key feature of mouse nomenclature is the Laboratory Registration Code or Lab Code, which is a code of usually up to three letters that identifies a particular institute, laboratory, or investigator that produced, and may hold stocks of, for example, a DNA marker, a mouse strain, or a mutation. Lab Codes are also used in naming chromosomal aberrations and transgenes. Lab Codes can be assigned by MGD or by the Institute of Laboratory Animal Research (ILAR) at http://www4.nas.edu/cls/afr.nsf.

The guidelines for transgenic nomenclature were originally developed by a committee sponsored by ILAR in 1992 and modified by the Nomenclature Committee in 1999 and 2000. Transgenic symbols should be submitted to MGD through the usual nomenclature submission form for new loci. These symbols can additionally be registered in Tbase: http://tbase.jax.org/.

The transgene symbol is made up of four parts:

- Tg, denoting transgene

- In parentheses, the official gene symbol of the inserted DNA

- The laboratory's line or founder designation or a serial number

- The Lab Code of the originating lab

For example, Tg(SV)7Bri is a transgene containing the SV40 large-T-antigen gene, the seventh transgenic line designated by the lab of Ralph Brinster.

The mouse strain on which the transgene is maintained should be named separately as in the Guidelines for Mouse Strain Nomenclature. In describing a transgenic mouse strain, the strain name should precede the transgene designation.

Mutations that are the result of gene targeting by homologous recombination in ES cells are given the symbol of the targeted gene, with a superscript consisting of three parts:

- The symbol tm to denote a targeted mutation

- A serial number from the laboratory of origin

- The Lab Code where the mutation was produced

For example, *Bmp4^{tm1Blh}* is the first targeted mutation of the bone morphogenetic protein 4 (*Bmp4*) gene produced in the laboratory of Brigid Hogan.

Intraperitoneal (IP) Injection

IP injection is the typical means of introducing most compounds, such as hormones and anesthetics, into the mouse.

MATERIALS

ANIMALS
Mouse

EQUIPMENT
Hypodermic needle, 26- or 30- gauge, 1/2 inch (smaller-size needle is recommended for young animals)
Syringe, 1–3 ml sterile disposable

PROCEDURE

1. Pick up the mouse by the scruff of its neck as close to the ears as possible. Be sure to take up enough skin so that the mouse cannot turn its head and bite you. Hold the tail by twisting it around your little finger (Fig. 3.1).

2. Use the hypodermic needle to pierce the skin and abdominal muscles to inject the solution into the intraperitoneal cavity, taking care to avoid the diaphragm and other internal organs. Wait briefly before withdrawing the needle so that liquid does not seep out (Fig. 3.1).

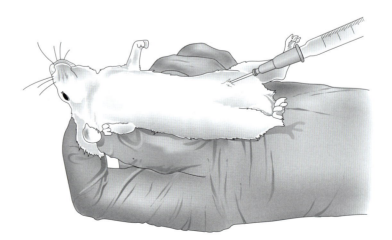

FIGURE 3.1. Method for intraperitoneal injection of a mouse.

COMMENTS

If by mistake the injection has been subcutaneous because the angle of needle entry into the body wall was not sufficient, a bleb will appear at the site of injection. If gonadotropins were injected subcutaneously, the mouse can be mated as usual, although the yield of embryos may be compromised. If an anesthetic was injected subcutaneously, wait longer than usual (two times as long) to make sure the mouse is anesthetized. An additional small dose of anesthetic may be required, but care should be taken due to potential overdose.

REFERENCES

Brinster R.L., Chen H.Y., Trumbauer M.E., Yagle M.K., and Palmiter R.D. 1985. Factors affecting the efficiency of introducing foreign DNA into mice by microinjecting eggs. *Proc. Natl. Acad. Sci.* **82:** 4438–4442.

Champlin A.K., Dorr D.L., and Gates A.H. 1973. Determining the stage of the estrous cycle in the mouse by the appearance of the vagina. *Biol. Reprod.* **8:** 491–494.

Foster H.L., Small J.D., and Fox J.G. 1981. *The mouse in biomedical research.* Vol. I. *History, genetics and wild mice.* Academic Press, New York.

———. 1982. *The mouse in biomedical research.* Vol. II. *Diseases.* Academic Press, New York.

———. 1983. *The mouse in biomedical research.* Vol. III. *Normative biology, immunology, and husbandry.* Academic Press, New York.

Gates A.H. 1971. Maximizing yield and developmental uniformity of eggs. In *Methods in mammalian embryology* (ed. J. C. Daniel), pp. 64–76. W.H. Freeman, San Francisco.

Grosschedl R., Weaver D., Baltimore D., and Costantini F. 1984. Introduction of a mu immunoglobulin gene into the mouse germ line: Specific expression in lymphoid cells and synthesis of functional antibody. *Cell* **38:** 647–658.

Hill A.C. and Stalley G.P. 1991. *Mycoplasma pulmonis* infection with regard to embryo freezing and hysterectomy derivation. *Lab. Anim. Sci.* **41:** 563–566.

Lyon M.F. and Searle A.G., eds. 1989. *Genetic variants and strains of the laboratory mouse,* 2nd edition. Oxford University Press, United Kingdom.

Silver L.M. 1995. *Mouse genetics: Concept and applications.* Oxford University Press, United Kingdom.

Taketo M., Schroeder A.C., Mobraaten L.E., Gunning K.B., Hanten G., Fox R.R., Roderick T.H., Stewart C.L., Lilly F., Hansen C.T., et al. 1991. FVB/N: An inbred mouse strain preferable for transgenic analyses. *Proc. Natl. Acad. Sci.* **88:** 2065–2069.

Tronik D., Dreyfus M., Babinet C., and Rougeon F. 1987. Regulated expression of the Ren-2 gene in transgenic mice derived from parental strains carrying only the Ren-1 gene. *EMBO J.* **6:** 983–987.

Whitten W.K. and Champlin A.K. 1978. Pheromones, estrus, ovulation and mating. In *Methods in mammalian reproduction* (ed. J.C. Daniel), pp. 403–417. Academic Press, New York.

Whittingham D.G. and Wood M.J. 1983. Reproductive physiology in the mouse. *Biomed. Res.* **111:** 137–164.

Recovery and In Vitro Culture of Preimplantation-stage Embryos

T HIS CHAPTER PROVIDES INFORMATION about the isolation and culture of preimplantation-stage mouse embryos that can be used for the introduction of new genetic information and a variety of other studies. It can therefore be used independently of Chapter 3, although readers will find the information on setting up a mouse colony and breeding mice very useful. This chapter contains an overview of culture media available for preimplantation-stage mouse embryos, describes important considerations for in vitro embryo culture, and provides protocols for media preparation and collection of all preimplantation stages.

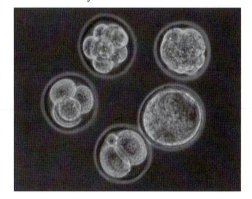

162 ■ CHAPTER 4

CONTENTS

CULTURE MEDIA FOR PREIMPLANTATION-STAGE EMBRYOS

History

The experimental research on preimplantation-stage mouse embryos became a reality in the mid-1950s after the pioneering work of Whitten (1956), who reported the development of eight-cell mouse embryos into blastocysts using Krebs-Ringer bicarbonate solution supplemented with glucose and bovine serum albumin (BSA). Following studies of factors affecting the implantation (McLaren and Mitchie 1956), the successful development and birth of mice after the transfer of in vitro-cultured blastocysts into the uteruses of pseudopregnant females was demonstrated by McLaren and Biggers (1958). Successful aggregation chimera formation and culture using 50% fetal bovine serum (FBS) and 50% Earle's salt solution containing 0.002% phenol red and lactic acid at 1.0 mg/ml was reported by Mintz (1967).

The microdrop culture method developed by R. Brinster in 1963 and still widely used today was central to progress in mouse embryo culture, because it allowed the development of subsequent assays and micromanipulations of preimplantation-stage embryos. The early modification of Krebs-Ringer bicarbonate used in this work included the reduction in the concentration of calcium from 2.54 mM to 1.71 mM, present in most modern culture media, to prevent the precipitation of the medium in the microdrops. Brinster's systematic study of culture media characteristics, effect of osmolarity changes, pH, amino acid composition, and energy sources established the foundation of preimplantation-stage mouse embryo culture. The results from these studies demonstrated that although the first three embryonic cleavage divisions were not supported by Whitten's original medium, addition of lactate supported the development from the two-cell stage (Brinster 1965a), and pyruvate rather than glucose was the primary energy source for mouse embryos, essential for their survival (Brinster 1965b,c). Later studies by Biggers et al. (1967) determined that the development of the mouse zygote to the two-cell stage had an absolute requirement for pyruvate. Glucose, however, was required to support the development of four-cell/eight-cell-stage embryos (Brinster and Thomson 1966). These studies led to the formulation of Brinster's Medium for Ovum Culture (BMOC), and the first basic medium published in table form was BMOC2 (Brinster 1968, 1969).

Several other media based on BMOC were subsequently developed, including modified Whitten's medium (Whitten 1971) and M16 medium, until now possibly the most widely used bicarbonate-buffered medium designed by Whittingham (1971). A variant of M16 medium, in which part of the bicarbonate buffer was replaced by HEPES buffer to maintain correct pH, was later developed for collecting embryos and for experiments in which the embryos are handled for prolonged periods outside the incubator during micromanipulation procedures such as microinjection (Quinn et al. 1982). Complete development to the blastocyst stage of the mouse zygote cultured in conventional embryo media such as M16 has been restricted to a few inbred strains (e.g., C3H) and F_1 hybrids (e.g., C57BL/10J x SJL/J, B6AF1, B6D2F1). Embryos from the majority of inbred and outbred mouse strains undergo cleavage arrest at the two-cell stage; for example, in SJL/J, C57BL/10J, 129/Rr, C3H x DBA, DBA, Swiss Webster, and CF1 strains (Whitten and Biggers 1968; Biggers 1971). This phenomenon, referred to as the "two-cell block," is mainly dependent on the strain of the oocytes and not on that of sperm (Goddard and Pratt 1983). The two-cell block could be overcome by the transfer of embryos to an oviduct either in situ or in organ culture (Whittingham and Biggers 1967), by injecting cytoplasm from F_1 hybrid embryos that do develop normally (Muggleton-Harris et al. 1982), or by including chelators of heavy metals such as EDTA to the media (Abramczuk et al. 1977). Although addition of EDTA to Whitten's medium allowed development to the blastocyst stage for C57BL/6 and ICR zygotes, it was less beneficial for the BALB/c strain (Abramczuk et al. 1977). In a more recent publication of Suzuki et al. (1996), EDTA did not affect the development of embryos derived from AKR/N, B10.Thy.1, and ddY, but significantly increased blastocyst development of embryos from ICR and PW/a strains. This study of in vitro development of mouse embryos from 55 different strains found considerable strain difference and confirmed the maternal effect on the two-cell block, as sperm from a single strain (ICR) was used for fertilization of oocytes from different strains.

For nearly 20 years, few changes were made in the culture conditions for mouse embryos; however, a renaissance of embryo culture research has occurred in the last decade largely because human embryo development in culture was far from optimal and contributed significantly to the embryonic loss observed during human in vitro fertilization (IVF). The development of second-generation embryo culture media has been characterized by a better understanding of physiological principles and metabolic requirements, leading to the improvement of embryo culture techniques. Most of the systematic studies used in the development of human embryo culture media have been done first on mouse embryos due to similarities of metabolic parameters between the two species. Mouse embryos cultured from the pronuclear to blastocyst stage provide a sensitive bioassay that is routinely used in human IVF clinics as a quality control test for medium components, batches of oil, culture medium, and equipment (Quinn et al. 1985b; Scott et al. 1993; Quinn and Horstman 1998). However, the results of such bioassays should be interpreted cautiously because they depend on such parameters as genetic background, age of the embryo at the recovery, and type of medium used. (See more about the quality control below, p. 174.)

Media Available for the Culture of Mouse Embryos

MTF

The rather simplistic composition of preimplantation embryo culture media based on Krebs-Ringer bicarbonate salt solution, supplemented with the energy sources pyruvate, lactate, and glucose, as well as protein in the form of serum albumin, is very different from oviduct and uterine fluids, which are complex in nature and contain many more components such as amino acids, vitamins, and growth factors (Leese 1988). Highly sensitive ultramicrofluorescent assays have been used to analyze metabolites in fluids and enzymatic activities in embryos (Gardner and Leese 1990). The mouse tubal fluid (MTF) medium was formulated using the ionic composition of M16, but it contained physiological concentrations of glucose, pyruvate, and lactate (see Table 4.1) (Gardner and Leese 1990). The main difference between MTF and M16 is the lactate concentration (4.79 and

TABLE 4.1. Composition of different media for the culture of mouse embryos (mM)

Component	BMOC2[1]	M16[2]	MTF[3]	CZB[4]	KSOM[5] (AA)	G1[6]	G2[6]
NaCl	94.88	94.66	114.2	81.62	95.00	85.16	85.16
KCl	4.78	4.78	4.78	4.83	2.5	5.5	5.5
KH_2PO_4	1.19	1.19	1.19	1.18	0.35	none	none
$MgSO_4 \cdot 7H_2O$	1.19	1.19	1.19	1.18	0.2	1.0	1.0
$CaCl_2 \cdot 2H_2O$	1.71	1.71	1.71	1.71	1.71	1.8	1.8
$NaHCO_3$	25.00	25.00	25.00	25.00	25.00	25.00	25.00
Na lactate	25.00	23.28	4.79	31.30	10.00	10.5	5.87
Na pyruvate	0.25	0.33	0.37	0.27	0.20	0.32	0.10
Glucose	5.56	5.56	3.40	0(5.56)	0.20	0.50	3.15
Glutamine	none	none	none	1.00	1.00	1.00	1.00
EDTA	none	none	none	0.11	0.01	0.01	none
Taurine	none	none	none	none	none	0.1	none
$NaH_2PO_4 \cdot 2H_2O$	none	none	none	none	none	0.5	0.5
BSA (mg/ml)	1.00	4.00	4.00	5.00	1.00	2.00	2.00
MEM NEAA[a]	none	none	none	none	(0.5x)	1x	1x
MEM EAA[b]	none	none	none	none	(0.5x)	none	1x(0.5x)

All media contain phenol red at 0.001–0.01 g/liter (*optional*), 0.06 g/liter (100 units/ml) penicillin G, potassium salt, 0.05 g/liter streptomycin sulfate.

[a]MEM NEAA, Minimum Essential Medium non-essential amino acids.

[b]MEM EAA, Minimum Essential Medium essential amino acids.

[1] Brinster (1968, 1969); [2]Whittingham (1971); [3]Gardner and Leese (1990); [4]Chatot et al. (1989), CZB should be supplemented with 5.56 mM glucose after 48 hours of culture. [5]Lawitts and Biggers (1993), KSOM with both NEAA and EAA is denoted KSOM-AA. [6]Barnes et al. (1995), later modifications of original G1/G2, denoted G1.2 and G2.2, contain reduced concentration of glutamine, EDTA and phosphate; its exact formulations are not disclosed for commercial reasons (Gardner et al. 1998).

23.3 mM, respectively). Culturing F_2 zygotes obtained by mating of (CBA/Ca x C57BL/6) F_1 mice for 4 days in a reduced lactate concentration significantly increased the embryos' cell number and fetal development, compared with embryos cultured in M16 medium (Gardner and Sakkas 1993). However, zygotes from outbred CF1 females, cultured in this medium, blocked at the two-cell stage, even in the presence of EDTA (Erbach et al. 1994). These studies suggested that embryo culture medium should reflect the constantly changing environment of the female reproductive tract at different stages of embryo development, synchronizing embryo requirements at certain developmental stages with their place in the female reproductive tract at that time. Moreover, the effect of other components of culture medium, such as amino acids and vitamins, should be further investigated because they appear to affect in vitro and subsequent in vivo mouse embryo development.

CZB

It has been generally accepted that preimplantation-stage embryos up to the eight-cell stage do not use glucose efficiently and require pyruvate and lactate as energy sources (Iyengar et al. 1983; Leese 1991). Moreover, in simple culture media, glucose is responsible for retardation or developmental arrest; the replacement of glucose with glutamine in BMOC2 medium allowed zygotes of a strain blocking at the two-cell stage to develop into blastocysts (Chatot et al. 1989). CZB medium, a modification of BMOC2 medium, contained an increased lactate/pyruvate ratio, 0.1 mM EDTA, and 1 mM glutamine as a replacement for glucose; it facilitated the development of the zygote to the blastocyst stage for several strains of mice that normally blocked at the two-cell stage (CF1 x B6SJLF1/J, DBA x B6SJLF1/J) (Chatot et al. 1990). However, successful development of these embryos from the morula to blastocyst stage in CZB medium required an exposure to 5.56 mM of glucose at 48 hours of culture (see Table 4.1) (Chatot et al. 1990). CZB medium was successfully used for development beyond the two-cell stage of such blocking strains of mice as CF1 and Swiss Webster (Chatot et al. 1989). More recently, CZB medium and its modifications have become the media of choice for embryo manipulations and culture in various mouse cloning experiments (Wakayama et al. 1998, 1999; Eggan et al. 2000, 2001; Humphreys et al. 2001; see also Chapter 13).

KSOM

Lawitts and Biggers (1991, 1992) used a sequential simplex optimization strategy that allowed optimization of several media components simultaneously and developed a SOM (Simplex Optimized Medium) that permitted outbred mouse zygotes (CF1) to overcome the two-cell block (Lawitts and Biggers 1991, 1992). The concentrations of NaCl and KCl were later increased, and the new medium was called KSOM (see Table 4.1) (Lawitts and Biggers 1993; Erbach et al. 1994). KSOM medium supported higher rates of cell division of the trophoblast cells and has resulted in higher yields of blastocyst development in an outbred mouse strain (CF1 x B6D2F1) when compared with its predecessor, SOM medium, and CZB medium (Erbach et al. 1994). KSOM medium has been used to grow zygotes to blastocysts and to support the development to term in such mouse strains as CF1, CD1, FVB, NOD, and C57BL/6 (Lawitts and Biggers 1993). A flushing holding medium (FHM) for handling embryos in the air has been developed based on KSOM medium, which contains HEPES buffer (20 mM) and decreased NaHCO$_3$ (4 mM) (see Table 4.3). Recent modification of KSOM supplemented with amino acids (see Amino Acids, below) is recommended for the culture of mouse embryos following IVF (Marschall et al. 1999; Sztein et al. 2000). KSOM supplemented with amino acids was also successfully used for the culture of reconstituted B6C3F1 oocytes following their enucleation and electrofusion (Liu et al. 2000).

Critical Components of Embryo Culture Media

Glucose

Significant research has been done regarding the role of glucose as an energy substrate in culture medium for mammalian preimplantation embryos. Glucose has a detrimental effect on the development of mouse zygotes during the first 48 hours of culture, yet it is required for blastocyst formation (Chatot et al. 1989, 1990). There is a growing tendency to remove glucose from human embryo culture media (Quinn 1995). However, glucose is present in the reproductive tract (Gardner and Leese 1990), and its complete removal from culture media does not duplicate physiological conditions.

It appears that phosphate plays a significant role in the adverse effects of glucose during in vitro culture, and that glucose is mainly a problem in the presence of a high concentration of phosphate (>0.35 mM) (Scott and Whittingham 1996; Quinn 1998). When human tubal fluid (HTF) medium, the first specific medium for human IVF (Quinn et al. 1985a), devoid of glucose and phosphate ions and containing exogenous protein in the form of BSA, was supplemented with 0.1 mM EDTA and 1 mM glutamine, it greatly improved in vitro development of mouse CF1 outbred and B6C3F1 hybrid embryos as well as human embryos; this medium is referred to as Basal XI HTF (Quinn 1995). Similarly, when KSOM medium was used (low phosphate concentration of 0.35 mM), the addition of glucose (to 5.56 mM) did not inhibit the development of outbred mouse zygotes (CF1 x B6D2F1) into hatching blastocysts (Summers et al. 1995). Even in the presence of a high level of phosphate (1.19 mM), glucose is not inhibitory to the cleavage-stage mouse embryo when nonessential amino acids and glutamine are added to the culture medium (Gardner and Lane 1996). Although it is possible to obtain blastocysts in the absence of glucose in culture medium, their postimplantation development is significantly impaired when compared with blastocysts cultured in the presence of glucose (Gardner and Lane 1996). The variable effects of glucose on preimplantation development may be due to the differences in the medium composition. Recent evidence suggests that the effects of glucose and phosphate are independent, and that glucose is not always an inhibitor of mouse preimplantation development in vitro (Biggers and McGinnis 2001).

Amino Acids

Early formulations of embryo culture media based on balanced salt solutions, lacking amino acids and supplemented only with carbohydrates and BSA, supported development to term, but embryos exhibited a delayed cleavage rate (Bowman and McLaren 1970) and reduced viability after transfer (Bowman and McLaren 1970). Originally, it had been reported that the development of a two-cell mouse embryo to the blastocyst stage had no absolute requirement for exogenous amino acids (Brinster 1965c). However, mouse embryos have the ability to take up amino acids (Brinster 1971), and amino acids are present at high levels in the fluids of the female reproductive tract (Miller and Schultz 1987). In vitro embryo development of different strains of mice was improved when culture medium was supplemented with specific amino acids (Mehta and Kiessling 1990; Gardner and Lane 1993, 1996; Ho et al. 1995). The addition of amino acids to modified MTF (Gardner and Lane 1993) and KSOM (Ho et al. 1995) media increased the rate of blastocyst formation, frequency of blastocyst hatching, and total cell number. Importantly, the addition of amino acids to KSOM medium enabled the blastocysts to synthesize mRNA of several types of proteins at in vivo levels (Ho et al. 1995).

Intensive studies of the role of amino acids in culture media has led to the updating of medium composition in recent years. The amino acids classified by Eagle (1959) as nonessential, i.e., not required for the development of somatic cells in culture, and an essential amino acid glutamine significantly reduced the duration of the first three cell cycles, resulting in increased rates of development and viability of mouse embryos (see Table 4.2) (Lane and Gardner 1994, 1997). Post-compaction, the nonessential amino acids stimulate blastocyst expansion and hatching. However, the addition of essential amino acids prior to the eight-cell stage inhibited development to the blastocyst stage (Lane and Gardner 1994, 1997) but stimulated the development of the inner cell mass (ICM) and subse-

TABLE 4.2. Concentration of amino acids used to supplement mouse embryo culture media

Amino acids	Concentration in Eagle's medium		1/2 Eagle's medium (mM)	
	mM	mg/liter	mM	mg/liter
Non-essential (NEAA)				
L-Alanine-HCl	0.1	8.9	0.05	4.45
L-Asparagine-H_2O	0.1	15.0	0.05	7.5
L-Aspartic acid	0.1	13.3	0.05	6.66
L-Glutamic acid	0.1	14.7	0.05	7.36
Glycine	0.1	7.5	0.05	3.75
L-Proline	0.1	11.5	0.05	5.76
L-Serine	0.1	10.5	0.05	5.26
Essential (EAA)				
L-Arginine-HCl	0.6	126.4	0.3	63.2
L-Cystine	0.1	24.0	0.05	12.02
L-Histidine-HCl-H_2O	0.2	41.9	0.1	20.96
L-Isoleucine	0.4	52.4	0.2	26.23
L-Leucine	0.4	52.4	0.2	26.24
L-Lysine-HCl	0.4	73.1	0.2	36.52
L-Methionine	0.1	14.9	0.05	7.46
L-Phenylalanine	0.2	33.0	0.1	16.52
L-Threonine	0.4	47.6	0.2	23.82
L-Tryptophan	0.05	10.2	0.025	5.11
L-Tyrosine	0.2	36.2	0.1	18.12
L-Valine	0.4	46.9	0.2	23.42

Both NEAA and EAA are added to KSOM medium at half the concentration used for the culture of human cell lines by Eagle (Eagle 1959; Ho et al. 1995; Biggers et al. 2000). Full Eagle's concentration (0.1 mM) of NEAA is used in both G1 and G2 media (Gardner and Lane 1997). Original full Eagle's concentration of EAA used in G2 medium should be reduced to half according to Lane et al. (2001). Glutamine, an essential amino acid, is present at 1 mM and can be substituted by more stable dipeptide L-alanyl-L-glutamine.

quent fetal development when added at the eight-cell stage (Lane and Gardner 1997).

The amino acid taurine is a major component of the free amino acid pool; it is present in mouse oocytes, embryos, and the female reproductive tract, and it has been shown to be beneficial to the development of mouse embryos (Schultz et al. 1981; Dumoulin et al. 1992). The effect of taurine may be due to its antioxidant properties and its ability to serve as an osmolyte or chelating agent (Li et al. 1993). It is present in several media used for human embryo culture; e.g., G1 (Barnes et al. 1995) and P1 (Carrillo et al. 1998).

The determination that amino acids are among key regulators of mammalian embryo development, together with earlier studies of embryo physiology, resulted in significant changes of medium composition regarding levels of carbohydrates and amino acids. The media are formulated according to the carbohydrate composition of reproductive tract fluids and take into account the metabolic requirements and changing physiology of the embryo. Two sequential media (G1 and G2) for the culture of the human zygote to the blastocyst stage have been formulated (Barnes et al. 1995). G1 medium was designed to support development to the eight-cell stage, and G2 the subsequent development to blastocysts (see Table 4.1). G1/G2 sequential media developed for human embryos were recently used successfully to culture mouse embryos obtained by adult somatic nuclear transfer (Munsie et al. 2000). Later modifications of G1 and G2 media, G1.2 and G2.2, contain reduced glutamine, EDTA, and phosphate concentrations and include specific vitamins in G2.2; the exact formulations of G1.2 and G2.2 (Scandinavian IVF Science/Vitrolife) have not been disclosed for commercial reasons (Gardner and Schoolcraft 1998; Gardner et al. 1998). Several other sequential media used for human embryo culture and tested on mouse embryos are commercially available; they include S1/S2 (Vitrolife, http://www.vitrolife.com); P1/Blastocyst media (Irvine Scientific, http://irvinesci.com); Enhance Day 1, Day 3, Day 5 HTF (Conception Technologies, http://conceptiontechnologies.com); and Quinn's Advantage Cleavage and Blastocyst Media (SAGE BioPharma, http://www.sagebiopharma.com).

The concentrations of essential amino acids used for embryo culture as described by Eagle are significantly higher than what the embryo would be exposed to in vivo (see Table 4.2). In addition, when amino acids are used at such concentrations, embryo-toxic ammonium is produced from amino acid metabolism and the spontaneous breakdown of these amino acids (Gardner and Lane 1993). The most recent study of the effect of essential amino acids on mouse embryo viability and ammonium production recommended reducing the concentration of essential amino acids to half of that described by Eagle (1959) in G2.2 medium following compaction (Lane et al. 2001), whereas nonessential amino acids were present at 0.1 mM (Eagle's concentrations) in both media. A more stable derivative of glutamine, alanyl-glutamine, used instead of glutamine at 1 mM, resulted in the production of significantly less ammonium (Lane et al. 2001).

The addition of both essential and nonessential amino acids to KSOM medium at one-half the concentration of Eagle's medium was shown to support the development of CF1 mouse zygotes to the late blastocyst stage, and preferentially to stimulate proliferation of cells in the ICM compared with trophectoderm (Biggers et al. 2000). Importantly, the effects of amino acids were studied in the absence of the exogenous nitrogen source, because BSA (common supplement in mouse embryo culture medium) was replaced with polyvinyl alcohol.

Recent Developments

It is well known that the culture of embryos in reduced volumes of medium and/or in groups increases blastocyst development, blastocyst cell number, and, most importantly, viability after transfer (Paria and Dey 1990; Lane and Gardner 1992). Mouse zygotes could develop to hatching blastocysts in the absence of exogenous protein when cultured at a density of 10 embryos per 10 µl of medium but not when embryonic density was 1 zygote per 10 µl (Quinn et al. 1993). Such beneficial effects may be due to the autocrine/paracrine factor(s) produced by the embryos and diluted in large volumes. Numerous growth factors are present in female reproductive tract fluids. The role of maternal and embryonic growth factors during preimplantation development has been demonstrated in many publications (for review, see Kane et al. 1997; Kaye 1997; Hardy and Spanos 2002).

Receptors and mRNA for several growth factors have been identified in preimplantation mouse embryos, such as platelet activating factor (PAF) and insulin-like growth factor II (IGF-II) (O'Neill 1997). Maternal leukemia inhibitory factor (LIF) is essential for embryo implantation in the mouse (Stewart et al. 1992). Recombinant LIF improves the development of murine blastocysts in culture (Fry 1992); more specifically, it has been shown to enhance mouse eight-cell development in vitro as seen by accelerated trophoblast outgrowth and rate of embryo hatching (Lavranos et al. 1995). A number of growth factors have been shown to promote blastocyst formation. For example, addition of heparin-binding epidermal growth factor (HB-EGF) or transforming growth factor-α (TGF-α) enhanced embryo development to the blastocyst stage (Paria and Dey 1990). Insulin, IGF-I, and IGF-II were shown to increase blastocyst number in the ICM (Harvey and Kaye 1990, 1992a,b). Treatment of mouse blastocysts with EGF increases the implantation rate (Morita et al. 1994). Thus, growth factors appear to play an important role in preimplantation embryo development, and their addition to culture media may further improve embryo viability in culture.

With recently improved culture conditions, human IVF clinics have a tendency to culture embryos for extended periods of time for later-stage embryo transfer with the potential advantage of the embryos' synchronization with the female tract, leading to an increased implantation rate and the ability to assess embryo development and viability further (Gardner and Lane 1997, 2000). However, evidence that in vitro embryo culture and manipulation can lead to reduced viability, aberrant growth, and developmental abnormalities should serve as a cautionary note (Bavister 1995). These defects are often linked to epigenetic changes in the embryo's genome that may be caused by in vitro culture. A recent example demonstrated the dramatic differential effect of culture conditions (Whitten's vs. KSOM medium with amino acids) on the expression of an imprinted H19 gene in preimplantation mouse embryos (Doherty et al. 2000). Preimplantation culture of mouse embryos in the presence of serum was shown to influence the regulation of growth-related imprinted genes such as Igf2, H19, and Grb10, leading to aberrant fetal growth and development (Khosla et al. 2001a). For a recent review of long-term effects of preimplantation embryo culture on gene expression and phenotype, see Khosla et al. (2001b).

Cultured embryos are exposed to atmospheric oxygen tension ($\sim$20% O_2) and are at risk of damage by free-radical production. This culture-induced enhanced oxidative damage due to the production of reactive oxygen species (ROS) is proposed to be one of the main contributory factors of developmental retardation

(Legge and Sellens 1991). The beneficial effect of EDTA on the cleavage-stage embryo is associated with its role as a chelator of heavy-metal ions, thereby reducing the production of ROS (Nasr-Esfahani et al. 1992). EDTA also has been reported to alter the metabolism of mouse embryos; i.e., it inhibits premature use of glycolysis. However, beneficial effects of EDTA seem to be confined to the cleavage-stage embryos. Culture of post-compaction-stage embryos with EDTA significantly reduces ICM and fetal development, compared to culture without EDTA (Gardner and Lane 1996; Gardner et al. 2000). The recent work of Lane and Gardner (2001) has shown EDTA to act as a chelator of divalent cations such as magnesium, to inhibit at least one glycolytic kinase, and to prevent a premature switch to glucose metabolism that is associated with developmental arrest. These results suggested that mouse embryos should be cultured in the presence of 10 μM EDTA for the first 48 hours, with EDTA omitted from the medium for post-compaction development (G1/G2 culture media and 5% O_2 were used). Inclusion of 10 μM EDTA in KSOM medium improved blastocyst formation and the hatching rate but reduced the blastocyst cell count, whereas reduced oxygen tension (5% O_2) showed a consistent beneficial effect on all developmental markers (Orsi and Leese 2001). Therefore, it was proposed by these authors that continuous culture of murine embryos should occur in an EDTA-free medium under 5% O_2 with supplementary protection against ROS provided by physiological chelators such as proteins and/or amino acids.

Successful embryo culture depends on multiple parameters, with the genetic background of the embryos and the medium composition being the most important. As demonstrated in the most recent publication of Kamjoo et al. (2002), when two mouse strains (MF1 and C57BL/6 x CBA) were cultured in two kinds of culture media (M16 and KSOM), they showed significantly different levels of apoptosis. The choice of the medium also depends on the actual purpose for which it is going to be used. For example, a relatively high concentration of glucose is needed for an optimum rate of IVF (Hoppe 1976; Fraser and Quinn 1981); however, media successfully used for mouse IVF such as modified Tyrode's (Fraser and Drury 1975) or HTF (Quinn et al. 1985a) do not support the development beyond the two-cell stage unless F_1 hybrids are used, because glucose inhibits the initial development of preimplantation mouse embryos (Chatot et al. 1989; Lawitts and Biggers 1991). In addition, optimal culture conditions depend on such parameters as oxygen and carbon dioxide concentration, stable environment, and incubation volume. (For more details, see General Considerations for In Vitro Embryo Culture, below.)

Significant progress in human embryo culture achieved in recent years, and largely neglected by researchers in mouse development, may be further validated with the advantage of genomic manipulation in the mouse and may possibly improve such areas as cloning, embryo and sperm cryopreservation, and IVF. Two recent publications show the significance of embryo culture in a field of mouse cloning. Evaluation of the blastocyst formation rate of mouse embryos cloned by somatic nuclear transfer (NT) in CZB medium as well as in sequential media G1/G2 and KSOM/G2 demonstrated that NT embryos are more sensitive to in vitro culture conditions than parthenogenetic control embryos, and that selection of the media can influence the preimplantation development of NT embryos (Heindryckx et al. 2001). The effect of culture conditions on early development of cloned mouse embryos studied by Chung et al. (2002) showed that the combination of Whitten's medium followed by KSOM resulted in an increased number of cells per blastocyst and fostered consistent production of blastocysts at a greater frequency than in CZB medium supplemented with glucose.

In summary, mammalian preimplantation embryo culture is a constantly developing field. The use of specific media for in vitro culture of preimplantation-stage embryos allowed studies of the early development and manipulation of the mammalian genome. Although M16 has been traditionally used in successful transgenic production for many years, at present it is hard to recommend one specific culture medium for all mouse embryo manipulations, although an exception may be made for KSOM medium supplemented with amino acids. The formulations of different preimplantation embryo culture media discussed above and the concentrations of amino acids used to supplement media are presented in Tables 4.1 and 4.2. Their preparation is described in Protocols 4.1–4.4.

GENERAL CONSIDERATIONS FOR IN VITRO EMBRYO CULTURE

The most important points for successful embryo culture are described in this section.

Medium Preparation

Ideally, use disposable sterile plastic containers (e.g., Nalge square medium bottles 2019-0250 or 50-ml conical tubes) and pipettes. If glassware is used for collecting and manipulating embryos, and for preparation and storage of culture medium, it should be designated solely for that use, absolutely free of detergent residues, and rinsed immediately after use at least six times in glass-distilled water before sterilization.

It is extremely important that the water used for making culture media be distilled at least twice in a still, in which the metal element is enclosed by glass (2x glass-distilled), or be purified by filtration (e.g., Milli-Q system producing 18 MΩ water) after an initial reverse osmosis process, and then stored in clean plastic containers. Water purified by filtration should be tested for endotoxin (see Chapter 8). Prolonged storage is not advised. It is also possible to use commercially available deionized water for tissue culture, embryo-tested or endotoxin-tested water, or water for human intravenous administration.

Use only chemicals of the highest grade possible. Test several batches of a particular component, for example, BSA and paraffin oil, and reserve one batch for culture use. Although embryo-tested products such as BSA (Sigma A3311), light mineral oil (Sigma M8410), and ES cell-qualified light mineral oil (Specialty Media ES-005-C) are ready to use, screening of new batches under actual laboratory conditions may be advisable. Double-washed, filtered, endotoxin- and embryo-tested light mineral (paraffin) oil suitable for human embryo culture (commercially available from Vitrolife, Irvine Scientific, SAGE BioPharma, etc.) is also a possible, albeit expensive, alternative. Paraffin oil sold by pharmacists for human medicinal use or light mineral oil may be used for embryo culture. It is not recommended to autoclave oil because this may increase its toxicity. If necessary, oil can be filtered through a 0.8-μm filter (see comments about the filtration below). The potential toxicity of batches of oil should be checked on spare embryos by culturing them in microdrops and assessing their development to the blastocyst stage or their viability following embryo transfer. (See more about quality control below, p. 174.) Prolonged storage of opened bottles of paraffin oil is not recommended because the oil may become toxic. Oil used for embryo culture may be washed by stirring with distilled water or BSA-free media, then centrifuged or allowed to separate, and decanted. For highly sensitive procedures, such as IVF, it is suggested that the oil be gassed by bubbling 5% CO_2 prior to use in embryo culture; alternatively, oil aliquots can be kept for several hours or overnight in the incubator with the caps loose.

Embryo culture media should not be stored for more than 1 week (maximum 2 weeks). Because fresh embryo culture medium has to be prepared weekly or biweekly and requires exact, time-consuming measurements, it is often convenient to prepare it from a series of stock solutions. Many stock salt solutions are stable for months at 4°C. Short-shelf-life stock solutions, such as pyruvate, lactate, bicarbonate, glucose, and glutamine, are prepared weekly and kept frozen or added fresh when media are made. All stock solutions may be stored for a few months when kept at –20°C or –70°C.

Protein-containing solutions can be sterilized by positive-pressure filtration; for example, through Millipore filters (Type GS, pore size 0.22 μm) or presterilized filter assemblies (Gelman Acrodisc, 0.2-μm pore size). Presterilized Nalgene filter units (Nalge, Type S 0.20-μm pore size, e.g., Nalgene 175-0020) are convenient for sterilizing up to 250 ml of solutions using a vacuum pump. However, caution should be used with all filters due to their potential embryotoxicity (Harrison et al. 1990). When filtering embryo culture media or any other reagents used with preimplantation-stage embryos, prewash the filter with distilled water or discard the first few milliliters of medium. That portion of the medium also can be used to measure the osmolarity, which should be within 10 milliosmoles of the calculated value.

It is generally recommended to gas freshly prepared embryo culture medium as well as the air space in a storage container with 5% CO_2 gas mixture and then re-gas before placing the opened container back at 4°C to maintain a physiological pH, because medium rapidly becomes alkaline outside an enriched CO_2 environment. Microdrop culture dishes (see Protocol 4.5) are usually made up several (at least 4–6) hours in advance (Chatot et al. 1989; Gardner and Lane 1996) or on the day prior to the experiment to gas-equilibrate overnight (Lawitts and Biggers 1993). This last option is absolutely required for such sensitive procedures as IVF.

Incubations are most often carried out in a humidified atmosphere of 5% CO_2, 95% air, regulated automatically. A small gassed and humidified container or vacuum desiccator humidified with sterile water can be placed inside a larger, conventional tissue culture incubator. Modular incubator chambers (MIC) (e.g., Billups-Rothenberg), which can be sealed, gassed, and placed at 37°C, as well as incubators with multiple chambers, may be convenient for use in embryo culture because they reduce fluctuations in temperature and carbon dioxide level caused by frequent door opening. Mouse embryo development in MICs at the same gas phase as the incubator (5% CO_2 in air) was significantly increased compared to the development in the main chamber (Gardner and Lane 1996).

Hoppe and Pitts (1973) reported that an atmosphere of 5% CO_2, 5% O_2, and 90% N_2 enhances the survival of early cleavage-stage embryos, and this gas phase is preferred by some laboratories (McGrath and Solter 1983). More recent studies have further supported this evidence. Even a transient 1-hour exposure to 20% oxygen reduced mouse embryo development in vitro (Pabon et al. 1989). Orsi and Leese (2001) showed that reducing oxygen tension from 20% to 5% enhanced mouse embryo development and increased cell number. Although mouse embryos can grow at elevated oxygen concentrations, more physiological low oxygen concentrations are beneficial. Carbon dioxide is required to maintain the pH of bicarbonate-buffered medium, and most often 5% CO_2 concentration is used; however, it is necessary to have 5.5–6.5% CO_2 to maintain the pH of medium containing 25 mM bicarbonate at 7.2–7.4 (Gardner and Lane 2000). Although mouse embryos could develop from the two-cell to the blastocyst stage in a wide pH range (5.9–7.8) (Brinster 1965), a more recent study has shown that a transient exposure of zygotes and two-cell-stage mouse embryos to medium with elevated pH significantly reduced their development to the blastocyst stage (Scott et al. 1993). It is advisable, therefore, to measure the pH of the medium under the actual laboratory conditions and to adjust the level of CO_2 if necessary to achieve the desired pH range of 7.2–7.4. Most recent publications regarding mouse embryo culture describe the use of 6% CO_2, 5% O_2, 89% N_2 gas mixtures with G1/G2 medium (Lane and Gardner 2001) and KSOM medium (Biggers and McGinnis 2001).

Embryo Collection and Culture

Prepare microdrop cultures (see Protocol 4.5) in plastic tissue culture dishes 4–6 hours in advance or even 1 day before the experiment. An oil overlay helps to stabilize the medium and minimize evaporation (which can lead to increased osmolarity), temperature, and pH changes caused by a loss of CO_2 when dishes are outside the incubator. Twenty- to 50-μl drops of medium are used, because it has been demonstrated that culture of embryos in reduced volumes and/or in groups significantly increases blastocyst development and cell number. The optimal embryo/incubation ratio for mouse embryo development in vitro is a group of 10 embryos in 20 μl (Lane and Gardner 1992). An open culture system in larger volumes without oil (e.g., using organ culture center well dishes, Falcon 35-3037) may be convenient for short-term culture of freshly collected embryos before and after manipulations as well as for short-range transportation (i.e., from one lab to another).

Cells, particularly oocytes, zygotes, and two-cell-stage embryos, are intolerant of pH and temperature fluctuations. Exposure of mouse zygotes to room temperature for just 5 minutes inhibited cleavage rate, and 10–15 minutes of exposure reduced the development to the blastocyst stage by half (Scott et al. 1993). Meiotic spindles of most mammals, including mouse, are very sensitive to temperature fluctuations (Pickering and Johnson 1987). Temperature fluctuations during in vitro manipulations of oocytes may disrupt spindles, contributing to abnormal chromosome distribution and failed or abnormal fertilization (Pickering et al. 1990; Almeida and Bolton 1995). Using polarized light microscopy of living human oocytes, a recent study showed that meiotic spindles of human oocytes are exquisitely sensitive to temperature alterations, and their integrity is important for normal fertilization and embryo development (Wang et al. 2001). Therefore, for experiments involving prolonged manipulation outside the incubator, as well as in vitro fertilization, it is advisable to maintain a constant temperature of 37°C by warming the stage of the microscope.

The medium used for collecting, manipulating, and culturing embryos contains antibiotics and is sterilized by filtration as described. Disposable sterile plasticware and freshly pulled embryo-manipulating pipettes (see p. 177) are usually used. If the embryos are to be returned to the oviduct or uterus after a short incubation period, further precautions against contamination are not required. However, if the embryos are to be incubated for more than 24 hours or if it is required by local regulations in specific pathogen-free facilities, sterile techniques are essential to avoid contamination with bacteria or yeast. In this case, use autoclaved instruments and sterile embryo-manipulating pipettes and set up the dissection microscope inside a laminar flow hood.

It has been recommended that embryos be collected using medium with a pH that is stable when exposed to air (Quinn et al. 1982, 1985b). Recent strong evidence for the use of HEPES-buffered culture media for all procedures with gametes and embryos outside the incubator, including micromanipulations, cryopreservation, and embryo transfer, is reported by Lane et al. (1999) and Lane and Gardner (2000). These works demonstrated that in the presence of phosphate-buffered medium, the internal pH of embryos can be dramatically altered, leading to severely compromised viability. Therefore, care should be taken to avoid pH fluctuations of the media during embryo manipulations and culture. Any medium used for embryo culture can be modified to include HEPES buffer for manipulations outside the incubator by replacing 20 mM $NaHCO_3$ with 20 mM HEPES at pH 7.4. The compositions of M2 (HEPES-buffered M16), FHM (HEPES-buffered KSOM), and another commonly used flushing medium, Dulbecco's phosphate-buffered saline (D-PBS), are shown in Table 4.3.

A minimum of time between sacrificing embryo donors and placing the embryos in culture dishes, as well as the proper removal of debris after embryo collection, contributes to successful embryo culture. Embryos should be washed through several drops of final equilibrated culture media when transferred from HEPES-buffered flushing medium to remove traces of HEPES.

Quality Control

Culture of mouse zygotes to the blastocyst stage is the most often recommended option for testing new batches of medium, its components, oil, BSA, filters, etc. The results of such tests should be interpreted with caution because they depend on multiple factors such as the genetic background of the mouse strain and the type of the medium used. Generally, outbred mouse embryos are more sensitive to the environmental factors than embryos of inbred strains or their F_1 hybrids, with two-cell block being one example of such sensitivity (see also p. 163) (Scott and Whittingham 1996; Gardner and Lane 2000). The phenomenon of two-cell block in certain embryo culture media, the influence of genetic makeup, and the chemical composition of the medium were discussed above (p. 170) in detail. The sensitivity of mouse embryos also depends on the age of the embryo at recovery, with zygotes being the most sensitive (Davidson et al. 1988a,b). Removal of the zona pellucida increases the sensitivity of the mouse embryo to culture conditions (Montoro et al. 1990), which makes optimal culture conditions necessary for successful aggregation chimera experiments (see Chapter 11).

Taking into consideration all of these factors, the zygotes are cultured in defined conditions. Blastocyst formation, assessed at a specific time (e.g., after 96

TABLE 4.3. Composition of media for embryo collection

Component	M2		FHM[1]		D-PBS[1]	
	mM	g/liter	mM	g/liter	mM	g/liter
NaCl	94.66	5.534	95.0	5.55	136.0	8.0
KCl	4.78	0.356	2.5	0.186	2.68	0.2
KH_2PO_4	1.19	0.162	0.35	0.0476	1.47	0.2
$Na_2HPO_4 \cdot 12H_2O$	none	none	none	none	8.1	2.89
$MgSO_4 \cdot 7H_2O$	1.19	0.293	0.20	0.0493	none	none
$MgCl_2 \cdot 6H_2O$	none	none	none	none	0.49	0.1
Na lactate	23.28	2.61	10.0	1.12	none	none
Na pyruvate	0.33	0.036	0.2	0.022	none	none
Glucose	5.56	1.00	0.2	0.036	none	none
Glutamine	none	none	1.0	0.146	none	none
EDTA	none	none	0.01	0.0038	none	none
$NaHCO_3$	4.15	0.349	4.0	0.336	none	none
$CaCl_2 \cdot 2H_2O$	1.71	0.25	1.71	0.251	0.9	0.133
HEPES	20.85	4.969	20.0	4.76	none	none
BSA				4.0	4.0	3.0

All media contain phenol red at 0.001–0.01 g/liter (*optional*), 0.06 g/liter (100 units/ml) penicillin G potassium salt, 0.05 g/liter streptomycin sulfate. pH is adjusted to 7.3–7.4.

Variation of D-PBS, sometimes called PB-1 supplemented with 1000 mg/liter of glucose (or dextrose) and 36 mg/liter of sodium pyruvate is also often used for embryo handling and in cryopreservation. Possible substitutions in D-PBS formulation: anhydrous $CaCl_2$ at 0.1 g/liter; $Na_2HPO_4 \cdot 7H_2O$ at 2.16 g/liter; anhydrous Na_2HPO_4 at 1.15 g/liter.

[1]Lawitts and Biggers (1993).

hours of culture), should reach more than 80%. However, this is a rather subjective outcome and a poor indicator of embryo quality and viability (Lane and Gardner 1996). A more sensitive and quantitative, but not very practical, parameter is the blastocyst's cell number (Scott et al. 1993). Although fetal formation is the ultimate assessment of embryo viability after in vitro culture, embryo transfer is not always practical and is affected by additional factors involved in the procedure. In this case, the ability of blastocysts to expand fully, hatch, and attach to the surface of a culture dish may provide the necessary outcome.

PREPARATION OF EMBRYO CULTURE MEDIA

Protocols 4.1–4.4 describe in detail the preparation of the most commonly used mouse embryo media: M16, M2, and KSOM. Other embryo media can be prepared in a similar way; their compositions are shown in Table 4.1.

Although M2 is used in all protocols for embryo collection and manipulation in air, it can be substituted with any other HEPES-buffered medium such as FHM (HEPES-KSOM) (see Table 4.3) or HEPES-CZB, ideally corresponding to the culture media used for the subsequent procedures of embryo manipulations and culture (M16, KSOM-AA, CZB, etc.).

Various mouse embryo culture media are available commercially from Sigma (M2 [M5910, M7167] and M16 [M1285, M7292] http://www.sigma-aldrich.com) and Specialty Media (M2 [MR-015, FHM [MR-024], KSOM-AA [MR-106, MR-121] http://www.specialtymedia.com). Sequential media for human embryo culture are manufactured by different companies such as Vitrolife, Irvine Scientific,

SAGE BioPharma, and Conception Technologies and can be used for mouse embryo culture.

MAKING PIPETTES FOR EMBRYO HANDLING

A mouth pipette assembly, consisting of an aspirator mouthpiece, tubing (e.g., latex 1/8″ ID, 1/32″ wall, VWR 62996-350), and Pasteur pipette or glass capillary pulled on a flame is probably the most commonly used device for all embryo handling (see Fig. 4.1A,B). Mouth-pieces are available from HPI Hospital Products, a division of MEDTECH International, 1501P-B4036-2 (flat) or as a part of an aspirator tube assembly from Drummond 2-000-0001 (round) http://www.drummondsci.com. Glass capillaries or Pasteur pipettes can be pulled on a flame to create a narrow opening. It is convenient to pull disposable glass Pasteur pipettes (e.g., VWR 14672-380), as they can be held securely in the hand when inserted into a 1000-μl tip connected to a tubing (Fig. 4.1A) (see also Protocols 4.6 and 4.7). If only the narrow part of the Pasteur pipette or hard glass capillary is used, a micropipette holder is necessary (e.g., part of Drummond Microcaps bulb assembly) (Fig. 4.1B). It is advisable to use a cotton plug inserted in a 1000-μl tip or Pasteur pipette and/or a small filter in a middle section of the tubing for both kinds of mouth pipettes.

A more hygienic alternative to a mouth pipette is a handheld assembly, consisting of a 4- to 5-cm piece of soft plastic tubing tied at one end and connected to a pulled capillary or narrow part of a Pasteur pipette. The inner diameter of the tubing should be approximately equal to a middle part of 200-μl tip that serves as a capillary holder (Fig. 4.1C). The system works well for all embryo-handling procedures except embryo transfer, which still requires a mouth pipette.

Drummond Microcaps (10009002, 10009003) use the same principle of moving micro quantities of liquid by squeezing the bulb (or plastic tubing as described above) connected to a capillary. The Ultramicro Accropet Pipettor (http://www.bel-art.com) provides one-handed operation for filling and dispensing of up to 0.2 ml by a knob rotation using micropipettes and capillary tubes. This device, as well as micrometer pneumatic or hydraulic systems similar to those used for microinjection (see Chapter 7), may serve as an alternative to mouth pipettes if necessary.

When using mouth- or hand-controlled pipettes for embryo handling, clean medium is usually pulled into the capillary before picking up embryos. This reduces the capillary action that draws the embryos into the pipette and may cause embryo losses. It also helps to avoid accidental expelling of air bubbles when embryos are released from the capillary. When using a mouth pipette, it is also possible to slow the capillary action by picking up a few air bubbles as described for embryo transfer (see Chapter 6, Fig. 6.1). Traditionally, light paraffin oil filled to just past the shoulder of the pipette is also used to reduce capillary action during embryo transfer. In any case, embryos should be kept near the tip of the pipette. When transferred from one medium to another, embryos are usually washed through several drops of medium; the medium remaining in the pipette is expelled and the pipette is filled from a new drop. When embryos are expelled into a drop, care should be taken not to push all the medium out of the pipette, which would cause bubble formation in the medium drop.

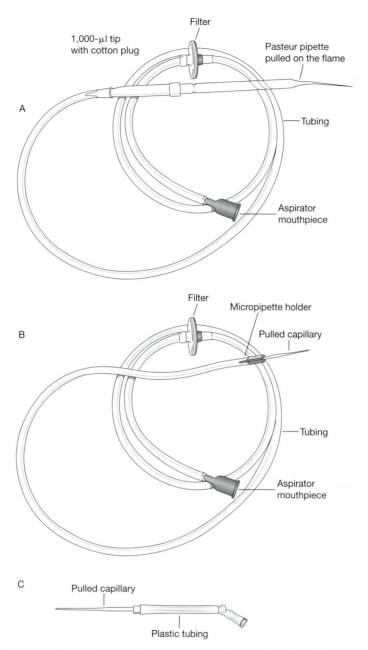

FIGURE 4.1. Suggested arrangements for embryo-handling pipettes. Mouth-controlled pipettes using pulled Pasteur pipette (*A*) or capillary (*B*). Hand-controlled pipette for embryo handling (*C*).

RECOVERING PREIMPLANTATION-STAGE EMBRYOS

In general, rigorous sterile techniques are not necessary when isolating zygotes and preimplantation embryos for micromanipulations. However, the working area should be clean, dust-free, and covered with fresh absorbent paper. The instruments can be autoclaved or sterilized by immersion in alcohol, followed by air-drying. Sterile techniques are necessary for long-term culture of embryos and if required by local SPF facilities regulations. Protocol 4.8 gives methods for humane sacrifice by cervical dislocation as well as for the dissection of female reproductive organs.

Zygotes may be collected several hours before they are to be injected (see Protocol 4.9). When mice are maintained on a light–dark cycle with the midpoint of the dark period at 12 midnight, it is convenient to begin dissections before noon. In case of superovulation, zygotes are collected 21–25 hours post-human chorionic gonadotropin (hCG) when the vaginal plug is checked in the morning after mating; i.e., on 0.5 dpc (see Chapters 2 and 3). Unfertilized oocytes enclosed in cumulus cells are collected exactly the same way as for zygotes but taking special care of the timing when it is done for IVF (see Chapter 14). If the recovery is done later in the day, when the cumulus (follicle) cells begin to fall off, the zygotes are somewhat more difficult to recover and need to be flushed from the oviduct as described for embryos of later stages in Protocol 4.10.

Two- to eight-cell embryos are present in the oviduct 20–60 hours postcoitum (p.c.). By this time, the embryos have lost their cumulus cells and can be flushed from the oviduct using a small amount of M2 medium. When donor females are superovulated, two-cell-stage embryos are collected on 1.5 dpc 45–48 hours post-hCG; eight-cell- to compacted morula-stage embryos are collected on 2.5 dpc from around 67–77 hours post-hCG. (Protocol 4.10 describes the collection of two-cell- to compacted morula-stage embryos.) Blastocysts can be flushed from the uterus between 3.5 and 4.5 dpc (Protocol 4.11).

Preparing M16 Culture Medium

M16 is a modified Krebs-Ringer bicarbonate solution and is very similar to Whitten's medium (Whitten 1971). For a full reference on M16, see Whittingham (1971).

MATERIALS

Compound	mM	Molecular weight	g/liter
NaCl	94.66	58.450	5.533[a]
KCl<!>	4.78	74.557	0.356
CaCl$_2 \cdot$2H$_2$O	1.71	147.200	0.252
KH$_2$PO$_4$	1.19	136.091	0.162
MgSO$_4 \cdot$7H$_2$O<!>	1.19	246.500	0.293
NaHCO$_3$	25.00	84.020	2.101
Sodium lactate	23.28	112.100	2.610 or 4.349 g of 60% syrup
Sodium pyruvate	0.33	110.000	0.036
Glucose	5.56	179.860	1.000
BSA			4.000
Penicillin G•potassium salt (final conc., 100 units/ml)			0.060
Streptomycin sulfate (final conc., 50 mg/ml)			0.050
Phenol red			0.010[c]
2x glass-distilled H$_2$O[b]			up to 1 liter

[a] Increase NaCl to 5.68 g/liter when CaCl$_2$ is omitted for Ca^{++}-free medium.
[b] See note on water under General Considerations, p. 171.
[c] The concentration of phenol red can be decreased to 0.0001–0.001 g/liter, because it may be embryotoxic.

CAUTION: See Appendix 2 for appropriate handling of materials marked with <!>.

PROCEDURE

1. Weigh the penicillin and streptomycin and dissolve them in a small volume of 2x distilled water.

2. Weigh the CaCl$_2$ and dissolve it in 2x distilled water.

3. Weigh the remaining compounds (except BSA and lactate) into a designated 1-liter volumetric flask and add ~500 ml of 2x distilled water. Allow to dissolve.

4. Add the penicillin, streptomycin, and CaCl$_2$ to a volumetric flask.

5. Weigh the lactate syrup into a designated 10-ml beaker and add it to the volumetric flask. Rinse the beaker several times with 2x distilled water; add the washings to the volumetric flask.

6. Determine the pH of the medium; it should be 7.2–7.4. If it is less, it can be adjusted using 0.2 M NaOH. Bring the volume to 1 liter. Gas the medium by bubbling with 5% CO_2, 95% air for 5 minutes to adjust the pH (this step can be omitted if the pH is 7.4).

7. Sprinkle BSA on top of the medium and allow it to dissolve slowly. Mix gently. Do not shake the medium, because it will froth and denature the protein.

8. Filter the medium through a Millipore filter into small sterile containers, gas the air space with 5% CO_2, 95% air, and cap tightly to maintain a pH of 7.2–7.4. Use positive pressure to filter to reduce foaming. Discard the first few milliliters that come through the filter.

9. Store at 4°C for up to 2 weeks. The osmolarity should be 288–292 mosmoles.

Preparing M2 Culture Medium

M2 is a modified Krebs-Ringer solution with HEPES buffer substituted for some of the bicarbonate (Quinn et al. 1982). It is used for collecting embryos and for handling them outside the incubator.

MATERIALS

Compound	mM	Molecular weight	g/liter
NaCl	94.66	58.450	5.533[a]
KCl<!>	4.78	74.557	0.356
$CaCl_2 \cdot 2H_2O$	1.71	147.200	0.252
KH_2PO_4	1.19	136.091	0.162
$MgSO_4 \cdot 7H_2O$<!>	1.19	246.500	0.293
$NaHCO_3$	4.15	84.020	0.349
HEPES	20.85	238.300	4.969
Sodium lactate	23.28	112.100	2.610 or 4.349 g of 60% syrup
Sodium pyruvate	0.33	110.000	0.036
Glucose	5.56	179.860	1.000
BSA			4.000
Penicillin G•potassium salt			0.060
Streptomycin sulfate			0.050
Phenol red			0.010[c]
2x Glass-distilled H_2O[b]			up to 1 liter

[a] Increase NaCl to 5.68 g/liter when $CaCl_2$ is omitted for Ca^{++}-free medium.
[b] See note on water under General Considerations, p. 171.
[c] The concentration of phenol red can be decreased to 0.0001–0.001 g/liter, as it may be embryo-toxic.

CAUTION: See Appendix 2 for appropriate handling of materials marked with <!>.

PROCEDURE

1. Weigh the HEPES and dissolve in 50–100 ml of 2x distilled water.

2. Adjust the pH to 7.4 with 0.2 M NaOH.

3. Weigh the penicillin and streptomycin and dissolve in 2x distilled water.

4. Weigh the $CaCl_2$ and dissolve in 2x distilled water.

5. Weigh the remaining components (except BSA and lactate) into a designated 1-liter volumetric flask and add 500 ml of 2x distilled water. Allow to dissolve.

6. Add the penicillin, streptomycin, HEPES, and $CaCl_2$ to the volumetric flask.

7. Weigh the lactate syrup into a designated 10-ml beaker and add to the volumetric flask. Rinse the beaker several times with 2x distilled H_2O, add washings to the volumetric flask, and adjust the volume to 1 liter.

8. Sprinkle BSA on top of the medium and allow to dissolve slowly. Mix gently. Do not shake the medium because it will froth and denature the protein.

9. If necessary, readjust the pH of the medium to 7.2–7.4 with 0.2 M NaOH.

10. Filter through a Millipore filter with positive pressure to reduce foaming. Discard the first few milliliters and aliquot into sterile containers.

11. Store at 4°C for up to 2 weeks. The osmolarity should be 285–287 mosmoles.

Preparing M2 and M16 Media from Concentrated Stocks

It is often more convenient to prepare concentrated stocks of the components of M2 and M16 media, but take note of the information below on storage.

MATERIALS

Stock A	Component	g/100 ml
(10x concentration)		
	NaCl	5.534
	KCl<!>	0.356
	KH_2PO_4	0.162
	$MgSO_4 \cdot 7H_2O$<!>	0.293
	sodium lactate	2.610
		or 4.349 g
		of 60% syrup
	glucose	1.000
	penicillin	0.060
	streptomycin	0.050

Stock B	Component	g/100 ml
(10x concentration)	$NaHCO_3$	2.101
	phenol red	0.010[a]

Stock C	Component	g/10 ml
(100x concentration)	sodium pyruvate	0.036

Stock D	Component	g/10 ml
(100x concentration)	$CaCl_2 \cdot 2H_2O$	0.252

Stock E	Component	g/100 ml
(10x concentration)	HEPES	5.958
	phenol red	0.010[a]

[a] The concentration of phenol red can be decreased to 0.0001–0.001 g/liter, because it may be embryotoxic.

CAUTION: See Appendix 2 for appropriate handling of materials marked with <!>.

PROCEDURE

Preparation of Stocks A, B, C, and D

1. Weigh the salts (except sodium lactate) into a designated volumetric flask and bring up to volume using 2x distilled water.

2. Weigh the sodium lactate into a designated 10-ml beaker.

3. Add the sodium lactate to the volumetric flask.

183

4. Rinse the beaker several times with 2x distilled water. Add the washings to the volumetric flask, and bring up to volume using 2x distilled water.

5. Filter all stocks through a Millipore filter. Convenient aliquots in sterile tubes can be made at this time according to the potential use (see Table 4.4).

Preparation of Stock E

1. Weigh the HEPES and phenol red into a designated beaker.

2. Add ~50 ml of 2x distilled water and allow to dissolve.

3. Adjust the pH to 7.4 with 0.2 M NaOH.

4. Pour into a 100-ml volumetric flask.

5. Rinse the beaker with 2x distilled water, add washings to the volumetric flask, and adjust to 100 ml.

6. Filter all stocks through a Millipore filter. Convenient aliquots in sterile tubes can be made at this time according to the potential use (see Table 4. 4).

TABLE 4.4. M2 and M16 media from concentrated stocks

Stock	10 ml	50 ml	100 ml	200 ml
M2 from concentrated stocks				
A (x10)	1.00	5.0	10.0	20.0
B (x10)	0.16	0.8	1.6	3.2
C (x100)	0.10	0.5	1.0	2.0
D (x100)	0.10	0.5	1.0	2.0
E (x10)	0.84	4.2	8.4	16.8
H_2O	7.80	39.0	78.0	156.0
BSA	40 mg	200 mg	400 mg	800 mg
M16 from concentrated stocks				
A (x10)	1.0	5.0	10.0	20.0
B (x10)	1.0	5.0	10.0	20.0
C (x100)	0.1	0.5	1.0	2.0
D (x100)	0.1	0.5	1.0	2.0
H_2O	7.8	39.0	78.0	156.0
BSA	40 mg	200 mg	400 mg	800 mg

STORAGE

If stored in a refrigerator at 4°C, stocks A, D, and E can be kept for up to 3 months. Stocks B and C must be changed every other week. All stocks stored frozen at –20°C or –70°C can be kept for longer periods, but once the 1x solution is made, it should not be stored for more than 1–2 weeks.

Preparation of M2 and M16 Media from Concentrated Stocks

1. Measure 2x distilled water accurately into a designated conical flask. A plastic 50-ml tube or a sterile culture bottle (e.g., Nalge square media bottles 2019-0250) may be used instead of a conical flask.

2. Measure stock solutions using plastic pipettes or tips. Leave the pipette in the conical flask.

3. Rinse the pipettes in the flask by sucking up water/medium mixture two or three times.

4. Measure the osmolarity of the medium (*optional*).

5. Gas the M16 medium by bubbling with 5% CO_2, 95% air for ~15 minutes to adjust the pH to 7.4 (*optional*).

6. Add the BSA to the medium to a final concentration of 4 mg/ml, allow to dissolve slowly, and mix gently. Do not shake the medium because it will froth and denature the protein.

7. If necessary, readjust the pH of M2 medium with 0.2 N NaOH to pH 7.2–7.4, using color standards.

8. Filter the medium through a Millipore filter into sterile plastic tubes. Gas the space in the tubes with M16 medium with 5% CO_2, 95% air for 30 seconds, and cap tightly to maintain a pH of 7.2–7.4.

9. Store at 4°C for 1–2 weeks.

Preparing KSOM Medium

KSOM medium was developed using the simplex optimization procedure; it contains comparatively low concentrations of NaCl, KCl, KH_2PO_4, lactate, and glucose. KSOM allows outbred zygotes to overcome two-cell block and it supports in vitro and in vivo development of various mouse strains (Lawitts and Biggers 1993; Erbach et al. 1994).

MATERIALS

Component	Final concentration		Concentrated stock	Stock volume for 100 ml
	mM	g/liter		
			A´ (10x) g/100 ml	10 ml
NaCl	95.0	5.55	5.55	
KCl<!>	2.50	0.186	0.186	
KH_2PO_4	0.35	0.0476	0.0476	
$MgSO_4 \cdot 7H_2O$<!>	0.20	0.0493	0.0493	
Glucose	0.20	0.036	0.036	
Penicillin G			0.060	0.060
Streptomycin			0.050	0.050
Sodium lactate	10.0	1.12 or 1.87 g of 60% syrup	1.12 or 1.87 g of 60% syrup	
			B´ (10x) g/100 ml	10 ml
$NaHCO_3$	25.0	2.10	2.10	
Phenol red			0.001	
			C´ (100x) g/10 ml	1 ml
Na pyruvate	0.20	0.022	0.022	
			D (100x) g/10 ml	1 ml
$CaCl_2 \cdot 2H_2O$	1.71	0.25	0.25	
			F (1000x) g/10ml	0.1 ml
EDTA	0.01	0.0038	0.038	
L-Glutamine	1.00	0.146	G (200x) 200 mM	0.5 ml
BSA (e.g., Sigma A3311)		1.000		100 mg

CAUTION: See Appendix 2 for appropriate handling of materials marked with <!>.

PROCEDURE

KSOM medium can be prepared from concentrated stocks as described above for M2 and M16 media. All stock solutions may be stored for a few months when kept at –20°C or –70°C. It is also possible to prepare 2x KSOM without the calcium chloride and BSA and then store it at –70°C for up to 3 months; 0.171 M calcium chloride and 100 mg/ml BSA stocks are frozen separately and added at the time of medium preparation (Biggers et al. 1997, 2000). The osmolarity of KSOM should be 256 mOsm. The prepared medium is filter-sterilized through a 0.2-μm Millipore filter. Aliquots in polypropylene tubes are stored at 4°C for 1–2 weeks.

COMMENTS

- L-Glutamine (200 mM; e.g., from Invitrogen Life Technologies 25030) can be substituted by a more stable dipeptide L-alanyl-L-glutamine (e.g., GlutaMAX supplement from Invitrogen Life Technologies 35050-061).

- If KSOM medium is supplemented with amino acids, stock solutions of minimal essential medium (MEM) essential amino acids (EAA) and nonessential amino acids (NEAA) can be used (e.g., Invitrogen Life Technologies: 100x [10 mM] NEAA 1140, 50x EAA 11130]). One-half of Eagle's concentrations are used, thus 0.5 ml of NEAA and 1 ml of EAA are added per 100 ml of KSOM medium (Ho et al. 1995; Biggers et al. 2000). See Table 4.2.

- The pH indicator phenol red, an optional but useful component of the medium, may be embryotoxic at high concentrations. It is sufficient to add 0.01 ml of 0.5% or 1% solution of phenol red to 100 ml of medium.

Setting Up Microdrop Cultures

Optimally, the microdrop culture should be set up several hours in advance or on a day before the experiment to permit temperature and gas equilibration.

MATERIALS

EQUIPMENT
5% CO_2, 5% O_2, and 90% N_2
Incubator, humidified at 37°C, 5% CO_2, 95% air
Micropipette and tips (Gilson P20 or P200)
Paraffin oil, embryo-tested light (e.g., Sigma M84510)
Pipette (5 ml) or Pasteur pipette
Pipettes for embryo handling (see Protocol 4.6 and Fig. 4.11)
Syringe, 1 ml with 26-gauge needle (*optional*)
Tissue culture dishes, plastic, 35-mm sterile

REAGENTS
M16 or other embryo culture medium

PROCEDURE

1. Use a 20-µl tip to dispense 20- to 40-µl drops of embryo culture medium in an array on the bottom of a 35-mm sterile plastic tissue culture dish (Fig. 4.2A). It is also convenient to dispense microdrops using a 1-ml syringe with a 26-gauge needle.

2. Immediately flood the dish with embryo-tested light paraffin oil from a 5-ml pipette or tube (Fig. 4.2B, C). The microdrops should not be exposed to the air without being covered with oil.

3. Place the dish in the incubator to equilibrate.

4. After the embryos' recovery and/or manipulations, use embryo-handling transfer pipettes (see Fig. 4.1 and Protocols 4.6 and 4.7) to place the embryos into the drops of medium under the oil (Fig. 4.2D). It is important to rinse the embryos from HEPES-buffered medium before placing them in culture. This can be done by transferring embryos from one drop of equilibrated culture medium to another.

5. Place the microdrop culture dish in the incubator. Take care to minimize its handling while exposed to the air.

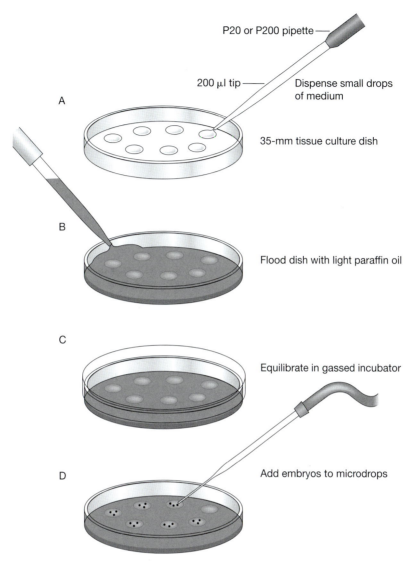

P20 or P200 pipette ——

200 μl tip ——

Dispense small drops of medium

A

35-mm tissue culture dish

B

Flood dish with light paraffin oil

C

Equilibrate in gassed incubator

D

Add embryos to microdrops

FIGURE 4.2. Setting up microdrop cultures. (*A*) Small drops of embryo culture medium are placed on the bottom of a 35-mm sterile plastic tissue culture dish using a 200-μl tip or syringe. (*B*, *C*) The dish is flooded with light paraffin oil to cover the drops of medium and placed in a 37°C gassed incubator to equilibrate. (*D*) Embryos are transferred into the microdrops.

Making Pipettes from Hard Glass Capillaries

MATERIALS

EQUIPMENT
Bunsen/alcohol burner (small) or microburner
Glass test tubes with metal caps (BDH 267/0040/07; 225/0014/08)
Hard glass capillary tubing (1.5-mm external diameter) or the narrow part of a glass
    Pasteur pipette cut off with a diamond-point pencil or abrasive stone (oilstone)
Micropipette holder (e.g., Drummond Microcaps (10009-002, 10009-003)
Oven for sterilization
Silicone tips for holding mouthpiece (Baxter Scientific P5066-14D)
Suction mouthpiece (HPI Hospital Products Med. Tech. 1501P-B4036-2 [flat]) or as a
    part of an aspirator tube assembly from Drummond 2-000-0001 (round) or Curtin
    Matheson Scientific 258616
Tubing (e.g., latex 1/8″ ID, 1/32″ wall)

PROCEDURE

1. Soften the glass tubing or Pasteur pipette by rotating it in a fine flame until the glass becomes soft. If necessary, a microburner can be constructed using a bent Pasteur pipette or 18- or 20-gauge needle connected to the gas supply.

2. Withdraw the glass from the heat and quickly pull both ends smoothly to produce a tube with an internal diameter of ~200–250 µm. Do not pull the glass tubing while it is still in the flame.

3. For a neat break, score the tubing with a diamond-point pencil or abrasive stone (oilstone) and bend gently to snap. Alternatively, pull on the cooled tubing until the two halves break apart. A narrow shaft ~2–3 cm long is optimal for a Pasteur pipette. It is important to achieve an evenly broken straight tip because it is difficult to handle embryos with a jagged capillary end.

4. Fire-polish the tip of the pipette by quickly touching the flame. This is generally recommended to minimize potential zona and oviduct/uterus damage during manipulations and embryo transfer but absolutely required for handling zona-free embryos. In addition, sharp edges of the pipette are easily caught by the plastic surface of the dish, causing the pipette to break, and they tend to collect more debris during embryo transfer.

5. If required, sterilize the pipettes by heating them in an oven using a glass tube with metal caps.

6. Mouth- or hand-controlled holders for the transfer pipettes can be constructed as shown in Figure 4.1 and described above.

Preparing Siliconized Pipettes

Although not essential, the routine use of siliconized pipettes minimizes the loss of embryos or embryonic tissues during transfer. However, do not use siliconized glass if oil is used as a braking agent in embryo-handling pipettes. Coating the pipette with 1% BSA immediately before use also helps to minimize stickiness during embryo transfer.

MATERIALS

EQUIPMENT
Beaker, large
Cotton wool
Fume hood
Kimwipes
Oven
Pasteur pipettes, glass

REAGENTS
Dimethyldichlorosilane<!> (Sigma D3879) or Repelcote (BDH 63216 6L)
HCl<!>, 1 mM
Water, distilled

CAUTION: See Appendix 2 for appropriate handling of materials marked with <!>.

PROCEDURE

1. Soak pipettes in 1 M HCl overnight. Rinse them three times in distilled water. Dry the pipettes in an oven at 100–180°C for 1 hour.

2. In a fume hood, fill a large beaker with Repelcote or dimethyldichlorosilane so that the shafts of the pipettes are completely immersed. Dip the pipettes into the silicone fluid. Remove and place on a slant on Kimwipes to dry.

3. Plug the pipettes with cotton wool and heat-sterilize at 180°C in suitable containers.

Opening the Abdominal Cavity and Locating Female Reproductive Organs

Here we describe the cervical dislocation method for humane sacrifice of a mouse. Local institutional regulations should be followed for this procedure.

MATERIALS

ANIMALS
Pregnant female mouse

EQUIPMENT
Absorbent paper
Forceps, watchmaker's #5, two pairs
Scissors (one pair fine scissors, one pair regular surgical scissors)

REAGENTS
Ethanol, 70% in a squeeze bottle

CAUTION: See Appendix 2 for appropriate handling of materials marked with <!>.

PROCEDURE

1. Place the mouse on top of a cage, so that it grips the bars with its front paws. Break its neck humanely by applying firm pressure at the base of the skull (Fig. 4.3) while at the same time pulling backward on the tail. Alternatively, a spatula, pencil, or cage card holder can be used to apply pressure to the base of the skull. The CO_2 inhalation method also may be used if necessary.

2. Lay the animal on its back on absorbent paper and soak it thoroughly in 70% ethanol from a squeeze bottle. This important step reduces the risk of contaminating the dissection with mouse hair.

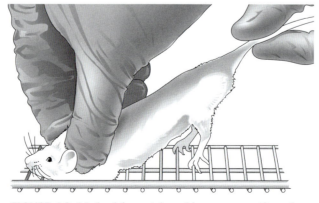

FIGURE 4.3. Method for quick and humane sacrifice of a mouse by cervical dislocation.

3. Pinch the skin and make a small lateral incision at the midline (the position is not critical) with regular surgical scissors (Fig. 4.4A). Holding the skin firmly above and below the incision, pull the skin toward the head and tail until the abdomen is completely exposed and the fur is well out of the way. Using the watchmaker's forceps and fine scissors, cut the body wall (peritoneum) as shown in Figure 4.4B. Push the coils of gut out of the way and locate the two horns of the uterus, the oviducts, and the ovaries (Fig. 4.4C).

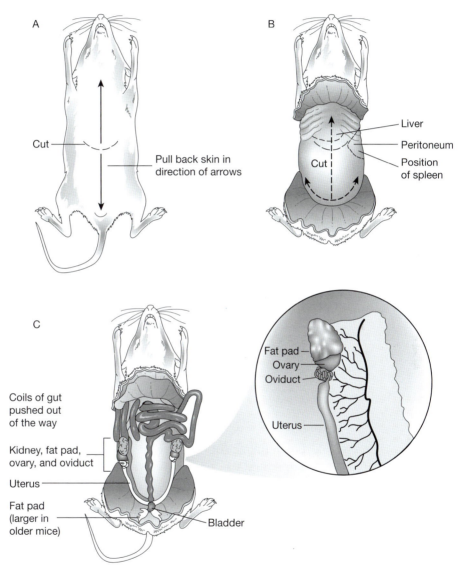

FIGURE 4.4. Dissection of reproductive organs of a female mouse. (*A*) The position of the small lateral incision in the skin is indicated by the dashed line. The skin is then pulled back in the direction of the solid arrows. (*B*) The body wall (peritoneum) is cut in the direction of the dashed arrows. (*C*) The alimentary tract displaced to reveal reproductive organs in the floor of the body cavity.

Collecting Zygotes and Removing Cumulus Cells with Hyaluronidase

MATERIALS

ANIMALS

Pregnant female mice sacrificed humanely (see Protocol 4.8)

EQUIPMENT

Embryo-handling pipette consisting of mouth or hand-held pipette assembly and pulled capillary (see Protocols 4.6 and 4.7)

Forceps

Forceps, watchmaker's #5, two pairs

Microdrop culture dishes (see Protocol 4.5)

Needles, 26-gauge (*optional*)

Organ culture dish (Falcon 3037) (*optional*)

Petri dishes, 35-mm, or embryological watch glasses

Scissors, fine

Stereomicroscope with transmitted and reflected or fiber optics (*optional*) illumination (preferably a ground-glass stage) with 20x and 40x magnification

REAGENTS

Hyaluronidase (e.g., Sigma H4272) stock solution in M2 medium (at room temperature or 37°C)

M2 medium at room temperature

PROCEDURE

1. Open the abdominal cavity as described above (see Protocol 4.8). Grasp the upper end of one of the uterine horns with fine forceps and gently pull the uterus, oviduct, ovary, and fat pad taut and away from the body cavity. This will reveal a fine membrane (the mesometrium), which connects the reproductive tract to the body wall and carries a prominent blood vessel. Poke a hole in the membrane close to the oviduct with the closed tips of a pair of fine forceps or scissors (Fig. 4.5A).

2. Pull the oviduct, ovary, and fat pad taut with fine forceps and cut between the oviduct and ovary with fine scissors as shown in Figure 4.5B. Do not be afraid to go close to the oviduct. Reposition the forceps and cut the uterus near the oviduct.

3. Transfer the oviduct and attached segment of uterus to a 35-mm petri dish or embryological watch glass containing M2 medium at room temperature. Oviducts from several mice can be collected in the same dish.

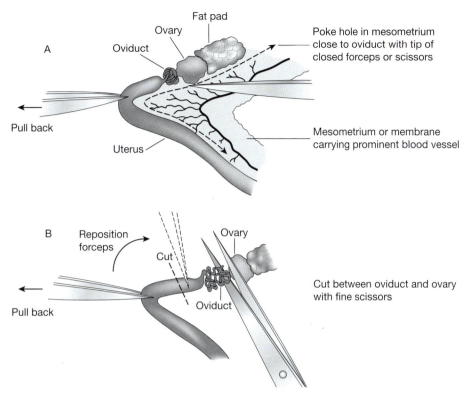

FIGURE 4.5. Dissection of the oviduct. (*A*) The ovary, oviduct, and end of the uterus are separated from the mesometrium with the closed tips of a pair of fine forceps. (*B*) A cut is made between the oviduct and the ovary. After repositioning the forceps, a second cut separates the oviduct from the uterus.

4. Newly ovulated oocytes, surrounded by cumulus cells, are found in the upper part of the oviduct (ampulla), which at this time (12 hours postovulation) is much enlarged. The fimbriated end of the oviduct (infundibulum) is also swollen during ovulation and can easily be located under 20x magnification in the stereomicroscope (Fig. 4.6A,B).

5. Transfer one oviduct at a time into another 35-mm petri dish containing hyaluronidase solution in M2 medium (~0.3 mg/ml) at room temperature (or 37°C) and view through the stereomicroscope at 20x or 40x magnification (Fig. 4.6B).

6. Use one pair of watchmaker's forceps to grasp the oviduct next to the swollen infundibulum and hold it firmly on the bottom of the dish. Use another pair of watchmaker's forceps or a 26-gauge needle to tear the oviduct close to where the zygotes are located (Fig. 4.6C,D), releasing the clutch of cumulus cells (Fig. 4.6E). If the zygotes do not flow out by themselves, use the forceps to push them out by gently squeezing the oviduct. If they stick to the outside of the oviduct, allow the oviduct to sit for several minutes in the hyaluronidase solution. Zygotes will be released as the digestion removes the sticky cumulus cells. If the zygotes stick to the forceps, simply lift the forceps out of the petri dish and they will be retained by the surface tension of the medium and will fall back to the bottom of the dish.

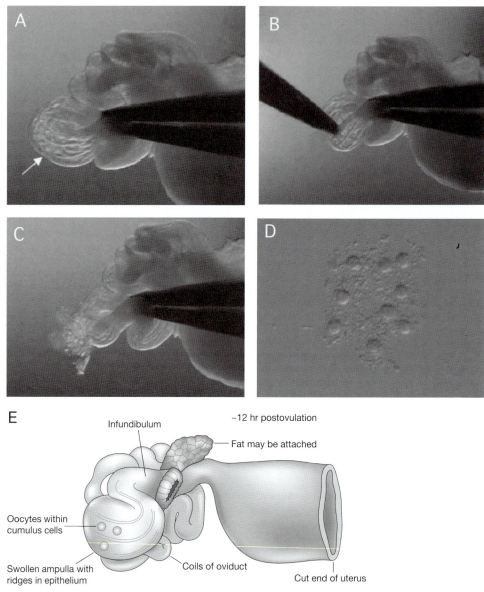

FIGURE 4.6. Isolation of zygotes from dissected oviduct. If the oviduct is removed soon after fertilization, the zygotes surrounded by cumulus (follicle) cells can be seen in the swollen upper part of the oviduct, the ampulla (A, E), and can be released by tearing the ampulla with fine forceps (B, C). (D) Cumulus-enclosed zygotes. (E) Diagram of procedure.

7. Allow the zygotes to incubate in the hyaluronidase solution for several minutes until the cumulus cells fall off. If necessary, pipette them up and down a few times, but do not leave them in the hyaluronidase solution for more than a few minutes after the cumulus cells are shed, because this may be harmful. Although ~0.3 mg/ml solution of hyaluronidase is usually recommended, more concentrated hyaluronidase solutions of 0.5–1 mg/ml that require a shorter incubation time of less than a minute may also be used.

8. Use pipettes to pick up the zygotes and transfer them to a petri dish containing several drops of fresh M2 medium to rinse off the hyaluronidase solution, cumulus cells, and debris. Then transfer the zygotes to a microdrop culture

dish, rinse through several drops of equilibrated medium, and keep at 37°C, 5% CO_2 until needed. An organ culture center well dish (Falcon 35-3037) with equilibrated embryo culture medium may be used as an alternative for short-term culture, i.e., before microinjection.

Collecting Two-cell- to Compacted Morula-stage Embryos

MATERIALS

ANIMALS
Pregnant female mice (20–60 hours p.c.) sacrificed humanely

EQUIPMENT
Embryo-handling pipette consisting of mouth or hand-held pipette assembly and pulled capillary

Flushing needle (either a 30- or 32-gauge hypodermic needle [end cut and/or ground to a blunt tip on an abrasive stone (oilstone) or sandpaper])

Forceps, fine

Forceps, watchmaker's #5, two pairs

Microdrop culture dish (see Protocol 4.5)

Organ culture dish (Falcon 3037) (*optional*)

Petri dishes (35-mm) or embryological watch glasses

Scissors, fine

Stereomicroscope with transmitted and reflected or fiber optics (*optional*) illumination (preferably a ground-glass stage) with 20x and 40x magnification

Syringe, 1-cc

REAGENTS
Ethanol<!>, 70%

M2 medium at room temperature

CAUTION: See Appendix 2 for appropriate handling of materials marked with <!>.

PROCEDURE

1. To reduce the risk of tearing the oviduct, cut the end of a 30- or 32-gauge hypodermic needle and grind it to a blunt tip (Fig. 4.7A). It is also possible just to grind the sharp tip of the needle without cutting it to create a smaller beveled tip that may prove useful for flushing the oviducts from very young females. Sterilize the needle by flushing it with 70% ethanol immediately before use.

2. Open the abdominal cavity as described above (see Protocol 4.8). Grasp the upper end of one of the uterine horns with fine forceps and gently pull the uterus, oviduct, ovary, and fat pad taut and away from the body cavity. This will reveal a fine membrane (the mesometrium), which connects the reproductive tract to the body wall and carries a prominent blood vessel. Poke a hole in the membrane close to the oviduct with the closed tips of a pair of fine forceps or scissors (see Fig. 4.5A).

3. Pull the oviduct, ovary, and fat pad taut with fine forceps and cut between the oviduct and ovary with fine scissors as shown in Figure 4.5B. Do not be

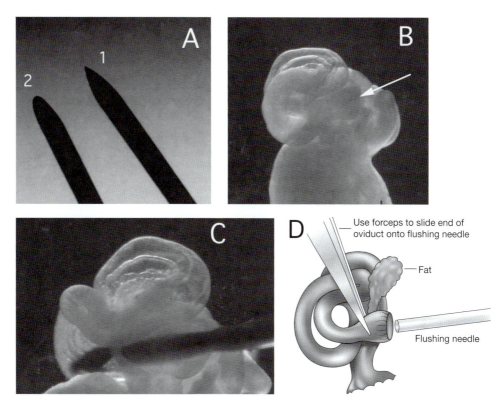

FIGURE 4.7. At cleavage stages, the embryos are recovered by flushing the oviduct with M2 medium using a flushing needle inserted into the end of the oviduct (infundibulum). (*A*) Flushing needle: 30-gauge needle (1) ground to a blunt tip (2). (*B*) Position of the infundibulum. (*C*) Flushing the oviduct through the infundibulum. (*D*) Diagram of the procedure.

afraid to go close to the oviduct. Reposition the forceps and cut the uterus near the oviduct, leaving at least 1 cm of the upper part of the uterus attached if the collection is taking place on 2.5 dpc.

4. Transfer the oviduct and attached segment of uterus to a 35-mm petri dish or embryological watch glass containing M2 medium at room temperature. Oviducts from several mice can be collected in the same dish. Place dish under stereomicroscope.

5. Test the syringe to be sure that it is free of air bubbles and that the M2 medium is flowing smoothly before inserting the needle.

6. The opening of the oviduct (infundibulum) at this time is no longer swollen and must be located within the coils of the oviduct (Fig. 4.7B,C). Use fine forceps to slide the end of the oviduct onto the flushing needle. Gently press the tip of the flushing needle against the bottom of the dish to hold it in place. Flush the oviduct with ~0.1 ml of M2 medium (Fig. 4.7D).

7. Use pipettes to pick up the embryos and wash them through several drops of fresh M2 medium to rinse off the debris.

8. Transfer the embryos to a microdrop culture dish, rinse through several drops of equilibrated medium, and keep at 37°C, 5% CO_2 until needed. An organ culture dish with equilibrated embryo culture medium may be used as an alternative for short-term incubation.

COMMENTS

- To prevent the oviduct from moving while locating the infundibulum, it can be placed in a very small drop of medium or even onto dry plastic if it is moved directly from a drop of M2.

- Because the tip of the flushing needle is blunt, it will not puncture the oviduct. Therefore, it is often possible to use it as a tool to press down the oviduct to the plastic and hold it in place inside the infundibulum while flushing.

- It is important to use only good-quality embryos for experiments and to distinguish them from delayed or fragmenting embryos. Examples of embryos collected at 1.5 and 2.5 dpc are presented in Figure 4.8. For examples of embryos collected at 0.5 dpc and 3.5 dpc, see Chapters 7 and 11.

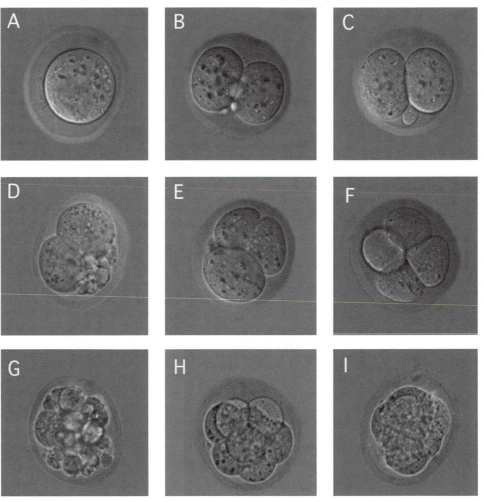

FIGURE 4.8. Examples of different-quality embryos that can be collected at 1.5 and 2.5 dpc: (*A*) unfertilized oocyte, (*B, C*) two-cell-stage embryos, (*D*) fragmenting two-cell-stage embryo, (*E*) three-cell-stage embryo, (*F*) four-cell-stage embryo, (*G*) fragmenting embryo, (*H*) eight-cell-stage embryo, (*I*) compacted morula.

Collecting Blastocysts

MATERIALS

ANIMALS

Pregnant female mice (3.5–4.5 dpc) sacrificed humanely (see Protocol 4.8)

EQUIPMENT

Embryo-handling pipette consisting of mouth or hand-held pipette assembly and pulled capillary (see Protocols 4.6 and 4.7)

Forceps, fine

Hypodermic needle, 26-gauge

Microdrop culture dish

Organ culture dish (Falcon 3037)(*optional*)

Scissors, fine

Stereomicroscope with transmitted and reflected or fiber optics (*optional*) illumination (preferably a ground-glass stage) with 20x and 40x magnification

Syringe, 1- or 2-ml

Tissue culture dishes, 35-mm sterile plastic or embryological watch glasses

REAGENTS

M2 medium

PROCEDURE

1. Open the abdominal cavity as described above (see Fig. 4.4). To remove the uterus, grasp it with fine forceps just above the cervix (located behind the bladder) (Fig. 4.9A) and cut across the cervix with fine scissors (Fig. 4.9B). Pull the uterus upward to stretch the mesometrium and use fine scissors to trim this membrane away close to the wall of the uterine horns (Fig. 4.9C). Then cut between the oviduct and the ovary, keeping the utero-tubal junction intact (Fig. 4.5B).

2. Place the uterus in a small volume of M2 medium in a 35-mm plastic tissue culture dish.

3. The next step depends on which way the uterus will be flushed:

Option 1 (from Cervix toward Oviduct)

- Because the utero-tubal junction acts as a valve, it should be cut lengthwise to allow flushing (Fig. 4.9D).

- Insert the needle into the cut cervix and slide it into the base of each horn to flush. Flush each horn with ~0.2 ml of M2 medium using a 26-gauge hypodermic needle and a 1- or 2-ml syringe (Fig. 4.9D).

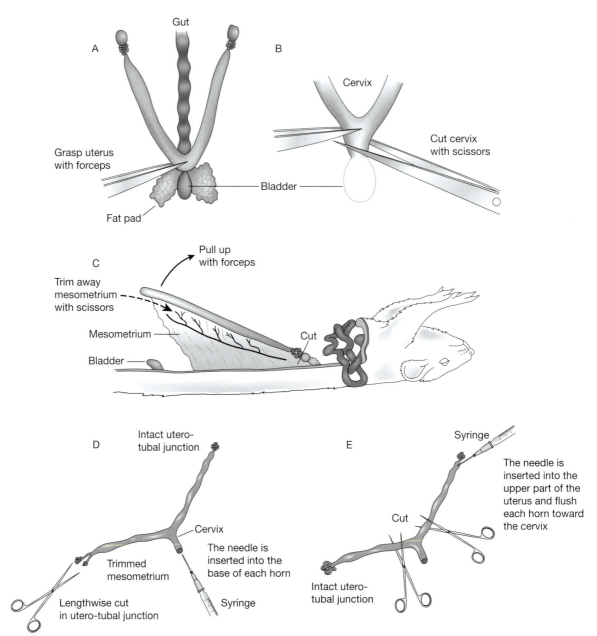

FIGURE 4.9. Dissection for flushing embryos from the uterus. (*A ,B*) The uterus is removed by cutting across the cervix. (*C*) The membrane (mesometrium) holding the uterus to the body wall is trimmed away, and the uterus is cut, keeping the junction with the oviduct intact. (*D*) The uterus is placed in a small volume of M2 medium, utero-tubal junction is cut lengthwise, and each horn is flushed from the cervix. (*E*) The uterus is cut near the cervix, the needle is inserted into the upper part of the uterus, and each horn is flushed toward the cervix.

Option 2 (toward Cervix)

- Cut each horn near the cervix.

- Insert a 26-gauge needle into the upper part of the uterus near utero-tubal junction and flush each horn toward the cervix (Fig. 4.9E). It is also possible to flush both oviduct and uterus through the infundibulum as described in Protocol 4.10.

4. Use a pipette to pick up the embryos and wash them through several drops of fresh M2 medium to rinse off the debris. Then transfer the embryos to a microdrop culture dish at 37°C, 5% CO_2 until needed.

COMMENTS

The yield of blastocysts is very low after they have hatched from the zona pellucida (at 4.5 dpc) and have attached to the uterus. The yield can sometimes be increased by leaving the utero-tubal junction intact and using Option 2 in step 3. Alternatively, inflate the uterine horns by injecting M2 medium from the cervix, and then cut the junction to release the fluid.

REFERENCES

Abramczuk J., Solter D., and Koprowski H. 1977. The beneficial effect of EDTA on development of mouse one-cell embryos in chemically defined medium. *Dev. Biol.* **61:** 378–383.

Almeida P.A. and Bolton V.N. 1995. The effect of temperature fluctuations on the cytoskeletal organisation and chromosomal constitution of the human oocyte. *Zygote* **3:** 357–365.

Barnes F.L., Crombie A., Gardner D.K., Kausche A., Lacham-Kaplan O., Suikkari A.M., Tiglias J., Wood C., and Trounson A.O. 1995. Blastocyst development and birth after in-vitro maturation of human primary oocytes, intracytoplasmic sperm injection and assisted hatching. *Hum. Reprod.* **10:** 3243–3247.

Bavister B.D. 1995. Culture of preimplantation embryos: Facts and artifacts. *Hum. Reprod. Update* **1:** 91–148.

Biggers J.D. 1971. Metabolism of mouse embryos. *J. Reprod. Fertil. Suppl.* **14:** 41–54.

Biggers J.D. and McGinnis L.K. 2001. Evidence that glucose is not always an inhibitor of mouse preimplantation development in vitro. *Hum. Reprod.* **16:** 153–163.

Biggers J.D., McGinnis L.K., and Raffin M. 2000. Amino acids and preimplantation development of the mouse in protein-free potassium simplex optimized medium. *Biol. Reprod.* **63:** 281–293.

Biggers J.D., Summers M.C., and McGinnis L.K. 1997. Polyvinyl alcohol and amino acids as substitutes for bovine serum albumin in culture media for mouse preimplantation embryos. *Hum. Reprod. Update* **3:** 125–135.

Biggers J.D., Whittingham D.G., and Donahue R.P. 1967. The pattern of energy metabolism in the mouse oocyte and zygote. *Proc. Natl. Acad. Sci.* **58:** 560–567.

Bowman P. and McLaren A. 1970. Cleavage rate of mouse embryos in vivo and in vitro. *J. Embryol. Exp. Morphol.* **24:** 203–207.

Brinster R.L. 1963. A method for in vitro cultivation of mouse ova from two-cell to blasocyst. *Exp. Cell Res.* **32:** 205–208.

———. 1965a. Lactate dehydrogenase activity in the preimplanted mouse embryo. *Biochim. Biophys. Acta.* **110:** 439–441.

———. 1965b. Studies on the development of mouse embryos in vitro. II. The effect of energy source. *J. Exp. Zool.* **158:** 59–68.

———. 1965c. Studies on the development of mouse embryos in vitro. IV. Interaction of energy sources. *J. Reprod. Fertil.* **10:** 227–240.

———. 1968. In vitro culture of mammalian embryos. *J. Anim. Sci.* **27:** 1–14.

———. 1969. Mammalian embryo culture. In *The mammalian oviduct* (ed. E.S.E. Hafez and R. Blandau), pp. 419–444. University of Chicago Press, Illinois.

———. 1971. Uptake and incorporation of amino acids by the preimplantation mouse embryo. *J. Reprod. Fertil.* **27:** 329–338.

Brinster R.L. and Thomson J.L. 1966. Development of eight-cell mouse embryos in vitro.

Exp. Cell. Res. **42:** 308–315.

Carrillo A.J., Lane B., Pridman D.D., Risch P.P., Pool T.B., Silverman I.H., and Cook C.L. 1998. Improved clinical outcomes for in vitro fertilization with delay of embryo transfer from 48 to 72 hours after oocyte retrieval: Use of glucose- and phosphate-free media. *Fertil. Steril.* **69:** 329–334.

Chatot C.L., Lewis J.L., Torres I., and Ziomek C.A. 1990. Development of 1-cell embryos from different strains of mice in CZB medium. *Biol. Reprod.* **42:** 432–440.

Chatot C.L., Ziomek C.A., Bavister B.D., Lewis J.L., and Torres I. 1989. An improved culture medium supports development of random-bred 1-cell mouse embryos in vitro. *J. Reprod. Fertil.* **86:** 679–688.

Chung Y.G., Mann M.R., Bartolomei M.S., and Latham K.E. 2002. Nuclear-cytoplasmic "tug of war" during cloning: Effects of somatic cell nuclei on culture medium preferences of preimplantation cloned mouse embryos. *Biol. Reprod.* **66:** 1178–1181.

Davidson A., Vermesh M., Lobo R.A., and Paulson R.J. 1988a. Mouse embryo culture as quality control for human in vitro fertilization: The one-cell versus the two-cell model. *Fertil. Steril.* **49:** 516–521.

———. 1988b. The temporal effects of changes in in vitro fertilization culture media on the one-cell mouse embryo system. *J. In Vitro Fert. Embryo. Transf.* **5:** 149–152.

Doherty A.S., Mann M.R., Tremblay K.D., Bartolomei M.S., and Schultz R.M. 2000. Differential effects of culture on imprinted H19 expression in the preimplantation mouse embryo. *Biol. Reprod.* **62:** 1526–1535.

Dumoulin J.C., Evers J.L., Bakker J.A., Bras M., Peters M.H., and Geraedts J.P. 1992. Temporal effects of taurine on mouse preimplantation development in vitro. *Hum. Reprod.* **7:** 403–407.

Eagle H. 1959. Amino acid metabolism in mammalian cell cultures. *Science* **130:** 432–437.

Eggan K., Akutsu H., Hoechedlinger K., Rideout W. 3rd, Yanagimachi R., and Jaenisch R. 2000. X-Chromosome inactivation in cloned mouse embryos. *Science* **290:** 1578–1581.

Eggan K., Akutsu H., Loring J., Jackson-Grusby L., Klemm M., Rideout W.M., 3rd, Yanagimachi R., and Jaenisch R. 2001. Hybrid vigor, fetal overgrowth, and viability of mice derived by nuclear cloning and tetraploid embryo complementation. *Proc. Natl. Acad. Sci.* **98:** 6209–6214.

Erbach G.T., Lawitts J.A., Papaioannou V.E., and Biggers J.D. 1994. Differential growth of the mouse preimplantation embryo in chemically defined media. *Biol. Reprod.* **50:** 1027–1033.

Fraser L.R. and Drury L.M. 1975. The relationship between sperm concentration and fertilization in vitro of mouse eggs. *Biol. Reprod.* **13:** 513–518.

Fraser L.R. and Quinn P.J. 1981. A glycolytic product is obligatory for initiation of the sperm acrosome reaction and whiplash motility required for fertilization in the mouse. *J. Reprod. Fertil.* **61:** 25–35.

Fry R.C. 1992. The effect of leukaemia inhibitory factor (LIF) on embryogenesis. *Reprod. Fertil. Dev.* **4:** 449–458.

Gardner D.K. and Lane M. 1993. Amino acids and ammonium regulate mouse embryo development in culture. *Biol. Reprod.* **48:** 377–385.

———. 1996. Alleviation of the '2-cell block' and development to the blastocyst of CF1 mouse embryos: Role of amino acids, EDTA and physical parameters. *Hum. Reprod.* **11:** 2703–2712.

———. 1997. Culture and selection of viable blastocysts: A feasible proposition for human IVF? *Hum. Reprod. Update* **3:** 367–382.

———. 2000. Embryo culture systems. In *Handbook of in vitro fertilization*, 2nd edition (ed. A. Trounson and D. Gardner), pp. 205–264. CRC Press, New York.

Gardner D.K. and Leese H.J. 1990. Concentrations of nutrients in mouse oviduct fluid and their effects on embryo development and metabolism in vitro. *J. Reprod. Fertil.* **88:** 361–368.

Gardner D.K. and Sakkas D. 1993. Mouse embryo cleavage, metabolism and viability: Role of medium composition. *Hum. Reprod.* **8:** 288–295.

Gardner D.K. and Schoolcraft W.B. 1998. Human embryo viability: What determines developmental potential, and can it be assessed? *J. Assist. Reprod. Genet.* **15:** 455–458.

Gardner D.K., Lane M.W., and Lane M. 2000. EDTA stimulates cleavage stage bovine embryo development in culture but inhibits blastocyst development and differentiation. *Mol. Reprod. Dev.* **57:** 256–261.

Gardner D.K., Vella P., Lane M., Wagley L., Schlenker T., and Schoolcraft W.B. 1998. Culture and transfer of human blastocysts increases implantation rates and reduces the need for multiple embryo transfers. *Fertil. Steril.* **69:** 84–88.

Goddard, M. J. and Pratt H.P. 1983. Control of events during early cleavage of the mouse embryo: An analysis of the '2-cell block'. *J. Embryol. Exp. Morphol.* **73:** 111–133.

Hardy K. and Spanos S. 2002. Growth factor expression and function in the human and mouse preimplantation embryo. *J. Endocrinol.* **172:** 221–236.

Harrison K.L., Sherrin D.A., Hawthorne T.A., Breen T.M., West G.A., and Wilson L.M. 1990. Embryotoxicity of micropore filters used in liquid sterilization. *J. In Vitro Fert. Embryo Transf.* **7:** 347–350.

Harvey M.B. and Kaye P.L. 1990. Insulin increases the cell number of the inner cell mass and stimulates morphological development of mouse blastocysts in vitro. *Development* **110:** 963–967.

———. 1992a. IGF-2 stimulates growth and metabolism of early mouse embryos. *Mech. Dev.* **38:** 169–173.

———. 1992b. Insulin-like growth factor-1 stimulates growth of mouse preimplantation embryos in vitro. *Mol. Reprod. Dev.* **31:** 195–199.

Heindryckx B., Rybouchkin A., Van Der Elst J., and Dhont M. 2001. Effect of culture media on in vitro development of cloned mouse embryos. *Cloning* **3:** 41–50.

Ho Y., Wigglesworth K., Eppigg J.J., and Schultz R.M. 1995. Preimplantation development of mouse embryos in KSOM: Augmentation by amino acids and analysis of gene expression. *Mol. Reprod. Dev.* **41:** 232-238.

Hoppe P.C. 1976. Glucose requirement for mouse sperm capacitation in vitro. *Biol. Reprod.* **15:** 39–45.

Hoppe P.C. and Pitts S. 1973. Fertilization in vitro and development of mouse ova. *Biol. Reprod.* **8:** 420–426.

Humpherys D., Eggan K., Akutsu H., Hochedlinger K., Riseout W.M. 3rd, Biniszkiewicz D., Yanagimachi R., and Jaenisch R. 2001. Epigenetic instability in ES cells and cloned mice. *Science* **293:** 95–97.

Iyengar M.R., Iyengar C.W., Chen H. Y., Brinster R.L., Bornslaeger E., and Schultz R.M. 1983. Expression of creatine kinase isoenzyme during oogenesis and embryogenesis in the mouse. *Dev. Biol.* **96:** 263–268.

Kamjoo M., Brison D.R., and Kimber S.J. 2002. Apoptosis in the preimplantation mouse embryo: Effect of strain difference and in vitro culture. *Mol. Reprod. Dev.* **61:** 67–77.

Kane M.T., Morgan P.M., and Coonan C. 1997. Peptide growth factors and preimplantation development. *Hum. Reprod. Update* **3:** 137–157.

Kaye P.L. 1997. Preimplantation growth factor physiology. *Rev. Reprod.* **2:** 121–127.

Khosla S., Dean W., Reik W., and Feil R. 2001a. Culture of preimplantation embryos and its long-term effects on gene expression and phenotype. *Hum. Reprod. Update* **7:** 419–427.

Khosla S., Dean W., Brown D., Reik W., and Feil R. 2001b. Culture of preimplantation mouse embryos affects fetal development and the expression of imprinted genes. *Biol. Reprod.* **64:** 918–926.

Lane M. and Gardner D.K. 1992. Effect of incubation volume and embryo density on the development and viability of mouse embryos in vitro. *Hum. Reprod.* **7:** 558–562.

———. 1994. Increase in postimplantation development of cultured mouse embryos by amino acids and induction of fetal retardation and exencephaly by ammonium ions. *J. Reprod. Fertil.* **102:** 305–312.

———. 1996. Selection of viable mouse blastocysts prior to transfer using a metabolic criterion. *Hum. Reprod.* **11:** 1975–1978.

———. 1997. Nonessential amino acids and glutamine decrease the time of the first three cleavage divisions and increase compaction of mouse zygotes in vitro. *J. Assist.*

Reprod. Genet. **14:** 398–403.

———. 2000. Regulation of ionic homeostasis by mammalian embryos. *Semin. Reprod. Med.* **18:** 195–204.

———. 2001. Inhibiting 3-phosphoglycerate kinase by EDTA stimulates the development of the cleavage stage mouse embryo. *Mol. Reprod. Dev.* **60:** 233–240.

Lane M., Hooper K., and Gardner D.K. 2001. Effect of essential amino acids on mouse embryo viability and ammonium production. *J. Assist. Reprod. Genet.* **18:** 519–525.

Lane M., Ludwig T.E., and Bayister B.D. 1999. Phosphate induced developmental arrest of hamster two-cell embryos is associated with disrupted ionic homeostasis. *Mol. Reprod. Dev.* **54:** 410–417.

Lavranos T.C., Rathjen P.D., and Seamark R.F. 1995. Trophic effects of myeloid leukaemia inhibitory factor (LIF) on mouse embryos. *J. Reprod. Fertil.* **105:** 331–338.

Lawitts J.A. and Biggers J.D. 1991. Optimization of mouse embryo culture media using simplex methods. *J. Reprod. Fertil.* **91:** 543–556.

———. 1992. Joint effects of sodium chloride, glutamine, and glucose in mouse preimplantation embryo culture media. *Mol. Reprod. Dev.* **31:** 189–194.

———. 1993. Culture of preimplantation embryos. *Methods Enzymol.* **225:** 153–164.

Leese H.J. 1988. The formation and function of oviduct fluid. *J. Reprod. Fertil.* **82:** 843–856.

———. 1991. Metabolism of the preimplantation mammalian embryo. *Oxf. Rev. Reprod. Biol.* **13:** 35–72.

Legge M. and Sellens M.H. 1991. Free radical scavengers ameliorate the 2-cell block in mouse embryo culture. *Hum. Reprod.* **6:** 867–871.

Li J., Foote R.H., and Simkin M. 1993. Development of rabbit zygotes cultured in protein-free medium with catalase, taurine, or superoxide dismutase. *Biol. Reprod.* **49:** 33–37.

Liu L., Oldenbourg R., Trimarchi J.R., and Keefe D.L. 2000. A reliable, noninvasive technique for spindle imaging and enucleation of mammalian oocytes. *Nat. Biotechnol.* **18:** 223–225.

Marschall S., Huffstadt U., Balling R., and Hrabe de Angelis M. 1999. Reliable recovery of inbred mouse lines using cryopreserved spermatozoa. *Mamm. Genome.* **10:** 773–776.

McGrath J. and Solter D. 1983. Nuclear transplantation in the mouse embryo by microsurgery and cell fusion. *Science* **220:** 1300–1302.

McLaren A. and Biggers J.D. 1958. Successful development and birth of mice cultivated in vitro as early embryos. *Nature* **182:** 877–878.

McLaren A. and Michie D. 1956. Studies on the transfer of fertilized mouse eggs to uterine foster-mothers. I. Factors affecting the implantation survival of native and transferred eggs. *J. Exp. Biol.* **33:** 394–416.

Mehta T.S. and Kiessling A.A. 1990. Development potential of mouse embryos conceived in vitro and cultured in ethylenediaminetetraacetic acid with or without amino acids or serum. *Biol. Reprod.* **43:** 600–606.

Miller J.G. and Schultz G.A. 1987. Amino acid content of preimplantation rabbit embryos and fluids of the reproductive tract. *Biol. Reprod.* **36:** 125–129.

Mintz B. 1967. Mammalian embryo culture. In *Methods in developmental biology* (ed. F.H. Wilt and N.K. Wessels), pp. 379–400. Cromwell, New York.

Montoro L., Subias E., Young P., Baccaro M., Swanson J., and Sueldo C. 1990. Detection of endotoxin in human in vitro fertilization by the zona-free mouse embryo assay. *Fertil. Steril.* **54:** 109–112.

Morita Y., Tsutsumi O., and Taketani Y. 1994. In vitro treatment of embryos with epidermal growth factor improves viability and increases the implantation rate of blastocysts transferred to recipient mice. *Am. J. Obstet. Gynecol.* **171:** 406–409.

Muggleton-Harris A., Whittingham D.G., and Wilson L. 1982. Cytoplasmic control of preimplantation development in vitro in the mouse. *Nature* **299:** 460–462.

Munsie M.J., Michalska A.E., O'Brien C.M., Trounson A.O., Pera M.F., and Mountford P.S. 2000. Isolation of pluripotent embryonic stem cells from reprogrammed adult mouse somatic cell nuclei. *Curr. Biol.* **10:** 989–992.

Nasr-Esfahani M.H., Winston N.J., and Johnson M.H. 1992. Effects of glucose, glutamine, ethylenediaminetetraacetic acid and oxygen tension on the concentration of reactive oxygen species and on development of the mouse preimplantation embryo in vitro. *J. Reprod. Fertil.* **96:** 219–231.

O'Neill C. 1997. Evidence for the requirement of autocrine growth factors for development of mouse preimplantation embryos in vitro. *Biol. Reprod.* **56:** 229–237.

Orsi N.M. and Leese H.J. 2001. Protection against reactive oxygen species during mouse preimplantation embryo development: Role of EDTA, oxygen tension, catalase, superoxide dismutase and pyruvate. *Mol. Reprod. Dev.* **59:** 44–53.

Pabon J.E., Jr., Findley W.E., and Gibbons W.E. 1989. The toxic effect of short exposures to the atmospheric oxygen concentration on early mouse embryonic development. *Fertil. Steril.* **51:** 896–900.

Paria B.C. and Dey S.K. 1990. Preimplantation embryo development in vitro: Cooperative interactions among embryos and role of growth factors. *Proc. Natl. Acad. Sci.* **87:** 4756–4760.

Pickering S.J. and Johnson M.H. 1987. The influence of cooling on the organization of the meiotic spindle of the mouse oocyte. *Hum. Reprod.* **2:** 207–216.

Pickering S.J., Braude P.R., Johnson M.H., Cant A., and Currie J. 1990. Transient cooling to room temperature can cause irreversible disruption of the meiotic spindle in the human oocyte. *Fertil. Steril.* **54:** 102–108.

Quinn P. 1995. Enhanced results in mouse and human embryo culture using a modified human tubal fluid medium lacking glucose and phosphate. *J. Assist. Reprod. Genet.* **12:** 97–105.

———. 1998. Glucose and phosphate–important or unimportant in culture media for embryos? *Fertil. Steril.* **70:** 782–783.

Quinn P. and Horstman F.C. 1998. Is the mouse a good model for the human with respect to the development of the preimplantation embryo in vitro? *Hum. Reprod.* (Suppl.) **4:** 173–183.

Quinn P., Barros C., and Whittingham D.G. 1982. Preservation of hamster oocytes to assay the fertilizing capacity of human spermatozoa. *J. Reprod. Fertil.* **66:** 161–168.

Quinn P., Hirayama T., and Marrs R.P. 1993. Cooperative interaction among mouse zygotes cultured in protein-free medium: Blastocyst development and hatching. In *Preimplantation embryo development* (ed. B.D. Bavister), p. 328. Springer-Verlag, New York.

Quinn P., Kerin J.F., and Warnes G.M. 1985a. Improved pregnancy rate in human in vitro fertilization with the use of a medium based on the composition of human tubal fluid. *Fertil Steril.* **44:** 493–498.

Quinn P., Warnes G.M, Kerin J.F., and Kirby C. 1985b. Culture factors affecting the success rate of in vitro fertilization and embryo transfer. *Ann. N.Y. Acad. Sci.* **442:** 195–204.

Schultz G.A., Kaye P.L., McKay D.J., and Johnson M.H. 1981. Endogenous amino acid pool sizes in mouse eggs and preimplantation embryos. *J. Reprod. Fertil.* **61:** 387–393.

Scott L. and Whittingham D.G. 1996. Influence of genetic background and media components on the development of mouse embryos in vitro. *Mol. Reprod. Dev.* **43:** 336–346.

Scott L.F., Sundaram S.G., and Smith S. 1993. The relevance and use of mouse embryo bioassays for quality control in an assisted reproductive technology program. *Fertil. Steril.* **60:** 559–568.

Stewart C.L., Kaspar P., Brunet L.J., Bhatt H., Gadi I., Kontgen F., and Abbodanzo S.J. 1992. Blastocyst implantation depends on maternal expression of leukaemia inhibitory factor. *Nature* **359:** 76–79.

Summers M.C., Bhatnagar P.R., Lawitts J.A., and Biggers J.D. 1995. Fertilization in vitro of mouse ova from inbred and outbred strains: Complete preimplantation embryo development in glucose-supplemented KSOM. *Biol. Reprod.* **53:** 431–437.

Suzuki O., Asano T., Yamamoto Y., Takano K., and Koura M. 1996. Development in vitro of preimplantation embryos from 55 mouse strains. *Reprod. Fertil. Dev.* **8:** 975–980.

Sztein J.M., Farley J.S., and Monbraaten L.E. 2000. In vitro fertilization with cryopreserved inbred mouse sperm. *Biol. Reprod.* **63:** 1774–1780.

Wakayama T., Perry A.C., Zuccotti M., Johnson K.R., and Yanagimachi R. 1998. Full-term development of mice from enucleated oocytes injected with cumulus cell nuclei.

Nature **394:** 369–374.

Wakayama T., Rodriguez I., Perry A.C., Yanagimachi R., and Mombaerts P. 1999. Mice cloned from embryonic stem cells. *Proc. Natl. Acad. Sci.* **96:** 14984–14989.

Wang W.H., Meng L., Hackett R.J., Oldenbourg R., and Keefe D.L. 2001. Limited recovery of meiotic spindles in living human oocytes after cooling-rewarming observed using polarized light microscopy. *Hum. Reprod.* **16:** 2374–2378.

Whitten W.K. 1956. Culture of tubal mouse ova. *Nature* **177:** 96.

———. 1971. Embryo medium. Nutrient requirements for the culture of preimplantation embryos in vitro. *Adv. Biosci.* **6:** 129–141.

Whitten W.K. and Biggers J.D. 1968. Complete development in vitro of the pre-implantation stages of the mouse in a simple chemically defined medium. *J. Reprod. Fertil.* **17:** 399–401.

Whittingham D.G. 1971. Culture of mouse ova. *J. Reprod. Fertil. Suppl.* **14:** 7–21.

Whittingham D.G. and Biggers J.D. 1967. Fallopian tube and early cleavage in the mouse. *Nature* **213:** 942–943.

CHAPTER *5*

Isolation, Culture, and Manipulation of Postimplantation Embryos

AFTER OVULATION AND FERTILIZATION, the preimplantation embryos undergo cleavage divisions and the initial steps of differentiation as they move down the oviduct toward the uterus. Upon reaching the uterus, the blastocyst-stage embryos emerge from their zonae pellucidae to implant at 4.5 dpc. Once the mouse embryo implants into the uterus, it becomes more difficult to manipulate. This chapter describes methods for dissecting postimplantation-stage mouse embryos and isolating specific tissues and cells for experimental analysis. In addition, roller and static culture systems are described to grow postimplantation mouse embryos in vitro for experimental manipulations and imaging outside the mother. Recently, gene transfer methods using nucleic acid electroporation into mouse embryos have been developed. A method is included to introduce gene expression constructs into postimplantation mouse embryos by electroporation.

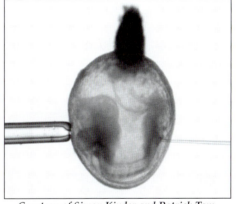

Courtesy of Simon Kinder and Patrick Tam,
Children's Medical Research Institute

CONTENTS

ISOLATING POSTIMPLANTATION EMBRYOS

Visualizing Early Embryo Implantation Sites

In the mouse, implantation of the blastocyst into the uterus occurs at ~4.5 dpc. This stage is one of the most difficult to study in the mouse because of the very small size of the embryo and its general inaccessibility in the uterus. Thus, methods to visualize the sites of blastocyst implantation help to facilitate studies of this critical step in mouse development. One of the earliest morphological changes considered prerequisite for the onset of the attachment reaction between the blastocyst trophectoderm and the uterine luminal epithelium is an increased vascular permeability in the uterine stromal bed at the site of blastocyst apposition (Paria et al. 2000, 2001). This increased localized uterine vascular permeability is easily detected in mice or rats by injecting a macromolecular blue dye solution intravenously a few minutes prior to sacrificing the animal (see Protocol 5.1). The dye binds with circulating proteins, and protein–dye conjugates accumulate in interstitial spaces only at the sites of increased vascular permeability. Distinct blue bands along the uterus demarcate the implantation sites, indicating that the

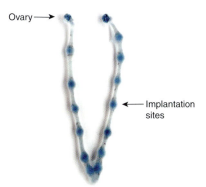

FIGURE 5.1. Dissected uterus from a pregnant female mouse at 6 days of gestation injected intravenously with a dye solution to visualize the embryo implantation sites, which are stained blue.

attachment process is in progress (Fig. 5.1). The visually marked implantation sites and unlabeled interimplantation regions can then be easily isolated for further study.

Isolating Postimplantation Embryos

Postimplantation development in the mouse occurs between 4.5 dpc and birth; i.e, ~80% of the gestational period. Thus, a comprehensive understanding of normal development requires the ability to isolate and analyze postimplantation-stage mouse embryos. In addition, mutations in many genes cause defects after implantation in the uterus, making it important to isolate embryos at these stages (Copp 1995). Postimplantation embryos require more complex media than preimplantation embryos if the cells are to remain viable and healthy during the dissection. Therefore, it is advisable to dissect the embryos in a tissue culture medium such as Dulbecco's modified Eagle's medium (DMEM) containing 10% fetal bovine serum (FBS). An important function of the serum is to stop the tissues from becoming too sticky; if necessary, the serum can be replaced with bovine serum albumin (BSA). HEPES buffer (~25 mM, pH 7.4) is added to maintain the pH during handling outside the incubator. Protocol 5.2 describes methods for isolating embryos at 5.5–8.5 dpc.

Isolating Extraembryonic Membranes

Extraembryonic tissues are essential for pre- and postimplantation mouse development. Defects in extraembryonic tissues can cause intrauterine growth restriction (IUGR) or fetal death and account for a significant amount of failed pregnancies in mammals (Cross et al. 1994). The molecular basis of these defects requires the ability to isolate and analyze these extraembryonic tissues. The best source of trophoblast tissue free of maternal contamination is the 7.5- to 8.5-dpc ectoplacental cone (see Fig. 5.7B, below) (Rossant and Croy 1985). For isolating the parietal yolk sac (PYS), visceral yolk sac (VYS), and amnion simultaneously, use mouse embryos at 13.5 dpc. The PYS begins to degenerate in older embryos and disappears by 15–16 dpc. Protocol 5.3 provides a method for isolating these membranes.

The VYS also can be used to determine the genotypes of certain types of ES cell-derived chimeras (Varlet et al. 1997). The mesoderm layer of the VYS is derived from the epiblast, whereas the endoderm layer of the VYS is derived from the visceral endoderm. Because ES cells contribute poorly to the visceral endoderm of the VYS in chimeras (Beddington and Robertson 1989), the genotype of the recipient embryo used to generate the chimera can be determined by analyzing the visceral endoderm of the VYS. VYS tissue layer separation (Protocol 5.3) provides pure visceral endoderm tissue to determine the genotypes of ES cell-derived chimeras (Varlet et al. 1997).

Separating Postimplantation Germ Layers

Germ layers are multipotent embryonic tissues that have the ability to differentiate, depending on intrinsic and extrinsic factors, into various tissue types. The ability to isolate and manipulate germ layers is therefore important for understanding the mechanisms that regulate their potency and ultimate lineage restriction. The basic steps involved in germ layer separation are illustrated in Figure 5.12, below (see Protocol 5.4). Although this figure shows the isolation of endoderm, mesoderm, and ectoderm from a late primitive-streak-stage embryo (~7.5 dpc), the principles of combining mechanical dissection and enzymatic methods of tissue separation apply to isolation of later tissues, such as neurectoderm, somites, or notochord (8.5–9.5 dpc), and to separation of the earlier bilaminar pre- and early primitive streak stages (5.5–6.5 dpc). In all cases, the embryo is first dissected into the region from which the tissues are to be separated. A 25 mM HEPES-buffered DMEM solution containing 10% FBS is used for the manipulation and storage of embryos. The dissected region is incubated in a mixture of 2.5% pancreatin and 0.5% trypsin in Ca++/Mg++-free Tyrode Ringer's saline or Ca++/Mg++-free PBS (pH 7.6–7.7) at 4°C. Small regions of the embryo require only a 10-minute incubation in the pancreatic/trypsin enzyme solution, whereas larger pieces from older embryos may require up to a 1-hour incubation. The duration of enzyme treatment must be established empirically. Enzymatic digestion is halted by returning the embryonic tissue to DMEM containing 10% FBS. Importantly, the digested embryonic regions are allowed to "rest" for 2 or more minutes, facilitating subsequent mechanical dissection. After 2 minutes in DMEM containing 10% FBS, the embryos can be subjected to further mechanical dissection, reducing the risk of the tissues tearing or falling to pieces.

Germ Layer Explant Recombination Culture

Germ layer explant recombination culture was initially established in the mouse using naïve epiblast and mesendoderm to define the tissues that can induce anterior neural fates in the epiblast of early wild-type mouse embryos (Ang and Rossant 1993). Protocol 5.5 is also useful for testing in vitro the inductive properties of germ layer fragments isolated from wild-type and mutant mouse embryos (Ang et al. 1994; Shawlot et al. 1999; Kimura et al. 2000). This procedure can be used for primitive-streak-stage embryos (6.5 and 7.5 dpc) and also for somite-stage embryos (8.5 and 9.5 dpc). The germ layer fragments are recombined and cultured in vitro in depression wells similar to those used to generate aggregation chimeras (see Chapter 11). This method has also been adapted to test the inductive properties of a variety of different embryonic regions (Shawlot et al. 1998; Tian et al. 2002).

Isolating Germ Cells from the Genital Ridge

Germ cells carry the genetic information from parent to offspring. Thus, the development of the mammalian germ line is one of the most interesting topics in developmental biology (McLaren 2001). Once the primordial germ cells have migrated into the genital ridges, they become relatively easy to isolate. Protocol 5.6 describes the isolation of germ cells from the genital ridges of fetal mice from 11.5 dpc onward. Because different embryos from the same mother can vary in their stage of development, it is advisable to classify them according to the morphology of their hind limb bud (Fig. 5.2). The aim of the EDTA treatment is to dissociate the germ cells from the stroma so that they can be released with minimal contamination by somatic cells. Isolated germ cells can then be used for subsequent analysis, culture, or transplantation.

Male and female genital ridges are morphologically distinct only by stage 6 (12.5 dpc) (Fig. 5.14) or later. In some situations, it may be necessary to know the genetic sex of the embryos younger than 12.5 dpc for germ cell isolation or other purposes. A simple marking system can be set up using a ubiquitously expressed X-linked GFP transgenic mouse line to provide an easy visual method for identifying male and female fetuses at any stage of postimplantation development (Hadjantonakis et al. 1998). GFP hemizygous ($X^{gfp}Y$) males are bred with wild-type females. All genetic female progeny will inherit the X-linked transgene ($X^{gfp}X$) and express GFP, whereas all genetic male progeny will not inherit the transgene (XY) and will be GFP negative. Thus, embryos from such a cross can be easily sorted by GFP fluorescence under the dissection microscope prior to genital ridge and germ cell isolation.

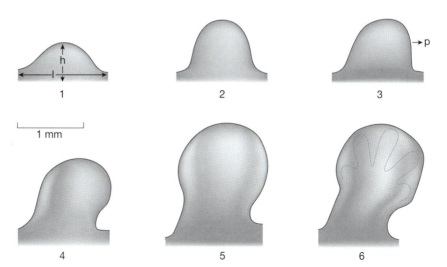

FIGURE 5.2. Morphology of hind limb bud from 10.5- to 12.5-dpc fetuses. Hind limb bud morphology is useful for correlating the differentiation of the genital ridges. Male and female genital ridges cannot be distinguished until stage 6 (12.5 dpc). (*1*) Length (l) > height (h); (*2*) h > l; (*3*) the limb bud is asymmetric, posterior (p) reentrant; (*4*) posterior and anterior reentrant; (*5*) limb bud is symmetric and has a circular outline; (*6*) the limb bud has an angulated outline and rays of toes are just visible.

CULTURING POSTIMPLANTATION EMBRYOS

Preparing Embryos

Embryos from 6.5 to 11.5 dpc are dissected from the uterus. All manipulations must be performed in medium containing 10% fetal bovine serum (FBS) to prevent the embryonic tissues from becoming unduly sticky. An appropriate medium for recovery and short-term storage of embryos on the bench is DMEM containing 10% FBS, buffered with HEPES (25 mM, pH 7.4). Reichert's membrane expands poorly in vitro and must therefore be reflected or removed before embryos are placed in culture. Removal of Reichert's membrane from a late primitive-streak-stage embryo is illustrated in Figure 5.10. The manipulation is basically the same for embryos of other stages.

Roller Culture of Postimplantation Embryos

Embryos from 6.5 to 11.5 dpc can be cultured in roller culture (see Protocol 5.7), although the older the embryo at the time of explantation, the shorter the period for which normal development can be sustained in vitro. Thus, embryos at these stages of postimplantation development can be isolated, experimentally manipulated, and then cultured for subsequent development outside the uterus (Kinder et al. 2001). The following are the principal features of "classical" whole embryo culture: (1) Rat serum is the major component of the medium, (2) cultures are carried out in small volumes of medium in tubes that are continually circulated on rollers to maintain constant equilibration of the medium with the gas phase, and (3) gas composition is varied according to the developmental stage of explanted embryos. Embryos develop normally under these conditions, with only slight retardation in growth, for 48 hours when explanted from early primitive streak to early somite stages. Thereafter, 36 hours or 24 hours is the limit of normal development in vitro. All instruments and receptacles used for preparing embryos and media must be scrupulously clean and not contaminated with toxic materials used for other embryological purposes (e.g., organic solvents used in histology, detergents). For more comprehensive protocols for whole embryo culture, see Beddington (1987), Cockroft (1990), and Sturm and Tam (1993).

Several different equipment setups have been used successfully for roller culture. A purpose-made miniroller apparatus inside an incubator is available from BTC Engineering. A miniroller marketed by Wheaton can be placed inside an incubator (e.g., Precision Scientific). Alternatively, tubes can be placed in a Bellco roller drum, which essentially is a wheel with holes in it that rotates at a variable speed. (A small hybridization incubator with increased rotation speed can also be used.) The incubator is maintained at 37–38°C and the rollers revolve at a constant speed of 30 rpm.

Static Culture of Postimplantation Embryos

The classic roller methods of postimplantation embryo culture have recently been modified into so-called "static" culture methods that do not require equipment to rotate the cultures, facilitating the direct visualization of the embryos during in vitro development. Protocols 5.8 and 5.9 describe these methods. Embryos at 6.0–7.0 dpc can be cultured satisfactorily in a static medium with the CO_2 and O_2 levels in the medium maintained in equilibrium that is achieved by diffusion from the ambient gaseous environment. Good development of prestreak to mid/late-streak embryos to the late allantoic bud/early head-fold stage can be achieved in about 65%, 75%, and 85% of 6.0-, 6.5-, and 7.0-dpc embryos, respectively. Embryos at 6.0–7.0 dpc develop more rigorously and are morphologically more comparable to embryos in vivo in static culture than roller culture. However, development beyond the early head-fold stage (equivalent to 8.0 dpc in vivo; Downs and Davies 1993) cannot be sustained as effectively by static culture and requires the use of a roller culture system, which allows more efficient gaseous exchange and better buffering capability. The static culture therefore is best limited to experiments that last for up to 48 (for 6.0-dpc embryos) to 24 hours (for 7.0-dpc embryos) of in vitro development.

The principal features of this culture method are: (1) The culture medium remains static throughout the experiment, (2) the gas composition usually remains unchanged during culture, (3) the culture medium contains a high content of rat or human cord serum in DMEM or serum only, and (4) cultures are carried out in small volumes of medium in dishes or wells of tissue culture chamber slides. The last feature is particularly advantageous if the experiment requires individual or small groups of three or four embryos to be distinguished from each other and/or whose development has to be monitored at regular intervals during the experiment. This method offers the most economic use of the precious serum-based culture medium and allows repetitive real-time observation of development with minimal handling of the embryo. Cultured embryos from pre-streak to late-streak stage tend to remain submerged at the bottom of the culture vessel and can be studied in situ under the dissecting or fluorescence microscope. However, embryos that develop beyond the late-streak stage start to accumulate fluid in the yolk sac and thus become buoyant in the serum-containing culture medium that is of a higher fluid density. More advanced embryos have to be taken out of the static culture to holding medium (e.g., PB1) for observation.

Static Culture Of Postimplantation Embryos for Imaging

Roller culture methods, although optimal for in vitro growth of postimplantation embryos, require constant motion, making direct observations of the embryos and cell movements and relationships difficult. Basically, the roller culture apparatus must be stopped and the embryo either visualized directly in the culture medium vessel or temporarily removed into a petri dish with holding medium for observation. It is currently not possible in a roller culture system to visualize the embryos continuously. Static methods for culturing postimplantation embryos eliminate the constant motion required by the roller culture methods that greatly facilitates continuous imaging of embryos. Recently, a protocol has been developed that allows static culture of early postimplantation embryos on a microscope stage, as described in Protocol 5.9. Embryos between 6.5 and 9.5 dpc can be cultured and imaged for 24 hours, with very little growth retardation. Although medium formulation and gas requirements are very similar to roller culture, great care must be taken to prevent evaporation. The quality of reagents and embryo preparation are also extremely important for obtaining healthy growth in culture.

INTRODUCING NUCLEIC ACIDS INTO POSTIMPLANTATION EMBRYOS BY ELECTROPORATION

Recently, methods have been devised to introduce gene constructs into the embryos of diverse species, including mice, by electroporation (Osumi and Inoue

2001; M. Takahashi et al., in prep.) (see Protocol 5.10). The introduction of genes into mouse embryos by electroporation is a convenient and efficient alternative to the generation of transgenic mice by pronuclear injection or by transfection with recombinant viruses. Gene constructs can be quickly tested for tissue-specific transcriptional activity or can be used to overexpress gene products. In addition, cell labeling by electroporation with lineage tracers such as lacZ and green fluorescent protein (GFP) can be used to monitor cellular behavior such as proliferation and migration. Recently, electroporation methods have also been used successfully to introduce double-stranded RNA (dsRNA) into preimplantation mouse embryos in which the zona pellucida had been weakened by transient acidified Tyrode's treatment (Grabarek et al. 2002). This method was used to generate gene knockdowns in preimplantation embryos by RNA interference (Fjose et al. 2001). Intriguingly, this suggests that it may be possible to use the electroporation method on preimplantation-stage embryos to introduce DNA constructs to generate transgenic mice.

Cultured postimplantation mouse embryos from 9.5 to 11.5 dpc (and earlier), or even subregions of embryos, can be readily electroporated with exogenous genes according to methods initially established for avian embryos (Muramatsu et al. 1997; Nakamura and Funahashi 2001). Expression can be localized to specific embryonic regions because negatively charged DNA will move toward a positively charged electrode that can be positioned according to the desired outcome. In principle, any region of the embryo can be a target for electroporation; however, vesicles and tubes (e.g., the neural tube) are very amenable for electroporation because they can be easily injected with DNA solutions. Although expression induced by electroporation is transient, it is often long enough to analyze effects, including the alteration of downstream gene expression and induction of subsequent phenotypes. The basic features of the electoporation method are to (1) construct gene expression vectors, (2) prepare a solution of purified plasmid DNA, (3) inject the DNA solution into regions in question, (4) apply square electrical pulses to move the DNA into tissues, and finally, (5) culture in vitro (Fig. 5.3).

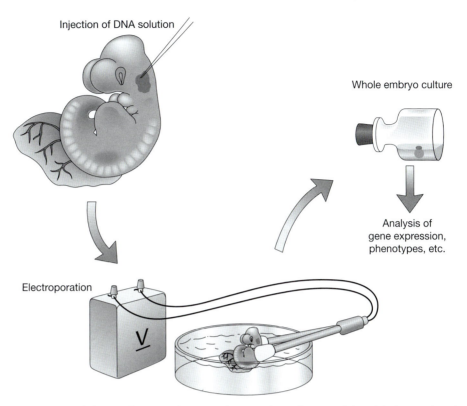

FIGURE 5.3. Scheme showing the procedures to introduce nucleic acids into subregions of postimplantation mouse embryos by electroporation for analysis.

Visualizing Early Embryo Implantation Sites by Dye Injection

This procedure was provided by S. K. Dey, Departments of Pediatrics and Cell Biology, Vanderbilt University Medical Center, Nashville, Tennessee.

MATERIALS

ANIMALS
Pregnant female mice (4.5–5.5 dpc)

EQUIPMENT
Bowl for warm water
Gauze sponges or paper towels
Hypodermic needle, 27-gauge, 1/2-inch
Syringe, 1-ml

REAGENTS
Anesthetic
Chicago Sky Blue 6B, also called Pontamine sky blue (Sigma C8679)

Dissolved in isotonic saline to provide a 1% solution (1 gm/100 ml saline). Filter the solution through a Whatman filter paper (Whatman International, Maidstone, United Kingdom, 1001110). Store in a glass bottle at room temperature.

PROCEDURE

1. Anesthetize the pregnant female mouse.

2. Rinse the tail three or four times with gauze sponges or paper towels soaked in warm water (60°C) with gentle pressure to dilate the tail veins.

3. Without further delay, inject 0.1 ml of the 1% blue dye solution through a tail vein using a 1-ml syringe fitted with a 27-gauge needle.

   a. The syringe filled with the dye solution should be free of air bubbles and the bevel of the needle should face up for ease of injection.

   b. Either of the two lateral tail veins can be used for injection. It is a good practice to start injection from the caudal end of the tail vein and then move rostrally up the vein in case of failures.

   c. During injection, the blue dye solution should move smoothly through the vein if the needle is inside the vein.

4. After the injection, place a piece of gauze or a paper towel with gentle pressure at the site of puncture to stop bleeding.

5. Sacrifice the animal 3 minutes after the dye injection. (See Protocol 4.8 for humane method of sacrifice.)

6. Dissect the uterus from the female (see Fig. 5.5A).

7. Implantation sites appear as distinct blue bands along the uterine horns (Fig. 5.1).

8. Separated implantation (blue bands) and interimplantation sites can be used for cellular, biochemical, and molecular biology analyses.

COMMENTS

- During the i.v. injection, if resistance is felt and there is local accumulation of the dye solution at the site of puncture, the injection should be terminated immediately and attempted at a second site rostral to the first site.

- Waiting too long after the dye injection to sacrifice animals decreases the demarcation between the blue bands. In the event blue bands are absent, the uterine horns can be flushed with medium to recover any unimplanted embryos.

MATERIALS

ANIMALS
Pregnant female mice sacrificed humanely

EQUIPMENT
Forceps, watchmaker's #5, two pairs
Scissors, fine
Stereomicroscope
Tissue culture dishes, sterile plastic
Tungsten needles (99.95% pure, 0.5-mm diameter) (Goodfellow Metals W005160 or Ernest F. Fullam 16210). See Figure 5.4 for sharpening tungsten needles.

REAGENTS
Dulbecco's modified Eagle's medium (DMEM)
Fetal bovine serum (FBS) or bovine serum albumin (BSA) (Sigma A9647)
HEPES buffer, 1 M (pH 7.4) (Sigma H0087)

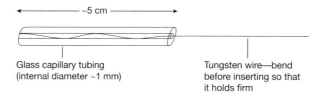

Glass capillary tubing
(internal diameter ~1 mm)

Tungsten wire—bend
before inserting so that
it holds firm

~5 cm

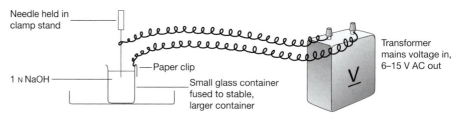

Needle held in
clamp stand

1 N NaOH

Paper clip

Small glass container
fused to stable,
larger container

Transformer
mains voltage in,
6–15 V AC out

V

FIGURE 5.4. Apparatus for sharpening tungsten needles by electrolysis. A very small piece of Plasticine can be attached to the tip of the blunt needle. Stop the electrolysis when it drops off.

PROCEDURE

Prestreak-stage (~5.5 dpc) Dissection

1. Open the abdominal cavity as described in Protocol 4.8, Figure 4.4. Cut one horn of the uterus below the oviduct and grasp the end firmly with the fine

forceps. Pull upward and separate the uterus from the mesometrium using the tips of the forceps or scissors (Fig. 5.5A). Then pull the uterus down tautly and out to one side.

2. Insert one tip of a pair of fine scissors (Fig. 5.5B) into the antimesometrial wall of the uterus near the cut end. Pointing the scissors slightly upward, "slide" them along, carefully cutting through the wall of the uterus. The decidua can now be "shelled" out of the uterus with forceps (Fig. 5.5C) and transferred to a plastic tissue culture dish containing medium composed of DMEM buffered with 25 mM HEPES (pH 7.4) containing 10% FBS. At this stage, the deciduum does not always separate cleanly from the endometrial stroma and is rather ragged.

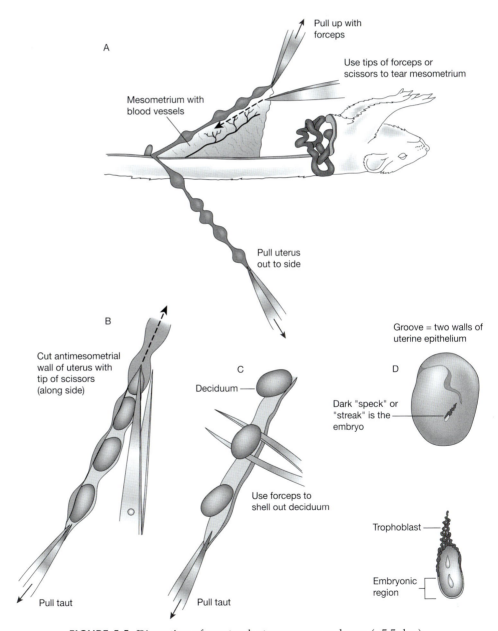

FIGURE 5.5. Dissection of prestreak-stage mouse embryos (~5.5 dpc).

3. Using transillumination, tease the deciduum apart with fine forceps. The two walls of the uterine epithelium have not yet fused completely and there is a distinct groove (see Fig. 5.5D). Locate the embryo, which appears as a dark speck or streak. The embryo can be "shelled out" with the tips of the forceps. The dark tissue is the trophoblast, and the embryonic region is transparent and very fragile (Fig. 5.5D).

Early Primitive-streak-stage (~6.5 dpc) Dissection

Either dissect the decidua from the uterus as described above or follow the procedure below.

1. Remove the uterus intact by cutting across the cervix and the two utero-tubal junctions and place it in a tissue culture dish containing medium composed of DMEM buffered with 25 mM HEPES (pH 7.4) containing 10% FBS. Cut into the individual swellings as shown in Figure 5.6A. The muscle layer is removed with forceps as shown in Figure 5.6B.

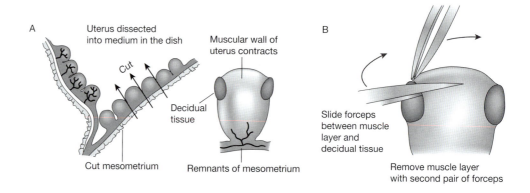

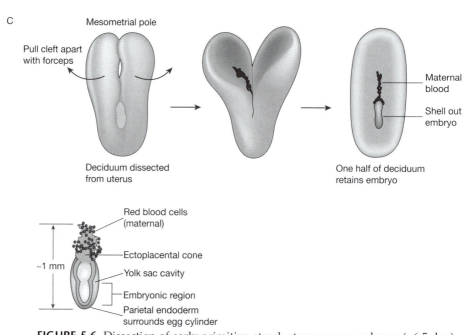

FIGURE 5.6. Dissection of early primitive-streak-stage mouse embryos (~6.5 dpc).

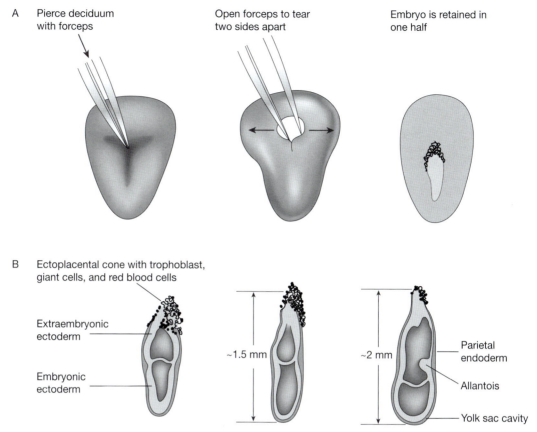

FIGURE 5.7. Dissection of late primitive-streak-stage mouse embryos (~7.5 dpc). (*A*) Technique for separating the deciduum to expose the embryo, which can then be shelled out with the tips of closed forceps. (*B*) Embryos are not synchronized in their development; more advanced embryos have a small allantois (*far right*).

2. Dissect the deciduum (Fig. 5.6C) to reveal the small embryo, which can be shelled out with the tips of closed forceps. The embryo can be dissected further into ectoplacental cone (trophoblast) and embryonic portions using tungsten needles in a scissor-like action. For details of making needles, see Figure 5.4, and for dissection techniques, see Snow (1978).

Late Primitive-streak-stage (~7.5 dpc) Dissection

Dissect the decidua and embryos essentially as described above and shown in Figure 5.7. At this stage, and earlier, the embryo can be separated into ectoderm, endoderm, and mesoderm by a combination of enzymatic digestion and mechanical dissection (see Protocol 5.4).

Early Neural-fold-stage (~8 dpc) Dissection

Dissect the embryo as shown in Figure 5.8A.

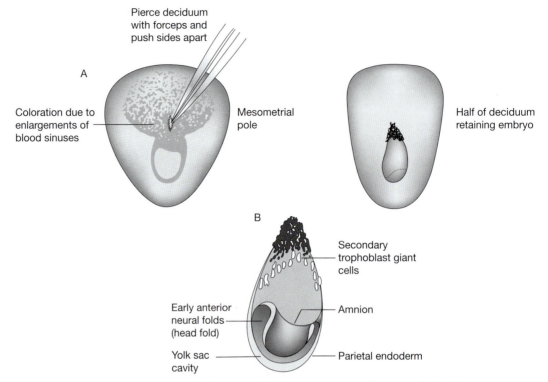

Pierce deciduum with forceps and push sides apart

A

Coloration due to enlargements of blood sinuses

Mesometrial pole

Half of deciduum retaining embryo

B

Secondary trophoblast giant cells

Early anterior neural folds (head fold)

Amnion

Yolk sac cavity

Parietal endoderm

FIGURE 5.8. Dissection of early neural-fold-stage mouse embryo (~8 dpc). (*A*) The technique for separating an embryo from deciduum is essentially as shown in Fig. 5.7. (*B*) The embryo now has a distinctive shape with large anterior neural folds and a deep neural groove.

Early Somite-stage (~8.5 dpc) Dissection

The following procedure is a simple method for dissection at this stage. In most embryos, the allantois has not yet fused with the chorion (Fig. 5.9B). More advanced embryos will have begun "turning," and in a few, the heart will be beating.

1. Use forceps to cut off the mesometrial one-third of the deciduum as shown in Figure 5.9A.

2. Pull the two halves of the deciduum apart with fine forceps and gently shell out the embryo.

3. If required, dissect and remove the extraembryonic membranes with forceps as shown in Figure 5.9B.

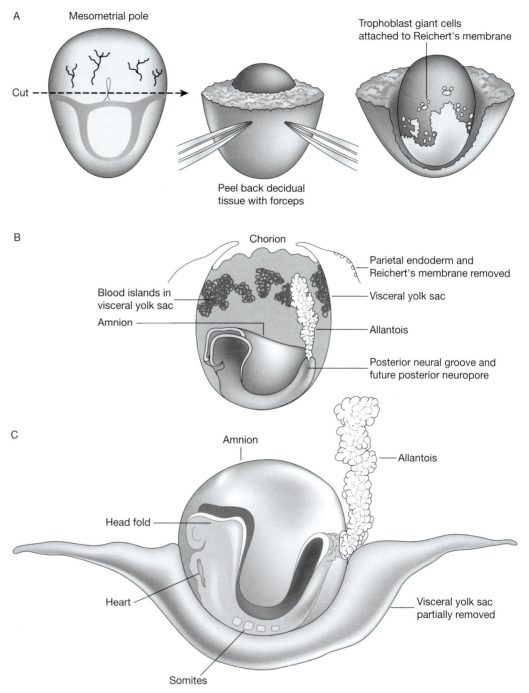

FIGURE 5.9. Dissection of early somite-stage embryos (~8.5 dpc).

Isolating Extraembryonic Membranes

MATERIALS

ANIMALS
Pregnant female mice (13.5 dpc) sacrificed humanely

EQUIPMENT
Forceps, fine, two pairs
Forceps, watchmaker's #5
Scissors, fine
Stereomicroscope
Tissue culture dishes, sterile plastic

REAGENTS
Dulbecco's modified Eagle's medium (DMEM) buffered with 2 mM HEPES (pH 7.4)
Dulbecco's modified Eagle's medium (DMEM) buffered with 25 mM HEPES (pH 7.4)
   containing 10% fetal bovine serum (FBS)
Hyaluronidase (Sigma H3884) (see Appendix 1) in DMEM (~300 μg/ml)
Methanol<!>/acetone<!> (1:1)
Pancreatin/trypsin enzyme solution (see Appendix 1)
Trypsin/EDTA solution (see Appendix 1)

CAUTION: See Appendix 2 for appropriate handling of materials marked with <!>.

PROCEDURE

1. Open the abdominal cavity as described in Protocol 4.8, Figure 4.4.

   a. Remove the uterus into a tissue culture dish with HEPES-buffered DMEM containing 10% FBS and cut transversely between implantations so that each conceptus is isolated (see Fig. 5.6A).

   b. Remove the muscle layer of the uterus (which in some cases retracts spontaneously) (Fig. 5.10A).

   c. Rotating the embryo, use the tips of two pairs of fine forceps in a "nipping" action to cut around the junction of Reichert's membrane and the placenta (Fig. 5.10B).

   d. Remove the membrane with parietal endoderm cells attached and transfer it to fresh serum-free medium.

   Some trophoblast cells will be attached to the membrane. The number of trophoblast cells will depend on the age and strain of the mouse. (C3H/He mice have very few trophoblast cells at 13.5 dpc.) By 15 days, the Reichert's membrane will be very thin and most of the parietal endoderm cells will have degenerated.

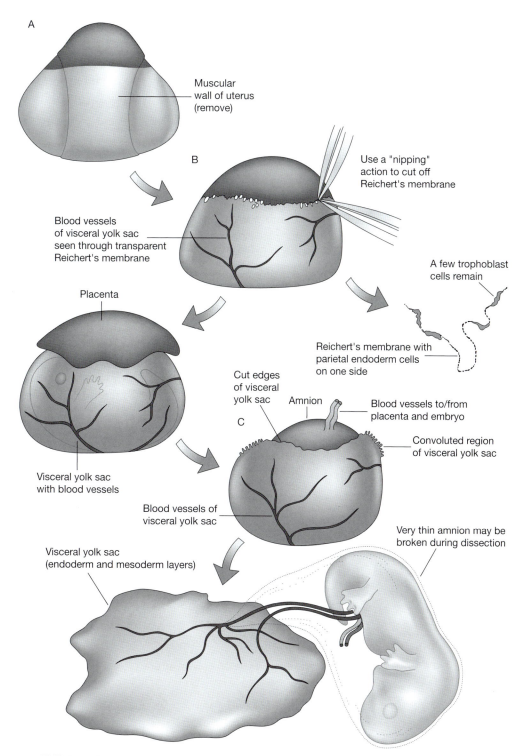

A

Muscular
wall of uterus
(remove)

B

Use a "nipping"
action to cut off
Reichert's membrane

Blood vessels
of visceral yolk sac
seen through transparent
Reichert's membrane

A few trophoblast
cells remain

Reichert's membrane with
parietal endoderm cells
on one side

Placenta

Cut edges
of visceral
yolk sac

Amnion

Blood vessels to/from
placenta and embryo

C

Convoluted region
of visceral yolk sac

Visceral yolk sac
with blood vessels

Blood vessels of
visceral yolk sac

Very thin amnion may be
broken during dissection

Visceral yolk sac
(endoderm and mesoderm layers)

FIGURE 5.10. Dissection of 13.5-dpc mouse embryo to recover extraembryonic tissues (parietal endoderm on Reichert's membrane, visceral yolk sac, and placenta).

2. To observe parietal endoderm cells in situ, attach the Reichert's membranes to the surface of a plastic tissue culture dish.

   a. Stab around the periphery of each membrane with the closed tips of a pair of forceps, thereby anchoring the membrane to the plastic (experiment with several brands of dishes to find the one with the softest surface).

   b. Rinse the Reichert's membranes gently in serum-free medium and, if required for use in immunocytochemistry, fix with a 1:1 mixture of methanol/acetone for 5 minutes at room temperature.

   c. Air-dry, and store at –70°C (Lane et al. 1983). (A 1:1 mixture of methanol and acetone will not dissolve the plastic tissue culture dish.)

3. After removing the parietal endoderm, use fine scissors to cut off the placenta and the proximal region of the visceral yolk sac (VYS). The VYS can now be separated from the embryo (surrounded by the amnion) (Fig. 5.10C).

   Note that clumps of visceral endoderm cells can be easily broken off from the convoluted region nearest to the placenta; if care is not taken, these cells can significantly contaminate the amnion or parietal yolk sacs dissected in the same tissue culture dish. (For discussion of the problem of cross-contamination, see Dziadek and Andrews 1983.)

4. Separate the epithelial layer of visceral endoderm from the underlying mesoderm (endothelial cells, blood islands, and fibroblasts) by incubating the VYS (from 9.5- to 13.5-dpc embryos) in pancreatin/trypsin enzyme solution.

   The time must be determined empirically for each batch of enzyme; 15 minutes to 1.5 hours at 4°C is usually required.

   a. Rinse the VYS in serum-free medium before starting the incubation.

   b. At the end of the incubation, transfer the VYS to HEPES-buffered DMEM containing 10% serum and tease the two layers of tissue apart with the watchmaker's forceps. The separate layers of tissue can be dissociated further (into single cells) by incubation in trypsin/EDTA solution.

COMMENT

The amnion becomes very "slimy" and difficult to handle toward the end of gestation. The copious glycosaminoglycan secretion can be removed by incubating the membrane briefly in hyaluronidase solution in DMEM (~300 μg/ml).

Separating Postimplantation Germ Layers

MATERIALS

ANIMALS

Late primitive-streak-stage embryos (~7.5 dpc)

EQUIPMENT

Forceps, watchmaker's #5, two pairs
Instrument tubes (Leitz 520142)
Microburner
Glass needles, siliconized, solid (1-mm external diameter; 0.5-mm internal diameter),
or tungsten needles (see Fig. 5.4).

Glass needles are prepared by first fusing the midregion of a Leitz thick-walled
glass capillary (Leitz 520119) and then pulling on a mechanical electrode puller.
The shaft of the needle should be ~1 cm long. In a fume hood, dip the tip of the
needle into Repelcote (BDH 63216 6L) or Sigmacote (Sigma SL-2). Introduce two
bends over a microburner or microforge: one 90° bend immediately behind the
shaft of the needle and one 130° bend 1 cm farther away from the needle point (Fig.
5.11). This produces a needle where the shaft runs parallel to the bottom of the dish
when held in the operator's hand.

Pasteur pipette, siliconized

This is hand-pulled over a microburner to obtain an internal diameter slightly
smaller than that of the embryonic region. Flame-polish the tip of the Pasteur
pipette by passing it rapidly through the flame of a microburner.

Stereomicroscope
Tissue culture dishes, 60-mm sterile plastic

REAGENTS

Dulbecco's modified Eagle's medium (DMEM) buffered with 25 mM HEPES
(pH 7.4)
Dulbecco's modified Eagle's medium (DMEM) buffered with 25 mM HEPES (pH 7.4)
containing 10% FBS
Pancreatin/trypsin enzyme solution (see Appendix 1)

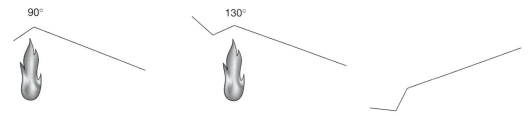

FIGURE 5.11. Bending glass needles over a microburner.

PROCEDURE

1. Dissect the embryo free of the deciduum into a plastic dish with HEPES-buffered DMEM containing 10% FBS, and remove the Reichert's membrane (Fig. 5.12A,B) as follows:

   a. Insert the closed tips of one pair of watchmaker's forceps into the yolk sac cavity (the cavity separating the parietal and visceral endoderm) in the vicinity of the ectoplacental cone, thereby pinning the embryo to the bottom of the plastic dish (Fig. 5.12A).

   b. Insert the closed tips of a second pair of watchmaker's forceps at the same spot and tear the membrane by drawing the second forceps toward the distal tip of the cylinder (Fig. 5.12B).

   c. Once released, the torn membrane will retract toward the ectoplacental cone, where it can be pinched off using the forceps like a pair of scissors.

2. Cut off the embryonic region using solid, siliconized glass or tungsten needles (Fig. 5.12C). Make cuts in the embryo by laying the shaft of the needle along the proposed line of the cut, pushing the needle shaft down against the bottom of the dish, and moving it from side to side using a cutting motion two or three times.

3. Rinse the embryonic region in serum-free medium. Then, incubate the embryonic region in pancreatic/trypsin enzyme solution at 4°C for 10–15 minutes and transfer it to a plastic dish of HEPES-buffered DMEM containing 10% FBS. Allow the embryo to "rest" for a few minutes.

4. Gently suck the embryo (distal tip first) into a siliconized Pasteur pipette to remove the endoderm layer (Fig. 5.12D). Use standard mouth-pipetting technique to aspirate the embryo in and out of the pipette two or three times. The endoderm will detach around the circumference of the embryo (Fig. 5.12E), but it may remain attached at the anterior of the primitive streak (the distal tip of the embryo). Cut the endoderm away from the embryo with a glass or tungsten needle.

5. If the mesoderm has not been detached already, insert the points of glass or tungsten needles under the mesoderm at the anterior aspect of the embryo.

   a. Stroke the embryo gently with the shafts of the needles (from anterior to posterior) to peel back the mesoderm layers on each side so that they fold back like two wings, attached to the ectoderm only along the primitive streak (Fig. 5.12F).

   b. Hold the embryo with one needle so that the primitive streak lies on the bottom of the dish.

   c. Cut off each mesoderm wing by making a single cut parallel to the primitive streak with another needle (Fig. 5.12F).

6. Hold the embryo with one needle so that the lateral side lies on the bottom of the dish. Using another needle, cut off the primitive streak region (Fig. 5.12G).

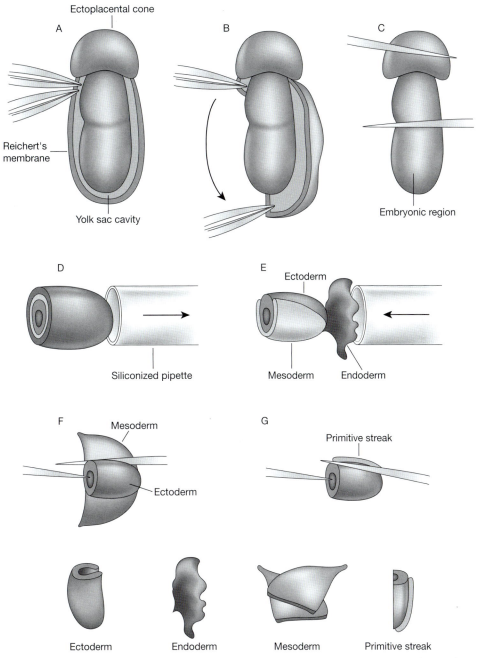

FIGURE 5.12. Method for separating ectoderm, endoderm, and mesoderm germ layers from a late primitive-streak-stage embryo.

Because the primitive streak is "J-shaped," the amount of ectoderm removed with the streak can be minimized by making two cuts—one cut parallel to the proximal distal axis of the embryo, which will remove two-thirds of the streak, and the other cut at right angles across the distal tip, which will cut off the anterior extreme of the streak. (*Note:* The primitive streak fraction will not be left intact using this procedure.)

Germ Layer Explant Recombination Culture

This procedure was provided by William Shawlot, Department of Genetics, Cell Biology, and Development, University of Minnesota, Minneapolis, Minnesota 55455.

MATERIALS

ANIMALS
Mouse embryos

EQUIPMENT
Aggregation needle (BLS, Hungary, http://www.bls-ltd.com DN-09)
Glass capillaries (World Precision Instruments TW100-4)
Incubator, humidified, 5% CO_2 at 37°C
#5 Micro dissecting tweezers (Roboz RS-4905)
Petri dishes, 3-cm bacteriological (Falcon 1008)
Pipette tip
Sharpening stone
Stereomicroscope
Transformer for electrolysis, 6–15 V AC out (Radio Shack)
Tungsten wire, 0.5-mm diameter (Goodfellow Corporation, Malvern, Pennsylvania) and holder for tungsten needles (Fine Science Tools 26016-12)

REAGENTS
Culture medium (DMEM containing 15% fetal calf serum [FCS], 2 mM glutamine, and 0.1 mM β-mercaptoethanol <!>)
Dulbecco's modified Eagle's medium (DMEM) containing 15% FCS
Mineral oil (Sigma M8410)
Pancreatin/trypsin solution (see Appendix 1)
Phosphate-buffered saline (PBS) containing 5% fetal bovine serum (FBS)
Sodium hydroxide (NaOH) <!>, 1 N

CAUTION: See Appendix 2 for appropriate handling of materials marked with <!>.

PROCEDURE

1. To prepare tungsten needles, cut a 4- to 5-cm length of wire and create a knife-like edge using a sharpening stone. Insert the sharpened wire into a needle holder for manipulation of embryos.

   Before each use, the tungsten needles can be re-sharpened by electrolysis in 1 N NaOH (see Fig. 5.4). Glass needles can be pulled on a commercial needle puller and used directly.

2. Dissect embryos from the uterus in PBS containing 5% FBS.

   a. For primitive-streak-stage embryos (6.5–7.5 dpc), deflect away the parietal endoderm layer from the embryonic portion using #5 micro dissecting tweezers.

   b. For older embryos, remove the yolk sac and the amnion and save the yolk sac for DNA isolation if necessary.

   c. Isolate the tissue fragment of interest using two needles. One needle is used to pin down the embryo and the second needle is used to cut away the tissue fragment.

   For fine isolation of tissues from 6.5-dpc embryos, glass needles generally work better. For cutting and separating germ layers from later-stage embryos, tungsten needles generally work better.

3. Rinse the embryo fragments in a petri dish containing PBS to remove serum that will interfere with enzymatic digestion.

   a. To separate the germ layers, place the tissue fragments (10–20) in 500 μl of pancreatin/trypsin solution for 5–10 minutes at 4°C.

   b. Stop the enzymatic reaction by transferring the tissue fragments into a clean dish of DMEM containing 15% FCS.

   c. Use a pipette tip containing a little bit of DMEM/serum to transfer the tissue fragments to prevent them from sticking to the pipette.

   d. Separate the germ layers using two tungsten or glass needles by gently working the tissue of interest free from the adjacent tissue layer.

   Some practice may be necessary. Isolated tissue fragments can be stored for short periods of time in DMEM/serum in the incubator while other germ layer fragments are being isolated.

4. Recombine the appropriate germ layer fragments in a 25-μl drop of culture medium. A depression well created with a darning needle in a 3-cm bacteriological petri dish will help keep the recombined tissue fragments together. Cover the medium droplets with mineral oil. Bacteriological plastic dishes are used to prevent the tissue fragments from sticking to the dish. Culture the explants for 1–2 days in 5% CO_2 at 37°C in a humidified tissue culture incubator. Recombinant tissue explants can then be processed for *lacZ* detection, RNA whole-mount in situ hybridization, or immunohistochemistry procedures (see Chapters 12 and 16).

Isolating Germ Cells from the Genital Ridge

Information for this protocol was provided by Anne McLaren, Wellcome/CRC Institute, University of Cambridge, Tennis Court Road, Cambridge CB2 IQR, United Kingdom.

MATERIALS

ANIMALS
Mouse embryos (12.5 dpc)

EQUIPMENT
Absorbent paper
Hypodermic needle, 26-gauge
Needle holder (Fisher 13-086; Gallenkamp DKD 430N)
Ophthalmic suture needles, half-curved #6 (Holborn Surgical and Veterinary Instruments E705; Anchor Products 1821-20)
Petri dishes, 35-mm plastic, or embryological watch glasses
Scissors, fine
Stereomicroscope
Syringe, 1-ml
Forceps, watchmaker's #5

REAGENTS
Dissection medium (prepare fresh): phosphate-buffered saline (PBS) complete (pH 7.2, see Appendix 1) or M2 medium (see Chapter 4)
Saline/EDTA solution plus glucose (see Appendix 1)

PROCEDURE

1. Dissect the embryos from their extraembryonic membranes (Fig. 5.10) and place them on absorbent paper.

    a. Use fine scissors to cut off the anterior half of the embryo just below the armpits (Fig. 5.13A).

    b. Make a cut along the ventral midline of the posterior half of the embryo (Fig. 5.13B) and scoop out the liver and intestines with the closed tips of the scissors.

2. Transfer the embryo fragment to a 35-mm plastic petri dish containing dissection medium and turn the fragment onto its back.

    a. Hold the embryo fragment with watchmaker's forceps and remove any remnants of intestines, etc., with a half-curved needle (Fig. 5.13C).

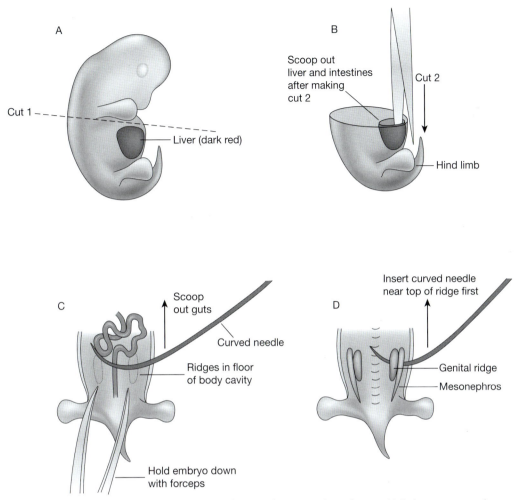

FIGURE 5.13. Dissection of genital ridges and mesonephros from a 12.5-dpc mouse embryo.

    b. The genital ridges lie on the dorsal wall of the embryo fragment, adjacent to the shield-like mesonephros. Slide the needle behind each genital ridge and mesonephros and cut them from the embryo (Fig. 5.13D).

3. Transfer the genital ridges and mesonephroi to fresh dissection medium with the watchmaker's forceps.

    a. Use the half-curved needle to cut the mesonephros from the genital ridge (Fig. 5.14C).

    From ~12.5 dpc (hind limb bud morphology stage 6), the ovaries and testes can be distinguished by their morphology. Testes are "striped" and larger than ovaries at the same stage (Fig. 5.14A), whereas the ovaries are "spotted" and smaller than the adjacent mesonephric shield (Fig. 5.14B).

4. Transfer the genital ridges to saline/EDTA solution plus glucose for ~15 minutes at room temperature.

    a. Return the genital ridges to the dissection medium.

    b. Holding each ridge with watchmaker's forceps, puncture the genital ridges with sharp stabs of a 26-gauge hypodermic needle (Fig. 5.14D).

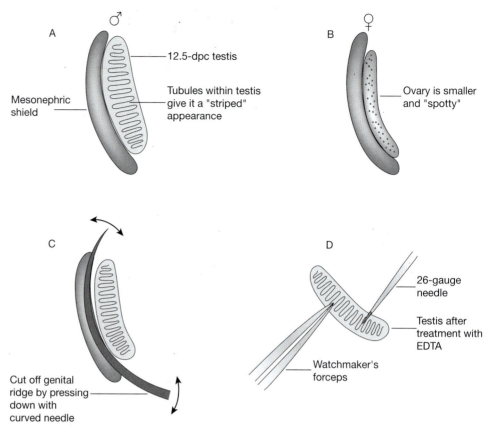

FIGURE 5.14. Technique for isolating germ cells from male and female genital ridges.

These punctures release the germ cells but few of the stromal cells. About 20 stabs are sufficient; additional puncturing of the genital ridges will release more germ cells but will also increase contamination with stromal cells. The germ cells can be recognized by their morphology, particularly when viewed with phase-contrast microscopy; they are large and have a smooth, often blebby outline (De Felici and McLaren 1983). It should be noted, however, that not all cells with this morphology are germ cells, and the population is by no means pure, as judged from alkaline phosphatase staining.

Roller Culture of Postimplantation Embryos

This procedure was provided by Patrick Tam, Children's Medical Research Institute, University of Sydney, Wentworthville, New South Wales, Australia.

MATERIALS

ANIMALS
Adult male rat
Mouse embryos (6.5–11.5 dpc)

EQUIPMENT
Bacteriological dish, plastic
Centrifuge
Forceps, large
Forceps, watchmaker's #5
Hypodermic needle, 19-gauge
Incubator, 37°C
Pasteur pipettes
Roller culture apparatus (e.g., BTC Engineering)
Roller tubes
Stereomicroscope
Syringe, 20-ml sterile
Syringe filter, 0.45-μm
Tube, 15-ml polypropylene, screw-capped (Corning 23519)
Tube, 50-ml polypropylene (Falcon)
Water bath, 56°C

REAGENTS
Dulbecco's modified Eagle's medium (DMEM)
Ether<!>
Gas mixture (see Fig. 5.15)
PB1 or M2 medium (see Chapter 4)
Rat serum
Silicon grease, high-vacuum

CAUTION: See Appendix 2 for appropriate handling of materials marked with <!>.

Preparing Medium

Up to the early somite stage, embryos can be grown in rat serum diluted 1:1 with DMEM, but older embryos grow better in 100% rat serum. Figure 5.16 illustrates the preparation of rat serum. Alternatively, rat serum can be purchased, but quality may vary among manufacturers and batches. Mouse serum can be used in combination with rat serum. The following mix is used for 7.5- to 11.5-dpc

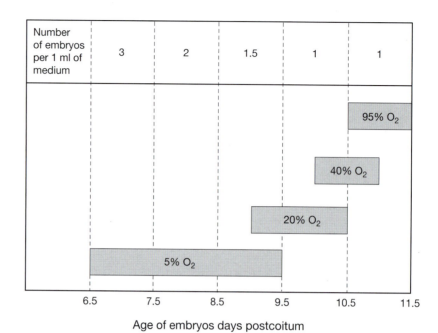

FIGURE 5.15. Percentage of oxygen required in gas phase for embryos of different developmental stages and number of embryos cultured per 1 ml of medium (all gas mixtures contain 5% CO_2 balanced with N_2).

embryos: 50% rat serum, 25% mouse serum, 25% DMEM. This mixture can work as well as 100% rat serum. Goat, sheep, and calf serum DO NOT support postimplantation mouse embryo development in vitro.

1. Withdraw blood slowly (to avoid hemolysis) from the dorsal aorta within the peritoneal cavity of an etherized male rat (Fig. 5.16A).

   a. Penetrate the dorsal aorta with a 19-gauge needle, attached to a sterile 20-ml syringe. Always insert the needle bevel downward to avoid blood spillage (Fig. 5.16B,C); 10–20 ml of blood can be withdrawn from a large rat.

   b. Once exsanguination is complete, euthanize the rat according to local institutional guidelines.

   There do not appear to be any marked strain differences in the quality of serum, but male serum has proved more consistent than female serum in supporting good embryonic development. This is presumably because the male is not subject to the cyclical hormonal and physiological changes associated with estrus. Older rats or retired breeders have a larger volume of blood that can be collected. Fatty serum is generally poor, and plasma is NEVER used.

   Ether is the anesthetic of choice because it readily evaporates from the serum without leaving toxic residues in the medium.

2. Remove the needle and decant the blood gently (again to avoid hemolysis of the red blood cells) into a 15-ml polypropylene, screw-capped centrifuge tube (Fig. 5.16D).

   a. Immediately spin the blood at 1200*g* at room temperature for 5 minutes (Fig. 5.16E). A white fibrin clot forms in the upper layer of plasma. Use large forceps to squeeze the clot to leave a stringy, white fibrin mass (Fig. 5.16E,F).

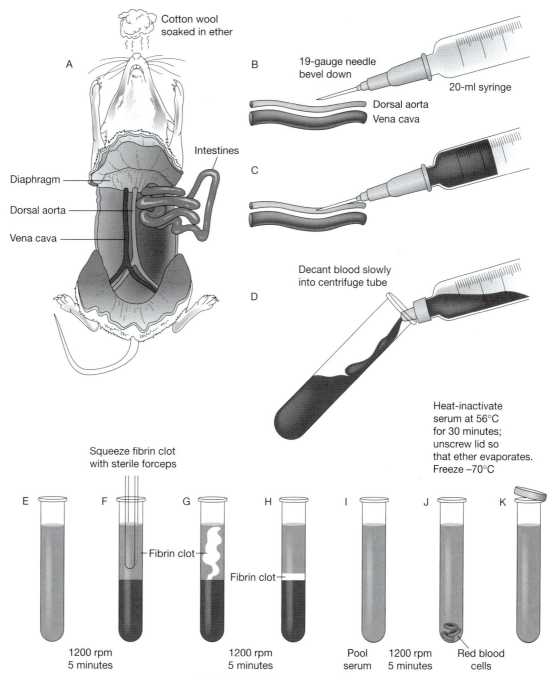

FIGURE 5.16. Preparation of rat serum.

b. Spin the tube again at 1200g for 5 minutes. There should now be an upper phase of clear serum separated from the red blood cells by the fibrin clot (Fig. 5.16H). If this does not happen, repeat the steps shown in Figure 5.16E–H.

3. Decant the serum with a sterile Pasteur pipette into a centrifuge tube. Serum from different rats can be pooled at this stage and stored on ice.

a. Spin the pooled serum a final time at about 1200g for 5 minutes to separate any remaining red blood cells (Fig. 5.16J).

b. Transfer the serum to a new tube (Fig. 5.16K). Aliquot the serum at this stage into convenient 5-ml volumes.

4. Heat-inactivate the serum in a water bath at 56°C for 30 minutes.

a. During this time, blow the gas mixture (see below) or air from a plugged Pasteur pipette over the surface of the serum to blow off the ether.

This is important because the ether is persistent and toxic to embryo growth.

b. Either use serum immediately or store at –20°C (store up to 4 months) or –80°C (store up to 6 months).

Preparing Embryos

1. Dissect the embryos into PB1 or M2 medium (25 to 30°C) in a plastic bacteriological dish.

a. Pin the embryo to the bottom of a plastic bacteriological dish by inserting the points of a pair of watchmaker's forceps into the yolk cavity in the region of the ectoplacental cone.

b. Insert the points of a second pair of watchmaker's forceps at the same spot and tear the membrane by drawing the points toward the distal tip of the cylinder. The membrane will retract toward the ectoplacental cone.

c. Pinch off the membrane by using the tips of the forceps like a pair of scissors.

Embryos that have been damaged in this process, most often incurring nicks in the endoderm layer, should be discarded.

2. Transfer the embryos into roller tubes using a hand-pulled siliconized Pasteur pipette whose internal diameter is slightly greater than that of the embryo.

Embryos older than 8.5 dpc should first be transferred to culture medium before placing them in roller tubes; this prevents carryover of an excessive volume of HEPES-buffered M2 (HEPES will interfere with the buffering of the medium with a CO_2 gas phase).

Embryo Culture

1. Mix the rat serum 1:1 with DMEM, and filter the mixture through a 0.45-μm syringe filter before use.

a. Aliquot the filtrate into 2- to 3-ml volumes in 50-ml polypropylene centrifuge tubes (Falcon). Let the tubes stand for at least 1 hour, with the lids ajar, in a 37°C tissue culture incubator containing 5% CO_2, 95% air. This will cause any remaining ether to evaporate.

b. Pre-equilibrate the medium with the appropriate gas mixture for at least 1 hour at 37°C before embryos are added. Gently blow gas through a plugged Pasteur pipette onto the surface of the medium; do not bubble through it. The preferred gas mixture for different embryonic stages and the recommended number of embryos per 1 ml of medium are shown in Figure 5.15.

2. Add the embryos to the culture medium.

a. Re-gas the tubes (again by blowing gas gently onto the surface of the medium), and seal tightly.

To ensure that the tubes remain gas-tight, apply a thin film of high-vacuum silicon grease to the screw thread at the top of the tube.

b. Place the tubes on a roller culture apparatus. To maintain the medium at the bottom of the tube, place 1-inch wedges under the fan box to keep the roller apparatus situated at a slight angle to the horizontal.

3. Re-gas the tubes every 12 hours with a gas mixture appropriate for the age of the embryos. The purpose-made miniroller apparatus from BTC Engineering permits constant gas flow. Change the medium every 24 hours and reduce the number of embryos per 1 ml of medium as recommended in Figure 5.15. Discard embryos that are visibly retarded in size or developmentally abnormal.

Static Culture of Postimplantation Embryos

This procedure was provided by Patrick Tam, Children's Medical Research Institute, University of Sydney, Wentworthville, New South Wales, Australia.

MATERIALS

ANIMALS
Embryos (6.0–7.0 dpc)

EQUIPMENT
Centrifuge, MSE benchtop
Culture dish, 60-mm
Filter paper
Eppendorf tubes, 1.5 ml
Incubator, 37°C
Pasteur pipette, sterile
Stereomicroscope
Tissue culture chamber slides (Lab-Tek Chamber Slides supplied by Nunc): 2-well slides (4812, 177429), each well holds 1.5 ml medium for 6-8 embryos; 4-well slides (4814, 177437), each well holds 1.0 ml medium for 4–5 embryos; 8-well slides (4818, 177435), each well holds 0.5 ml medium for 1–3 embryos.

> The wells of chambered slides offer superior optical quality for imaging. The design of the well also reduces the area of the gas–medium interface, thus reducing the loss of medium by evaporation, and increases the depth of the medium for the same volume of medium placed in the culture dish. These factors may contribute to better development of embryos in the wells.

Tissue culture dishes (35 x 10-mm dish, e.g., Corning 25000, holds 2.0 ml of medium for 8–10 embryos)
Water bath, 56°C

REAGENTS
Culture medium
    a. DR75 medium: 3:1 v/v mixture of rat serum and Dulbecco's modified Eagle medium (DMEM)
    b. DRH medium: 2:1:1 v/v mixture of rat serum, human cord serum, and DMEM
    c. Rat serum only
Gas mixture <!>: 5% CO_2, 5% O_2, 90% N_2 or 5% CO_2 in air

CAUTION: See Appendix 2 for appropriate handling of materials marked with <!>.

PROCEDURE

Preparing the Culture Medium

The culture medium is made up of a mixture of serum and DMEM or 100% serum. The serum component of the medium is either rat serum or a mixture of rat and human cord serum. (See below for preparation of human serum.) The three culture medium formulations listed above can be used for static cultures. DR75 medium gives a more consistent culture result than pure rat serum, which may show variable embryotrophic activity between batches. DRH medium produces better results than DR75 medium in culturing 6.0- to 6.5-dpc embryos.

1. Prepare rat serum as described in Protocol 5.7.

2. Prepare the human cord serum.

   a. Fresh cord blood is collected by attending nurses in the maternity ward from placentae of caesarean deliveries. It is essential, for biohazard concerns, that the mother has been screened for potential infectious agents such as hepatitis and AIDS viruses.

   b. The blood is drawn into syringes from the umbilical vessels and transferred to a non-heparinized sterile plastic culture tube (15-ml tubes, 17 x 100-mm style) and kept on ice while being transported to the laboratory.

   c. Centrifuge the partially clotted blood at 3600 rpm in a MSE benchtop centrifuge using a swinging-bucket rotor.

   d. Aspirate the straw-colored serum (the supernatant) using a sterile Pasteur pipette and dispense in 1-ml aliquots to eppendorf tubes. Store frozen at –20°C until use.

3. Before adding the serum to the medium, thaw the serum and heat-inactivate it for 30–35 minutes at 56°C.

Setting up the Static Culture

1. Equilibrate the final mixture in the culture incubator under the appropriate gas mixture for 30–60 minutes before use.

2. Add the embryos to the culture dish, taking care not to carry over excessive HEPES-buffered medium.

3. Place the culture dishes or chamber slides in a 60-mm culture dish, which is humidified by moist filter papers placed in the dish. Keep the whole assembly in an incubator gassed with 5% CO_2 in air.

   Alternatively, the dishes and chamber slides can be kept inside a tightly sealed plastic container filled with a gas mixture of 5% CO_2, 5% O_2, and 90% N_2 or 5% CO_2 in air. No differences are discerned in embryos grown under these different gaseous conditions.

4. Transfer embryos to fresh medium after 24 hours of culture if the experiment is extended more than 24 hours.

Static Culture of Postimplantation Embryos for Imaging

This procedure was provided by Mary Dickinson, Elizabeth Jones, David A. Crotty, and Scott Fraser, Division of Biology, California Institute of Technology, Pasadena, California 91125.

MATERIALS

ANIMALS
Embryos (6.5–9.5 dpc)

EQUIPMENT
Bubbler

> The inlet of the gas consists of a tube with a fritted tip of coarse porosity. This is kept submerged under water, and the gas outlet is above the water level. The bubbler holds ~150 ml of water (Fig. 5.17).

Dental wax (Surgident Periphery Wax, Heraeus Kulzer)
Hood, 37°C
Incubator, 37°C, 5% CO_2
Inverted microscope, with time-lapse imaging capability, heater box

> Constructed around the microscope stage and optics with cardboard (4 mm thick) covered by a thermal insulation (Reflectix Co., Markleville, Indiana; 5/16″ thick, foil-foil insulation). Temperature is controlled using heaters (Lyon Electric Company 115-20), originally designed for chick incubators (see Fig. 5.17).

Lab-Tek Chambers (Nalge Nunc 155380)
Stereomicroscope
Syringe filter, 0.2 xm
Teflon tape

> The chambers are modified by soldering a small hole into the lid and inserting a tubing connector (Cole-Parmer 6365-44). This is used to attach the inlet gas tube to the chamber.

Transfer pipette

REAGENTS
Culture medium

> 1 ml of DMEM/F12 (GIBCO 11330032)
> 1 ml of heat-inactivated rat serum
> 10 µl of 100x Pen-Strep (Sigma P0781)
> 10 µl of 1 M HEPES buffer solution (pH 7.4) (Irvine Scientific 9319)

> Rat serum is prepared as described in Protocol 5.7 with the following exception. The blood is not collected with a syringe as described, but rather using a Vacutainer Safety-lok butterfly needle (Becton-Dickinson 367283) and Vacutainer blood collection tubes (Becton-Dickinson 366512).

Dissection medium
    45 ml of DMEM/F12 (GIBCO 11330032)
    4.5 ml of heat-inactivated fetal bovine serum (FBS) (GIBCO 16140063)
    0.5 ml 100x Pen-Strep solution (Sigma P 0781)
Ethanol, 70% <!>
Gas mixture: 5% CO_2<!>, balance air
Hair
Mineral oil (Sigma M8410)

CAUTION: See Appendix 2 for appropriate handling of materials marked with <!>.

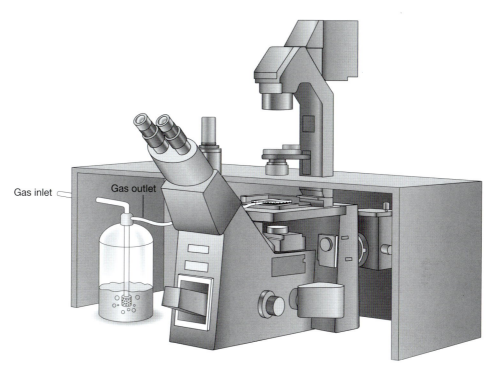

FIGURE 5.17. Static culture system for time-lapse imaging of postimplantation mouse embryos. The microscope and static culture are housed in a 37°C heater box. The gas mixture for the culture is humidified using a bubbler.

PROCEDURE

1. Prepare the dissection medium the night before culture by mixing the ingredients listed above. Prior to dissection, warm the dissecting medium to 37°C.

2. Prepare the medium the night before culture and store at 4°C. Filter using a 0.2-μm filter to sterilize. Prior to use in culture, allow the medium to equilibrate in a 37°C 5% CO_2 incubator for 1 hour.

3. Collect embryos at the appropriate stage in a 37°C hood following the procedure for isolating postimplantation embryos. The dissecting medium must be kept warm during dissection.

Yolk sacs can be left intact for 8.5-dpc embryos, but are removed for 9.5-dpc embryos.

4. After dissection, transfer three embryos to one well of a Lab-Tek chamber with a minimal amount of the dissection medium, using a transfer pipette.

5. It is necessary to immobilize 8.5-dpc embryos because they are very buoyant. This can be achieved by tying a hair (sterilized in 70% ethanol and then washed in culture medium) around the decidua and fixing the ends of the hair to the bottom of the chamber using a small amount of dental wax. Younger embryos are less buoyant and 9.5-dpc embryos are imaged without yolk sacs, thus they do not require immobilization.

6. Add 2 ml of culture medium to the culture chamber. For 8.5-dpc and younger embryos, cover the medium with as thin a layer of mineral oil as possible. Cultures of 9.5-dpc embryos should NOT be covered with mineral oil.

7. Place the embryos in a 37°C tissue culture incubator in an atmosphere containing 5% CO_2 for 1 hour, to equilibrate.

8. Seal the chamber by wrapping Teflon tape around the lid.

9. Set up the inlet gas to pass through the bubbler before being delivered to the chamber. Set the flow rate as low as the regulator can operate.

10. The microscope, bubbler, and heater box should be prewarmed to 37°C to prevent drifts in the optics. Place the culture chamber within the heater box, on the microscope stage (Fig. 5.17).

11. Image the embryos using bright-field or fluorescence illumination on an inverted microscope. During the imaging process the setup should be monitored for changes in temperature or focus, and to ensure that gas flow remains constant.

Electroporation

This protocol was provided by Noriko Osumi, Department of Developmental Neurobiology, Tokohu University Graduate School of Medicine, Sendai, Japan.

MATERIALS

BIOLOGICAL MOLECULES

Purified DNA solution of expression vectors

Expression vectors available for mammalian cells can also be used for embryo electroporation. The pCAX (CMV + beta-actin) and pEFX (EF-1 promoter) vectors are frequently used (Osumi and Inoue 2001; Takahashi et al. 2002). Plasmid DNA must be very pure and free of contaminants (e.g., use CsCl to isolate plasmid DNA). Dissolve the DNA in PBS on the day of the experiment or the day before (if it is highly concentrated), and adjust to its final concentration (up to 5 mg/ml; depending on your purpose). Fast Green (0.05%; Sigma F7258) can be added to facilitate visualization during the injection process. To monitor the tissues that have been electroporated with the gene construct, a GFP vector is often co-electroporated.

EQUIPMENT

Electrodes

Chamber-type (e.g., CUY 524) or forceps-type (CUY 650-P5) electrodes (Fig. 5.18) are available from Unique Medical Imada (Miyagi, Japan). The distance between the two electrodes is an important factor and is varied according to embryonic stages and experimental purposes (e.g., 20-mm distance for 9.5- to 10.5-dpc embryos). If an electroporation chamber is used, sterilize it with ethanol, rinse several times with Tyrode Ringer's solution, and fill with Tyrode Ringer's solution just before electroporation. After each electroporation, the chamber should be rinsed several times to remove degraded materials in the solution that may be toxic to embryos. Forceps-type electrodes are used in a 100-mm petri dish filled with Tyrode Ringer's solution, so that the solution does not need to be replaced after each electroporation.

Electroporator (Electro-Square Porator CUY21, NEPA Gene, Japan)

Injection needles

Pulled from glass capillaries (e.g., outer diameter = 1 mm, inner diameter = 0.58 mm, length = 100 mm: B100-58-10; Sutter Instrument, Novato, California) to make a fine open tip.

Petri dishes, 60 mm

REAGENTS

Tyrode Ringer's solution

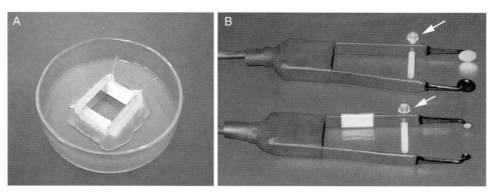

FIGURE 5.18. Chamber-type (*A*) and forceps-type (*B*) electrodes used for electroporation. The distance between the two electrodes of the forceps-type electrode is adjustable by a screw (*arrows*).

PROCEDURE

1. Preculture 9.5- to 11.5-dpc embryos as described above with the yolk sac opened (see Protocols 5.2 and 5.8) for 1.5–2 hours prior to electroporation.

2. Place the embryo in a petri dish with Tyrode Ringer's solution at room temperature.

3. Inject 0.25 µl of plasmid DNA solution into the target area of the embryo with a fine glass needle (Fig. 5.19A).

4. Adjust the position of the embryo so that the tissue to be electroporated is at a right angle against the positive electrode to which DNA moves, and apply square pulses using forceps-type electrodes (e.g., 50 msec, 5 pulses at 1-second intervals, 70V, five times).

5. Culture the electroporated embryo (see Protocol 5.8) for subsequent analysis (Fig. 5.19B,C).

COMMENTS

- Opening the yolk sac after the 30-somite stage enhances diffusion of oxygen directly into the embryo and/or via the yolk sac blood circulation, thereby compensating for the loss of function of the chorioallantoic placenta in vitro. In addition, opening the yolk sac will facilitate access to the target areas for injection. For detailed procedures for opening the yolk sac, see Osumi and Inoue (2001).

- Conditions for electroporation (voltage, duration, number of pulses, etc.) should be optimized empirically according to the specific purposes of the experiment.

- Inhibition experiments can be performed by using dominant-negative gene constructs.

- Phenotypes of mutant embryos can be rescued by overexpression of the wild-type gene.

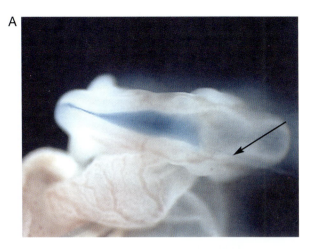

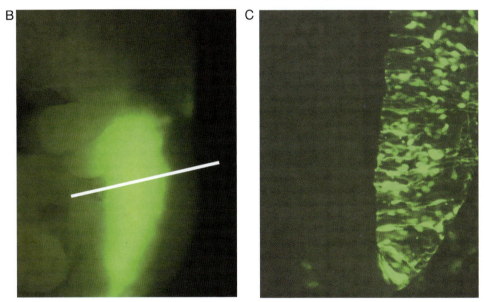

FIGURE 5.19. (*A*) Injection of a DNA solution (*blue*) into the cavity of the neural tube of the embryo using a fine glass needle (*arrow*). Electroporation is performed immediately after DNA injection. (*B*) 30 hours after electroporation. GFP (*green*) is expressed in the targeted hindbrain region of the cultured embryo. (*C*) Transverse section cut at the level of the line shown in panel *B*. GFP is efficiently expressed on the electroporated side of the neural tube.

REFERENCES

Ang S.L. and Rossant J. 1993. Anterior mesendoderm induces mouse Engrailed genes in explant cultures. *Development* **118:** 139–149.

Ang S.L., Conlon R.A., Jin O., and Rossant J. 1994. Positive and negative signals from mesoderm regulate the expression of mouse Otx2 in ectoderm explants. *Development* **120:** 2979–2989.

Beddington R.S. 1987. Isolation, culture, and manipulation of post-implantation mouse embryos. In *Mammalian development: A practical approach* (ed. M. Monk), pp. 43–69. IRL Press at Oxford University Press, England.

Beddington R.S. and Robertson E.J. 1989. An assessment of the developmental potential of embryonic stem cells in the midgestation mouse embryo. *Development* **105:** 733–737.

Cockroft D.L. 1990. Dissection and culture of postimplantation embryos. In *Postimplantation mammalian embryos: A practical approach* (ed. A.J. Copp and D.L. Cockroft), pp. 15–40. IRL Press at Oxford University Press, England.

Copp A.J. 1995. Death before birth: Clues from gene knockouts and mutations. *Trends Genet.* **11**: 87–93.

Cross J.C., Werb Z., and Fisher S.J. 1994. Implantation and the placenta: Key pieces of the development puzzle. *Science* **266**: 1508–1518.

De Felici M. and McLaren A. 1983. In vitro culture of mouse primordial germ cells. *Exp. Cell Res.* **144**: 417–427.

Downs K.M. and Davies T. 1993. Staging of gastrulating mouse embryos by morphological landmarks in the dissecting microscope. *Development* **118**: 1255–1266.

Dziadek M.A. and Andrews G.K. 1983. Tissue specificity of alpha-fetoprotein messenger RNA expression during mouse embryogenesis. *EMBO J.* **2**: 549–554.

Fjose A., Ellingsen S., Wargelius A., and Seo H.C. 2001. RNA interference: Mechanisms and applications. *Biotechnol. Annu. Rev.* **7**: 31–57.

Grabarek J.B., Plusa B., Glover D.M., Zernicka-Goetz M. 2002. Efficient delivery of dsRNA into zona-enclosed mouse oocytes and preimplantation embryos by electroporation. *Genesis* **32**: 269–276.

Hadjantonakis A.K., Gertsenstein M., Ikawa M., Okabe M., and Nagy A. 1998. Non-invasive sexing of preimplantation stage mammalian embryos. *Nat. Genet.* **19**: 220–222.

Kimura C., Yoshinaga K., Tian E., Suzuki M., Aizawa S., and Matsuo I. 2000. Visceral endoderm mediates forebrain development by suppressing posteriorizing signals. *Dev. Biol.* **225**: 304–321.

Kinder S.J., Tsang T.E., Wakamiya M., Sasaki H., Behringer R.R., Nagy A., and Tam P.P. 2001. The organizer of the mouse gastrula is composed of a dynamic population of progenitor cells for the axial mesoderm. *Development* **128**: 3623–3634.

Lane E.B., Hogan B.L., Kurkinen M., and Garrels J.I. 1983. Co-expression of vimentin and cytokeratins in parietal endoderm cells of early mouse embryo. *Nature* **303**: 701–704.

McLaren A. 2001. Mammalian germ cells: Birth, sex, and immortality. *Cell Struct. Funct.* **26**: 119–122.

Muramatsu T., Mizutani Y., Ohmori Y., and Okumura J. 1997. Comparison of three non-viral transfection methods for foreign gene expression in early chicken embryos in ovo. *Biochem. Biophys. Res. Commun.* **230**: 376–380.

Nakamura H. and Funahashi J. 2001. Introduction of DNA into chick embryos by in ovo electroporation. *Methods* **24**: 43–48.

Osumi N. and Inoue T. 2001. Gene transfer into cultured mammalian embryos by electroporation. *Methods* **24**: 35–42.

Paria B.C., Song H., and Dey S.K. 2001. Implantation: Molecular basis of embryo-uterine dialogue. *Int. J. Dev. Biol.* **45**: 597–605.

Paria B.C., Lim H., Das S.K., Reese J., and Dey S.K. 2000. Molecular signaling in uterine receptivity for implantation. *Semin. Cell. Dev. Biol.* **11**: 67–76.

Rossant J. and Croy B.A. 1985. Genetic identification of tissue of origin of cellular populations within the mouse placenta. *J. Embryol. Exp. Morphol.* **86**: 177–189.

Shawlot W., Deng J.M., and Behringer R.R. 1998. Expression of the mouse cerberus-related gene, Cerr1, suggests a role in anterior neural induction and somitogenesis. *Proc. Natl. Acad. Sci.* **95**: 6198–6203.

Shawlot W., Wakamiya M., Kwan K.M., Kania A., Jessell T.M., and Behringer R.R. 1999. Lim1 is required in both primitive streak-derived tissues and visceral endoderm for head formation in the mouse. *Development* **126**: 4925–4932.

Snow M.H.L. 1978. Techniques for separating early embryonic tissues. In *Methods in mammalian reproduction* (ed. J. C. Daniel), pp. 167–178. Academic Press, New York.

Sturm K. and Tam P.P.L. 1993. Isolation and culture of whole postimplantation embryos and germ layer derivatives. *Methods Enzymol.* **225**: 164–190.

Tian E., Kimura C., Takeda N., Aizawa S., and Matsuo I. 2002. Otx2 is required to respond to signals from anterior neural ridge for forebrain specification. *Dev. Biol.* **242**: 204–223.

Varlet I., Collignon J., and Robertson E.J. 1997. Nodal expression in the primitive endoderm is required for specification of the anterior axis during mouse gastrulation. *Development* **124**: 1033–1044.

CHAPTER 6

Surgical Procedures

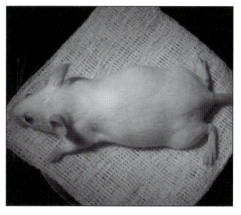

BASIC SURGICAL PROCEDURES described in this chapter are required for a variety of experiments used in production and analysis of transgenic and chimeric mice. These procedures include transfer of preimplantation-stage mouse embryos into the oviduct and uterus of pseudopregnant recipients, vasectomy and ovariectomy, Caesarean section and fostering, tissue transplantation, and various biopsy techniques. This chapter contains general brief guidelines for mouse surgeries followed by protocols. The exact details of anesthesia, antiseptic procedures, and post-surgical care are omitted because they depend on local rules and regulation and vary greatly among different facilities.

CONTENTS

GENERAL GUIDELINES

Essential components of successful surgery include aseptic surgical technique, use of the appropriate surgical instruments, administration of appropriate anesthesia, and postoperative care. All surgical protocols should be approved by the local animal care committee in accordance with national and local regulations. Mouse surgery can be performed in a procedure room contiguous with the animal room or in a designated area of the research lab that is clean and well lit. Countertops and microscopes used during the surgery should be disinfected (e.g., with quaternary ammonium or 70% alcohol). Additional procedures (e.g., use of laminar flow hood) may be necessary when the surgery is performed in a specific-pathogen-free (SPF) facility (see Chapter 3).

Surgical-grade stainless steel instruments can be purchased from companies such as Roboz Surgical Instrument (http://www.roboz.com), Fine Science Tools (http://www.finescience.com), and World Precision Instruments (http://www.wpiinc.com). Instruments are sterilized before the surgery (e.g., using steam or dry heat). Between procedures on different animals, the tips of the instruments can be disinfected in 70% alcohol or sterilized in a glass bead sterilizer (available from Inotech Biosystems International, http://www.inotechintl.com). It is important to handle surgical instruments carefully. After each use they should be cleaned, air-dried, and inspected. Delicate instruments with sharp points (e.g., Dumont forceps, fine scissors) should be protected by covering them with protective caps (e.g., disposable pipette tips or flexible rubber tubing). Specifically, the tips of fine forceps should be finely ground to a single point using an oilstone or fine sandpaper; otherwise, it will be difficult to grasp tissues with them.

A Ketamine/Xylazine mixture and Avertin (tribromoethanol) are two different types of anesthetics that are most commonly used for mouse surgeries. Although respiration and cardiovascular function are decreased during Ketamine/Xylazine anesthesia, this combination is considered to be very reliable for mouse surgery (Erhardt et al. 1984). The anticholinergic preanesthetic atropine can be used to reduce bronchial and salivary secretions (caused by Ketamine) and bradycardia (caused by Xylazine) (Nowrouzian et al. 1981; Magoon et al. 1988).

Avertin is a short-acting anesthetic that has rapid induction and recovery. Its use has been questioned due to apparent abdominal adhesions and acute peritoneal inflammation (Zeller et al. 1998). However, in a rigorous study, Papaioannou and Fox (1993) confirmed the efficacy, safety, and suitability of Avertin for mouse surgeries when minimal precautions (storage in the dark at 4°C) to prevent its decomposition were used. The correct dose of Avertin may vary with preparations and should be re-determined each time a new stock is prepared (see Appendix 1 for more details).

Inhalation anesthetics (halothane or isofluorane) in mouse surgery are often considered superior to injectable anesthetics due to their safety, efficacy, and short recovery time. However, they require the use of special equipment, supply of oxygen, and gas scavenging system, and, generally, these requirements limit the position of the mouse during surgery. In addition to general anesthesia, local anesthetics such as bupivacaine, injected subcutaneously, may provide prolonged intraoperative anesthesia and extended postoperative analgesia (Grant et al. 1994).

After the mouse is anesthetized, the fur may be removed from the incision site (5–10 mm wide) using electric clippers, scalpel blades, or scissors if this is required by local regulations; alternatively, carefully brush the fur away from the exposed skin, which is disinfected with 70% alcohol and/or surgical iodine. It is advisable to moisten the mouse's eyes with sterile phosphate-buffered saline (PBS) or ophthalmic ointment to prevent corneal drying during anesthesia.

To access organs within the body cavity, make an initial incision through the skin with heavy scissors to expose the body wall muscle. Loosen the skin from the muscle by tearing or cutting the intervening connective tissue. Then make an incision in the body wall muscle with fine scissors to gain access to the internal organs. The incision should be located very near the organ that is to be manipulated. As an example, different approaches can be used to access both sides of the female reproductive tract. Two dorsal incisions less than 1 cm long can be made about 1 cm to the left and to the right of the vertebral column below the level of the last rib. Alternatively, a single 1-cm incision made along the dorsal midline or perpendicular to the vertebrate column allows one to reach both sides of the tract by moving the skin to either side of the mouse.

Surgical drapes are often used when organs are exteriorized from and returned to the body cavity. Abdominal muscles are stitched using absorbable sutures. A variety of stitches can be used. Typically, double knots are used for a single stitch. Skin incisions may be closed with sutures, tissue adhesives, or wound clips. If wound clips are used, they must be placed very securely using a wound clip applier; otherwise, the animal might pull them out. The wound clips may be removed using a wound clip remover, but this should not be done earlier than 1 week after surgery (optional).

Postsurgical animals are transferred to a temperature-controlled warming plate or blanket or clean warm cage under a 50W light bulb (taking care to cover their eyes) and returned to the animal room when they are able to walk around the cage. Consult your animal care committee for specific postoperative care requirements in your institution.

These brief guidelines can be applied to all mouse surgical procedures described below; details regarding the exact antiseptic procedures and postsurgical care are often omitted from the protocol because they are largely based on local rules and regulations.

VASECTOMY AND OVARIECTOMY

Vasectomized males are bred with females to produce pseudopregnant recipients for oviduct and uterine transfers. Females in estrus are selected as described in Chapter 3. After mating with a sterile male, the female's reproductive tract becomes receptive for transferred embryos, even though her own unfertilized oocytes degenerate. A newly plugged pseudopregnant female will not resume her natural cycle for ~11 days. Vasectomized males can also be purchased from various animal suppliers (see Appendix 3). The vasectomy procedure is described in Protocol 6.1.

Protocol 6.2 describes the surgical removal of the ovaries, in combination with the administration of progesterone (Depo-Provera), which causes the arrest of blastocyst development. The viability of the embryos is maintained for up to 10 days following ovariectomy. This technique has been described in detail by Bergstrom (1978). Delayed blastocysts can be used to generate ES cell lines (Robertson 1987).

EMBRYO TRANSFER

Embryos from the one-cell to the blastocyst stage (0.5–3.5 dpc) can be transferred into the reproductive tract of a pseudopregnant recipient to complete their development. Embryos from the one-cell stage to the blastocyst stage (0.5–3.5 dpc) can be transferred into the ampullae of 0.5-dpc pseudopregnant recipients in an oviduct transfer (Protocol 6.3), whereas 3.5-dpc blastocysts are transferred into the uterine horns of 2.5-dpc (or 3.5-dpc) pseudopregnant recipients in a uterine transfer (Protocol 6.4). A 2.5-dpc pseudopregnant female is more optimal for uterine transfer than 3.5 dpc, because the manipulated embryo has sufficient time to catch up developmentally.

The oviduct transfer procedure is generally thought to be most efficient when performed with embryos enclosed within a zona pellucida, but this relates only to one- and two-cell-stage embryos (Nichols and Gardner 1989). This study suggested that zona-free embryos should be cultured to the morula stage prior to oviduct transfer. It is possible to obtain equal efficiency when transferring zona-free blastocysts into the oviduct or into the uterus (K. Vintersten, pers. comm.). Thus, in case there is a shortage of 2.5-dpc pseudopregnant recipients, 3.5-dpc blastocysts can be transferred into the oviduct of a 0.5-dpc pseudopregnant female. Alternatively, 3.5-dpc blastocysts can be cultured an additional night to 4.5 dpc and transferred into the uterus of a 2.5-dpc or oviduct of 0.5-dpc pseudopregnant female.

Oviduct Transfer

Oviduct transfer of mouse embryos was first reported by Tarkowski (1959). Whittingham (1968) applied a well-described method for the rat (Noyes and Dickman 1961) to the mouse, and it is the basis of the procedure described in Protocol 6.3.

More than 75% of unmanipulated zygotes or two-cell-stage embryos should develop to normal fetuses when transferred into the oviducts of 0.5-dpc recipients. In the case of zygotes injected with DNA, the rate of successful development will vary greatly between constructs; ~20–30% of injected zygotes should give rise to pups. With such a yield, the number of injected zygotes to be transferred should be about 20–25.

Uterine Transfer

The method described for uterine transfer in Protocol 6.4 is based on extensive work by McLaren and Mitchie (1956), which resulted in the first successful development and birth of in vitro-cultured mouse embryos (McLaren and Biggers 1958). A detailed description of similar methods is given by Mintz (1967) and Rafferty (1970). A success rate of at least 75% fetuses or pups should be expected for zona-enclosed unmanipulated blastocysts (C57BL/6).

Uterine transfer is most often performed for chimera production. The yield of ES cell-derived chimeras is highly dependent on the particular ES cell line and clone. Some ES cell clones have low developmental potential and produce mice with low chimerism. Other ES cell clones may cause decreased embryo viability. Typically, 40–60% of transferred manipulated embryos should develop to term. Usually between 10 and 15 embryos are transferred into 1 foster mother, and not more than 8 embryos per uterine horn if unilateral transfers are done. The success rate for zona-free CD1/ICR host blastocysts is somewhat lower, and 16–20 embryos per recipient are usually transferred. In case of recipient shortage, up to 24 manipulated zona-free embryos may be transferred per recipient, split between both uterine horns.

Recipient Females

Regardless of the stage of the transferred embryos, the pups are delivered according to the stage of the foster mother. The reason for transferring embryos into a female mouse at an earlier age of pseudopregnancy is to give the embryo time to "catch up" in its development before being exposed to conditions favorable for implantation. In general, for both procedures, a high percentage (>75%) of unmanipulated embryos should develop to term. The outcome is highly variable for manipulated embryos and largely depends on injected DNA or introduced ES cells. Ideally, enough embryos should be transferred to give a litter size of 5–7 pups. If there are only 1 or 2 embryos in the uterus, they may grow too big to be born without being damaged. Additionally, some mothers may not take care of small litters. If the litters are too large (more than 10), then a few of the pups may grow up small, with a risk of being sterile. Provided 2 foster female mice have been given embryos of exactly the same genotype and/or composition, they can be placed in the same cage after oviduct or uterine embryo transfer. The foster mothers will subsequently help each other raise the joint litter.

Female recipient mice should be at least 6 weeks of age and weigh 25–35 grams. F_1 hybrids (e.g., C57BL/6 x CBA or C57BL/6 x DBA/2) or outbred females, such as Swiss Webster, ICR, CD1, or MF1 mice, make good foster moth-

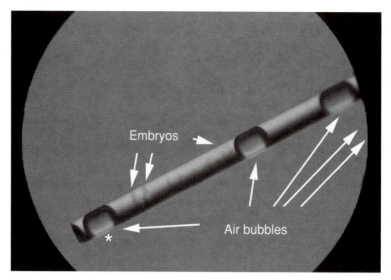

FIGURE 6.1. Embryo transfer pipette showing arrangement of air bubbles, medium, and embryos. Monitor the position of the air bubbles during the transfer to ensure that all of the embryos have been injected. The air bubble marked by the asterisk is only used for oviduct transfer.

ers. It is advisable to have excess pseudopregnant females available. Females are mated to vasectomized males as described in Chapter 3, and the day of the vaginal plug is considered to be 0.5 day of pseudopregnancy. Use 0.5-day pseudopregnant females for oviduct transfer. Use 2.5- to 3.5-day pseudopregnant females for uterine transfer.

Technical Aspects of Embryo Transfer

Transfer pipettes (pulled hard glass capillary or Pasteur pipette) are made as described in Chapter 4 (Fig. 4.1A, B) and connected to an aspirator mouthpiece with tubing. If a pulled Pasteur pipette is used, the narrow part of it should be 2–3 cm in length. The external diameter of the transfer pipette should be around 200 µm; i.e., with an internal diameter just larger than 1 embryo and smaller than 2 (Fig. 6.1). The pipette should have a clean, perpendicular break. In addition, it is advisable to have the tip of the pipette flame-polished or polished on the microforge in the same way as a holding pipette (see Chapter 7) to minimize damage to the oviduct and uterus. Sharp edges also tend to collect debris that might block the pipette opening. A bend can be made about 1 cm from the end of the pipette used for uterine transfer to facilitate judging how far the pipette has been inserted into the uterus. Prepare several transfer pipettes before starting the surgery.

It is important to reduce the capillary action in the pipette in order to have very good control moving minimal volumes of liquid. One way to do this is to fill the pipette with clean medium just past the shoulder of the pipette prior to picking embryos to reduce capillary action. Another simple way to inhibit capillary action and achieve good control is to load the pipette with a few air bubbles by touching a drop of medium with the tip of the pipette and quickly withdrawing it (Fig. 6.1). Air bubbles in the pipette also help show movement of the medium during embryo transfer. An alternative way to see movements is to use blue-colored Affigel beads (BioRad 153-7302). The traditional method of filling the transfer pipette with light paraffin oil to just past the shoulder of the pipette also reduces capillary action. The viscosity of the oil allows greater control when picking up or expelling the embryos. The use of oil with zona-free embryos is not recommended because the embryos might stick to it. As an alternative to using oil, the other end of the pipette can be melted down to a narrow opening as is done in a braking pipette used for implantation under the kidney capsule (see Protocol 6.6, p. 274).

In any case, load the embryos into the pipette near the tip and close to each other in a minimal amount of liquid at a short distance (around 5–7 mm) from the most distal air bubble in the pipette; place an additional air bubble at the tip of the pipette if oviduct transfer is being done (asterisk in Fig. 6.1). Two or three air bubbles placed into the oviduct indicate successful embryo transfer; optimally, no air bubbles or just one air bubble should be transferred into the uterus because they could potentially interfere with implantation. To reduce the stickiness of the pipette, the capillary pipette used for embryo transfer may be soaked in 1% bovine serum albumin (BSA) prior to loading embryos. This procedure is especially useful when handling embryos lacking a zona pellucida.

The number of embryos to be transferred at a time and placed in M2 or other HEPES-buffered medium should be prepared for no more than one or two recipients anesthetized at that time.

CAESAREAN SECTION AND FOSTERING

Caesarean section is necessary if the recipient of an embryo transfer or any pregnant mouse has not given birth by the delivery time normal for that particular strain. This often occurs when only one or two embryos are present and they have grown too big to be born naturally (see Recipient Females section, p. 255). In addition, Caesarean section is useful for collecting pups at 18.5 dpc if a transgenic or mutant mouse dies at birth to ensure that all pups are recovered for analysis. This technique is described in Protocol 6.5. Successfully fostering pups obtained by Caesarean section is a skill that requires practice but can be achieved in a variety of ways, as described in Protocol 6.5. See also Chapter 3.

TISSUE TRANSPLANTATION

Protocols 6.6 and 6.7 are used to determine in vivo the developmental potential of adult or embryonic tissues, including early-stage embryos, somites, and tail bud, as well as ES cells. The resulting teratomas and teratocarcinomas are screened by histology and markers to assess the differentiation of the resulting tissues.

TISSUE BIOPSY

The technique in Protocol 6.8 describes the surgical procedure for obtaining liver tissue for analysis. Similar procedures for splenectomy, nephrectomy, and castration are given in Protocols 6.9, 6.10, and 6.11.

Blood collection for analysis can be performed by tail bleeding, as described in Protocol 6.12.

Vasectomy for Generation of Sterile Males

Two methods are given. In the first, the vas deferens is accessed through the abdominal wall. The second, less invasive method accesses the vas deferens through the scrotal sac. In both methods mice are ready for mating after ~10–14 days. For novices, it is recommended that sterility of the vasectomized males be tested. Breed vasectomized males with fertile females to obtain plugs for timed matings. The plugged females should not become pregnant.

MATERIALS

ANIMALS
Male mice (at least 2 months of age from any strain with good breeding performance)

EQUIPMENT
Alcohol or Bunsen burner (*optional*)
Animal clippers (*optional*)
Clean cage
Fiber optic illuminator (*optional*)
Forceps, fine blunt
Forceps, watchmaker's #5, two pairs
    (one pair designated for the flame if cauterization is used)
Hypodermic needle, 26-gauge, 1/2-inch
Scissors, fine dissection
Surgical needle, curved (e.g., size 10, triangular, pointed)
Suture, surgical silk (size 5-0)
Syringe, 1-ml
Tissues
Warming plate, heating pad, or 50W light bulb
Weight scale
Wound clips and applier (*optional*)

REAGENTS
Anesthetic (see Appendix 1)
Ethanol, 70% <!>

CAUTION: See Appendix 2 for appropriate handling of materials marked with <!>.

PROCEDURE

Method 1

1. Weigh and anesthetize the mouse by intraperitoneal (IP) injection (see Chapter 3, p.158).

2. Place the mouse on its back to expose the abdomen. Shave the abdomen of the mouse, if required by local rules and regulations. Use a squirt bottle to apply 70% ethanol to the abdomen and wipe with tissues.

3. To access the body cavity, cut the skin with fine dissection scissors, making a 1.5-cm transverse incision at a point level with the top of the legs. Make a similar-sized transverse incision in the body wall; put one stitch through the body wall on one side of the incision with a curved surgical needle and leave a piece of silk suture in place (this helps to find the body wall later). Both testes can be reached through the single incision (Fig. 6.2A).

4. Use fine blunt forceps to gently grasp and pull out one of the testicular fat pads (Fig. 6.2A). The associated testis, vas deferens, and epididymis will come with the fat pad. The vas deferens lies underneath the testis and can be recognized by a blood vessel running along one side. Hold the vas deferens with one pair of watchmaker's forceps (Fig. 6.2B) and cut it with fine scissors (Fig. 6.2C) or cauterize with the red-hot tips of a second pair of watchmaker's forceps such that a portion (~1 cm) of the vas deferens is removed (Fig. 6.2D).

5. Pick up the fat pad with blunt forceps and carefully place the testis back inside the body cavity.

6. Repeat steps 4 and 5 on the other testis. It is advisable to keep the removed portions of each vas deferens on a paper towel next to the mouse to make sure that both vasa deferentia of each mouse were removed.

7. Sew up the body wall with two or three stitches. Then sew up the skin or clip the skin together with wound clips.

8. At the end of the procedure, place the mouse in a clean cage and keep it warm under a 50W light bulb (taking care to cover the eyes) or by placing the cage on a warming plate or heating pad until the mouse recovers from the injected anesthetic. Follow local animal care committee regulations for more details.

Method 2

1. Weigh and anesthetize the mouse by IP injection (see Chapter 3, p.158).
2. Push both testes down into the scrotal sacs by gently applying pressure to the abdomen.
3. Make a 10-mm incision through the skin along the midline of the scrotal sac (see Fig. 6.2E).
4. Locate the midline wall between the testes sacs under the covering membranes of the scrotal sacs. It will appear as a light whitish line.
5. Make a 5-mm incision in the testes membrane close to the left side of the midline wall.
6. Carefully push the testis to the left. The vas deferens will appear between the testis and the midline wall as a bright white tubule with a single blood vessel.
7. Using forceps, pull the vas deferens out while holding the testis in place. Hold the vas deferens with one pair of forceps and cauterize it with the red-hot tips of a second pair of forceps or cut with fine scissors such that the portion (~1 cm) of the vas deferens in the loop is removed (Fig. 6.2B–D).

8. Repeat steps 5–7 on the other testis. It is advisable to keep the removed portions of each vas deferens on a paper towel next to the mouse to make sure that both sides were removed.

9. Sew up or clip the skin together with wound clips.

10. At the end of the procedure, place the mouse in a clean cage and keep it warm under a 50W light bulb (taking care to cover the eyes) or by placing the cage on a warming plate or heating pad until the mouse recovers from the injected anesthetic. Follow local animal care committee regulations for more details.

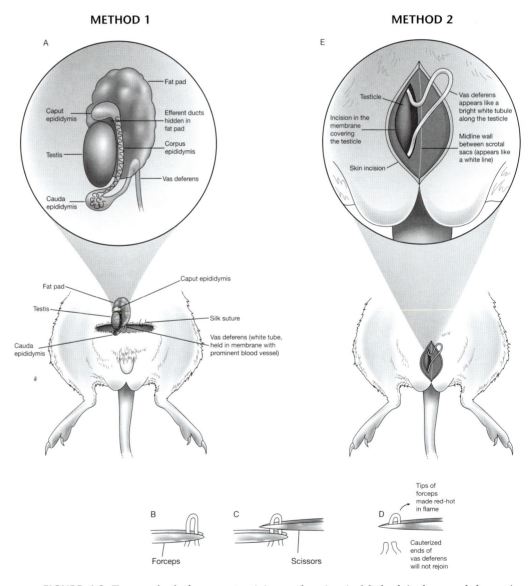

FIGURE 6.2. Two methods for vasectomizing male mice. in Method 1, the vas deferens is accessed through an abdominal incision (*A*) and cut (*C*) or cauterized (*D*). In Method 2, both vasa deferentia are accessed through a small incision in the scrotal sac without retracting the testis and cauterized or cut the same way as in Method 1 (*B, C, D*).

Ovariectomy for Induction of Blastocyst Implantation Delay

This procedure is performed during the afternoon of the third day of pregnancy (see Comments) to ensure that the morulae have moved sufficiently into the oviducts and will be less likely to be damaged during the ovariectomy.

MATERIALS

ANIMALS
Pregnant female mouse (2.5 dpc)

EQUIPMENT
Animal clippers (*optional*)
Clean cage
Dissecting microscope, low-power binocular (*optional*)
Fiber optic illuminator (*optional*)
Forceps, blunt fine
Forceps, watchmaker's #5, two pairs
Hypodermic needle, 26-gauge, 1/2 inch
Needle, curved surgical (e.g., size 10, triangular, pointed)
Scissors, fine dissection
Spring scissors (*optional*)
Suture, surgical silk (size 5-0)
Syringe, 1-ml
Tissue
Warming plate, heating pad, or 50W light bulb
Weight scale
Wound clips and applier (*optional*)

REAGENTS
Anesthetic (see Appendix 1)
Depo-Provera (6α-methyl-17α-hydroxy-progesterone acetate) solution (10 mg/ml in sterile saline) (Sigma M1629)
Ethanol, 70% <!>

CAUTION: See Appendix 2 for appropriate handling of materials marked with <!>.

PROCEDURE

1. Weigh and anesthetize the mouse by IP injection (see Chapter 3, p. 158).

2. Place the mouse on its abdomen for a dorsal incision, or on its side for a lateral incision. Use a squirt bottle to apply 70% ethanol to the back or side and wipe with tissues. Remove fur if local regulations require.

3. Use dissection scissors to make a small incision through the skin of the back of the mouse. Make a small incision over the left ovarian fat pad, grasp the

fat pad with the fine blunt forceps, and pull the ovary and oviduct out of the mouse.

4. Use two pairs of watchmaker's forceps to tear the bursal membrane surrounding the left ovary and oviducts. Alternatively, use fine spring scissors to cut the bursal membrane and expose the ovary. Tie the intact ovary and ovarian blood vessel by slipping a loop of surgical silk suture between the oviduct and the ovary. Use fine scissors to excise the ovary distal to the suture.

5. Using fine blunt forceps, replace the uterine horn into the body cavity. Close the body wall incision with a single stitch (*optional*).

6. Repeat steps 4 and 5 on the right ovary.

7. Close the incision in the skin with sutures or wound clips.

8. While the female is still anesthetized, use a 1-ml syringe with a 26-gauge needle to inject subcutaneously 1 mg of Depo-Provera in 0.1 ml of sterile saline into its flank.

9. At the end of the procedure, place the mouse in a clean cage and keep it warm under a 50W light bulb (taking care to cover the eyes) or by placing the cage on a warming plate or heating pad until the mouse recovers from an injected anesthetic. Follow local animal care committee regulations for more details.

COMMENTS

Schedule for collection of delayed blastocysts:
1. Day 0: Cage female with fertile male.
2. Day 0.5: Check plug.
3. Day 2.5: Perform ovariectomy and administer 1 mg of Depo-Provera.
4. Day 6.5–10.5: Recover delayed blastocysts by flushing uterine horns as described in Protocol 4.11.

Oviduct Transfer

The Whittingham (1968) method, which applied a well-described method for the rat (Noyes and Dickman 1961) to the mouse, is the basis of the procedure described below. It is best to practice this procedure first on a cadaver and then on an anesthetized 0.5-dpc pseudopregnant mouse using a dye solution or blue Affigel beads rather than embryos to gain experience in finding the opening of the oviduct (infundibulum). The position of the infundibulum is relatively invariant from mouse to mouse, and, with a little practice, the procedure will become routine.

MATERIALS

ANIMALS
Mouse embryos (0.5–3.5 dpc)
Pseudopregnant female recipient mice at 0.5 dpc

EQUIPMENT
Fiber optic illuminator (*very useful*)
Forceps, blunt fine with serrated tips
Forceps, sharp
Forceps, watchmaker's #5, two pairs
Hypodermic needle, 26-gauge 1/2-inch
Lid of 9-cm plastic petri dish, glass plate, or paper towel
Light bulb, 50W
Needle, curved surgical (e.g., size 10, triangular, pointed) (*optional*)
Plasticine (*optional*)
Scissors, fine dissection
Serrefine clamp or baby Dieffenbach clip (1.5 inch or smaller) (e.g., Roboz Surgical Instrument RS7440; Weiss B950B, or Fine Science Tools 18050-35)
Stereomicroscopes (ideally one for the surgery and one for loading the embryo transfer pipette) with transmitted and reflected light
Suture, surgical silk (size 5-0) (*optional*)
Syringe, 1-ml
Tissues (several rolled into small swabs are useful for soaking up any blood)
Transfer pipettes and mouth pipette assembly (see Chapter 4 and Fig. 4.1A,B)
Warming plate or heating pad
Weight scale
Wound clips and applier

REAGENTS
Anesthetic (see Appendix 1)
Beads, blue Affigel (BioRad 153-7302)
Epinephrine (*optional*)
Ethanol, 70% <!>
M2 medium or other HEPES-buffered medium at room temperature

CAUTION: See Appendix 2 for appropriate handling of materials marked with <!>.

PROCEDURE

1. Weigh and anesthetize the recipient mouse by IP injection (see Chapter 3).

2. Place the mouse on the lid of a 9-cm petri dish, a paper towel, or other supporting material so that it can be easily lifted onto the microscope stage.

3. Load a transfer pipette with embryos. Because they will be outside the incubator for several minutes, transfer embryos into M2 or other HEPES-buffered medium before loading the transfer pipette.

    a. Take up a small amount of M2 medium into the transfer pipette, then a small air bubble, then M2 medium, and then a second air bubble; repeat until good control is reached with reduction of capillary action (see Fig. 6.1).

    b. Draw up the embryos in a minimal (about 5–7 mm) volume of M2 medium (Fig. 6.1). Place an additional small air bubble at the tip of the pipette.

    c. Store the transfer pipette (still in the mouth-pipetting device) by pressing the side of pipette into a piece of Plasticine stuck to the base of a stereomicroscope and leave it there until ready to place the embryos in the oviduct. It is also possible to rest the pipette on a tube rack or any other support next to the microscope, making sure that the tip of the pipette does not touch anything.
    BE CAREFUL NOT TO DISTURB THE PIPETTE.

4. Expose the recipient's reproductive tract.

    a. Wipe the back of the mouse with tissues soaked in 70% ethanol. Remove fur if local legislation requires. Make a small incision in the skin with fine dissection scissors, along the dorsal midline, at the level below the last rib (Fig. 6.3A,B). Wipe the incision with 70% ethanol-soaked tissue to remove any loose hairs. As discussed above, both oviducts may also be approached from a single incision made at the dorsal midline perpendicular to the vertebral column at the level below the last rib, or from transverse incisions about 1 cm to the sides of the spinal cord.

    b. Slide the skin around until the incision is over the ovary (orange) or fat pad (white), both of which are visible through the body wall (Fig. 6.3C). Then pick up the body wall with watchmaker's forceps and make a small incision just over the ovary with fine dissection scissors. It is also possible to pierce the body wall using sharp forceps. Stretch the incision with the scissors (or forceps) to stop any bleeding (Fig. 6.3D). With a curved surgical needle, thread a piece of surgical silk suture through the body wall so that it will be easy to locate later (*optional*).

5. Transfer the embryos.

    a. Use blunt fine forceps to pick up the ovarian fat pad and pull out the attached left ovary, oviduct, and uterus (Fig. 6.3E). Clip a Serrefine clamp onto the fat pad and lay it down away from you over the middle of the back, so that the oviduct and ovary remain outside the body cavity (Fig. 6.3F).

b. Gently pick up the petri dish with the mouse and place it on the stage of a stereomicroscope with the mouse's head to the left.

c. Using the stereomicroscope, find the opening to the oviduct (infundibulum) and the swollen ampulla located underneath the bursa (a thin transparent membrane containing blood vessels that surrounds the oviduct and ovary). Arrange the mouse, oviduct, etc., so that the pipette can enter easily. For right-handed people, it is most convenient to have the head to the left and the ovary pulled away from you, toward the right side of the mouse and held outside the body cavity with a Serrefine clamp. For left-handed people, reverse the position of the mouse. Place a drop of epinephrine on the bursa to reduce subsequent bleeding, which can obscure the opening of the oviduct and potentially clog the transfer pipette (*optional*). With two pairs of sharp watchmaker's forceps, tear a hole in the bursa over the infundibulum. Alternatively, cut the bursa with fine spring scissors. Be careful not to break any large blood vessels. Locate the infundibulum (Fig. 6.3G).

d. Use blunt fine forceps to cradle the infundibulum gently and then insert the prepared transfer pipette into the oviduct opening (Fig. 6.3H). Blow on the transfer pipette until air bubbles or blue beads (if practicing) have entered the ampulla. Air bubbles or blue beads visible inside the oviduct indicate successful transfer.

e. Unclip the Serrefine clamp and remove the mouse from the stereomicroscope. Use blunt fine forceps to pick up the fat pad and place the uterus, oviduct, and ovary back inside the body cavity. Sew up the body wall with one or two stitches (*optional*) and close the skin with wound clips.

6. Repeat steps 3, 4, and 5 to transfer additional embryos to the right oviduct, if desired.

7. At the end of the procedure, place the mouse in a clean cage and keep it warm under a 50W light bulb (taking care to cover the eyes) or by placing the cage on a warming plate or heating pad until the mouse recovers from an injected anesthetic. Follow local animal care committee regulations for more details.

COMMENTS

- It is advisable to check whether ovulation has occurred as expected right after exposing ovaries and before transferring the embryos. Bloody fluid and small clots on ovaries should be readily visible even without a microscope to confirm recent ovulation. Ruptured graafian follicles may be seen under the microscope. Swollen ampulla and naturally ovulated unfertilized oocytes of the recipient surrounded by a cumulus mass should also be visible. If signs of ovulation are questionable, a different recipient should be used if available.

- An alternative procedure for embryo transfer into ampullae through a tear in the oviduct wall is described by Nakagata (1992). The oviduct wall may be punctured through the bursa using a 30-gauge needle (or acupuncture needle), eliminating the need to open the bursa to access the infundibulum. The fine blood vessels of the oviduct can serve as landmarks for the puncture, which may be difficult to see.

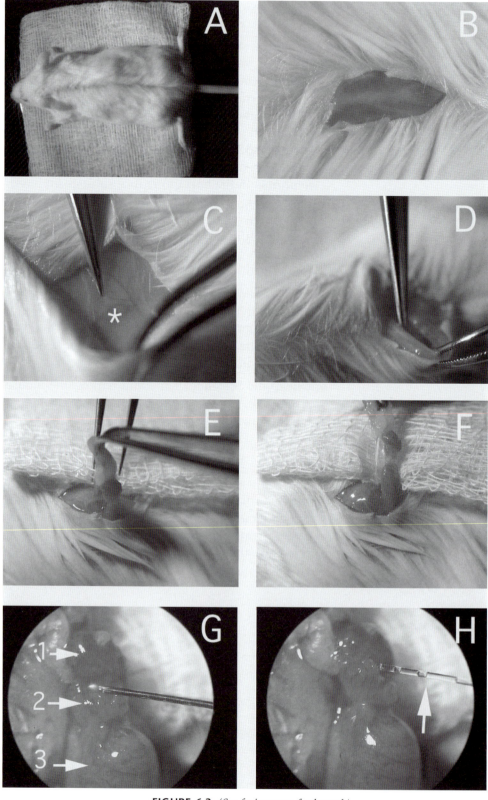

FIGURE 6.3. *(See facing page for legend.)*

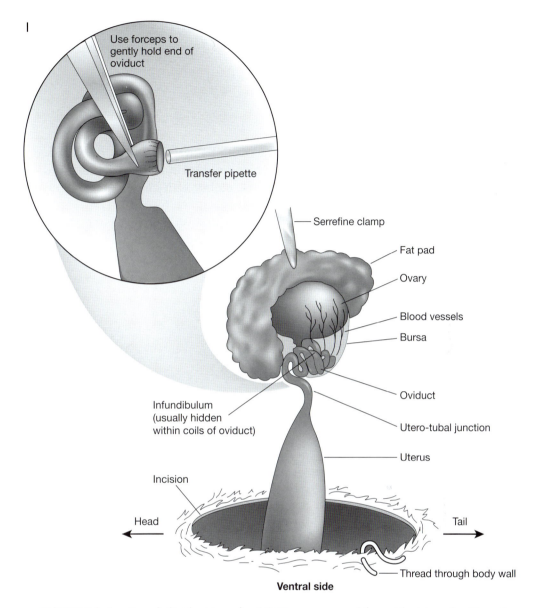

FIGURE 6.3. (*continued*) Oviduct transfer. (*A*) Mouse prepared for oviduct transfer. Place the anesthetized mouse on a 9-cm petri dish lid for ease of handling. The orientation shown here gives easy access to the oviduct for transfer of embryos by the technique described. (*B*) Make a small incision along the dorsal midline. (*C*) The ovary (*orange*) or fat pad (*white*) is visible through the body wall (marked with the star). (*D*) Cut the body wall over the ovary with fine dissection scissors or pierce using sharp watchmaker's forceps and stretch with the scissors (or forceps) to stop the bleeding. (*E,F*) The ovarian fat pad with attached ovary, oviduct, and uterus is picked and pulled out. Ovary (1), oviduct (2), and the proximal end of the uterus (3) are held outside the body by means of a Serrefine clamp attached to the fat pad above the ovary. (*G*) The ovary and the oviduct are surrounded by a thin transparent membrane (bursa) containing blood vessels. Make a small incision in the bursa (avoiding the blood vessels) and locate the end of the oviduct (infundibulum). (*H*) Insert the tip of the transfer pipette with embryos into the infundibulum. The edge of the infundibulum may be held very gently with fine forceps. (*I*) Diagram showing the technique.

Uterine Transfer

The method described below is based on extensive work by McLaren and Mitchie (1956), which resulted in the first successful development and birth of in vitro-cultured mouse embryos (McLaren and Biggers 1958). A detailed description of similar methods is given by Mintz (1967) and Rafferty (1970). It is best to practice this procedure first on a cadaver and then on an anesthetized 2.5-dpc pseudopregnant mouse using blue Affigel beads rather than embryos.

MATERIALS

ANIMALS AND EMBRYOS
Mouse blastocysts (3.5 dpc)
Pseudopregnant female recipient mice at 2.5 dpc

EQUIPMENT
Clean cage
Fiber optic illuminator (*very useful*)
Forceps, fine blunt with serrated tips
Forceps, watchmaker's #5, two pairs
Hypodermic needle, 26-gauge 1/2-inch
Lid of 9-cm plastic petri dish, glass plate, or paper towel
Needle, curved surgical (e.g., size 10, triangular, pointed)
Pipettes, transfer and mouth pipette assembly (see Chapter 4 and Fig. 4.1A, B, p. 177)
Plasticine (*optional*)
Scissors, fine dissection
Serrefine clamp or baby Dieffenbach clip (4 cm or smaller) (Roboz Surgical Instruments RS7440; Weiss B950B, or Fine Science Tools 18050-35)
Stereomicroscopes (ideally one for the surgery and one for loading the embryo transfer pipette) with transmitted and reflected light
Suture, surgical silk (size 5-0)
Syringe, 1-ml
Tissues
Warming plate, heating pad, or 50W light bulb
Weight scale
Wound clips and applier

REAGENTS
Anesthetic (see Appendix 1)
Beads, blue Affigel (BioRad 153-7302)
Dulbecco's modified Eagle's medium (DMEM) with HEPES, M2 or any other HEPES-buffered medium
Epinephrine (*optional*)
Ethanol, 70% <!>
Paraffin oil, embryo-tested light (*optional*)

CAUTION: See Appendix 2 for appropriate handling of materials marked with <!>.

PROCEDURE

1. Weigh and anesthetize the recipient mouse by IP injection (see Chapter 3).

2. Place the mouse on the lid of a petri dish (or other supporting material) so that it can be lifted onto the microscope stage easily.

3. Load a transfer pipette with embryos (see Fig. 6.1).

   a. Take up a small amount of M2 medium into the transfer pipette, then a small air bubble, then M2 medium, and then a second air bubble. Repeat until good control is reached with reduction of capillary action.

   b. Draw up the blastocysts in a minimal volume of medium (filling about 5–7 mm of the pipette).

   c. Store the transfer pipette (still in the mouth-pipetting device) by pressing its side into a piece of Plasticine stuck to the base of a stereomicroscope and leave it there until ready to place the embryos in the uterus. It is also possible to rest the pipette on a tube rack or any other support next to the microscope, making sure that the tip of the pipette does not touch anything.

   BE CAREFUL NOT TO DISTURB THE PIPETTE.

4. Expose the uterus.

   a. Wipe the back of the recipient mouse with tissues soaked in 70% ethanol and then make a single small longitudinal incision (less than 1 cm) in the skin with fine dissection scissors in the midline at the level of the last rib (Fig. 6.4A,B). Remove fur if required by local legislation. Wipe the incision with 70% ethanol-soaked tissue to remove any loose hairs.

   b. Slide the skin to the left or right until the incision is over the ovary (orange-pink) or fat pad (white), both of which are visible through the body wall (Fig. 6.4C). Then pick up the body wall with watchmaker's forceps and make a small incision (avoiding larger blood vessels) just over the ovary with fine dissection scissors. It is also possible to pierce the body wall using sharp forceps. Stretch the incision with the scissors to stop any bleeding. With a curved surgical needle, thread a piece of surgical silk suture through the body wall of the mouse so that the body wall will be easy to locate later (*optional*).

5. Transfer the embryos.

   a. Use blunt fine forceps to pick up the ovarian fat pad and pull out the attached left ovary, oviduct, and upper part of the uterus. Clip a Serrefine clamp onto the fat pad and lay it down over the middle of the back, so that the ovary, oviduct, and uterus remain outside the body wall (Fig. 6.4D).

   b. Gently pick up the mouse and place it on the stage of a stereomicroscope. This procedure is easier if the mouse is laid out initially on the lid of a 9-cm petri dish or paper towel.

   c. Hold the top of the uterus gently with blunt fine forceps and use a 26- or 30-gauge 1/2-inch needle with the bevel facing up to make a hole in the uterus a few millimeters down from the utero-tubal junction (Fig. 6.4E). Avoid small blood vessels in the uterine wall. If bleeding occurs, use tissue to remove the blood. Make sure that the needle has entered the uterine

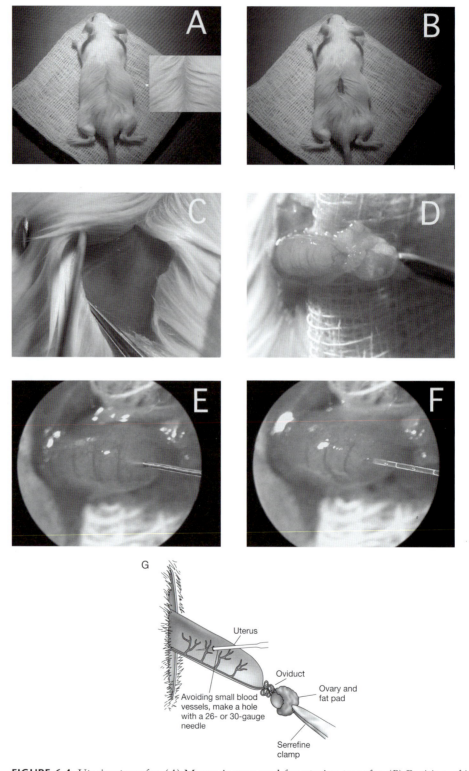

FIGURE 6.4. Uterine transfer. (*A*) Mouse is prepared for uterine transfer. (*B*) Position of incision for uterine transfers. (*C*) The ovary (*orange*) and fat pad (*white*) are visible through the body wall. (*D*) Ovary, oviduct, and the proximal end of the uterus are held outside the body by means of a Serrefine clamp attached to the fat pad above the ovary. (*E*) A 30- or 26-gauge needle is used to make a hole through the uterine wall into the lumen. (*F*) After removing the needle, insert the transfer pipette with embryos into the hole. The uterus may be held very gently with fine forceps. (*G*) Diagram showing the technique.

lumen and has not become lodged in the wall of the uterus. To test whether the needle has entered the lumen, pull it out slightly. If it slides easily, the needle has penetrated the lumen. Do not move the needle too much or the wall of the uterus may be lacerated. Watch the angle and depth of the needle's penetration and keep it parallel to the horn.

d. Keeping an eye on the hole made by the needle, pull out the needle and insert ~5 mm of the prepared transfer pipette containing the blastocysts into the hole (Fig. 6.4F). Blow gently on the transfer pipette until the air bubble or blue beads closest to the blastocysts are at the tip of the pipette and all of the blastocysts have been expelled. Watch the movement of air bubbles or blue beads (if practicing) in the pipette and remove the pipette when the first bubble reaches the opening in the uterus. Transferring too many bubbles into the uterus should be avoided because they might interfere with implantation.

> If there is no flow, do not force it. Withdraw the pipette, check under the microscope to see if it is clogged, expel the embryos, and reload the same or a new pipette.

e. Unclip the Serrefine clamp and remove the mouse from the stage of the stereomicroscope. Use blunt fine forceps to pick up the fat pad and place the uterus, oviduct, and ovary back inside the body cavity. Sew up the body wall with one or two stitches (*optional*) and close the skin with wound clips.

6. Repeat steps 3, 4, and 5 on the right side of the mouse, if desired.

7. At the end of the procedure, place the mouse in a clean cage and keep it warm under a 50W light bulb (taking care to cover the eyes) or by placing the cage on a warming plate or heating pad until the mouse recovers from an injected anesthetic. Follow local animal care committee regulations for more details.

COMMENT

A modified method of uterine transfer performed from the oviduct side of the utero-tubal junction was recently reported by Chin and Wang (2001). This method prevents the embryos from flowing out of the punctured hole because the utero-tubal junction provides a natural barrier. A 100% implantation rate has been reported using this method (Chin and Wang 2001).

Caesarean Section and Fostering

Caesarean section is required if the recipient of an embryo transfer or any pregnant mouse has not given birth by the delivery time normal for the particular strain.

MATERIALS

ANIMALS
Foster mother
Pregnant female mouse

EQUIPMENT
Desktop lamp or warming plate
Forceps, blunt fine
Forceps, watchmaker's #5, one pair
Scissors, fine dissection
Tissues

REAGENTS
Ethanol, 70% <!>

CAUTION: See Appendix 2 for appropriate handling of materials marked with <!>.

PROCEDURE

1. If a suitable foster mother is available, sacrifice the pregnant female (see Fig. 4.3). Place the pregnant female on her back to expose the abdomen. Use a squirt bottle to apply 70% ethanol to the abdomen and wipe with tissues.

2. Quickly open the abdomen with fine dissection scissors and dissect out the uterus onto a paper towel. Carefully cut through the uterine wall to expose the embryos within their fetal membranes. Dissect the pups from the yolk sac and amnion and cut the umbilical cord. Wipe away the amniotic fluid and secretions, particularly from the mouth and nostrils, with tissues.

3. Gently pinch the body of the pup with blunt fine forceps to stimulate breathing. If the pups squeak when pinched, they will usually be okay. Place the pups on moist tissues under a desktop lamp or on a 37°C warming plate to keep them warm until they become pink, move vigorously, and squeak when their tails are gently pinched.

4. The next step is to persuade the foster mother to accept the pups as her own and to take care of them. An ideal foster mother is a female who has successfully reared one or more litters of her own and has given birth the same day

or on the previous day or two. If a supply of breeding females is not routinely available, it is a good idea to set up natural matings 1–2 days earlier or at the same time that the recipient females are mated. Use foster mothers of a different coat color or clip the tails of the remaining natural pups, so that the foster pups can be distinguished later. Remove the mother from the cage and reduce the size of the original litter so that the final size is 6–8 pups, depending on the strain. If necessary, remove the entire natural litter and replace it with pups to be fostered. Mix the foster pups with bedding or nesting material or remaining pups of the natural litter. Mix pups and bedding together before returning the mother to the original cage. It is very important to minimize the temperature difference between the two sets of pups and to keep pups to be fostered warm.

Transplantation of Tissues under the Kidney Capsule

This protocol is based on information provided by Davor Solter (Max-Planck-Institute of Immunobiology, D-79108 Freiburg-Zähringen, Germany).

MATERIALS

ANIMALS AND TISSUE
Adult or embryonic tissue to be transplanted
Recipient histocompatible (or immunodeficient) mouse

EQUIPMENT
Anesthetic (see Appendix 1)
Clean cage
Forceps, 26-mm diameter Desmarres chalazion (Shields/Dina C62-2810)
Forceps, blunt fine
Forceps, watchmaker's #5
Hypodermic needle, 26-gauge 1/2-inch
Needle, curved surgical (size 10, triangular, pointed)
Pipette, braking

> This pipette is used if a small piece of tissue is to be transplanted. To make a braking pipette, pull out a glass capillary (transfer pipette) using a microburner to give a very fine open tip. Insert the end of the capillary with the fine tip into a larger piece of glass tubing using, for example, a Drummond micropipette tip or sealing wax (see Fig. 6.5). Pull out the other end of the capillary to the desired dimensions. Basically, the resistance to the flow of liquid will be inversely proportional to the diameter of the fine end of the capillary.

Scissors, fine dissection
Stereomicroscope
Suture, surgical silk (size 5-0)
Syringe, 1-ml
Tissues
Warming plate, heating pad, or 50W light bulb
Weight scale
Wound clips and applier

REAGENTS
Ethanol, 70% <!>
Sterile isotonic or phosphate-buffered saline (PBS) (pH 7.2; see Appendix 1)

CAUTION: See Appendix 2 for appropriate handling of materials marked with <!>.

PROCEDURE

1. Weigh and anesthetize the recipient mouse.

2. Wipe the back of the recipient mouse with 70% ethanol. Shave fur if required by local legislation. Then make an incision ~1-cm long as for embryo transfer (see p. 254). Slide the incision to one side and cut the body wall just above the level of the ovary. Use blunt fine forceps to pull out the kidney by its fat pad. Immobilize the kidney in Desmarres chalazion forceps (Fig. 6.5). Allow the surface of the kidney to air-dry for a few minutes. (This enables the capsule to be picked up with watchmaker's forceps.)

3. Use watchmaker's forceps to make a small transverse tear in the exposed capsule membrane and then moisten the capsule with sterile saline or PBS. Use moistened watchmaker's forceps to make a pocket underneath the capsule and then use forceps or a pipette to insert the tissue to be grown. The function of the braking pipette is to prevent the backflow of blood into the pipette, which could result in the pipette opening becoming clogged.

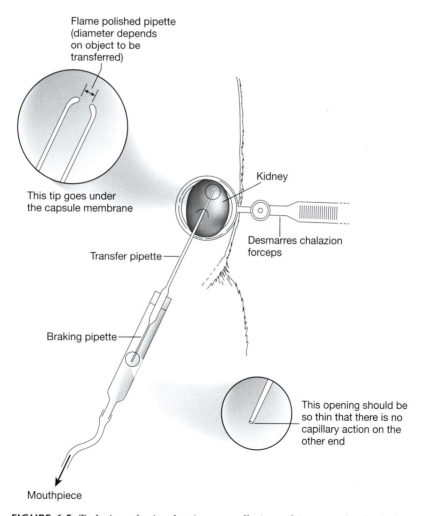

FIGURE 6.5. Technique for implanting a small piece of tissue under the kidney capsule using a braking pipette.

4. Push the tissue underneath the capsule as far away from the tear as possible. Release the kidney from the Desmarres chalazion forceps and replace it in the body cavity with blunt fine forceps. Sew up the body wall with one or two stitches and close the skin with wound clips.

5. At the end of the procedure, place the mouse in a clean cage and keep it warm under a 50W light bulb (taking care to cover the eyes) or by placing the cage on a warming plate or heating pad until the mouse recovers from an injected anesthetic. Follow local animal care committee regulations for more details.

Subcutaneous Injection of ES Cells

This procedure is based on Robertson (1987).

MATERIALS

ANIMALS AND TISSUE
Embryonic stem (ES) cells (~2 x 10^6 cells per mouse)
Recipients, histocompatible (or immunodeficient) mouse

EQUIPMENT
Anesthetic (optional)
Hypodermic needle, 26-gauge 1/2 inch
Syringe, 1 cc
Weight scale (*optional*)
Weight scale (*optional*)

PROCEDURE

1. Prepare a single cell suspension of ES cells (see Chapter 8, Trypsinization).

2. Weigh and anesthetize the recipient mouse. This step is optional, as the procedure may be done without anesthesia.

3. Inject the ES cells (~2 x 10^6 cells per mouse) subcutaneously into the flanks of the mouse.

4. Collect the resulting tumors 6 weeks after the transplantation for analysis (see Chapter 16).

Partial Hepatectomy

MATERIALS

ANIMALS
Mouse

EQUIPMENT
Animal clippers
Clean cage
Fiber optic illuminator (*optional*)
Forceps, blunt
Forceps, watchmaker's #5, two pairs
Needle, 26-gauge, 1/2 inch
Needle, curved surgical (e.g., size 10, triangular, pointed)
Scissors, dissection
Suture, surgical silk (size 5-0)
Syringe, 1-ml
Tissues
Warming plate, heating pad, or 50W light bulb
Weight scale
Wound clips and applier

REAGENTS
Anesthetic (see Appendix 1)
Ethanol, 70% <!>

CAUTION: See Appendix 2 for appropriate handling of materials marked with <!>.

PROCEDURE

1. Weigh and anesthetize the mouse.

2. Shave the upper abdomen of the mouse and wipe the skin with 70% ethanol to remove loose hairs.

3. Use dissection scissors to make a transverse incision ~1.5-cm long in the skin just below the sternum. Wipe the skin again with 70% ethanol. Pick up the body wall with watchmaker's forceps and make a transverse incision, also just below the sternum. Avoid cutting through the two major blood vessels (~1 cm apart) that run longitudinally in the body wall, on either side of the midline. Although cutting the blood vessels will not endanger the animal's life, considerable bleeding may result and will obscure the incision.

4. Hold the body wall with one pair of watchmaker's forceps, and then with a pair of blunt forceps, reach inside the body cavity and gently pick up the

edge of the top lobe of the liver. Gently bring the lobe of the liver outside the body wall and then, holding the forceps horizontally, grasp the liver with the blunt forceps. Ease the lobe of the liver farther outside the mouse until approximately two-thirds of the liver is exposed. Then thread a piece of surgical silk under the lobe of the liver and tie it off securely to constrict the blood vessels and reduce bleeding when the liver is cut. It is possible to pre-tie the silk as a large loop and slip the liver lobe through the loop. Use scissors to cut the lobe distal to the suture. The liver lobe can now be processed as needed.

5. Sew up the body wall with one or two stitches and close the skin with wound clips.

6. At the end of the procedure, place the mouse in a clean cage and keep it warm under a 50W light bulb (taking care to cover the eyes) or by placing the cage on a warming plate or heating pad until the mouse recovers from an injected anesthetic. Follow local animal care committee regulations for more details.

Splenectomy

MATERIALS

ANIMALS
Mouse

EQUIPMENT
Animal clippers
Clean cage
Fiber optic illuminator (*optional*)
Forceps, blunt
Forceps, watchmaker's #5, two pairs
Needle, 26-gauge, 1/2 inch
Needle, curved surgical (e.g., size 10, triangular, pointed)
Scissors, dissection
Suture, surgical silk (size 5-0)
Syringe, 1-ml
Tissues
Warming plate, heating pad, or 50W light bulb
Weight scale
Wound clips and applier

REAGENTS
Anesthetic (see Appendix 1)
Ethanol, 70% <!>

CAUTION: See Appendix 2 for appropriate handling of materials marked with <!>.

PROCEDURE

1. Weigh and anesthetize the mouse by IP injection (see Chapter 3, p. 158).

2. Shave the left side of the lower back of the mouse and wipe the skin with 70% ethanol to remove loose hairs.

3. Use dissection scissors to make a small transverse incision in the skin just to the left of the spinal cord at the level just below the last rib. Wipe the skin with 70% ethanol. Make a transverse incision in the body wall. Use blunt forceps to gently pull out the spleen. Cut the gastro-splenic ligament, which is at one end of the spleen, with dissection scissors. A major artery and vein run together through the fatty tissue to the middle of the spleen. Tie these vessels together with surgical silk suture. Cut the vessels distal to the knot and remove the entire spleen.

4. Sew up the body wall with one or two stitches and close the skin with wound clips.

5. At the end of the procedure, place the mouse in a clean cage and keep it warm under a 50W light bulb (taking care to cover the eyes) or by placing the cage on a warming plate or heating pad until the mouse recovers from an injected anesthetic. Follow local animal care committee regulations for more details.

Nephrectomy

MATERIALS

ANIMALS
Mouse

EQUIPMENT
Animal clippers
Clean cage
Dissection scissors
Fiber optic illuminator (*optional*)
Forceps, blunt
Forceps, watchmaker's #5, two pairs
Needle, 26-gauge, 1/2 inch
Needle, curved surgical (e.g., size 10, triangular, pointed)
Suture, surgical silk (size 5-0)
Syringe, 1-ml
Tissues
Warming plate, heating pad, or 50W light bulb
Weight scale
Wound clips and applier

REAGENTS
Anesthetic (see Appendix 1)
Ethanol, 70% <!>

CAUTION: See Appendix 2 for appropriate handling of materials marked with <!>.

PROCEDURE

1. Weigh and anesthetize the mouse by IP injection (see Chapter 3).

2. Shave the left or right side of the mouse at the level of the middle of the back and wipe the skin with 70% ethanol to remove loose hairs.

3. Use dissection scissors to make a longitudinal incision in the skin 0.5–1.0 cm long just to the left of the spine at the level of the last rib. Wipe the skin with 70% ethanol. Make a 0.5-cm longitudinal incision in the body wall, and with a surgical needle, draw a piece of surgical silk suture through the dorsal body wall.

4. Locate the kidney. Gently squeeze the mouse to push the kidney through the body wall. Alternatively, try to pull the kidney out by grasping the fatty tissue at the base of the kidney with blunt forceps. Pull a piece of suture under the kidney and securely tie off the renal artery, renal vein, and ureter. Use dis-

section scissors to cut off most of the kidney, leaving a small piece of kidney behind to keep the knot from slipping off.

5. Sew up the body wall with one or two stitches and close the skin with wound clips.

6. At the end of the procedure, place the mouse in a clean cage and keep it warm under a 50W light bulb (taking care to cover the eyes) or by placing the cage on a warming plate or heating pad until the mouse recovers from an injected anesthetic. Follow local animal care committee regulations for more details.

Castration

Castration is used to eliminate testicular hormones or to obtain testes for analysis without sacrificing the male. One or both testes can be removed depending on the experiment. Castration of male mice is performed in a similar manner as vasectomy (see Protocol 6.1). Access to the testes can be achieved using the approach of Protocol 6.1, Method 1 or 2.

MATERIALS

ANIMALS
Male mice

EQUIPMENT
Alcohol or Bunsen burner (*optional*)
Anesthetic (see Appendix 1)
Animal clippers (*optional*)
Fiber optic illuminator (*optional*)
Forceps, fine blunt
Forceps, watchmaker's #5, two pairs
Hypodermic needle, 26-gauge, 1/2-inch
Needle, curved surgical (e.g., size 10, triangular, pointed)
Scissors, coarse
Scissors, fine dissection
Suture, surgical silk (size 5-0)
Syringe, 1-ml
Tissues
Weight scale
Wound clips and applier (*optional*)

REAGENTS
Ethanol, 70% <!>

CAUTION: See Appendix 2 for appropriate handling of materials marked with <!>.

PROCEDURE

1. Weigh and anesthetize the mouse by IP injection (see Chapter 3, p. 158).
2. Push both testes down into the scrotal sacs by gently applying pressure to the abdomen.
3. Make an ~1-cm incision through the skin along the midline of the scrotal sac (see Fig. 6.2E).

4. Locate the midline wall between the testes sacs under the covering membranes. It will appear as a light whitish line.

5. Make a 5-mm incision in the membrane on the left side of the membrane.

6. Carefully push the testis out.

7. With blunt forceps, grasp the fat pad which adheres to the testis. Do not pull it out too far, or it will be difficult to place it back in the scrotal sac again. Locate the vas deferens with the prominent blood vessel running along it. While holding the testis in place, cauterize the vas deferens and the blood vessel with the red-hot tips of a pair of forceps. Dissect the testis away from the fat pad and remove it. Push back the fat pad into the scrotal sac.

8. Repeat steps 4–7 on the other testis if needed.

9. Sew up or clip the skin with wound clips. It is possible but not necessary to stitch the scrotal sac membrane.

10. At the end of the procedure, place the mouse in a clean cage and keep it warm under a 50W light bulb (taking care to cover the eyes) or by placing the cage on a warming plate or heating pad until the mouse recovers from an injected anesthetic. Follow local animal care committee regulations for more details. Wound clips may be removed after 10–14 days.

Blood Collection by Tail Bleeding

MATERIALS

ANIMALS
Mouse

EQUIPMENT
Lamp with 100W bulb
Microfuge tube, 1.5-ml
Microhematocrit capillary tube, heparinized
Razor blade
Slide warmer

REAGENTS
Anesthetic (depending on amount of blood to be collected; Ketamine/Xylazine or Avertin can be used)
Isofluorane (~0.5 ml in a cotton ball placed in a covered 1-liter beaker)
Heparin (*optional*)
Phosphate-buffered saline (PBS) (see Appendix 1)

PROCEDURE

1. Anesthetize the mouse by placing it in a covered 1-liter beaker containing isofluorane. Remove the mouse when it stops moving and falls over when the beaker is shaken. It will remain unconscious for ~1 minute.

2. Use a razor blade to cut off ~2 cm of the mouse's tail.

3. *To obtain up to 75 µl of blood:* Use a heparinized microhematocrit capillary tube held horizontally and touch the heparinized end to the bleeding tip of the mouse's tail, allowing the tube to fill by capillary action.

   To obtain 100–200 µl of blood: To inhibit clotting, place a small drop of heparin (~1000 U/ml) in a 1.5-ml microfuge tube and shake the tube to coat the inside surface with heparin. The mouse will jerk its tail when it starts to regain consciousness, so hold the mouse by the scruff of its neck (Fig. 3.1) and quickly insert the cut tail into the heparin-treated microfuge tube and hold it there until the desired volume of blood is collected. More blood may be recovered by gently stroking the tail toward the cut end, although this can alter the plasma/blood cell composition of the collected blood. If the mouse is particularly calm, let it stand on top of a cage after it recovers from the anesthesia and hold its tail loosely in the collection tube. The blood can also be collected in a tube containing PBS and the blood cells recovered by centrifugation.

To obtain 200–500 µl of blood: Anesthetize the mouse using Ketamine/Xylazine or Avertin (a lighter dose than would be required for major surgery). Place the mouse on a 37°C slide warmer to keep it warm and cut its tail (as described above). Position the cut tail inside a heparin-treated tube or a tube containing PBS. Up to 0.5 ml can be collected from a 20-gram mouse without endangering its life. If the blood clots and stops flowing, cut off another segment of tail.

REFERENCES

Bergstrom S. 1978. Experimentally delayed transplantation. In *Methods in mammalian reproduction* (ed. J.C. Daniel), pp. 419–435. Academic Press, New York.

Chin H.J. and Wang C.K. 2001. Utero-tubal transfer of mouse embryos. *Genesis* **30:** 77–81.

Erhardt W., Hebestedt A., Aschenbrenner G., Pichotka B., and Blumel G. 1984. A comparative study with various anesthetics in mice (pentobarbitone, ketamine-xylazine, carfentanyl-etomidate). *Res. Exp. Med.* **184:** 159–169.

Grant G.J., Vermeulen K., Langerman L., Zakpwski M., and Turndorf H. 1994. Prolonged analgesia with liposomal bupivacaine in a mouse model. *Reg. Anesth.* **19:** 264–269.

Magoon K.E., Hsu W.H., and Hembrough F.B. 1988. The influence of atropine on the cardiopulmonary effects of a xylazine-ketamine combination in dogs. *Arch. Int. Pharmacodyn. Ther.* **293:** 143–153.

McLaren A. and Biggers J.D. 1958. Successful development and birth of mice cultivated in vitro as early embryos. *Nature* **182:** 877–878.

McLaren A. and Michie D. 1956. Studies on the transfer of fertilized mouse eggs to uterine foster-mothers. I. Factors affecting the implantation and survival of native and transferred eggs. *J. Exp. Biol.* **33:** 394–416.

Mintz B. 1967. Mammalian embryo culture. In *Methods in developmental biology* (ed. E.H. Wilt and N.K. Wessels), pp. 379–400. Cromwell, New York.

Nakagata N. 1992. Embryo transfer through the wall of the fallopian tube in mice. *Jikken Dobutsu* **41:** 387–388.

Nichols J. and Gardner R.L. 1989. Effect of damage to the zona pellucida on development of preimplantation embryos in the mouse. *Hum. Reprod.* **4:** 180–187.

Nowrouzian I., Schels H.F., Ghodsian I., and Karimi H. 1981. Evaluation of the anaesthetic properties of ketamine and a ketamine/xylazine/atropine combination in sheep. *Vet. Rec.* **108:** 354–356.

Noyes R.W. and Dickman Z. 1961. Survival of ova transferred into the oviduct of the rat. *Fertil. Steril.* **12:** 67–79.

Papaioannou V.E. and Fox J.G. 1993. Efficacy of tribromoethanol anesthesia in mice. *Lab. Anim. Sci.* **43:** 189–192.

Rafferty R.A. 1970. *Methods in experimental embryology of the mouse.* Johns Hopkins Press, Baltimore.

Robertson E.J. 1987. Embryo-derived cell lines. In *Teratocarcinoma and embryonic stem cells: A practical approach* (ed. E.J. Robertson). IRL Press, Oxford.

Tarkowski A.K. 1959. Experimental studies on regulation in the development of isolated blastomeres of mouse eggs. *Acta. Theriol.* **3:** 191–287.

Whittingham D.G. 1968. Fertilization of mouse eggs in vitro. *Nature* **220:** 592–593.

Zeller W., Meier G., Burki K., and Panoussis B. 1998. Adverse effects of tribromoethanol as used in the production of transgenic mice. *Lab. Anim.* **32:** 407–413.

CHAPTER 7

Production of Transgenic Mice

Techniques developed during the past few decades have permitted the introduction of new genetic material into the mouse germ line. These techniques have revolutionized mouse genetics by making it possible to transfer any cloned gene, normal or mutant, into the germ line or to mutate genes randomly or specifically by random insertion of gene trap vectors or homologous recombination in embryonic stem cells, respectively. In this chapter, we discuss methods for introducing exogenous genes directly into the pronucleus of mouse zygotes. Protocols are given not only for the microinjection procedure itself, but also for DNA purification and production of glass capillaries, which are essential prerequisites for a successful outcome of these experiments. In addition, the use of large DNA constructs, such as BACs and YACs, is presented.

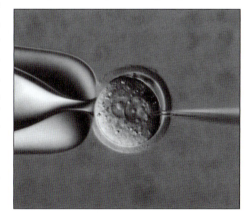

CONTENTS

APPLICATIONS OF TRANSGENIC MOUSE TECHNOLOGY

The use of transgenic mice in biomedical research is now widespread, and numerous applications have been developed. Several general reviews have been published, including those by Palmiter and Brinster (1986), Jaenisch (1988), and Hanahan (1989). The most common uses of transgenic mice are (1) for studies of tissue-specific and developmental-stage-specific gene regulation and (2) for experiments of the phenotypic effects of transgene expression. Most cloned genes introduced into the mouse germ line have shown appropriate tissue-specific and stage-specific patterns of expression, despite their integration into ectopic sites in the host genome. Therefore, gene transfer into the mouse embryo has provided an important experimental assay for defining the *cis*-acting DNA sequences that dictate specific patterns of transcription in the developing animal. In addition, the characterization of regulatory sequences through such studies has provided tools for the second category of experiments in which the expression of gene products of interest is directed to specific sites in the developing mouse.

Applications involving the phenotypic effects of transgene expression include studies of genes encoding hormones (Palmiter et al. 1982; Hammer et al. 1985), oncogenes (Brinster et al. 1984; Stewart et al. 1984), developmental regulatory genes (Wolgemuth et al. 1989; Kessel and Gruss 1990), immune system genes (Storb et al. 1984), and viral genes (Chisari et al. 1985; Berns 1991; Skowronski et al. 1993). In general, transgenic mice expressing novel genes represent gain-of-function mutations. It is also possible to create loss-of-function mutations by random insertional mutation (for review, see Woychik and Alagraman 1998) and, in one case, by gene targeting (Brinster et al. 1989). Dominant negative mutants have been obtained through expression of some transgenes (Stacey et al. 1988; Metsaranta et al. 1992; Peters et al. 1994), and, in some cases, the expression of transgenes encoding antisense RNA has been successful in inhibiting the expression of endogenous genes (for reviews, see Sokol and Murray 1996 and Erickson 1999). Recently, double-stranded RNA (RNA interference or RNAi) has been reported to knock down gene expression in mice (Wianny and Zernicka-Goetz 2000; McCaffrey et al. 2002).

Other uses of transgenesis in mice include

- Cell lineage ablation (Palmiter et al. 1987; Borrelli et al. 1989; Heyman et al. 1989)

- The production of new cell lineage markers (Lo et al. 1987; Beddington et al. 1989)

- Studies of somatic and meiotic recombination (Murti et al. 1992)

- Introduction of substrates for mutagen testing (Gossen et al. 1989; Kohler et al. 1991)

- Testing the ability of cloned genes to complement mutations in the mouse (Readhead et al. 1987; Stacey et al. 1987; Probst et al. 1998; Zhu et al. 2000)

- The development of methods for protein production in large animals (Pittius et al. 1988; Archibald et al. 1990; Niemann et al. 1999)

In addition to a need for controlling transgene expression by developmental or cell-type-specific regulatory elements, there has been an increasing need for temporal/inducible regulation (Lewandoski 2001). There are currently three systems dominating this area: tetracycline, RU-486, and the tamoxifen-inducible systems. The tetracycline system is composed of activator and responder transgenes. The activator transgene directs the expression of a fusion protein composed of the bacterial tetracycline repressor (TetR) protein and the herpes simplex virus VP16 transcription activation domain. TetR can bind both tetracycline and a 19-bp sequence called the tet operator (*tetO*) that is not represented in the mouse genome (Gossen et al. 1995). The responder is the *tetO* sequence combined with a minimal promoter ligated to a gene of interest. The expression of the engineered activator transcription factor does not cause ectopic activation of endogenous genes, but rather activates *tetO* transgenes that are dependent on the presence or absence of tetracycline (tet-ON and tet-OFF systems, respectively) and can be administered or withdrawn at any time during the life of the animals (Shin et al. 1999).

The tamoxifen- and RU-486-inducible systems are based on different principles and require another popular genetic switch system—the Cre site-specific DNA recombinase. The Cre recombinase of the P1 phage recombines DNA between two 34-bp consensus sequences called *loxP* sites. The *loxP* site is not represented in the mouse genome; therefore, it can be placed in transgenes to achieve postintegrational transgene modifications mediated by the Cre protein. Currently, many powerful "genetic switch" designs have been developed, in which the site-specific recombinases play critical roles. These are discussed in more detail in Chapter 9, where possible designs are given for Cre recombinase-conditional transgenes, knockouts, and genome alterations. We mention the Cre/*loxP* system here in the context of the tamoxifen- and RU-486-inducible systems and the requirement for the *loxP*-containing target transgene features. The usual way of introducing the Cre recombinase into cells to perform postintegrational modification on a transgene is to express it from another transgene with the required specificity. The site of action of the recombinase is in the nucleus, and additional regulation of its action is provided by its fusion with one or two hormone-binding domains of the progesterone or estrogen receptor. The estrogen receptor domain contains a mutation that dramatically reduces endogenous estrogen binding but allows binding of one of its "artificial" analogs, tamoxifen. Likewise, the progesterone domain contains a mutation that allows binding of RU-486 but not progesterone. In both systems, the consequence of the binding of the exogenously administered ligand to its domain is the translocation of the Cre protein to the nucleus, where it can perform its function.

Transgenes with *loxP* sites can be used as targets for Cre recombinase (Grieshammer et al. 1998). Because transgenes usually integrate as multiple copies in head-to-tail tandem arrays, the site of integration will have many *loxP* sites. More complex integrations are possible, leading to a situation with some of the *loxP* sites in reverse orientation relative to the others. This can lead to chromosome destruction upon Cre expression (Lewandoski and Martin 1997). Because mice do not tolerate monosomy except for the X chromosome, the result is embryonic lethality. Thus, if using such a system, single-copy transgene integration may be better. The ideal is to aim for the production of single-copy transgenes, which may be accomplished more easily by ES cell-mediated transgenesis (see Chapter 9) than by pronuclear injection.

In the early years of using transgenic techniques, a by-product of the generation of transgenic mice was the generation of random insertional mutations with a wide variety of phenotypic effects (Woychik et al. 1985; Woychik and Alagramam 1998). Approximately 5–15% of transgenic lines, produced by any of the available methods (see below), carry either recessive embryonic lethal or viable, but phenotypically visible, insertional mutations. The integrated transgene "tags" the mutated gene and allows it to be cloned molecularly and characterized, permitting the identification of potentially new genes with important developmental functions (for reviews, see Jaenisch 1988; Gridley 1991; Meisler 1992). Frequently, however, the transgene causes severe genomic disturbance at the site of integration, making the characterization tedious. However, the availability of the mouse genome sequence might ease these difficulties and make the insertion site characterization more appealing. Methods for cloning of DNA flanking the insertion loci are discussed in Chapter 12. More efficient complementary approaches of gene trapping are alternative strategies used to identify new genes on the basis of their expression patterns during development, in in vitro differentiation assays, and by sequence tags (Gossler et al. 1989; Friedrich and Soriano 1991; Skarnes et al. 1992; Hicks et al. 1997; Zambrowicz et al. 1998; Wiles et al. 2000). These methods are ES cell-based and are discussed in Chapter 9.

GENE TRANSFER INTO THE MOUSE GENOME BY PRONUCLEAR MICROINJECTION

The microinjection of DNA directly into the pronuclei of fertilized zygotes is the most extensively and successfully used method of gene transfer in the mouse. In pioneering experiments, Lin (1966) used a micromanipulation system to show that mouse zygotes could survive the insertion of a fine glass needle into their pronuclei. In addition, the injection of macromolecules such as bovine serum albumin (BSA) into zygotes and their transfer into recipient pseudopregnant females yielded viable mice. Lin's technique was ahead of its time, because recombinant DNA technologies had not yet been born. At the end of 1980, Gordon et al. (1980) reported the generation of transgenic mice by pronuclear injection of DNA. Within the next year, six other reports documented the integration, germ-line transmission, and expression of genes in transgenic mice (Brinster et al. 1981; Costantini and Lacy 1981; Gordon and Ruddle 1981; Harbers et al. 1981; E. Wagner et al. 1981; T. Wagner et al. 1981). Today this technique is well established, and has become a standard procedure in many laboratories around the world. More recently, the use of larger DNA fragments is becoming more and more popular (Copeland et al. 2001). Yeast artificial chromosomes (YACs) (Schedl et al. 1992; Peterson et al. 1997), P1-derived artificial chromosomes (PACs) (Sternberg 1992), and bacterial artificial chromosome (BAC) DNA (Antoch et al. 1997; Nielsen et al. 1997; Yang et al. 1997) can all be used for the generation of transgenic mice. These methods result in the stable chromosomal integration of the foreign DNA in 10–40% of the resulting mice. For a comprehensive review of these methods, see Girlado (2001).

It is important to keep in mind with any kind of pronuclear DNA injection that integration takes place, with a single exception (Brinster et al. 1989), in a random manner. Hence, it is not possible to predict the integration site of the transgene, and the number of copies of the transgene integrated is difficult to control. Thus, each of the resulting founder animals will have a unique site, copy number, and pattern of transgene integration. Microinjection of either linear or circular DNA most frequently results in the integration of multiple copies. This is thought to occur by rapid circularization of linear molecules after injection into the pronucleus, followed by homologous recombination between several circular molecules and integration into the genome (Brinster et al. 1981, 1982). The structure of the ends of a linear molecule appears to have little effect on either the frequency of integration or the structure of the integrated molecules. Regardless of the type of ends a linear DNA fragment carries, the injected molecules will most frequently integrate as a head-to-tail array (Brinster et al. 1981; Costantini and Lacy 1981). In most cases, the integration will take place at only one site; however, occasionally separate integration sites on different chromosomes can occur (Lacy et al. 1983). If integration occurs at more than one chromosomal location in a single founder, two different lines of transgenic mice can be established from their progeny. Integration will generally take place shortly after microinjection and before the first cleavage. However, in ~20–30% of founder mice, the integration is delayed and takes place only after the first cleavage, creating mosaics (Wilkie et al. 1986). Therefore, only a subset of the somatic and germ cells carry the transgene. When founders that are mosaic are bred with nontransgenics, less than 50% of the progeny inherit the transgene. The number of copies of the foreign DNA sequence that integrates ranges from one to several hundred and can be somewhat controlled by the concentration of the DNA injected into the zygote (Ellis et al. 1997).

Although the methods described in this section involve the introduction of transgenes directly into a zygote or early embryo, it should be noted that transgenes can also be introduced by viral infection of preimplantation embryos (Soriano et al. 1986; Lois et al. 2002) or by using embryonic stem (ES) cells (see Chapters 9 and 10). Gene transfer into ES cells has been performed using calcium phosphate-mediated DNA transformation (Gossler et al. 1986), retroviral infection (Robertson et al. 1986), and lipofection (Strauss and Jaenisch 1992). ES cells are most useful when it is necessary to select for rare integration events (e.g., gene-targeting studies); however, it may be desirable in some instances to introduce a transgene into ES cells and then produce chimeric mice instead of microinjecting the gene into mouse zygotes. These instances include cases in which an in vitro ES cell differentiation assay has been developed to verify the transgene expression specificity prior to the introduction into the mouse genome (Ding et al. 2001) or to complement a mutation in ES cells and to test for complementation in chimeric mice (Simon et al. 1992).

New genetic material may also be introduced into the mouse embryo by nuclear transplantation or by intracytoplasmic sperm injection (ICSI; Perry et al. 1999, 2001) (see Chapter 13), although at the moment this approach has more importance in mammalian species other than the mouse (Keefer et al. 2001; Zakhartchenko et al. 2001).

DESIGNING TRANSGENES

Effects of Prokaryotic Vector Sequences

Although prokaryotic cloning vector sequences have no apparent effect on the integration frequency of microinjected genes, they can severely inhibit the expression of genes introduced into the mouse germ line (see, e.g., Chada et al. 1985; Krumlauf et al. 1985; Townes et al. 1985; Hammer et al. 1987 reviewed in Palmiter and Brinster 1986; Rusconi 1991). Therefore, gene constructs are purified away from plasmid vector sequences before introduction into the mouse germ line to eliminate the potential influence of the vector sequences. Several cloning vectors are available containing multiple cloning sites flanked by two sites for the restriction enzyme *Not*I, which is represented less frequently in eukaryotic DNA and can usually be used to excise the gene construct (Lathe et al. 1987; Taylor et al. 1992). Alternatively, pBluescript (Stratagene) has a multiple cloning site flanked by *Bss*HII sites, which also occur infrequently in eukaryotic DNA. The construct should be designed in such a way that co-migrating bands of gene construct fragment and vector backbone DNA are easily separated on an agarose gel.

The injection of very large pieces of DNA, including BACs and YACs, has become increasingly popular (Giraldo and Montoliu 2001). YACs exist as linear molecules, whereas BACs exist as circular molecules. Their large size can make certain manipulations difficult. For example, it is typically not possible to find suitable restriction sites that will release a BAC insert from vector sequences. Fortunately, the expression of genes within BAC and YAC DNA does not appear to be significantly influenced by vector sequences (Jakobovits et al. 1993; Mendez et al. 1997; Kaufman 1999). Many BAC cloning vectors contain unique sequences to linearize the DNA (e.g., *Isce*I). Although linearization of BACs facilitates the integration of an intact construct, it also increases DNA viscosity, which can make the microinjection process more difficult. BACs can also be microinjected as intact circular molecules to generate transgenic mice (Probst et al. 1998). It is thought that a random double-strand break occurs, facilitating integration into the genome. This approach carries the risk of random linearization within constructs prior to integration, making subsequent analyses more complex. Because vector sequences appear to have a minimal effect on BAC and YAC transgene expression, it may be desirable to retain vector sequences as tags for genotyping purposes.

Length of DNA Construct

The length of a DNA construct used to produce transgenic mice has so far been limited only by cloning and handling considerations: A 50-kb bacteriophage λ clone (Costantini and Lacy 1981), a 60-kb cosmid insert (Taylor et al. 1992), and a 70-kb fragment produced by in vitro ligation of two cosmid inserts (Strouboulis et al. 1992) have been introduced intact. Large sequences can also be reconstructed in vivo through co-injection of two or more overlapping fragments, which undergo efficient homologous recombination with each other before (or during) integration into the genome (Palmiter et al. 1985; Pieper et al. 1992; Verma-Kurvari et al. 1996). As discussed above, the use of larger DNA fragments for the generation of transgenic mice has become more widespread. Large BAC (Nielsen et al. 1997; Zhu et al. 2000) and PAC (Duff et al. 2000) DNAs of several hundred kilobases have been successfully used, as have large YAC DNAs over 1000 kb (Lamb et al. 1999).

Co-injection of Two (or More) Transgenes

Several procedures can be used when two (or more) different genes are introduced into the same mouse. Two different transgenic lines can be produced and then bred together; the two genes can be linked into a single construct before microinjection; or, more simply, the two genes can be mixed and co-injected into mouse zygotes. In most of the resulting transgenics, the two gene constructs will be co-integrated at the same site in the genome (Behringer et al. 1989). Transgenic mice carrying one or the other gene construct will also be obtained in co-injection experiments.

It is important that the final concentration of the DNA solution be kept at the "normal" level, even when two separate DNA molecules are mixed in the same injection solution. Each construct will then be present in the solution at half the normal concentration. If the sizes of the two constructs vary considerably, care should be taken to calculate the mixture in such a way that each fragment contributes with approximately equal numbers of molecules per volume.

Functional transgenes can also be generated directly via homologous recombination between microinjected DNA fragments (Palmiter et al. 1985; Pieper et al. 1992; Verma-Kurvari et al. 1996) or bacteriophage P1 clones (Wagner et al. 1996) with high frequency and accuracy, which allows the efficient generation of relatively large transgenes.

Distinguishing Expression of Transgenes and Endogenous Genes

If it is important to distinguish the products of the transgene from endogenous RNAs or proteins, the following options can be used. Mouse genes can be marked by insertion of an oligonucleotide into an untranslated region (Peschon et al. 1987; Shi et al. 1989), a "mini-gene" encoding a shortened RNA can be engineered (Krumlauf et al. 1985), or one allele of a mouse gene can be introduced into a strain carrying a different allele (see, e.g., Tronik et al. 1987). Many times, genes from heterologous species, including humans, have been used. In these cases, species-specific probes can be used. Fusion of a short epitope/peptide tag (myc, FLAG, HA) with the protein of interest can facilitate the recognition of the transgene-encoded protein through epitope-specific antibodies (Swierczynski et al. 1996; Previtali et al. 2000). One should consider, however, that these short extensions of the protein may or may not interfere with its function. Interestingly, there are cases where a large extension, such as a fusion between the protein of interest and a reporter gene (alkaline phosphatase [Muller et al. 1998]) of green fluorescent protein (GFP) (Van Roessel and Brand 2002), retain the function of both components, and the reporter can serve as a convenient readout of the level of expression or the cellular and subcellular localization of the fusion protein. It is also possible to link transgene and reporter expression transcriptionally without creating a fusion protein between them. A viral internal ribosomal entry site (IRES) placed between their full coding regions allows independent translation of the two proteins from a single mRNA (Jang et al. 1988). However, in this situation, the IRES-linked reporter activity has been uncoupled posttranscriptionally from the primary transgene product and, thus, may not fully mirror transgene expression.

Strategies for Identifying Regulatory Sequences

As discussed above, most transgenic mice are produced for studying the control of gene expression and the effects of transgene expression in the intact animal, and for marking cell lineages. To study the control of gene expression, mice are produced that contain the entire gene of interest (if it is of manageable size), as well as several kilobases each of 5´- and 3´-flanking DNA (Storb et al. 1984). If the gene is expressed appropriately in transgenic mice, smaller DNA fragments with 5´, 3´, or intragenic regions removed can be tested. If the gene is not expressed appropriately, it may be regulated by more distant sequences that were absent from the initial construct. In some cases, upstream or downstream regulatory regions can be tentatively localized by testing for DNase-hypersensitive sites (see, e.g., Grosveld et al. 1987). A complementary approach is to identify a BAC that contains the gene of interest and to generate BAC transgenic mice. In many cases, correctly regulated expression will be obtained, indicating that the regulatory elements for gene expression reside within the BAC. Deletion mapping using transgenic mice can then be used with smaller fragments of the BAC. The availability of the mouse, human, and other genomic sequences, and the ability to perform comparisons to identify conserved noncoding sequences, have become increasingly important for pinpointing candidate regulatory elements (Onyango et al. 2000; Amid et al. 2001).

Reporter Genes

In defining the regulatory sequences of a gene, it is often advantageous to generate constructs that contain a "reporter" gene. Reporter genes that have been used successfully in transgenic mice include the *Escherichia coli lacZ* gene (Goring et al. 1987), the *E. coli* chloramphenicol acetyltransferase (*CAT*) gene (see, e.g., Overbeek et al. 1985), the firefly luciferase gene (Lira et al. 1990; Lee et al. 1992), the heat-resistant human placental alkaline phosphatase (hPAP) gene (DePrimo et al. 1996), and the green fluorescence protein (*GFP*) gene (Obake et al. 1997), or its derivatives. The *lacZ* gene is particularly useful for studies of tissue- or position-specific gene expression in the mouse embryo because whole embryos up to mid-gestation can be stained to detect β-galactosidase (β-Gal) activity visually (see Chapter 16). At later stages, β-Gal activity can be detected by staining fixed frozen tissue sections. In addition, there are many variations of the *lacZ* gene which produce β-Gal variants that can localize to different subcellular regions (including the cytoplasm, nucleus, and axons). Similar to *lacZ*, the human placental alkaline phosphatase gene (*ALPP*) is an excellent reporter for detecting expression in whole-mount embryos or in tissue sections at a cellular level (see Chapter 16). Alkaline phosphatase is slightly more sensitive than *lacZ*, but the stain does not penetrate into deeper tissues as easily. The *CAT* and the luciferase genes (whose expression is normally detected by enzyme assays of tissue extracts) are useful reporter genes because the assays to detect them are easy, sensitive, and quantitative. However, these assays do not facilitate visual detection of spatial patterns of expression.

One of the most exciting reporter genes recently introduced into transgenic mice is GFP and its variants. The gene is derived from the jellyfish, *Aequorea victoria*. Its protein product retains its light-emitting property when proper excitation light is applied in heterologous systems such as the mouse. Through intensive mutagenesis, derivatives have been identified with altered emission intensity and with maximum excitation and emission wavelengths. The most commonly used variant is the enhanced GFP (EGFP), which has an excitation maximum of 488 nm with emission maximum at 507-nm wavelength. Enhanced variants are also available for cyan and yellow emission. When these reporters are used, it is not necessary to fix cells for visualization. Live cells can be observed, facilitating new types of experiments not previously possible. The different reporters are discussed in Chapter 16.

Using Previously Identified Tissue-specific Regulatory Sequences in Transgene Constructs

A variety of tissue-specific genes have been used to produce constructs expressed in transgenic mice. In many of these cases, the correct pattern of gene expression has been obtained (Palmiter and Brinster 1986). These tissue-specific genes are too diverse and the results too complex to catalog here. When designing a transgene to express a certain protein in a specific pattern in the mouse, we suggest a literature database search (e.g., Medline) using key words "transgenic," "mouse," and the tissue or cell type of interest. In addition, the TBASE database for transgenic animals can be searched (see http://tbase.jax.org). An ever-increasing list of transgenic lines with tissue/cell-type-specific expression patterns is given in the Cre recombinase transgenic database http://www.mshri.on.ca/nagy/.

In evaluating the literature, it is important to distinguish between experiments in which an intact gene has been expressed in a correct pattern and experiments in which regulatory sequences from one gene have been used to direct the expression of a different gene (such as a reporter gene). In the former experiments, the 5′-, intronic, or 3′-flanking sequences may or may not be sufficient to confer the same pattern of expression to a reporter gene. In the latter experiments, it is also possible that substitution of a different "reporter gene" will alter the pattern of expression. In some cases, unexpected patterns of expression can be generated by producing novel fusion genes (see, e.g., Behringer et al. 1988; Russo et al. 1988; Al Shawi et al. 1991; Trudel et al. 1991).

Expression of cDNAs and the Role of Introns in Transgene Expression

It is often easier to use a cDNA clone to provide the coding sequences for the desired gene product when constructing fusion genes for expression in transgenic mice because they are usually smaller than genomic sequences. However, many cases show that the frequency and levels of gene expression obtained with cDNA-based constructs are often dramatically lower than those obtained when genomic sequences, including introns and exons, are used (Brinster et al. 1988). In some cases, this may be due to the presence of enhancers in the introns, but this may not be the only explanation. The addition of heterologous introns to cDNA-based constructs can give sizable increases in expression levels without altering the tissue specificity of expression (Choi et al. 1991; Palmiter et al. 1991). However, this may be gene-specific. Therefore, when expression levels are important, it is advisable (1) to use genomic sequences, rather than a cDNA-based fusion gene, whenever the gene is of a manageable size; (2) to construct a minigene that combines several introns and exons with parts of the cDNA (Krumlauf et al. 1985; Hammer et al. 1987); or (3) to introduce exogenous exons (Choi et al. 1991; Palmiter et al. 1991) into cDNA-based constructs. The transcript from these constructs should be checked carefully, because it is possible that cryptic splice acceptor or donor sites are present in the cDNA, which could result in splicing out of essential coding regions from the mRNA.

Using "Housekeeping Gene" Promoters to Direct Ubiquitous Expression

It is sometimes useful to express a gene product with a ubiquitous or widespread tissue distribution, rather than a tissue-specific pattern, in a transgenic mouse line. Although the ideal regulatory sequences, which would produce a level of expression similar to that of a reporter gene in all cells and at all developmental stages, have not been identified, several fusion genes have yielded fairly widespread expression in the developing embryo and/or adult mouse. These include constructs containing a β-actin promoter (Balling et al. 1989; Beddington et al. 1989), a cytomegalovirus (CMV) enhancer–chicken β-actin promoter combination (Niwa et al. 1991), the mouse metallothionein promoter (Palmiter et al. 1983; Iwamoto et al. 1991), the mouse *Hmgcr* promoter (Mehtali et al. 1990; Tam and Tan 1992), the histone H4 promoter (Choi et al. 1991), and the promoter of the *Rosa26* locus (Kisseberth et al. 1999).

The expression directed by all of the above regulatory elements is very dependent on the integration site. Therefore, the investigator should be prepared to screen through multiple transgenic mouse lines to obtain the desired expression pattern. ES cell-mediated transgenesis (see Chapter 9) provides an alternative to screen efficiently for integrations that result in widespread and high-level transgene expression (Hadjantonakis et al. 1998).

Factors Affecting the Efficiency of Gene Transfer

DNA molecules isolated with high quantity and purity can be integrated into the mouse genome by microinjection into the pronucleus of a zygote. Although the frequencies of integration obtained with different DNA fragments or constructs may vary severalfold under seemingly identical conditions, there is no evidence that this is an effect of the DNA sequence or base composition. Furthermore, although very large transgenes (i.e., >30 kb) may be more difficult to isolate or construct, the length of the injected DNA does not appear to be a limiting factor in the frequency of obtaining transgenic mice.

Several properties of the injected DNA preparation have been observed to affect its frequency of integration and/or the structure of the integrated DNA sequences. These include

- *The linear or circular form of the DNA.* Transgenic mice are generally produced by microinjection of linear DNA molecules, which integrate into the genome at a higher frequency than circular DNA (Brinster et al. 1985). The possibility of injecting circular plasmids may be exploited when transient expression (e.g., a Cre-expressing plasmid) without integration is desired. In cases of large DNA constructs, such as BAC and YAC DNA, practice has shown that linear and circular DNA integrate with equal efficiency (Camper and Saunders 2000). If circular DNA is injected, linearization will occur randomly before integration takes place. This process carries the risk that the DNA will break at a point where the transgene function may be disrupted.

- *The DNA concentration.* The DNA concentration is one of the most important factors influencing the efficiency of the pronuclear microinjection method. Low concentrations yield a low rate of integration. High concentrations, on the other hand, are toxic to the zygote. The optimal concentration for transgene integration is 1.5–2 ng/µl, which should be determined as accurately as possible. Previously, it was thought that there was no correlation between concentration and integration copy number. Lately, however, it has been shown that by reducing the concentration to around 0.1 ng/µl, a larger proportion of the founder animals show a single-copy integration pattern. However, the efficiency of generating founders using a reduced DNA concentration is significantly reduced (Ellis et al. 1997).

- *The purity of the DNA.* DNA samples for microinjection should be absolutely free of contaminants (e.g., traces of phenol, ethanol, or enzymes) that might harm the zygotes. The transgenic method requires a single cell, the zygote, to produce a viable mouse. This is in contrast to tissue culture transfections, where a loss of cells in the experiment can be tolerated.

- *The buffer in which the DNA is dissolved.* The composition, quality, and purity of the injection buffer are of major importance because the smallest particles will inevitably clog the injection needle, making the injection procedure very difficult. Usually molecular cloning methods routinely use TE buffers containing 1 mM EDTA. This concentration of EDTA is not compatible with mouse zygote viability (Brinster et al. 1985). Therefore, the buffers used for DNA injection into mouse embryos use an EDTA concentration of 0.1–0.3 mM. Large DNA constructs have additional requirements on the buffer composition (see Protocol 7.7).

- *The handling of and general conditions surrounding both embryo donor and recipient mice.* These conditions are of major importance for the successful outcome of microinjection experiments. For further information about optimal husbandry conditions, see Foster et al. (1983).

The number of animals needed per gene transfer experiment depends on several factors: age and quality of the embryo donors and stud males, physical parameters in the animal facility and the microinjection lab, embryo culture conditions, DNA purity, DNA concentration, and the manual skills of the operator. Generally, 20–30% of the injected oocytes will immediately lyse due to mechanical damage, although a >90% survival rate can be achieved if injection conditions are optimized. On average, about 30% of the surviving embryos will develop to term, and about 20% of those will carry the transgene. Therefore, if all conditions are met, 100–150 zygotes need to be injected to produce 3–6 founder animals.

The efficiency of transgenic mouse production when using BAC or YAC DNA is not significantly lower than for plasmid-based DNA fragments, but the variability between individual constructs is increased. These variations can most probably be attributed not to the large DNA size, but more likely to a higher level of copurified contaminants in the DNA solution than in the case of smaller constructs (Giraldo and Montoliu 2001). It is also possible that the increased dosage of genes residing on the BACs or YACs may be incompatible with embryo viability.

General Considerations about Large DNA Constructs

Although the basic technique for microinjection of large DNA constructs is similar to those used for the short DNA segments, there are some special requirements. The size of the DNA molecule may be a hundredfold that of a "standard" gene construct. With increased size, the risk of damaging the DNA increases dramatically. Shearing of the DNA can occur during purification, dilution, and microinjection. To prevent such damage, it is important to handle the DNA solution with the utmost care. The preparation of BAC/PAC/YAC DNA is somewhat more tedious than for plasmids (see Protocols 7.1–7.4). The inner diameter of the microinjection needle tip should be slightly larger than for small gene constructs, and the injection pressure should be kept as low as possible while still allowing the proper injection of the pronucleus. Polyamines are added to the microinjection buffer to protect the DNA from shearing. A comprehensive review of BAC/YAC/PAC transgenesis, comparing all three applications, is given by Giraldo and Montoliu (2001).

ISOLATION AND PURIFICATION OF DNA

General Considerations

One of the most essential factors affecting the successful outcome of pronuclear microinjections is the purity of the DNA. The slightest trace of contaminating agents will be harmful to the zygotes, which will either lyse shortly after injection or stop embryonic development at later preimplantation or postimplantation stages. Because the purification procedure is so important and practically every laboratory uses different protocols, we discuss each step in detail and present possible alternatives. Whatever the specific protocols, the basic steps involved are as follows:

1. Propagation of bacteria containing the desired plasmid.

2. Harvest and lysis of bacterial cells (see page 303).

3. Removal of bacterial cell components/plasmid purification (see Protocols 7.1, 7.2, and 7.4).

4. Digestion of the plasmid with restriction enzymes, followed by heat inactivation (see page 305).

5. Separation of the fragment from the vector backbone on an agarose gel (see page 306).

6. Isolation of the gene construct insert (see page 306).

7. Removal of any contaminants from the DNA (see page 306 and Protocols 7.1, 7.3, and 7.5).

8. Determination of the DNA concentration (see page 308).

9. Dilution to the correct concentration for injection.

Contaminants and Methods for Their Removal

It is essential to remove the following contaminants prior to injection:

1. *Bacterial cell contents.* Bacterial RNA and chromosomal DNA contaminate the DNA after lysis of *E. coli* for recovery of plasmid DNA.

2. *Endotoxins.* Gram-negative bacteria (such as *E. coli*) release small amounts of endotoxins to their surroundings when they are growing and substantially larger amounts during plasmid preparation.

3. *Ethidium bromide.* EtBr is a fluorescent dye that intercalates in the DNA structure and emits an orange-colored fluorescence upon UV illumination. EtBr always contaminates DNA solutions after plasmid purification over cesium chloride (CsCl) gradients, and after staining DNA in agarose gels. EtBr is toxic, a very powerful mutagen, and a potential carcinogen. Gels and solutions containing EtBr should be handled with caution and disposed of appropriately. Store EtBr protected from light.

4. *CsCl.* This is introduced into the DNA solution by the use of CsCl gradient centrifugation for plasmid preparation.

5. *Enzymes.* These are used for restriction digestion of plasmid DNA for the release of gene construct insert.

6. *Phenol and chloroform.* Residual organic compounds may contaminate the DNA solution after extraction with these agents.

7. *Salts.* These are added to the DNA during alcohol precipitation.

8. *Agarose.* The DNA solution may contain traces of agarose after enzymatic digestion of agarose slices.

Several standard procedures have been established for the effective removal of contaminants. Detailed protocols are given in Sambrook and Russell (2001). Here, we list methods for the removal of each compound listed above.

1. *CsCl gradient centrifugation.* CsCl gradients can be used to remove bacterial cell contents, bacterial RNA, and chromosomal DNA; 2x CsCl gradient centrifugation is necessary for the removal of endotoxins.

2. *Heat inactivation of enzymes.* Appropriate heat inactivation of restriction enzymes should be performed after restriction digestion of plasmid DNA for the release of transgenic insert.

3. *Phenol chloroform extraction.* Organic solvents are commonly used to remove proteins and enzymes.

4. *Ethanol precipitation combined with ethanol washes and extensive dialysis.* This procedure can be used to remove agarose, CsCl, and organic solvents (phenol, chloroform). Precipitation is usually performed in the presence of salts, which in the case of large-molecular-weight DNA molecules is more efficient with ammonium acetate, compared with sodium acetate.

5. *Extraction with water-saturated* n-*butanol.* This procedure can be used to remove EtBr.

6. *Dialysis.* Dialysis against a large volume of TAE buffer or injection buffer for a minimum of 16 hours (24–48 hours is optimal) can be used to rid traces of salts, CsCl, and organic solvents.

7. *Electroelution.* This method is used to release DNA from agarose gels.

8. *Gelase.* Gelase can be used as an alternative to electroelution for releasing DNA from agarose gels.

9. *Ion-exchange chromatography.* The DNA is bound to a matrix, washed in low-salt concentrations, and subsequently eluted in high-salt concentrations. This

process yields DNA or RNA of a purity and biological activity equivalent to two rounds of purification in CsCl gradients. Residual salts must be removed by precipitation, ethanol washes, and dialysis. The chromatography technique removes cellular debris, proteins, carbohydrates, small metabolites, bacterial chromosomal DNA, EtBr, organic solvents (phenol, chloroform), enzymes, agarose, dyes, and low-molecular-weight impurities. However, some types of ion-exchange columns will not separate DNA from RNA. Endotoxins are also not always removed, because the negative charge on the toxins may react with the matrix resins (Cotten et al. 1994).

10. *Silica matrix columns.* The technique is based on a simple bind–wash–elute procedure. Nucleic acids are adsorbed to the silica gel membrane in the presence of high concentrations of chaotropic salts, which remove water from hydrated molecules in solution. Polysaccharides and proteins do not adsorb and are removed. After a wash step, pure nucleic acids are eluted under low-salt conditions in small volumes. This method yields a pure DNA solution, which in most cases is directly suitable for microinjection. Silica matrix columns can be used to remove carbohydrates, proteins, RNA, small DNA fragments, enzymes, organic solvents (phenol, chloroform), agarose, salts, oils, and detergents.

Commercially Available Kits That Can Be Recommended for Plasmid Preparation and DNA Purification

DNA can be purified for microinjection by several methods. The classic preparation is a time-consuming and tedious procedure that includes the use of CsCl gradient centrifugation for plasmid DNA preparation, recovery of DNA from agarose either through electroelution or gelase, extraction of DNA with organic solvents, removal of EtBr with water-saturated *n*-butanol, ethanol precipitation and several washes, and, finally, extended dialysis. This protocol produces an ultrapure DNA, well suited for microinjection. In recent years, more and more commercially available kits have appeared both for plasmid DNA preparation and for the purification of DNA for microinjection. The efficiency and suitability of these kits may vary not only between manufacturers, but also between batches of kits (L. Montoliu and T. Saunders, pers. comm.). It is impossible to give advice on the choice between existing kits; new products of increasingly good quality are constantly being launched on the market. Table 7.1 summarizes kits that have given good results for mammalian transgenesis.

Preparation of Standard-sized Plasmid DNA from Bacterial Cultures

The plasmid-containing bacterial culture should be grown up in selection medium to an optimal density before the cells are harvested. Traditionally, harvested bacterial pellets have been lysed either with sodium dodecyl sulfate (SDS) or with alkaline SDS, boiling lysis; the choice of method depends on the strain of *E. coli* (Sambrook and Russell 2001).

Bacterial cell contents, RNA, and chromosomal DNA have often been removed from the crude DNA preparation by buoyant density centrifugation in gradients of CsCl-EtBr (Sambrook and Russell 2001). This method is very effi-

TABLE 7.1 Kits for plasmid preparation and DNA purification

	Mode of action	Suitable for	Contaminants removal	Maximum DNA size
Qiagen Endo Free Plasmid Kit (12362)	Modified alkaline lysis procedure for preparation of plasmid DNA from bacterial cultures. Gravity-flow, anion-exchange filtration through special resins. Contains a highly recommended additional endotoxin removal step.	Plasmid purification	This system can be used instead of CsCl centrifugation and phenol/chloroform extraction. Cellular debris, proteins, carbohydrates, small metabolites, bacterial chromosomal DNA, RNA, dyes, and low-molecular-weight impurities are removed.	Plasmids up to 150 kb can be purified. However, plasmids larger than 45–50 kb may give a reduced elution efficiency.
Elutip-D, (Schleicher & Schuell, 462615)	Ion-exchange chromatography, containing a matrix similar to RPC-5. A prefilter is used if DNA is recovered from standard agarose gels to remove agarose particles.	Can be used as a step in DNA fragment purification. The eluted DNA requires further purification by precipitation and dialysis to remove excess salts.	Because both DNA and RNA bind to the matrix, these cartridges cannot be used to separate DNA from RNA efficiently.	The largest fragment size that can be purified is 50 kb.
BIO 101 GENECLEAN Spin Turbo (1102-)	DNA is bound to a glass silica column in high concentrations of chaotropic salts and is eluted in a low salt concentration or water.	Final purification of DNA for microinjection.	These columns can be used to remove proteins, carbohydrates, RNA, DNA smaller then 200 bp, residual organic compounds that may be left after phenol/chloroform extractions, salts, agarose and EtBr enzymes, oils, and detergents.	The authors have used these columns for DNA fragments up to 20 kb with great success. According to the manufacturer, they can be used for fragments up to 300 kb.
QIAquick (28704) and QIAEX II (20021) gel extraction kits	DNA purification with spin columns containing a silica-gel membrane. DNA binds to the membrane in the presence of high salt concentrations, and is eluted in Tris buffer or water.	Both kits can be used for extraction of DNA from both standard and LMP agarose gels.	These columns remove salts, enzymes, agarose, dyes, ethidium bromide, oils, and detergents.	QIAquick can only be used for fragments that are a maximum of 10 kb, whereas QIAEX II is suitable for fragments up to 50 kb.
NucleoBond AX 500 Tip (Clontech 4003-1).	NucleoBond kits are based on an anion-exchange system, where a microporous, hydrophilic silica bead resin is packed inside a polypropylene column (tip).	High-copy and low-copy plasmids, cosmids, BACs, and PACs grown from bacterial cultures can be purified using these kits.	The resulting plasmid DNA is as pure as plasmids isolated by traditional CsCl procedures.	PACs and BACs.

cient, and separation depends on different amounts of EtBr that can be bound to linear and closed-circular DNA molecules. However, the protocol is time-consuming, requires expensive equipment, and introduces additional toxic contaminants into the DNA preparation that have to be removed in additional purification steps. Discontinuous CsCl gradient centrifugation (where three solutions containing different concentrations of CsCl are layered in the centrifuge tube) can be performed faster, but the resolution of plasmid DNA from bacterial DNA is not as good as with continuous gradients. Therefore, this latter method cannot be recommended for purification of plasmid DNA for microinjection purposes.

Lately, several commercial kits have appeared on the market (see Table 7.1), allowing all three steps of plasmid purification in one protocol: lysis, removal of cellular components, and separation of plasmid DNA from bacterial DNA. These purification kits rely on ion-exchange chromatography to separate plasmid DNA from cellular nucleic acid. They contain disposable chromatography columns with different matrices, such as glass, diatomaceous earth, or anionic resins such as DEAE or QAE. Traditionalists still believe that plasmids purified by banding in CsCl-EtBr gradients are the purest and best DNAs for microinjection, although data suggest that the two methods are equally suitable. We recommend starting with a commercial kit. If problems arise that could be connected to DNA impurities, then the CsCl gradient method can be used as an alternative.

The isolation of BAC DNA can be performed with plasmid purification kits especially designed for large-molecular-weight DNA (see Protocol 7.2). The extremely large size of YAC DNA makes the isolation more troublesome. Some plasmid purification kits may also be able to handle YACs, but the recommended method is the use of agarose plugs and pulsed-field gel electrophoresis (PFGE) (see Protocol 7.4). Unfortunately, YACs are generally of a similar size as yeast chromosomes, which may result in co-migrating or closely migrating endogenous yeast chromosomes. Although some studies indicate that the co-integration of contaminating endogenous yeast chromosomes does not seem to have an overt effect on the expression of YAC transgenes, it is always preferable to microinject YAC DNA samples free of contaminating yeast chromosomes (Giraldo 2001). This problem can be solved by the use of so-called yeast window strains. Each of these strains has a defined alteration in its karyotype that allows the isolation of relatively pure YAC DNA (Hamer 1995).

Digestion of Plasmid DNA for Release of Genetic Construct Fragment

In most cases, it is strongly recommended to separate the transgenic fragment from as much of the vector backbone as possible before injection. The plasmid should be incubated at appropriate temperatures, depending on the restriction enzymes used. On the following day, residual enzymes should be inactivated with the appropriate heat treatment. It is important to choose restriction enzymes in such a way that the resulting fragments have significantly different sizes to aid the separation of co-migrating bands on an agarose gel. If necessary, the vector backbone can be cut at additional points to create smaller fragments. It is recommended to start with a large amount (at least 100 µg) of plasmid DNA to recover enough (high concentration) purified fragment for injection. A small aliquot of the digest should be run on an electrophoresis gel to verify complete digestion and DNA integrity. The bands may be stained in EtBr, and then visualized under UV light. A "beard-shaped" smear of bands, which indicates DNA degradation, and additional bands, which could indicate inappropriate cutting, are not acceptable.

Gel Separation

The digested plasmid DNA can be loaded into an agarose gel, where the genetic construct fragment can easily be separated from the vector backbone by electrophoresis. TAE (not TBE!) buffer should be used both in the agarose gel and as the running buffer. Several types of agarose can be used for this purpose, and the choice mainly depends on the DNA size. Standard agarose gels are generally prepared to a concentration of 0.8–1.5%. Small DNA fragments (<500 bp) are better resolved in a higher agarose concentration of 1.5–2%. Fragments above 10–15 kb are better separated in lower agarose concentrations. PFGE should be used for DNA molecules above 50 kb. The quality of the agarose is a critical factor for the successful recovery of high-quality DNA. Therefore, we strongly recommend using only the highest possible grade, ultrapure agarose. Traditionally, the gel is stained either before or afterward with EtBr to detect the DNA under UV light. This is the most sensitive method, and undoubtedly the most suitable for DNA quantification (see p. 308) and for determining DNA integrity (see p. 307). However, EtBr is toxic, highly mutagenic, deleterious for embryos, and must be completely removed from the DNA before injection. The visualization also requires the exposure of the DNA to UV light, which potentially can damage the DNA structure. For separation of fragments of different sizes, where the quality of the digest has previously been determined, it is not necessary to use such a sensitive detection system as EtBr, since the migrating bands will be sufficiently visible also with other methods. One simple, reliable, and nontoxic staining method that allows the visualization of stained DNA under ambient light uses methylene blue (Flores et al. 1992). As an alternative, it is also possible to stain only parts of the gel in EtBr and to retrieve clean DNA by using the stained parts as markers.

Recovery of DNA Fragments from Agarose Gels

Several methods exist for recovering DNA from agarose gels (see Table 7.2). The choice will depend on the DNA size, and to some extent, personal preference.

Fragment Purification

Numerous protocols exist for the final purification of the transgenic fragment. The variations range from a few simple steps to very complex and time-consuming procedures. Expected contaminants in the DNA sample to be purified and individual experience are factors that will influence the method of choice. Additionally, a protocol that gives excellent results in one laboratory may not be the optimal one in other investigators' hands. We recommend first to try the simple procedure described in Protocol 7.1. If problems occur that are suspected of originating from impure DNA, the classic but more tedious procedure of CsCl gradient centrifugation and extensive dialysis may be used. Independently of the purification method, there are certain criteria that are important to keep in mind:

- All traces of residual cellular components, bacterial DNA and RNA, organic solvents, ethanol, salts, agarose, and EtBr have to be removed.

TABLE 7.2. Methods for recovering DNA from agarose gels

Electroelution with Biotrap	Biotrap BT1000 electro separation system (Schleicher & Schuell). DNA fragments between 14 and 15,000 bp can be eluted.
Electroelution with dialysis bags	Can be used for a large range of DNA fragment sizes (Sambrook).
Anion exchange or silica matrix chromatography	Commercial kits such as Elutip-D (Schleicher & Schuell), NACS Life Technologies, GeneClean BIO 101, Q-Biogene.
Enzymatic extraction	(Agarase, Gelase) extraction from LMP agarose gels. The gel slice containing the DNA transgenic fragment is digested with the enzyme Gelase or Agarase, which hydrolyzes the agarose to disaccharides.
Pulsed-field gel electrophoresis	Large DNA fragments are resolved by PFGE, and recovered from the gel by enzymatic (agarose) treatment. PFGE can be used to separate DNA fragments of up to 6 Mb.

- Care should be taken to keep the DNA fragment intact (see below).

- All reagents used for the purification steps should be of the highest possible quality.

- Sterile water from Sigma or Gibco, or 18 Ohm Milli-Q water, should be used for buffer preparations. Sigma water for embryo transfer (W1503) is recommended for the preparation of injection buffer.

- The microinjection buffer should be filtered through a 0.22-micron filter before use (see p. 309 and Protocols 7.6 and 7.7).

- Several methods have been used for purification and concentration of YAC DNA for microinjection. Many investigators use a second standard gel electrophoresis after the initial PFGE run (Schedl et al. 1993), followed by agarase treatment to recover the DNA. A second possibility is dialysis in sucrose (Gaensler et al. 1993). Ultrafiltration units have recently become more frequently used (see Protocol 7.5). This is the fastest and most convenient method, although great care should be taken not to damage DNA during the filtration. BAC DNA can be purified with the same methods as used for YACs, or with some commercial kits (see Protocol 7.3).

PREPARING DNA FOR MICROINJECTION

Determining and Maintaining High DNA Quality

DNA should be handled with care during all steps of purification, dilution, and storage. Larger molecules (>100 kb) are especially at high risk of breakage if not handled very carefully, but even smaller molecules of 50 kb can be damaged by vigorous handling. Sheared, nicked, or degraded DNA is unsuitable for microinjection. Because the purification process is a time-consuming process to repeat, it is worthwhile to prevent such problems from arising in the first place.

Special consideration should be taken regarding the following points:

- Use wide-bore pipettes, or cut off the tips of narrow pipettes to widen the opening.

- When mixing DNA-containing solutions, pipette slowly and no more than absolutely necessary.

- Avoid vortexing and vigorous shaking; use stirring instead when possible.

- Avoid exposing plasmid DNA to prolonged denaturing conditions.

- Avoid both overdigestion and partial digestion of plasmid DNA by restriction enzymes.

- Dissolve DNA very slowly and carefully after alcohol precipitation. Do not use extensive heating. This is one of the most likely steps during which DNA damage can occur.

- Keep the DNA concentration high.

- Add polyamines to very large-molecular-weight DNA solutions (see Protocol 7.7).

- Minimize the exposure of DNA to UV light.

- Avoid loading more than 200 ng of DNA per lane in standard agarose gels.

- Always use sterilized pipettes and tubes.

- Always use TAE buffer. Do not use TBE buffer for preparation of DNA for microinjection.

The integrity of the DNA should be determined before final dilution for injection. This is best done by loading small aliquots onto a standard agarose gel and running the electrophoresis at low speed. The quality of the bands should be strictly monitored, and pictures with varying exposure should be evaluated. The correct size of the fragment should be confirmed. Any degradation that appears as a smearing of the band into the low-molecular-weight range should be taken seriously. (These smears can more easily be seen on slightly overexposed photographs.) BAC DNA quality and integrity can be best assessed by PFGE of *Not*I-digested BAC DNA (T. Saunders, pers. comm.).

Determining Concentration of DNA for Injection

Brinster et al. (1985) found that optimal DNA integration efficiency (20–40%) is achieved with 1 µg/ml or higher concentrations of linear DNA. However, at high DNA concentrations (>10 µg/ml), embryo survival decreases significantly. Optimal numbers of transgenic mice were produced when DNA was injected at a concentration of 1–2 µg/ml, corresponding to 200–400 molecules/pl of 5-kb DNA fragment.

Previously, it was thought that little, if any, correlation existed between the concentration of DNA injected and the number of copies that integrate (Brinster 1985). Recently, however, it has been shown that a careful reduction of DNA concentration can increase the single-copy integration ratio (Ellis et al. 1997). Although efficiencies are clearly lower, a significantly increased proportion of produced founders will show a single-copy integration pattern. This approach is an alternative for experiments where single-copy integration is important, and in such cases, the lower efficiencies can be accepted. In this case, for standard constructs up to 20–30 kb, a concentration of 0.25–0.5 ng/µl should be used, and for large BACs, the concentration can be further reduced to 0.1 ng/µl.

It is very important to determine accurately the concentration of the purified DNA solution. Inaccurate determinations result in a low frequency of transgenics, or a low survival rate for injected zygotes. Optical density measurements indicate the maximum DNA concentration in a sample but can give falsely high values if contaminated with RNA or other UV-absorbing substances. It is important to use an absolutely reliable DNA concentration or the injection buffer as a reference. A more accurate measurement can be performed by the use of fluorometry; however, this assay gives no indication of the integrity of the DNA. Therefore, it is important to also run an agarose gel with one or more known volumes of the DNA solution in some lanes (estimated to yield 100–500 ng) and a series of dilutions of a standard DNA of known concentration (e.g., a linearized, CsCl-purified plasmid whose concentration is known or bacteriophage λ DNA digested with *Hin*dIII) in several other lanes as controls. The intensities of the "unknown" DNA bands are compared to the standard or control DNA bands to estimate sample DNA concentration. Due to the much larger size of BAC/YAC transgenes, the number of DNA molecules per volume is much lower than for a standard-sized construct when prepared at the same concentration. In contrast to what could be expected, this does not seem to have an effect on integration efficiencies. Therefore, the concentration of large-molecular-weight DNA solutions should be kept to the standard 1–2 ng/µl.

Filtering the DNA Solution?

If the injection buffer is prefiltered, the DNA is carefully prepared according to the above recommendations (see Protocols 7.1–7.5), and the concentration of the stock DNA solution after purification is kept to at least 50 µg/ml, problems with clogged needles during microinjection can be avoided to a large extent. If problems still occur, the already diluted DNA solution may be filtered through a 0.22-micron filter (Millipore Ultrafree-MC) that has been prewashed with water/injection buffer). However, the DNA solution should not be filtered if the construct is larger than 5 kb. Another alternative may be to centrifuge the solution at maximum speed for 10 minutes, then carefully remove the upper half of the solution for injection. Note that hard centrifugation should not be performed with large constructs, such as YACs and BACs.

Storage of Prepared DNA

Stock as well as diluted DNA ready for injection can be stored at 4°C. If it is stored for more than 2 weeks, it should be checked on a gel to assure lack of degradation. Stocks may also be stored at –20°C. Dilutions should be made fresh and not refrozen. BAC and YAC DNA should be prepared as close to injection as possible, stored at 4°C, and not frozen.

PREPARATION OF TECHNICAL EQUIPMENT AND ZYGOTES

Choice of Mouse Strain

The choice of genetic background of the zygote should be considered carefully (see page 150, in Chapter 3). In most cases, a cross between two F_1 animals, such as (C57BL/6 x CBA)F_1, (C57BL/6 x SJL)F_1, or (C57BL/6 x DBA/2)F_1, is used because these combinations have proven to provide large numbers of good-quality embryos. We have used embryos obtained by crossing CD1 outbred females with C57BL/6 inbred males, and this combination has worked very efficiently for us. In these cases, however, the genetic background of the founder animals will be mixed. It has been shown that transgene expression can be modulated by the genetic background of the donor embryo (Chisari et al. 1989; Harris et al. 1988). If an inbred background is required, it is possible to use, for example, C57BL/6 or FVB/N zygotes, although the microinjection process will be more difficult and the viability of the embryos is lower. Another alternative is the use of outbred (CD1 or ICR) zygotes, which generally are cheaper to purchase; ICR CCD-1 are reasonable in superovulation but they lyse more easily than hybrid strains (see Chapter 3). Embryo donor females should be superovulated (see Chapter 3, and Protocol 3.1) to increase the number of zygotes obtained per mouse.

Preparation of Zygotes

Great care should be taken when isolating and preparing zygotes for injection (see Protocol 4.9). The time spent outside the incubator, and the length of hyaluronidase treatment, should be kept to a minimum, because zygotes are sensitive to temperature fluctuations.

The zygote quality should be assessed, and only those zygotes that show a normal morphology should be selected for injection. An initial sorting can be carried out under a dissection microscope. Morphological signs indicating good and poor quality are shown in Figure 7. 1.

Theoretically, two polar bodies should be visible. However, occasionally one of the polar bodies is divided or fragmented, and often the zygote is turned in such a way that the polar bodies are difficult to detect under a low-magnification view. Therefore, zygotes that seem to lack polar bodies or seem to have three polar bodies should not be discarded at this stage. However, the possibility of screening for high-quality zygotes under the dissecting microscope is to a large extent dependent on the quality of the optics and the experience of the operator.

Making Holding Pipettes

The holding pipette serves the purpose of gently but firmly holding the zygote in position during injection. Several commercial companies offer prefabricated and sterile-packed holding capillaries of high quality. Homemade pipettes are of at least equal quality, and can be slightly adjusted in shape and size to suit personal preferences. A decision as to whether to use commercial or homemade capillaries will mainly be based on the quantity needed and the price/time ratio in terms of their fabrication.

Making a good holding pipette, as described in Protocol 7.8, takes time and patience, but it is well worth the effort because it can save considerable time when performing microinjection. The shape and the inner diameter of the holding pipette are of great importance. The opening should be absolutely smooth so that the zygote is not damaged. Neither a too-large nor a too-small opening is comfortable for the zygote. If the opening is too small, the zygote will easily turn around on its axis when poked by the injection needle, so that the pronucleus moves out of the focal plane. However, if the opening is too large, the zygote can easily be sucked into the holding pipette or become greatly misshapen during the catching procedure.

There is another problem associated with holding pipettes with a too-large outside diameter. When such a pipette is resting on the microscope slide surface,

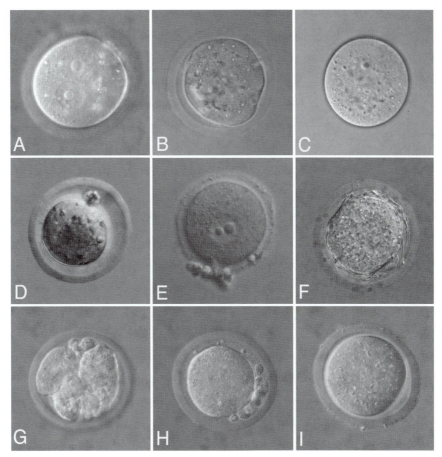

FIGURE 7.1. Morphology of zygotes. (*A*) Normal morphology. (*B*) Zygote close to division. The cytoplasm shows a slightly uneven form; the two pronuclei are large and located close to each other, in the middle of the cytoplasm. (*C*) Zygote without zona pellucida. (*D*) Abnormal size and "shrunken" appearance. Optimally, a clear but not too large previtelline space should be seen. (*E*) Cumulus cells still attached to the zona pellucida. This will make microinjection difficult, because the cells are very sticky and prohibit an easy orientation of the zygote on the holding capillary. (*F*) Severe polyspermy (sperm trapped under the zona pellucida can best be visualized under dark-field illumination). (*G*) Fragmentation of the cytoplasm. (*H*) Fragmentation of the polar bodies. (*I*) Unfertilized zygote.

the opening onto which the zygote is sucked will inevitably be positioned farther away from the slide surface. The zygote, which is smaller, will therefore have to "jump up" and off the slide surface. This scenario usually results in a zygote position that is not adequately fixed so that it easily rotates during injection, and the optimal focusing possibility of the pronuclei is lost. The shape of an ideal holding pipette is illustrated in Figure 7.2.

Some investigators prefer to make a bend on the holding pipette, very close to the tip. This angle may help in the correct horizontal positioning of the capillary at the bottom of the injection chamber (Fig. 7.3). However, if the angle at which the capillary enters the injection chamber can be kept low (<20°), such bending will not be necessary. If the outer diameter is correct, the holding capil-

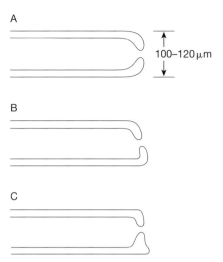

FIGURE 7.2. Holding pipette for microinjection. (*A*) Correct symmetrical shape of the tip; (*B,C*) incorrect shape of the tip.

lary can be positioned resting on the surface of the injection chamber and slightly pushed down on the glass until it rests horizontally.

Making Injection Pipettes

The quality of the injection pipette has a major influence not only on the efficiency with which injections can be performed, but even more for the survival rate of the zygotes. It is worthwhile to optimize the production procedure. Injection needles should be regarded as disposable items. Accumulation of debris on the outer or inner surface is a common problem; these pipettes should be disposed of immediately.

Injection pipettes can be purchased from commercial companies. However, they may also be produced using either a horizontal or vertical pipette puller (see Protocol 7.9). In either case, use thin-walled glass capillary tubing with an outside diameter compatible with the device into which the pipette will be inserted for injection (typically 1 mm or 1.5 mm). Tubing containing an internal glass filament is convenient because it allows the pipette to be filled through the blunt end by capillary action. Glass capillaries from most suppliers can be used directly out of the package without cleaning or sterilization. However, keep the package of unused capillaries closed so that no dust accumulates. Avoid touching the ends and middle section of the unused capillaries with fingers when preparing pipettes. This caution is important to avoid debris from entering the pipette when the DNA solution is filled into it, and to avoid traces of fat and debris from

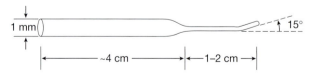

FIGURE 7.3. Holding pipette for microinjection. The pipette is shown after introducing a bend so that it can be used in the depression slide system illustrated in Fig. 7.5.

fingerprints burning onto the tip of the capillary when it is pulled. Do not use powdered gloves to handle pipettes; any powder entering the pipette will inevitably clog the pipette tip!

Pipettes should be pulled the same day, or only a few days before they are to be used in order to avoid dust accumulation, which also leads to clogging. They can be stored until use in petri dishes, supported by two ridges of Plasticine or rolled-up masking tape. Use a mechanical pipette puller to pull a piece of 10- to 15-cm-long glass capillary tubing so that it has a suitable tip for injection. A good injection pipette should have a tip with a diameter of less than 1 μm, which is too small to be seen clearly with a light microscope. If the hole in the end of the pipette can be seen in the microscope, it is probably too large. Pipettes with large tips are difficult to insert into the pronucleus and tend to lyse many of the zygotes; however, if the tip is too small, the hole will clog very easily. The best way to learn what a good pipette looks like is to use various pipettes to determine which ones work best. It is possible to pull the pipette with the correct opening (more than 0.5 but less than 1.5 micron); if this turns out to be difficult, the tip can be chipped on a glass bead on the microforge or against the holding pipette. Some investigators prefer to grind or bevel the tip at an angle in a silicon carbide slurry stirring on a stir plate (Gundersen et al. 1993). This technique yields a pipette tip with a larger opening that allows the fluid to flow out of the pipette more rapidly. A larger opening is required for BAC and YAC DNA to avoid shredding of the DNA.

Another important pipette parameter is the degree of taper in the region near the tip of the pipette, which can be measured by the diameter of the pipette at a fixed distance from the tip. At a point of 50 μm from the tip of the pipette, the diameter should be ~10–15 μm or less. The distance between the shoulder and tip of the pipette is not critical, but it should be at least 5–8 mm. Mann and McMahon (1993) have shown 8 mm to be optimal. If this distance is much shorter, it may be difficult to position the tip near the bottom of the injection chamber because the pipette shoulder will interfere with the injection. However, if the distance is too long, the needle will lose stability and will bend easily when moved through the thicker oil phase of the injection chamber and when injection is attempted.

The optimal shape of these needles is achieved by using relatively moderate heat values, combined with rather hard pull strength and velocity and a low time delay. When using an electronic puller with a built-in test program, it is very important to run this before the first capillary is pulled to avoid burning up the filament. On the Sutter P97, this is called a ramp test. The value obtained from the ramp test will determine the maximum of heat that can be used for the particular filament and glass combination. The exact positioning and quality of the heating filament will have a great influence on the shape of the pulled needle. Therefore, it is important to check that the filament is not damaged and that it is exactly positioned before use.

Some investigators prefer to treat the injection pipettes with silicon (DePamphilis et al. 1988). This technique may be useful when attempting to inject zygotes of "difficult strains." It is important that the pipette enter the embryo perpendicularly and not obliquely to make sure the membrane is punctured and not torn. When using cavity slides as injection chambers, it is critical to bend the injection needle so that the tip is horizontal.

Microinjection Setup

There are many ways to set up a microinjection workstation (Protocol 7.10). Some microscope manufacturers provide complete setups including all imaginable components in an almost "ready-to-use" configuration. This alternative allows one to get started in a very speedy way and generally provides excellent state-of-the-art equipment, but it is expensive. Other alternatives can be used to achieve an equally good injection environment, but the assembly of the individual components is not always very easily optimized.

The basic components of any microinjection workstation are an inverted microscope in the middle with two micromanipulators on each side (see Fig. 7.4). The micromanipulators allow the operator to perform very precise movements during injection. In addition, the control of pressure must be regulated, both in the injection needle and in the capillary that holds the zygote in place. It is vital for successful injection and embryo survival to avoid any vibrations on the microscope stage. If the laboratory environment is free of vibrations, a heavy table will be sufficient. However, in most cases, a sturdy marble table, or even a pneumatic antivibration table, may be needed.

Microscope

Microinjection is most easily performed using an inverted microscope. Several microscope models specifically useful for microinjection are now available from Leica, Nikon, Zeiss, and Olympus (see Chapter 17). Microinjection of zygote

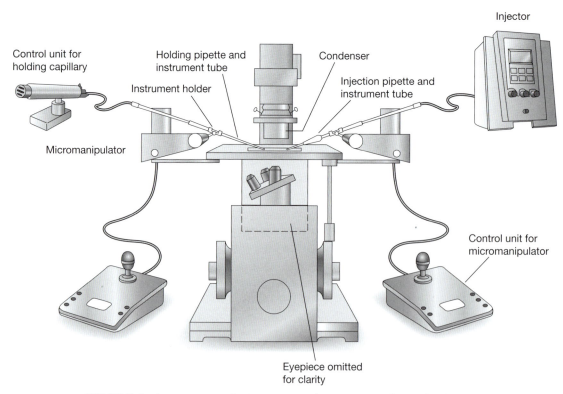

FIGURE 7.4. Arrangement of microscope and micromanipulators for the pronuclear injection method.

pronuclei is most easily performed at a magnification of 400–500x, usually achieved with a 40x objective and 10–12.5x eyepieces. In addition, a low-power objective (2.5–5x) is used for transferring zygotes into and out of the chamber.

Pronuclei can be best seen using Nomarski differential interference contrast (DIC) optics or Hoffman Modulation Contrast optics. DIC optics are preferable because of the better resolution and sharp visualization of the outline of the pronucleus. However, Hoffman optics are generally somewhat cheaper and may for this reason be an alternative to consider. DIC optics must be used with glass injection chambers, whereas Hoffman optics can be used with disposable plastic tissue culture dishes. Phase contrast is not sufficient in visualizing pronuclei; however, bright-field optics are sufficient for the low-power objective.

Micromanipulators

Two micromanipulators are needed for the movement of the holding and injection capillaries, respectively. Commercially available micromanipulators, such as Leica, Eppendorf, or Narishige, are all suitable, and the choice of brand is a matter of personal preference and some practical considerations. For detailed discussions, see Chapter 17.

Pressure Control in the Holding Pipette

The control of the holding capillary (capture and release of the zygotes) can be achieved by using an air-, oil-, or water-filled system (e.g., Eppendorf CellTram air), by simple mouth pipetting, or by using a homemade system consisting of a 5- to 10-cc syringe connected to the capillary holder through the tubing with the option of an oil- or air-filled capillary. Commercial control units (e.g., Eppendorf, Sutter, Narishige) should be used with the supplied air-filled tubing, and the holding capillary/tubing/control unit set up according to the manufacturer's instructions. In some cases, one might experience too fast a movement in the capillary due to a different inner diameter of the glass. In this case, it may be helpful to fill the capillary with light oil. If a custom-made micrometer screw-controlled unit is used, it is in most cases advisable to use thick Tygon tubing (Tygon R3603) and to fill the complete set (syringe, tubing and holding capillary) with oil to achieve a precise movement.

The simplest and most efficient method for controlling the holding pipette is by mouth pipetting. The only disadvantage is the difficulty for the operator to talk during injection, which may make teaching more difficult (local regulations may also forbid the use of mouth pipetting).

The choice of oil viscosity for the oil-filled alternative will depend on the inner diameter of both the connecting tubing and the capillary. As a general rule, the heavier the oil and the smaller the inner diameter of the tubing/capillary, the slower the motion will be.

Pressure Control in the Injection Pipette

The flow of DNA in the injection pipette is best controlled by an automatic injector (e.g., Eppendorf FemtoJet; see Chapter 17). These devices guarantee a precise and easily controlled injection process. It is also possible to use a simple 10- or 50-ml glass syringe lightly greased with paraffin oil connected through the micromanipulator handle to the injection pipette, with thick Tygon tubing with a tubing adapter and Luer lock to enable relief of the pressure when necessary. This is a cheap and useful alternative but is recommended only for experienced operators. The risk of "backflow" into the injection pipette is high, and the injection process is much more difficult to control. In either case, an absolutely airtight connection has to be established between the control device and injection pipette. If the Eppendorf FemtoJet is used, the pressure constant (Pc) should be set to around 10, which allows a slow constant flow of DNA out of the injection pipette tip. This flow largely prevents DNA from clogging the pipette tip, and it blocks any backflow of medium into the tip, which could dilute the injected DNA. The injection pressure (Pi) should initially be set to 30–40. After the first injection, this value should be adjusted in such a way that the pronucleus clearly swells up within one second, but not faster. If the pressure is set too low, injection will take too long and may harm the zygote. If, on the other hand, the pressure is set too high, the DNA blows up the pronucleus too fast, which could lead to lethal damage to the zygote. The injection should take place in "Manual" mode, where the injection time is adjusted for each pronucleus, and kept until a clear swelling is achieved.

As a possible alternative, an oil-filled system can be used for controlling the flow of DNA in the injection pipette. This system consists of a micrometer screw connected to a syringe, for example, the Narishige microinjector IM-9B or the Eppendorf CellTram Vario. In this case, a filament-free pipette is first loaded with oil from the back side, and then with the DNA solution from the front by applying suction with the control device. The pipette is then moved into the injection chamber, and slow positive pressure is achieved by twisting the control unit knob. This system, if correctly used, prevents any backflow of media into the injection pipette, but it is not possible to control the pressure as accurately as with an electronic injector. If the pressure is set too high, the result will inevitably be an increased rate of zygote lysis.

Injection Chambers

The injection takes place in an injection chamber, which serves as a microenvironment in which zygotes can survive for a short period of time. The chamber consists of a small drop of M2 medium covered by light paraffin oil. The injection pipette is inserted from the right-hand side, the holding capillary from the left-hand side, and the zygotes take place in the middle (see Fig. 7.5). There are several alternatives for setting up such a chamber:

- *Plastic petri dish with glass bottom (Willco wells GWst-3522)* (Fig. 7.6). M2 medium is placed in a special plastic dish with very low walls and a large, flat inner well with glass bottom, and then covered with oil. These dishes are very convenient to use. They also have the advantage of being disposable, thus eliminating the need for cleaning and the associated risk of toxic contamination from residual detergents. However, these dishes do not allow very shallow angles when inserting holding and injection pipettes (see Protocol 7.9, Comments), a factor that may contribute to increased lysing rates. An inverted lid of a normal plastic tissue culture dish can be used in a similar fashion if the microscope is equipped with Hoffman optics.

- *The depression slide injection chamber* (Fig. 7.7). This can be used with any optical system. A drop of M2 medium is placed in the center of the depression of

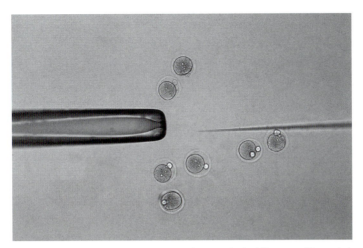

FIGURE 7.5. Microinjection chamber with holding capillary, injection pipette, and zygotes ready for injection.

a glass depression slide and covered with light paraffin oil. These slides can be reused after rinsing with plenty of distilled water followed by 70% ethanol and then drying.

- *Metal frame/glass slide injection chamber* (Fig. 7.8). For these chambers, a metal frame with an outside shape exactly matching a microscope slide is used. In the center of the metal, a rectangular hole is cut with the approximate size of 10 x 20 mm. The frame is covered with a thin layer of high-vacuum grease (Dow Corning 698) and pressed onto a glass slide. A drop of M2 is then placed in the middle of the frame, onto the glass, and covered with light paraffin oil.

Factors to consider include the following:

- Depending on the glass used for the microinjection chamber, the M2 drop may float out very widely to form a large thin drop, which is not deep enough for harboring the holding pipette and zygotes. If this happens, initially a very small microdrop of M2 should be placed on the glass. Then cover the drop with oil. With a mouth pipette and drawn-out glass capillary, add additional M2 to the drop until the desired volume has been reached. The oil will keep the medium in shape, allowing a higher drop to develop. Alternatively, the slide/coverslip can be siliconized (Sigmacote, Sigma) to achieve a less flattened drop.

- If the drop of medium tends to become very high, the DIC optics may have difficulties with the light bending in a way that may result in less-than-optimal

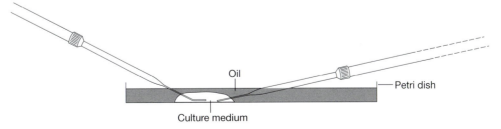

FIGURE 7.6. Petri dish injection chamber.

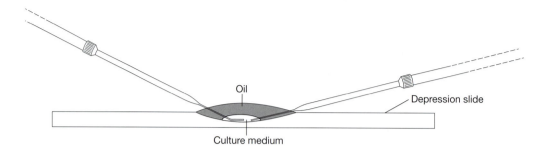

FIGURE 7.7. Depression slide injection chamber.

visualization of the pronuclei. In this case, the drop should be flattened out prior to adding the oil by inserting a glass capillary and making a swift round movement.

• Regardless of which injection chamber is used, great care should be taken to keep oil from spilling into the inverted optics. If oil should drop onto the objectives, it should immediately be removed carefully, and the objective cleaned. If oil has entered lower parts of the optics, the microscope should not be used further, and a professional cleaning should be performed as soon as possible to avoid severe damage.

MICROINJECTION

Timing of Pronuclear Injection

The pronuclei of fertilized mouse oocytes can be most easily injected when they are at their maximum size, just before the nuclear membrane disappears prior to the first cleavage. The pronuclei swell progressively during the one-cell stage, and they are in an optimal state for injection for a period of ~3–5 hours. The tim-

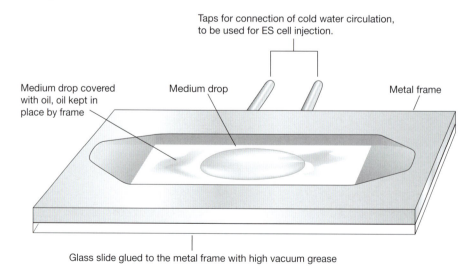

FIGURE 7.8. Metal frame injection chamber.

ing of the zygote's developmental stage can be controlled by careful adjustment of the time of human chorionic gonadotropin (hCG) administration, the light cycle in the animal facility, and time of embryo collection. It is important to keep in mind, however, that the effects of all these variables on embryonic development highly depend on the strain or hybrid cross used. For more details, see Chapter 3. Typically, on a 5 a.m. to 7 p.m. light cycle, if the hCG is administered at noon the day prior to injection and the zygotes are collected around 10 a.m., injection will be optimal between noon and 4 p.m. The best timing for hormone administration, light cycle, and embryo collection should be determined empirically. If the embryo development is not advanced enough, the pronuclei will be very small, peripheric, and difficult to inject. In this case, hCG should be given earlier, and/or the light cycle should be set back. On the other hand, if the embryos are at too late a stage of development, the pronuclei will already be fused, and the first division will be taking place. The solution here is to give the hormones later and/or set the light cycle forward.

Microinjection of Mouse Zygotes

The microinjection procedure is described in detail in Protocol 7.11. The zygotes are captured and held in position by the holding capillary, and the injection pipette is carefully inserted into one of the pronuclei. DNA is injected until a clear swelling of the pronucleus can be observed, and then the injection pipette is removed. Once the microinjection setup is well adjusted, this process can be performed rather swiftly; however, it may take considerable time to adjust all parameters in such a way that zygotes can be optimally injected. These adjustments should be performed with attention to detail, as they will greatly contribute to the survival rate of injected zygotes. As a general rule, one should concentrate on injecting with great care rather then trying to increase the speed of the micromanipulation. The aim should be to achieve a high survival rate and a high rate of founder animals, and this goal is better reached by injection focusing on quality rather than quantity. Therefore, we do not discuss the number of oocytes that can be injected within a certain amount of time, as this is greatly influenced by the experience of the operator.

After Microinjection

Zygotes that have survived microinjection can be transferred the same day into the oviducts of 0.5-dpc pseudopregnant female mice (see Chapter 6). It is also possible to culture them overnight in vitro to the two-cell stage and then transfer them in the same way as one-cell-stage zygotes. Keep in mind that in vitro culture may decrease embryo viability. However, in vitro culture can be performed in case there is a lack of recipients, or in other special cases. The overnight culture allows the injection of more zygotes in one session and can serve as a quality control judged by the number of zygotes developing to healthy two-cell-stage embryos. This approach could also be useful when the viability of the embryos is expected to be reduced due to very early transgene expression, and it therefore is desirable to monitor their development. Approximately 30% of microinjected zygotes transferred into the oviduct will develop to term. To avoid very small litters, which may not be cared for by the foster mother, it is advisable to transfer 20–30 microinjected zygotes into each pseudopregnant recipient. It is possible to transfer all oocytes into one oviduct of the pseudopregnant recipient; however, transferring half the embryos into each of the two oviducts is highly recommended.

ESTABLISHMENT AND MAINTENANCE OF TRANSGENIC MOUSE LINES

Transgenic mice that develop from injected zygotes are termed "founders." Founders can be screened for the presence of the transgene by analysis of tissues obtained by biopsy. Founders that carry the transgene are then bred to establish transgenic lines. Alternatively, or in parallel, founders can be bred immediately to obtain transgenic progeny that are then used to analyze transgene expression. To develop a transgenic line from a male founder, the founder is bred with several nontransgenic females. There is no need to keep the male with the females until the pups are born; instead, he can be mated to a new female as soon as the first one has successfully been mated. In this way, the male founder can sire many litters in a short period of time. To develop a transgenic line from a female founder, the founder is bred with a nontransgenic male. Accordingly, it is necessary to wait until the female founder has given birth and raised at least one litter containing transgenic progeny to establish a line. To ensure the establishment of a transgenic line, both female and male founders should not be sacrificed until transgenic progeny have been identified. Although the majority of founders will transmit the transgene to 50% of their offspring, ~20–30% of founders are mosaics and will transmit the transgene at a lower frequency, e.g., 5–10% (Wilkie et al. 1986), or not at all.

A founder derived from an inbred strain is typically bred with mice of the same inbred strain to maintain the transgene on a specific inbred background. If the founder is an F_2 hybrid, or if it is not essential to maintain the transgene on an inbred background, it is more efficient to breed the founder with F_1 hybrid or outbred mice, which often have a higher reproduction rate and make better mothers than inbred mice.

A simple and productive way to maintain a transgenic line is to breed hemizygous transgenic males with nontransgenic females. The transgenic progeny can be identified by the presence or absence of the transgene. A transgenic line can also be maintained by breeding hemizygous transgenic females with nontransgenic males, but this is not as productive. Maintaining the transgene in the hemizygous state minimizes complications caused by insertional mutations or by gradual loss of fertility frequently observed with inbreeding.

Despite the above concerns, if desired, homozygous transgenic mice can be produced by setting up hemizygous intercrosses. One quarter of the progeny from such a cross should be homozygous for the transgene, one-half hemizygous, and one-quarter nontransgenic (for assays to distinguish homozygous from hemizygous animals, see Chapter 12). Approximately 5–15% of the random DNA integration events in transgenic mice produce recessive lethal mutations; consequently, not all transgenic lines can be maintained with the transgene in the homozygous state.

Simple, Reliable Steps for DNA Fragment Isolation and Purification

MATERIALS

EQUIPMENT

BioTrap electroelution device (Schleicher & Schuell, BioRad BT1000)
Electrophoresis chamber suitable for the BioTrap device
GENECLEAN Turbo columns (BIO 101, 1102)
Microcentrifuge tubes
Qiagene Endo Free Plasmid kit (12362)
Scalpel blade, clean

REAGENTS

Bacterial culture containing plasmid with transgenic construct
Distilled water
GENECLEAN Turbo salt (BIO 101, 1102)
GENECLEAN Turbo Wash solution (BIO 101, 1102)
Methylene Blue solution, 0.1% in 0.5 M sodium acetate <!> (pH 5.2)
Microinjection buffer (see Protocol 7.6)
Plasmid DNA (100 µg) containing transgenic construct
Restriction enzymes
TAE buffer<!> (see Appendix 1)
  1x solution: 40 mM Tris acetate, 1 mM EDTA (pH 8.0)

CAUTION: See Appendix 2 for appropriate handling of materials marked with <!>.

PROCEDURE

1. Prepare plasmid DNA from the bacterial culture using the Qiagene Endo Free Plasmid kit and following the manufacturer's instructions. Aim to isolate at least 100 µg of DNA.

2. Cut the plasmid DNA with the relevant restriction enzymes.

3. Heat-inactivate the restriction enzymes.

4. Run the plasmid digest on a standard agarose gel in TAE buffer.

5. Stain the gel by immersing it in 0.1% methylene blue solution for 20 minutes.

6. Destain the gel in distilled water for 30 minutes, changing the water every 10 minutes.

7. View the gel under short-wave UV illumination and then excise the gel fragment containing the transgenic insert with a clean scalpel.

8. Assemble the BioTrap device with the relevant filters according to the manufacturer's instructions. Take care to place the glycerine-stored filters in the right orientation. Keep the elution chamber to the smallest possible size.

9. Place the gel fragments in the BioTrap device. Fill the device with TEA buffer and place it in a conventional electrophoresis chamber.

10. Electroelute the DNA in TAE buffer for 4–5 hours at 150 V.

11. Carefully remove the eluted DNA from the elution chamber.

12. Calculate the amount of eluted DNA, assuming a loss of 25% during elution.

13. Divide the DNA solution into microcentrifuge tubes, with a maximum of 10 μg in each.

14. Add 5 volumes of GENECLEAN TurboSalt to each tube.

15. Mix well, and add the solution from each tube to a separate GENECLEAN Turbo cartridge.

16. Let the DNA bind to the matrix in the filter at 37°C for 10 minutes.

17. Centrifuge the cartridges for a few seconds at 14,000g, then remove the flowthrough.

18. Add 500 μl of TurboWash solution to each cartridge.

19. Centrifuge the cartridges for a few seconds at 14,000g, then remove the flowthrough.

20. Repeat steps 9 and 10.

21. Centrifuge the cartridges for 4 minutes at 14,000g, then move them to clean catch tubes.

22. Add 30 μl of microinjection buffer, and let the DNA elute at 37°C for 10 minutes.

23. Centrifuge the cartridges for 10 seconds at 14,000g and recover the eluted DNA.

COMMENTS

Additional Steps That May Improve DNA Quality:

- **Elutip-D-minicolumn** (Schleicher & Schuell 27370) or **NACS** columns. Should be followed by ethanol precipitation and dialysis to remove the high salt concentrations.

- **Isopropanol precipitation** followed by several 70% ethanol washes and extensive dialysis.

- **Phenol/chloroform extraction** (2–3x) with chloroform containing 5% isoamyl alcohol, followed by 2x chloroform extractions, ethanol precipitation, several washes with 70% ethanol, and extensive dialysis.

- **Dialysis** for 12–72 hours against 2 liters of injection buffer. The buffer should be changed several times during the dialysis.

- **Filtering** through 0.22-micron filter (prewashed with water/injection buffer) can be useful for smaller constructs, but should not be done when DNA molecules larger than 5 kb are purified (see p. 309).

Isolating BAC DNA from Bacterial Cultures Using the NucleoBond System

This protocol was kindly provided by Dr. Thom Saunders, University of Michigan, Transgenic Animal Model Core, Ann Arbor, Michigan 48109-0674.

MATERIALS

EQUIPMENT
NucleoBond AX-500 Tip (Clontech 4003-1)
NucleoBond Folded Filters (Clontech 4062-1)
Tube, 50 ml

REAGENTS
Bacterial culture pellet, from 100- to 300-ml bacterial culture
NucleoBond Buffer Set I (Clontech 4040-1)

Buffers S1 and S3 in the buffer set should be stored at 4°C; buffers S2, N2, N3, and N5 at room temperature

PROCEDURE

1. Carefully resuspend a bacterial cell pellet from a maximum 300 ml of bacterial culture in 12 ml of Buffer S1 in a 50-ml tube.

2. Add 12 ml of Buffer S2. Mix gently by inverting the tube, and incubate at room temperature for 5 minutes. Do not vortex.

3. Add 12 ml of Buffer S3 and mix gently by inverting six to eight times until a homogeneous suspension is formed. Incubate the mixture on ice for 10 minutes.

4. Put the folded NucleoBond filter onto a 50-ml tube, wet it with 0.5–1.0 ml of water, and fill it with the cooled lysate from step 3. Collect the cleared flowthough in the 50-ml tube.

   To get the maximum recovery of DNA, rinse the filter paper with an additional 1.5 ml of water.

5. Equilibrate an AX-500 cartridge with 5 ml of Buffer N2.

6. Load the cleared flowthough from the filter paper onto the equilibrated AX-500 cartridge.

7. Wash the cartridge with 2 x 12 ml of Buffer N3.

8. Elute DNA with 6 ml of Buffer N5.

   If this elution step is repeated one additional time, up to 30% more DNA can be isolated.

COMMENTS

- Take care that the supernatant of step 5 is clear! Any precipitate will clog the cartridge and prevent the adsorption of nucleic acids.

- To prevent lower-than-expected yields of plasmid DNA due to incomplete bacterial lysis, the amounts of the buffers are increased (compared to the manufacturer's protocol) to reduce viscosity and to promote diffusion. A minimum of 4.0 ml of each of Buffers S1, S2, and S3 should be used per 100 ml of culture.

- It has been reported in some cases that the use of a high-GC-content type of elution buffer helps to increase the yields of very large plasmids such as BACs, regardless of their GC content. However, use of this formamide buffer can lead to complications in the precipitation step unless care is taken to prevent salt precipitation (room temperature propanol precipitation and centrifugation are required), formamide removal (a second propanol precipitation is required), and propanol removal (an additional ethanol precipitation is required). It is important to control the duration of this denaturing step thoroughly. The following addition of potassium acetate is a crucial step.

Purification of BAC DNA Prepared with the NucleoBond System

This protocol was kindly provided by Dr. Thom Saunders, University of Michigan, Transgenic Animal Model Core, Ann Arbor, Michigan 48109-0674.

MATERIALS

EQUIPMENT
Centrifuge

REAGENTS
BAC DNA prepared by the NucleoBond system
Ethanol, 70% <!>
Isopropanol <!>
Microinjection buffer (see Protocol 7.7)

CAUTION: See Appendix 2 for appropriate handling of materials marked with <!>.

PROCEDURE

1. Add 0.7 volume of room-temperature isopropanol to the DNA solution from Protocol 7.2.

2. Centrifuge at >12,000g at 4°C for 10–20 minutes.

3. Wash the pellet with 70% ethanol.

4. Air-dry the pellet for 5 minutes, then dissolve it in microinjection buffer. Store at 4°C until use.

Large-scale Preparation of Agarose Plugs of Yeast DNA

This protocol was kindly provided by Dr. Lluis Montoliu, Centro Nacional de Biotechnologia (CNB-CSIC), Campus de Cantoblanco, 28049 Madrid, Spain.

MATERIALS

EQUIPMENT

Agarose block formers (plug molds). Seal bottom with tape.
Centrifuge
Flask, 1-liter
Hematocytometer
Water bath, 37–40°C

REAGENTS

EDTA, 50 mM (pH 8.0)

Overnight YAC culture

Selection medium for YACs (i.e., AHC or drop-out medium, SD-W-U, e.g., Sigma Y1251 plus the relevant drop-out medium supplement depending on the yeast strain)

Solution I
~1 M Sorbitol (Merck, autoclaved)
20 mM EDTA (pH 8.0) (autoclaved)
14 mM β-Mercaptoethanol<!> (Merck)
2 mg/ml Zymolyase-20T<!> (ICN 320921)
Sterile water
Prepare fresh, do not store.

Solution II
~1 M Sorbitol (Merck, autoclaved)
20 mM EDTA (pH 8.0) (autoclaved)
2% SeaPlaque GTG agarose (FMC 50112) 0.2 g
14 mM β-Mercaptoethanol<!> (Merck)
Sterile water
Prepare fresh. Melt the Sorbitol plus EDTA and agarose in the microwave. Equilibrate in a water bath down to 37–40°C. Add the β-mercaptoethanol and keep the solution at 37–40°C until use.

Solution III
~1 M Sorbitol (Merck, autoclaved)
20 mM EDTA (pH 8.0) (autoclaved)
10 mM Tris-HCl (pH 7.5)<!> (autoclaved)
14 mM β-Mercaptoethanol<!> (Merck)
2 mg/ml Zymolyase-20T<!> (ICN 320921)
Sterile water
Prepare fresh.

Solution IV
1% Lithium dodecyl sulfate<!> (Sigma L4632)

100 mM EDTA (pH 8.0) (autoclaved)
10 mM Tris-HCl (pH 8.0)<!> (autoclaved)
Sterile water
Filter-sterilize (0.22 micron). Store at room temperature.
100% NDS Buffer
Mix 350 ml of water with 93 g of EDTA and 0.6 g of Tris <!> base. Equilibrate to pH >8.0 with solid NaOH <!> pellets. Add 5 g of N-laurylsarcosine (Sigma) pre-dissolved in 50 ml of water. Equilibrate to pH 9.0 with 10 M NaOH and bring the final volume to 500 ml with water. Filter-sterilize (0.22 micron). Store at 4°C.

CAUTION: See Appendix 2 for appropriate handling of materials marked with <!>.

PROCEDURE

1. Inoculate 200 ml of selection medium for YACs (i.e., AHC or drop-out medium, SD-W-U) with 1 ml of an overnight culture in 1-liter flasks. Let the culture grow at 30°C with vigorous shaking until saturation (1–2 days).

2. Count the number of yeast cells with a hematocytometer.

3. Place the cells in a 50-ml tube and centrifuge at 600g for 5 minutes at room temperature. Discard the medium.

4. Wash the cells by resuspending them in 50 mM EDTA (pH 8.0). Use 40 ml of this solution per 100 ml of original medium. Spin down the cells as in step 2. Repeat this washing step one more time with 10–20 ml of 50 mM EDTA (pH 8.0). Discard the medium.

5. Weigh the cell pellet (assuming a density of 1 g/ml).

6. Warm the pellet at 37–40°C for 30 seconds. Immediately add enough pre-warmed Solution I to give a final concentration of 8×10^9 yeast cells/ml. Carefully resuspend the cells. The volume of liquid added should be kept as small as possible. If the cell counting turns out to be difficult, resuspend the cells with a volume of buffer equal or similar to the volume of cells.

7. Immediately add an equal volume of prewarmed (37–40°C) Solution II. Mix quickly and pipette 80-μl aliquots into agarose block formers (plug mold) previously bottom-sealed with tape and placed on ice. This will make a final concentration of 4×10^9 yeast cells/ml of agarose plug. Proceed as quickly as possible and avoid trapping air bubbles in the plugs. Shake the tube with cells and agarose every 20 seconds in the water bath (37–40°C) to prevent solidification too early, until pipetting is finished.

8. Chill on ice for 10 minutes until the agarose plugs solidify.

9. Transfer the agarose plugs into Solution III for spheroplasting, using 8 ml of solution per milliliter of plug. Incubate at 37°C for 2–3 hours with gentle agitation.

   This is one of the most important steps determining the quality and yield of the DNA. An inefficient spheroplasting step will result in very poor yield. Correspondingly, overdigesting the sample or using enzymatic batches of lower quality might result in DNA degradation and/or bad electrophoretic mobilities.

10. Decant Solution III and replace it with an identical volume of Solution IV (8 ml of solution per ml of plug). Incubate at 37°C with gentle agitation for at least 1 hour. Replace the buffer with fresh Solution IV and continue incubating overnight at 37°C with gentle agitation.

11. The next day, decant the buffer and wash the agarose plugs with 20% NDS buffer. Use 8 ml of buffer per milliliter of plug. Proceed for 2 hours with gentle agitation at room temperature. Repeat the wash one or two times. Agarose plugs can be loaded directly onto PFGE gels or stored in this buffer at 4°C.

   For better resolution and PFGE running conditions, it is recommended to equilibrate the agarose plugs with running buffer or TE (pH 8.0) before loading the gels. Equilibration is performed with at least four consecutive washes of 30 minutes each in excess of buffer. Alternatively, plugs can also be stored in sterile 0.5 M EDTA (pH 8.0) for longer periods (months) at 4°C.

Purification of YAC DNA with Filtration Units

This protocol was kindly provided by Dr.Lluis Montoliu, Centro Nacional de Biotechnologia (CNB-CSIC), Campus de Cantoblanco, 28049 Madrid, Spain.

MATERIALS

EQUIPMENT
Eppendorf tube, 1.5-ml
Gel comb with preparative slot in center
Millipore dialysis filter (Millipore VMWP02500, pore size 0.05 micron)
Millipore ultrafiltration unit (Millipore Ultrafree MC 30,000 NMWL UFC3 TTK 00)
Petri dish
Pipette, cut off, yellow-tipped
Pulsed-field gel electrophoresis equipment
Ruler
Scalpel
UV light<!>
Water bath, 40°C
Water bath or hot block, 40°C, 65°C

REAGENTS
Agarose gel
Agarose gel, LMP (NuSieve GTG, FMC) in 1x TAE
Agarose plugs (see Protocol 7.4)
Equilibration buffer
    10 mM Bis-Tris-HCl (pH 6.5)<!>
    0.1 mM EDTA
    100 mM NaCl
    0.030 mM Spermine<!> (Sigma, tetrachloride, S1141)
    0.070 mM Spermidine <!> (Sigma, trihydrochloride, S2501)
Ethidium bromide (EtBr) staining solution
    Add 50 µl of 10 mg/ml EtBr <!> stock solution per 1 liter of running buffer.
Gelase (Epicentre, or Agarase from NEB)
Lambda DNA multimers (NEB or Boehringer Mannheim)
Microinjection buffer (see Protocol 7.7)
SeaPlaque GTG (FMC) LMP
TAE buffer, 0.5x, 1.0x (see Appendix 1)
    1x solution: 40 mM Tris acetate<!>, 1 mM EDTA (pH 8.0)

CAUTION: See Appendix 2 for appropriate handling of materials marked with <!>.

PROCEDURE

1. Pour a 1% SeaPlaque GTG (FMC) LMP PFGE gel using a comb with a preparative slot in the center.

2. Load the agarose plugs vertically and consecutively into the preparative slot (6–9 blocks, depending on the size). Include marker lanes on both sides with a very small slice of the same batch of agarose plugs, and marker lanes with λ DNA multimers.

3. After the gel has run, cover all slots with 1% SeaPlaque GTG (FMC) LMP, let it solidify, and start the gel with appropriate running conditions to ensure optimal resolution in the desired size range.

   A wide range of chromosomal sizes (50–2000 kb) can be recovered by using 0.5x TAE buffer at 10°C, running at 180 V with a pulse of 30 seconds for 12 hours followed by a pulse of 60 seconds during 15 hours. Avoid using too much running buffer within the PFGE tank (not more than 2 mm over the surface of the gel) to improve the quality of the chromosomal separation.

4. After the gel has run, cut off marker lanes plus a small part of the preparative lane on either side and stain these two external parts with EtBr staining solution. The central part of the gel containing most of the preparative slot remains unstained in running buffer.

5. Mark the locations of both the YAC chromosomal bands and the additional bands on the gel slices under UV light by cutting a nick with a scalpel.

6. Put the stained and marked parts next to the preparative central lane and carefully cut out the YAC-containing agarose slice using the marked nicks and a ruler as a guide. Then, remove the YAC DNA-containing slice. Try to produce an agarose slice not thicker than 5–6 mm.

7. Equilibrate the gel slices, excluding the YAC DNA slice, in 1x TAE buffer three times for 30 minutes each.

8. Equilibrate the gel slice containing YAC DNA in equilibration buffer for a minimum of 2 hours.

9. Transfer the gel slices onto a sterile surface, and carefully remove all drops of buffer with tissue paper.

10. Weigh the gel slice in a sterile Eppendorf tube.

11. Melt the slice by placing the tube in a hot block or water bath for 10 minutes at 65°C.

12. Centrifuge the tube for 5 seconds.

13. Add 4–8 units of Gelase per 100 mg of gel slice. It is important to prewarm the enzyme to room temperature to avoid immediate solidification of the gel upon contact. Use equilibration buffer if agarase is used.

14. Place the tube in a 40°C water bath. After 5 minutes, gently pipette up and down two or three times with a cut-off yellow-tipped pipette. Proceed with the digestion for 2–3 hours at 40°C, mixing the sample gently every hour.

15. Chill the tube on ice for 5–10 minutes and check for the completeness of the agarose-gel digest.

This is a very important step. The appearance of a pale brown or opaque cloud within the tube clearly indicates that the digestion has not been completed. In this case, go back to step 13 and perform a second incubation with additional enzyme.

16. Centrifuge the digest at maximum speed for 20 minutes.

17. Transfer up to 400 μl of the digested agarose into the upper reservoir of a Millipore ultrafiltration centrifuge for 2 minutes at 6000 rpm. Continue with the centrifugation step (in rounds of 2 minutes) until about 320 μl has passed through the membrane.

18. Incubate the tubes at 4°C for a few hours. Resuspend the YAC DNA (possibly attached to the surface of the membrane) by pipetting up and down with a cut-off yellow tip (maximum 2–3 times) very carefully and slowly.

    This is the most risky step in which it is possible to shear/break the YAC–DNA preparation.

19. Add 40 ml of microinjection buffer to a petri dish. Carefully place a Millipore dialysis filter floating on the buffer surface with the glossy side up.

20. Carefully spot the YAC DNA liquid solution onto the center of the filter. Dialyze for 2–3 hours. Allow the dialysis to proceed quietly without any shaking or movement. Do not let the dialysis proceed for more than 3 hours.

21. Pipette off the solution, and transfer the YAC DNA solution into a sterile Eppendorf tube. This is the stock YAC DNA solution. Recoveries between 50 and 70% of the original volume are normal. Store at 4°C, do not freeze, and do not centrifuge more than very briefly.

22. Check the concentration of YAC DNA on an agarose gel (see p. 308). Do not allow the samples to enter more than 2 cm into the gel matrix.

    This procedure should yield highly concentrated YAC–DNA preparations (20–60 ng/μl).

23. The integrity of YAC DNA can be checked by loading an aliquot of the preparation into a PFGE or, when available, by analyzing the molecules at the level of electron microscopy.

Preparing Injection Buffer for Standard-sized DNA

MATERIALS

EQUIPMENT
Filter, 0.22-micron (Whatman 6809-1102)

REAGENTS
ddH$_2$O, ultrapure (Sigma W1530)
EDTA, 0.5 M (Sigma E7889)
HCl<!>, 1 M
Tris-HCl<!>, 1 M (Sigma T2663)

CAUTION: See Appendix 2 for appropriate handling of materials marked with <!>.

PROCEDURE

1. Prepare a 5 mM Tris, 0.1 mM EDTA solution with ultrapure water.
2. Adjust the pH to 7.4 with 1 M HCl.
3. Filter through a prewashed 0.22-micron filter.
4. Store at 4°C or at –20°C.

Preparing Injection Buffer for BAC/YAC DNA

This protocol was kindly provided by Dr. Lluis Montoliu, Centro Nacional de Biotechnologia (CNB-CSIC), Campus de Cantoblanco, 28049 Madrid, Spain.

MATERIALS

EQUIPMENT
Falcon tube, 50 ml

REAGENTS
ddH$_2$O, sterile
EDTA, 0.5 M (pH 8.0)
NaCl, 5 M
Spermidine<!> (Sigma, trihydrochloride, S2501)
Spermine<!> (Sigma, tetrahydrochloride, S1141)
Tris-HCl<!>, 1 M (pH 7.5)

CAUTION: See Appendix 2 for appropriate handling of materials marked with <!>.

PROCEDURE

1. For the 1000x polyamine mix, dissolve the spermine and spermidine together in sterile ddH$_2$O so that the end concentration is 30 mM spermine and 70 mM spermidine.

2. Filter-sterilize (0.22 micron) the mixture and store at –20°C.

3. For the basic injection buffer, add 0.5 ml of 1 M Tris-HCl to a plastic disposable 50-ml Falcon tube.

4. Add 10 µl of 0.5 M EDTA.

5. Add 1 ml of 5 M NaCl.

6. Add sterile water up to 50 ml.

7. Aliquot, filter-sterilize, and store the mixture at 4°C.

8. To prepare ready-to-use injection buffer, add 50 µl of polyamine mix to 50 ml of basic injection buffer. Mix and use directly.

    Polyamine mix is stable at –20°C for several months. The microinjection buffer without polyamines added is stable at 4°C for several months. The ready-to-use buffer (polyamines added) should be prepared fresh for each experiment and should not be stored.

Making Holding Pipettes

MATERIALS

EQUIPMENT

Borosilicate glass capillary (e.g., Leica 520119)
Bunsen burner (small) or micro burner or commercial capillary puller (Sutter, etc.)
Diamond-point pencil or oil stone
Microforge (Bachofer, Alcatel, Narishige, De Fonbrune) (*optional*)
Petri dishes, large
Plasticine or rolled-up pieces of masking tape for temporary storage of holding
    pipettes
Stereomicroscope with micrometer scale in the eyepiece

PROCEDURE

1. Hold a 10-cm piece of glass capillary tubing with both hands and heat a small region at the center by rotating it in the flame of a microburner. The trick to obtaining a fine diameter is to draw out the capillary to ~1 cm while it is still in the flame, and then quickly remove the tubing from the flame and give it a second pull to draw out a thin region 5–10 cm long. Take care to pull absolutely straight, so that the two resulting pipettes are not bent in the area where the glass has been melted. The outside diameter of the drawn-out region of the pipette should be ~80–120 μm (i.e., close to the diameter of the zona pellucida). Alternatively, program a commerical capillary puller to produce the desired shape capillary.

2. Break the pipette ~2–3 cm from the shoulder of the pipette. Make sure to achieve a perfectly flush tip by scoring the glass lightly either with a diamond-point pencil or on an oilstone at the desired position; bend until the glass breaks at this point. If the tip shows the slightest sign of being uneven, chipped, or not at exactly a 90° angle, the pipette will not be useful and should be discarded.

3. Clamp the pipette into the microforge.

   a. Position the drawn tip of the pipette so that it exactly opposes the heating filament from all angles.

   b. Heat the filament and position it very close to the tip of the pipette until the glass begins to melt. Use an eyepiece with a micrometer to observe the inside diameter of the pipette as it melts. Allow it to shrink to a diameter of ~15 μm. The diameter of the opening should be sufficient to hold the embryo but not so large as to allow its aspiration into the shaft of the pipette; ~20% of the diameter of the embryo is generally a good size.

   c. Once the pipette reaches the desired diameter, quickly turn off the current through the filament, or increase the distance between pipette tip and filament. If the filament has the correct temperature and is positioned close enough to the pipette, it should take only 2–3 seconds to melt the glass.

   d. The hole in the pipette tip should be straight, not pointing to one side, and the end should be smooth and perpendicular to the long axis of the pipette (Fig. 7.2). If the tip is crooked and uneven, or if the internal diameter is too small, break off a few millimeters of glass and melt the tip again.

4. If an angle is desired at the pipette tip, position the pipette horizontally in the microforge so that the drawn-out region protrudes over the heating filament.

   a. Move the filament so that it almost touches the pipette at a point 2–3 mm from the tip of the pipette.

   b. Carefully heat the filament until the pipette starts to soften at the point being heated; allow the pipette to bend under its own weight until it is at a 15° angle (see Fig. 7.3).

   c. Quickly turn off heat to the filament to stop the bending. This bend allows the end of the pipette to be positioned horizontally in the injection chamber.

5. Place the ready-made holding capillaries in petri dishes with a thin thread of Plasticine or rolled-up masking tape. The capillaries can be stored for a longer time as long as they are kept in a dust-free environment.

FIRE-POLISHING

If a microforge is not available, it is possible to fire-polish the end of the pipette by touching it to the flame of a microburner for the shortest possible time. Only the very tip of the capillary should be inserted into the side of the flame, but not the top of the flame, which is too hot. However, this technique requires a lot of practice to be successful.

    One holding pipette usually lasts for an entire microinjection session; however, it cannot be reused indefinitely because it tends to become dirty. Some investigators prefer to use the capillaries without melting the tip to a narrow opening. In these cases, the control of the movement in the holding capillary has to be very precisely adjustable so as not to damage the zygotes.

Making Injection Pipettes

MATERIALS

EQUIPMENT

Forceps

Glass capillary tubing with an internal glass filament (e.g. W-P Instruments TW100F; Clark Electromedical Instruments GC 100TF-15). Alternatively, commercially available injection needles can be used (Eppendorf FemtotipII)

Mechanical pipette puller (e.g., Sutter P-97)

Petri dishes for temporary storage of pipettes

Plasticine or rolled-up pieces of masking tape for temporary storage of pipettes

Through filament, 3-mm (Science Bioproducts FT 330B)

PROCEDURE

Determination of Useful Settings with Sutter Puller P-97

1. Use a 3-mm through filament. Insert a capillary in the puller using clean forceps.

2. Run a ramp test.

3. Set the Heat to the ramp value −10.

4. Set the Pull to 100.

5. Set the Velocity to 150.

6. Set the Time to 100.

7. Pull a capillary with these approximate settings.

8. Evaluate the shape of the tip, and determine which parameters need to be changed.

9. Adjust the settings of Heat, Pull, Velocity, and Time with 10 increments. Keep a record of how the shape of the needle changes with each modification.

COMMENTS

The parameters can be varied in a way that the resulting pipettes are close to perfect. However, this procedure takes time and requires a great deal of patience. Good advice about settings for the Sutter puller can be found at http://www.sutter.com/. The following guidelines can aid in the process, but it is important to keep in mind that all parameters interact with each other. The results of changing one parameter can be influenced by the values of others.

• Increasing the Heat will result in a longer and finer taper and smaller tip diameter.

• Increasing the Pull will result in a longer taper, with less influence on the tip diameter.

- Increasing the Velocity will result in a finer tip diameter.
- Increasing the Time will result in a shorter taper, more rapidly decreasing diameter.
- Changing the pressure requires some additional programming steps. Increasing the pressure will result in a shorter taper and wider tip diameter.

Microinjection Setup

MATERIALS

EQUIPMENT

Antivibration table (see Chapter 17)

Control unit for holding pipette (see Chapter 17). An alternative is Tygon tubing and a mouth pipette device.

Holding pipettes

Injection chamber

Injection pipettes

Injector (see Chapter 17). An alternative is a 10-ml glass syringe (10 cc, Becton Dickinson 2590,6458) and Tygon tubing.

Inverted microscope with DIC or Hoffman optics, 2.5x or 5x, and a 32x or 40x objective, long-working-distance condenser (see Chapter 17)

Microloader (Eppendorf long specifically for Femtotips)

Micromanipulators (see Chapter 17)

Needles, 5-cm-long, 26-gauge, for loading holding pipette

Tygon tubing (*optional*)

REAGENTS

DNA solution, 1–2 ng/μl

M2 and M16 culture media (see Chapter 4)

Paraffin oil, light (e.g., Merck 1.07161)

PROCEDURE

1. Set up the microinjection workplace on the antivibration table (see Fig. 7.4). Place the inverted microscope in the middle and the two micromanipulators on each side, directed toward the working area. Assemble the electronic or hydraulic manipulators onto the microscope stage, and secure the mechanical Leitz manipulators onto a base plate (see Chapter 17).

2. Attach the injector to the right-hand-side micromanipulator handle, following the manufacturer's instructions (see Protocol 11, Comments). Alternatively, injection can be performed by attaching thick Tygon tubing to the micromanipulator pipette and connecting the other end to a glass syringe. This method is only recommended in the case that no injector is available, because the injection process will be less precise and requires higher manual skills to perform.

3. Set up an injection chamber and place it on the microscope stage (see pp. 316–318). Use low power (2.5–5x) to get a full view of the microinjection drop.

4. If using a mechanical control unit for the holding pipette, follow the manufacturer's instructions for filling it with oil or leaving it filled with air. If using

a mouth pipette device to control the holding pipette, fill the pipette with paraffin oil as follows: Fill a syringe with light paraffin oil, and attach a needle. Insert the needle from the back side into the holding pipette. Push in the oil by a continuous pressure, until the oil enters through the tip of the holding pipette. Make sure no air bubbles are left in the pipette.

5. Assemble the holding pipette into the handle of the left-hand-side micromanipulator. Insert the holding pipette into the drop, and adjust it so that it lies horizontally on the bottom of the chamber and enters the drop at a straight 90° angle.

6. Load the DNA solution into the microinjection pipette by using a very long and thin microloader (Eppendorf Microloader), inserted through the wide back end. As an alternative, fill the injection pipette by dipping the blunt end into the DNA solution and letting the DNA enter the tip of the pipette by capillary force.

7. Assemble the injection pipette into the handle of the right-hand-side micromanipulator. Switch on the injector and follow the manufacturer's instructions about specific settings to create a slow constant flow of DNA through the tip of the pipette. Alternatively, if a syringe is used instead of an injector, apply light pressure to the syringe.

8. Insert the injection pipette into the microinjection drop, and adjust the pipette so that it lies horizontally on the bottom of the chamber and enters the microinjection drop at a straight 90° angle, exactly parallel to the holding pipette (see Figs. 7.9 and 7.10).

A

B

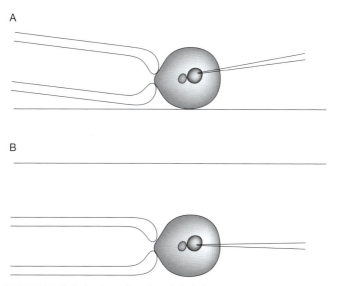

FIGURE 7.9. Injection chamber. (*A*) Side view; (*B*) top view.

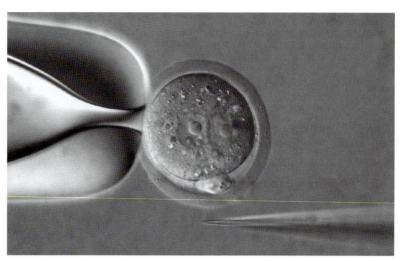

FIGURE 7.10. Positioning of the injection pipette at a 6 o'clock position, where the Z level is adjusted until both the pronucleus and pipette tip are in focus.

COMMENTS

- It is important to adjust both the holding and injection pipettes in such a way that they face each other at exactly a 180° angle. Otherwise, it will be difficult to keep the zygote in position during injection.

- If the holding pipette is of the straight type, without an angle at the tip, it should be lowered into the drop until it starts sliding forward on the surface of the injection chamber. If the outer diameter of the holder is correct, the zygote will be captured firmly (see Fig. 7.2).

- The injection pipette should be inserted into the injection chamber at as low an angle as possible relative to the surface of the slide: 5° is optimal, 10° is good, 15° is often too steep, because the exact targeting of the pronucleus will be more difficult, and the mechanical damage increased, with a higher lysing rate as a result.

- The speed with which the holding pipette catches and releases the oocyte can be adjusted by loading oil into the holding pipette. The higher the viscosity of the oil, and the smaller the inner diameter of the holding pipette tip, the slower the motion will be.

Microinjection of Mouse Zygotes

MATERIALS

EQUIPMENT
Microinjection setup as described in Protocol 7.10

REAGENTS
Fertilized mouse embryos 0.5 dpc (see Chapter 3)
KSOM, M16 or other embryo culture medium

PROCEDURE

1. Use an embryo transfer pipette to transfer a group of fertilized oocytes into the injection chamber. The number of zygotes to be moved into the microinjection drop should be determined by the skills of the injector and quality of the setup. Do not attempt to work with more zygotes than can be injected within 20–30 minutes.

2. Examine the zygotes under high power, making sure that two pronuclei are visible and that the morphology is good (see Fig. 7.1). Discard all zygotes that appear abnormal.

3. To ensure that the injection pipette is not closed at the tip or clogged, place the tip of the injection pipette close to (but not touching) a zygote in the same horizontal plane as the midplane of the zygote (i.e., in the same focal plane, on high power). Apply pressure using the regulator of the injector.

    If the pipette is open, a stream of DNA will move the zygote away from the tip of the injection pipette.

    If the pipette tip is closed or clogged, flush DNA with high power through the injection pipette by using the "Clear" function on the FemtoJet. Repeat the test. If the tip is still not open, tip it carefully on the holding pipette and so break up to a larger tip diameter. If the diameter becomes too large, or the tip is still not open, discard the pipette and use a new one.

4. To prepare a zygote for injection, place the tip of the holding pipette next to the zygote and suck it onto the end of the pipette by applying a negative pressure to the pressure control unit. Focus the microscope to locate the pronuclei.

    A pronucleus can be most easily injected if it is located in the zygote hemisphere closest to the injection pipette. The pronucleus should also be as close as possible to the central axis of the holding pipette; if it is far from this axis, the zygote will tend to rotate when the injection pipette is pushed toward the pronucleus.

If it is necessary to reorient the zygote to place the pronucleus in a better position, release the zygote from the holding pipette, use the injection pipette and/or the holding pipette to rotate it slightly, and then suck the zygote back onto the holding pipette.

5. When satisfied with the position of the zygote, give the syringe controlling the holding pipette an extra twist to be sure the zygote is held firmly. The zona pellucida should be seen being pulled slightly into the opening of the pipette, but the zygote itself should not be deformed. Either of the two pronuclei may be injected.

6. Refocus on the pronucleus to be injected, making sure that its borders can be seen sharply (the focus is set to the midplane of the pronucleus).

   a. Bring the tip of the injection pipette into the same focal plane as the midplane of the pronucleus.

   b. Move the injection pipette to the same y-axis position as the targeted pronucleus (either 6 o'clock or 12 o'clock of the embryo) and adjust the height of the pipette so that the tip of the pipette appears completely sharp (without changing the focus!). This is an important step that allows the pipette to target the pronucleus exactly (see Fig. 7.10).

7. Move the injection pipette to a 3 o'clock position without changing its vertical level. Push the injection pipette through the zona pellucida, into the cytoplasm, and toward the pronucleus. Make sure that both the tip of the pipette and the outline of the pronucleus remain in focus; if the zygote moves and the pronucleus goes out of focus, the pipette will not hit the pronucleus.

   a. Continue pushing the pipette forward, entering into the pronucleus. Avoid touching the nucleoli as they are very sticky and will adhere to the pipette.

   b. When the tip of the pipette appears to be inside the pronucleus, apply injection pressure through the injector.

8. If the pronucleus swells visibly (see Figs. 7.11 and 7.12), it has been successfully injected! *Quickly* pull the pipette out of the zygote.

   A pipette pulled out slowly frequently will remain attached to nuclear components (perhaps the nuclear membrane or chromosomes). Also, flow of the solution may disturb the plasma membrane if the withdrawal is slow.

9. If the pronucleus does not swell, the pipette has become clogged or has not punctured the oocyte plasma membrane.

   a. If a small round "bubble" forms around the tip of the pipette (see Figs. 7.11 and 7.12), then the pipette has not punctured the plasma membrane. The plasma membrane is very elastic and can be pushed far back into the zygote, even into the pronucleus, without being pierced. In this case, try pushing the pipette right through the pronucleus and out the other side; then pull back on the pipette slightly so that the tip is again inside the pronucleus. This maneuver frequently moves the pipette through the plasma membrane and into the pronucleus.

   b. Another sign that the pipette has actually pierced the membrane is that at the point of entry the membrane will be roughly perpendicular to the wall of the pipette (Figs. 7.11B and 7.12B), whereas if the membrane has not been pierced, it will appear to be indented (Figs. 7.11A and 7.12A).

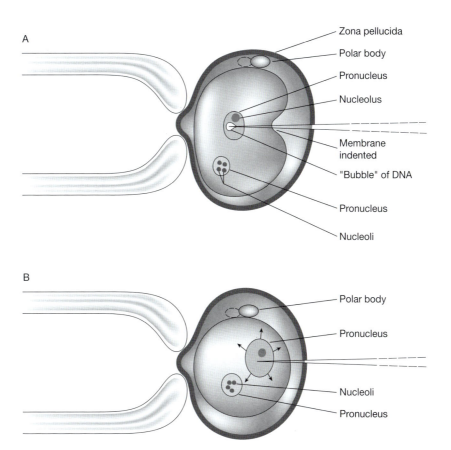

A

Zona pellucida
Polar body
Pronucleus
Nucleolus
Membrane indented
"Bubble" of DNA
Pronucleus
Nucleoli

B

Polar body
Pronucleus
Nucleoli
Pronucleus

FIGURE 7.11. Microinjection into the pronucleus. (*A*) Incomplete penetration of the zygote plasma membrane. The plasma membrane is pushed into the pronucleus by the microinjection pipette. The DNA solution expelled from the pipette forms a visible bubble of plasma membrane within the pronucleus, but the pronucleus does not swell. When the pipette is removed, the DNA will flow back out of the zygote. (*B*) Successful injection of the pronucleus. The plasma membrane has been pierced and the zygote returns to a spherical shape. The tip of the pipette remains inside the pronucleus, and DNA expelled from the pipette causes the pronucleus to swell visibly.

10. It is difficult, if not impossible, to accurately control the volume of DNA solution introduced into the pronucleus. Most investigators estimate that 1–2 picoliters (pl) is injected, but the fraction of DNA remaining in the nucleus is unknown. The size of the pronucleus varies from zygote to zygote, and the injected volume has to be adjusted accordingly. As a guideline, injection should be continued until a clear increase in pronucleus size has been achieved.

11. Cytoplasmic granules flowing out of the oocyte after removal of the injection pipette are a clear sign that the zygote will soon lyse. In this case, or if nuclear components are sticking to the tip of the injection pipette after injection, the oocyte should be discarded. If the zygote appears to be intact and successfully injected, it should be sorted into the group of "good zygotes," and another zygote should be picked up for injection.

12. The same injection pipette can be used as long as it continues to inject successfully. Switch to a new injection pipette if (a) you are unable to get into the

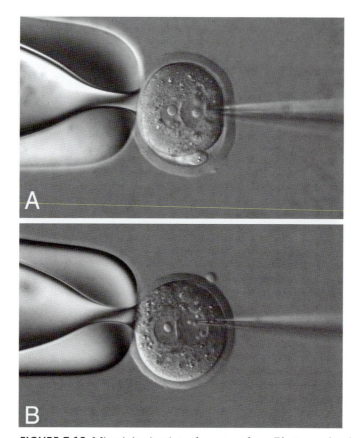

FIGURE 7.12. Microinjection into the pronucleus. Photographs showing the same features as Fig. 7.11. (*A*) Incomplete penetration of the zygote plasma membrane. (*B*) Complete penetration.

pronuclei of several zygotes, even though the pronuclei can be seen clearly; (b) two oocytes in succession lyse immediately after injection; (c) the tip of the pipette becomes visibly "dirty," or nuclear contents stick to the pipette; (d) the tip of the pipette breaks and appears to be more than 1 μm in diameter; or (e) the pipette clogs and cannot be cleaned by flushing through with high pressure ("Clean" function on the FemtoJet).

13. When all of the zygotes in the chamber have been injected, they should immediately be moved back into either M16 or KSOM medium and incubated at 37°C. A new group of zygotes can be transferred into the injection chamber, and the injection procedure continued until all zygotes are injected.

14. Some injected zygotes will inevitably lyse due to the mechanical damage caused by the injection procedure. Lysing will take place 5–30 minutes after completed injection. These lysed zygotes can be easily distinguished from healthy ones. Lysed zygotes appear translucent, fill out the whole zona pellucida, and give a lightweight impression (they will swim up more easily if blown upon with media). Healthy zygotes have a distinct space between the plasma membrane and zona pellucida, and the cytoplasm appears compact and evenly shaped. Typically, about 75% of the zygotes survive the injection.

COMMENTS

- It is of utmost importance to carefully select only good-quality zygotes for injection (see Fig. 7.1). There should be no signs of sperm under the zona pellucida. The oocyte should not fill out the whole zona pellucida (there should be a marked previtelline space), but it should also not appear "shrunken." Two pronuclei should be clearly visible. No fragmentation should be apparent.

- If only one pronucleus can be detected, the oocyte is most likely unfertilized. If more then two pronuclei can be detected, the oocyte is polyspermic (fertilized by multiple sperm). The absence of a second polar body is not always a reliable sign of failed fertilization, because occasionally polar bodies are fragmented, making them difficult to count accurately. However, the oocyte itself should not show any signs of fragmentation.

- The quality assessment can be performed while injecting, but care should be taken to sort the normal and successfully injected oocytes from abnormal or unsuccessfully injected ones immediately, since this discrimination will be impossible to do at a later stage. Electronically controlled micromanipulators can aid the sorting of good- and low-quality embryos by making use of programmed positions of the holding pipette. In this case, set the position 1 far to one side of the microinjection drop, and the position 2 to the other. Each injected zygote can then be automatically moved to the corresponding group.

- To minimize the risk of the zygote turning in the holding pipette's grip during injection, it is important to make sure that the targeted pronucleus lies approximately in the midplane of the cytoplasm. If this is the case, the outline of the pronucleus will be sharp at the same time as the zona pellucida appears as the most "crisp" with most details visible in it.

- Most modern electronic injectors (such as the FemtoJet) have an "Automatic" and a "Manual" injection mode. The difference here is simply the mode of injection in relation to the release button or foot pedal action. In case of the "Automatic" mode, the length of time during which injection is triggered is preset and equal each time. In the "Manual" mode, however, the duration of the injection is determined individually; as long as the button or foot pedal is held, injection proceeds. Because pronuclei can vary significantly in size, the "Manual" mode should be used, so that the appropriate pronuclear swelling can be achieved for each individual oocyte.

- The settings of injection pressure and constant flow pressure have to be determined empirically and are dependent on the inner diameter of the injection pipette. As basic settings for the FemtoJet, P_c can be set to 10–15, and P_i set to 40–50. If during injection the pronucleus swells very fast, resulting in nucleoli pressed out into the cytoplasm, the pressure should be reduced. If, on the other hand, the pronucleus swells very slowly, the pressure should be increased.

- The microinjection setup is virtually the same whether normal-sized or large constructs are injected. However, it is important to keep in mind the higher viscosity of high-molecular-weight DNA solutions, and the danger of shearing the DNA when pushing it through a very small pipette tip. The injection pressure should be kept as low as possible, and if the injection pipette is clogged, it should immediately be changed. The inner diameter of the injection pipette should be slightly increased to allow an easy flow of DNA without increasing the injection pressure. This can most easily be achieved by "ticking" the tip of the injection pipette onto the holding capillary until it breaks at a larger dimension. If the new diameter is broken too widely, the injection pipette should be changed. The lysing rate will grow proportionally with the tip diameter, but this is a drawback one should accept when working with large constructs, rather than risking the shearing of the DNA by pushing it through a very small pipette tip.

- It is easiest to target the larger of the two pronuclei and/or the one that is closest to the injection pipette. If the embryos are well timed, the pronuclei should both appear in the center of the embryo, both be large and clearly visible. The advice to inject the male pronucleus is outdated. At the optimal time for injection, it is often difficult to tell which pronucleus is the male and which is the female. The only criterion for the choice should be the ease of injection: The pronucleus that is largest and/or nearest to the injection pipette should be targeted.

- If the pronuclei are small, the embryo has not yet reached the optimal time point for injection. It may help to bring the embryos back to the incubator for an hour and start the injection process later. In case the embryos are left for too long at 37°C before the injection is started, the two pronuclei will have fused and injection will be impossible. The developmental stage of the oocytes highly correlates with the timing of hCG injection and the light cycle in the animal room, so that the injection time point easily can be influenced by the superovulation protocol (see Chapter 3).

- The microinjection chamber should be removed from the microscope stage immediately after each injection session to prevent any accidental spillage of oil or medium into the optics.

- Microinjection is not a trivial process. Allow a considerable learning time before high efficiencies can be achieved. Embryos are living material, which should be handled with care. It is not sufficient for them barely to survive the manipulation and in vitro culture; they should be treated in such a way that after replacement in vivo they can develop into healthy mice. This is an important picture to keep in mind during all embryo micromanipulations, culturing, and handling.

TROUBLESHOOTING GUIDE

The process of creating genetically altered mice through pronuclear injection is very sensitive. Numerous potential problems may arise, eventually leading to either a low number of pups born or no founders among the offspring. This troubleshooting guide gives an overview of the most common difficulties, their possible underlying causes, and suggestions for solutions.

1. **Low plugging rate.**
   - The stud males are too young or too old. Optimal age is 2–6 months.
   - The stud males have been mated too often. They should not be mated with superovulated females more than twice per week, and they should be allowed at least 1 day rest between matings.
   - The batch of PMSG/hCG used is of low quality, or the hormones may have been prepared incorrectly. Hormones should be kept in frozen stocks, and not allowed to be thawed longer than 1 hour prior to use.
   - The light–dark cycle is not constant from day to day. It is important that the cycle be kept exactly the same each day, with absolutely no variations. No entry into mouse rooms should be allowed after the start of the dark period.
   - Environmental problems in the animal facility. High noise levels— especially noise that is not of a constant nature—vibrations, temperature fluctuations, and suboptimal humidity should be avoided as much as possible.

2. **Poor yield of zygotes.**
   - Problems with the light cycle, hormones, or environment; see point 1.
   - Suboptimal age of donor females. Depending on the strain or hybrid cross used, it may be necessary to use older or younger females that respond better to the hormone regimen.
   - Incorrect hormone dosage. The dose of PMSG/hCG required for superovulation varies between strains and hybrid crosses, and also with the age of the females. It is therefore important to determine the optimal dosage empirically.

3. **Few or no zygotes display two pronuclei.**
   - The females were not actually plugged.
   - The oocytes are not fertilized. This may be due to the use of too-old or overworked males, or too-young females.
   - The zygotes were fertilized very recently. In this case, the pronuclei will become visible after a few hours in in vitro culture. Administer the hormones at an earlier time.
   - The dissecting or injection microscope is poorly adjusted, the optics are incorrectly aligned, and/or the objectives are not clean.
   - The injection microscope is not used with DIC or Hoffman optics, or plastic dishes are used in combination with DIC optics.
   - See also point 5.

4. The pronuclei are too small to inject easily.

- Injection has been attempted too soon after fertilization. Administer the hormones earlier, or inject later in the day.

- Some strains have small pronuclei that are difficult to inject. In case flexibility in the choice of genetic background is acceptable, try a different strain or hybrid combination.

5. The pronuclei cannot be seen properly.

- See point 4.

- Small oil bubbles may have collected under the surface of the medium drop in the microinjection chamber, which alters the image. This often happens when oocytes are moved in and out of the drop. The oil drops can be removed with the mouth pipette while focusing on them at low-power magnification, or a new chamber should be set up.

- The medium drop in the injection chamber may not be completely covered by oil. This is most likely to happen if the medium drop is relatively high and the oil cover thin. Also in this case, the image will be distorted. Add more oil on top of the medium drop.

- DIC or Hoffman optics are not properly aligned and adjusted. Ask the microscope manufacturer for help with accurate optimization.

- The eyepiece or the objective is dirty (clean carefully using ethanol and water with dust-free soft paper lens tissue). Take good care never to let any oil drop down from the microinjection chamber onto the objectives! For this reason, it is important to remove the microinjection chamber from the microscope stage immediately after each injection session.

6. Strands of nuclear material stick to the injection pipette after withdrawal from the pronucleus.

- The injection pipette has touched the nucleoli. Nucleoli resemble small bright round spots within the pronucleus. They are extremely sticky and will immediately stick to the injection pipette if touched. Try to avoid touching the nucleoli during injection.

- The pipette is dirty and should be replaced. Extremely small particles, which cannot be seen even under high magnification, make the outside surface of the injection pipette sticky, which in turn results in an increased risk of the tip touching and attaching to nuclear material.

- The injection pipette has been pulled out too slowly. Try to pull out faster after the pronuclear swelling.

- One possible method for reducing this problem is to siliconize the pipettes before use with Sigmacote (Sigma).

7. The injection pipette clogs repeatedly.

- Flush with the "Clean" button if an electronic injector is used.

- The opening in the pipette is too small. Alter the settings on the pipette puller, or enlarge the opening by chipping on the holding capillary.

- The injection pressure is too low. Increase the constant flow pressure.

- Impurities are present in the DNA solution. Alter the purification method.

8. **The injection pipette or the holding pipette drifts when the joystick is not being touched** (using Leitz micromanipulator).

   - Reduce the movement ratio by twisting the joystick collar in the appropriate direction.

   - The joystick is too far out of position. Reposition the pipette with the positioning screws on the manipulator, resetting the joystick to a middle position.

9. **The tip of the injection pipette moves during injection.**

   - The tip of the pipette is not symmetrical: Adjust the pipette puller so that the capillary is centered inside the heating filament.

   - The silicone rubber tubing inside the Leitz micromanipulator pipette holder is worn out, and should be replaced.

10. **Too many zygotes lyse after injection.**

    - The tip of the injection pipette is too wide. Replace the pipette. Adjust the settings on the puller.

    - The angle between the long axis of the pipette and the plane of the microscope stage is too great. Try to enter the injection pipette in as low an angle as possible.

    - The injection pipette is dirty. Replace it.

    - The DNA is impure. Improve the purification protocol.

    - The DNA concentration is too high. Reduce the concentration to half.

    - The injection pipette is moved in the y axis when inserted into the zygote. Take more care not to move the injection pipette other than straight into the embryo.

    - The injected volume is too high. Stop injection as soon as the pronucleus has expanded.

    - The injection pipette is pulled out too slowly. Try to pull out the pipette with a swift movement.

    - The injection and/or constant flow pressure is too high. Reduce the DNA flow in the pipette.

    - Microenvironmental problems such as extensive vibration on the microscope stage, too low a temperature in the microinjection room, or suboptimal medium composition may also contribute to reduced embryo survival.

11. **The plasma membrane or pronucleus is difficult to penetrate.**

    - The injection pipette is dull. Use a sharper pipette, or try tapping the pipette against the holding capillary to break off a small piece at the tip.

    - The injection pipette is dirty. Replace it.

    - Try to push the injection pipette through the pronucleus into the distal cytoplasm, then pull back so that the tip of the pipette is inside the pronucleus.

    - Reorient the zygote, and position a larger pronucleus in line with the center of the holding pipette.

    - When inserting the pipette, aim exactly at the middle of the pronucleus.

- Make sure that the z axis of the injection pipette lies exactly in the midplane of the pronucleus (see point 6).

12. **The zygotes do not divide (if cultured) after injection.**

    - The zygotes were not fertilized. Make sure only to inject oocytes with two visible pronuclei.

    - The zygotes were fertilized by more then one spermatozoon each. Screen away any zygotes that display three or more pronuclei.

    - The culture medium is of poor quality, or the incubator is set to the wrong temperature/CO_2 concentration, or the humidity in the incubator is not sufficient. Check the culture conditions by culturing uninjected zygotes. Make sure that the water reservoir in the incubator is filled.

    - The zygotes were maintained for too long at room temperature during injection. The temperature in the injection room may be too low, or the microscope stage suffers from vibrations.

    - The zygotes were mechanically damaged by the injection (see point 9, above).

    - The DNA contains toxic impurities. Compare injection of a different DNA or a buffer control.

    - The injected DNA concentration or volume was too high.

    - Transient expression of the injected gene is toxic to the zygote.

13. **The pregnancy rate among pseudopregnant recipients is low, or the percentage development to term is low.**

    - See point 12.

    - The females were not actually plugged, or the plug date was misrecorded. Foster mothers for oviduct transfer should be plugged on the same day as transfer is done, or on the following day if injected zygotes are cultured overnight.

    - The zygotes/embryos were not successfully implanted into the oviduct.

    - The females were overly traumatized by surgery or anesthetic.

    - The reproductive organs were damaged or traumatized during the transfer procedure. Try to avoid touching the ovary, and handle the oviduct with great care during transfer.

    - A large proportion of the zygotes were damaged by the injection procedure or DNA preparation (see point 11, above).

    - Noise, smell, vibration, etc. in the animal facility may affect pregnancy rates (see Chapter 3, p. 154).

14. **The proportion of founders among the pups born is low.**

    - The volume of injected DNA was too low; increase the degree of pronuclear swelling.

    - The DNA concentration was lower than estimated; remeasure or increase the concentration.

    - The DNA was diluted by backflow of culture medium into the injection pipette; make sure that the constant flow pressure is adequate. If injection is performed by use of a syringe, squeeze on the plunger once before entering the injection pipette into the zygote.

- The DNA was not injected into the pronucleus because (a) the pronuclear membrane was not penetrated, resulting in DNA deposition in the cytoplasm, or (b) neither the pronuclear membrane nor the oolemma was penetrated, resulting in DNA deposition outside the zygote plasma membrane. Use sharper pipettes. Adjust the optics if the pronucleus is difficult to see. Make sure only to transfer zygotes where a clear pronuclear swelling can be detected.

- The vasectomized males were not correctly vasectomized, so that the foster mother carried her own fertilized embryos in addition to the injected zygotes. For example, if the CD1 outbred strain is used both as vasectomized males and foster females, any dark coat color can be used for the injected zygotes. If albino mice are detected in the litters, the vasectomized males must be suspected of not being properly vasectomized. Vasectomized males may be tested for sterility before being used (see Chapter 6).

- The expression of the transgene is toxic to the developing fetus.

15. **The transgenic founder mouse breeds but does not transmit the transgene to the progeny as expected.**

- The founder is not transgenic, but a false positive.

- The founder mouse is mosaic, with a few or no transgenic cells contributing to the germ line. Breed further litters, use other founders, or produce additional founders.

- The expression of the transgene is toxic to developing F_1 mice, but the founder survived because of mosaicism. If the founder is male, timed matings may be set up, and embryos analyzed at various developmental stages.

16. **The F_1 offspring from a founder show segregating integration and/or expression patterns.**

- The founder was mosaic, with two or more separate integration patterns. As a general rule, all F_1 offspring from a founder mouse should be carefully analyzed not only with PCR, but also with means to determine integration patterns.

REFERENCES

Al Shawi R., Burke J., Wallace H., Jones C., Harrison S., Buxton D., Maley S., Chandley A., and Bishop J.O. 1991. The herpes simplex virus type 1 thymidine kinase is expressed in the testes of transgenic mice under the control of a cryptic promoter. *Mol. Cell. Biol.* **11:** 4207–4216.

Amid C., Bahr A., Mujica A., Sampson N., Bikar S.E., Winterpacht A., Zabel B., Hankeln T., and Schmidt E.R. 2001. Comparative genomic sequencing reveals a strikingly similar architecture of a conserved syntenic region on human chromosome 11p15.3 (including gene ST5) and mouse chromosome 7. *Cytogenet. Cell Genet.* **93:** 284–290.

Antoch M.P., Song E.J., Chang A.M., Vitaterna M.H., Zhao Y., Wilsbacher L.D., Sangoram A.M., King D.P., Pinto L.H., and Takahashi J.S. 1997. Functional identification of the mouse circadian Clock gene by transgenic BAC rescue. *Cell* **89:** 655–667.

Archibald A.L., McClenaghan M., Hornsey V., Simons J.P., and Clark A.J. 1990. High-level expression of biologically active human alpha 1-antitrypsin in the milk of transgenic mice. *Proc. Natl. Acad. Sci.* **87:** 5178–5182.

Balling R., Mutter G., Gruss P., and Kessel M. 1989. Craniofacial abnormalities induced by ectopic expression of the homeobox gene Hox-1.1 in transgenic mice. *Cell* **58:** 337–347.

Beddington R.S., Morgernstern J., Land H., and Hogan A. 1989. An in situ transgenic enzyme marker for the midgestation mouse embryo and the visualization of inner cell mass clones during early organogenesis. *Development* **106:** 37–46.

Behringer R.R., Peschon J.J., Messing A., Gartside C.L., Hauschka S.D., Palmiter R.D., and Brinster R.L. 1988. Heart and bone tumors in transgenic mice. *Proc. Natl. Acad. Sci.* **85:** 2648–2652.

Behringer R.R., Ryan T.M., Reilly M.P., Asa-kura T., Palmiter R.D., Brinster R.L., and Townes T.M. 1989. Synthesis of functional human hemoglobin in transgenic mice. *Science* **245:** 971–973.

Berns A. 1991. Tumorigenesis in transgenic mice: Identification and characterization of synergizing oncogenes. *J. Cell. Biochem.* **47:** 130–135.

Borrelli E., Heyman R.A., Arias C., Sawchenko P.E., and Evans R.M. 1989. Transgenic mice with inducible dwarfism. *Nature* **339:** 538–541.

Brinster R.L., Allen J.M., Behringer R.R., Gelinas R.E., and Palmiter R.D. 1988. Introns increase transcriptional efficiency in transgenic mice. *Proc. Natl. Acad. Sci.* **85:** 836–840.

Brinster R.L., Chen N.Y., Trumbauer M.E., Yagle M.K., and Palmiter R.D. 1985. Factors affecting the efficiency of introducing foreign DNA into mice by microinjecting eggs. *Proc. Natl. Acad. Sci.* **82:** 4438–4442.

Brinster R.L., Chen H.Y., Warren R., Sarthy A., and Palmiter R.D. 1982. Regulation of metallothionein–thymidine kinase fusion plasmids injected into mouse eggs. *Nature* **296:** 39–42.

Brinster R.L., Braun R.E., Lo D., Avarbock M.R., Oram F., and Palmiter R.D. 1989. Targeted correction of a major histocompatibility class II E alpha gene by DNA microinjected into mouse eggs. *Proc. Natl. Acad. Sci.* **86:** 7087–7091.

Brinster R.L., Chen H.Y., Messing A., van Dyke T., Levine A.J., and Palmiter R.D. 1984. Transgenic mice harbouring SV40 T-antigen genes develop characteristic brain tumors. *Cell* **37:** 367–379.

Brinster R.L., Chen H.Y., Trumbauer M., Senear A.W., Warren R., and Palmiter R.D. 1981. Somatic expression of herpes thymidine kinase in mice following injection of a fusion gene into eggs. *Cell* **27:** 223–231.

Camper S.A. and Saunders T.L. 2000. Transgenic rescue of mutant phenotypes using large DNA fragments. In *Genetic manipulation of receptor expression and function* (ed. D. Accili.), pp. 1–22. John Wiley, New York.

Chada K., Magram J., Raphael K., Radice G., Lacy E., and Costantini F. 1985. Specific expression of a foreign beta-globin gene in erythroid cells of transgenic mice. *Nature* **314:** 377–380.

Chisari F.V., Pinkert C.A., Milich D.R., Filippi P., McLachlan A., Palmiter R.D., and Brinster R.L. 1985. A transgenic mouse model of the chronic hepatitis B surface antigen carrier state. *Science* **230:** 1157–1160.

Chisari F.V., Klopchin K., Moriyama T., Pasquinelli C., Dunsford H.A., Sell S., Pinkert C.A., Brinster R.L., and Palmiter R.D. 1989. Molecular pathogenesis of hepatocellular carcinoma in hepatitis B virus transgenic mice. *Cell* **59:** 1145–1156.

Choi T., Huang M., Gorman C., and Jaenisch R. 1991. A generic intron increases gene expression in transgenic mice. *Mol. Cell. Biol.* **11:** 3070–3074.

Copeland N.G., Jenkins N.A., and Court D.L. 2001. Recombineering: A powerful new tool for mouse functional genomics. *Nat. Rev. Genet.* **2:** 769–779.

Costantini F. and Lacy E. 1981. Introduction of a rabbit beta-globin gene into the mouse germ line. *Nature* **294:** 92–94.

Cotten M., Baker A., Saltik M., Wagner E., and Buschle M. 1994. Lipopolysaccharide is a frequent contaminant of plasmid DNA preparations and can be toxic to primary human cells in the presence of adenovirus. *Gene Ther.* **1:** 239–246.

DePamphilis M.L., Herman S.A., Martinez-Salas E., Chalifour L.E., Wirak D.O., Cupo D.Y., and Miranda M. 1988. Microinjecting DNA into mouse ova to study DNA replication and gene expression and to produce transgenic animals. *Biotechniques* **6:** 662–680.

DePrimo S.E., Stambrook P.J., and Stringer J.R. 1996. Human placental alkaline phosphatase as a histochemical marker of gene expression in transgenic mice. *Transgenic Res.* **5:** 459–466.

Ding H., Roncari L., Shannon P., Wu X., Lau N., Karaskova J., Gutmann D.H., Squire J.A., Nagy A., and Guha A. 2001. Astrocyte-specific expression of activated p21-ras results in malignant astrocytoma formation in a transgenic mouse model of human gliomas. *Cancer Res.* **61:** 3826–3836.

Duff K., Knight H., Refolo L.M., Sanders S., Yu X., Picciano M., Malester B., Hutton M., Adamson J., Goedert M., Burki K., and Davies P. 2000. Characterization of pathology in transgenic mice over-expressing human genomic and cDNA tau transgenes. *Neurobiol. Dis.* **7:** 87–98.

Ellis J., Pasceri P., Tan-Un K.C., Wu X., Harper A., Fraser P., and Grosveld F. 1997. Evaluation of beta-globin gene therapy constructs in single copy transgenic mice. *Nucleic Acids Res.* **25:** 1296–1302.

Erickson R.P. 1999. Antisense transgenics in animals. *Methods* **18:** 304–310.

Flores N., Valle F., Bolivar F., and Merino F. 1992. Recovery of DNA from agarose gels stained with methylene blue. *BioTechniques* **13:** 203–205.

Foster, H.L., Small J.D., and Fox J.G. 1983. *The mouse in biomedical research.* Academic Press, New York.

Friedrich G. and Soriano P. 1991. Promoter traps in embryonic stem cells: A genetic screen to identify and mutate developmental genes in mice. *Genes Dev.* **5:** 1513–1523.

Gaensler K.M., Kitamura M., and Kan Y.W. 1993. Germ-line transmission and developmental regulation of a 150-kb yeast artificial chromosome containing the human beta-globin locus in transgenic mice. *Proc. Natl. Acad. Sci.* **90:** 11381–11385.

Giraldo P. and Montoliu L. 2001. Size matters: Use of YACs, BACs and PACs in transgenic animals. *Transgenic Res.* **10:** 83–103.

Gordon J.W. and Ruddle F.H. 1981. Integration and stable germ line transmission of genes injected into mouse pronuclei. *Science* **214:** 1244–1246.

Gordon J.W., Scangos G.A., Plotkin D.J., Barbosa J.A., and Ruddle F.H. 1980. Genetic transformation of mouse embryos by microinjection of purified DNA. *Proc. Natl. Acad. Sci.* **77:** 7380–7384.

Goring D.R., Rossant J., Clapoff S., Breitman M.L., and Tsui L.C. 1987. In situ detection of β-galactosidase in lenses of transgenic mice with a gamma-crystallin/lacZ gene. *Science* **235:** 456–458.

Gossen J.A., de Leeuw W.J., Tan C.H., Zwarthoff E.C., Berends F., Lohman P.H., Knook D.L., and Vijg J. 1989. Efficient rescue of integrated shuttle vectors from transgenic mice: A model for studying mutations in vivo. *Proc. Natl. Acad. Sci.* **86:** 7971–7975.

Gossen M., Freundlieb S., Bender G., Muller G., Hillen W., and Bujard H. 1995. Transcriptional activation by tetracyclines in mammalian cells. *Science* **268:** 1766–1769.

Gossler A., Joyner A.L., Rossant J., and Skarnes W.C. 1989. Mouse embryonic stem cells and reporter constructs to detect developmentally regulated genes. *Science* **244:** 463–465.

Gossler A., Doetschman T., Korn R., Serfling E., and Kemler R. 1986. Transgenesis by means of blastocyst-derived embryonic stem cell lines. *Proc. Natl. Acad. Sci.* **83:** 9065–9069.

Gridley T. 1991. Insertional versus targeted mutagenesis in mice. *New Biol.* **3:** 1025–1034.

Grieshammer U., Lewandoski M., Prevette D., Oppenheim R.W., and Martin G.R. 1998. Muscle-specific cell ablation conditional upon Cre-mediated DNA recombination in transgenic mice leads to massive spinal and cranial motoneuron loss. *Dev. Biol.* **197:** 234–247.

Grosschedl R., Weaver D., Baltimore D., and Costantini F. 1984. Introduction of an immunoglobulin gene into the mouse germ line: Specific expression in lymphoid cells and synthesis of functional antibody. *Cell* **38:** 647–658.

Grosveld F., van Assendelft G.B., Greaves D.R., and Kollias G. 1987. Position-independent, high-level expression of the human beta-globin gene in transgenic mice. *Cell* **51:** 975–985.

Gundersen K., Hanley T.A., and Merlie J.P. 1993. Transgenic embryo yield is increased by a simple, inexpensive micropipet treatment. *BioTechniques* **14:** 412–414.

Harbers K., Jahner D., and Jaenisch R. 1981. Microinjection of cloned retroviral genomes into mouse zygotes: Integration and expression in the animal. *Nature* **293:** 540–542.

Hadjantonakis A.K., Gertsenstein M., Ikawa M., Okabe M., and Nagy A. 1998. Generating green fluorescent mice by germline transmission of green fluorescent ES cells. *Mech. Dev.* **76:** 79–90.

Hamer L., Johnston M., and Green E.D. 1995. Isolation of yeast artificial chromosomes free of endogenous yeast chromosomes: Construction of alternate hosts with defined karyotypic alterations. *Proc. Natl. Acad. Sci.* **92:** 11706–11710.

Hammer R.E., Brinster R.L., Rosenfeld M.G., Evans R.M., and Mayo K.E. 1985. Expression of human growth hormone-releasing factor in transgenic mice results in increased somatic growth. *Nature* **315:** 413–416.

Hammer R.E., Krumlauf R., Camper S.A., Brinster R.L., and Tilghman S.M. 1987. Diversity of alpha-fetoprotein gene expression in mice is generated by a combination of separate enhancer elements. *Science* **235:** 53–58.

Hanahan D. 1989. Transgenic mice as probes into complex systems. *Science* **246:** 1265–1275.

Harris A.W., Pinkert C.A., Crawford M., Langdon W.Y., Brinster R.L. and Adams J.M. 1988. The E mu-myc transgenic mouse. A model for high-incidence spontaneous lymphoma and leukemia of early B cells. *J. Exp. Med.* **167:** 353–371.

Heyman R.A., Borrelli E., Lesley J., Anderson D., Richman D.D., Baird S.M., Hyman R., and Evans R.M. 1989. Thymidine kinase obliteration: Creation of transgenic mice with controlled immune deficiency. *Proc. Natl. Acad. Sci.* **86:** 2698–2702.

Hicks G.G., Shi E.G., Li X.M., Li C.H., Pawlak M., and Ruley H.E. 1997. Functional genomics in mice by tagged sequence mutagenesis. *Nat. Genet.* **16:** 338–344.

Iwamoto T., Takahashi M., Ito M., Hamatani K., Ohbayashi M., Wajjwalku W., Isobe K., and Nakashima I. 1991. Aberrant melanogenesis and melanocytic tumour development in transgenic mice that carry a metallo-thionein/ret fusion gene. *EMBO J.* **10:** 3167–3175.

Jaenisch R. 1988. Transgenic animals. *Science* **240:** 1468–1474.

Jakobovits A., Moore A.L., Green L.I., Vergara G.J., Maynard-Currie C.E., Austin H.A., and Klapholz S. 1993. Germ-line transmission and expression of a human-derived yeast artificial chromosome. *Nature* **362:** 255–258.

Jang S.K., Krausslich H.G., Nicklin M.J., Duke G.M., Palmenberg A.C., and Wimmer E. 1988. A segment of the 5′ nontranslated region of encephalomyocarditis virus RNA directs internal entry of ribosomes during in vitro translation. *J. Virol.* **62:** 2636–2643.

Kaufman R.M., Pham C.T., and Ley T.J. 1999. Transgenic analysis of a 100-kb human beta-globin cluster-containing DNA fragment propagated as a bacterial artificial chromosome. *Blood* **94:** 3178–3184.

Keefer C.L., Baldassarre H., Keyston R., Wang B., Bhatia B., Bilodeau A.S., Zhou J.F., Leduc M., Downey B.R., Lazaris A., and Karatzas C.N. 2001. Generation of dwarf goat (*Capra hircus*) clones following nuclear transfer with transfected and nontransfected fetal fibroblasts and in vitro-matured oocytes. *Biol. Reprod.* **64:** 849–856.

Kessel M. and Gruss P. 1990. Murine developmental control genes. *Science* **249:** 374–379.

Kisseberth W.C., Brettingen N.T., Lohse J.K., and Sandgren E.P. 1999. Ubiquitous expression of marker transgenes in mice and rats. *Dev. Biol.* **214:** 128–138.

Kohler S.W., Provost G.S., Fieck A., Kretz P.L., Bullock W.O., Sorge J.A., Putman D.L., and Short J.M. 1991. Spectra of spontaneous and mutagen-induced mutations in the lacI gene in transgenic mice. *Proc. Natl. Acad. Sci.* **88:** 7958–7962.

Krumlauf R., Hammer R.E., Tilghman S.M., and Brinster R.L. 1985. Developmental regulation of alpha-fetoprotein genes in transgenic mice. *Mol. Cell Biol.* **5:** 1639–1648.

Lacy E., Roberts S., Evans E.P., Burtenshaw M.D., and Costantini F.D. 1983. A foreign beta-globin gene in transgenic mice: Integration at abnormal chromosomal positions and expression in inappropriate tissues. *Cell* **34:** 343–358.

Lamb B.T., Bardel K.A., Kulnane L.S., Anderson J.J., Holtz G., Wagner S.L., Sisodia S.S., and Hoeger E.J. 1999. Amyloid production and deposition in mutant amyloid precursor protein and presenilin-1 yeast artificial chromosome transgenic mice. *Nat. Neurosci.* **2:** 695–697.

Lathe R., Vilotte J.L., and Clark A.J. 1987. Plasmid and bacteriophage vectors for excision of intact inserts. *Gene* **57:** 193–201.

Lee K.J., Ross R.S., Rockman H.A., Harris A.N., OBrien T.X., van B.M., Shubeita H.E., Kandolf R., Brem G., Price J., Evans S.M., Zhu H., Franz W.M., and Chien K.R. 1992. Myosin light chain-2 luciferase transgenic mice reveal distinct regulatory programs for cardiac and skeletal muscle specific expression of a single contractile protein gene. *J. Biol. Chem.* **267:** 15875–15885.

Lewandoski M. 2001. Conditional control of gene expression in the mouse. *Nat. Rev. Genet.* **2:** 743–755.

Lewandoski M. and Martin G.R. 1997. Cre-mediated chromosome loss in mice. *Nat. Genet.* **17:** 223–225.

Lin T.P. 1966. Microinjection of mouse eggs. *Science* **151:** 333–337.

Lira S.A., Kinloch R.A., Mortillo S., and Wassarman P.M. 1990. An upstream region of the mouse ZP3 gene directs expression of firefly luciferase specifically to growing oocytes in transgenic mice. *Proc. Natl. Acad. Sci.* **87:** 7215–7219.

Lo C.W., Coulling M., and Kirby C. 1987. Tracking of mouse cell lineage using microinjected DNA sequences: Analyses using genomic Southern blotting and tissue-section in situ hybridizations. *Differentiation* **35:** 37–44.

Lois C., Hong E.J., Pease S., Brown E.J., and Baltimore D. 2002. Germline transmission and tissue-specific expression of transgenes delivered by lentiviral vectors. *Science* **295:** 862–872.

Mann J.R. and McMahon A.P. 1993. Factors influencing frequency production of transgenic mice. *Methods Enzymol.* **225:** 771–781.

McCaffrey A.P., Meuse L., Pham T.T., Conklin D.S., Hannon G.J., and Kay M.A. 2002. RNA interference in adult mice. *Nature* **418:** 38–39.

Mehtali M., LeMeur M., and Lathe R. 1990. The methylation-free status of a housekeeping transgene is lost at high copy number. *Gene* **91:** 179–184.

Meisler M.H. 1992. Insertional mutation of 'classical' and novel genes in transgenic mice. *Trends Genet.* **8:** 341–344.

Mendez M.J., Green L.L., Corvalan J.R., Jia X.C., Maynard-Currie C.E., Yang X.D., Gallo M.L., Louie D.M., Lee D.V., Erickson K.L., et al. 1997. Functional transplant of megabase human immunoglobulin loci recapitulates human antibody response in mice. *Nat. Genet.* **15:** 146–156.

Metsaranta M., Garofalo S., Decker G., Rintala M., de Crombrugghe B., and Vuorio E. 1992. Chondrodysplasia in transgenic mice harboring a 15-amino acid deletion in the triple helical domain of pro alpha 1(II) collagen chain. *J. Cell Biol.* **118:** 203–212.

Muller H., Dai G., and Soares M.J. 1998. Placental lactogen-I (PL-I) target tissues identified with an alkaline phosphatase-PL-I fusion protein. *J. Histochem. Cytochem.* **46:** 737–743.

Murti J.R., Bumbulis M., and Schimenti J.C. 1992. High-frequency germ line gene conversion in transgenic mice. *Mol. Cell. Biol.* **12:** 2545–2552.

Nielsen L.B., McCormick S.P., Pierotti V., Tam C., Gunn M.D., Shizuya H., and Young S.G. 1997. Human apolipoprotein B transgenic mice generated with 207- and 145-kilobase pair bacterial artificial chromosomes. Evidence that a distant 5′-element confers appropriate transgene expression in the intestine. *J. Biol. Chem.* **272:** 29752–29758.

Niemann H., Halter R., Carnwath J.W., Herrmann D., Lemme E., and Paul D. 1999. Expression of human blood clotting factor VIII in the mammary gland of transgenic sheep. *Transgenic. Res.* **8:** 237–247.

Niwa H., Yamamura K., and Miyazaki J. 1991. Efficient selection for high-expression transfectants with a novel eukaryotic vector. *Gene* **108:** 193–200.

Okabe M., Ikawa M., Kominami K., Nakanishi T., and Nishimune Y. 1997. "Green mice" as a source of ubiquitous green cells. *FEBS Lett.* **407:** 313–319.

Onyango P., Miller W., Lehoczky J., Leung C.T., Birren B., Wheelan S., Dewar K., and Feinberg A.P. 2000. Sequence and comparative analysis of the mouse 1-megabase region orthologous to the human 11p15 imprinted domain. *Genome Res.* **10:** 1697–1710.

Overbeek P.A., Chepelinsky A., Khillan J.S., Piatigorsky J., and Westphal H. 1985. Lens-specific expression and developmental regulation of the bacterial chloramphenicol acetyltransferase gene driven by the murine alpha-crystallin promoter in transgenic mice. *Proc. Natl. Acad. Sci.* **82:** 7815–7819.

Palmiter R.D. and Brinster R.L. 1986. Germ-line transformation of mice. *Annu. Rev. Genet.* **20:** 465–499.

Palmiter R.D., Hammer R.E., and Brinster R.L. 1985. Expression of growth hormone genes in transgenic mice. *Banbury Rep.* **20:** 123–132.

Palmiter R.D., Norstedt G., Gelinas R.E., Hammer R.E., and Brinster R.L. 1983. Metallothionein-human GH fusion genes stimulate growth of mice. *Science* **222:** 809–814.

Palmiter R.D., Sandgren E.P., Avarbock M.R., Allen D.D., and Brinster R.L. 1991. Heterologous introns can enhance expression of transgenes in mice. *Proc. Natl. Acad. Sci.* **88:** 478–482.

Palmiter R.D., Behringer R.R., Quaife C.J., Maxwell F., Maxwell I.H., and Brinster R.L. 1987. Cell lineage ablation in transgenic mice by cell-specific expression of a toxin gene. *Cell* **50:** 435–443.

Palmiter R.D., Brinster R.L., Hammer R.E., Trumbauer M.E., Rosenfeld M.G., Birnberg N.C., and Evans R.M. 1982. Dramatic growth of mice that develop from eggs microinjected with metallothionein-growth hormone fusion genes. *Nature* **300:** 611–615.

Perry A.C., Wakayama T., Kishikawa H., Kasai T., Okabe M., Toyoda Y., and Yanagimachi R. 1999. Mammalian transgenesis by intracytoplasmic sperm injection. *Science* **284:** 1180–1183.

Perry A.C., Rothman A., de las Heras J.I., Feinstein P., Mombaerts P., Cooke H.J., and Wakayama T. 2001. Efficient metaphase II transgenesis with different transgene archetypes. *Nat. Biotechnol.* **19:** 1071–1073.

Peschon J.J., Behringer R.R., Brinster R.L., and Palmiter R.D. 1987. Spermatid-specific expression of protamine 1 in transgenic mice. *Proc. Natl. Acad. Sci.* **84:** 5316–5319.

Peters K., Werner S., Liao X., Wert S., Whitsett J., and Williams L. 1994. Targeted expression of a dominant negative FGF receptor blocks branching morphogenesis and epithelial differentiation of the mouse lung. *EMBO J.* **13:** 3296–3301.

Peterson K.R., Clegg C.H., Li Q., and Stamatoyannopoulos G. 1997. Production of transgenic mice with yeast artificial chromosomes. *Trends. Genet.* **13:** 61–66.

Pfeifer A., Ikawa M., Dayn Y., and Verma I.M. 2002. Transgenesis by lentiviral vectors: Lack of gene silencing in mammalian embryonic stem cells and preimplantation embryos. *Proc. Natl. Acad. Sci.* **99:** 2140–2145.

Pieper F.R., de Wit I., Pronk A.C., Kooiman P.M., Strijker R., Krimpenfort P.J., Nuyens J.H., and de Boer H.A. 1992. Efficient generation of functional transgenes by homologous recombination in murine zygotes. *Nucleic Acids Res.* **20:** 1259–1264.

Pittius C.W., Hennighausen L., Lee E., Westphal H., Nicols E., Vitale J., and Gordon K. 1988. A milk protein gene promoter directs the expression of human tissue plasminogen activator cDNA to the mammary gland in transgenic mice. *Proc. Natl. Acad. Sci.* **85:** 5874–5878.

Previtali S.C., Quattrini A., Fasolini M., Panzeri M.C., Villa A., Filbin M.T., Li W., Chiu S.Y., Messing A., Wrabetz L., and Feltri M.L. 2000. Epitope-tagged P(0) glycoprotein causes Charcot-Marie-Tooth-like neuropathy in transgenic mice. *J. Cell Biol.* **151:** 1035–1046.

Probst F.J., Fridell R.A., Raphael Y., Saunders T.L., Wang A, Liang Y., Morell R.J., Touchman J.W., Lyons R.H., Noben-Trauth K., Friedman T.B., and Camper S.A. 1998. Correction of deafness in shaker-2 mice by an unconventional myosin in a BAC transgene. *Science* **280:** 1444–1447.

Readhead C., Popko B., Takahashi N., Shine H.D., Saavedra R.A., Sidman R.L., and Hood L. 1987. Expression of a myelin basic protein gene in transgenic shiverer mice: Correction of the dysmyelinating phenotype. *Cell* **48:** 703–712.

Robertson E., Bradley A., Kuehn M., and Evans M. 1986. Germ-line transmission of genes introduced into cultured pluripotential cells by retroviral vector. *Nature* **323:** 445–448.

Rusconi S. 1991. Transgenic regulation in laboratory animals. *Experientia* **47:** 866–877.

Russo A.F., Crenshaw E.B., III, Lira S.A., Simmons D.M., Swanson L.W., and Rosenfeld M.G. 1988. Neuronal expression of chimeric genes in transgenic mice. *Neuron* **1:** 311–320.

Sambrook J. and Russell D. 2001. *Molecular cloning: A laboratory manual*, 2nd edition. Cold Spring Harbor Laboratory Press, Cold Spring Harbor, New York.

Schedl A., Beermann F., Thies E., Montoliu L., Kelsey G., and Schutz G. 1992. Transgenic mice generated by pronuclear injection of a yeast artificial chromosome. *Nucleic Acids Res.* **20:** 3073–3077.

Schedl A., Larin Z., Montoliu L., Thies E., Kelsey G., Lehrach H., and Schutz G. 1993. A method for the generation of YAC transgenic mice by pronuclear microinjection. *Nucleic Acids Res.* **21:** 4783–4787.

Shi Y., Son H.J., Shahan K., Rodriguez M., Costantini F., and Derman E. 1989. Silent genes in the mouse major urinary protein gene family. *Proc. Natl. Acad. Sci.* **86:** 4584 –4588.

Shin M.K., Levorse J.M., Ingram R.S., and Tilghman S.M. 1999. The temporal requirement for endothelin receptor-B signalling during neural crest development. *Nature* **402:** 496–501.

Simon M.C., Pevny L., Wiles M.V., Keller G., Costantini F., and Orkin S.H. 1992. Rescue of erythroid development in gene targeted GATA-1- mouse embryonic stem cells. *Nat. Genet.* **1:** 92–98.

Skarnes W.C., Auerbach B.A., and Joyner A.L. 1992. A gene trap approach in mouse embryonic stem cells: The lacZ reported is activated by splicing, reflects endogenous gene expression, and is mutagenic in mice. *Genes Dev.* **6:** 903–918.

Skowronski J., Parks D., and Mariani R. 1993. Altered T cell activation and development in transgenic mice expressing the HIV-1 nef gene. *EMBO J.* **12:** 703–713.

Sokol D.L. and Murray J.D. 1996. Antisense and ribozyme constructs in transgenic animals. *Transgenic Res.* **5:** 363–371.

Soriano P., Cone R.D., Mulligan R.C., and Jaenisch R. 1986. Tissue-specific and ectopic expression of genes introduced into transgenic mice by retroviruses. *Science* **234:** 1409–1413.

Stacey A., Mulligan R., and Jaenisch R. 1987. Rescue of type I collagen-deficient phenotype by retroviral-vector-mediated transfer of human pro alpha 1(I) collagen gene into Mov-13 cells. *J. Virol.* **61:** 2549–2554.

Stacey A., Bateman J., Choi T., Mascara T., Cole W., and Jaenisch R. 1988. Perinatal lethal osteogenesis imperfecta in transgenic mice bearing an engineered mutant pro-alpha 1(I) collagen gene. *Nature* **332:** 131–136.

Sternberg N.L. 1992. Cloning high molecular weight DNA fragments by the bacteriophage P1 system. *Trends Genet.* **8:** 11–16.

Stewart T.A., Pattengale P.K. and Leder P. 1984. Spontaneous mammary adenocarcinomas in transgenic mice that carry and express MTV/myc fusion genes. *Cell* **38:** 627–637.

Storb U., O'Brien R.L., McMullen M.D., Gollahon K.A., and Brinster R.L. 1984. High expression of cloned immunoglobulin kappa gene in transgenic mice is restricted to B lymphocytes. *Nature* **310:** 238–241.

Strauss W.M., and Jaenisch R. 1992. Molecular complementation of a collagen mutation in mammalian cells using yeast artificial chromosomes. *EMBO J.* **11:** 417–422.

Strouboulis J., Dillon N., and Grosveld F. 1992. Developmental regulation of a complete 70-kb human beta-globin locus in transgenic mice. *Genes Dev.* **6:** 1857–1864.

Swierczynski S.L., Siddhanti S.R., Tuttle J.S., and Blackshear P.J. 1996. Nonmyristoylated MARCKS complements some but not all of the developmental defects associated with MARCKS deficiency in mice. *Dev. Biol.* **179:** 135–147.

Tam P.P.L. and Tan S.-S. 1992. The somito-genetic potential of cells in the primitive streak and the tail bud of the organogenesis-stage mouse embryo. *Development* **115:** 703–715.

Taylor L.D., Carmack C.E., Schramm S.R., Mashayekh R., Higgins K.M., Kuo C.C., Woodhouse C., Kay R.M., and Lonberg N. 1992. A transgenic mouse that expresses a diversity of human sequence heavy and light chain immunoglobulins. *Nucleic Acids Res.* **20:** 6287–6295.

Townes T.M., Lingrel J.B., Chen H.Y., Brinster R.L., and Palmiter R.D. 1985. Erythroid-specific expression of human beta-globin genes in transgenic mice. *EMBO J.* **4:** 1715–1723.

Tronik D., Dreyfus M., Babinet C., and Rougeon F. 1987. Regulated expression of the Ren-2 gene in transgenic mice derived from parental strains carrying only the Ren-1 gene. *EMBO J.* **6:** 983–987.

Trudel M., D'Agati V., and Costantini F. 1991. C-myc as an inducer of polycystic kidney disease in transgenic mice. *Kidney Int.* **39:** 665–671.

van Roessel P. and Brand A.H. 2002. Imaging into the future: Visualizing gene expression and protein interactions with fluorescent proteins. *Nat. Cell Biol.* **4:** 15–20.

Verma-Kurvari S., Savage T., Gowan K., and Johnson J.E. 1996. Lineage-specific regulation of the neural differentiation gene MASH1. *Dev. Biol.* **180:** 605–617.

Wagner E.F., Stewart T.A., and Mintz B. 1981. The human beta-globin gene and a functional viral thymidine kinase gene in developing mice. *Proc. Natl. Acad. Sci.* **78:** 5016–5020.

Wagner S.D., Gross G., Cook G.P., Davies S.L., and Neuberger M.S. 1996. Antibody expression from the core region of the human IgH locus reconstructed in transgenic mice using bacteriophage P1 clones. *Genomics* **35:** 405–414.

Wagner T.E., Hoppe P.C., Jollick J.D., Scholl D.R., Hodinka R.L., and Gault J.B. 1981. Microinjection of a rabbit beta-globin gene into zygotes and its subsequent expression in adult mice and their offspring. *Proc. Natl. Acad. Sci.* 78: 6376–6380.

Wianny F. and Zernicka-Goetz M. 2000. Specific interference with gene function by double-stranded RNA in early mouse development. *Nat. Cell. Biol.* **2:** 70–75.

Wiles M.V., Vauti F., Otte J., Fuchtbauer E.M., Ruiz P., Fuchtbauer A., Arnold H.H., Lehrach H., Metz T., von Melchner H., and Wurst W. 2000. Establishment of a gene-trap sequence tag library to generate mutant mice from embryonic stem cells. *Nat. Genet.* **24:** 13–14.

Wilkie T.M., Brinster R.L., and Palmiter R.D. 1986. Germline and somatic mosaicism in transgenic mice. *Dev. Biol.* **118:** 9–18.

Wolgemuth D.J., Behringer R.R., Mostoller M.P., Brinster R.L., and Palmiter R.D. 1989. Transgenic mice overexpressing the mouse homoeobox-containing gene Hox-1.4 exhibit abnormal gut development. *Nature* **337:** 464–467.

Woychik R.P. and Alagramam K. 1998. Insertional mutagenesis in transgenic mice generated by the pronuclear microinjection procedure. *Int. J. Dev. Biol.* **42:** 1009–1017.

Woychik R.P., Stewart T.A., Davis L.G., D'Eustachio P., and Leder P. 1985. An inherited limb deformity created by insertional mutagenesis in a transgenic mouse. *Nature* **318:** 36–40.

Yang X.W., Model P., and Heintz N. 1997. Homologous recombination based modification in *Escherichia coli* and germline transmission in transgenic mice of a bacterial artificial chromosome. *Nat. Biotechnol.* **15:** 859–865.

Zakhartchenko W., Mueller S., Alberio R., Schernthaner W., Stojkovic M., Wenigerkind H., Wanke R., Lassnig C., Mueller M., Wolf E., and Brem G. 2001. Nuclear transfer in cattle with non-transfected and transfected fetal or cloned transgenic fetal and postnatal fibroblasts. *Mol. Reprod. Dev.* **60:** 362–369.

Zambrowicz B.P., Friedrich G.A., Buxton E.C., Lilleberg S.L., Person C., and Sands A.T. 1998. Disruption and sequence identification of 2,000 genes in mouse embryonic stem cells. *Nature* **392:** 608–611.

Zhu Y., Jong M.C., Frazer K.A., Gong E., Krauss R.M., Cheng J.F., Boffelli D., and Rubin E.M. 2000. Genomic interval engineering of mice identifies a novel modulator of triglyceride production. *Proc. Natl. Acad. Sci.* **97:** 1137–1142.

Isolation and Culture of Blastocyst-derived Stem Cell Lines

T HE ISOLATION AND GENETIC MANIPULATION of embryonic stem (ES) cells represents one of the most important and far-reaching achievements in mammalian developmental biology. Following a significant foundation laid down by research on multipotent embryonic carcinoma (EC) cells, ES cells were first derived from blastocysts in culture by Evans and Kaufman (1981) and Martin (1981). Soon thereafter, Bradley et al. (1984) showed that ES cells were capable of contributing to many different tissues in chimeras generated by blastocyst injection, including the germ line. ES cells are typically used as vehicles for modifying the mouse genome. However, ES cells can also be used for chimera studies (see Chapter 11). Several recently derived and characterized ES cell lines (Eggan et al. 2001) have made the production of completely ES-cell-derived animals very efficient. This approach can be used directly to generate and analyze heterozygous and potentially homozygous mutants (Carmeliet et al. 1996; Eggan et al. 2002). ES cells can also be manipulated in vitro to generate many differentiated cell types that may be used in the future for cell-based therapies of disease. Recently, an additional type of stem cell, trophoblast stem (TS) cell (Tanaka et al. 1998), has been isolated from blastocysts. These cells hold promise for providing new insights into trophoblast differentiation and placental biology. It seems reasonable to expect that primitive endoderm stem cell lines will soon also be generated from blastocysts, which then completes the effort of establishing permanent cell lines from all three developmental distinct lineages of the blastocyst.

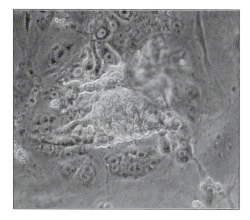

CONTENTS

ACQUIRING ES CELL LINES

There are several ways to obtain ES cell lines in your laboratory: (1) requesting already established and characterized ES cell lines from the laboratory in which they were generated, (2) purchasing ES cell lines from a commercial source, or (3) establishing and characterizing your own ES cell lines from blastocyst-stage embryos (see Protocol 8.5). Those who decide to generate their own ES cell lines should note that genetic background can significantly influence the ease of ES cell establishment and developmental potential.

Most ES cell lines have been established from the 129 inbred mouse strain. Interestingly, substrains of 129 are genetically very diverse (Simpson et al. 1997; Threadgill et al. 1997), because these substrains were generated by crosses with other inbred strains to introduce new traits. This is why, for example, the alleles of the albino (*c*) and the pink-eyed dilution (*p*) loci are highly variable and why the tremendous coat color variation exists among 129 substrains. Most 129 strain mice are homozygous for the white-bellied agouti allele (A^w/A^w) of the *Agouti* locus. The 129 substrain that appears to contain the least genetic contamination from other strains is 129S6/SvEvTac maintained by Taconic, and originally derived from The Jackson Laboratory. ES cell lines such as CCE, D3, and AB1 are derived from agouti-pigmented 129 substrains (Doetschman 1985; Robertson 1986; McMahon and Bradley 1990). Some ES cell lines like E14 (Hooper et al. 1987) and its subclone IB10 are derived from the light cream-colored substrain, 129P2/Ola. This substrain is homozygous for the chinchilla (c^{ch}) allele of the *c* locus and pink-eyed dilution (*p*) allele of the *p* locus. The R1 ES cell line (Nagy et al. 1993) is derived from the F_1 hybrid of two 129 substrains (129X1/SvJ and 129S1/SV-+^{p}+$^{Tyr-c}$Kitl^{Sl-J}/+). R1 ES cells are heterozygous at the *c* and *p* loci (*C/c,P/p*) and generate unique coat color variation in the F_2 generation after germ-line transmission.

C57BL/6 (B6) ES cell lines have been generated in several laboratories, and there are reports of successful germ-line transmission of targeted alterations from such cells (Ledermann and Burki 1991). However, anecdotally, it has been suggested that B6 ES cell lines are less stable in terms of maintaining germ-line potential through the sometimes multiple rounds of genetic alterations that occur in current gene-targeting strategies. The success rate of establishing ES cell lines from F_1 or mixed hybrid genetic backgrounds, especially those that contain the 129 strain, is usually very high. For laboratories with experience with ES cell culture and manipulations, establishing new ES lines is not difficult. Conditions and detailed protocols (e.g., Protocol 8.5) to derive ES-cell lines from blastocysts are well established. For laboratories that are just beginning to use ES-cell-based technologies, it is more prudent to acquire established and well-characterized ES cell lines.

Table 8.1 is a partial list of individuals who will provide germ-line-competent ES cell lines to the scientific community. Certain conditions governing their use, and/or the use of mouse lines derived from them, may be attached to some of these ES cell lines. Fibroblast feeders are also required for the maintenance of most of the ES cell lines. Some, but not all, laboratories also provide STO fibroblasts (see below) along with their ES cell lines. ES cells can also be obtained from commercial sources, including Taconic (http://www.taconic.com) and Specialty Media (http://www.specialtymedia.com).

CULTURING ES CELLS

Mycoplasma and Mouse Antibody Production Testing

ES cell lines and feeder cells should be tested for the presence of pathogens. Pathogen-contaminated ES cell lines may have reduced germ-line potential and can introduce infections into your mouse colony (see Chapter 3). It is very important to keep in mind that pathogens can also be transmitted from feeder cells to ES cells, so it is essential to test batches of primary mouse embryonic fibroblasts (MEF) if their pathogen-free status is unknown (Nicklas and Weiss 2000).

Mycoplasma is a common pathogen that can infect ES cell and feeder cell lines. A simple test for *Mycoplasma* involves staining cultures with the UV fluorescent dye Hoechst 33258 (Chen 1977). The presence of DNA from *Mycoplasma* (also yeast and slowly growing bacteria) will be revealed by "bead-like," extranuclear staining associated with the cytoplasm of the cells. Subconfluent or sparse cultures on tissue culture dishes with cells grown for at least two passages in antibiotic-free medium are rinsed in saline and then fixed at room temperature for 15–30 minutes in two changes of Carnoy's fixative (100% methanol/acetic acid,

TABLE 8.1. Sources of germ-line-competent ES cell lines

Allan Bradley, The Sanger Centre, Wellcome Trust Genome Campus, Hinxton, Cambridge, CB10 1SA
ES and STO cell lines:
http://www.imgen.bcm.tmc.edu/molgen/labs/bradley/cell.htm
AB1 (origin: 129S7/SvEvBrd)
AB2.1 (origin: 129S7/SvEvBrd, *Hprt* deficient)
SNL 76/6 feeder cell line (*neo^r*)
Rudolf Jaenisch, The Whitehead Institute, 9 Cambridge Center, Cambridge MA 02139.
ES cell line:
J1 (origin: 129S4/SvJae)
Andras Nagy, Samuel Lunenfeld Research Institute, Mt. Sinai Hospital, Toronto, Canada M5G 1X5
ES cell lines: http://www.mshri.on.ca/nagy
R1 (origin: 129X1/SvJ x 129S1/SV-+^{p}+$^{Tyr-c}$Kitl^{Sl-J}/+)
R1-fluorescent protein-marked ES cell lines
Elizabeth Robertson, Department of Molecular and Cellular Biology, Harvard University, 16 Divinity Ave., Cambridge, MA 02138
ES cell line:
CCE (origin: 129S6/SvEv)
Rosa26-marked ES cell lines
Colin Stewart, Center for Cancer Research , National Cancer Institute, FCRDC, Frederick, MD 21702
ES cell line:
W9.5 (origin: 129X1/SvJ)

3:1). If the cultures are confluent, it will be difficult to see any staining associated with the edges of the cells. The Hoechst dye is dissolved in Hanks' buffered saline (pH 7.0) at a concentration of 0.05 μg/ml (diluted from a 1 mg/ml stock). The dye is added to the cells for 10 minutes to overnight followed by two washes with distilled water. A coverslip is mounted with phosphate-buffered saline (PBS) or water, and the cells are viewed with a 40x objective lens under UV fluorescence. To learn what positive staining looks like without having to grow *Mycoplasma*-positive cells, a kit of control slides with fixed cells positive and negative for *Mycoplasma* can be obtained from Bionique Testing Laboratory (http://www.bionique.com). This company also provides a testing service. Alternatively, a PCR kit is available from Stratagene (302007) for *Mycoplasma* testing. However, it may be necessary to use multiple primers for detection of all possible *Mycoplasma* subtypes.

Keep in mind that there are several subtypes of *Mycoplasma* with significantly different growth rates, and that some ES cell lines are more sensitive to these infections than others. A slow-growing *Mycoplasma* subtype may be difficult to detect unless the supernatant is cultured under optimal conditions for up to 3 weeks. This is best performed in specially equipped laboratories. The widely used E14-1 ES cell line (Hooper et al. 1987) is contaminated with *Mycoplasma hominis*, and is a good example of a case where the infection is often missed due to the relatively difficult detection of the pathogen. Interestingly, when E14-1 ES cells are treated with G418 during selection, *M. hominis* is eliminated.

Mouse antibody production (MAP) testing is probably the most sensitive and specific way to detect and identify murine viral contaminants. It is offered by Charles River Laboratories (http://www.criver.com) and performed by inoculation of Virus Antibody Free (VAF) animals with samples to be tested and consequent assay of serum samples for the presence of virus-specific antibodies. MAP testing is recommended for checking newly acquired established ES cell lines from other laboratories or newly derived ES cell lines that will be used for making chimeras in specific-pathogen-free (SPF) facilities. A PCR-based alternative to MAP testing has been developed at the University of Missouri (http://www.radil.missouri.edu). The cost of this test is lower than commercial MAP testing, and the results are achieved much faster.

Culture Conditions for ES Cells

To maintain ES cells at their full developmental potential, optimal growth conditions should be provided in culture. If ES cells are cultured inappropriately, they progressively acquire genetic lesions (e.g., aneuploidy) that will compromise their germ-line potential. If suboptimal culture conditions exist, variants are selected that have undergone chromosomal rearrangements and/or mutations that increase their growth rate and decrease their pluripotency. Suboptimal conditions include those in which some nutrient or growth factor is in limited supply and a culture regimen that involves leaving the cells too long at high density (see Fig. 8.1), which favors the differentiation of cells into endoderm-like cells.

ES cells should be cultured at a reasonably high density, and the cells should be split at 1:3 to 1:6 every 2–3 days (see Protocol 8.3). This is preferable to subculturing the cells to very low densities, even if it means replacing the medium daily. In addition, because clumps of ES cells tend to differentiate into endoderm, it is very important to dissociate the cultures into single cells after trypsinization. A typical cell culture is shown in Figure 8.1. Note that the cells grow in tightly packed groups with "smooth" outlines. If the cells start to spread onto the substratum or form colonies with "rough" endoderm on the surface, the cultures are suboptimal (Fig. 8.1F).

To minimize the danger of accumulating chromosomal anomalies in ES cells, vials of stock cells should be frozen (see Protocol 8.4) as soon as possible after a new cell line is derived or a bona fide pluripotent ES cell line capable of colonizing the germ line is acquired from another laboratory. Keep a record of the number of passages that the cells undergo after they are thawed. As a general rule, always go back to the original pool of ES cells with each experiment, and do not grow cells for more than three or four passages before beginning an experiment. Before starting a project to target a mutation by homologous recombination, it is prudent to test that the stock cells are still capable of colonizing the germ line after culture under the conditions of your laboratory. If the cell line does not give a high rate of germ-line transmission, one option is to derive 5–10 subclones from single cells and to test these for germ-line transmission. Evidence from several laboratories suggests that this is a reliable way of recovering germ-line-competent subclones and may be less work than deriving a completely new cell line. Once genetically altered cells have been selected, it is important to reduce the time that they spend in culture before introducing them into embryos.

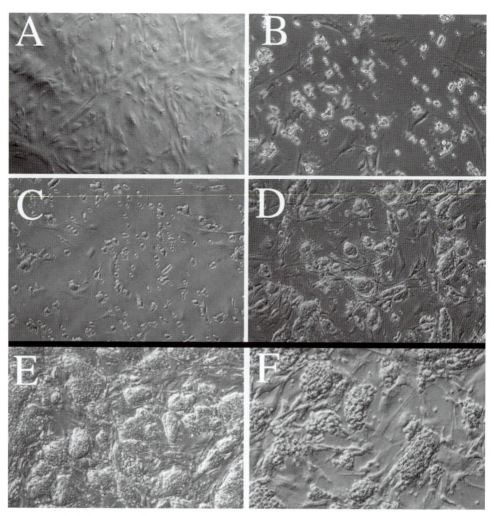

FIGURE 8.1. Culturing embryonic stem cells. (*A–D*) Stages of optimal ES cell maintenance. (*A*) Feeder cells before, (*B*) 4 hours, (*C*) 1 day, (*D*) 2 days after plating, respectively. (*E*) and (*F*) are examples of overgrown and differentiating cultures. Note the smooth outline of the colonies of densely packed cells (*D*). Differentiated cells adhere very tightly to the surface of the dish or form a rough endoderm-like layer on the surface of the colonies.

Culture Medium

The medium most frequently used for ES cell culture is Dulbecco's modified Eagle's medium (DMEM) (high glucose, 4500 mg/liter). If a cell line is obtained from another laboratory, it is advisable to use the same culture medium and supplements used by that laboratory because medium requirements can vary between different ES cell lines. DMEM can be purchased as a powder (e.g., Invitrogen Life Technologies 12100) requiring reconstitution or as a ready-made solution in 500-ml or 1-liter bottles. High-quality liquid medium can be obtained from commercial sources, such as GIBCO products from Invitrogen Life Technologies: http://www.invitrogen.com (e.g., KO-DMEM medium 10829-018) or Specialty Media: http://www.specialtymedia.com (e.g., SLM-220B). Evidence suggests that it may be better to use 2.2 g/liter of bicarbonate instead of the standard 3.7 g/liter because this amount is better suited to a 5% CO_2 environment and brings the osmolality to 290 mOsm/liter, which is close to that of serum (P. Soriano, pers. comm.).

DMEM is a bicarbonate-buffered culture medium designed to maintain a pH of 7.2–7.4 in an atmosphere of 5% CO_2, 95% air. Thus, it is a good idea to monitor the concentration of CO_2 regularly and directly in the incubator, rather than relying on the incubator gauges, to ensure accurate pH. This can be measured using a simple apparatus, the FYRITE test kit from Bacharach (http://www.bacharach-inc.com). It is essential to always keep the incubator humidified. This is easily done using trays of water.

If the medium is reconstituted from a powder, use care to ensure that it is made correctly and use filtered MilliQ-water or distilled water free of bacterial endotoxins. If filtered MilliQ-water is the water source, make sure to change filters regularly, because bacteria may grow on the resin and release endotoxins into the water. A simple test for the presence of endotoxin is based on the gelling of protein present in a *Limulus* (horseshoe crab) amebocyte lysate and is available from Associates of Cape Cod, Inc. Endotoxin-free distilled water can also be purchased commercially.

Once the medium has been made, it should be dispensed into glass bottles that have been cleaned with a tissue culture-grade detergent and rinsed thoroughly with endotoxin-free distilled water. Medium can also be stored in disposable plastic tissue culture bottles. Medium should be stored in the dark at 4°C. Tissue culture media have a limited shelf life even at 4°C, particularly if they contain glutamine (which oxidizes to glutamic acid) and serum (which contains glutaminase). If complete medium is stored for longer than 2 weeks, supplement it with additional 1–2 mM L-glutamine because this amino acid is unstable. Glutamax medium from Invitrogen Life Technologies (10566) contains L-glutamine in a stabilized dipeptide form and can be used without extra supplements. Do not store media near a fluorescent light because this will lead to the breakdown of vitamins.

Disposable plasticware should be used for routine procedures such as subculturing and freezing of cells. If this is not possible, then all pipettes and containers should be cleaned carefully to avoid traces of detergent residue. It is advisable to use glassware designated for tissue culture only to minimize the risk of contamination with residual reagents. Cells attach efficiently only to the surfaces of plastic tissue culture dishes. To avoid confusion, clearly label bacteriological-grade dishes. Plastic tissue culture dishes or inserts that are resistant to the chemicals used in histological and electron microscopic procedures (e.g., Falcon 3006 Optical) are also available.

When medium, trypsin/EDTA, and Ca^{++}/Mg^{++}-free PBS are made in the laboratory, be sure to use endotoxin-free water and tissue culture-grade reagents. Although it is more expensive to purchase these reagents ready-made from a company that tests for *Mycoplasma*, endotoxin, and cell toxicity, this can eliminate potential problems.

Before use, the ES cell medium should be supplemented with the following:

1. Glutamine to 2 mM from a 100x concentrated stock (200 mM) (e.g., 25030-081 or Glutamax 35050-061, from Invitrogen Life Technologies), stored as aliquots at –20°C.

2. MEM nonessential amino acids to 0.1 mM from a 100x concentrated stock (10 mM; e.g., Invitrogen Life Technologies 11140-050), stored at 4°C.

3. 0.1 mM β-mercaptoethanol (βME) (tissue culture grade, Sigma M7522) or 0.15 mM monothioglycerol (tissue culture grade, Sigma M6145). Monothioglycerol

is less volatile than βME), stored as aliquots at –20°C. *Caution: Miscalculations that increase the βME concentration in the ES cell culture medium are guaranteed to cause ES cell culture failure.*

4. Sodium pyruvate (*optional*) to 1 mM from a 100x concentrated stock (e.g., Invitrogen Life Technologies 11360-070), stored at 4°C.

5. Gentamicin to 50 μg/ml (50 mg/ml stock, Sigma G1522), or penicillin to 50 U/ml and streptomycin to 50 μg/ml, or no antibiotics, particularly if cells are to be tested for *Mycoplasma* (see p. 361). A 100x stock solution of penicillin–streptomycin is stored at –20°C (e.g., Invitrogen Life Technologies 15070-063).

6. 15% Fetal bovine serum (FBS) (see below).

7. Leukemia inhibitory factor (LIF), 500–1000 units/ml (see below).

Serum

FBS is used to supplement the DMEM for ES cell culture. It is very important to identify a batch of serum that gives optimal growth, and, if possible, to reserve enough of this batch to supply the laboratory's culture needs for some time, to avoid repeated testing. Small quantities of specific serum batches can be obtained from different suppliers (e.g., HyClone, Gemini, Invitrogen Life Technologies, Sigma, Specialty Media, Wisent) and tested for plating efficiency. Trypsinize a plate of exponentially growing ES cells and seed them at low density (~1 x 10^3 cells per 6-cm plate) without feeder cells (see below) in medium supplemented with each serum sample at 15%. Set up five or six plates per sample and include a dish in which 30% serum is added to test for toxicity at high concentrations and another dish with 10% serum for better judgment of small differences in quality between batches. Refeed the cultures daily. After 7–10 days, when colonies of ES cells are visible to the naked eye, remove the medium, wash the ES cell colonies with PBS, and stain them with 2% methylene blue for 2–5 minutes. Check for colony morphology as well as plating efficiency (number of colonies per cells plated). The plating efficiency should be between 5% and 10%. Once a batch of serum has been selected, it should be used as a standard to test new serum batches. Many laboratories have found that heat inactivation of the FBS is unnecessary. The heat treatment may destroy important components of the serum that are needed for optimal ES cell culture conditions.

Different ES cell lines may behave differently in the same FBS. If cells are received from another laboratory where one source of serum was successfully used, it is advisable to obtain the same batch. If this batch is no longer available from the supplier, ask the originating lab for information about the second-best batch used in their last serum screen.

Most laboratories use the less expensive newborn calf serum (not necessarily selected batches) for culturing primary MEF and STO cell lines. However, some laboratories believe that serum quality is very important for STO cell culture and use the same serum used for ES cell culture to maintain their STO cell lines.

Leukemia Inhibitory Factor

Leukemia inhibitory factor (LIF; also known as DIA, differentiation inhibiting factor) is a secreted polypeptide cytokine that inhibits the spontaneous differentiation of ES cells (Smith et al. 1988; Williams et al. 1988). It is one of the active components of Buffalo rat-conditioned medium used to grow ES cells (Smith and Hooper 1987) and is produced by both primary MEF and STO cells. It is possible to substitute MEF or STO cells completely with purified LIF. Typically, if ES cells are "weaned off" feeder cells and cultured on gelatinized plates in medium supplemented with LIF, they go through a so-called "crisis" in which the colonies are flatter and there is extensive differentiation at the margins. However, in most cases, good morphology reappears after a few passages, and sublines can be derived with normal morphology under these conditions. Indeed, some laboratories are successfully using "gelatin-trained" ES cells that retain good germ-line potential if they are kept in medium supplemented with LIF. Currently, CHEMICON International (http://www.chemicon.com) is the exclusive manufacturer and supplier of LIF under the name of ESGRO (ESG-1107). The observation that LIF-deficient fibroblast cells are unable to maintain ES cells in an undifferentiated state is clear evidence that LIF is the crucial factor required for pluripotency. It is worth noting, however, that MEF and STO cells are probably producing other factors that enhance the survival or growth of ES cells in addition to LIF. Most laboratories continue to use feeder cells to culture ES cells but may also, in some cases, supplement the medium with LIF. LIF should be added to a concentration of 500–1000 units/ml when culturing ES cells with MEF feeders or 1000 units/ml when using STO cell feeders.

Trypsin/EDTA

Use a mixture of trypsin and EDTA in Ca^{++}/Mg^{++}-free PBS to detach cells from the tissue culture dishes and to dissociate them from one another. For routine culturing of primary MEF or STO mouse fibroblast cell lines, use a final concentration of 0.05% trypsin/0.02% (0.53 mM) EDTA. This can be prepared by diluting a 10x stock of 0.5% trypsin/5.3 mM EDTA from a supplier, such as Invitrogen Life Technologies (15400-054), with 1x Dulbecco's PBS. A solution of 0.05% trypsin/ 0.53 mM EDTA may also be purchased ready-made (e.g., Invitrogen Life Technologies 25300-054). The diluted solution can be dispensed into convenient amounts (e.g., 5–10 ml) and stored frozen at –20°C.

For routine subculture of ES cell lines, dissociating ES colonies after selection, and deriving ES cells de novo, a more concentrated trypsin/EDTA solution is recommended. The final concentration is 0.25% trypsin/0.04% (1 mM) EDTA in Hank's buffered saline (or Tris-buffered saline). This can be purchased as 1x solution from Invitrogen Life Technologies (25200-056). The higher trypsin/EDTA concentration (some laboratories use as high as 0.5% trypsin) helps to dissociate the closely packed colonies of ES cells into single cells. It is also possible to prepare trypsin solution by rehydrating lyophilized powder (e.g., Bacto Trypsin 1:250, Difco, Becton Dickinson 215320 http://www.bd.com) in saline/EDTA to the desired concentration (see Appendix 1).

FEEDER CELLS

To maintain the undifferentiated state of ES cells, two kinds of feeder cells are most commonly used: primary cultures of MEFs and the STO mouse fibroblast cell line. Two methods are used to prepare mitotically inactive feeder cells: treatment with mitomycin C and γ-irradiation (see Protocol 8.2). Mitomycin C crosslinks DNA and blocks cell proliferation. γ-Irradiation avoids the risk of carrying over mitomycin C from the feeder cells to the ES cells and can be more convenient and less labor-intensive than mitomycin C treatment if a radiation source is readily available.

Primary Mouse Embryonic Fibroblasts

The advantage of using MEFs is that, as primary cells, they represent a very potent, reliable, and reproducible source of feeder cells. MEFs cannot be maintained indefinitely in culture, so there is no temptation to use them beyond the time when they may have lost their growth-enhancing activity (see below for STO cells). The disadvantage of using MEFs is that they do have a limited life span in culture and must be replenished continuously from frozen stocks. However, this is not a great problem and only requires the establishment of a stock of MEFs once a year or less often (see Protocol 8.1). Keep in mind that MEFs from wild-type mice cannot be used as feeder cells during drug selection (e.g., G418, puromycin, hygromycin). However, this can be overcome by performing the drug selection on gelatinized plates and using medium supplemented with LIF. It is also possible to use MEFs derived from a mouse strain in which the drug-resistance gene is constitutively expressed from a transgene or gene-targeted mutation. Gene knockouts that are homozygous viabile and fertile are advantageous because all embryos derived from a homozygous parent will express the drug-resistance gene. Homozygous *neo* transgenic animals are available through The Jackson Laboratory, such as the line C57BL/6J-TgN(pPGKneobpA)3Ems (002356). Another useful mouse line recently developed in the laboratory of Rudolf Jaenisch is also available from The Jackson Laboratory as TgN(DR4)1Jae (003208). This line was prepared by intercrossing three different strains (Tucker et al. 1997), one bearing resistance genes for neomycin and puromycin, the second bearing the gene for hygromycin resistance, and the third carrying a spontaneous mutant for the *Hprt* gene.

STO Fibroblasts

STO cells were derived from *S*IM mice and are a *t*hioguanine- and *o*uabain-resistant subline of fibroblasts isolated by Dr. Alan Bernstein (Bernstein et al. 1976). They have been widely used as feeder cells for the culture of embryonic carcinoma (EC) and ES cells. The advantage of using a continuous cell line instead of MEFs for preparing feeders is that the cells are easily grown and do not require a lengthy and tedious method to replenish frozen stocks. However, great care must be taken to maintain the STO cells under optimal growth conditions and to guard against the selection of more rapidly growing variants, which may occur if stock cultures are allowed to become dense. A routine should be established for thawing frozen stocks after a fixed number of passages. STO cells that have been stably transfected with a *neo*ʳ vector and also with a LIF expression vector (SNL or STO/N/L cells) are available from Allan Bradley (The Sanger Centre, Wellcome Trust Genome Campus, Hinxton) and Elizabeth Robertson (Biological Labs, Harvard University, Cambridge, Massachusetts). The expression of the *neo*ʳ gene allows the cells to be used as feeder cells during the G418 selection process.

STO cells are routinely cultured in DMEM with 7–10% newborn calf serum on gelatinized tissue culture dishes (see Protocol 8.2).

COUNTING CHROMOSOMES IN ES CELLS

Aneuploidy in ES cells is the major cause of failure in obtaining contributions to all tissues of chimeras, including the germ line. In a study (Longo et al. 1997), the majority of ES cell clones that harbored 50–100% euploid metaphases did transmit through the germ line; in contrast, none of the ES cell clones with more than 50% chromosomally abnormal metaphases transmitted through the germ line. Euploid ES cell clones cultured in vitro for more than 20 passages rapidly became severely aneuploid. This result correlated closely with the percentage of chimerism and with the number of ES cell–embryo chimeras.

It was found that chromosomal abnormalities occurred rather frequently in ES cells during gene targeting. Cells having an abnormal number of chromosomes, in particular trisomy 8, were found in three independently derived ES cell lines (Liu et al. 1997). This abnormality conferred a selective growth advantage on these cells and led to depletion and eventual loss of normal ES cells during consecutive passages. In comparison with parental ES cells, ES cells with trisomy 8 rarely contributed to the germ line.

The subjective choice of clones with a high probability of contributing to the germ line may be based on cell morphology and growth rate. The cells should grow in dense, three-dimensional colonies, having distinct edges with a minimal number of flattened colonies and fibroblast-like outgrowths. Clones with unusually high or low growth rates should be avoided if possible. A simple and fast method of chromosome counting from metaphase spreads may be helpful when many ES cell clones need to be tested (see Protocol 8.6). ES cell clones showing euploidy on 70–80% of spreads are used for chimera production.

It is important to remember, however, that the chromosome counting method will not show such chromosomal aberrations as translocations or spontaneous mutations, and clones with good euploid count may still perform poorly. The only reliable way to determine the developmental capacity of ES cells is to introduce them into the embryonic environment and screen for germ-line potential.

IN VITRO DIFFERENTIATION OF ES CELLS

Pluripotent ES cells can develop into many types of differentiated tissues if they are placed back into a differentiating environment. This can occur in vivo when the ES cells are injected into or aggregated with an embryo, or in vitro if their culture conditions are modified to induce differentiation. There are an increasing number of differentiating culture conditions that can bias the differentiation of ES cells into desired cell types, such as ventricular cardiomyocytes, dopaminergic neurons, pancreatic β cells, hematopoietic cells, endothelial cells, and adipocytes (Kawasaki et al. 2000; Turksen 2002). Determining the mechanisms that control ES cell differentiation into therapeutically important cell types is a quickly growing area of research. Knowledge gained from these studies may eventually lead to the use of stem cells to repair specific damaged tissues. Many times ES cell differentiation proceeds through an intermediate stage called the embryoid body (EB). EBs are round structures composed of ES cells that have undergone some of the initial stages of differentiation. EBs can then be manipulated further to generate more differentiated cell types. Protocol 8.7 provides a method to differentiate ES cells into EB.

TROPHOBLAST STEM CELLS

Three different cell types can be found in later-stage blastocysts: outside epithelium called trophectoderm, inside primitive ectoderm, and primitive endoderm forming inner cell mass (ICM) (see Chapter 2 for details). The developmental potential of ES cells represents the primitive ectoderm lineage, which gives rise to the embryo proper and some extraembryonic tissues. Recently, culture conditions have been established for a second blastocyst-derived cell line, trophoblast stem (TS) cells. These cells retain the developmental potential of the trophectoderm, and therefore contribute only to the trophoblast lineages of the placenta (Tanaka et al. 1998). The availability of TS cells makes possible new studies of trophoblast differentiation and placental function. Technologies to modify TS cells genetically are currently being developed and could lead to the creation of a system that can selectively introduce genetically altered cellular components into the placenta of a developing embryo. Established TS cell lines are available from Janet Rossant, Samuel Lunenfeld Research Institute, Mt. Sinai Hospital, Toronto, Canada M5G 1X5.

Derivation and Culture of TS Cell Lines

The culture conditions for TS cell establishment and maintenance are a little more complex than those of ES cells (see Protocol 8.8). On the other hand, under these conditions they are more stable and easier to establish from different genetic backgrounds than ES cells (see Protocol 8.9). Because LIF is essential for ES cells, fibroblast growth factor-4 (FGF4) and heparin are critical components in the medium for TS cells and are necessary to maintain the cells in an undifferentiated state. In addition, there are some still essential unidentified compounds, because the maintenance of TS cells also requires either MEF feeders or MEF-conditioned medium. TS cell cultures contain a certain percentage (5–10%) of differentiated cells, even in the best conditions. Since they are more trypsin-resistant and often postmitotic, their relative levels do not increase after each passage. For more details, see Kunath et al. (2001).

Preparing Mouse Embryo Fibroblasts

Although the authors use the procedure outlined below, a slightly different protocol for preparing primary cultures is described by Abbondanzo et al. (1993).

MATERIALS

EMBRYOS
Mouse embryos (15.5–16.5 dpc) dissected in a 10-cm sterile plastic petri dish (For the choice of strain, see Mouse Embryo Fibroblasts, p. 368).

EQUIPMENT
Cell culture dishes, 150-mm
Centrifuge
Conical tubes, plastic, 15- and 50-ml sterile screw-cap
Cryovials
Forceps and scissors for dissection (sterilize prior to use by autoclaving or by dipping in alcohol)
Glass beads (3–5 mm in diameter, autoclaved), sterile
Magnetic stirrer in a 37°C warm room or large incubator at 37°C
Microscope, inverted
Petri dishes, 10-cm, sterile
Stir bars (1–2 inch, autoclaved), sterile
Surgical blades, sterile

REAGENTS
Dulbecco's modified Eagle's medium (DMEM) supplemented with 10% newborn calf serum or fetal bovine serum (FBS) (see p. 366) Trypan Blue (Flow Labs 16-910-49)
Trypsin/EDTA (see p. 367)

PROCEDURE

1. Aseptically dissect mouse embryos at 15.5–16.5 dpc (see Chapter 5) from one or two pregnant mice (about 8 embryos per litter). Place the embryos in a 10-cm sterile plastic petri dish containing sterile PBS. The dish does not have to be made of tissue culture-grade plastic.

2. Remove the embryo's limbs, and scoop out and discard the internal organs. Leave the lower head, but remove the brain or upper part of the head containing the brain.

3. Place the carcasses into a 50-ml screw-capped sterile conical tube containing sterile PBS. Rinse them three or more times with sterile DMEM without serum (*optional*).

4. Place the embryos into a 10-cm petri dish and remove the medium.

5. Mince the embryos into very small pieces with a sterile surgical blade or scissors.

6. Place the minced embryos into a 50-ml screw-capped tube containing ~10 ml of trypsin/EDTA in PBS or similar buffer. Add 5 ml of sterile glass beads and a stir bar and close the tube.

7. Incubate at 37°C for 30 minutes with stirring.

8. Add 10 ml of trypsin/EDTA and incubate at 37°C for another 30 minutes with stirring.

9. Repeat step 8 (final volume 30 ml+).

10. Decant the cell suspension into two 50-ml tubes each containing 3 ml of FBS.

11. Wash the screw-capped tube twice with DMEM + 10% FBS and add the washings to the tubes prepared in step 10.

12. Centrifuge at 1000 rpm (200g) for 5 minutes and resuspend the pellet in 50 ml of DMEM plus 10% FBS.

13. Count the number of viable nucleated cells using Trypan Blue. (~5 x 10^7 to 5 x 10^8 cells can be expected from 10 fetuses.)

14. Plate 5 x 10^6 cells per 150-mm tissue culture dish containing DMEM plus 10% FBS.

15. Change the medium the next day and allow the cells to grow until the dishes are confluent. Many cell types will be seen, such as nerve, cartilage, and fibroblast, but except for the latter, they will not survive the subculture.

16. Split the cells 1:6 and allow them to grow to confluency. When confluent, freeze the MEF stocks by following Protocol 8.4 given for ES cell freezing and test for *Mycoplasma* and mouse pathogens (see p. 361).

    For laboratories processing large volumes of cells, each 150-mm dish can be frozen in one cryovial (see Protocol 8.4). There is no hard-and- fast rule as to the number of passages the MEF cells can sustain before their growth rate declines; however, the number of cells should be similar from vial to vial.

17. To start the preparation of feeders for ES cells, thaw one vial of frozen MEF in 5 or 6 x 150-mm sterile plastic tissue culture dish containing DMEM plus 10% FBS (see Protocol 8.4).

    In about 3 days, when the cells are confluent, three options are possible:

    a. split the cells if necessary one more time before mitomycin C treatment or irradiation (see Protocol 8.2);

    b. treat with mitomycin C or irradiate and use directly as feeders for ES cell culture;

    c. treat with mitomycin C or irradiate and freeze in cryovials for later use.

COMMENT

MEF and STO cells grow well in DMEM supplemented with just 10% serum. However, many laboratories also add 0.1 mM βME and additional 0.2 mM L-glutamine to the medium. As an antioxidant, βME prevents cell damage by free radicals and has been shown to increase plating efficiencies with primary cultures (Oshima 1978).

Preparing Feeder Cell Layers from STO or MEF Cells

MATERIALS

EQUIPMENT
Centrifuge
Incubator, humidified 37°C 5% CO_2, 95% air
Microscope, inverted
Tissue culture dishes, various sizes
Tissue culture dishes, pretreated with gelatin for STO cells
For γ irradiation:
    Gammacell 40 Exactor (e.g., MDS Nordion, http://www.mds.nordion.com)

REAGENTS
DMEM supplemented with 10% newborn calf serum or FBS
    Gelatin (Swine Skin type II, Sigma G2500) (0.1% in water, autoclaved)
    Incubator, humidified 37°C 5% CO_2, 95% air
    Phosphate-buffered saline (PBS), Ca^{++}/Mg^{++}-free (see Appendix 1)
    Trypsin/EDTA (see p. 367)
For mitomycin C treatment:
Mitomycin C<!> stock (0.5–1.0 mg/ml in PBS or glass-distilled water)
Store for 1 week at 4°C protected from light. If stored frozen, check that it is all in
solution after thawing (Sigma M0503). Use at 10 μg/ml.
STO or MEF cells

CAUTION: See Appendix 2 for appropriate handling of materials marked with <!>.

PROCEDURE

Mitomycin C Treatment

1. Treat confluent STO or MEF cells in tissue culture dishes with DMEM plus 10% FBS and 10 μg/ml mitomycin C.

2. Return the dishes to the humidified incubator at 37°C with 5% CO_2 for 2–3 hours.

3. Wash the dishes extensively with several changes of PBS and collect the cells by trypsinization. Pellet the cells by low-speed centrifugation (1000 rpm for 5 minutes).

4. Remove the supernatant and resuspend the pellet with fresh DMEM plus 10% FBS.

5. Count the cells, and dilute them to give a final cell density of 2×10^5/ml. For example, one confluent 150-mm MEF plate can generate approximately the following numbers of feeder plates:

> 5 x 100-mm plates (10 ml each)
>
> 12 x 60-mm plates (5 ml each)
>
> 25 x 35-mm plates (2 ml each)
>
> 25 x 4-well or 4 or 5 x 24-well plates (0.5 ml/well)
>
> 6 x 96-well plates (0.2 ml/well)

6. Plate the MEF cells directly onto tissue culture dishes (see Comments below). STO cells are plated onto tissue culture dishes pretreated with gelatin. This is achieved by flooding each dish with a 0.1% solution of gelatin, and then completely removing it. There is no need for drying the plates before use.

Treatment by γ-Irradiation

1. Expose confluent cells to 3,000–10,000 rads of γ-irradiation.

2. Harvest the cells by trypsinization. Pellet the cells by low-speed centrifugation (1000 rpm for 5 minutes).

3. Count the cells, and plate them on tissue culture-grade dishes.

   Irradiated cells can be frozen at a concentration of 5×10^6 cells/ml. After thawing, 1 ml of this suspension is plated on four or five 6-cm dishes.

COMMENTS

- Generally, there is no need to gelatinize the dishes for MEF cells because healthy fibroblast cells will attach to the surface of tissue culture-grade plates; however, irradiated MEF may last longer if plated on gelatinized plates.

- It is important to plate the fibroblasts at the correct density to ensure that a confluent uniform monolayer is produced. A density of $5 \times 10^4/cm^2$ is suggested for STO cells.

- Feeder plates may be used for up to 1 week after they are made, and the medium should be replaced with ES cell culture medium immediately before use. Check for the integrity of the monolayer before use.

Passage of ES Cells

ES cells should be split at 1:3 to 1:7 every 2–3 days depending on their growth rate when they reach 70% confluency. They should never be allowed to grow past 90% confluency, but rather they should form tightly packed colonies not touching each other. Medium should be replaced daily. Because clumps of ES cells tend to differentiate into endoderm, it is very important to dissociate the cultures into single cells after trypsinization.

MATERIALS

CELLS
ES cells (subconfluent 60-mm plate, ready for the passage)
STO or MEF feeder plates

EQUIPMENT
Centrifuge
Incubator, humidified 37°C, 5% CO_2, 95% air
Microscope, inverted
Pipettes
Test tubes, sterile plastic with cap (e.g., 14-ml)
Tissue culture dishes

REAGENTS
ES cell culture medium with FBS and supplements (ES-DMEM, as described on p. 365–366)
Phosphate-buffered saline (PBS), Ca^{++}/Mg^{++}-free (see Appendix 1)
Trypsin/EDTA solution (see p. 367)

PROCEDURE

1. When the culture reaches ~70% confluence, aspirate the medium, rinse the plate with PBS, and add 0.5–1 ml of trypsin.

2. Place the plate in an incubator at 37°C for 5 minutes. During this time, the cell clumps should be lifted off and disaggregated.

3. Stop the action of trypsin by adding 4 ml of ES-DMEM to the plate.

4. Transfer the cell suspension into a 14-ml sterile test tube and spin the cells down at 1000 rpm for 5 minutes. Aspirate the supernatant.

5. Gently resuspend the pellet in fresh ES-DMEM, and seed the cells onto new feeder plates in the ratio of 1:3 to 1:7 (~1 x 10^6 cells per 60-mm plate, 2 x 10^6 cells per 100-mm plate). Tilt the plate a couple of times both in the X and Y directions to distribute the cells evenly. Then place the plates in the incubator.

6. Repeat steps 1–5 every other day for maintenance, keeping track of the passage number.

Freezing and Thawing of ES Cells Using Cryovials

It is important to freeze ES cell stocks as soon as possible to reduce the time that they are in culture. Keep a careful record of the number of times cells are passaged and the location of the cryovials. Usually ES cells are frozen at about 2×10^6 to 5×10^6 cells/ml (~4 or 5 vials from a 100-mm dish). This protocol is for cryovials containing this number of cells that usually need to be thawed on a 60-mm feeder plate. When thawing cells, it is important to bring them to 37°C as quickly as possible.

MATERIALS

CELLS
Cryovial with frozen ES cells (for thawing procedure)
ES cells (subconfluent plate ready to be frozen)
STO or MEF feeder plate (60 mm)

EQUIPMENT
Centrifuge
Cryovials (e.g., Nalge 5000-0012), labeled
Cryo container (e.g., Nalge 5100-0001)
Incubator, humidified 37°C, 5% CO_2, 95% air
Microscope, inverted
Pipettes
Test tubes, sterile plastic with cap (e.g., 14-ml)
Tissue culture dishes
Water bath, 37°C, with floating device for cryovials

REAGENTS
DMEM containing 20–25% FBS and 10% dimethyl sulfoxide (DMSO) <!> (e.g., Sigma D2650, D5879), cooled on ice
ES cell culture medium with FBS and supplements (ES-DMEM) (see pp. 365–366)
Phosphate-buffered saline (PBS), Ca^{++}/Mg^{++}-free (see Appendix 1)
Trypsin/EDTA solution (see p. 367)

CAUTION: See Appendix 2 for appropriate handling of materials marked with <!>.

PROCEDURE

Freezing

1. Change the medium 2–3 hours before freezing ES cells (*optional*).

2. Prepare the freezing medium and cool it on ice. Freezing medium can be prepared by adding 10% FBS and 10% DMSO to ES-DMEM culture medium

already containing 15% FBS.

> It is also possible to use 2x freezing medium that is added to the cell suspension in equal volume. It consists of 20% DMSO, 20% FBS, and 60% ES-DMEM (already containing 15% FBS).

3. Remove the medium from the dish with ES cells, add PBS to rinse the cells, aspirate PBS, and add trypsin/EDTA solution (e.g., 2 ml per 100-mm dish). Place the plate in an incubator at 37°C for 5 minutes.

4. Stop the action of trypsin by adding 4 ml of ES-DMEM to the plate. Transfer the cell suspension into a 14-ml sterile test tube and pellet the cells down at 1000 rpm for 5 minutes. Aspirate the supernatant.

5. Gently resuspend the pellet in the appropriate volume of cold freezing medium to reach the concentration of 2×10^6 to 5×10^6 cells/ml (vial) and pipette them into clearly labeled cryovials.

6. Put the vials into a cryo container and immediately place the container in a –70°C freezer overnight. Next day cryovials can be transferred to liquid nitrogen for the storage.

COMMENTS

It is important to freeze the cells slowly at ~1°C per minute to –70°C before storing them in liquid nitrogen. This prevents the formation of ice crystals inside the cells. This slow cooling rate can be achieved in several ways. For example, freezing containers holding cryovials made by Nalge Nunc (5100-001) can be purchased from various suppliers (e.g., Sigma C1562, Fisher 15-350-50). Alternatively, make a simple container from two pieces of Styrofoam; holes made with a heated metal spatula (or similar instrument) should be placed so that there is ~1-inch thickness of Styrofoam around each cryovial in all directions. It is also possible to use a Styrofoam tube rack with the lid.

Thawing

1. Thaw the vial of ES cells by quickly warming it in a 37°C water bath up to the point when ice crystals almost disappear.

2. Aseptically transfer the cell suspension into a 14-ml tube using the pipette filled with ES-DMEM (e.g., 4 ml) to dilute the DMSO slowly.

3. Pellet the cells by centrifuging at 1000 rpm for 5 minutes and then aspirate the supernatant.

4. Gently resuspend the pellet in fresh ES-DMEM, plate on a 6-cm feeder plate, tilt the plate a couple of times in both the X and Y directions to distribute the cells evenly, and then place the plate in the incubator.

5. The next day, aspirate the medium to remove floating dead cells, and add fresh medium. If the correct procedure was used, cells should be ready for passage in 2–3 days.

De Novo Isolation of ES Cell Lines from Blastocysts

The starting material for de novo isolation of stem cell lines can either be normal 3.5-dpc expanded blastocysts or "delayed" blastocysts. Delayed blastocysts are usually collected 4–6 days after ovariectomy (see Chapter 6). For both groups of blastocysts, tissue culture procedures are identical, the only difference being the timing of the first disaggregation, because delayed blastocysts will initially grow more slowly.

MATERIALS

EMBRYOS AND CELLS
Expanded blastocysts (3.5 dpc) or "delayed blastocysts"
Feeder layer (STO or mouse embryonic fibroblast cells, MEFs)

EQUIPMENT
Microscope, inverted and dissecting
Pasteur pipettes, finely drawn, attached to tubing and mouthpiece (see Chapter 4) or Gilson P2 pipetman
Tissue culture dishes, 10-mm well (4 x 10-mm well plates are ideal, e.g., Nalge Nunc International multidish 4-well 176740); also 35-, 60-, and 100-mm dishes

REAGENTS
ES cell culture medium (ES-DMEM, see pp. 365–366)
Light paraffin oil (e.g., embryo-tested mineral oil from Sigma M8410, or ES cell-qualified light mineral oil from Specialty Media, ES-005-C)(also see Chapter 4)
M2 medium (see Chapter 4) or Dulbecco's modified Eagle's medium (DMEM) plus 10% FBS and 25 mM HEPES (pH 7.4)
Phosphate-buffered saline (PBS), Ca^{++}/Mg^{++}-free (see Appendix 1)
Trypsin/EDTA (see p. 367)

PROCEDURE

1. Flush the embryos from the uterine horns (see Chapter 4) with either M2 medium or DMEM plus 10% serum and 25 mM HEPES (pH 7.4) and place them individually into 10-mm well tissue culture dishes containing a preformed feeder layer of either STO or MEF and 0.5 ml of ES cell culture medium (see pp. 365–366).

The first stage of embryo culture can also be performed in microdrops of ES medium without feeder cells incubated under light paraffin oil or on gelatinized four-well plates. After 1–2 days of culture, the embryos hatch from the zona pellucida and attach to the surface of the tissue culture dish with spreading of the trophoblast cells. Shortly after embryo attachment, the inner cell mass (ICM) becomes readily distinguishable and can be seen to grow rapidly over the next 2 days. Normally after 1–2 days, the ICM is considerably enlarged. However, embryos within a group may vary considerably, so that daily monitoring of individual embryos is essential. However, it is important to avoid extensive exposure of the embryos to room temperature and normal atmosphere, since environmental changes may trigger differentiation. Examples of the progressive changes in the morphology of cultured blastocysts are given in Figure 8.2A–D.

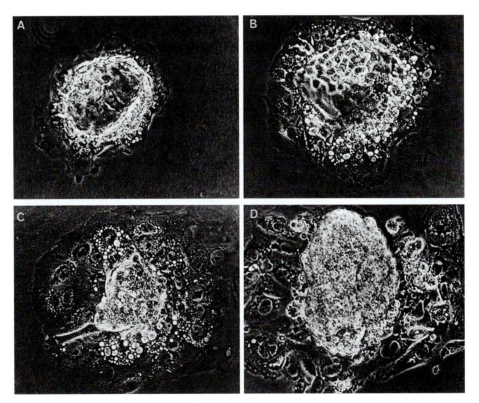

FIGURE 8.2. Progressive changes in the morphology of cultured blastocysts. (*A*) A 48-hr culture; (*B*) 72-hr culture; (*C*) 96-hr culture; (*D*) 120-hr culture. When the inner-cell-mass-derived component has attained approximately the size and morphology illustrated in *D*, it is suitable for isolation and disaggregation as described in the text.

2. When the ICM-derived clump has reached the stage illustrated in Figure 8.2D (a stage normally reached 4–5 days after explantation into culture), dislodge it from the underlying sheet of trophoblast cells using a finely drawn Pasteur pipette. Clumps that have an outer layer of endoderm have differentiated too much and are less likely to generate ES cells. Wash the clump of cells with two changes of Ca++/Mg++-free PBS. Use a finely drawn pipette to transfer the cells to a microdrop of 0.25% trypsin/0.2% EDTA in PBS or similar buffer under light paraffin oil. An alternative for this and the next steps is to use a Gilson P2 pipetman to pick the outgrowths and transfer them into 25- to 30-μl trypsin drops or in a 96-well plate.

3. Incubate the microdrop at 37°C for 3–4 minutes. Fill another finely drawn Pasteur pipette with serum-containing medium (the diameter of the pipette end should be no greater than the cell clump). Use the filled pipette to disaggregate the ICM clump gently into smaller cellular aggregates of three or four cells.

   It is NOT advisable to attempt to reduce the ICM-derived clump to a single-cell suspension.

4. Transfer the disaggregated contents of the microdrop into a fresh 10-mm feeder cell tissue culture well. Inspect the individual cultures daily, but only for a brief time. Generally, after 2 days, primary colonies of cells will become readily visible and will have one of several different morphologies:

   a. trophoblast-like cells

   b. epithelium-like cells

   c. endoderm-like cells

   d. ES cell-like cells (see Comments, below)

5. If clumps with ES cell morphology dominate the culture, start passing the cells by the regular trypsinization described for maintenance (see above). If this is not the case, the ES cell-like clump(s) should be physically separated from the differentiated counterparts as described in steps 2–5.

6. When ES cell-like colonies dominate the 10-mm well, the cells should be gradually expanded to 35-mm, through 60-mm and then 100-mm plates. It is recommended that newly derived ES cells be expanded to one or two 100-mm plates before they are frozen for storage (see Protocol 8.4).

COMMENTS

- *Trophoblast-like cells:* In nearly 100% of cases, areas of trophoblast-like cells rapidly become apparent. These cells are morphologically identical to those seen in the initial blastocyst outgrowths. Examples of the morphology of these trophoblast giant cells are given in Figure 8.3. In addition, colonies are frequently located during the early stages of tissue culture that closely resemble pluripotent stem cells but, in fact, turn out to give rise to exclusively trophoblast-like cells over the next 2–3 days (Fig. 8.3A–C).

- *Epithelium-like cells:* Occasionally, colonies of a very distinct and easily recognizable phenotype will form. These cells are very slow growing and form discrete patches on the feeder layer. The constituent cells pack together to give a flat pavement, epithelium-like structure, often with a very marked, highly refractile edge to the cell colony. An example of a typical colony is given in Figure 8.3D.

- *Endoderm-like cells:* Areas of rounded, refractile, loosely attached cells grow in a few cultures. These appear to be similar to the endodermal cell type that forms when stem cells are encouraged to differentiate in culture.

- *Stem-cell-like cells:* These are cells that grow progressively and maintain a stable ES-cell-like phenotype (Fig. 8.4). Discrete colonies of a stem-cell morphology are located and marked. Those colonies that fail to show any differentiation but retain cells of an exclusively ES phenotype are removed selectively after up to 7–8 days of tissue culture. These colonies are then dissociated in microdrops of trypsin/EDTA (see steps 2 and 3, above) and passaged into fresh, small feeder wells to keep the cell density high. In successful cultures, small nests of stem cells appear within 2–3 days of subculture. Depending on the relative rate of growth, these cultures are expanded 3–5 days later by trypsinizing the whole well and transferring its contents onto a larger feeder dish. The experimental stages of the technique and the timings of the various disaggregation events are summarized diagrammatically in Figure 8.5.

- An established stem cell line requires careful subculture at 2- to 3-day intervals, depending on the cell growth rate. If care is not taken when passaging to dissociate the cells, the colonies may form clumps, which differentiate extensively to form large quantities of endoderm. To improve the viability of the cells, subconfluent cultures are generally refed 2–3 hours before passage. The cells are washed thoroughly with two changes of PBS, and a small amount of 0.25% trypsin/0.2% EDTA in PBS or similar buffer is added to the tissue culture dish. After incubation at 37°C for 3–4 minutes, the cells can be seen to detach from the dish. The cell clumps are broken up to give a suspension of small clumps by pipetting vigorously with a Pasteur pipette. After adding serum-containing medium to the cell suspension, the cells are reseeded onto fresh feeder wells.

- The overall efficiency of ES cell derivation varies even for permissive strains. It may be improved by using knockout serum replacement (Invitrogen Life Technologies 10828-028) instead of serum and extra LIF (up to 2000 units/ml) (A. Nagy lab). The addition of MEK1 inhibitor (Parke-Davis 098059; New England Biolabs 9900L) facilitates the derivation of ES cells (A. Smith lab) because it enhances the effect of LIF on self-renewal and further reduces differentiation in ES cells (Burdon et al. 1999).

- The sex and karyotype of all newly established ES cell lines need to be determined. It is preferable to have a male ES cell line because XY ES cells can convert the sex of the host embryo in chimera production, and chimeric males can produce more offspring than females. The transmission of the ES cell genome through the germ line of chimeras depends on euploidy of ES cells (Longo et al. 1997). It is also important to test newly established ES cells for *Mycoplasma* and murine antibody production (See *Mycoplasma* and MAP testing, pp. 361–363).

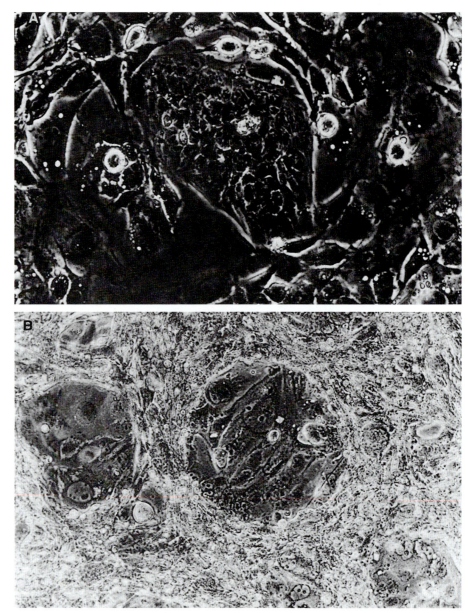

FIGURE 8.3. (*See facing page for parts C and D.*) Morphology of primary cell colonies following dissociation of the inner cell mass. (*A,B,C*) Examples of "trophoblast-like" colonies. These colonies may first be apparent 1–2 days after passage as small nests of cells that have a superficially "stem-like-cell" morphology (*A*). With further culture, the cells alter in appearance and flatten and spread to give areas composed of a monolayer of giant cells. (*B,C*) Although embryo-derived cells of this phenotype grow well in initial cultures, they fail to proliferate with long-term culture unless specific conditions are used (Tanaka et al. 1998). (*D*) "Epithelial-like" cells. This cell type forms discrete colonies that grow comparatively slowly. If cultured for long periods (2 weeks), large, flat colonies composed of a monolayer of cells will result.

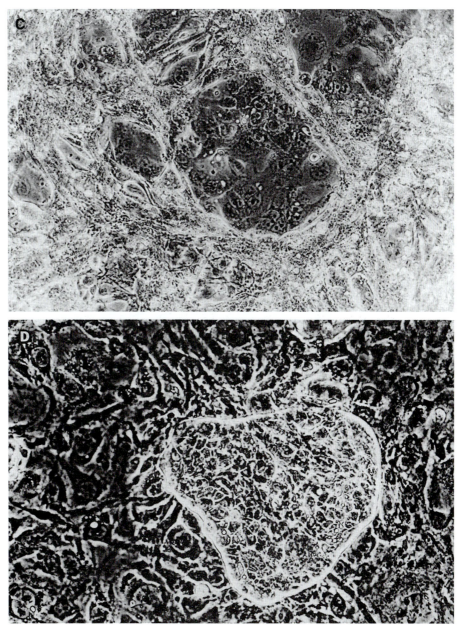

FIGURE 8.3. (*See facing page for legend.*)

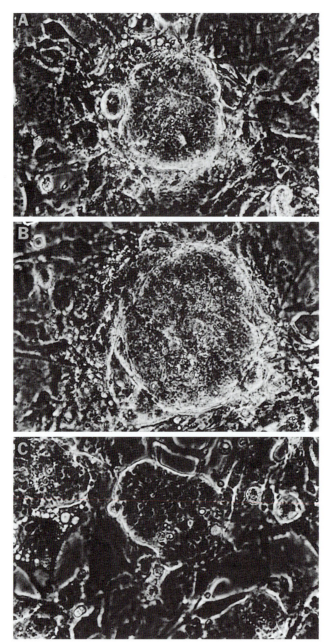

FIGURE 8.4. Morphology of embryonic stem cell colonies. (*A*) Appearance of a colony of stem cells 2 days after disaggregation of the inner cell mass. (*B*) Same colony 2 days later. Note that the colony remains composed of a homogeneous population of stem cells and that no overt cellular differentiation has occurred. Stem cells are comparatively small, typically have a large clear nucleus containing one or more prominent nucleoli, and are tightly packed within the multilayered primary colony. (*C*) The colony shown above in *B* was subcultured (as described in the text) into a fresh feeder well. Within 2 days, numerous small nests of stem cells (illustrated) appear in the culture.

DAY

0

5–6

5–6

9

12–13

14

16

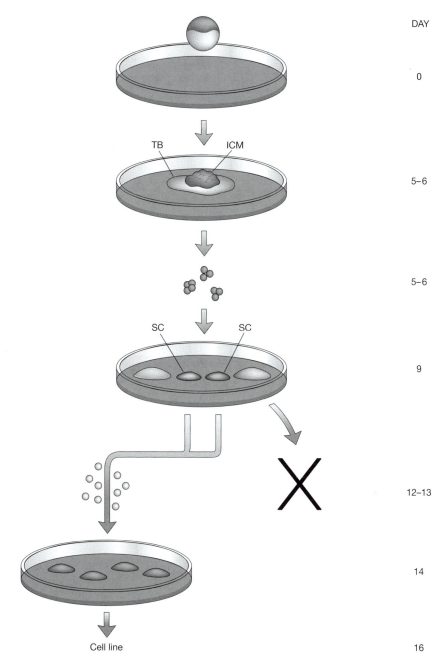

FIGURE 8.5. Summary of the procedure for deriving embryonic stem cell lines. Blastocysts are recovered and placed into wells with feeders (day 0). By day 5–6 of culture, the ICM cells have proliferated to generate a "blastocyst outgrowth." The blastocyst outgrowth is dislodged from the trophoblast (TB) cells and dissociated into cell clumps that are transferred to another well with feeders. On day 9, primary colonies are examined and classified according to morphology. Colonies that resemble stem cells (SC) are individually removed, dissociated into a small clump of cells, and reseeded into wells with feeders on day 12–13. Two days later (day 14), the wells are inspected for the presence of ES cell colonies, and wells containing stem cells are subcultured (day 16) to generate permanent ES cell lines.

Counting Chromosomes in ES Cells

Analysis of mouse karyotype by chromosome banding is given in Chapter 12, Protocol 12.6.

MATERIALS

CELLS
ES cells in a 35- or 60-mm dish, subconfluent culture

EQUIPMENT
Centrifuge
Centrifuge tube, 15-ml conical
Microscope with 100x oil immersion and 40x phase-contrast or fluorescence
  objectives
Microscope slides, precleaned (e.g., Fisherbrand Superfrost 12-550-12)

REAGENTS
Colcemid, 10 µg/ml stock (e.g., Invitrogen Life Technologies 15210–012)
ES cell medium (see Culture Medium, p. 364)
Fixative
  Three parts methanol <!> to 1 part glacial acetic acid <!>. Make up fresh each time
  and keep at 4°C before use. Use a fume hood and dispose of properly into a sol-
  vent waste!
  Giemsa stain (e.g., Invitrogen Life Technologies 10092-013)
  Gurr's buffer tablets (pH 6.8) (e.g., Invitrogen Life Technologies 10582-013 or
  BDH 33-193)
Hypotonic solution
  0.075 M potassium chloride <!> (0.559 g of KCl in 100 ml of water). Prewarm at
  37°C before use.
Mounting medium with DAPI <!> (Vectashield from Vector H-1200)
Phosphate-buffered saline (PBS) without calcium and magnesium
0.25%Trypsin/1 mM EDTA (e.g., Invitrogen Life Technologies 25200-056)

CAUTION: See Appendix 2 for appropriate handling of materials marked with <!>.

PROCEDURE

1. Culture actively growing ES cells (i.e., 24 or 48 hours after 1:5 passage on
   gelatinized plate) in the presence of Colcemid (0.05–1 µg/ml) for 1 hour.

2. Remove the medium, wash with PBS, add trypsin, and incubate for 3–5 minutes.

3. Harvest the cells into a 15-ml conical centrifuge tube, centrifuge for 5 minutes at 1000 rpm, aspirate the supernatant, and flick the pellet. Resuspend the cells in 1.5 ml of fresh medium.

4. Add 10 ml of warmed (37°C) KCl solution slowly, drop by drop, along the wall of the tube, flicking the tube gently at the same time. Invert several times, and incubate at 37°C for 15–20 minutes.

   The time in KCl is crucial. If it is too short, the chromosomes will be too tightly packed with cytoplasm present around metaphases; if too long, the chromosomes will not be contained in their appropriate group.

5. Add two or three drops of ice-cold fixative, invert the tube, and centrifuge for 5 minutes to pellet the cells.

6. Aspirate the supernatant, leaving ~1 ml behind. Flick the pellet.

7. Using vortexing, add up to 10 ml of ice-cold fresh fixative drop by drop using vortexing to disperse cells thoroughly in the fixative. Centrifuge for 5 minutes and then aspirate the supernatant, leaving ~ 1 ml behind. Flick the pellet.

8. Repeat step 7 at least one more time or, better, two or three times, for a total of three or four fixations.

9. Slides are best made within 3 hours of the final fixation. They can be made later if the cells are kept in fixative at 4°C or –20°C and resuspended in fresh fixative before making spreads.

10. After final fixation, resuspend the cells in a small volume of fixative (~0.5 ml) that can be diluted later if the suspension is too dense. Several slides can be made from each sample.

11. Drop the cell suspension onto a glass slide from a 10-cm distance. Air-dry at room temperature. Alternatively, make spreads by slowly dispersing 60–80 μl of cell suspension on a wet slide in humid conditions; place a few drops of fixative on top of the slide before air-drying it.

12. Scan the first slide with a 40x objective. If necessary, dilute the suspension or pellet the cells and resuspend them in a smaller volume.

13. Add a drop of mounting medium with DAPI to each slide, and seal with a cover slip. Chromosomes can be seen under UV light. Alternatively, stain slides for up to 10 minutes with freshly made Giemsa stain (2.5 ml of Giemsa stain in 47.5 ml of Gurr's pH 6.8 buffer). Rinse with water until clear. Air-dry.

14. Count the chromosomes using the 100x oil immersion objective. Look for spreads in widely separated fields. Ignore spreads with fewer than 39 chromosomes. A total of 20–40 good spreads should give reasonable sampling. ES cell clones with a minimum 70% of spreads containing 40 chromosomes are acceptable. It is possible occasionally to get germ-line transmission through a female chimera that has lost the Y chromosome. If targeted ES cells are aneuploid, they may be subcloned, and a subclone with euploid count introduced into chimeras.

Differentiating ES Cells into Embryoid Bodies

Non-tissue culture grade plastic is used for ES cell differentiation in suspension culture to reduce the number of cells adhering to the surface. An alternative strategy of ES cell differentiation in hanging drop culture (Wobus et al. 2002) may give more uniform development of EBs and better control of their density.

MATERIALS

CELLS
ES cells

EQUIPMENT
Falcon tube, 50-ml
Microscope, inverted
Petri dishes, bacterial-grade
Pipette

REAGENTS
ES cell culture medium with FBS and supplements but *without LIF* (ES-DMEM w/o LIF, p. 365–366)
Phosphate-buffered saline (PBS), Ca^{++}/Mg^{++}-free (see Appendix 1)
Trypsin/EDTA (see p. 367)

METHOD

Day 0

1. Trypsinize the cells until the colonies lift off.

2. Add 5 ml of PBS to the dish, pipette gently. Try to keep the loosely connected clumps of cells together using gentle handling.

3. Transfer the cells to a 50-ml tube containing 20 ml of PBS. Let the colonies settle to the bottom of the tube.

4. Aspirate the trypsin/PBS from the tube. Repeat the washing step.

5. Add ES-DMEM **without LIF** to the tube. Then directly plate the cells into bacterial-grade petri dishes—e.g., 5×10^5 to 10×10^5 cells per 100-mm dish. This is a very general guideline. Different ES cell lines may have different plating efficiency and would require different cell density.

Day 2

6. It is possible to feed the cells directly in the dish by carefully tilting the plate and removing only a portion of the medium. Alternatively, pipette the cell clumps and the medium into 50-ml tubes. Let the clumps settle to the bottom of the tubes, aspirate the old medium, add fresh ES cell culture medium **without LIF**, and return the cells to the plate.

Day 4

7. Replace the medium as in day 2. At this stage of the differentiation, the aggregates are called simple EBs.

Days 7–14

8. Change the medium every 2–3 days.

By day 15, a high percentage of the EBs will have developed fluid-filled cavities. These structures are called cystic EBs.

Culturing TS Cell Lines

This protocol was provided by the laboratory of Janet Rossant, Samuel Lunenfeld Research Institute, Mount Sinai Hospital, Toronto, Ontario M5G 1X5, Canada.

MATERIALS

CELLS
TS cells
MEF feeder cells

CULTURE MEDIA

FGF4 stock solution (1000x, 25 µg/ml). Human recombinant fibroblast growth factor-4 (FGF4; Sigma F2278, 25 µg). Resuspend lyophilized FGF4 in its vial with 1.0 ml of phosphate-buffered saline (PBS)/0.1% (w/v) bovine serum albumin (BSA). Mix well, aliquot 100 µl, and freeze at –80°C. Thaw each aliquot as needed and store it at 4°C. Do not refreeze.

PBS/0.1%(w/v) BSA fraction V (10 mg, Sigma A3311) in PBS without Ca++ /Mg++ (10 ml). Resuspend 10 mg of BSA fraction V in 10 ml of PBS without Ca++ /Mg++ to a final concentration of 0.1% (w/v). Filter through a 0.45-µm filter, aliquot, and store at –80°C.

Heparin stock solution (1000x, 1.0 mg/ml) (Sigma H3149, 10,000 units). Resuspend in PBS and store at –80°C. The stock can be also prepared as 10,000x (10 mg/ml) solution and used to make batches of 1000x.

TS medium (650ml) is prepared by adding the following reagents to 500 ml of RPMI medium 1640 (e.g., Invitrogen 61870 or 11875)

    FBS (e.g., CanSera, CS-COB-500, may be substituted by other suppliers) 130 ml (20% final)

    penicillin and streptomycin (100x stock, e.g., from Invitrogen 15070-063) (50 µg/ml each final)

    sodium pyruvate (100 mM stock, e.g., Invitrogen Life Technologies 11360) 6.5 ml (1 mM final)

    β-mercaptoethanol <!>(10 mM stock, e.g., Sigma M7522) 6.5 ml (100 µM final)

    L-glutamine (200 mM stock, e.g., Invitrogen 25030 or 35050) 6.5 ml (2 mM final)

TS+F4H medium: Add 10 µl of 1000x FGF4 (25 ng/ml) and 10 µl of 1000x heparin (1 µg/ml) into 10 ml of TS medium.

Feeder-conditioned medium (feeder-CM): Prepare mitomycin-treated mouse embryonic fibroblasts (MEF) (see Protocol 8.2) and culture in TS medium for 72 hours. Collect the medium. Spin to remove the floating cells and debris, filter (0.45 µm), and store at –20°C in aliquots. Thaw each aliquot as needed and store it at 4°C; do not refreeze. Use the MEF to prepare two more batches of feeder-CM, then discard the cells. MEF are used up to 10 days after mitomycin treatment.

70cond medium: Add 3 ml of TS medium into 7 ml of feeder-CM.

70cond+F4H medium: Add 10 µl of 1000x FGF4 (25 ng/ml) and 10 µl of 1000x heparin (1 µg/ml) into 10 ml of 70cond medium.

EQUIPMENT
Incubator, 37°C, 5% CO_2, 95% air
Microscope, inverted
Tissue culture dishes

REAGENTS
0.1% Trypsin/EDTA (see p. 367)

CAUTION: See Appendix 2 for appropriate handling of materials marked with <!>.

PROCEDURE

1. To culture TS cells on MEF:

   a. Culture the TS cells in TS + F4H medium on half the density of MEF as usually used for ES cells (i.e., 1×10^5 cells/ml), in a standard tissue culture incubator (37°C, 5% CO_2).

   b. Change the medium every other day and passage the cells (1:10–1:20) every 4th day or when the culture has reached ~80% confluency. Passaging TS cells at higher densities may lead to precocious differentiation.

   c. Trypsinize the cells to very small clumps with some single cells. A complete single-cell suspension is not required, and it may even be detrimental to the culture. It is usually sufficient to trypsinize for 3–4 minutes at 37°C with some pipetting up and down.

2. To culture TS cells on tissue culture plastic:

   a. Culture the TS cells in 70cond + F4H medium. TS cells grow well on standard tissue culture dishes without MEF if the medium is supplemented with feeder-CM. The dishes do not need to be gelatin-coated.

   b. Feed and passage the cells as described above.

3. TS cells are frozen and thawed using procedures similar to those for ES cells (Protocol 8.4), except that a higher serum content is used. A 2x freezing medium is freshly prepared and kept on ice before being added to the cell suspension in equal volume: 2x freezing medium is 50% FBS, 20% DMSO<!>, 30% TS medium (already containing 20% FBS).

COMMENT

When switching from MEF to feeder-CM it may be desirable to remove the MEF immediately. The different adherence rates of MEF (fast) and TS cells (slow) can be used to obtain a pure TS cell population. Passage the cells to a new plate and incubate the culture for 1 hour at 37°C/5% CO_2. Remove the supernatant and plate onto another dish. This population of cells should consist almost entirely of TS cells. Such differential plating enriches for stem cells at the expense of differentiated trophoblast cells.

Derivation of TS Cell Lines from Blastocysts

The derivation of TS cell lines from 3.5-dpc mouse blastocysts is similar to the derivation of embryonic stem (ES) cell lines. However, the success rate is considerably higher, and less expertise is required to recognize pluripotent TS cell colonies. This protocol is from the lab of Janet Rossant, Samuel Lunenfeld Research Institute, Mount Sinai Hospital, Toronto, Ontario M5G 1X5, Canada.

MATERIALS

EMBRYOS AND CELLS

Expanded blastocysts (3.5 dpc)

MEF feeder layer

EQUIPMENT

Incubator, 37°C, 5% CO_2, 95% air

Microscope, inverted and dissecting

Pasteur pipettes, finely drawn, attached to tubing and mouthpiece (see Chapter 4) or Gilson P2 pipetman

Tissue culture dishes, 10-mm-well (4 x 10-mm well plates are ideal, e.g., Nalge Nunc International multidish 4-well 176740)

REAGENTS

70cond + 1.5x F4H (see Protocol 8.8)

M2 medium (see Chapter 4) or Dulbecco's modified Eagle's medium (DMEM) plus 10% serum and 25 mM HEPES (pH 7.4)

Phosphate-buffered saline (PBS), Ca++/Mg++-free (see Appendix 1)

Trypsin/EDTA (see p. 367)

TS + F4H medium (see Protocol 8.8).

PROCEDURE

1. Set up matings (natural or superovulated) between the mice of interest.

2. Prepare 4-well plates with MEF (at half the density used to culture ES cells and for the preparation of feeder-CM; i.e., 1×10^5 cells/ml) in TS medium the day before flushing the embryos.

3. Replace the TS medium with TS + F4H medium (500 µl per well) in the morning of flushing the embryos (**Day 1**).

4. Collect 3.5-dpc blastocysts under sterile conditions, place one blastocyst per well in the four-well plates containing TS + F4H medium, and culture at 37°C and 5% CO_2.

5. The blastocysts should hatch and attach to the wells in 24–36 hours (**Day 2**).

6. On **Day 3**, a small outgrowth is formed from each embryo. Feed each culture with TS + F4H medium (500 μl).

7. On **Day 4**, the outgrowth is usually disaggregated; however, this depends on its size. It is smaller than the size of the outgrowth disaggregated for ES cell line derivation. The ideal size for TS cell line derivation is illustrated in Figure 8.6AB. Larger outgrowths will also work, but with less efficiency.

8. Once suitable outgrowths have been chosen, they may be disaggregated directly in the wells in which they were cultured. Remove the medium and wash the cells with PBS (500 μl). Aspirate the PBS, add 0.1% trypsin/EDTA (100 μl), and incubate for 5 minutes at 37°C, 5% CO_2. Using a P2 pipetteman or a drawn Pasteur pipette, disaggregate the clump by pipetting up and down vigorously until the outgrowth is reduced to small clumps of cells. Immediately stop the trypsin reaction by adding 70cond+1.5x F4H (400 μl) and return the cultures to the incubator.

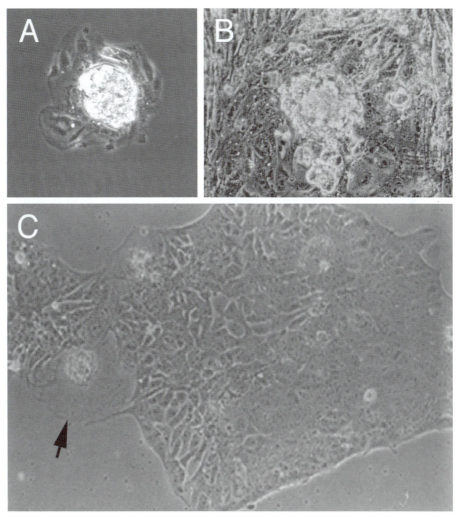

FIGURE 8.6. Trophoblast stem cells. Blastocyst outgrowth on gelatinized plate (*A*) and on fibroblast feeders (*B*). (*C*) Typical TS cell colony on gelatinized plate. The arrow shows a differentiated cell into a giant trophoblast cell.

9. Change the medium (70cond + 1.5 x F4H) 16 hours after the disaggregation.

10. On **Day 6** feed each culture (70cond + F4H, 500 ml) and continue to refeed every 2 days.

11. Between **Days 7 and 11** (this time period is highly variable), TS cell colonies will begin to appear. They appear as flat, epithelial sheets with a distinctive colony boundary (Fig. 8.6C).

12. Continue to feed the cultures until TS cell colonies get sufficiently large to cover about 50% of the well. Some differentiation will be observed at the edges of the colonies. This is normal. These differentiated cells are most often giant cells as well as other unidentified cell types that may be between a stem cell and giant cell phenotype.

13. Passage the half-confluent well of TS cells to a 6-well plate or 35-mm dish with MEF. Aspirate the medium and wash with PBS. Aspirate the PBS, add trypsin/EDTA (100 µl), and incubate for 5 minutes at 37°C, 5% CO_2. Stop the trypsin reaction by adding TS + 1.5 x F4H (400 µl) and pipetting up and down to get a near-single-cell suspension. Transfer the cells to a 6-well plate or 35-mm dish with TS + 1.5 x F4H medium (2.5 ml) on MEF. This first passage is crucial because it is the most likely time for the culture to differentiate.

14. Change the medium 16 hours after passage (TS + 1.5 x F4H).

15. Feed the cells every 2 days (TS + F4H). After five or six more passages on MEF, TS cells may be cultured without them in the presence of 70cond + F4H.

REFERENCES

Abbondanzo S.J., Gadi I., and Stewart C.L. 1993. Derivation of embryonic stem cell lines. *Methods Enzymol.* **225:** 803–823.

Bernstein A., MacCormick R., and Martin G.S. 1976. Transformation-defective mutants of avian sarcoma viruses: The genetic relationship between conditional and nonconditional mutants. *Virology* **70:** 206–209.

Bradley A., Evans M., Kaufman M.H., and Robertson E. 1984. Formation of germ-line chimaeras from embryo-derived teratocarcinoma cell lines. *Nature* **309:** 255–256.

Burdon T., Stracey C., Chambers I., Nichols J., and Smith A. 1999. Suppression of SHP-2 and ERK signalling promotes self-renewal of mouse embryonic stem cells. *Dev. Biol.* **210:** 30–43.

Carmeliet P., Ferreira V., Breier G., Pollefeyt S., Kieckens L., Gertsenstein M., Fahrig M., Vandenhoeck A., Harpal K., Eberhardt C., Dechercq C., Pawling J., Moons L., Collen D., Risau W., and Nagy A. 1996. Abnormal blood vessel development and lethality in embryos lacking a single VEGF allele. *Nature* **380:** 435–439.

Chen T.R. 1977. In situ detection of mycoplasma contamination in cell cultures by fluorescent Hoechst 33258 stain. *Exp. Cell Res.* **104:** 255–262.

Doetschman T.C., Eistetter H., Katz M., Schmidt W., and Kemler R. 1985. The in vitro development of blastocyst-derived embryonic stem cell lines: Formation of visceral yolk sac, blood islands, and myocardium. *J. Embryol. Exp. Morphol.* **87:** 27–45.

Eggan K., Akutsu H., Loring J., Jackson-Grusby L., Klemm M., Rideout W.M., 3rd, Yanagimachi R., and Jaenisch R. 2001. Hybrid vigor, fetal overgrowth, and viability of mice derived by nuclear cloning and tetraploid embryo complementation. *Proc. Natl. Acad. Sci.* **98:** 6209–6214.

Eggan K., Rode A., Jentsch I., Samuel C., Hennek T., Tintrup H., Zevnik B., Erwin J., Loring J., and Jackson-Grusby L., et al. 2002. Male and female mice derived from the same embryonic stem cell clone by tetraploid embryo complementation. *Nat. Biotechnol.* **20:** 455–459.

Evans M.J. and Kaufman M.H. 1981. Establishment in culture of pluripotential cells from mouse embryos. *Nature* **292:** 154–156.

Hooper M., Hardy K., Handyside A., Hunter S., and Monk M. 1987. HPRT-deficient (Lesch-Nyhan) mouse embryos derived from germline colonization by cultured cells. *Nature* **326:** 292–295.

Kawasaki H., Mizuseki K., Nishikawa S., Kaneko S., Kuwana Y., Nakanishi S., Nishikawa S.I., and Sasai Y. 2000. Induction of midbrain dopaminergic neurons from ES cells by stromal cell-derived inducing activity. *Neuron* **28:** 31–40.

Kunath T., Strumpf D., Rossant J., and Tanaka S. 2001. Trophoblast stem cells. In *Stem cell biology.* (eds. D.R. Marshak, R.L. Gardner, and D. Gottlieb). Cold Spring Harbor Laboratory Press, pp. 267–287.

Ledermann B. and Burki K. 1991. Establishment of a germ-line competent C57BL/6 embryonic stem cell line. *Exp. Cell. Res.* **197:** 254–258.

Liu X., Wu H., Loring J., Hormuzdi S., Disteche C.M., Bornstein P., and Jaenisch R. 1997. Trisomy eight in ES cells is a common potential problem in gene targeting and interferes with germ line transmission. *Dev. Dyn.* **209:** 85–91.

Longo L., Bygrave A., Grosveld F.G., and Pandolfi P.P. 1997. The chromosome make-up of mouse embryonic stem cells is predictive of somatic and germ cell chimaerism. *Transgenic Res.* **6:** 321–328.

Martin G.R. 1981. Isolation of a pluripotent cell line from early mouse embryos cultured in medium conditioned by teratocarcinoma stem cells. *Proc. Natl. Acad. Sci.* **78:** 7634–7638.

McMahon A.P. and Bradley A. 1990. The wnt-1 (int-1) proto-oncogene is required for development of a large region of the mouse brain. *Cell* **62:** 1073–1085.

Nagy A., Rossant J., Nagy R., Abramow-Newerly W., and Roder J.C. 1993. Derivation of completely cell culture-derived mice from early-passage embryonic stem cells. *Proc. Natl. Acad. Sci.* **90:** 8424–8428.

Nicklas W. and Weiss J. 2000. Survey of embryonic stem cells for murine infective agents. *Comp. Med.* **50:** 410–411.

Oshima R. 1978. Stimulation of the clonal growth and differentiation of feeder layer dependent mouse embryonal carcinoma cells by beta-mercaptoethanol. *Differentiation* **11:** 149–155.

Robertson E., Bradley A., Kuehn M., and Evans M. 1986. Germ-line transmission of gene introduced into cultured pluripotential cells by retroviral vector. *Nature* **323:** 445–448.

Simpson E.M., Linder C.C., Sargent E.E., Davisson M.T., Mobraaten L.E., and Sharp J.J. 1997. Genetic variation among 129 substrains and its importance for targeted mutagenesis in mice. *Nat. Genet.* **16:** 19–27.

Smith A.G. and Hooper M.L. 1987. Buffalo rat liver cells produce a diffusible activity which inhibits the differentiation of murine embryonal carcinoma and embryonic stem cells. *Dev. Biol.* **121:** 1–9.

Smith A.G., Heath J.K., Donaldson D.D., Wong G.G., Moreau J., Stahl M., and Rogers D. 1988. Inhibition of pluripotential embryonic stem cell differentiation by purified polypeptides. *Nature* **336:** 688–690.

Tanaka S., Kunath T., Hadjantonakis A.K., Nagy A., and Rossant J. 1998. Promotion of trophoblast stem cell proliferation by FGF4. *Science* **282:** 2072–2075.

Threadgill D.W., Yee D., Matin A., Nadeau J.H., and Magnuson T. 1997. Genealogy of the 129 inbred strains: 129/SvJ is a contaminated inbred strain. *Mamm. Genome* **8:** 390–393.

Tucker K.L., Wang Y., Dausman J., and Jaenisch R. 1997. A transgenic mouse strain expressing four drug-selectable marker genes. *Nucleic Acids Res.* **25:** 3745–3746.

Turksen K., ed. 2002. *Embryonic stem cells: Methods and protocols. Methods in molecular biology,* vol. 185. Humana Press, Totowa, New Jersey.

Williams R.L., Hilton D.J., Pease S., Willson T.A., Stewart C.L., Gearing D.P., Wagner E.F., Metcalf D., Nicola N.A., and Gough N.M. 1988. Myeloid leukemia inhibitory factor maintains the developmental potential of embryonic stem cells. *Nature* **336:** 684–687.

Wobus A. M., Guan K., Yang H.-T., and Boheler K.R. 2002. Embryonic stem cells as a model to study cardiac, skeletal muscle, and vascular smooth muscle cell differentiation. *Methods Mol. Bio.* **185:** 127–156.

Vector Designs for Embryonic Stem Cell-based Transgenesis and Genome Alterations

EVERAL UNIQUE PROPERTIES of mouse embryonic stem cells make them a very efficient vehicle for introducing genetic alterations into this species. Transgenesis, homologous recombination-based gene targeting, heterologous site-specific recombinases, positive and negative selectable markers, reporters, and the availability of the mouse genome sequence have created an arsenal of tools that allow tailoring of the mouse genome at a level never before imagined. We are in the position of being able to copy almost any genetic change linked to a specific human disease, from point mutations through small deletions to large specific chromosomal aberrations. This ability makes mouse genetics a powerful

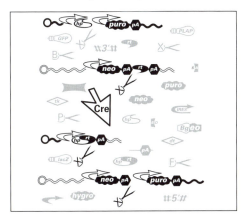

approach for addressing important gene functions and identifies the mouse as the ideal system to model human disease. This chapter introduces the basic principles in the most commonly used genetic alteration strategies and points out the importance of combining these tools in new ways to increase the success of understanding the tremendous complexity in the interplay among genes.

GLOSSARY OF ELEMENTS

Altering the mouse genome to affect a specific gene function uses a small number of genetic and protein elements. These are combined with each other in different designs to achieve much larger numbers of genetic changes.

Genomic and Gene Elements

Promoters: Promoters were originally defined as sites on the DNA where the RNA polymerase recognizes a specific signal that allows it to bind tightly as a prerequisite for transcriptional initiation. Today this definition is broader, and it also includes *cis*-regulatory elements responsible for controlling the transcriptional machinery and determining the level and specificity of transcription. Some embryonic stem (ES) cell-based gene/genome alterations, like gene traps or "promoterless" gene targeting, use endogenous promoters ( ), but almost all of them use exogenous promoters ( ) built into the vector and introduced into the cells. For reliable ES cell expression, the promoter of the mouse Pgk-1 gene is the choice for driving the expression of selectable markers in most applications.

Exon; expressed region ( ): A section of the genomic DNA that codes for a protein, or part of a protein. Such sections are separated by noncoding regions (introns).

Intron; intragenic region ( ): Noncoding sequence that separates coding sequences (exons).

Polyadenylation signal ( ): The polyadenylation signal specifies cleavage of the transcript and addition of a poly(A) tail to the resulting 3′ end. This process is essential for the transcript synthesized by RNA polymerase II to develop into mature mRNA.

Splice donor (Sᴰ) and **Splice acceptor (sᴬ)**, respectively: RNA splicing acts on primary transcript to remove introns and join exons. The introns usually contain clear signals for splicing. Most introns start at the 5′ end from the sequence GU and end at the 3′ end with the sequence AG, and are referred to as the splice donor (SD) and splice acceptor (SA) sites, respectively. These sequences at the two sites, however, are not sufficient to signal the presence of an intron. Another important sequence, called the branch site and located 20–50 bases upstream of the acceptor site, is also required. The branch site consensus sequence is CU(A/G)A(C/U), where A is conserved in all genes (McKeown 1993).

IRES, Internal Ribosome Entry Site (IRES): These sequences allow the initiation of translation in a cap-independent manner (Mountford and Smith 1995). This type of translation initiation was first demonstrated on picornavirus (encephalomyocarditis virus [EMCV]) RNAs along with other viral messages (Jang et al. 1988, 1989). The IRES sequence, which comprises ~500 base pairs, allows the construction of artificial bicistronic transgenes or gene alterations containing two (or more) coding regions for proteins. To achieve translation for a second protein in transgenes, an IRES sequence must be placed between the two coding regions. In addition, the second coding region must use the original ATG of the IRES where the cap-independent translation starts. Keep in mind that both the cap-dependent and cap-independent translation initiation sites must be optimized (Mizuguchi et al. 2000; Hennecke et al. 2001).

Arms of homology: Arms of homology are essential components of gene targeting in ES cells. They are placed in the target vector flanking a positive selectable marker gene and have the same nucleotide sequence as two segments of the gene of interest with equivalent orientation. Relative to the orientation of the gene of interest, we distinguish the two arms as the 5′ arm (⛏5′⛏) and the 3′ arm (⛏3′⛏). Sequence between these two segments of the gene will be replaced by the positive selectable marker of the vector, if the integration occurs through homologous recombination. The length of the homology arms affects the recombination frequency (Thomas and Capecchi 1987; Hasty et al. 1991a). To obtain a reasonable targeting frequency, the homology of the two arms should be at least 6000–8000 base pairs total. The length of the arms can be uneven, but a length shorter than 1000 base pairs is not recommended for any arm.

The genomic DNA for the arm should be cloned from the same inbred strain from which the ES cell line was derived. Isogenic DNA conditions can further increase the targeting frequency (te Riele et al. 1992).

Selectable Markers

CAUTION: See Appendix 2 for appropriate handling of materials marked with <!>.

Selectable markers are essential elements of almost all ES cell-based genome manipulation strategies, due to the fact that the stable transfection efficiency of ES cells is relatively low (~0.1%), except in cases where the integration is virus-mediated. Therefore, those cells in which integration has occurred (positive selection) or in which the result of genome alteration was loss of a selectable marker (negative selection) must be identified and selected for. The typical markers used in ES cell-based genome alteration strategies are described below.

Positive Selectable Markers

The positive selectable markers are built into most of the constructs described below to identify cells that have picked up DNA transiently or integrated it into its genome. There are four markers that are frequently used in different ES cell-based applications:

neo; <u>ne</u>omycin phosphotransferase or aminoglycoside 3′-phosphotransferase gene (neo): This is the most generally applied selection marker of *E. coli* origin. The selective drug is G418 <!>, an aminoglycosidic antibiotic that inhibits protein synthesis.

The *neo* gene allows production of aminoglycoside 3′-phosphotransferase, which inactivates G418 by phosphorylation. The *neo*-resistance gene exists in the ES cell technology in two different "flavors"—a wild-type gene and a point mutant, which has severalfold lower activity than the former. For certain applications one is a better fit than the other. For example, if *neo* assists the trapping of genes with an extremely low expression level, the wild-type *neo* should be used. However, the mutant *neo* must be used when producing homozygous ES cells with a high concentration of G418.

G418 solutions are stable for 3 months if kept at –20°C. Working concentrations with ES cells are 100–400 µg/ml active component. ES cells can tolerate up to 2–5 mg/ml concentration if the resistance gene expression is high enough.

puro; <u>puro</u>mycin resistance gene (puro): Puromycin<!> is an aminonucleoside antibiotic produced by *Streptomyces alboniger*. It specifically inhibits peptidyl transfer on ribosomes, therefore it inhibits the growth of various insect and animal cells, including ES cells. The expression of the *pac* gene (puromycin-*N*-acetyltransferase from *S. alboniger*) confers puromycin resistance to transfected cells. The suggested concentration for ES cell selection is 1 µg/ml. Stock solutions can be made in water (5–50 mg/ml) and stored at –20°C.

hygro; <u>hygro</u>mycin resistance gene, hph^r (hygro): Hygromycin B <!>, an aminoglycosidic antibiotic produced by *Streptomyces hygroscopicus*, is used for the selection and maintenance of prokaryotic and eukaryotic cells transfected with the hygromycin-resistance gene of *E. coli* origin. Hygromycin B kills cells by inhibiting protein synthesis. The resistance gene encodes for a kinase that inactivates hygromycin B through phosphorylation.

A final concentration of 150 µg/ml hygromycin B is recommended with ES cells for this selectable marker. The sensitive cells are dead after 4–5 days. They seem to remain attached to the clones of resistant cells, so it is very hard to see the ES colonies at the early phase of selection. Clones may be picked a week after selection has been started.

hprt; <u>h</u>ypoxanthine <u>p</u>hospho<u>r</u>ibosyl<u>t</u>ransferase (hp): The *hprt*-encoded enzyme converts free purine bases into the corresponding nucleoside phosphate and thus makes them available for the synthesis of nucleic acids. It is relatively easy to derive *hprt*-deficient ES cells from male ES cell lines due to the X-chromosome-linked nature of the gene.

The deficient cells cannot survive hypoxanthine, aminopterin, thymidine (HAT) medium selection (Szybalski 1992). If a functional *hprt* gene is reintroduced into the cells, only the cells that become transgenic will survive. A bipartite minigene has also been created from the *hprt* gene by splitting it with an intron (Ramirez-Solis et al. 1995). This gene can be separated into a 5′ and a 3′ half. The halves are not functional by themselves. However, they gain function if the 5′ part (*⟨hp⟩*) immediately precedes the 3′ part (*⟨rt⟩*). The bipartite nature of this gene has assisted many genome manipulation strategies.

COMMENT

The most generally applied positive selectable marker is the neomycin-resistance gene. Usually when *neo* is no longer available, because the cells are already resistant, hygromycin- or puromycin-resistance genes are used. Each antibiotic has a different time course of selection. Puromycin selection is the fastest, with the sensitive cells dying within 3 days. The neomycin selection takes about 1 week to complete, and hygromycin selection is the slowest, 7–10 days.

Negative Selectable Markers

The negative selectable markers are used in applications where the loss of the gene is the desired event, such as creating genomic deletions and enriching for homologous recombination-mediated insertions. There are three markers that are frequently used in different ES cell-based applications:

HSV-*tk*, herpes simplex virus-1 thymidine kinase gene (⟨tk⟩): Unlike cellular thymidine kinases, HSV-TK is nonspecific; it uses pyrimidine as well as purine nucleosides or analogs as substrates. Ganciclovir (GCV) <!> is an acyclic analog of 2′-deoxyguanosine. GCV can be phosphorylated to GCV-monophosphate (GCV-MP) by HSV-TK. GCV-MP is further converted to the diphosphate and triphosphate (GCV-TP) forms by host kinases. GCV-TP lacks essential residues for DNA chain elongation.

Incorporation of GCV-TP into DNA by polymerases causes premature termination of DNA synthesis and subsequent cell death. FIAU (5-iodo-2′-fluoro-2′deoxy-1-β-D-arabino-furanosyl-uracil) is an analog that can be used instead of ganciclovir.

The use of HSV-*tk* requires some cautions. Expression of HSV-*tk* in germ cells is known to cause sterility (Wilkie et al. 1991). Therefore, ES cells transgenic for HSV-*tk* are usually not germ-line competent. A truncated HSV-*tk* gene, however, does not cause this problem (Salomon et al. 1995). Both ganciclovir and FIAU selection produce the bystander killing effect, in which the dying nonresistant cells can transfer their toxic products to the nearby resistant cell, causing them to die as well. Sparse plating of cells for selection and use of FIAU, which has less bystander killing effect, are necessary to get this selection working properly.

***dt*, diphtheria toxin (⟨dt⟩):** Diphtheria toxin from *Corynebacterium diphtheriae* <!> is so poisonous to cells that a single molecule may be enough to kill (Honjo et al. 1969). Diphtheria toxin causes the inactivation of elongation factor-2 (EF-2) and inhibition of chain elongation in protein synthesis. This selection has been frequently used for negative selection in gene targeting and for conditional transgenesis in ablation studies.

hprt; **h**ypoxanthine **p**hospho**r**ibosyl**t**ransferase (hprt): This enzyme, which is used as a positive selectable marker in *hprt*-deficient cells, can also be used as a negative selectable marker. *hprt*-deficient cells are sensitive to medium containing HAT, but resistant to medium containing 6-thioguanine (6TG). If they become positive for *hprt*, the resistance/sensitive relation changes; the cells survive HAT but die in 6TG. Therefore, 6TG can be used as a negatively selecting agent.

Reporters

LacZ; β-**galactosidase gene from** *E. coli* (lacZ) This is a very convenient reporter gene for use in transgenic mice. It is frequently used in genome/gene alterations to visualize gene expression or tagged cells. More details are given in Chapter 16.

GFP; **G**reen **F**luorescent **P**rotein gene (GFP): This gene was isolated from the jellyfish, *Aequorea victoria*. GFP's natural role in jellyfish is to transduce the blue chemiluminescence of the protein aequorin into green fluorescent light by energy transfer. In a short period of time, the *GFP* gene has become a useful tool in many fields of life science research by allowing the observation of cell behavior, gene expression, and metabolic processes in live cells. Because GFP tolerates amino- and carboxy-terminal fusion to a broad variety of proteins, it also facilitates the studies of protein trafficking and protein–protein interactions. It is also gaining more and more importance in ES cell-based gene and genome manipulations in the mouse. Several mutants with spectral and strength variations in light emission have been isolated. The expression of enhanced cyan, green, and yellow variants has been shown to be compatible with the ES cell-based gene/genome alteration technologies (Hadjantonakis et al. 2002).

hPLAP, **H**uman **P**lacental **A**lkaline **P**hosphatase gene (hPLAP): Human placental alkaline phosphatase is heat stable, unlike other AP proteins typically expressed by cells. With simple heat treatment (see Chapter 16), the endogenous alkaline phosphatase activity can be eliminated while the *hPLAP* remains active. This reporter is similar to *lacZ* in convenience and the mechanism of detection. It is at least as sensitive as *lacZ* and is superior for labeling axon processes in the nervous system.

Site-specific Recombinases

The site-specific recombinases catalyze the recombination between two consensus DNA sequences. If these sites are properly designed into transgene(s) or targeted locus/loci and the recombinase is expressed in the same cell, the site-specific recombination can further modify the targeted loci or alter the integrated transgene. This postinsertional modification can result in new function or loss of function in the locus.

Cre Recombinase

Cre recombinase of the bacteriophage P1 is the most widely used recombinase in combination with transgenesis and gene targeting in the mouse (Nagy 2000). This prokaryotic enzyme was first shown to work in the mouse by Lakso et al. (1992). Cre protein catalyzes recombination between two 34-bp-long *loxP* recognition sites: The *loxP* sequence has a unique structure in which palindromic repeats of 13-bp sequences flank an 8-bp core sequence. The asymmetry of the core sequence confers orientation on the *loxP* site. The Cre-mediated strand cleavage and exchange between *lox* sites occurs following the first bases and before the last base of the 8-bp cores.

loxP sequence (✂):

```
5´-ATAACTTCGTATA gcatacat TATACGAAGTTAT-3´
3´-TATTGAAGCATAT cgtatgta ATATGCTTCAATA-5´
   ------------------>          <----------------------
   palindrome repeat    core    palindrome repeat
```

Although the recombination efficiency is generally quite sensitive to any alterations in the *lox* sequence, operational mutant versions have been identified (Albert et al. 1995; Lee and Saito 1998). Recombination between the same, homotypic pairs of *lox* sites can occur efficiently in vivo, whereas recombination between heterotypic pairs occurs much less efficiently.

Commonly used mutant *lox* sites include:

lox66	. . . TATACGAA*CGGTA*	(3´ palindrome sequence mutant)
lox71	*TACCG*TTCGTATA . . .	(5´ palindrome sequence mutant)
lox FAS	*tacctttc*	(core segment variant)
lox511	g*t*atacat	(core segment mutant - single)
lox2272	g*g*atac*tt*	(core segment mutant - double)
lox2372	g*g*atac*ct*	(core segment mutant - double)

(The mutant nucleotides are underlined and italicized.)

COMMENT

Note that the *loxP* site, reading from the 3´ strand, contains two, not-the-same-frame ATG sites, which could interfere with proper translation initiation of a coding region if the *loxP* site is placed in front and its orientation is inverted.

Flp Recombinase

The second most popular recombinase is Flp recombinase from yeast (*Saccharomyces cerevisiae*). Its mechanism of action is similar to Cre/*loxP*, but so far it appears to be less efficient. However, a recently developed enhanced Flp (Flpe) has improved the efficiency of this recombinase, bringing it closer to the efficiency of Cre (Buchholz et al. 1998; Rodriguez et al. 2000). The 34-bp consensus recombination site of Flp is called *FRT*. *FRT* has the same structure as *loxP*, but has a different sequence.

FRT sequence (⊢<):

```
5´-GAAGTTCCTATAC tttctaga GAATAGGAACTTC-3´
3´-CTTCAAGGATATG aaagatct CTTATCCTTGAAG-5´
    --------------------→                    ←--------------------
    palindrome repeat      core      palindrome repeat
```

Similar to Cre, the Flp recombinase is also extremely sensitive to changes in the *FRT* site. Several mutants, however, have been identified (Schlake and Bode 1994) with efficient recombination between homotypic sites but with a severely compromised recombination rate between the heterotypic combinations and between mutant and wild-type *FRT* sites (Seibler et al. 1998).

Mutant *FRT* sites are:

F3 - tt*caaat*a (core segment mutant)
F5 - tt*caaaag* (core segment mutant)

(The mutant nucleotides are underlined and italicized.)

φC31 Integrase 3

The φC31 integrase from the *Streptomyces* phage φC31 was recently demonstrated to function in human cells (Groth et al. 2000). It carries out a site-specific recombination at the ttg triplet (see below) between the attPP´ and attBB´ sites. Because both sites have unique sequences flanking the ttg recombination site, the recombination does not recreate either of the original recognition sites. The resulting hybrid sites, on the other hand, are no longer substrates for the integrase. The recombination can occur only once. The operational level of φC31 expression seems to be compatible with germ-line transmission of ES cells (Belteki et al., in prep.). This system is, however, still at the early phase of characterization in ES cell technology, but it certainly holds promise for becoming an additional recombinase system with a special use by taking advantage of its unique property of "once-and-no-more" action.

52-bp short form of *attBB´* site (⊟<):

```
5´-TGCGGGTGCCAGGGCGTGCCC ttg GGCTCCCCGGGCGCGTACTCCACCTCAC-3´
3´-ACGCCCACGGTCCCGCACGGG aac CCGAGGGGCCCGCGCATGAGGTGGAGTG-5´
```

51-bp short form of *attPP´* site (⊟<):

```
5´-GTGCCCCAACTGGGGTAACCT ttg AGTTCTCTCAGTTGGGGGCGTAGGGTC-3´
3´-CACGGGGTTGACCCCATTGGA aac TCAAGAGAGTCAACCCCCGCATCCCAG-5´
```

Inducible Systems

Parallel to the increasing understanding of transcriptional regulation of prokaryotic and eukaryotic systems, several artificial inducible gene expression systems have been designed in the mouse and other eukaryotic model systems. All of these systems represent a particular solution for the same concept: A small molecule, which can be easily administered to and taken up by embryos or animals, is essential for the action of an artificial transcription factor and responder gene (tetracycline system) or a "DNA-active" protein and responder gene (e.g., Cre recombinase), where the components of the systems have been pre-engineered into the animals by transgenesis.

Tamoxifen- and RU-486-inducible Recombinase Expression

Estrogen or a progesterone receptor ligand-binding domain in the presence of the ligand translocates the receptor to the nucleus. The domain and certain mutant forms retain this property in chimeric protein setups as well. On the basis of this phenomenon, the Cre recombinase is fused to mutant ligand-binding domains, which have lost their ability to bind endogenous estrogen or progesterone, but still bind tamoxifen (an estrogen antagonist) or RU-486 (a synthetic steroid), respectively. The nuclear translocation of the resulting chimeric Cre molecule becomes dependent on the presence of these ligands (Brocard et al. 1998; Kellendonk et al. 1999). Both systems have provided a certain level of inducibility (Danielian et al. 1998; Guo et al. 2002), but neither of them is the ideal one, with the result of no Cre activity at the noninduced stage and complete Cre action at the induced stage.

Tetracycline-inducible Gene Expression

The tetracycline-inducible gene expression system is more flexible than the tamoxifen- or RU-486-inducible Cre recombinase systems, since the latter is not reversible after the Cre recombinase produces the required alteration on the target. The tetracycline-inducible gene expression system can be turned on and off, depending on the tetracycline status in the cells, and is not necessarily connected to a recombinase system. It uses the DNA-binding domain of the bacterial *tet*-repressor protein and a strong transcriptional activator domain (VP16 from the herpesvirus), which are fused together. Such a heterologous protein can bind to the tetracycline operator element and activate transcription, depending on the presence of tetracycline (Gossen et al. 1995).

Reporters to Detect Recombinase Action

Testing of the efficiency and specificity of any recombinase system is highly recommended before complex experiments are designed and initiated. Several transgenic recombinase activity reporter mouse lines (i.e., a transgenic line expressing a reporter gene in response to recombinase-mediated recombination) have been developed. All are based on the same principle: a single-copy, close to the ubiquitously expressing promoter followed by a recombinase recognition site-flanked transcriptional STOP region, and then the coding region of a reporter gene. The reporter initiates expression under the control of the promoter only after recombinase-mediated excision removes the STOP region. For both Cre and Flp, such a conditional *lacZ* reporter construct has been introduced into the ROSA26 gene-trap integration site (Mao et al. 1999; Soriano 1999; Awatramani et al. 2001). Other systems (Z/AP and Z/EG) use two reporters: Cells express *lacZ* before Cre excision and heat-resistant human placental alkaline phosphatase for Z/AP (Lobe et al. 1999) and enhanced green fluorescent protein for Z/EG

(Novak et al. 2000) after excision. This behavior was achieved by random and single-copy integration of a transgene by ES cell-mediated transgenesis (see Protocol 9.1).

ES CELL-MEDIATED TRANSGENESIS

Those who have followed the development of pronucleus injection-based transgenic and ES cell-mediated genome alteration technologies over the last two decades might believe that DNA injection is used for generating random insertional transgenes and ES cells are used for the remainder of alterations. This was certainly true for a long time, until the ES cell technologies were simplified and obtained wide acceptance. ES cell-mediated random insertional transgenesis is not only an alternative for producing transgenic mice, but it also has its own advantage when there is a need for low copy number (single copy) integration. In addition, it also allows checking of expression of the transgene before a decision is made to introduce the transgenic cells in vivo (see Protocol 9.1 and corresponding comments).

Example A

A positive selectable marker incorporated into the constructs is the usual solution for transgene introduction into the cells. In these cases, the selectable marker is removable after integration, for example, flanked by *loxP* sites.

Example B

If the transgene is expected to be expressed in ES cells as well, the selectable marker can be connected to the gene of interest through an IRES sequence that couples the expression of the two coding regions. The strength of *neo* resistance may reflect the level of expression for the gene of interest.

Example C

In most cases, co-electroporation of the selectable gene with the nonselectable gene is also possible. In this example, the expression of the gene of interest is coupled with GFP expression using the IRES sequence between them. A puromycin expressor transgene provides a fast and efficient selection for cells that have picked up DNA. If the DNA concentration of the two constructs used for electroporation is biased toward the gene of interest (20:2 μg), a large proportion of the puromycin-surviving colonies will be transgenic for the gene of interest as well. If the puromycin selection is stopped after 3–4 days, it is easy to identify transgenic lines in which the selectable marker is expressed only transiently and does not form a stable integrant transgene.

Example D

The Cre recombinase can be used to activate the expression of the gene of interest in "conditional" transgenic settings. In this transgene, the promoter drives a *loxP*-flanked β*geo*+pA gene (Friedrich and Soriano 1991), where the *loxP* sites have the same orientation. This flanked region is followed by the coding region of the gene of interest with its own polyadenylation signal. This transgene expresses the β*geo*, which gives an easy readout for the specificity and level of expression in addition to providing a selectable marker to identify transgenic clones. The gene of interest becomes expressed only if the *loxP*-flanked region is removed by Cre recombinase-mediated excision. In these cells, the β*geo* gene will no longer be expressed. The Cre recombinase could be produced by another transgene with a given specificity. The expression status of the gene of interest is dependent on whether the Cre recombinase has ever been expressed and acted in an ancestor of a cell.

If such a transgene is introduced into ES cells, there are several issues that require attention. If it is properly established in animals via germ-line transmission, it is one of the most versatile transgenes possible. Crossing with any of the Cre recombinase transgenic lines can activate the expression of the gene of interest. Therefore, a huge variety of specificity is available after creating one single conditional transgenic line. The proper establishment of the conditional transgenic line is, however, tricky. First the transgene has to be single-copy integrant and expressed in a ubiquitous, or at least in a widespread, manner. Even with a presumably ubiquitous promoter, such as β-actin, the random integration very rarely gives widespread expression. A large number of integration sites have to be screened for expression to find an acceptable overall expressor transgene.

The construct given for this example provides an easy screen if it is introduced into ES cells. The transgenic cells can be cloned on the basis of *neo* resistance, the expression of the transgene can be determined from the nature and strength of *lacZ* expression with a simple histochemical staining, and the ES cell-mediated transgenesis favors the single-copy integration. There is a reasonable likelihood that a single-copy, strong, and overall expressor transgenic ES cell line can be identified among one or two hundred colonies screened. Protocol 9.1 gives further details for this procedure.

Example E

This construct is a more sophisticated version of Example D and contains one more step. This conditional transgene has an additional IRES-reporter attached to the coding region of the gene of interest, which can give an easy visual read-out if the *loxP*-flanked β*geo* is removed and the transgene is activated. Instead of GFP, other reporters (except *lacZ*) or selectable markers (other than *neo*) can be attached, depending on the actual needs of the experiment.

GENE TARGETING

Mouse ES cells have provided a large number of cells for targeting with specially designed DNA vectors to select for low-frequency events such as homologous recombination between the ES cell genome and homologous sequences placed in the (target) vector. A properly designed target vector can create a null mutation (gene knockout) in the gene under investigation (Capecchi 1989). It is difficult to estimate the number of genes that have been knocked out so far in the mouse, but the number certainly exceeds 5000 at the time of publication of this manual.

The importance and impact of the information gained from these studies on our understanding of gene function in normal development and in disease processes cannot be emphasized enough. We have also learned the limitation of creating null alleles, because many genes have multiple functions used in different developmental stages. The combination of these new tools with the "traditional" gene-targeting methods has reached an extremely sophisticated level, resulting in the ability to copy almost any genetic change linked to a specific human disease, from point mutations through small deletions to large specific chromosomal aberrations.

Example A

The basic concept in building a target vector to knock out a gene is to use two segments of homologous sequences to a gene of interest that flank a part of the gene essential for function, e.g., a coding exon. In the target vector, a positive selectable marker is placed between the homology arms, e.g., the *neo* gene. Upon homologous recombination between the arms of the vector and the corresponding genomic regions, the positive selectable marker will replace the functionally essential segment and create a "null" allele. Frequently, but not always, a negative selectable marker is placed outside of one or both of the homology arms, e.g., a HSV-*tk* or *dt* gene. If homologous recombination occurs at the arms, the negative selectable marker is not introduced into the genome; therefore, the cell will survive the negative selection. In the case of random insertion, which is the majority of cases, the negative selectable marker may also integrate and the cell dies. This positive/negative selection was first introduced by the Capecchi lab (Mansour et al. 1988) to enrich for homologous recombination.

Example B

In the "classical" target vector depicted in Example A, the *neo* gene with its promoter remains in the targeted allele. There are concerns that the presence of a selectable marker expression cassette can have an effect on the expression of the neighboring genes (Kim et al. 1992; Fiering et al. 1993; Braun et al. 1994; Olson et al. 1996). The interference that this cassette represents can lead to misinterpretation of a phenotype. For removal of the selectable marker, a classical targeting vector has to be just slightly modified: *loxP* or *FRT* sites must be placed around the selectable marker. By removing the selectable marker from the targeted allele, the cells become G418-sensitive, and they can be the subject of a second targeting with the original target vector to create homozygous knockout ES cells (Rossant and Nagy 1995; Abuin and Bradley 1996; Nagy 1996). The *neo*-selectable marker can be removed in vivo by a simple crossing of an animal transmitting the targeted allele with a transgenic partner expressing the Cre recombinase at an early stage (e.g., preimplantation stage) of embryonic development. If the Cre recombinase is expressed properly in the double transgenic offspring, the *neo*-selectable marker is removed from the targeted allele. After this modification, this allele is structurally different from the original targeted allele, and this difference can be utilized in further studies, for example, in chimera studies (see Chapter 11).

Example C

Replacement of the expression of the gene of interest with the expression of a reporter is also a favored strategy. In this case, the reporter *lacZ* (Le Mouellic et al. 1990) or GFP (Godwin et al. 1998) provides a convenient readout for the location and behavior of the cells harboring the targeted allele(s).

Example D

If the gene of interest is expressed in ES cells, a target vector can use the endogenous expression to drive the selectable marker by placing the marker in an expressible position (Schwartzberg et al. 1990). Usually this position is the first coding exon, in frame with the endogenous ATG or replacing the endogenous ATG with the marker's ATG 3′ from the transcriptional initiation. This is a very efficient strategy, since most of the random integrations do not provide expression for the positive selectable marker and the cells do not survive the selection. There is no need for negative selection in this strategy.

Example E

This example shows the fusion protein between *lacZ* and *neo*; β*geo* provides a combination of Examples C and D. In this strategy, the endogenous promoter-driven β*geo* gene provides LacZ activity and G418 resistance for the targeted cells.

Example F

For some ES cell-expressed genes, the first coding region is not always available for insertion of the selectable marker to create a promoterless target vector as detailed in Examples D and E. The IRES sequence provides a solution for expression of the marker (Jeannotte et al. 1991) from an internal exon or artificial exon created at the site of insertion within the gene.

Example G

Homologous recombination occurs between the gene of interest and the target vector somewhere within the homology arms. This construct offers a way of introducing specific subtle changes into the locus. The positive selectable marker, however, has to be positioned properly (for example, into an intron) and made removable by a recombinase system. Targeting with such a construct can create, e.g., a point mutation, and leaves behind an intron positioned in the *loxP* or *FRT* site after the removal of the selectable marker. It is unlikely that this short recognition site interferes with the proper expression of the altered allele; therefore, the effect of subtle change can be studied. Be aware, however, that not all of the homologous recombination events introduce the designed change into the targeted allele, since the recombination in the mutant arm could occur between the mutation and the positive selectable marker. In this case, only the marker is introduced into an intron. Frequently, the intronic insertion of the *neo* cassette severely decreases the level of mRNA produced by the allele (Meyers et al. 1998; Nagy et al. 1998). The *neo* sequence could also introduce cryptic splice sites that could lead to truncation of the protein encoded by the gene of interest. Nevertheless, intronic insertion of the *neo* gene could also provide very informative "knock down" alleles for furthering the understanding of gene function (Meyers et al. 1998; Nagy et al. 1998).

Example H

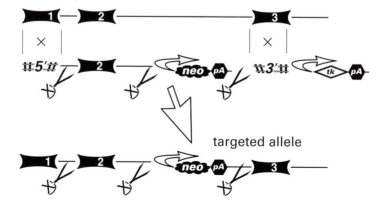

targeted allele

The Cre/*loxP* recombinase system combined with gene targeting offers one of the most powerful genetic dissections of a complex phenotype—the conditional, cell-type- or lineage-specific gene knockouts. The target vector designed for this purpose is very similar to the one shown in Example G, except that no alteration was made in any of the coding exons. An additional (third) *loxP* site is introduced into an intron in a way that *loxP* sites not only flank the selectable marker, but also flank an essential coding exon in an arrangement depicted in the construct presented in this example (Gu et al. 1994). In a targeted allele, none of the coding exons is disturbed; the new sequence insertion occurs only in the introns. The selectable marker insertion, however, could be quite deleterious for proper functioning of the allele. When a Cre expression vector is transiently expressed in ES cells heterozygous for this allele, both total and partial excision could occur among the three *loxP* sites, resulting in three scenarios as follows:

1. Complete excision

2. Type I excision

3. Type II excision

Complete and Type II excisions are the desired outcomes. These two, in fact, lose G418 resistance; therefore, they are relatively easy to identify, even without the assistance of a selection marker, among several hundred colonies randomly picked after transient Cre expression. The Complete excision outcome provides a null allele for the gene of interest, whereas the Type II excision allele is expected to be a wild-type equivalent functional allele. After germ-line transmission, the functionality of the Type II allele, however, is dependent on the expression of a Cre recombinase in the cells. If Cre recombinase is expressed in a cell, it removes the *loxP*-flanked essential exon from this allele, rendering it nonfunctional. A simple breeding system can be designed to place a null allele, a conditional allele, and a cell type/tissue-specifically expressed Cre recombinase transgene together in the same embryo or animal. This individual will represent a cell type/tissue-specific gene knockout, in which the specificity is defined by the Cre recombinase expression.

Example I

Critical, and perhaps the most important part in the previous example (H), is obtaining the Type II excision, which is the conditional targeted allele. The combination of the Cre and Flp recombinase in the same strategy offers a more straightforward and safer way of obtaining that allele. The selectable marker is flanked by *FRT* sites, and two *loxP* sites are used around an essential exon. Then transient expression of the *Flpe* gene (Schaft et al. 2001) in ES cells heterozygous for the targeted allele removes the selectable marker only, leaving the *loxP*-flanked exon and an intronic *FRT* site behind. Another alternative is pronuclear injection of a Flpe expression vector (Schaft et al. 2001) into zygotes harboring the targeted allele.

Example J

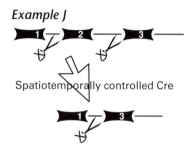

If a conditional allele has been created by flanking an essential exon or region of the gene of interest with *loxP* sites, the possibility arises of controlling the knock-out not only in a spatial manner, but also in a temporal manner. This result can be achieved by using inducible Cre recombinase transgenic systems in combination with the conditional knockout allele. Tamoxifen (Danielian et al. 1998; Guo et al. 2002), RU486 (Wunderlich et al. 2001), and tetracycline-inducible (Utomo et al. 1999) Cre transgenic lines have been established with different lineage/cell type-specificity for Cre expression, and the inducibility of Cre action has been demonstrated.

There are two caveats, so far, worth mentioning here. One is that none of the existing inducible systems provides the ideal inducibility conditions—zero Cre activity at the noninduced stage and total (100%) excision activity at the induced stage. Partial excision creates mosaicism in the target cell type, which could complicate the phenotype analysis. The other caveat is related to the time required to clear a protein from a cell by the site-specific recombinase-assisted conditional knockout approach. There are several layers of delay present in the systems. The Cre transcription starts first, followed by translation and accumulation of the protein; then Cre enzyme catalyzes recombination at the target *loxP* sites. This event eliminates the transcriptional supply of the protein of interest, but the previously made mRNA is still in the cell for a while to provide a template for translation. Then, after decline of translation, the protein has to be degraded. Depending on the expression dynamics of Cre and the target gene and the stability of the messages and protein, this delay can vary from several hours to days, which could limit addressing fast developmental processes with a conditional knockout approach (Nagy 2000).

Nevertheless, the ability to generate site-specific somatic mutations in a spatiotemporally controlled manner is one of the most powerful genetic tools for dissecting multiple gene functions. The conditional site-specific recombination system allows the analysis of knockout phenotypes that cannot be addressed by conventional gene targeting.

GENE AND PROMOTER TRAPS

Typically, DNA introduced into ES cells integrates into the genome randomly. The integration can occur outside or inside of an endogenous gene. Specially designed vectors, in the latter case, may not only disturb the gene, but also can report its transcriptional activity and/or even some properties of the original protein products (Gossler et al. 1989). These "gene sensor" vectors are called gene trap vectors. There are several research centers around the globe that are running large-scale gene trap programs with the goal of eventually trapping all of the genes of the mouse genome (Stanford et al. 2001). Because the gene trap insertions are highly mutagenic, this method not only reports the existence of the trapped gene and its important features (for example, sequence and expression pattern), but also immediately provides mutations for genetic studies.

Example A

Most of the vectors introduced into ES cells deliver a positive selectable marker gene, which simply makes the isolation of clones with transgenic insertion possible. In the classical gene trap vector, the selectable marker is separated from the "trapper" part as an independently expressed transgene. The trapping is based on the phenomenon that splice acceptors (SA) function in a heterologous gene environment; therefore, if a SA–reporter complex (i.e., SA–*lacZ*) integrates into an intron of a gene, the reporter will be spliced into the mRNA transcribed by endogenous regulation of the locus. If the translation from this hybrid mRNA finds the code of the reporter in-frame, a fusion protein is produced with β-galactosidase at the carboxyl terminus. This enzyme is quite tolerant to terminal extensions; therefore, in most cases, it remains functional and its activity can be detected by a simple histochemical staining (see Chapter 16). This gives a convenient readout for the expression regulation of the trapped gene. In addition, the hybrid mRNA can be found on the basis of the known *lacZ* sequence, and the overhanging sequence of the trapped gene origin reveals its identity. During the more than 10 years of gene-trapping history, the vectors used have been going through a diversification process, thus creating alternative solutions for practically the same task.

Example B

If trapping aims at genes that are expressed in ES cells, the vector can be further simplified by using β*geo*, the fusion gene between *lacZ* and *neo* (Friedrich and Soriano 1991).

Example C

An elegantly modified version of the construct introduced in Example B is when a transmembrane domain (TM) is placed right after the splice acceptor site, in front of β*geo* (Skarnes et al. 1995). If a trapped gene does not encode for secretion of a transmembrane protein, the fusion protein between the amino terminus of the gene and the β*geo* will not have a signal peptide, which would protect the TM-containing protein, sequestering it in the lumen of the endoplasmic reticulum. This results in the abolishment of β-gal activity. When a signal peptide is added by the trapped gene, the fusion protein will be translocated into the cytosol, where β-gal activity can be detected. Such a trap vector preferentially detects genes encoding secreted factors or transmembrane proteins, such as receptors.

Example D

Products of reporters can be localized differently in cells, and some can be used to visualize certain cellular compartments better than others. For example, human placental alkaline phosphatase is preferentially allocated to axons of neurons, whereas βgal stays in the neuron cell bodies. The above double-reporter gene-trap vector was applied in a large-scale identification of genes involved in defining connections in the mouse brain (Leighton et al. 2001).

Example E

If the splice acceptor is left out from the trap vector, its "gene sensing" ability will be restricted to insertions into exons. This approach is called a promoter trap and is obviously less efficient than a gene trap. The extremely high mutagenic activity of a promoter trap, however, could be an advantage.

HOMOLOGOUS RECOMBINATION—REPLACEABLES

In the previous examples, there was always a "little" constraint; the positive selectable marker was either left behind in the modified locus or was removable by a site-specific recombinase after the targeted allele or transgenic insertion was identified. Even the latter method left behind at least a recognition site for the recombinase. Here we describe two concepts for pure replacements using homologous recombination-based integration. The advantage of these strategies is their high purity; their disadvantage is generally low efficiency.

Example A

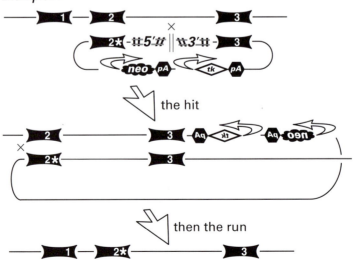

The "hit-and-run" strategy was originally designed for introducing point mutations into genes of interest in a pure way, without leaving any extra sequence behind (Hasty et al. 1991b). It is based on a special property of an insertion target vector, where the selectable marker is not placed between the arms of homology, but connects them from outside. Upon homologous recombination (Hit), the entire vector inserts into a single site, creating a duplication of the genomic sequence determined by the arms of homology. The arms of homology form one of the duplicons, which have the same orientation and flank the selectable marker. Therefore, after integration, the genome could loop, the duplicons could pair, and homologous recombination could occur (Run). A positive selectable marker (*neo*) helps the identification of the Hit and a negative selectable marker (HSV-*tk*) is used to find the Run. If a point mutation was engineered into one of the homology arms of the vector, the Run event could leave the mutation behind. However, it depends on the site of homologous recombination between the duplicons. The Run event must occur 5′ from the mutant site.

Example B

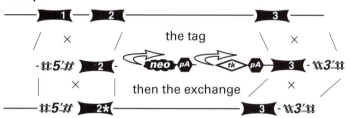

The concept of the Tag and Exchange strategy is similar to that of the Hit and Run. However, two targeting steps are used (Askew et al. 1993). The first target vector inserts a positive and a negative selectable marker to an intron of a gene of interest (Tag). The positive selectable marker is used in this step to identify the targeted clone, which then becomes the subject of the second targeting with a very special target vector. This vector is just a piece of genomic DNA bridging over the site of selectable markers placed by the first targeting. Homologous recombination of this simple vector removes the selectable markers from the locus and restores the original wild-type status (Exchange). This event can be selected with the negative selectable marker. If a subtle change, e.g., a point mutation, is engineered into one of the homology arms of the second vector, the Exchange event could leave that change behind, depending on the site of homologous recombination between the tagged locus and the second vector in the mutant arm. In addition to using this method to introduce point mutations into a gene, a further extension of this idea exists. Practically anything can be placed into the second vector, which then replaces the selectable marker complex. The replacement is pure; no constraints are attached.

SITE-SPECIFIC RECOMBINATION-MEDIATED INSERTIONS

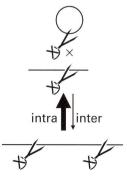

Another group of methods is the development of site-specific insertion into the genome by using recombinases. These are the recombinase-mediated site-specific integrations or recombination-mediated cassette exchanges (RMCE). Both Flp and Cre recombinases and the ϕC31 integrase work intra- and intermolecularly. In the previous examples, the intramolecular actions of these enzymes were shown. One of the categories of the intermolecular usage of these proteins is to mediate site-specific insertions. A common intrinsic problem with Cre and Flp in this respect is that they act between two homotypic sites, and the insertion of a small molecule, such as a plasmid, into the genome creates two recombinase recognition sites flanking the inserts. These sites are "easy" targets for a subsequent intramolecular excision. The intramolecular excision is much more efficient than the intermolecular insertion; therefore, the equilibrium favors the excision to the insertion stage. Several smart designs have been developed to select for the latter, unfavored stage or to make the insertion the favored one.

Example A

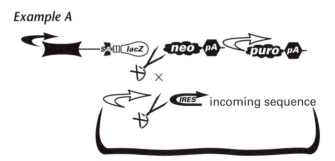

This example shows a way of modifying a gene trap vector, which allows a Cre recombinase-mediated insertion of any cDNA or any other sequence into the gene trap insertion site (modified after Hardouin and Nagy 2000). The puromycin selection assists the isolation of random integration into the genome, and the *SA–lacZ* (note that no pA is added after *lacZ*!) is the sensor for the gene trap event. A *loxP*-promoterless *neo*-pA segment is placed between these two cassettes. The *neo* is not expressed; therefore, the cell line is G418-sensitive. Then a Cre recombinase expression vector and a circular incoming sequence (plasmid) are introduced into the cells. The plasmid consists of a *loxP* site, an IRES sequence, the incoming cDNA, the plasmid backbone, and a promoter. In the insertion stage, the plasmid delivers a promoter in front of *neo*, and the cells become G418-resistant. The other "business side" of the plasmid inserts the IRES incoming sequence (e.g., an IRES-cDNA) after the *lacZ* coding region, providing the trapped regulation to the incoming sequence. Because the excision stage and the intact gene trap insertion are G418-sensitive, the only *neo* selection survivors are the cells with insertions.

Example B

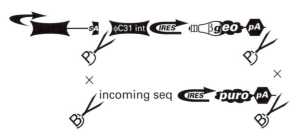

As mentioned earlier, φC31 is unique among the recombinases. It recombines between two heterotypic sites (*attPP′* and *attBB′*), and the resulting recombined sites are no longer substrates for the enzyme. This property can be used for an efficient replacement of a target sequence with an incoming segment. The gene trap insertion in this example (modified from G. Belteki and A. Nagy, in prep.) contains an *attPP′*-flanked region harboring the coding region of the φC31 integrase and connected to a β*geo* reporter/selectable marker. This latter detects the gene trap events. The incoming sequence is flanked with the other site, *attBB′*, and within the flanked region it is followed by an IRES-puro-pA. There is no need for transient expression on φC31 to induce the replacement because the cells produce it already. Upon replacement (cassette exchange), the active recognition sites for the integrase are destroyed and the φC31 integrase and the β*geo* are replaced by the incoming sequence and IRES-puro construct. The cells become G418-sensitive and puromycin-resistant.

Example C

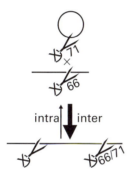

Significant effort has been made to identify mutant variants for both *lox* and FRT sites, which could help to increase the stability of the insertion stage after Cre- or Flp-mediated cassette exchange (Schlake and Bode 1994; Lee and Saito 1998; Siegel et al. 2001). For example, *lox66* and *lox71* are mutants at the 3′ and 5′ ends of the palindrome arms, respectively. They recombine with each other relatively well, but their recombination results in a wild-type *loxP* site and a double-mutant *lox 66/71* site, which then recombine with each other at much lower frequency (Araki et al. 1997).

Example D

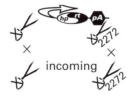

Another combination of *lox* sites, *loxP* and *lox2272*, seems to be an excellent combination for RMCE (Kolb 2001). The *hprt* gene can be used as selection marker only if the ES cells are HPRT-deficient. The advantage of this marker in a cassette exchange project is that it can be both positively and negatively selected. Positive selection assists the insertion of the *lox*-site-flanked region into the genome, for example, by homologous recombination into a gene of interest, whereas negative selection against the inserted *hprt* assists the identification of the RMCE.

Alternatively, a separate positive (e.g., *neo*) or negative (e.g., HSV-*tk*) selectable marker can be placed between the *loxP* and *lox2272* sites.

INDUCED SITE-SPECIFIC CHROMOSOMAL ABERRATIONS

As the distance between the recognition sites of the recombinases increases, or if these sites are placed interchromosomally, the frequency of recombination drastically decreases. Luckily, the number of ES cells and the general superior efficiency of the Cre recombinase can compensate for this decrease. With well-designed positive selectable markers, depending on the positioning and orientation of the *loxP* sites, large chromosomal deletions, duplications, inversions, and translocations can be identified (Yu and Bradley 2001).

Example A: Deletion

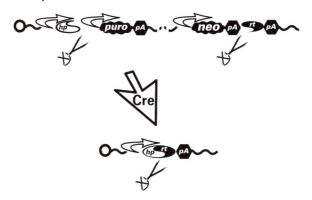

Creating deletion of a large chromosomal region requires independent targeting of the breakpoints with *loxP* sites at the same orientation and selectable markers assisting both the targeting and the subsequent deletion detection. In the strategy presented here (designed by Allan Bradley's laboratory), an *hprt*-deficient ES cell line is used. One breakpoint of the planned deletion can be targeted with puromycin and the other with neomycin selectable markers. The target vectors deliver one of each halves of the bipartite *hprt* minigene. It is important to position the minigene halves outside the area that is to be deleted. In this case, if deletion occurs between the *loxP* sites, the bipartite halves will be brought together and the *hprt* minigene function will be restored; the cells will lose G418 and puromycin resistance but will gain HAT resistance.

Example B: Inversion

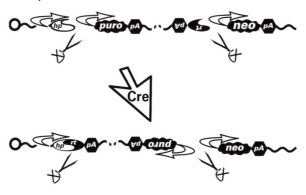

Similar to making a deletion, independent targeting of the breakpoints is required here using *hprt*-deficient ES cells. In this case, however, the *loxP* sites are inverted. One breakpoint of the planned inversion can be targeted with puromycin and the other with neomycin. The target vectors deliver one of each of the halves of the bipartite *hprt* minigene. It is important to position the halves either both at the proximal side or both at the distal side of the *loxP* sites relative to the centromere. In this case, if recombination occurs between the inverted *loxP* sites, the bipartite halves will be brought together and the *hprt* minigene function will be restored; the cells become triple resistant to G418, puromycin, and HAT.

Example C: Translocation

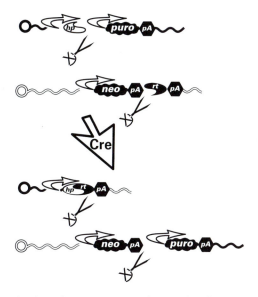

The breakpoints are independently targeted in *hprt*-deficient ES cells. It is important to place the *loxP* sites at the same orientation relative to the centromeres of their chromosomes (otherwise dicentric and acentric chromosomes will form after recombination). One breakpoint of the planned translocation can be targeted with puromycin and the other with neomycin. The target vectors deliver one of each of the halves of the bipartite *hprt* minigene. It is also critical to position the halves so that one of them is at the proximal side and the other is at the distal side of their corresponding *loxP* sites relative to the centromere. In this case, if recombination occurs between the *loxP* sites, the bipartite halves will be brought together into a functional *hprt* minigene in one of the translocation chromosomes, whereas the other chromosome will comprise both *neo* and puromycin genes. Therefore, the cells become triple resistant to G418, puromycin, and HAT.

CREATING HOMOZYGOUS MUTANT ES CELL LINES

ES cells alone are able to support the development of the mouse embryo or an adult if the placenta and the primitive endoderm-derived extraembryonic membranes are provided by a tetraploid embryo (see Chapter 11). The superior developmental potential of these cultured cells allows the derivation of mutants directly from the tissue culture dish and in this way saves money, resources, and time for the investigator. Concerning the gene knockouts (targeting), there are several means of creating homozygous mutant ES cell lines. Five of these methods are described in the following examples.

Example A

For mutations that are already introduced into the mouse, the most straightforward way of generating homozygous mutant ES cells is to derive new ES cell lines from homozygous embryos. In most cases, this requires generation of ES cells from embryos of a segregating population (i.e., F_2 generation), where only a portion of the embryos have the desired genotype. After establishment, the genotype of the cell should be determined by Southern blot or reliable PCR (see Chapter 12). If the inbred genetic background is not a requirement, using a heterozygous genetic background is recommended; this will decrease the risk of picking up additional phenotypes due to genetic and/or epigenetic damages associated with culturing of the cells.

Example B

Historically, the first method (Mortensen et al. 1991; Sawai et al. 1991) for generating deficient ES cell lines was two independent, consecutive targetings of ES cells using two different positive selectable markers.

Example C

The ability to remove the *loxP*-flanked selectable marker from a targeted allele by transient expression of the Cre recombinase in ES cells makes the original target vector reusable for a second round of targeting (Abuin and Bradley 1996). This method does not require building a second target vector as needed in Example B, but additional electroporation and cloning steps are needed.

Example D

An efficient and still not completely understood way to make homozygous ES cells for targeted mutations is simply to increase the concentration of the G418 for a second round of selection performed on the already G418-resistant cells heterozygous for a targeted allele (Mortensen et al. 1992). In the effective high concentration (6–20x of the original G418 concentration used to identify the primary targeted cells), almost all of the cells die. A few (30–100 per electroporation) survive and form colonies. The effective concentration is different from gene to gene and should be determined experimentally. Genotyping reveals that most of these cells are homozygous for the targeted allele and contain two copies of the G418 resistance gene.

For a long time, the underlying mechanism behind this phenomenon was unknown. Recently, however, it was shown that from the three possible mechanisms—gene conversion (local event), mitotic recombination (regional event), or duplication of the entire chromosome (chromosomal event)—the chromosomal event is highly prevalent (Lefebvre et al. 2001). As a consequence, cells are made homozygous not only for the targeted gene, but also for every gene on the corresponding chromosome.

Producing homozygous ES cells with a high concentration of G418 works with most genes. One might want to be careful with this method, however, if the gene of interest is located on an imprinted chromosome. This method will result in uniparental disomy for that chromosome, which could result in a phenotype by itself.

It is also important to know that the selection works only if the mutant *neo* (Yenofsky et al. 1990) gene is used for targeting. The wild-type *neo* renders the cells so G418 resistant that they survive even 5–10 mg/ml, a concentration of the drug that starts to have general toxicity on cells, regardless of their level of resistance.

High G418 selection can be used to identify homozygous ES cell lines when the mechanism is Cre recombinase-mediated mitotic recombination. The prerequisite for this method is placing *loxP* sites in both homologous chromosomes at homotopic sites. If the *neo*-containing allele of the gene of interest is distal from the *loxP* site and Cre-mediated recombination occurs in the G_2 phase of the cell cycle, chromatin segregation could result in a homozygous targeted allele (Koike et al. 2002). In this case, only the chromosome arm distal from the *loxP* site will be homozygous.

Example E

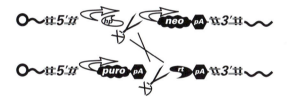

An elegant but heavily selectable marker-intensive design is using Cre recombinase-mediated mitotic recombination as well (Liu et al. 2002). Similar to the strategy above, a single *loxP* site is placed homotopically on both homologous chromosomes. The bipartite *hprt* minigene is separated between the chromosomes, but when the recombinase recombines between them during the G_2 phase, a functional *hprt* gene is formed. Furthermore, in *hprt*-positive cells that resulted from X segregation during cleavage, the chromosome arm distal from the *loxP* site is homozygous. This technology, perhaps, will have limited applications in homozygous knockouts, but could gain importance in generating loss of heterozygosity after random mutagenesis.

In Vitro Screen to Obtain Widespread, Transgenic Expression in the Mouse

This protocol corresponds thematically to the topics described in this chapter. It uses several protocols detailed in Chapter 10, and thus the general information in Chapter 10 should be consulted before attempting to execute Protocol 9.1.

This protocol details the steps for creating widespread (ubiquitous-like) transgenic expression by taking advantage of the special features of ES cell-mediated transgenesis. This type of transgene expression may seem to be a simple task, but the reality is that even the promoter of the most accepted ubiquitously expressed gene can express a transgene from a random integration in far from the expected manner. Most of these transgenes are expressed in a mosaic or tissue-restricted fashion, depending on the site of integration and the copy number of the transgene. The number of copies seems to affect the stability of expression negatively. ES cell-mediated transgenesis is not exempted from these phenomena. But the advantage of using ES cells is that transgenesis results in mostly a single-copy or low-copy-number integration. Furthermore, at the cellular level, ES cell-mediated transgenesis can easily provide several hundred different integration sites in a single experiment. The in vitro expression of these clones can be assayed before a decision is made about introducing the candidate cell line(s) into the mouse. Therefore, the transgenic vector should have an easy readout for the nature of transgene expression.

One vector that is perfect for this task is the Cre excision-conditional transgene vector already introduced above (ES Cell-mediated Transgenesis, Example D, p. 409).

Upon introduction into ES cells, such transgenes will expess β*geo* only, which confers G418 resistance on the cells, thus allowing the identification of transgenic colonies. In addition, staining the cells for *lacZ* will provide an easy characterization of the expression in undifferentiated and in vitro-differentiated ES cells.

Concerning the promoter of choice, the combination of a cytomegalovirus (CMV) enhancer and a chicken β-actin promoter with the first intron (Niwa et al. 1991) has provided good results in several projects aiming at widespread expression of the transgene (Okabe et al. 1997; Hadjantonakis et al. 1998). During the production of Z/AP (Lobe et al. 1999) and Z/EG (Novak et al. 2000), Cre-excision reporter lines have resulted, and very useful plasmids (pCALL and pCALL2) have been built that require only a single cloning step to build any conditional transgene vector.

Protocol 9.1 is a "super" protocol in the sense that other protocols are used as individual steps. Therefore, the Materials section names the protocols used in this procedure.

MATERIALS

For equipment, reagents, cells, and DNA, please refer to Protocols 10.1, 10.2, 10.5, 11.5 (or 11.10B), and 16.15.

PROCEDURE

1. Execute Protocol 10.1—electroporating DNA into ES cells and selecting for G418-resistant colonies. Performing step 9 is optional; apply step 10 to all of the plates instead of one, and ignore step 11, because negative selection is not being dealt with in this case.

2. Execute Protocol 10.2—isolating individual ES cell colonies by picking. Harvest a total of about 200 colonies and grow them up in four replica plates. Freeze two of the four plates (see Protocol 10.3), and expand the remaining two into four 96-well plates again. From this passage onward, use gelatinized plates because these cells will be used for *lacZ* staining and DNA preparation. Use ES cell culture medium with LIF to keep the cells undifferentiated.

3. Stain one plate for *lacZ* expression (follow Protocol 16.15; see comment on staining cultured cells).

4. Evaluate *lacZ* expression for the level of expression and the overall (mosaic) aspect in each well.

5. Withdraw LIF from the ES cell medium and culture cells at low density for 5–8 days. The cells will differentiate during this period of time.

6. Perform steps 3 and 4 on the differentiated cells.

7. Evaluate the two staining results together. Select the cell lines that have given the most overall type of *lacZ* expression. If a high level of expression is also desired, consider the level of expression as a criterion as well.

8. Prepare DNA from the best 15–25 lines and process them for Southern blot analyses to determine the copy number of the transgenes (see Chapter 12).

9. Select three or four ES cell lines with the single-copy transgene and best expression evaluation. Introduce them into the chimeras and obtain germ-line transmission (see Chapter 11).

10. After germ-line transmission, evaluate the *lacZ* expression at the stage of interest following Protocol 16.15. Usually 50–75% of the transgenic lines generated will give satisfying overall expression.

11. Activate the expression of the genes of interest in the offspring by crossing them with a general deletor or tissue-specific Cre transgenic animals.

COMMENTS

- Do not overstain in step 3! Stop the staining reaction when only about 10–15 % of the colonies can be scored as strong expressors.
- An alternative method to activate the expression of the gene of interest is injection of a circular Cre expressor plasmid into the pronucleus of transgenic zygotes (Araki et al. 1995).

- With a slight modification, this protocol can be used to screen in vitro for cell-type-specific expression of a transgene introduced into ES cells. The difference in this approach is to use a cell-type-specific promoter in a slightly different construct, where the *loxP*-flanked β*geo*–gene-of-interest–pA is followed by an independently expressed promoter–puro–pA construct.

  If ES cells are not supposed to express the transgene, due to the cell-type specificity, the transgenic ES cells will be G418-sensitive and puromycin-resistant. Staining for *lacZ* can also be used to confirm the lack of expression from the specific promoter in undifferentiated ES cells. The other criterion is the detection of *lacZ* expression in in vitro-differentiated cells into the cell type of promoter specificity. These cells can also be selected with G418 since they will be resistant. Using cellular markers specific for the cell type of interest can further characterize the identity of the G418-resistant cells. Clones passing this complex criterion system are then selected for introduction into the mouse. This procedure significantly enriches for transgenic lines with the required cell-type-specificity expression.

- For additional comments, see Protocols 10.1, 10.2, 10.5, 11.5 (or 11.10B), and 16.15, and for general considerations refer to Chapter 12.

REFERENCES

Abuin A. and Bradley A. 1996. Recycling selectable markers in mouse embryonic stem cells. *Mol. Cell. Biol.* **16:** 1851–1856.

Albert H., Dale E. C., Lee E., and Ow D. W. 1995. Site-specific integration of DNA into wild-type and mutant lox sites placed in the plant genome. *Plant J.* **7:** 649–659.

Araki K., Araki M., Miyazaki J., and Vassalli P. 1995. Site-specific recombination of a transgene in fertilized eggs by transient expression of Cre recombinase. *Proc. Natl. Acad. Sci.* **92:** 160–164.

Araki K., Araki M., and Yamamura K. 1997. Targeted integration of DNA using mutant lox sites in embryonic stem cells. *Nucleic Acids Res.* **25:** 868–872.

Askew G.R., Doetschman T., and Lingrel J.B. 1993. Site-directed point mutations in embryonic stem cells: A gene-targeting tag-and-exchange strategy. *Mol. Cell. Biol.* **13:** 4115–4124.

Awatramani R., Soriano P., Mai J.J., and Dymecki S. 2001. An Flp indicator mouse expressing alkaline phosphatase from the ROSA26 locus. *Nat. Genet.* **29:** 257–259.

Braun T., Bober E., Rudnicki M.A., Jaenisch R., and Arnold H.H. 1994. MyoD expression marks the onset of skeletal myogenesis in Myf-5 mutant mice. *Development* **120:** 3083–3092.

Brocard J., Feil R., Chambon P., and Metzger D. 1998. A chimeric Cre recombinase inducible by synthetic, but not by natural ligands of the glucocorticoid receptor. *Nucleic Acids Res.* **26:** 4086–4090.

Buchholz F., Angrand P.O., and Stewart A.F. 1998. Improved properties of FLP recombinase evolved by cycling mutagenesis. *Nat. Biotechnol.* **16:** 657–662.

Capecchi M.R. 1989. Altering the genome by homologous recombination. *Science* **244:** 1288–1292.

Danielian P.S., Muccino D., Rowitch D.H., Michael S.K., and McMahon A.P. 1998. Modification of gene activity in mouse embryos in utero by a tamoxifen- inducible form of Cre recombinase. *Curr. Biol.* **8:** 1323–1326.

Fiering S., Kim C.G., Epner E.M., and Groudine M. 1993. An "in-out" strategy using gene targeting and FLP recombinase for the functional dissection of complex DNA regulatory elements: Analysis of the β-globin locus control region. *Proc. Natl. Acad. Sci.* **90:** 8469–8473.

Friedrich G. and Soriano P. 1991. Promoter traps in embryonic stem cells: A genetic screen to identify and mutate developmental genes in mice. *Genes Dev.* **5:** 1513–1523.

Godwin A.R., Stadler H.S., Nakamura K., and Capecchi M.R. 1998. Detection of targeted *GFP-Hox* gene fusions during mouse embryogenesis. *Proc. Natl. Acad. Sci.* **95:** 13042–13047.

Gossen M., Freundlieb S., Bender G., Muller G., Hillen W., and Bujard H. 1995. Transcriptional activation by tetracyclines in mammalian cells. *Science* **268:** 1766–1769.

Gossler A., Joyner A.L., Rossant J., and Skarnes W.C. 1989. Mouse embryonic stem cells and reporter constructs to detect developmentally regulated genes. *Science* **244:** 463–465.

Groth A.C., Olivares E.C., Thyagarajan B., and Calos M.P. 2000. A phage integrase directs efficient site-specific integration in human cells. *Proc. Natl. Acad. Sci.* **97:** 5995–6000.

Gu H., Marth J.D., Orban P.C., Mossmann H., and Rajewsky K. 1994. Deletion of a DNA polymerase beta gene segment in T cells using cell type-specific gene targeting. *Science* **265:** 103–106.

Guo C., Yang W., and Lobe C.G. 2002. A Cre recombinase transgene with mosaic, widespread tamoxifen-inducible action. *Genesis* **32:** 8–18.

Hadjantonakis A.K., MacMaster S., and Nagy A. 2002. Embryonic stem cells and mice expressing different GFP variants for multiple non-invasive reporter usage within a single animal. *BMC Biotechnology* **2:** 11.

Hadjantonakis A.K., Gertsenstein M., Ikawa M., Okabe M., and Nagy A. 1998. Generating green fluorescent mice by germline transmission of green fluorescent ES cells. *Mech. Dev.* **76:** 79–90.

Hardouin N. and Nagy A. 2000. Gene-trap-based target site for cre-mediated transgenic insertion. *Genesis* **26:** 245–252.

Hasty P., Rivera-Perez J., and Bradley A. 1991a. The length of homology required for gene targeting in embryonic stem cells. *Mol. Cell. Biol.* **11:** 5586–5591.

Hasty P., Ramirez-Solis R., Krumlauf R., and Bradley A. 1991b. Introduction of a subtle mutation into the Hox-2.6 locus in embryonic stem cells. *Nature* **350:** 243–246.

Hennecke M., Kwissa M., Metzger K., Oumard A., Kroger A., Schirmbeck R., Reimann J., and Hauser H. 2001. Composition and arrangement of genes define the strength of IRES-driven translation in bicistronic mRNAs. *Nucleic. Acids. Res.* **29:** 3327–3334.

Honjo T., Nishizuka Y., and Hayaishi O. 1969. Adenosine diphosphoribosylation of aminoacyl transferase II by diphtheria toxin. *Cold Spring Harbor Symp. Quant. Biol.* **34:** 603–608.

Jang S.K., Davies M.V., Kaufman R.J., and Wimmer E. 1989. Initiation of protein synthesis by internal entry of ribosomes into the 5′ nontranslated region of encephalomyocarditis virus RNA in vivo. *J. Virol.* **63:** 1651–1660.

Jang S.K., Krausslich H.G., Nicklin M.J., Duke G.M., Palmenberg A.C., and Wimmer E. 1988. A segment of the 5′ nontranslated region of encephalomyocarditis virus RNA directs internal entry of ribosomes during in vitro translation. *J. Virol.* **62:** 2636–2643.

Jeannotte L., Ruiz J.C., and Robertson E.J. 1991. Low level of Hox1.3 gene expression does not preclude the use of promoterless vectors to generate a targeted gene disruption. off. *Mol. Cell Biol.* **11:** 5578–5585.

Kellendonk C., Tronche F., Casanova E., Anlag K., Opherk C., and Schutz G. 1999. Inducible site-specific recombination in the brain. *J. Mol. Biol.* **285:** 175–182.

Kim C.G., Epner E.M., Forrester W.C., and Groudine M. 1992. Inactivation of the human beta-globin gene by targeted insertion into the beta-globin locus control region. *Genes Dev.* **6:** 928–938.

Koike H., Horie K., Fukuyama H., Kondoh G., Nagata S., and Takeda J. 2002. Efficient biallelic mutagenesis with Cre/loxP-mediated inter-chromosomal recombination. *EMBO Rep.* **3:** 433–437.

Kolb A.F. 2001. Selection-marker-free modification of the murine beta-casein gene using a lox2272 [correction of lox2722] site. *Anal. Biochem.* **290:** 260–271.

Lakso M., Sauer B., Mosinger B., Jr., Lee E. J., Manning R.W., Yu S.H., Mulder K.L., and Westphal H. 1992. Targeted oncogene activation by site-specific recombination in transgenic mice. *Proc. Natl. Acad. Sci* . **89:** 6232–6236.

Lee G. and Saito I. 1998. Role of nucleotide sequences of loxP spacer region in Cre-mediated recombination. *Gene* **216:** 55–65.

Lefebvre L., Dionne N., Karaskova J., Squire J.A., and Nagy A. 2001. Selection for transgene homozygosity in embryonic stem cells results in extensive loss of heterozygosity. *Nat. Genet.* **27:** 257–258.

Leighton P.A., Mitchell K.J., Goodrich L.V., Lu X., Pinson K., Scherz P., Skarnes W.C., and Tessier-Lavigne M. 2001. Defining brain wiring patterns and mechanisms through gene trapping in mice. *Nature* **410:** 174–179.

Le Mouellic H., Lallemand Y., and Brulet P. 1990. Targeted replacement of the homeobox gene Hox-3.1 by the *Escherichia coli* lacZ in mouse chimeric embryos. *Proc. Natl. Acad. Sci.* **87:** 4712–4716.

Lobe C.G., Koop K.E., Kreppner W., Lomeli H., Gertsenstein M., and Nagy A. 1999. Z/AP, a double reporter for cre-mediated recombination. *Dev. Biol.* **208:** 281–292.

Mansour S.L., Thomas K.R., and Capecchi M.R. 1988. Disruption of the proto-oncogene int-2 in mouse embryo-derived stem cells: A general strategy for targeting mutations to non-selectable genes. *Nature* **336:** 348–352.

Mao X., Fujiwara Y., and Orkin S.H. 1999. Improved reporter strain for monitoring Cre recombinase-mediated DNA excisions in mice. *Proc. Natl. Acad. Sci.* **96:** 5037–5042.

McKeown M. 1993. The role of small nuclear RNAs in RNA splicing. *Curr. Opin. Cell Biol.* **5:** 448–454.

Meyers E.N., Lewandoski M., and Martin G.R. 1998. An Fgf8 mutant allelic series generated by Cre- and Flp-mediated recombination. *Nat. Genet.* **18:** 136–141.

Mizuguchi H., Xu Z., Ishii-Watabe A., Uchida E., and Hayakawa T. 2000. IRES-dependent second gene expression is significantly lower than cap-dependent first gene expression in a bicistronic vector. *Mol. Ther.* **1:** 376–382.

Mortensen R.M., Conner D.A., Chao S., Geisterfer-Lowrance A.A., and Seidman J.G. 1992. Production of homozygous mutant ES cells with a single targeting construct. *Mol. Cell. Biol.* **12:** 2391–2395.

Mortensen R.M., Zubiaur M., Neer E.J., and Seidman J.G. 1991. Embryonic stem cells lacking a functional inhibitory G-protein subunit (alpha i2) produced by gene targeting of both alleles. *Proc. Natl. Acad. Sci.* **88:** 7036–7040.

Mountford P.S. and Smith A.G. 1995. Internal ribosome entry sites and dicistronic RNAs in mammalian transgenesis. *Trends Genet.* **11:** 179–184.

Nagy A. 1996. Engineering the mouse genome. In *Mammalian development* (ed. P. Lonai), pp. 339-371. Harwood Academic Publishers, Amsterdam.

———. 2000. Cre recombinase: The universal reagent for genome tailoring. *Genesis* **26:** 99–109.

Nagy A., Moens C., Ivanyi E., Pawling J., Gertsenstein M., Hadjantonakis A.K., Pirity M., and Rossant J. 1998. Dissecting the role of N-myc in development using a single targeting vector to generate a series of alleles. *Curr. Biol.* **8:** 661–664.

Niwa H., Yamamura K., and Miyazaki J. 1991. Efficient selection for high-expression transfectants with a novel eukaryotic vector. *Gene* **108:** 193–199.

Novak A., Guo C., Yang W., Nagy A., and Lobe C.G. 2000. Z/EG, a double reporter mouse line that expresses enhanced green fluorescent protein upon Cre-mediated excision. *Genesis* **28:** 147–155.

Okabe M., Ikawa M., Kominami K., Nakanishi T., and Nishimune Y. 1997. "Green mice" as a source of ubiquitous green cells. *FEBS Lett.* **407:** 313–319.

Olson E.N., Arnold H.H., Rigby P.W. and Wold B.J. 1996. Know your neighbors: Three phenotypes in null mutants of the myogenic bHLH gene MRF4. *Cell* **85:** 1–4.

Ramirez-Solis R., Liu P., and Bradley A. 1995. Chromosome engineering in mice. *Nature* **378:** 720–724.

Rodriguez C.I., Buchholz F., Galloway J., Sequerra R., Kasper J., Ayala R., Stewart A.F., and Dymecki S.M. 2000. High-efficiency deleter mice show that FLPe is an alternative to Cre- loxP. *Nat. Genet.* **25:** 139–140.

Rossant J. and Nagy A. 1995. Genome engineering: The new mouse genetics. *Nat. Med.* **1:** 592–594.

Salomon B., Maury S., Loubiere L., Caruso M., Onclercq R., and Klatzmann D. 1995. A truncated herpes simplex virus thymidine kinase phosphorylates thymidine and nucleoside analogs and does not cause sterility in transgenic mice. *Mol. Cell. Biol.* **15:** 5322–5328.

Sawai S., Shimono A., Hanaoka K., and Kondoh H. 1991. Embryonic lethality resulting from disruption of both N-myc alleles in mouse zygotes. *New Biol.* **3:** 861–869.

Schaft J., Ashery-Padan R., van der Hoeven F., Gruss P., and Stewart A.F. 2001. Efficient FLP recombination in mouse ES cells and oocytes. *Genesis* **31:** 6–10.

Schlake T. and Bode J. 1994. Use of mutated FLP recognition target (FRT) sites for the exchange of expression cassettes at defined chromosomal loci. *Biochemistry* **33:** 12746–12751.

Schwartzberg P.L., Robertson E.J., and Goff S.P. 1990. Targeted gene disruption of the endogenous c-abl locus by homologous recombination with DNA encoding a selectable fusion protein. *Proc. Natl. Acad. Sci.* **87:** 3210–3214.

Seibler J., Schubeler D., Fiering S., Groudine M., and Bode J. 1998. DNA cassette exchange in ES cells mediated by Flp recombinase: An efficient strategy for repeated modification of tagged loci by marker-free constructs. *Biochemistry* **37:** 6229-6234.

Siegel R. W., Jain R. and Bradbury A. 2001. Using an in vivo phagemid system to identify non-compatible loxP sequences. *FEBS Lett.* **499:** 147–153.

Skarnes W.C., Moss J.E., Hurtley S.M., and Beddington R.S. 1995. Capturing genes encoding membrane and secreted proteins important for mouse development. *Proc. Natl. Acad. Sci.* **92:** 6592–6596.

Soriano P. 1999. Generalized lacZ expression with the ROSA26 Cre reporter strain. *Nat. Genet.* **21:** 70–71.

Stanford W.L., Cohn J.B., and Cordes S.P. 2001. Gene-trap mutagenesis: Past, present and beyond. *Nat. Rev. Genet.* **2:** 756–768.

Szybalski W. 1992. Use of the HPRT gene and the HAT selection technique in DNA-mediated transformation of mammalian cells: First steps toward developing hybridoma techniques and gene therapy. *BioEssays* **14:** 495–500.

te Riele H., Maandag E.R., and Berns A. 1992. Highly efficient gene targeting in embryonic stem cells through homologous recombination with isogenic DNA constructs. *Proc. Natl. Acad. Sci.* **89:** 5128–5132.

Thomas K.R. and Capecchi M.R. 1987. Site-directed mutagenesis by gene targeting in mouse embryo-derived stem cells. *Cell* **51:** 503–512.

Utomo A.R., Nikitin A.Y., and Lee W.H. 1999. Temporal, spatial, and cell type-specific control of Cre-mediated DNA recombination in transgenic mice. *Nat. Biotechnol.* **17:** 1091–1096.

Wilkie T.M., Braun R.E., Ehrman W.J., Palmiter R.D., and Hammer R.E. 1991. Germ-line intrachromosomal recombination restores fertility in transgenic MyK-103 male mice. *Genes Dev.* **5:** 38–48.

Wunderlich F.T., Wildner H., Rajewsky K., and Edenhofer F. 2001. New variants of inducible Cre recombinase: A novel mutant of Cre-PR fusion protein exhibits enhanced sensitivity and an expanded range of inducibility. *Nucleic Acids Res.* **29** E47.

Yenofsky R.L., Fine M., and Pellow J.W. 1990. A mutant neomycin phosphotransferase II gene reduces the resistance of transformants to antibiotic selection pressure. *Proc. Natl. Acad. Sci.* **87:** 3435–3439.

Yu Y. and Bradley A. 2001. Engineering chromosomal rearrangements in mice. *Nat. Rev. Genet.* **2:** 780-790.

10

Introduction of Foreign DNA into Embryonic Stem Cells

I N THE PREVIOUS CHAPTER we gave several examples of smart DNA designs for performing special and unique changes or functions in the mouse genome upon integration. One might ask why embryonic stem (ES) cells are needed for mediation, since there are other, more direct ways for introducing these vectors into the mouse. The answer is "the number." Most of those specific functions that we are expecting to obtain from the vectors have specific requirements for site or mode of integration (which are very rare events), such as homologous recombination, gene-trap insertion, or expression-permissive sites for single-copy transgenes. Therefore, one of the most important components in the success

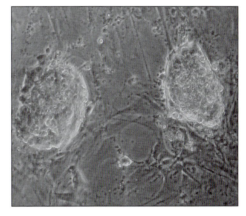

of ES cells in the new genome manipulation technologies is the number. One 10-cm tissue culture plate of ES cells contains more than 10 million cells, which are simultaneously the subject of genetic alterations in a single experiment. The other important component is that ES cells can survive in antibiotics (Gossler et al. 1986) if they express bacterial antibiotic-resistance genes, such as neomycin phosphotransferase, puromycin-, or hygromycin-resistance genes. Last, after all of these manipulations of transfection, antibiotic selection, and frozen storage, the genetically altered sublines of cells retain the germ-line compatibility of the original line in the majority of cases.

GENERAL CONSIDERATIONS

The most widely used method to alter the genome of embryonic stem (ES) cells is to introduce a specifically designed DNA fragment using electroporation. The DNA will then integrate into the genome of ES cells. Because the integration of the exogenous DNA occurs only in a fraction of the cells treated, the common component of all of these vectors is a positive selectable marker gene. This is usually an antibiotic-resistance gene, which keeps the cells alive after integration when the culture is treated with the corresponding antibiotic drug. During such treatment, all sensitive cells die, and the resistant cells grow into individual colonies of the appropriate size for subcloning by picking. Then the subclones are expanded, replica-plated, frozen, and used as a source for genomic DNA preparation for genotyping.

DNA PURIFICATION FOR INTRODUCTION INTO THE ES CELL GENOME

It is critical to use high-quality DNA, free of contaminating chemicals. The DNA should be purified either on a Qiagen column (e.g., Qiagen 12143) or by CsCl centrifugation. DNA is then linearized, extracted with phenol/chloroform/isoamyl alcohol (25:24:1) (e.g., Invitrogen Life Technologies, 15593-031), precipitated with ethanol, and dissolved in sterile phosphate-buffered saline (PBS) or water at a concentration of 1 mg/ml. The completion of the restriction digestion to linearize the DNA should be assessed by agarose gel electrophoresis. When using Qiagen purification, it is important to follow the manufacturer's instructions precisely. In particular, do not grow bacterial colonies too densely in a very rich culture medium. This may result in carbohydrate copurifying with the DNA.

ELECTROPORATING DNA INTO ES CELLS AND SELECTION METHODS

Electroporation is one of the most efficient methods for introducing DNA into ES cells. However, pilot studies should be performed to optimize the conditions for each DNA construct. Protocol 10.1 is only a guide, and the steps that should be varied in the optimization process are indicated. The selection method described here, as an example, is one of the most complex. It is based on the positive-negative selection technique devised by Mansour et al. (1988). Similar principles can be applied with other selection systems. This one is based on using targeting constructs in which the bacterial neomycin-resistance (*neo*^r) gene disrupts the coding sequence of the mouse gene. In addition, the herpes virus thymidine kinase (HSV-*tk*) gene is placed at one or both ends of the targeting construct so that it will be eliminated during homologous recombination, but usually not during random insertion into the chromosomes (see Chapter 9). The antibiotic G418 is used to select for cells in which the DNA construct containing the *neo*^r gene has been integrated, either randomly or by homologous recombination. The effective concentration of G418 first slows down cell proliferation, but kills cells after 3–5 days. The nucleoside analog ganciclovir or FIAU (1-[2′-deoxy-2′-fluoro-β-D-arabinofuranosyl]-5-iodouracil) is converted by the HSV-*tk* to a cytotoxic derivative. DNA that has integrated by homologous recombination will have lost HSV-*tk* and will be resistant to the drug, whereas cells that have incorporated the DNA randomly are likely to retain the HSV-*tk* gene. Thus, cells containing random integrations into a chromosomal location that allows expression of the HSV-*tk* gene will be killed. The use of FIAU is recommended above the use of ganciclovir, because there is a concern that the latter has some degree of bystander killing effect in which the death of sensitive cells could eliminate the nearby resistant cells. For further details about positive-negative selection, see Mansour et al. (1988).

ISOLATING INDIVIDUAL ES CELL COLONIES BY PICKING, REPLICA-PLATING, AND FREEZING ES CELL LINES FOR THE TIME PERIOD OF GENOTYPING

Assuming that the colonies are visible to the naked eye after ~8–10 days of culture in selective medium, different protocols can be used for isolating individual colonies and expanding them to a sufficient number to allow isolation and screening of DNA.

- Colonies can be picked using a finely drawn Pasteur pipette or automatic pipettor and placed in microdrops of trypsin the same way as is done for de novo ES cell derivation (see Protocol 8.5).

- Picking, replicating, and freezing ES cell colonies and analyzing DNA using 96-well plates and a multichannel pipettor, as devised by Allan Bradley, is possibly the most widely used procedure that potentially allows automation in such high-throughput functional analysis technologies as gene trapping. This method is described in detail in Protocols 10.2, 10.3, 10.4, and 10.5.

- The original method of picking, replicating, and freezing ES cell colonies was based on 24-well plates. The procedure is now only used when a relatively small number of colonies need to be processed because it is more labor-intensive. Cells from individual colonies can be frozen directly in 24-well plates, expanded to a larger surface, and frozen individually in cryovials. The preparation of DNA from ES cells grown in 24-well plates is described in Protocol 10.6; this method is often used to confirm the targeting event in potentially positive clones identified by a first screening in 96-well plates.

- An alternative rapid PCR (polymerase chain reaction) method for identifying targeted colonies, developed by Philippe Soriano, and described in Protocol

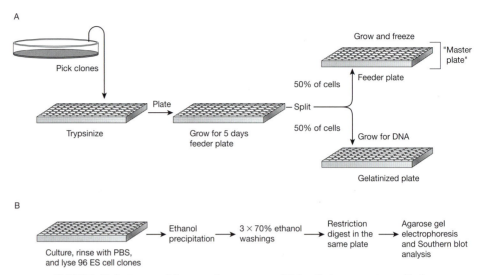

FIGURE 10.1. Protocol for rapid screening of ES cell clones in 96-well plates.

10.7, is based on DNA analysis of a small part of colonies pooled directly from the selection plates. It allows obtaining the PCR result before picking colonies and saves time because only positive colonies are expanded.

RAPID PREPARATION OF DNA FROM CELLS IN 96-WELL TISSUE CULTURE DISHES

It is extremely important to reduce the time during which ES cells are in culture between selection and injection into blastocysts. Picking, culturing, freezing, extracting, and analyzing DNA from individual colonies grown in 24-well tissue culture dishes and individual dishes can be very time-consuming. Allan Bradley and colleagues (Ramirez-Solis et al. 1992, 1993) have devised a rapid method using 96-well tissue culture dishes (Fig. 10.1).

Preparation of DNA from ES cells grown to confluence in gelatinized 96-well tissue culture dishes, and analysis of the DNA by "mini-Southern," are carried out using Protocol 10.4. With this method, the cells are lysed directly in each well and nucleic acids are precipitated in the dish, to which the DNA remains bound. The DNA is then cut by restriction enzymes while still remaining in the well of the tissue culture dish. This protocol has been tested for several restriction enzymes. Not all enzymes give complete DNA restriction; therefore, a pilot reaction with the enzyme of choice should be performed before a large screen is started. As a guide, Table 10.1 contains a list of enzymes evaluated during several years in our laboratories. When handling a large number of tissue culture dishes, be sure to label dish bottoms and lids to avoid confusion.

TABLE 10.1. Rating of restriction enzymes for ability to digest genomic ES cell DNA in 96-well plates

Good performance	*Apa*I , *Asp*718, *Avr*II, *Bam*HI, *Bgl*l, *Bgl*II, *Dra*III, *Drd*I, *Eco*RV, *Hind*III, *Hpd*I, *Nco*I, *Pst*I, *Pvu*II, *Sca*I, *Sst*I , *Stu*I, *Xba*I
Variable performance	*Eco*RI, *Kpn*I, *Sac*I
Bad performance	*Bsp*DI, *Hind*II, *Sac*I, *Sal*I, *Sma*I, *Xho*I, *Xmn*I

Electroporating DNA into ES Cells and Selection Methods

MATERIALS

CELLS AND DNA

ES cells (exponentially growing)

Linear DNA, 25 µg electroporation (circular DNA for transient expression of a transgene, e.g., Cre recombinase)

EQUIPMENT

Centrifuge

Electroporation apparatus (e.g., Bio-Rad Gene Pulser II apparatus 165-2105 or 165-2106, or MicroPulser 165-2100) equipped with a capacitance extender PLUS module (165-2108)

Electroporation cuvettes, 0.4-cm electrode gap sterile (e.g., Bio-Rad 165-2088)

Incubator, 37°C, 5% CO_2, 95% air

Microscope, inverted

Pasteur pipettes

Tissue culture dishes, 10-cm, containing either G418-resistant STO or MEF feeder cells

Tissue culture dishes, 10-cm, gelatinized

REAGENTS

Electroporation buffer (Specialty Media ES-003-D) (*optional*, but gives better recovery of electroporated ES cells when used instead of PBS)

ES cell culture medium with 15% fetal bovine serum (FBS) and supplements (ES-DMEM) (see Chapter 8 for details)

Gelatin, 0.1% solution (e.g., Sigma G2500) in tissue culture-grade water

G418 (e.g., Geneticin, Invitrogen Life Technologies, in liquid 10131, in powder 11811; Sigma G9516) to 200 µg/ml ES-DMEM final concentration of the active ingredient. Note that batches vary in purity.

G418 containing ES-DMEM (above) plus (final concentration 2×10^{-6} M; e.g., ganciclovir, Sigma G2566; or Cymevene, Syntex Pharmaceuticals BN 52304) (or 0.2 µM FIAU)(e.g., Moravek Biochemical MC251, http://www.moravek.com)

Phosphate-buffered saline (PBS), Ca^{++}/Mg^{++}-free (see Appendix 1)

Trypsin/EDTA (see p. 367)

PROCEDURE

1. Begin with an exponentially growing culture of early-passage ES cells. Change the medium ~3 hours before harvesting the cells.

2. Trypsinize the cells, add ES cell culture medium with serum (ES-DMEM) to inhibit the action of the trypsin, and pipette the cells up and down to produce a single-cell suspension.

3. Centrifuge the cells gently to pellet (200*g* for 5 minutes), wash once with PBS or electroporation buffer, and resuspend as a single-cell suspension at a density of ~1 x 10^7 cells/ml in Ca^{++}/Mg^{++}-free PBS or electroporation buffer.

4. For each electroporation, place 0.8 ml of the cell suspension in a 0.4-cm-wide sterile cuvette (e.g., Bio-Rad 165-2088). Add 25 µg of linear DNA (circular for transient expression) and mix well. Let the mixture stand at room temperature for 5 minutes (*optional*).

5. Apply a single pulse to the cells. Conditions for a typical single pulse are 230–250 V, 500 µF. Alternatively, 800 V, 3 µF has been used successfully. Conditions should be tested. Approximately 50% cell death is to be expected with optimal transfection efficiency when PBS is used.

6. After applying the pulse, allow the cells to stand at room temperature for 5 minutes. (Some laboratories incubate "on ice.") Be sure to rinse the cuvette to recover all of the cells.

7. Meanwhile, change the medium on 10-cm tissue culture dishes of either mitotically inactivated STO or *neo*^{r} MEF feeder cells, or if you select in feeder-free conditions, prepare gelatinized plates (see Comments, below). If using the positive-negative selection technique of Mansour et al. (1988), divide the 0.8-ml sample from the cuvette from the electroporation equally among four tissue culture dishes.

8. In all cases, allow the cells to recover for ~24 hours in nonselective medium, i.e., ES-DMEM without any added selection agents.

9. After 24 hours, refeed one tissue culture dish with nonselective medium. This dish serves as a control to assess the number of cells that survived electroporation.

10. Add ES cell culture medium (ES-DMEM) containing G418 (200 µg/ml final concentration of the active ingredient) to a second tissue culture dish to determine the fraction of cells transformed by integration of the *neo*^{r} gene.

11. Feed the remaining dishes with medium containing G418 plus ganciclovir (2 x 10^{-6} M) (or 0.2 µM FIAU; McMahon and Bradley 1990) to enrich for cells that contain a functional *neo*^{r} gene but have lost the HSV-*tk* gene as a result of targeted integration. There is no need to set up controls from the remaining cuvettes.

12. Incubate the cells in culture for ~8–10 days, changing the medium every day.

COMMENTS

- Different ES cell lines are not equally sensitive to G418, and different G418 batches are not equally potent. Therefore, it is advisable to test the toxicity of the G418 beforehand to find the lowest dose giving 100% killing of nontransfected cells within 5 days.

- Ideally, ~50–100 colonies should become visible in the double-selected plate after 10 days in culture. If significantly fewer colonies are obtained, then electroporation conditions or the construct's design can be varied in an attempt to increase the overall efficiency. Parameters that may be varied include:

  1. The use of a double pulse instead of a single pulse from the electroporation apparatus, or changing the voltage;

2. the amount of DNA (10–100 μg) added to the cuvette;

3. the promoter used to drive the *neo*^r gene;

4. the length of DNA in the targeting construct and the region of the gene chosen.

- If colonies are to be screened by Southern blotting of DNA, the number of colonies that can be easily screened in one experiment is ~300, and a final electroporation should be performed to generate this number, based on the results of the pilot experiments described above.

- The same protocol can be applied for targeting with *neo*-positive selection without *tk*-negative selection or with diphtheria toxin (dt)-based negative selection, which does not require a selective agent. Obviously, the selection against *tk* should be left out.

- To gelatinize the tissue culture plate, flood the plates with a 0.1% solution of gelatin for a few seconds. Remove the gelatin completely, and add ES cell culture medium supplemented with leukemia inhibitory factor (LIF) (see Chapter 8) to the plates. This treatment increases the adhesiveness of the ES cells to the dish.

- If the selection for vector integration is based on puromycin, the killing of sensitive, non-transgenic cells is faster, and usually takes about 3 days. It is advisable, however, to wait with picking (see Protocol 10.2) for an additional 4–5 days, until the colonies reach an acceptable size.

| **Isolating Individual ES Cell Colonies by Picking**

MATERIALS

CELLS

ES cells under the selection after DNA electroporation
neo-resistant STO or MEF feeder cells

EQUIPMENT

Incubator, 37°C, 5% CO_2, 95% air

Inverted phase-contrast microscope with a low-power objective lens, or a dissection stereomicroscope

Micropipettor, 20-μl automatic (e.g., Gilson), fitted with a sterile disposable tip

Micropipettor, 200-μl automatic multichannel (8 or 12 channel for 96-well plates) (e.g., Costar 4888, 4880) fitted with sterile disposable tips

Multichannel aspirator system (*optional*) (e.g., Inotech Biosystems Vacuset IV-520)

Tissue culture dish (uncoated), 96-well U- or V-bottomed, containing 30-μl microdrops of trypsin/EDTA solution or similar solution (see p. 439) in every other row

Tissue culture dishes, 96-well, flat-bottomed

Each well should contain either 1×10^4 STO feeder cells or $1.5–2.0 \times 10^4$ MEF feeder cells in 0.1 ml of ES cell culture medium (ES-DMEM) containing G418. If the selection was performed without feeder cells, prepare gelatinized 96-well tissue culture dishes instead.

REAGENTS

ES cell culture medium (ES-DMEM) supplemented with the effective concentration of G418

ES cell culture medium (ES-DMEM)

Phosphate-buffered saline (PBS), Ca^{++}, Mg^{++}-containing (see Appendix 1)

Trypsin/EDTA (p. 367)

PROCEDURE

1. Prepare sets of 96-well tissue culture dishes containing *neo*-resistant STO or MEF feeder cells or gelatinized 96-well tissue culture dishes with LIF-supplemented ES cell culture medium. Also prepare a 96-well U- or V-bottomed tissue culture dish (uncoated) with 30 μl of 0.25% trypsin/0.04% EDTA buffer in every other row.

2. Remove the culture medium from the dish containing the ES cells and rinse the ES cells once with Ca^{++}/Mg^{++}-containing PBS. Add just enough PBS to cover the surface of the tissue culture dish with a thin film of Ca^{++}/Mg^{++}-containing PBS buffer (~2 ml).

3. This step can be carried out either inside a tissue culture hood or on the open bench, depending on local requirements. Using an inverted microscope or stereomicroscope and an automatic pipettor, draw up ~8–10 µl of trypsin/EDTA buffer into a sterile disposable pipette tip. Hold the tip of the pipette very close to a single colony of ES cells and dispense a small amount of the trypsin/EDTA buffer over the colony. After a short time, draw the colony into the pipette tip and transfer it into the next well of the trypsin/EDTA-containing 96-well plate.

4. Repeat this step until the six rows (48 colonies) have been filled up. Place the plate at 37°C for 5 minutes.

5. With the multichannel micropipettor set to 40 µl, pipette the cells gently up and down to disperse the colony, starting with the first row. Seed the cells into a corresponding row of a flat-bottomed 96-well tissue culture dish.

6. After all 48 colonies have been harvested in this way, collect 48 more colonies to fill up the 96-well tissue culture plate with cells.

   It is important not to let the dish dry out and not to leave it without medium for extended periods of time. Thus, if many colonies are to be picked from one dish, it is best to pick them in several sessions, replacing the PBS with complete medium between sessions so that the remaining colonies can recover.

7. After harvesting the required number of colonies, incubate the tissue culture dishes for ~3–4 days, changing the medium on the day after picking and on days 3 and 4. By day 4 after picking, almost-confluent cultures should be present in most of the wells.

8. Expand each colony so that samples can be taken for DNA analysis. Store the remainder of the cells frozen while the DNA is screened. It is generally recommended to have two replica plates for DNA preparation and one or two replica plates as a frozen stock.

9. Rinse each well with PBS, trypsinize the cells, and split them into corresponding wells of two new dishes. Label each daughter plate clearly for later identification. When these cells are almost confluent, freeze one replica plate (see Protocol 10.3) and passage the other one again. Alternatively, passage both plates at the same time, and, when the final four plates reach confluence, freeze two plates and prepare DNA from the remaining two.

COMMENTS

- When replacing the ES cell medium in the selection plate before picking, use a Ca^{++}/Mg^{++}-containing PBS. If Ca^{++} and Mg^{++} are omitted, the cell–cell connections loosen much faster, resulting in an involuntary detachment of the colonies.

- Picking from feeders is slightly more difficult than from the gelatinized plate, because the ES cell clumps are strongly attached to the feeder layer. Picking could lift a sheet of feeder cells, which makes the procedure troublesome. Therefore, cutting the feeders around the ES cell colony using the pipette tip is recommended before lifting the clump.

- When only every other row of the trypsin-containing plate is used (48 wells), it is easier to keep track of the position of the next empty well during picking.

- Concerning the G418 selection, it is important to maintain it for 3–4 days after picking. It happens, especially when "low-end" G418 concentrations are used, that some nonresistant cells survive the regular phase of selection if they expressed *neo* transiently or were cross-fed by nearby G418 feeder cells. These false-positive colonies die in G418-containing medium shortly after picking. This extension of G418 selection also helps to kill *neo*-sensitive cells that often survive in mixed colonies, in which the resistant cells could cross-feed the sensitive cells.

- The advantage of picking into 96-wells is the high speed of the procedure and the large number of colonies that can be managed at a time. The drawback is that the amount of DNA obtained per well is small, usually only enough for one Southern blot analysis. In addition, not all of the restriction enzymes work properly with the crude DNA prepared from the plate (Table 10.1). These limitations have to be taken into consideration at the targeting or transgenic vector design stage (see Chapter 9).

- To avoid the above-mentioned problems, some laboratories grow the cells in 24-well plates. The protocol for this procedure is practically the same as for 96-well plate picking, except that the transfer of the cells after the trypsinization step is done into feeder-prepared or gelatinized 24-well plates.

- Because the presence of feeder cells may interfere with reliable genotyping of the ES cells, it is important to pass the cells on feeder-free, gelatinized plates for a couple of passages before DNA preparation.

Freezing ES Cells In 96-Well Plates

MATERIALS

CELLS
ES cells, in 96-well dish containing ES cells confluent in most of the wells

EQUIPMENT
Multichannel micropipettor and aspirator system
Styrofoam box, large enough to accommodate the 96-well plates but small enough to fit into a –70°C freezer
Tissue culture dish, 96-well

REAGENTS
ES cell culture medium with 15% FBS and supplements (ES-DMEM)
2x Freezing medium
    60% ES-DMEM (containing 15% FBS)
    20% fetal bovine serum
    20% dimethyl sulfoxide<!>
    Prepare fresh and cool on ice.
Mineral oil, cold, sterile (e.g., Sigma M8410, or Specialty Media ES-005-C)
Trypsin/EDTA solution (0.25% trypsin/0.04% EDTA)

CAUTION: See Appendix 2 for appropriate handling of materials marked with <!>.

PROCEDURE

1. Rinse the cells with PBS and trypsinize the cells of an 80% confluent 96-well plate by adding 25 µl of trypsin/EDTA per well. Stop the action of the trypsin with 70 µl of ES cell culture medium. The final volume of the cell suspension should be 100 µl per well.

2. Working quickly on ice, aliquot 100 µl of 2x freezing medium into each well. Pipette the cells up and down several times to get a homogeneous suspension. Alternatively, transfer the cell suspension into the new V-bottomed 96-well plate already containing 100 µl of 2x freezing medium and keep on ice.

3. Add 50 µl of cold sterile mineral oil on top of each well.

4. Wrap the plates in Parafilm. Place the plates in a precooled Styrofoam box, and store in a –70°C freezer, preferably not longer than 2–3 months, until ready for thawing and expansion. It is also possible to use foil to wrap the plates before placing them in a –70°C freezer.

COMMENTS

- It is important to keep in mind that the 96-well plate-based frozen storage at –70°C has a shelf life of 2–3 months. Analysis of the clones (e.g., genotyping) must be accomplished during this period, and then the selected clones must be expanded and stored properly in cryovials in liquid nitrogen for further studies.

- Some emerging technologies, which require the storage of large numbers of ES sublines (libraries), might benefit from a longer 96-well-based frozen storage. For such applications, Continental Lab Products manufactures special 96-well plates (2600 mini), with breakoff capbands (VWR/Falcon 352117), allowing the removal of individual wells. These plates can be conveniently stored long term in –150°C freezers (Sanyo SY1155TN) with 96-well plate racks (Sanyo CDP12LR).

Thawing ES Cells from a 96-Well Plate

After the candidate clones are identified by genomic Southern or PCR, the method of rescue of the cells from the frozen 96-well plates is very important. As we noted before, cell storage at –70°C should not be longer than 2–3 months.

MATERIALS

CELLS
ES cells, identified clones

EQUIPMENT
Cryovials
Feeder plates, 4- or 24-well, containing ES cell culture medium with serum and supplements (see pp. 436–437)
Incubator, 37°C, 5% CO_2, 95% air

REAGENTS
ES cell culture medium
Ethanol, 70%<!>

CAUTION: See Appendix 2 for appropriate handling of materials marked with <!>.

PROCEDURE

1. Remove the plate containing identified clones from the freezer. Unwrap the plate and warm it quickly (place in the incubator).

2. When ice crystals have almost disappeared, wipe the outside of the plate with 70% ethanol.

3. Transfer the contents of the wells into the wells of newly prepared 4- or 24-well plates with feeders.

4. Rinse the original wells of the 96-well plate with more ES cell culture medium and transfer to the same wells. Change the medium after overnight culture and daily thereafter.

5. Passage the cells to a larger plate (e.g., 35 mm or directly to 60 mm) when they reach 70–80% confluency, leaving some cells behind to grow for the secondary DNA analysis. If the cells do not reach confluency in a few days, but form few large colonies in a well, they can be trypsinized, broken into smaller cell clumps, and plated back on the same plate.

6. Passage 1:5 every other day (see Protocol 8.3). Freeze the cells in cryovials (see Protocol 8.4).

7. Prepare DNA from ES cells confluent in 4- or 24-wells as described in Protocol 10.6 to confirm that the thawed clone is indeed targeted.

8. Once targeting is confirmed, ES cell clones can be thawed for chromosome counting (Protocol 8.6), and the best ones can be used for making chimeric animals (see Chapter 11).

Rapid Preparation of DNA from Cells in 96-Well Tissue Culture Dishes

This protocol was originally established by Ramirez-Solis et al. (1992).

MATERIALS

CELLS

Confluent ES cells in 96-well plate

EQUIPMENT

Gel electrophoresis and DNA transfer equipment

A 6-inch x 10-inch 1% agarose gel with three 33-tooth combs spaced 3.3 inches apart gives enough space for 96 samples plus 1 molecular-weight marker lane for every comb.

Incubator, humidifed, 37°C, 5% CO_2, 95% air

Multichannel pipettor (Fisher 21-233)

Paper towels

Plastic container (Tupperware-type)

Tissue culture dish, gelatinized 96-well, containing confluent ES cells

REAGENTS

Ethanol, 70% <!>

Gel electrophoresis loading buffer

Lysis buffer

10 mM Tris<!> (pH 7.5)

10 mM EDTA

10 mM NaCl

0.5% (w/v) sarcosyl

Add 1 mg/ml Proteinase K just before use.

NaCl, 5 M, and absolute ethanol <!> mixture

Add 150 µl of 5 M NaCl to 10 ml of cold absolute ethanol. (Prepare fresh.)

Phosphate-buffered saline (PBS) (see Appendix 1)

Restriction enzyme digestion mixture

1x restriction buffer

1 mM spermidine <!>

100 µg/ml bovine serum albumin (BSA)

50–100 µg/ml RNase

10 units of each restriction enzyme/sample

TAE, 1x

CAUTION: See Appendix 2 for appropriate handling of materials marked with <!>.

PROCEDURE

1. Allow the ES cells to grow in the gelatinized wells until they turn the culture medium orange or yellow in 1 day (4–5 days after plating cells).

2. Once the cells are ready for DNA extraction, rinse the wells twice with PBS and add 50 µl of lysis buffer to each well. Add 1 mg/ml Proteinase K to the lysis buffer just prior to use.

3. Incubate the dishes in a humid atmosphere at 60°C overnight (e.g., by placing the plates in a closed plastic Tupperware-type container with wet paper towels in a conventional 60°C oven).

4. The next day, carefully add 100 µl of a mixture of NaCl and ethanol (150 µl of 5 M NaCl to 10 ml of cold absolute ethanol) to each well using a multichannel pipettor.

5. Allow the 96-well dish to stand at room temperature on the bench for 30 minutes without mixing. The nucleic acids precipitate as a filamentous network.

   It is useful to check the precipitate between each of the subsequent washes.

6. Invert the dish carefully to discard the NaCl/ethanol solution; the nucleic acids will remain attached to the dish. Blot the excess liquid on paper towels.

7. Wash the nucleic acids in the dish three times by gently adding 150 µl of 70% ethanol per well with the multichannel pipettor. Discard the alcohol by inverting the dish after each wash. At this point, DNA can be stored in 70% ethanol at –20°C.

8. After the final 70% ethanol rinse, invert the dish and allow it to dry on the bench for 10–15 minutes. The DNA is ready to be cut with restriction enzymes.

9. Prepare the restriction digestion mixture.

10. Add 30 µl of restriction digestion mixture to each well with a multichannel pipettor. Mix the contents of the well using the pipette tip and incubate the reaction in a humidified incubator with 5% CO_2 at 37°C overnight.

11. Add gel electrophoresis loading buffer to the samples, perform conventional electrophoresis, and DNA transfer to blotting membranes.

    Gel electrophoresis in 1x TAE at 80 V for 4–5 hours gives a good separation in the 1- to 10-kb range.

Preparation of DNA from Cells in 24-Well Tissue Culture Plates

This protocol is modified, with permission, from Wurst and Joyner (1993).

MATERIALS

CELLS
Confluent ES cell cultures in 24- or 4-well tissue culture plates

EQUIPMENT
Microfuge tubes, 1.5-ml
Micropipettor, 200–1000 µl automatic, with a disposable tip or glass rid
Rocking platform or orbital shaker

REAGENTS
Ethanol, 70%<!>
Isopropanol<!>
Lysis buffer:
    100 mM Tris-HCl<!> (pH 8.5)
    5 mM EDTA
    0.2% SDS<!>
    200 mM NaCl
    100 µg/ml Proteinase K (20 mg/ml Proteinase K stock is stored at –20° and added
      to the lysis buffer prior to use).
Phosphate-buffered saline (PBS) (see Appendix 1)
TE Buffer (10 mM Tris-HCl [pH 8.0], 1 mM EDTA)
Optional
100% Ethanol<!> (kept at –20°C)
Phenol<!>:chloroform<!>:isoamyl alcohol<!> (25:24:1)
Sodium acetate (3 M, pH 5.2)

CAUTION: See Appendix 2 for appropriate handling of materials marked with <!>.

PROCEDURE

1. Aspirate the medium from each well containing confluent ES cells, rinse the wells once with cold PBS, add 200–500 µl of lysis buffer per well, and incubate for 4–16 hours at 55°C in humid conditions.

   Optional: Harvest cells by trypsinization (see Protocol 8.3), transfer cell suspension to microfuge tubes, wash with cold PBS, add 200–500 µl of lysis buffer per sample, and incubate for 4–16 hours at 55°C. It is also possible to collect cells by adding the lysis buffer directly to the cells. After the incubation, add equal volumes of phenol/chloroform/isoamyl alcohol to the tube and vortex for 1 minute. Centrifuge at 14,000 rpm for 5 minutes. Carefully transfer the top layer (water phase) to the new tube and add 1/10 volume of 3 M sodium acetate and 3 volumes of 100% ethanol (–20°C). Centrifuge at 14,000 rpm for 10 minutes, remove supernatant, and continue with step 4.

2. After the incubation, shake the plate on a shaker for 15 minutes, add an equal volume of isopropanol per well, and shake until DNA precipitate becomes visible (15–30 minutes).

3. Remove the DNA using the disposable tip of a micropipettor or a glass rod and place the DNA into eppendorf tubes. Spin briefly to pellet.

4. Wash the DNA pellet two times with 70% ethanol and dry for 10 minutes at room temperature. Resuspend the DNA pellet in 50–100 μl of TE buffer, and incubate for 3–4 hours at 55°C to dissolve. Store at 4°C.

Genotyping ES Cell Colonies Prior to Picking

This PCR-based method has been developed by Philippe Soriano http://www.fhcrc.org/labs/soriano/protocols/pcres.html for identifying targeted colonies prior to picking.

MATERIALS

CELLS
ES colonies in G418 selection

EQUIPMENT
Capillary tubes, finely drawn
Eppendorf tube, 0.6-ml
Incubator, 5% CO_2 at 37°C
PCR equipment

REAGENTS
Light mineral oil (Sigma M3516)
Lysis buffer
    1x PCR buffer
    1.7 μM SDS<!>
    50 μg/ml Proteinase K
Oligonucleotides
    These are usually 20-mers, with a GC content of ~50% and a melting temperature
       between 55°C and 60°C.
1x PCR buffer (Gittschier buffer)
    16.6 mM ammonium sulfate<!>
    67 mM Tris (pH 8.8)<!>
    6.7 mM $MgCl_2$<!>
    5 mM β-mercaptoethanol<!>
    6.7 μM EDTA
PCR components
    1x PCR buffer (10x stock)
    1 mM each dNTP (Pharmacia; stock of the mix 10 mM)
    10% dimethyl sulfoxide (DMSO)<!> (Aldrich)
    80 μg/ml bovine serum albumin (BSA) (nuclease-free; Boehringer Mannheim BM
       711 454; stock 1.6 mg/ml)
    0.1 μM each oligonucleotide primer
    5 μl of DNA in PCR lysis buffer (do not forget positive and negative controls)
    2.5 units (0.25 μl) of AmpliTaq polymerase (Perkin Elmer)

CAUTION: See Appendix 2 for appropriate handling of materials marked with <!>.

PROCEDURE

1. Pick the cells as follows:

   a. Pick part of the colonies at ~days 8–10 of G418 selection (the day when G418 is first applied is counted as day 0).

   b. Aspirate the medium, invert the dish, and mark the colonies with a colored marker under a stereomicroscope. Care should be taken to identify only undifferentiated colonies, and different-colored markers can be used to identify clones in different pools.

   c. Cover the cells on the dish with 10 ml of PBS.

   d. Using a finely drawn capillary tube, pick approximately one-fifth of each colony. Repeat this process with the same capillary for the remaining colonies in the pool (usually ~25 colonies). Even though the same capillary tube is used throughout this process, cross-contamination between colonies is not a problem.

   e. Transfer the picked colonies to a 0.6-ml eppendorf tube.

   f. Centrifuge the cells for 30 seconds to pellet them and resuspend the pellet in 12 µl of PCR lysis buffer.

Once a PCR-positive pool has been identified, individual colonies can be picked as described above. At this point, use a different capillary tube when picking each colony. If care is taken, it is also possible to pick just a little from each colony and transfer cells directly to a tube containing 12 µl of PCR lysis buffer with minimal PBS carryover. This way, it is not necessary to pellet and resuspend the cells.

2. Lyse the cells in PCR lysis buffer at 37°C for 1 hour and then heat-inactivate the Proteinase K for 10 minutes at 85°C. Pellet the cells.

3. PCR amplifications are usually performed in a 25-µl final volume. Make a stock mixture with everything except DNA and dispense 20 µl of this mixture per tube before adding DNA. Once DNA is added, overlay with light mineral oil. Reactions include the PCR components listed in the Materials (above).

4. Optimize PCR conditions using a mock construct diluted with genomic DNA. Perform test PCRs in a 25-µl final volume, using 100 ng of genomic DNA (e.g., DNA extracted from the tail of a wild-type mouse), and dilutions of mock plasmid. Aim for a UV-visible product after 40 cycles using 1 fg of plasmid DNA, and for an extension of 2–3 kb. PCR extension times are based on 700 bp/minute. Normal PCR conditions are 93°C for 2 minutes, then 40 cycles at 93°C for 30 seconds, 55°C for 30 seconds, and 65°C (for 2-kb extension) for 3 minutes. Note the lower temperature of extension due to the presence of DMSO.

REFERENCES

Gossler A., Doetschman T., Korn R., Serfling E., and Kemler R. 1986. Transgenesis by means of blastocyst-derived embryonic stem cell lines. *Proc. Natl. Acad. Sci.* **83:** 9065–9069.

Mansour S.L., Thomas K.R., and Capecchi M.R. 1988. Disruption of the proto-oncogene int-2 in mouse embryo-derived stem cells: A general strategy for targeting mutations to non-selectable genes. *Nature* **336:** 348–352.

McMahon A.P. and Bradley A. 1990. The Wnt-1 (int-1) proto-oncogene is required for development of a large region of the mouse brain. *Cell* **62:** 1073–1085.

Miller S.A., Dykes D.D., and Polesky H.F. 1988. A simple salting out procedure for extracting DNA from human nucleated cells. *Nucleic Acids Res.* **16:** 1215.

Ramirez-Solis R., Davis A.C., and Bradley A. 1993. Gene targeting in embryonic stem cells. *Methods Enzymol.* **225:** 855–878.

Ramirez-Solis R., Rivera-Perez J., Wallace J.D., Wims M., Zheng H., Bradley A. 1992. Genomic DNA microextraction: A method to screen numerous samples. *Anal. Biochem.* **2:** 331–335.

Wurst W. and Joyner A.L. 1993. Production of targeted embryonic stem cell clones. In *Gene targeting: A practical approach,* 1st Edition (ed. A.L. Joyner). pp. 33–61. IRL Press at Oxford University Press, New York.

Production of Chimeras

THE CONCEPT OF CHIMERAS HAS BEEN a part of human culture since the prehistoric ages. The Chimera of Arezzo (see inset), which is said to date from the Etruscan period (5th century B.C.), demonstrates the basic principle of a creature composed of parts of several individuals. A more scientific definition is that a chimera is an individual whose cells originate from more than one embryo. Chimerism is a naturally occurring phenomenon, the best example of which is twin calves in which a prenatally joined circulation creates chimeric hematopoiesis. Human bone marrow and organ transplant recipients are also chimeras according to this definition. In this chapter, we describe the two basic

methodologies for creating chimeras, primarily for obtaining germ-line transmission from ES cells. We also discuss the important features of the various types of chimeras. In addition, we point out the typical applications of these unique animals.

CONTENTS

GENERAL CONSIDERATIONS

The first experimental mouse chimeras were created more than 40 years ago by Andrzej Tarkowski (Tarkowski 1961), who removed the zona pellucida from two genetically distinct eight-cell-stage embryos and placed them in physical contact with each other during in vitro culture. These aggregates resulted in viable live chimeras after transfer to surrogate mothers. Richard Gardner (Gardner 1968) subsequently showed that chimeras could also be created by injecting blastocyst cells directly into the cavity of blastocysts. From the 1960s to the 1980s, many important biological questions about embryonic and organ development were addressed using chimeras as a research tool. With the advent of embryonic stem (ES) cells, chimeras gained additional importance. In mice, they have become the vehicles for transmitting the ES cell genome in vivo. Two predominant methods have been developed for introducing ES cells into preimplantation-stage embryos: the so-called injection chimeras (Gardner 1968) and aggregation chimeras (Wagner et al. 1985; Nagy et al. 1990; Wood et al. 1993). Both techniques result in efficient germ-line transmission. The developmental potential (and restriction) of ES cells has become more and more evident through the study of their somatic contribution in ES cell–embryo chimeras (Beddington and Robertson 1989).

An even deeper understanding was gained when ES cells were aggregated with developmentally compromised tetraploid mouse embryos (Nagy et al. 1990). These experiments clearly showed that ES cells retain the full potential of the primitive ectoderm lineage and that ES cells can give rise to all the lineages of the embryo proper, and some tissues of the extraembryonic membranes, including the amnion and mesoderm layer of the yolk sac. Interestingly, they retain the developmental restrictions of their presumptive origin as well; in vivo they are unable to contribute to the primitive endoderm and the trophectoderm lineages, and therefore they do not contribute to the extraembryonic endoderm and the trophoblast of the placenta (Nagy et al. 1990).

Recently, conditions have been identified for the derivation of trophoblast stem (TS) cells (Tanaka et al. 1998), which are permanent stem cell lines derived from the trophectoderm of the blastocyst. As with ES cells, TS cells also retain the developmental potential and restriction of their origin. These advances have made it possible to use diploid embryos, tetraploid embryos, ES cells, and TS cells in different combinations to create chimeras. These components all represent unique developmental potential and restriction, which determine the allocations of their derivatives in the resulting chimeras. Thus, the newest generations of chimeras have become very powerful tools for analyzing complex phenotypes (Nagy and Rossant 2001).

The injection of embryonic cells directly into the cavity of blastocysts is one of the fundamental methods for generating chimeras. ES cells can also be injected into blastocysts, and this is probably the most common method for introducing genetic alterations performed in ES cells into the mouse by producing germ-line-transmitting chimeras (Bradley et al. 1984). Chimeras can also be generated by aggregation of embryonic cells with morula-stage embryos. Although ES cells are typically established from the blastocyst stage, they are still capable of integrating one day earlier into the eight-cell-stage embryo. By taking advantage of this property, a relatively simple way of introducing ES cells back into the embryonic environment has been developed (Nagy and Rossant 1999). Thus, ES cells can also be aggregated with morula-stage embryos to generate chimeras. This procedure is technically less demanding and less expensive than the blastocyst injection method. Consequently, aggregation chimeras have become an important alternative to ES cell injection chimeras for transmitting the ES cell genome into mice (Wood et al. 1993).

These two procedures have several similarities, but they also have major differences, advantages, and disadvantages. The injection technique requires expensive equipment (an inverted microscope with phase-contrast optics, micromanipulators, etc.), whereas the aggregation method can be performed simply using a good stereo dissection microscope. Injection is usually a slower process, not allowing more than 50–100 blastocysts to be injected during 1 day, whereas 200 or more embryos can easily be aggregated in the same amount of time. On the other hand, aggregated embryos must be kept in culture for a much longer period of time than is the case for injected blastocysts. This is an important factor, because the slightest suboptimal culture conditions may have a major influence on the developmental potential of the chimeric embryos. When injection is used, individual ES cells are carefully selected and screened by morphological criteria. This selection cannot be performed with the aggregation technique because clumps of ES cells are selected under lower magnification. In cases where ES cell quality is high, both techniques generally work equally efficiently in terms of producing highly chimeric embryos for ES cell lines known to work with the aggregation technique. However, the careful selection of ES cells during the injection technique may in some cases contribute to a higher efficiency when slightly lesser quality ES cell clones are used. Finally, during the aggregation method, the zona pellucida is removed from the embryos, which makes them more sensitive to in vitro conditions. The resulting cultured embryos are fragile and sticky, and may require some experience in their handling and transfer. Regardless of the techniques used, the resulting chimeric embryos are placed into the reproductive tract of pseudopregnant female mice. If high chimerism has been achieved in the embryos, the likelihood is great that the germ cells will also be at least partially ES-cell-derived. By breeding such chimeras to wild-type mice, it is possible to establish stable germ-line transmission of the transgene or targeted allele.

The genetic background of the ES cells should differ from the host embryo in two important ways: (1) The coat-color markers of the host blastocyst should differ from those of the ES cell line to make coat-color chimerism distinguishable and (2) the host embryo's background should be matched to the ES cells in such a way that provides the ES cells a high probability of contributing to the germ line of the resulting chimera. Traditionally, the most widely used ES cell lines (Simpson et al. 1997) are derived from one of the 129 substrains (or an F_1 cross between two of the 129 substrains) (see Table 11.1). For these ES cell lines, C57BL/6 host blastocysts have proven to be a very suitable choice (Table 11.2). The coat color of chimeras is easily distinguishable, and the ES cells have an optimal developmental advantage, which makes the creation of highly chimeric animals very efficient. When ES cell lines derived from other strains are used, the optimal host strain has to be determined in each particular case. For example, BALB/c blastocysts have been shown to work well in combination with C57BL/6-derived ES cells (Lemckert et al. 1997). It is important to note that the genetic background combination that works well in the case of ES cell injection may not necessarily be the same when the ES cells are aggregated with morula-stage embryos (see Table 11.2). More recently, a superior developmental potency of F_1 hybrid ES lines has been demonstrated (Eggan et al. 2001). These cells allow highly efficient survival of completely ES cell-derived animals made by ES cell injection into tetraploid embryos. The use of these cells for genetic manipulations holds promise for higher efficiency of obtaining germ-line transmission, even after extensive genetic manipulations in vitro. However, using F_1 hybrid ES cells eliminates the possibility of immediately obtaining the genetic alteration(s) on an inbred background. A series of back-crossings with an inbred strain is required to generate a uniform genetic background among the animals.

The number of cells injected into the blastocyst is generally recommended to be around 15. This number corresponds fairly well to the number of primitive ectoderm cells present in the inner cell mass (ICM) of the host blastocyst at the

TABLE 11.1. Origin, JAX stock numbers, and phenotypes of some commonly used ES cell lines of 129 origin

129 Strains	Stock number (TJL)	ES Cell line	Appearance
129P2/OlaHsd	n.a.	E14TG2a	pink-eyed, light-bellied chinchilla
		HM-1 ($Hprt^{b-m1}$)	
129P3/JEms	000690	mEMS32	pink-eyed, light-bellied chinchilla
129X1/SvJ	000691	RW-4	pink-eyed, light-bellied, light chinchilla
		PJ1-5	or albino
129X1 x 129S1	000691 (X1)	R1 ($+^{Kitl-Slj}$)	light-bellied agouti
	000090 (S1)		
129S1/Sv-+^{P} +$^{Tyr-c}$	000090	W9.5 ($+^{Kitl-Slj}$)	light-bellied agouti
$Kitl^{Sl-J/+}$		CJ7 ($+^{Kitl-Slj}$)	
The origin of			
129S1/SvImJ ($+^{Kitl-Slj}$)	002448		
129S2/SvPas	n.a.	D3	light-bellied agouti
129S4/SvJae	n.a.	J1	light-bellied agouti
129S4/SvJaeSor	n.a.	AK7	light-bellied agouti
129S6/SvEv	n.a.	EK.CCE	light-bellied agouti
		CP-1	
129S6/SvEvTac	n.a.	TC1	light-bellied agouti
		W4	
129S7/SvEvBrd-$Hprt^{b-m2}$	n.a.	AB1 ($+^{Hprt-bm2}$)	light-bellied agouti
		AB2.1 ($Hprt^{b-m2}$)	

Adapted from Simpson (1997) and courtesy of Carol Cutler Linder, The Jackson Lab. n.a. indicates not applicable.

time of injection. For the aggregation method, a loosely connected clump of 8–15 ES cells is usually aggregated with a morula-stage embryo. However, the number of ES cells combined with the host embryo should not be taken as an absolute rule because the optimal ratio varies between different genetic backgrounds of both host embryo and ES cell line. For this reason, the optimal number should be determined experimentally. If too many cells are introduced, the result will often be reduced embryo viability. On the other hand, with too few cells, the degree of chimerism may be reduced.

The majority of ES cell lines commonly used have an XY (male) genotype. Half of the host blastocysts used for injection or aggregation are male (XY), whereas the other half are female (XX). When XY ES cells are introduced into XY embryos, the resulting embryo will develop as a male. If, however, XY ES cells are introduced into XX embryos, the resulting embryo will contain both XY and XX cells. If the XY (ES cell) contribution to the developing gonad is high, the chimera will typically develop as a male. However, if the XY contribution to the gonad is too low, the chimera will develop either as a female or a hermaphrodite

TABLE 11.2. The favored genetic background combinations result in a high probability of germ-line transmission

Method of introducing ES cells	Genetic background of ES cells		
	129	C57BL/6	BALB/c
Injection	C57BL/6 (Wood et al. 1993)	BALB/c (Lemckert et al. 1997) or C57BL/6-Tyr-c (Schuster-Gossler et al. 2001)	C57BL/6 (Dinkel et al. 1999)
Aggregation	CD1, ICR (Wood et al. 1993)[a]	no report	no report

[a]Some laboratories are successfully using C57BL/6 or C57BL/6 x DBA, C57BL/6 x CBA F_1 hybrid morulae as host embryos for aggregations.

(possessing both male and female reproductive organs). Hermaphrodites may occasionally be identified at weaning age as having abnormal external genitalia. However, more often these animals are recognized as sterile after unsuccessful breeding. In cases of complete sex conversion of the XX host embryo by the XY ES cells, all of the germ cells of the resulting male chimera will be ES-cell-derived because XX germ cells cannot mature into spermatozoa. These chimeras will produce ES-derived offspring exclusively. However, if the host embryo was XY, some of the germ cells will be ES-cell-derived and others will be derived from the host embryo. Such a chimera will produce both ES- and host-derived offspring. Interestingly, female chimeras with a high ES cell contribution, as judged by coat color, can be obtained from XY ES cell lines. Such females may be capable of germ-line transmission when bred with wild-type males. It is believed that in these cases the Y chromosome has been lost from a portion of the XY ES cells (XO) used to generate the chimera. Generally, female chimeras with a low proportion of ES-cell-derived coat color are not contributing to the germ line and are usually discarded.

TYPES OF CHIMERAS

Diploid Embryo–Diploid Embryo Aggregation Chimeras

Here, typically two eight-cell- to morula-stage embryos are aggregated (see Protocol 11.10A). Both embryos have the potential to contribute to all of the lineages of the entire conceptus, including all extraembryonic membranes (Table 11.3). These types of chimeras were extensively studied for many years after the first reports of their production (Tarkowski 1961; Mintz 1962). In the current era of widespread induced mutations in the mouse, these chimeras have gained renewed importance for analyzing mutant phenotypes. Usually, eight-cell- to morula-stage embryos from a heterozygous mutant intercross are aggregated with wild-type embryos to analyze the behavior of mutant cells in the presence of wild-type cells. This is a very informative approach for addressing questions about the cell-autonomous versus non-cell-autonomous nature of a mutant phenotype and the abnormal behavior of mutant cells (Rossant and Spence 1998). A chimera also creates a situation where wild-type cells can rescue early defects of mutant embryos, allowing the embryo to develop further and so reveal other mutant phenotypes (Tanaka et al. 1997). Many times, a lack or biased contribution of mutant cells to certain lineages or tissues is observed in chimeras. This should be interpreted with caution. Cell lineage compartments of the developing embryo are also competitive fields between the two chimera components. A lack

TABLE 11.3. Contribution of the components to different embryonic lineages in different types of chimeras

Chimera components		Lineage composition			
A	B	embryo proper	yolk sac mesoderm	yolk sac endoderm	placenta's trophoblast
Diploid embryo	diploid embryo	AB	AB	AB	AB
Tetraploid embryo	diploid embryo	B	B	AB	AB
ES cells	diploid embryo	AB	AB	B	B
ES cells	tetraploid embryo	A	A	B	B

of contribution of cells of one genotype does not necessarily indicate an inability for differentiation. This situation can also suggest that cells of one genotype are out-competed by the other (Lindahl et al. 1998).

The genotyping of these chimeras, when segregating populations (i.e., −/−, +/−, and +/+) are used as one of the components, also requires careful consideration. Because there are cells from the wild-type component "contaminating" the chimeric tissue samples, it can be difficult to distinguish the heterozygous–wild-type and the homozygous mutant–wild-type chimeras. One way of circumventing this problem is to use two distinguishable mutant alleles to create transheterozygous mutant embryos (Fig. 11.1) (Rivera-Perez et al. 1995). The presence of both mutant alleles and the wild-type allele in a chimera indicates that it is mutant–wild type. Because gene targeting with removable selectable markers is becoming a common practice (see Chapter 9), the production of two distinguishable null alleles in a gene does not necessarily mean extra work. The selectable marker-in and marker-out alleles of the targeted gene provide the necessary reagents for easy genotyping for a diploid embryo–diploid embryo experiment.

ES Cell–Diploid Embryo Aggregation and Injection Chimeras

In these chimeras (Table 11.3), the diploid embryo component contributes to all parts of the conceptus. The ES cell component, however, will contribute to the embryo proper, amnion, yolk sac mesoderm, allantois, and allantoic component of the placenta, but not to the trophoblast or yolk sac endoderm. This important feature of ES cell–diploid embryo chimeras stems from the in vivo developmental limitation of ES cells. They will not contribute to the trophoblast of the placenta and extraembryonic endoderm, including the yolk sac endoderm; therefore, these components are derived exclusively from the host embryo. This property makes these chimeras a useful tool for separating certain extraembryonic phenotypes from phenotypes in the embryo proper. (See Protocol 11.10B for assembling aggregates between ES cells and diploid embryos, and Protocols 11.1–11.5 for blastocyst injection of ES cells.)

Thus, the mutant ES-cell-derived component of the chimeric embryo proper can be studied when the trophoblast and extraembryonic endoderm are nonchimeric wild type (Ciruna et al. 1997; Duncan et al. 1997). The reverse combination, in which wild-type ES cells are injected into mutant embryos, can also be very informative. In this scenario, the ES cell derivatives may rescue some or all of the components of the embryo proper phenotype, but the primary extraem-

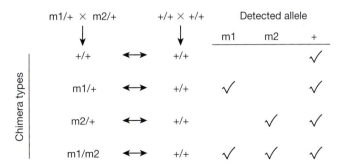

FIGURE 11.1. Genotyping of segregating components of the chimeras by using distinguishable mutant (m1, m2) alleles.

bryonic mutant genotype remains intact (Varlet et al. 1997). The genotyping of chimeras generated by the injection of wild-type ES cells into mutant blastocysts can also benefit from the use of two distinguishable mutant alleles. In ES cell–diploid embryo chimeras, one can also take advantage of the purely embryo-derived nature of the yolk sac endoderm to genotype the host embryo. The isolation of the endoderm from the bilayered yolk sac is a relatively simple procedure (see Protocol 5.3).

Diploid Embryo–Tetraploid Embryo Aggregation Chimeras

This combination also creates chimeras with unique features. The contribution of the tetraploid component can only be observed during prenatal development. Tetraploid cells are found in most of the chimeras only in the trophoblast of the placenta and in the endoderm layer of the yolk sac (Table 11.3) (Tarkowski et al. 1977). The nature of this restriction is not the inability of the tetraploid cells to contribute to the primitive ectoderm derivatives, but rather that they are simply selected against in this highly proliferating lineage. (See Protocol 11.9 for production of tetraploid embryos and Protocol 11.10C for assembling aggregates between diploid and tetraploid embryos.)

The typical use of these chimeras in phenotypic analysis is in instances when the induced mutation has an extraembryonic phenotype (Guillemot et al. 1994; Duncan et al. 1997; Yamamoto et al. 1998; Adams et al. 2000). These chimeras create a situation in which wild-type tetraploid cells can contribute to the defective trophoblast and primitive endoderm derivatives or, alternatively, mutant cells to these lineages if the diploid embryo is wild type and the tetraploid embryo is mutant (Damert et al. 2002). Interestingly, tetraploid cells contribute to the chimeric yolk sac endoderm in a patchy, rather than "salt and pepper," manner, which provides a unique opportunity to study gene function in the primitive endoderm derivatives of diploid embryo–tetraploid embryo chimeras, including the visceral endoderm of the yolk sac (Damert et al. 2002).

ES Cell–Tetraploid Embryo Aggregation and Injection Chimeras

The complementing developmental potentials of ES cells and tetraploid embryos became obvious when these components were used separately with diploid embryo partners in chimeras. The tetraploid embryos contribute to the trophoblast lineage of the placenta and the extraembryonic endoderm, including the endoderm layer of the yolk sac (Tarkowski et al. 1977), whereas the ES cells contribute only to the mesoderm layer of the yolk sac, the amnion, the embryo proper, and the allantois/umbilical cord. This perfectly appearing complementation strongly invited the experiment for testing whether tetraploid embryos and ES cells can complement each other when they form the two components of chimeras. The expectation was correct, and in such chimeras the tetraploid cells can take over all the lineages of the trophectoderm and primitive endoderm derivatives, while the ES cells exclusively colonize the primitive ectoderm derivatives, including the yolk sac mesoderm and the entire embryo proper (see Table 11.3) (Nagy et al. 1990). (See Protocol 11.10D for assembling aggregates between ES cells and tetraploid embryos.)

Using this strategy, live-born mice can be generated that are completely derived from ES cells (Nagy et al. 1993; Eggan et al. 2001). Most of the commonly used 129 and other inbred ES cell lines can support the development to term of completely ES-cell-derived embryos, which develop with extraembryonic membranes provided by the tetraploid embryo component. The efficiency of producing completely ES-cell-derived newborns or later-stage embryos varies between these ES lines. With inbred ES cell lines in exceptional cases, it can reach 15% of the chimeric embryos transferred. However, these newborns die shortly after delivery or Caesarean section due to their inability to breathe (Nagy et al. 1990). Recently, experimental evidence suggests that the inbred status of ES cell lines may contribute to these survival problems. Chimeras generated using noninbred ES cell lines (e.g., R1 or F_1(C57BL/6 x 129) and other hybrids can survive beyond term and develop into fertile and completely normal animals (Nagy et al. 1993; Eggan et al. 2001, 2002).

There are several unique applications for ES cell–tetraploid embryo chimeras. The efficiency of obtaining completely ES-cell-derived embryos can serve as a fast and stringent measure of the general developmental potential of newly established ES cell lines. In addition, if serial genetic alterations are planned in ES cells before introducing them into the mouse, it is advisable to test the developmental normality of the intermediate cell lines and use the best lines for the next alteration. ES cell–tetraploid embryo chimeras are a quick way to evaluate these intermediate cell lines by simply measuring how efficiently they support completely ES-cell-derived development.

These chimeras also provide a unique opportunity to access in vivo phenotypes of targeted mutations directly from ES cells, when ES cells homozygous for a targeted allele are produced (Carmeliet et al. 1996). Many induced mutations have both embryonic and extraembryonic phenotypes. In most of these cases, it is not obvious which embryonic phenotype is primary and which is secondary (e.g., a consequence of a defect in extraembryonic development). The most common application of the ES cell–tetraploid embryo chimeras is the dissection of the embryonic versus extraembryonic phenotype of a mutation. If ES cells homozygous for a mutant allele are combined with wild-type tetraploid embryos, the trophectoderm and primitive endoderm lineages will be wild type, supporting a mutant embryo proper (Carmeliet et al. 1996; Barbacci et al. 1999). If wild-type ES cells are used with mutant tetraploid embryos, the mutant/wild-type scenario is the opposite (Damert et al. 2002).

PRODUCTION OF ES CELL INJECTION CHIMERAS

Preparing Blastocysts for Injection

It is essential to use the highest possible quality blastocysts for injection, otherwise the injection process may become very difficult and the developmental potential of transferred blastocysts may be severely compromised. Blastocysts chosen for injection should be timed such that only well-expanded, but not overexpanded or hatched, embryos are used. Small blastocysts are difficult to inject without disturbing the ICM, whereas over-expanded blastocysts are often difficult to penetrate. Examples of optimal and suboptimal blastocysts are shown in Figure 11.2.

It is possible to influence the developmental stage of the blastocysts by (1) adjusting the light cycle in the animal room, (2) optimizing the timing of hor-

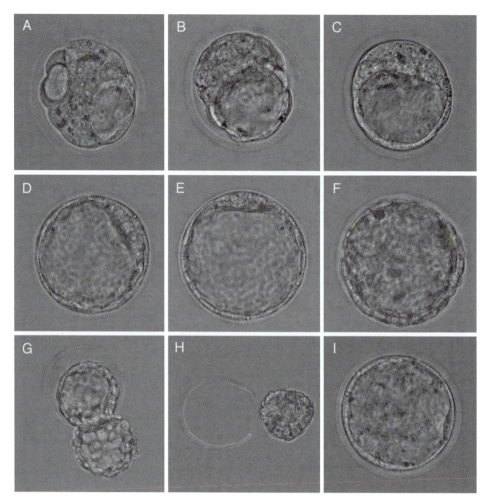

FIGURE 11.2. Developmental stage and quality of blastocysts for injection. (*A*) Very early blastocyst-stage embryo with the blastocyst cavity just starting to form; too early to inject. (*B*) Early blastocysts with small cavity; too early to inject. (*C*) Growing blastocyst; not yet large enough to inject. (*D*) Almost fully expanded blastocyst; possible to inject. (*E*) Fully expanded blastocyst; optimal to inject. (*F*) Overexpanded blastocyst (trophoblast layer slightly thicker; shape not completely spheric; very difficult to inject). (*G*) Hatching blastocyst (not possible to inject). (*H*) Hatched blastocyst (not possible to inject). (I) Well-expanded blastocyst, which is turned in a way that the inner cell mass cannot be seen (turn around before injection is attempted; see Protocol 11.5, point 4).

mone treatment if superovulated (see Chapter 3) animals are used, or (3) starting the injections earlier or later during the day. If the blastocyst cavity is too small, the light cycle should be adjusted so that the dark cycle starts earlier, the hormones are given earlier, or the injections are started later. If the blastocysts are over-expanded, the opposite strategy should be used (for further details, see Chapter 3). Protocol 11.5 describes injection of blastocysts.

In most people's hands, embryos collected from naturally mated mice tend to be of higher quality than embryos from superovulated females. However, if the timing of the hormone treatment is optimized, superovulated females can also be used. Lower quality is usually due to the fact that embryos collected from superovulated females tend to be slightly more advanced than those from naturally mated animals (while kept under the same light cycle). This problem can be solved by collecting 2.5-dpc morulae from the oviduct and culturing them overnight to the blastocyst stage instead of using the usual collection of 3.5-dpc blastocysts. The short in vitro culture will slow down the development of these embryos slightly, and in many cases may provide a good alternative to natural matings. However, it should be pointed out that this strategy should only be used if absolutely optimal culture conditions can be provided.

Great care should be taken to handle the embryos as cautiously as possible both during recovery from the uterus/oviduct and following manipulation. The time spent outside the incubator should be kept to a minimum, and the embryos should be carefully washed with KSOM-AA or other embryo culture medium (Chapter 4) when transferred from HEPES-buffered medium to culture conditions in a 5% CO_2 incubator.

Although blastocysts are somewhat more robust than one- and two-cell-stage embryos, and will better survive suboptimal temperatures/culture conditions, one must keep in mind that survival alone is not enough; these embryos should be handled and cultured correctly so that they can develop to healthy pups.

Preparing ES Cells for Injection

ES cells used for injection should show an optimal undifferentiated morphology in culture (see Chapter 8). The cells should be growing exponentially and should be harvested at subconfluent density. If frozen cells are to be used, thaw the vial 2–3 days prior to injection. The medium should be changed the next day, and cells should be passaged to avoid having them grow too densely. On the day of injection, the ES cells should be prepared an hour before the embryos are ready to be injected. After preparation, the cells can be used for 3–4 hours. It is important to keep the feeder-dependent ES cells growing on good-quality mouse embryonic fibroblasts (MEF) at all possible times to minimize the risk of differentiation. During injection, however, MEF can be disturbing and should therefore be removed. This can be done by plating out the ES cells on gelatin for the last passage before injection. A better alternative is to perform thorough preplating, as described in Protocol 11.1.

Preparing the Injection and Holding Pipettes

The quality of the tools, particularly the injection pipette, is very important for successful blastocyst injection (Bradley and Robertson 1986; Papaioannou and Johson 1993). Both holding pipettes and injection needles are commercially available from several vendors (e.g., Eppendorf VacuTip; TransferTipES). Generally these capillaries are of high quality, but they are expensive. The injection and holding pipettes can be reused in consecutive experiments if they are stored overnight in a trypsin solution and rinsed well. Both the holding capillaries and injection pipettes can be custom-made in the laboratory. Homemade capillaries are generally of superior quality, the size can be adjusted to personal preference, and they are considerably more cost-effective than purchased premade pipettes. The preparation of holding capillaries is described in Protocol 7.8.

A micropipette puller (e.g., Sutter Instruments) is required for the preparation of injection needles. A good injection pipette should have a tip with an inner diameter ranging between 10 and 15 μm. Single ES cells from different lines may vary considerably in size; therefore, it is convenient to be able to choose the size most suitable in each case. Too small an inner diameter will harm the cells by squeezing them in the capillary, whereas too large an outer diameter will impose too much physical damage on the blastocyst during injection. The length and shape of the taper are of great importance. ES cell needles should be as long as possible and have a slowly increasing diameter. The aim here is to achieve a taper length with practically equal diameter, long enough to hold 50–60 ES cells. If the tip diameter increases too fast, the ES cells will easily mix with media and/or oil, which greatly hinders efficient injection.

Both before and after preparation of injection needles, glass capillaries should be kept as dust-free as possible. Dust inside the injection needle results in a less smooth movement of the contents and a high risk of small oil bubbles attaching to the wall. If these oil bubbles are suddenly released and are injected together with the ES cells, this is deleterious for the blastocyst.

When using an electronic puller with a built-in test program, it is very important to run the program before the first capillary is pulled, to avoid burning up the filament. On the Sutter P97, this test is called a "ramp test." The value from the ramp test will determine the maximum heat that can be used for the particular filament and glass combination. (See Protocol 11.2 for details.)

Capillary pullers have several individually variable settings, which determine the shape of the pulled needle. On modern electronic pullers, these settings often interact with each other in a way that makes the determination of optimal values complicated. (See Protocol 11.2, Comments section, for a discussion of the effects of changing these settings.) The optimal shape of ES cell injection needles is achieved by using relatively high heat values, combined with a rather weak pull, low velocity, and possibly a high time delay.

Preparing the Tips of ES Cell Injection Needles

There are two methods for creating an injection needle with a sharp tip that allows the easy penetrance of the blastocyst: (1) Break the capillary either with a razor blade or with the help of a microforge, and then grind the tip to a flat and sharp opening. Optionally, a very sharp spike can be created on the tip by melting it on a microforge. (2) Manually break the tip with a razor blade to obtain an arched cut. Both procedures will result in a pointed and sharp tip if performed properly. However, the shape of the tip will differ, the first option resulting in a straight-cut opening (Fig. 11.3A), and the second option producing an arched cut (Fig. 11.3B). An arched cut is preferred by many laboratories, because it makes it easier to pick up the ES cells and pack them closely in the needle with a minimum amount of medium between individual cells (Fig. 11.3C). Commercially produced injection needles are either ground or "spiked." For homemade needles, the cut tip is recommended, because it will, with a bit of practice, provide the highest possible quality injection needles. (See Protocol 11.3 for this technique.) If a piezo drive (e.g., Primetech, Ibaraki, Japan) is used for injection of ES cells into blastocysts, a flat-tipped injection needle is necessary (Eggan et al. 2001).

Assembling the Microinjection Setup

The basic microscope/micromanipulator setup (see Chapter 7) can easily be modified for blastocyst injection. The same inverted microscope with two micromanipulators, one on each side of the microscope stage, can be used. Phase-contrast optics is recommended rather than differential interference contrast (DIC). These two optic systems provide completely different images: DIC gives a high-

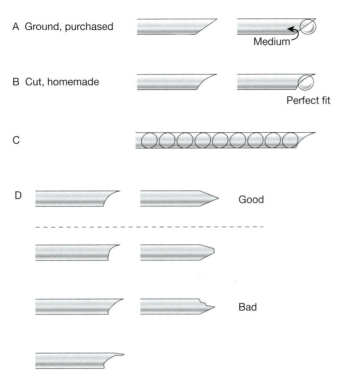

FIGURE 11.3. Shape of ES cell injection needles. (*A*) Ground tip. (*B*) Cut tip. (*C*) ES cells loaded into the injection needle. (*D*) Correct and incorrect shapes of cut tips.

resolution view, which can detect the membrane of a pronucleus inside a zygote. Phase-contrast optics, on the other hand, provides a better image of the outside surface of an object, forming a clear light "halo" around three-dimensional shapes. This makes it possible to select those individual ES cells that are most suitable for injection (see Protocol 11.5). As for pronuclear injection, a low-power magnification with a 5x or 10x objective of any type is sufficient for arranging the holding/injection capillaries, blastocysts, and ES cells. A high-power (20x) phase-contrast objective is required for the actual injection process.

As with pronuclear injection, the holding capillary should be controlled either with a micrometer screw-type control unit (e.g., Eppendorf CellTram air) or a simple mouth pipetting device (see Chapter 7). The injection needle, however, should be connected to a control device that offers a very fine movement both in and out of the capillary. This can be achieved with a control unit similar to that used for the holding capillary, but in this case filled with oil (e.g., Eppendorf CellTram Oil Vario). The manufacturers' recommendations about tubings, connections, and oil filling should always be followed if commercial control units are used! As an alternative, a 10-cc glass syringe connected to the instrument holder through thick, air-filled Tygon tubing can be used; however, in this case the injection needle is filled with oil.

ES cells tend to stick to each other and to all available surfaces when kept at room temperature or above. To make it easier to collect cells from the injection chamber, it is advisable to cool the microscope stage. Several commercial cooling stages are currently available (see Chapter 17), and many mechanical workshops have manufactured custom-made cooling stages for individual laboratories. The principle in most cases is simple: A metal frame is fitted to the microscope stage. This frame is hollow, and cooled water is circulated inside. The water is cooled to a temperature that will maintain 10–14°C in the injection chamber (see Fig. 7.8). However, other systems are also available (see Chapter 17).

The movement of holding and injection capillaries is controlled by the micromanipulator on the same side of the microscope stage (see Fig. 7.4). However, it is practical to control the suction/release of the capillaries with the opposite hand. The reason for the crossed control will be evident when injection is performed—it is difficult to control and move a capillary at the same time with the same hand.

PRODUCTION OF AGGREGATION CHIMERAS

The production of aggregation chimeras can be divided into several subprocedures following each other in a logical order. This order should minimize stress on the embryos. Aggregation plates are prepared in advance (see Protocol 11.7) and equilibrated in an incubator while embryos and ES cells are processed. The next critical step is the preparation of the collected embryos for aggregation; i.e., zona pellucida removal (Protocol 11.8). This step is followed by the actual aggregation procedure (Protocols 11.10 A,B,C,D) and overnight culture. The last step is the transfer of the aggregates into pseudopregnant females as described in Chapter 6. If ES cells are aggregated with diploid or tetraploid embryos, the cells are thawed and cultured in advance before being prepared for the experiment (Protocol 11.6).

TRANSFER OF CHIMERIC BLASTOCYSTS

The chimeric blastocysts should be transferred either to the uterus of a 2.5-dpc recipient or to the oviduct of a 0.5-dpc pseudopregnant recipient. Ideally, blastocysts are divided into groups of 5–7, and each group is returned to a single uterine horn, resulting in 10–14 embryos per recipient. The success rate for zona-free blastocysts (CD1/ICR) is somewhat lower, and 16–20 embryos per recipient are usually transferred. In case of recipient shortage, up to 24 manipulated zona-free embryos may be transferred per recipient. For details of embryo transfer, see Chapter 6.

The pups are usually born at 19.5 dpc, according to the schedule of the mother (and not that of the embryos). Hence, 3.5-dpc blastocysts, which are transferred to the oviduct of 0.5-dpc or the uterus of 2.5-dpc recipients, will implant with a 3-day or 1-day delay, respectively. The pregnancy rate for unmanipulated blastocysts after transfer should approach 90% if the procedure is carried out correctly and the recipients are maintained in an optimized and pathogen-free environment.

For ES cell chimeras, the embryo implantation rate is highly dependent on the developmental potential of the cell line. ES cells of very high potency can support normal development in embryos, which are highly chimeric, whereas less potent ES cells will result in either low chimerism or reduced embryo viability. On average, 40–60% of transferred chimeric embryos will develop into healthy newborns. The transfer of too few embryos or a reduced number of developing embryos due to poor ES cell quality will often result in only a few (2 or 3) fetuses. These fetuses often grow too big, which causes difficulties at birth. In this case, the mother will generally not deliver on 19.5 dpc, and, if there is no intervention, the pups will die during the delayed birth. This situation can be solved by delivery of the pups by Caesarean section (see Protocol 6.5).

BREEDING OF CHIMERIC MICE PRODUCED BY INJECTION OR AGGREGATION WITH ES CELLS

Because most ES cell lines are male, there should be a distortion of the sex ratio toward male mice among the chimeras resulting from either injection or aggregation (Robertson et al. 1986). It is worth test-breeding all chimeras with 50% or more ES-derived coat color for germ-line transmission; however, germ-line transmission can be obtained with males with significantly less than 50% ES-derived coat color. If the host embryo is of the C57BL/6 genotype and the ES cells are derived from any of the agouti-pigmented 129 substrains, chimeric males are bred with nonagouti inbred (e.g., C57BL/6 *a/a; C/C*) or outbred females (e.g., CF1 *a/a; c/c* or Black Swiss *a/a; C/C*). If the host embryo is albino, albino females should be used for the test breeding, with pigmented offspring indicating germ-line transmission. Chimeras generated using ES cells from an agouti-pigmented substrain of 129 and albino blastocysts can be recognized at birth because they have pigmented eyes. Furthermore, when these chimeras are test-bred with albino mates, germ-line transmission can be recognized at birth because the germ-line pups have pigmented eyes. Depending on the experiment, it may be useful to have such a color-marking system.

The degree of chimerism can easily be detected in the pups at the time when the fur becomes visible (around 10 days after birth). Depending on the coat-color markers of both ES cells and the host embryo, different color combinations can be seen. To understand the coat-color combinations that may arise in the most commonly used chimera combinations, it is important to keep in mind that the genotypes of both the melanocytes (where melanin is produced) and the hair follicle cells (which may influence the production of melanin in the melanocytes) will contribute to the resulting color of the hair.

A chimera between a C57BL/6 host embryo and an agouti-pigmented substrain of 129 has two colors: black (C57BL/6) and agouti (129). However, a chimera between C57BL/6 and 129P2/Ola will have three colors: black (C57BL/6), agouti, and chinchilla (creamy) (129P2/Ola). The 129P2/Ola mouse is chinchilla colored. The agouti pigmentation arises in this case when one of either (but not both) melanocytes or hair follicle cells is derived from 129P2/Ola, and the other from C57BL/6. If both hair follicle cells and melanocytes are derived from the 129P2/Ola ES cells, a chinchilla coat color will result.

The choice of host mouse strain depends on a variety of factors, including availability of different mouse strains and the background strain in which the mutation will be maintained. There is evidence that the penetrance and phenotype of a developmental mutation can vary on different genetic backgrounds; this should be taken into consideration when designing experiments and interpreting results (Threadgill et al. 1995). A good strategy would be initially to breed the mutation with nonagouti inbred or outbred mice to assess germ-line transmission. Confirmed chimeras with germ-line transmission of 129 ES cells can then be mated to 129 females. The mutation will immediately be on a 129 inbred background, but, because there is no coat-color marker to indicate germ-line transmission, all of the offspring must be screened by DNA analysis. A disadvantage with the 129 background is the often poor intrinsic breeding performance of this strain. Because the C57BL/6 strain is generally considered to be the "gold standard," many laboratories start backcrossing germ-line F_1 progeny to C57BL/6 to generate a congenic strain. If F_1 hybrid ES cells were used, the immediate production of an inbred background of mutations or transgenes cannot be achieved. In this case, backcrossing to an inbred strain is the only way to make the genetic background homogeneous.

Only one allele in ES cells is usually mutated by homologous recombination, and consequently only 50% of the offspring "fathered" by the ES cell compartment of a chimera will inherit the mutation. Therefore, DNA is prepared from tails of these offspring and screened by Southern analysis or polymerase chain reaction (PCR) (as described in Chapter 12). For PCR, DNA from an earpunch or small piece of toe (if toe clipping is allowed in the facility) can be used as well. Heterozygous male and female mice are then crossed to obtain an expected yield of 50% heterozygous, 25% homozygous wild-type, and 25% homozygous mutant embryos or offspring. If the homozygous mutants die shortly after birth, they are often eaten by the mothers and therefore missed. To avoid this loss, Caesarean section can be performed on day 18.5 of pregnancy (see Protocol 6.5). If no heterozygous animals are found among the offspring of a germ-line (coat color)-transmitting chimera, there are two possibilities: the targeted ES cell line contained contaminating nontargeted cells or the targeted gene is haploinsufficient (Carmeliet et al. 1996; Bi et al. 2001) or imprinted.

Preparation of ES Cells for Injection

MATERIALS

CELLS
Subconfluent ES cells grown on 6-cm tissue culture dish with MEF feeder layer

EQUIPMENT
Centrifuge
Incubator, 37°C
Pipette
Tubes, 15-ml (Falcon)

REAGENTS
ES cell culture medium (see Chapter 8)
ES cell injection medium
    Add 200 µl of 1 M HEPES to 10 ml of ES medium; keep on ice. Store up to 7 days at 4°C.
Phosphate-buffered saline (PBS) without Ca^{++} or Mg^{++}
Trypsin, 0.25% in PBS with 0.2% EDTA

PROCEDURE

1. Remove the medium and wash the cells once with PBS.

2. Add 1 ml of trypsin, and incubate the cells for 10 minutes at 37°C.

3. Stop the trypsin reaction by adding 4 ml of ES cell culture medium. Resuspend the cells by pipetting them gently up and down three or four times. Make sure to produce a single-cell suspension, but do not pipette more than necessary.

4. Place the suspension in a tissue culture dish and incubate the cells at 37°C for 10–15 minutes.

5. Most of the MEF feeders will be loosely attached to the plastic surface, whereas the ES cells will be in suspension. Move the dish very carefully, taking care not to disturb these layers! Very carefully tilt the dish, and collect all medium containing the ES cells, leaving the MEFs attached to the dish.

6. Centrifuge the cell suspension at 1000 rpm for 5 minutes.

7. Resuspend the cells thoroughly in 500 µl of ES cell injection medium, and add the mixture to a 15-ml Falcon tube.

8. Place the tube on ice for 30 minutes.

9. Carefully remove 3/4 of the medium, taking care not to disturb the cells that have settled at the bottom of the tube. This step allows the removal of dead cells floating in the medium.

469

10. Resuspend the cells in new cold ES cell injection medium. The cells are now ready for injection.

COMMENTS

Once ES cells have been prepared, they should be kept on ice at all times until injection. The cells will settle at the bottom of the tube and should be carefully resuspended in the existing medium immediately prior to removal of cells for injection. This preparation will keep the cells in a single-cell suspension, preventing them from forming aggregates and preventing the pH from changing, even though the cells are kept outside the incubator. The cells should be used within 3–4 hours. If a longer injection time is planned, new cells should be prepared at a later time.

Determining Useful Settings with the Sutter Puller P-97

MATERIALS

EQUIPMENT
Forceps

Glass, borosilicate glass capillary without inner filament (e.g., Clark Electromedical Instruments GC120T-15)

Puller (e.g., Sutter P-97) with 3-mm through filament (Science Bioproducts FT330B)

PROCEDURE

1. Use a 3-mm through filament.

2. Insert a capillary in the puller using clean forceps.

3. Run a ramp test.

4. Set the Heat to the ramp value +20.

5. Set the Pull to 50.

6. Set the Velocity to 75.

7. Set the Time to 25.

8. Pull a capillary with approximately these settings.

9. Evaluate the shape of the tip, and determine which parameters need to be changed.

10. Adjust the settings of Heat, Pull, Velocity, and Time by increments of 10.

    Keep a record of how the shape of the needle changes with each modification.

COMMENTS

The parameters can be optimized to achieve injection needles with a shape exactly according to personal preference. However, this procedure takes time and requires a great deal of patience. The following guidelines can aid in the process, but it is important to keep in mind that all parameters interact with each other. The results of changing one parameter can be influenced by values of the others.

- Increasing the Heat will result in a longer and finer taper, and smaller tip diameter.
- Increasing the Pull will result in a longer taper, with less influence on the tip diameter.
- Increasing the Velocity will result in a finer tip diameter.
- Increasing the Time will result in a shorter taper, more rapidly decreasing diameter.
- The pressure is usually preset to 500. Changing the pressure requires some additional programming steps. Increasing the pressure will result in a shorter taper and wider tip diameter. (Additional information and advice on settings can be found at http://www.sutter.com.)

Breaking the Tips of ES Cell Injection Needles

MATERIALS

EQUIPMENT
Pulled injection needles
Scalpel, disposable
Silicon tubing, very thick (external diameter 15–20 mm)
Stereo dissecting microscope (20x magnification)

PROCEDURE

1. Cut an ~3-cm-long piece of thick silicon rubber tubing longitudinally to produce two half-round pieces. Place one tubing piece with the rounded surface upward under a stereomicroscope.

2. Place the pulled injection needle across the top of the tubing at about a 45° angle (see Fig. 11.4A). Hold a sharp and clean scalpel in the other hand, again at about a 45° angle to the rubber tubing (see Fig. 11.4B). Because the glass needle is very thin at the end, it will stick slightly to the rubber.

3. Position the ultrathin tip of the capillary on the top part of the rubber in such a way that the desired diameter section of the capillary is not touching the rubber (this is achieved by the rounded shape of the tubing) (see Fig. 11.4C).

4. Apply a very gentle touch with the scalpel to the glass needle at the optimum diameter position. The glass will in many cases break in an arched shape. If not, the cut can be repeated at a slightly further upward position.

COMMENTS

The tip shape should be evenly rounded on both sides, the bend should not be too steep and also not too shallow, and the tip should be absolutely sharp (see Fig. 11.3D). This procedure requires patience when first attempted. In the beginning, most needles will break suboptimally. However, with some practice, it is possible to break almost every needle into a good shape. Do not give up! It is worth the effort to learn this technique.

Some investigators prefer to put a 30° angle bend in the injection and holding pipettes using a microforge so that the pipette ends can be held parallel to the bottom of the injection chamber. However, if the angle of the capillaries as they enter the injection chamber is kept low, it is not necessary to perform this procedure.

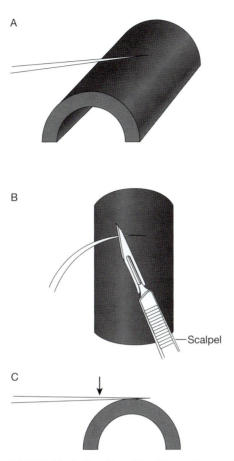

FIGURE 11.4. Breaking ES cell injection needle tips. (*A*) Hold the capillary in the left hand and place it on the rubber support (silicon tubing cut in half). (*B*) Because the glass is very thin, the tip will stick slightly to the rubber. Making use of this slight resistance, give the glass a bend. Hold a sharp and clean scalpel in the right hand. (*C*) Break the glass with a smooth touch at the position where it just starts to lift off the bent surface of the rubber (*arrow*).

Assembling the Microinjection Setup

MATERIALS

EQUIPMENT

Controlled pressure device for holding capillary (CellTram Air, Eppendorf)
Controlled pressure device for injection needle (Cell Tram Oil Vario, Eppendorf)
Cooling stage for the microscope (see Chapter 17 and Fig. 7.8)
Holding pipette (see Protocol 7.8) or Eppendorf VacuTip
Injection chamber (see Chapter 7)
Injection pipette (see above)
Micromanipulators, right- and left-hand, with two instrument holders (see Chapter 17)
Micrometer syringe, high-precision (Stoelting 51218M) for use with a 1-ml Hamilton syringe as an alternative to control device for holding capillary
Microscope, inverted fixed-stage (see Chapter 17)
Paraffin oil, light (Fisher O121-1; BDH 29436)
Polyethylene tubing, small diameter, thick-walled flexible (e.g., 0.76-mm inside diameter, 1.22-mm outside diameter) (Clay Adams PE-60) or thick Tygon tubing from any supplier, 3.2 x 6.4 mm
Silicone oil, low-viscosity (BDH 630054T; Aldrich 14,615.3) high-viscosity (Fluka DC200, 85422, 1000 mPa.s)
Syringe, 10-cc glass, connected to thick (3.2 x 6.4 mm) Tygon tubing as an alternative to control device for injection needle
Transfer TipES (Eppendorf)

PROCEDURE

1. Place the holding pipette in the left-hand micromanipulator. Connect the instrument holder to a control unit (e.g., CellTram Air) with the manufacturer's tubing. Position the control unit within easy reach of your right hand. Alternatively, connect the instrument holder with thick silicon tubing to a mouth pipette.

2. When an oil-filled control unit is used, it is important to remove *all* air bubbles from the system before proceeding.

   Failure to remove all air bubbles will cause cells to rush in and out of the pipette, or cause no movement of the meniscus at all when the control unit is operated.

3. If an Eppendorf TransferTipES is used as an injection needle in combination with CellTram oil, leave the injection needle empty.

   If other brands, or homemade needles and/or control devices other than CellTram are used, the injection needle may need to be filled with oil to adjust the fine control (see Comments, below). Fill with oil a 10-cc syringe fitted with a needle of the same dimension as the inner diameter of the glass capillary used for pulling the injection needle. Insert the needle into the back side of the injection capillary and squeeze the plunger until oil has completely filled the injection needle (a small oil drop should be visible at the tip). Withdraw the needle slowly and make sure that no air bubbles are trapped inside the injection needle.

4. Place the (oil-filled) injection pipette in the right-hand micromanipulator instrument holder, and connect it to a control unit (e.g., CellTram Oil Vario). Position the control unit on the left-hand side of the microscope stage. Alternatively, use a glass syringe connected to the instrument holder with thick Tygon tubing to control the injection needle. In this case, fill the injection needle with heavy silicon oil to compensate for the wide diameter of the tubing.

5. Align the instruments in the injection chamber according to Figure 7.5. Take care to align both holding and injection capillaries so that they oppose each other in a straight line.

COMMENTS

- All connections must be absolutely tight. Any leaking oil or air will inevitably result in uncontrolled movements.
- Any oil used in the needle must be tested for possible embryo toxicity (see Chapter 4). Heavy oil will slow down the movement in the injection needle. Use heavy oil if the control appears to be too fast. A lighter oil should be chosen if the movement is too slow.

Injecting Blastocysts

MATERIALS

EMBRYOS AND CELLS
Blastocyst-stage embryos (see Chapter 4)
ES cells prepared according to Protocol 11.1

EQUIPMENT
Holding pipettes (see Chapter 7)
Humidified incubator, at 37°C, 5% CO_2, 95% air
Injection chamber (see Chapter 7)
Injection pipette (see Fig. 11.3)
Microscope/micromanipulator setup (see Chapters 7 and/or 17)
Mineral oil, embryo-tested light
Transfer pipettes and mouth pipette device (see Chapter 4)

REAGENTS
M16 or KSOM-AA medium (see Chapter 4)

PROCEDURE

1. Precool the injection chamber (see chamber description in Fig. 7.8) on the microscope stage.

2. Use a transfer pipette to transfer the expanded blastocysts in groups of ~10 into the precooled injection chamber. Then, introduce a few hundred ES cells (a "cloud" of cells, filling, at most, half the drop) into the injection chamber and allow them to settle on the bottom.

3. Using high-power magnification, select individual cells carefully on the basis of size (small, compared with the feeder cells) and shape (uniformly round, compared with more ragged or "rough" feeder cells) (see Fig. 11.5A). Draw ~10–15 cells into the injection pipette and position them near the tip in a minimal amount of medium (see Comments).

4. Immobilize a single blastocyst by applying suction to the holding pipette and move it toward the center of the microscope field. Position the embryo with the inner cell mass (ICM) at either the 6 or 12 o'clock position. If the ICM is difficult to visualize, it is most probably located on the side nearest or farthest away from the optics. In this case, the blastocyst should be turned around with the help of the injection needle until the ICM can clearly be seen (see Comments, below).

5. Align the tip of the injection needle in the same focal plane as the midpoint, or equator, of the blastocyst. Touch the end of the injection pipette gently to

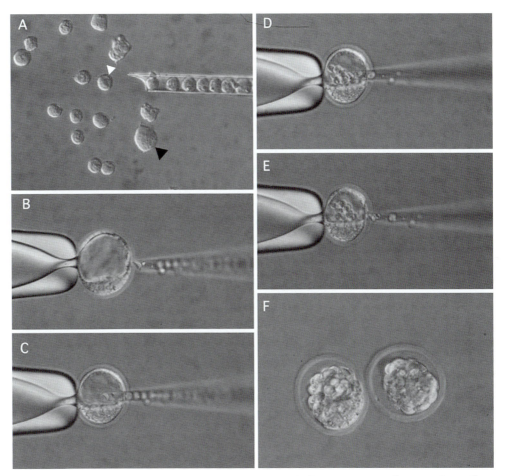

FIGURE 11.5. Procedure for injection of blastocysts. (*A*) The injection needle is used to collect individual cells, which are positioned near the needle tip (*black arrow*, fibroblast; *white arrow*, ES cell). (*B*) Immobilize a blastocyst on a holding pipette so that the inner cell mass is positioned at 6 or 12 o'clock. The tip of the injection needle is brought into the same focal plane as the mid-point, or equator, of the blastocyst. Touch the end of the injection pipette gently to the surface of the embryo. (*C*) With a single, swift, continuous movement, introduce the needle into the blastocyst cavity. (*D*) Release the cells (*arrow*) slowly into the blastocyst cavity. (*E*) Slowly withdraw the needle. (*F*) The blastocyst collapses.

the surface of the embryo (Fig. 11.5B). Be careful not to damage the zona pellucida. This will make a small indentation in the embryo, which will give an indication of the position of the tip of the pipette relative to the surface of the blastocyst. The tip of the pipette can be moved into an "equatorial" position on a trial-and-error basis.

6. With a single, swift, continuous movement, introduce the loaded injection pipette into the blastocyst cavity. Aim to insert the injection needle at a junction between two trophoblast cells. This will minimize the damage to the embryo and will make successful penetration much easier (Fig. 11.5C). Take care not to touch the ICM with the injection needle. If the first attempt to penetrate the trophoblast layer is unsuccessful, and the blastocyst is not collapsed, try to insert the needle *exactly* at the same position again, but this time with a swifter movement. The blastocyst should be discarded if it collapses without successful injection (see Comments, below).

7. Slowly expel the cells (Fig. 11.5D) inside the blastocyst cavity. Take care not to insert any oil bubbles or lysed (dark) cells into the blastocyst.

8. Withdraw the injection pipette slowly (Fig. 11.5E). If the pressure is high inside the blastocyst cavity, injected cells may be pushed out while the needle is being withdrawn (see Comments, below, for suggestions for preventing this problem). The blastocyst will collapse once the pipette is removed, resulting in the cells' coming into close contact with the surface of the ICM (Fig. 11.5F).

9. Remove the injected embryos periodically and incubate them in a humidified incubator with 5% CO_2 at 37°C in microdrops of M16 or KSOM-AA medium. The blastocysts will re-expand after 1–3 hours in culture. In most instances, the injected cells can no longer be seen in the blastocyst cavity after re-expansion, but will have already integrated into the ICM.

COMMENTS

- Only well-expanded blastocysts should be used. Injection of partially expanded embryos can result in damage to the ICM.

- It is possible to inject expanded blastocysts at a rate of 20–30 blastocysts per hour if the injection needle used is of optimal quality and the operator has a great deal of experience. However, the quality of injections is of much greater importance than the final quantity injected. Occasionally, blastocyst injection can be hampered by the accumulation of cellular debris around the tip or inside the injection pipette. This will make the pipette tip blunt and the inside of the needle sticky. Replace the injection pipette if the debris cannot be removed either by purging the pipette with light paraffin oil, by touching the needle tip to the oil–medium interface, or by carefully rubbing it against the holding pipette. The shape, size, and quality of the injection needle determine the number of cells that can be picked up at one time. High-quality needles will allow a large number of cells to be tightly packed, permitting the injection of several blastocysts consecutively. However, care should be taken not to insert too many cells in the needle, because they may mix with medium and oil drops.

- Blastocysts can easily be brought into the right position by turning them around with the injection needle. If the blastocyst needs to be turned in a clockwise or counterclockwise direction, the injection needle should be kept at the equatorial level of the embryo and carefully touched to the zona pellucida at the same time as it is moved. If, on the other hand, the blastocyst needs to be turned in the direction toward or away from the optic plane, the injection needle should be moved down to the bottom of the injection chamber. Placing the needle at a 6 o'clock position, and then swiftly moving it to a 12 o'clock position under the blastocyst can nicely turn around the embryo (see Fig. 11.6).

- It is important to remember that high pressure already exists in the blastocyst cavity before injection. This pressure is particularly high in over-expanded embryos, and often in blastocysts collected from superovulated females. The pressure increases additional-

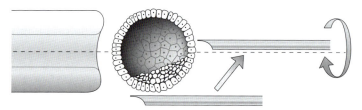

FIGURE 11.6. Turning the blastocyst with the help of the injection needle.

ly as ES cells are injected. When the needle is removed, the cells might be squeezed out again, with the fluid escaping from the blastocyst cavity. The excess pressure can be released either before or after the injection of the cells:

1. Insert the injection needle through the trophoblast layer, and then withdraw it again to the position where the tip of the needle is still between two trophoblast cells. Wait until the blastocyst visibly reduces in size (1 second). Insert the injection needle again and expel the cells. Withdraw the needle after injection.

2. Insert the injection needle, expel the cells, and withdraw the needle to the position with the tip still in the trophoblast layer. Wait until the pressure is equilibrated, and only then withdraw the needle completely. According to our experience, alternative 1 is generally suitable for embryos collected from superovulated females, whereas alternative 2 can be used for any well-expanded blastocyst.

• Recently some laboratories have found that the piezo-equipped injection system makes the blastocyst penetration of the pipette significantly easier (Eggan et al. 2001; Kawase et al. 2001).

• Some laboratories add DNase I to the injection medium to reduce problems with stickiness.

TROUBLESHOOTING FOR ES CELL INJECTION

1. **The cells are difficult to handle; they are sticky and/or fragile.**
   • The trypsin treatment was carried out for too long a time, the cells were resuspended too vigorously, or the cells were not kept on ice.
   • The microscope stage is too warm; use a cooling stage.

2. **The blastocysts are collapsing when the injection needle is being inserted.**
   • The injection needle is blunt: Use only needles of optimal quality.
   • The blastocyst is too small (not fully expanded) or too large (over-expanded, close to hatching): Use blastocyst at the right stage. To change the developmental stage, the light cycle in the animal facility can be changed (see Chapter 3).
   • The culture medium is suboptimal; check the quality of the medium by culturing C57BL/6 embryos from the one-cell stage to blastocysts. At least 70–75% should reach the hatching stage.

3. **The injection needle seems to be sticky, with debris hanging on the tip.**
   • During collection, dead cells and feeders may attach to the tip of the needle. Carefully "tap" on the joystick (mechanical manipulator) or the capillary holder (electronic and hydraulic manipulator) with your finger to shake off the debris. If this does not help, move the needle swiftly through the oil surface and down into the medium drop again.

4. **The injection needle seems to be sticky, with debris on the inside wall.**
   • ES cells are lysing inside the injection needle. Take care not to pick up cells that are too large and may get squeezed. Choose only healthy-looking cells with shiny, round morphologies.
   • Change the needle; it cannot be cleaned on the inside.

5. **The blastocysts do not stay in position during injection.**
   • The holding capillary is not straight, not evenly shaped, or the inner diameter is too small. Use only perfect-quality holding capillaries.
   • The holding capillary and/or injection needle are not inserted into the injection chamber in a straight line with each other. Reposition the capillaries.
   • The injection needle is not positioned exactly at the equator of the blastocyst when injection is attempted. Reposition the injection needle tip accurately.

6. **The ES cells "pop out" of the blastocyst cavity immediately after injection.**
   - The pressure inside the blastocyst is too high. See Comments, above, for solving the problem.
   - The blastocysts are over-expanded and close to hatching. Use earlier-stage embryos.
7. **The tip of the injection needle moves during injection.**
   - See Chapter 7, Troubleshooting, point 9.

Preparing ES Cells for Aggregation

For details of ES cell culture, see Chapter 8.

MATERIALS

CELLS
Subconfluent ES cells grown on 6-cm tissue culture dish with MEF feeder layer

EQUIPMENT
Inverted or dissecting microscope
Tissue culture dishes, gelatinized, feeder-free

REAGENTS
ES cell culture medium (see Chapter 8)
Gelatin, 0.1% solution of (e.g., Sigma G2500) in tissue culture-grade water (see Chapter 8)
Phosphate-buffered saline, Ca^{++}/Mg^{++}-free (see Appendix 1)
Trypsin, 0.25% in PBS with 0.2% EDTA

PROCEDURE

1. Passage a subconfluent plate of ES cells 1 or 2 days prior to the aggregation onto feeder-free gelatinized tissue culture plates (see Chapter 8). ES cells are preplated to deplete feeder cells (see Protocol 11.1) and seeded more sparsely than usual (e.g., using 1:10–1:50 dilutions) on a few gelatinized plates to create colonies of 5–15 cells for the aggregation experiment 1 or 2 days later. Growth for 24 hours is enough for most clones, but 48 hours may be necessary for slower-growing ES cell clones.

2. On the day of the aggregation experiment, after preparation of the aggregation plate (see Protocol 11.7) and the zona-free embryos (see Protocol 11.8), remove the medium from the ES cell plates prepared for aggregation.

3. Wash the cells with Ca^{++}/Mg^{++}-free PBS and then aspirate the PBS from the plate.

    (*Optional*) Wash the cells with trypsin to help loosen up the cells. This step minimizes the amount of trypsin needed in the next step.

4. Add a minimal amount of trypsin to just cover the cells (e.g., 0.5 ml for a 60-mm plate). Leave the plate at room temperature or put it in the incubator for 1–2 minutes. Gently swirl the plate and watch the cells under the microscope as they are lifting up from the bottom of the dish.

5. When about one-third of the colonies are floating, knock the plate against the microscope stage several times to get the majority of the colonies separated from the bottom of the dish.

6. Stop the action of the trypsin by adding 3–4 ml of ES cell culture medium. The cells should form loosely connected clumps, as shown in Figure 11.7.

COMMENTS

A subconfluent culture of ES cells on feeders also can be used. It should be trypsinized lightly for a very short time (30 seconds to 1 minute), the dish tapped gently on the microscope stage, and small clumps removed into a new plate with ES cell medium. If necessary, large clumps can be gently resuspended in trypsin using a 1-ml pipette or 1000-µl tip to get clumps of the right size (8–15 cells).

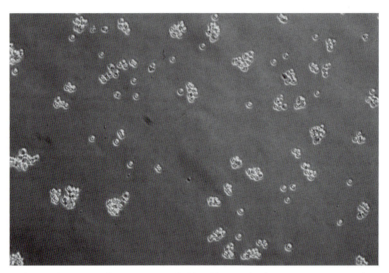

FIGURE 11.7. Briefly trypsinized ES cells, ready for aggregation with preimplantation-stage embryos.

| **Preparing the Aggregation Plate**

MATERIALS

EQUIPMENT

Aggregation needle (DN-09, BLS, Hungary, http://www.bls-ltd.com)
Dissecting microscope
Incubator, humidified, at 37°C, 5% CO_2
Syringe, 1-cc, with 26-gauge needle
Tissue culture dishes, plastic, 35-mm (e.g., Falcon 35-3001)

REAGENTS

Embryo-tested light mineral oil (e.g., Sigma M8410)
70% Ethanol<!>
KSOM-AA medium or other embryo culture medium (see Comments)

CAUTION: See Appendix 2 for appropriate handling of materials marked with <!>.

PROCEDURE

1. Place microdrops of KSOM-AA medium into a 35-mm tissue culture dish in four rows (Fig. 11.8A); the first and fourth rows should contain three drops and the remaining rows five. The easiest and fastest way of creating drops is using a medium-filled 1-cc syringe with a 26-gauge needle. The drops should be ~3 mm in diameter or 10–15 μl.

2. Cover the drops *immediately* with embryo-tested light mineral oil. It is essential that the drops are completely submerged beneath the oil; however, do not overfill the dish with oil.

3. Wipe the aggregation needle with 70% ethanol. Then create deep depressions in the drops (e.g., six in each microdrop) of rows 2–4 by pressing the aggregation needle into the plastic (Fig. 11.8B,C). Perform this step under the dissecting microscope to ensure that the depression is sufficiently deep, smooth, and positioned approximately halfway between the center and the periphery of the drop. Avoid the use of the center of the drops, because if air bubbles form in the medium during further procedures, they can hinder the view of the central region. It is advisable to equilibrate the aggregation plate in the incubator for at least 30 minutes before making the depressions.

4. Prepare a few aggregation plates the same way, each with 40–60 depressions (depending on the expected number of embryos to be aggregated). Place the plates into the incubator to equilibrate with CO_2 and temperature for at least several hours. If possible, it is recommended to prepare the plates 1 day before they are to be used for aggregation.

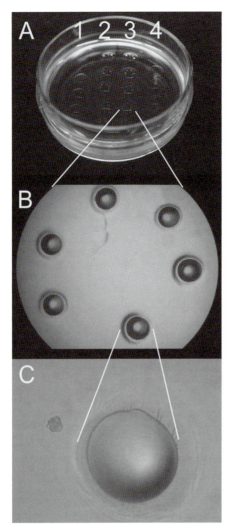

FIGURE 11.8. Aggregation plate. (*A*) Rows of microdrops under oil. (*B*) Arrangement of depression wells in a microdrop. (*C*) A depression well at high magnification. A zona-free embryo is placed outside the well to assist judging the diameter.

COMMENT

The quality of embryo culture conditions is a major factor in the production of aggregation chimeras because embryos are cultured overnight or, in the case of tetraploid embryos, for 48 hours before being transferred into pseudopregnant females (see more about preimplantation embryo culture in Chapter 4). Both M16 and KSOM media have been successfully used for many years for the production of aggregation chimeras. KSOM medium supplemented with amino acids (KSOM-AA) has been used in most recent experiments. Possibly other preimplantation embryo culture media can be effectively used as well.

Removal of Zona Pellucida

The removal of the zona pellucida of preimplantation mouse embryos using an acidified Tyrode's solution has become popular because of its simplicity. R. Gwatkin (1964) showed that the zona pellucida of mouse blastocysts could be dissolved by exposure to acidified media. This treatment was initially reported to be incompatible with embryo viability (Bowman and McLaren 1969). However, Brun and Psychoyos (1972) subsequently developed a protocol to remove the zona pellucida of rat blastocysts and retain viability using a buffer with a pH of 3.7. The current method of using acidified Tyrode's solution (pH 2.5) to remove the zona pellucida of mouse embryos was developed by Nicolson and colleagues (1975).

MATERIALS

EMBRYOS

Preimplantation-stage embryos

For morula aggregation, embryos are collected at 2.5 dpc as described in Chapter 4. If tetraploid embryos are used, two-cell-stage embryos are collected at 1.5 dpc, fused as described in Protocol 11.9, cultured overnight, and used for aggregations at the three- and four-cell stage.

EQUIPMENT

Aggregation plates, equilibrated (see Protocol 11.7)
Dissecting microscope (see Comment)
Incubator, humidified, at 37°C, 5% CO_2
Petri dishes, plastic (see Comment)
Pipette assembly, mouth or hand-held, for embryo manipulation (Chapter 4)

REAGENTS

Acidic Tyrode's solution (e.g., Sigma T1788) at room temperature (see Appendix 1)
M2 or other HEPES-buffered medium (see Chapter 4)
KSOM-AA medium or other embryo culture medium

PROCEDURE

1. Place two drops of M2 medium and several drops of acidic Tyrode's solution (kept at room temperature) in a plastic petri dish.

   A fixed drop arrangement (see Fig. 11.9A) helps in maneuvering the embryos.

2. Transfer the embryos of the same genotype into one of the M2 drops (Fig. 11.9B). Pick up 20–50 embryos (the number depends on the speed of manipulations) with as little medium as possible and wash them through one drop of acidic Tyrode's solution. Then transfer the embryos to a fresh drop of acidic Tyrode's solution.

3. Move the embryos up and down using the very tip of the pipette while observing zona dissolution (Fig. 11.9C). Transfer the embryos into the fresh

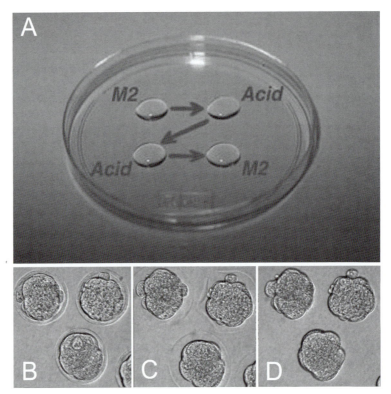

FIGURE 11.9. Removal of the zona pellucida. (*A*) Dish with acid Tyrode's solution and M2 medium drops. (*B*) Embryos with intact zona pellucida. (*C*) Zona pellucida dissolving. (*D*) Embryos without zona pellucida.

drop of M2 *immediately* after the zonae dissolve. Do not allow the embryos to touch each other (Fig. 11.9D).

4. Repeat the procedure with the remaining embryos. Use new drops of acidic Tyrode's solution for each group of embryos; the same drops can be used if their volume is 100–200 μl. It is very important to transfer minimal volumes of solutions between drops because acid diluted with M2 medium will not work efficiently, and acid in the culture medium may damage embryos.

5. Wash the embryos through a few drops of M2 medium and then through several drops of KSOM-AA medium before placing them individually into the aggregation plates.

COMMENTS

• The embryos are sticky during the zona removal in acidic Tyrode's solution. The use of a bacteriological-grade plastic petri dish or the lid of a tissue culture plate is essential to prevent embryos from sticking to the dish. In addition, working slightly below room temperature and constantly moving the embryos further decreases the sticking.

• It is useful to fire-polish the tip of finely drawn Pasteur pipettes when handling embryos without zonae because they are fragile and easily damaged by sharp edges of the capillary (see Chapter 4).

• The use of frosted instead of transparent glass in the stage of the dissecting microscope provides a better view of the zona pellucida during its removal.

Production of Tetraploid Embryos

Recently, tetraploid mouse embryos have become an important component of both embryo–embryo and ES cell–embryo chimeras. Tetraploid embryo production is an essential part of these two types of chimeras. The production of tetraploid embryos by electrofusion was first described by Kubiak and Tarkowski (1985). The procedure described here is derived from that publication but contains modifications and fine-tuning for the available instruments (Nagy and Rossant 1999).

MATERIALS

EMBRYOS

Two-cell embryos collected at 1.5 dpc as described in Chapter 4

EQUIPMENT

CF-150B pulse generator (BLS, Hungary, http://www.bls-ltd.com)

Dissecting microscope(s) (see Comments)

Electrode-chamber with 250-μm gap; for example, GSS-250 (BLS, Hungary, http://www.bls-ltd.com)

GSS-500 and GSS-1000 electrode chambers with 500- or 1000-μm gap distance, respectively, may also be used, and are available from the same company.

Incubator, humidified, at 37°C, 5% CO_2

Microdrop culture dish (see Chapter 4)

Petri dishes, 100-mm plastic (e.g., Fisher 08-757-13A)

Pipette assembly, mouth or hand-held, for embryo manipulating

REAGENTS

Mannitol, 0.3 M (Sigma M4125)

Dissolve in ultrapure water, add 0.3% bovine serum albumin (BSA; Sigma A3311), and filter through a 0.22-μm Millipore filter. Store in aliquots at –20°C. Use a fresh aliquot for each experiment (see Appendix 1).

M2 or other HEPES-buffered medium (see Chapter 4)

KSOM-AA medium or other embryo culture medium

PROCEDURE

1. Flush two-cell-stage (1.5-dpc) embryos from the oviducts of naturally mated or superovulated females as described in Chapter 4. Usually up to 200 embryos can be fused right after collection, using HEPES-buffered medium, within 15 minutes. If more than 200 embryos were collected, some of them should be placed in a microdrop culture plate until ready for the fusion in order to avoid prolonged exposure to HEPES-buffered medium.

TABLE 11.4. Suggested impulse parameters for the CF-150B fusion instrument and different electrodes

Electrode	Voltage	Duration	Repeat number	AC voltage
GSS-250	30 V	40 μsec	2	2 V
GSS-500	50 V	35 μsec	2	2 V
GSS-1000	147 V	26 μsec	2	2 V

2. Set the pulse generator to nonelectrolyte fusion and the effective DC and zero AC voltages and the effective pulse duration. The values of these two parameters are dependent on the electrode chamber used. (See Table 11.4 for suggested values.) These values should be fine-tuned for the individual fusion instrument and electrode used.

3. Place a 100-mm petri dish containing the electrode chamber under a dissecting microscope and connect the cables to the pulse generator (see Fig. 11.10A).

4. Place two large drops of M2 medium (drops 1 and 4) and two drops of mannitol solution (drops 2 and 3) in the dish as shown in Figure 11.10A. Drop 3 should be eccentric with respect to the electrodes, to avoid blocking vision by accidental air bubbles.

5. Place all of the embryos in drop 1 (the M2 drop).

6. Rinse 30–35 embryos through the mannitol drop 2 and place them in drop 3 (mannitol) between the electrodes. Slowly increase the AC field by turning the corresponding knob. This will correctly orient (see Fig. 11.10B) most of the embryos. Gently push or pick up those that are not oriented and let them fall back again. The AC field can also be set up in advance to the desired minimal voltage (usually 1–2 V), which positions the plane of blastomere contact parallel to the electrodes upon placing embryos between electrodes.

   CAUTION: Setting too high a voltage of the AC field may lyse the embryos.

7. When all of the embryos are properly oriented, push the trigger to apply the set fusion pulse.

8. Transfer the embryos into drop 4, the M2 drop.

9. Repeat steps 6–8 with the remaining embryos, taking no longer than 15 minutes. Change the mannitol drop in the electrode chamber to a fresh one if more embryos need to be processed.

10. Wash the embryos through several drops of KSOM-AA medium and place them into microdrops under oil in the incubator (37°C, 5% CO_2).

11. At 30–45 minutes later, examine the embryos and separate those that have undergone complete fusion of the two blastomeres (see embryo #3 in Fig. 11.10C) to a new microdrop. This step is very important because the fusion occurs at the late two-cell stage. If not examined during this time period, fused embryos will cleave again, and at that point, they will be indistinguishable from diploid embryos that are not fused.

12. Culture the fused embryos for 24 hours in the incubator. During this period the healthy embryos will cleave twice; therefore, expect to see four-cell-stage tetraploid embryos. This is the stage at which the tetraploid embryos are

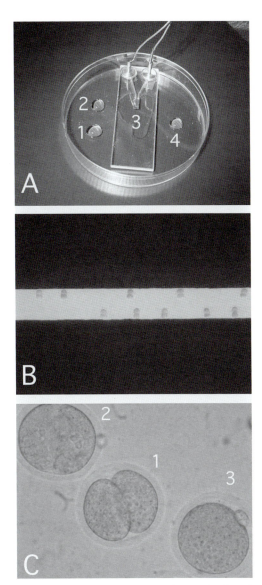

FIGURE 11.10. Production of tetraploid embryos. (*A*) Electrofusion chamber with drops of mannitol (drops 2 and 3) and M2 medium (drops 1 and 4). (*B*) Embryos lined up between the electrodes. (*C*) Two-cell-stage embryos undergoing fusion; (*1*) embryos immediately after pulse applied, (*2*) embryos undergoing fusion of two blastomeres, and (*3*) fused two-cell-stage embryo.

used for aggregation. For injection chimeras, culture for an additional 24 hours or more. During this time the tetraploid embryos will develop to the blastocyst stage.

COMMENTS

- The developmental timing through the preimplantation stages remains unchanged with the tetraploid embryos. Therefore, the compaction of tetraploid embryos starts at the four-cell stage.
- If the fusion parameters are optimized, the efficiency of fusion should be above 90%. A second pulse can be applied to unfused embryos.

- The development rate to the four-cell stage should be at least 80% 24 hours after fusion when optimal culture conditions are used. A small proportion of embryos may stay at the "one-cell" stage and cannot be used. Two-cell-stage embryos are delayed and may be used later in the day when they develop to the four-cell stage.

- Because embryos from some mouse strains exhibit the "two-cell block" phenomenon in certain culture media (such as M16), it is important to collect the embryos at the late two-cell stage when using these media; however, this is not a concern when KSOM-AA medium is used.

- If possible, it is convenient to use two microscopes next to each other during fusion: one for groups of embryos in M2 and mannitol washes, the other for the electrode chamber.

Assembling Aggregates between Diploid Embryos

MATERIALS

EMBRYOS
2.5-dpc embryos (Protocol 11.8)

EQUIPMENT
Aggregation plates (Protocol 11.7)
Incubator, humidified, at 37°C, 5% CO_2
KSOM-AA medium or other embryo culture medium
Microdrop or organ culture dish (Falcon 35-3037)
Pipette assembly, mouth or hand-held, for embryo manipulation
Stereo dissecting microscope

PROCEDURE

1. Prepare the aggregation plates (Protocol 11.7) at least several hours or 1 day before the experiment.

2. Collect eight-cell-stage embryos of the first component (genotype) of the chimeras as described in Chapter 4. It is convenient to place collected embryos before zona removal in a microdrop or organ culture dish in the incubator until ready for the next steps.

3. Collect eight-cell-stage embryos of the second component (genotype) of the chimeras. Place them in the incubator for short-term culture using KSOM-AA or other embryo culture medium or remove the zona pellucida immediately.

4. Remove the zona pellucida (Protocol 11.8).

5. Place zona-free embryos individually into each of the depressions in the aggregation plate (e.g., six depressions in each microdrop as shown in Fig. 11.11A).

6. Repeat steps 4 and 5 for the embryos of the other genotype. Make sure that embryos of the second group are physically attached to the embryos of the first group when they are placed individually into the depressions. The embryo aggregates should look like those shown in Figure 11.11B.

7. Culture the aggregates in a humidified incubator with 5% CO_2 at 37°C for 24 hours.

8. After overnight culture, all of the aggregates should form a single embryo at the late morula or early blastocyst stage as it is shown for diploid embryo–ES cell aggregation in Figure 11.12D.

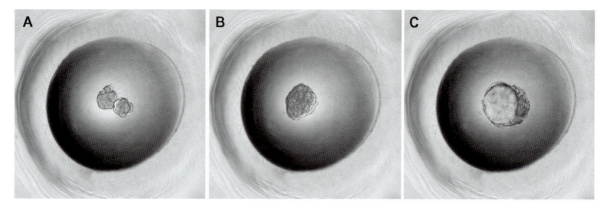

FIGURE 11.11. Aggregation of the diploid embryos. (*A*) The two eight-cell-stage embryos placed into a depression well. (*B*) Chimeric morula after 12 hours in culture. (*C*) Expanding chimeric blastocyst after 24 hours in culture.

COMMENTS

- Depending on the method of embryo production (natural matings or superovulation) and/or embryo genotype, the quality of the embryos may vary. For maximum efficiency, only perfect eight-cell- to morula-stage embryos should be used for aggregation. If the culture conditions were optimal and the embryos were minimally stressed, all of the aggregates will develop properly and can be transferred into pseudopregnant females after 24 hours in culture.
- Compacted morulae and blastocysts are transferred to the uterus of pseudopregnant 2.5-dpc recipient female mice as described in Chapter 6. Embryos without zonae are more

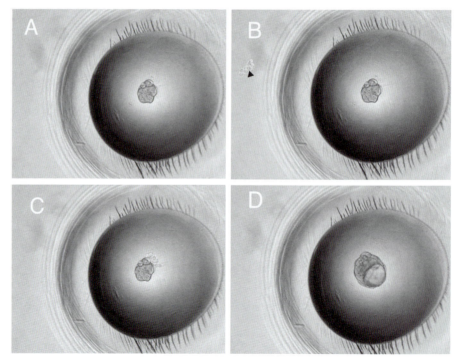

FIGURE 11.12. Aggregation of ES cells with diploid embryos. (*A*) Zona-free diploid embryo in a depression well. (*B*) Diploid embryo in a depression well, with a clump of ES cells (*arrow*) outside the well. (*C*) Clump of ES cells in a depression well, with a diploid embryo. (*D*) Aggregated blastocyst after overnight culture.

fragile and sticky and should be handled with care. In case of a shortage of recipients, morulae can be cultured for another night and expanded blastocysts transferred the following day into 2.5-dpc foster mothers; 3.5-dpc recipients also work but are less efficient than 2.5 dpc. It is also possible to transfer embryos into the oviduct of 0.5-dpc recipients. (See Chapter 6 for more details.)

Assembling Aggregates between ES Cells and Diploid Embryos

MATERIALS

EMBRYOS AND CELLS
ES cells (Protocol 11.6)
2.5-dpc embryos (Protocol 11.8)

EQUIPMENT
Aggregation plates (Protocol 11.7)
Dissecting microscope
Incubator, humidified, at 37°C, 5% CO_2
KSOM-AA medium or other embryo culture medium
Organ culture dish (Falcon 35-3037)
Pipette assembly, mouth or hand-held, for embryo manipulation
Tissue culture plate, gelatinized

PROCEDURE

1. Prepare ES cells by plating them 1 or 2 days prior to the aggregation onto a gelatinized tissue culture plate (see Chapter 8 and Protocol 11.6 for details).

2. Prepare the aggregation plates (Protocol 11.7) at least several hours or 1 day before the experiment.

3. Collect eight-cell- to morula-stage embryos as described in Chapter 4. If necessary, it is possible to place collected embryos before zona removal in the microdrop or organ culture dish in the incubator until you are ready for the next steps. Otherwise, place the collected embryos in HEPES-buffered medium and move on to step 4.

4. Remove the zona pellucida (Protocol 11.8).

5. Wash the embryos through several drops of equilibrated culture medium and place zona-free embryos individually into the small depressions of the aggregation plate. Place only one embryo into each depression (Fig. 11.12A).

6. Prepare ES cells for aggregation (Protocol 11.6).

7. Pick 100–200 clumps of loosely connected cells and transfer them into a large drop of KSOM-AA or equivalent embryo culture medium. Then pick from this drop again, but now try to enrich for the clumps containing the right number (8–15) of cells and transfer the clumps to the first and fourth rows of the aggregation plate that do not contain depressions (Fig. 11.8A). It is also possible to transfer clumps of the right size in a minimal volume of medium directly into the depression-free microdrops.

8. Perform the last selection for the best candidate ES cell clumps and transfer 6–10 of those into the middle of the drops containing the embryos in rows 2 and 3 (Figs. 11.8A and 11.12B).

9. Place one clump (after the final selection), touching each embryo sitting inside the depressions of the aggregation plate (Fig. 11.12C).

10. When the assembly of the aggregates is finished, carefully remove unused ES cell clumps from the middle of the drops.

11. Culture the aggregates in a humidified incubator with 5% CO_2 at 37°C for 24 hours.

12. After overnight culture, all of the aggregates should form a single embryo at the late-morula or early-blastocyst stage (Fig. 11.12D) and can be transferred to the uterus of a pseudopregnant female mouse (see Chapter 6).

COMMENTS

It is also possible to put zona-free embryos into the microdrops not inside, but next to, the depressions at step 5 of the protocol. In this case, clumps of ES cells are placed into depressions before the embryos, making it easier to see the clump size better. Both methods work equally well.

Assembling Aggregates between Diploid and Tetraploid Embryos

The procedure for producing diploid embryo–tetraploid embryo chimeras requires the timed combination of four-cell-stage tetraploid embryo production (Protocol 11.9) and the execution of the procedure described for diploid embryo–diploid embryo aggregation (Protocol 11.10A). Therefore, to produce this kind of chimera, perform the following procedure in an orderly manner.

MATERIALS

EMBRYOS
Diploid 2.5-dpc embryos (Protocol 11.8)
Tetraploid embryos (Protocol 11.9)

EQUIPMENT
Aggregation plates (Protocol 11.7)
Incubator, humidified, at 37°C, 5% CO_2
Organ culture dish (Falcon 35-3037)
Pipette assembly, mouth or hand-held, for embryo manipulation
Stereo dissecting microscope

PROCEDURE

1. Prepare a microdrop culture plate several hours or 1 day before tetraploid embryo production.

2. Collect two-cell-stage embryos at 1.5 dpc, fuse them as described in Protocol 11.9, and culture them overnight to produce four-cell-stage tetraploid embryos for aggregation with diploid embryos the next day.

3. Prepare the aggregation plates (Protocol 11.7) at least several hours or 1 day before the aggregation experiment.

4. Collect eight-cell-stage embryos as described in Chapter 4. It is convenient to place collected embryos before zona removal in a microdrop or organ culture dish in the incubator until ready for the next steps. Alternatively, remove zona pellucida immediately.

5. Remove zonae pellucidae from diploid embryos (Protocol 11.8).

6. Place the zona-free diploid embryos individually into each of the depressions in the aggregation plate (e.g., six depressions in each microdrop as shown in Fig. 11.11A).

7. Repeat steps 4 and 5 for tetraploid embryos. Make sure that the tetraploid embryos are physically attached to the embryos of the first group when they are placed individually into the depressions (as shown for diploid embryo aggregation in Fig. 11.11B).

8. Culture the aggregates in a humidified incubator with 5% CO_2 at 37°C for 24 hours.

9. After overnight culture, all of the aggregates should form a single embryo at the late-morula or early-blastocyst stage (as shown for diploid embryo–ES cell aggregation in Fig. 11.12D) and can be transferred into the uterus of 2.5-dpc pseudopregnant females (see Chapter 6).

COMMENTS

- Although the tetraploid component frequently contributes to the extraembryonic membranes, the tetraploid cells are at a disadvantage even here. This results in less than 50% contribution from this compartment in almost all the chimeras.

- Occasionally it happens that a small contribution of tetraploid cells is observed in the embryo proper. This tetraploid "contamination" affects less than 10% of the embryos. This is a small portion of the embryos, but it is important to keep in mind, and to design the experiment in a way that makes these embryos easily recognizable. Otherwise, this contamination might make phenotypic interpretations difficult.

Assembling Aggregates between ES Cells and Tetraploid Embryos

The procedure for producing ES cell–tetraploid embryo chimeras requires the timed combination of four-cell-stage tetraploid embryo production and the execution of the procedure described for ES cell–diploid embryo aggregation (Protocol 11.10B) with the replacement of the diploid embryos with tetraploid embryos. Therefore, for production of this kind of chimera, perform the following procedure in an orderly manner.

MATERIALS

EMBRYOS
ES cells (Protocol 11.6)
Tetraploid embryos (Protocols 11.8 and 11.9)

EQUIPMENT
Aggregation plates (Protocols 11.8 and 11.7)
Incubator, humidified, at 37°C, 5% CO_2
Organ culture dish (Falcon 35-3037)
Pipette assembly, mouth or hand-held, for embryo manipulation
Stereo dissecting microscope
Tissue culture plate, gelatinized

REAGENTS
KSOM-AA medium or other embryo culture medium

PROCEDURE

1. Prepare ES cells by plating them 1 or 2 days prior to the aggregation onto a gelatinized tissue culture plate (for details, see Chapter 8 and Protocol 11.6). Prepare a microdrop culture plate several hours or 1 day before tetraploid embryo production.

2. Collect two-cell-stage embryos at 1.5 dpc, fuse them as described in Protocol 11.9, and culture them overnight to produce four-cell-stage tetraploid embryos for aggregation with ES cells the next day.

3. Prepare the aggregation plates (Protocol 11.7) at least several hours or 1 day before the aggregation experiment.

4. Remove the zonae pellucidae from the tetraploid embryos (Protocol 11.8).

5. Place the zona-free tetraploid embryos individually into each of the depressions in the aggregation plate (e.g., six depressions in each microdrop; see Comments below).

6. Prepare ES cells for aggregation (Protocol 11.6).

7. Transfer several clumps of the right size (10–15 cells) in a minimal volume of medium directly into the depression-free microdrops (follow steps in Protocol 11.10B).

8. Place one clump of 10–15 cells touching each tetraploid embryo in the depressions of the aggregation plate. Similar to aggregation with diploid embryos, it is also possible to put zona-free embryos in the microdrops next to the depressions. If this method is used, clumps of ES cells are placed into the depressions before the embryos. Both methods work equally well.

9. Culture the aggregates in a humidified incubator with 5% CO_2 at 37°C for 24 hours.

10. After overnight culture, all of the aggregates should form a single embryo at the late-morula or early-blastocyst stage and can be transferred to the uterus of a pseudopregnant female mouse (see Chapter 6).

COMMENTS

- It is more efficient to use two tetraploid embryos with a clump of ES cells in one aggregate when the experiment allows (i.e., embryos of the same genotype in sufficient number are used). When two tetraploid embryos are aggregated with a clump of ES cells, care should be taken that all three components touch each other so they form one embryo after overnight culture (Fig. 11.13).

- Occasionally it happens that a small contribution of tetraploid cells is observed in the embryo proper. This tetraploid "contamination" affects less than 10% of the embryos. It is a small portion of the embryos, but it is important to keep in mind and to design the experiment in a way that makes these embryos easily recognizable. Otherwise this contamination may make phenotypic interpretations difficult.

- Some laboratories culture tetraploid embryos into the blastocyst stage and successfully use these embryos for ES cell injection (Protocol 11.5), for deriving complete ES-cell embryos or animals (Wang et al. 1997).

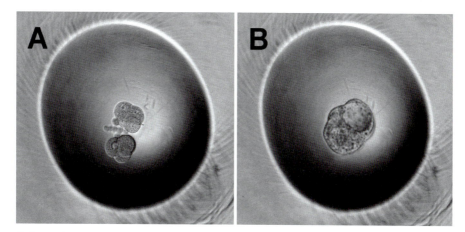

FIGURE 11.13. Aggregation of tetraploid embryos with ES cells using two tetraploid embryos. (*A*) Two tetraploid embryos are sandwiching a clump of ES cells. (*B*) The aggregate develops into a blastocyst after overnight culture.

Disaggregrating Cleavage-stage Embryos and the Inner Cell Mass of Blastocysts into Individual Cells

This procedure is adapted from Ziomek and Johnson (1980).

MATERIALS

EMBRYOS

Embryos that have been allowed to recover for ~1 hour after removing the zona pellucida (see Protocol 11.8) or ICMs isolated by immunosurgery (see Protocol 11.12)

EQUIPMENT

Culture dishes, bacteriological plastic
Incubator, humidified, at 37°C, 5% CO_2, 95% air
Pipette, flame-polished glass

REAGENTS

Calcium-free M16 medium plus 6 mg/ml bovine serum albumin (BSA), rather than the usual M16 medium plus 4 mg/ml BSA (see Chapter 4)
M16 medium (see Chapter 4)
Paraffin oil, embryo-tested light (see Chapter 4)

PROCEDURE

1. Transfer the ICM or embryos without zonae to microdrops of calcium-free M16 medium plus 6 mg/ml BSA under light paraffin oil (see Protocol 4.5). Incubate in a humidified incubator with 5% CO_2 at 37°C for 10–15 minutes. Disaggregate the blastomeres or ICM cells by pipetting them through a flame-polished glass pipette.

2. Remove the blastomeres or ICM cells from the calcium-free M16 medium as soon as possible. The isolated blastomeres and ICM cells are very sticky, so place them individually or in small groups in single drops of conventional M16 medium in petri dishes to which they will not adhere (e.g., bacteriological plastic culture dishes).

COMMENTS

- Cell death may be reduced by transferring the embryos into M16 medium containing calcium before disaggregating the blastomeres by pipetting (C.F. Graham, pers. comm.).
- It is also possible to incubate zona-free embryos in 0.05% trypsin, 0.53 mM EDTA solution for 2–5 minutes at room temperature and dissect them using a pulled glass capillary. Zona-free embryos should recover in a microdrop culture for about 30 minutes before trypsin treatment (C. Chazaud, pers. comm.).

Immunosurgery: Isolating the Inner Cell Mass of Blastocysts

This technique was first described by Solter and Knowles (1975) for selectively killing the outer trophectoderm (TE) of the blastocyst, sparing the inner cell mass (ICM). A schematic representation is shown in Figure 11.14 .

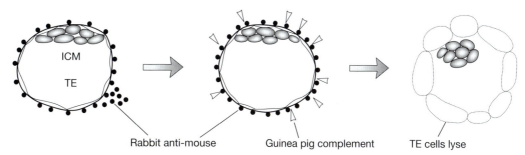

Rabbit anti-mouse   Guinea pig complement   TE cells lyse

FIGURE 11.14. Immunosurgery. Schematic representation of the technique of immunosurgery for isolating inner cell masses. Blastocysts are incubated with rabbit anti-mouse serum (*closed circles*), washed thoroughly, and exposed to guinea pig complement (*triangles*). Only the outer trophectoderm (TE) cells are lysed; the ICM cells are protected from exposure to the rabbit antibodies by the tight permeability seal of the trophectoderm.

MATERIALS

ANIMALS
Guinea pig serum (see step 2)
Mouse spleen or tissue culture cells (see step 1)
Pregnant female mice (3.5–4.5 dpc)
Rabbit

EQUIPMENT
Incubator, humidified, at 37°C, 5% CO_2, 95% air
Pasteur pipette, finely drawn (*optional*)

REAGENTS
Agar (Difco Noble) or agarose (Sigma A6013; Calbiochem 121852)
Dulbecco's modified Eagle's medium (DMEM) or similar medium
Guinea pig serum

> The serum can be either purchased (e.g., Murex Diagnostics or GIBCO) as freeze-dried guinea pig serum or prepared from fresh guinea pig blood obtained by cardiac puncture. However, the source of the guinea pig appears to be important. It is best to use animals kept under specific-pathogen-free (SPF) conditions because they are less likely to have developed natural antibodies against bacterial carbohydrates that may cross-react with carbohydrates on the surface of mouse cells. If the serum batch is toxic to the mouse cells, it is sometimes possible to remove the endogenous antibodies by preadsorbing the serum with agar (Cohen and Schlesinger 1970). Rat serum can be used instead.

HEPES-buffered DMEM (20 mM) plus 10% heat-inactivated fetal bovine serum (FBS)
M2 medium (see Chapter 4)
Rabbit anti-mouse serum (prepared in advance; Rockland Immunochemicals)

PROCEDURE

1. To prepare rabbit anti-mouse serum, bleed a rabbit 10 days after three injections at 14-day intervals of 4×10^8 mouse cells (e.g., spleen cells or mouse tissue culture cells from any strain of mouse). Allow the blood to clot, and collect the serum. Heat-inactivate the serum by placing at 56°C for 30 minutes to destroy complement and store in aliquots at –70°C.

2. For guinea pig complement, it is sufficient to use guinea pig serum rather than purified complement. Dilute serum 1:3 with DMEM to 3 ml and add 80 mg of agar or agarose. Let the diluted serum stand on ice for 30–60 minutes with occasional shaking and then centrifuge at 4°C to pellet the agar. Store the supernatant in aliquots at –70°C. This serum can usually be used without further dilution.

3. Collect blastocysts as described previously (see Protocol 4.11). Remove the zona pellucida (Protocol 11.8) either before adding the anti-mouse serum or after lysing the trophectoderm cells with complement. Antibodies and complement can pass through the zona pellucida. By leaving the zona removal step to last, it is possible to obtain very clean ICMs free of trophectoderm debris.

4. Transfer the blastocysts in a few microliters of medium (e.g., M2 medium or 20 mM HEPES-buffered DMEM plus 10% heat-inactivated FBS) to a much larger volume (e.g., 3 ml) of DMEM containing rabbit anti-mouse serum diluted 1:30 to 1:100 (the optimal dilution should be determined in preliminary tests).

5. Incubate the blastocysts in a humidified incubator with 5% CO_2 at 37°C for 10 minutes. Antibodies will bind to the outer TE cells, but they are prevented by the zonular tight junctions between these cells from penetrating the blastocyst cavity and binding to the ICM cells.

6. Wash the blastocysts in two changes of 20 mM HEPES-buffered DMEM plus 10% heat-inactivated FBS.

7. Add guinea pig serum (diluted 1:3 to 1:50) as prepared in step 2 and incubate in a humidified incubator with 5% CO_2 at 37°C for 30 minutes. The outer TE cells should begin to lyse within a few minutes.

8. Wash the ICMs and transfer them to culture medium. The dead TE cells can be removed by gently pipetting the ICMs in a finely drawn Pasteur pipette, but this removal is not necessary for subsequent development.

COMMENT

In vitro growth and differentiation of isolated ICMs have been described previously (Solter and Knowles 1975; Hogan and Tilly 1978a,b; Wiley et al. 1978). The immunosurgery process can be repeated on the ICMs that have differentiated a continuous outer layer of endoderm (Hogan and Tilly 1977).

TROUBLESHOOTING FOR EMBRYO TRANSFER, CHIMERA PRODUCTION, AND BREEDING FOR GERM-LINE TRANSMISSION

1. **The plugging rate of donor females is low.**
   - See Chapter 7, Troubleshooting, point 1.
   - The stud males have been mated too often. Stud males can be naturally mated with two adult females 2–3 days in a row. However, after such a mating period, the studs should be given 3–4 days' rest before the next time they are mated.

2. **The number of embryos isolated per donor female is low.**
   - See Chapter 7, Troubleshooting, point 2.
   - The donor females are of suboptimal age. Naturally mated animals should be at least 8 weeks old, and at the most 16 weeks old.

3. **The pregnancy rate after blastocyst transfer is low.**
   - See Chapter 7, troubleshooting point 13.

4. **Very few or no pups are born after blastocyst transfer.**
   - The embryo transfer surgery was not performed properly.
   - The environment in the animal room is suboptimal (see Chapter 7, Troubleshooting, point 1).
   - The developmental potential of the ES cells is low. Embryos have incorporated many ES cells and do not survive. Perform karyotyping or chromosome counting of the clone (see Protocol 8.6). Check the undifferentiated morphology of the ES cells in culture. Use a different clone. Make sure that the ES cell culture conditions are optimal (see Chapter 8).
   - Too many ES cells have been injected or aggregated. Do not use more than 15 cells per embryo.
   - Suboptimal culture conditions. Make sure that the culture medium is optimal, and that the incubator maintains 37°C and 5% CO_2.

5. **Many pups are born, but very few or none of them are chimeric.**
   - Too few ES cells have been injected or aggregated. Keep the number of cells close to 15 per embryo or increase this number.
   - The developmental potential of the ES cells is low. The embryo has selected host cells against ES cells. In this case, use another ES cell clone.

6. **The sex distribution among good chimeras is not biased toward males, and most good chimeras are female.**
   - If the ES cell line used is XY, many cells may have lost the Y chromosome during culture and are XO. Females with very high ES-derived coat color can potentially contribute to the germ line and should be bred; however, the chances for germ-line transmission are greatly reduced. If possible, injections/aggregations should be repeated with a different clone.

7. **The chimeras do not produce any offspring at all.**
   - Incomplete sex conversion resulting in incomplete differentiation or hermaphroditism. Use other chimeras (produce more chimeras).
   - Lethal or severe phenotype caused by a targeted mutation on the X chromosome. Mutation of an imprinted autosomal gene can also cause heterozygous lethality or mutant phenotype.
   - Haploinsufficient phenotype.

8. **The chimeras do not give germ-line transmission.**
   - The developmental potential of ES cells is low (see points 2 and 3).

9. **The chimeras produce offspring derived from the ES cells according to the coat color, but the right genotype cannot be identified.**
   - The wrong ES cell clone was used.

- Wild-type ES cells contaminated the clone.
- The introduced genome alteration or transgene is heterozygous (or hemizygous) lethal.
- The genetic alteration cell-autonomously interferes with germ cell formation.
- The targeted gene is uniparentally expressed (imprinted) by the parental genome equivalent to the gender of germ-line chimeras.

REFERENCES

Adams R.H., Porras A., Alonso G., Jones M., Vintersten K., Panelli S., Valladares A., Perez L., Klein R., and Nebreda A.R. 2000. Essential role of p38α MAP kinase in placental but not embryonic cardiovascular development. *Mol. Cell* **6:** 109–116.

Barbacci E., Reber M., Ott M.O., Breillat C., Huetz F., and Cereghini S. 1999. Variant hepatocyte nuclear factor 1 is required for visceral endoderm specification. *Development* **126:** 4795–4805.

Beddington R.S. and Robertson E.J. 1989. An assessment of the developmental potential of embryonic stem cells in the midgestation mouse embryo. *Development* **105:** 733–737.

Bi W., Huang W., Whitworth D.J., Deng J.M., Zhang Z., Behringer R.R., and de Crombrugghe B. 2001. Haploinsufficiency of Sox9 results in defective cartilage primordia and premature skeletal mineralization. *Proc. Natl. Acad. Sci.* **98:** 6698–6703.

Bowman P. and McLaren A. 1969. The reaction of the mouse blastocyst and its zona pellucida to pH changes in vitro. *J. Reprod. Fertil.* **18:** 139–140.

Bradley A. and Robertson E. 1986. Embryo-derived stem cells: A tool for elucidating the developmental genetics of the mouse. *Curr. Top. Dev. Biol.* **20:** 357–371.

Bradley A., Evans M., Kaufman M.H., and Robertson E. 1984. Formation of germ-line chimaeras from embryo-derived teratocarcinoma cell lines. *Nature* **309:** 255–256.

Brun J.L. and Psychoyos A. 1972. Dissolution of the rat zona pellucida by acidified media and blastocyst viability. *J. Reprod. Fertil.* **30:** 489–491.

Carmeliet P., Ferreira V., Breier G., Pollefeyt S., Kieckens L., Gertsenstein M., Fahrig M., Vandenhoeck A., Harpal K., Eberhardt C., Declercq C., Pawling J., Moone L., Collen D., Risaw W., and Nagy A. 1996. Abnormal blood vessel development and lethality in embryos lacking a single VEGF allele. *Nature* **380:** 435–439.

Ciruna B.G., Schwartz L., Harpal K., Yamaguchi T.P., and Rossant J. 1997. Chimeric analysis of fibroblast growth factor receptor-1 (Fgfr1) function: A role for FGFR1 in morphogenetic movement through the primitive streak. *Development* **124:** 2829–2841.

Cohen A. and Schlesinger M. 1970. Absorption of guinea pig serum with agar. *Transplantation* **10:** 130–132.

Damert A., Miquerol L., Gertsenstein M., Risau W., and Nagy A. 2002. Insufficient VEGF-A activity in yolk sac endoderm compromises hematopoietic and endothelial differentiation. *Development* **129:** 1881–1892.

Dinkel A., Aicher W.K., Wamatz K., Burki K., Eibel H., and Ledermann B. 1999. Efficient generation of transgenic BALB/c mice using BALB/c embryonic stem cells. *J. Immunol. Methods* **223:** 255–260.

Duncan S.A., Nagy A., and Chan W. 1997. Murine gastrulation requires HNF-4 regulated gene expression in the visceral endoderm: Tetraploid rescue of Hnf-4(–/–) embryos. *Development* **124:** 279–287.

Eggan K., Akutsu H., Loring J., Jackson-Grusby L., Klemm M., Rideout W.M., 3rd, Yanagimachi R., and Jaenisch R. 2001. Hybrid vigor, fetal overgrowth, and viability of mice derived by nuclear cloning and tetraploid embryo complementation. *Proc. Natl. Acad. Sci.* **98:** 6209–6214.

Eggan K., Rode A., Jentsch I., Samuel C., Hennek T., Tintrup H., Zevnik B., Erwin J., Loring J., Jackson-Grusby L., Speicher M.R., Kuehn R., and Jaenisch R. 2002. Male and female mice derived from the same embryonic stem cell clone by tetraploid embryo complementation. *Nat. Biotechnol.* **20:** 455–459.

Gardner R.L. 1968. Mouse chimeras obtained by the injection of cells into the blastocyst. *Nature* **220:** 596–597.

Guillemot F., Nagy A., Auerbach A., Rossant J., and Joyner A.L. 1994. Essential role of Mash-2 in extraembryonic development. *Nature* **371:** 333–336.

Gwatkin R.B.L. 1964. Effect of enzymes and acidity on the zona pellucida of the mouse egg before and after fertilization. *J. Reprod. Fert.* **7:** 99.

Hogan B. and Tilly R. 1978a. In vitro development of inner cell masses isolated immunosurgically from mouse blastocysts. I. Inner cell masses from 3.5-day p.c. blastocysts incubated for 24 h before immunosurgery. *J. Embryol. Exp. Morphol.* **45:** 93–105.

———. 1978b. In vitro development of inner cell masses isolated immunosurgically from mouse blastocysts. II. Inner cell masses from 3.5- to 4.0-day p.c. blastocysts. *J. Embryol. Exp. Morphol.* **45:** 107–121.

———. 1977. In vitro culture and differentiation on normal mouse blastocysts. *Nature* **265:** 626–629.

Kawase Y., Iwata T., Watanabe M., Kamada N., Ueda O., and Suzuki H. 2001. Application of the piezo-micromanipulator for injection of embryonic stem cells into mouse blastocysts. *Contemp. Top. Lab. Anim. Sci.* **40:** 31–34.

Kubiak J.Z. and Tarkowski A.K. 1985. Electrofusion of mouse blastomeres. *Exp. Cell. Res.* **157:** 561–566.

Lemckert F.A., Sedgwick J.D., and Korner H. 1997. Gene targeting in C57BL/6 ES cells. Successful germ line transmission using recipient BALB/c blastocysts developmentally matured in vitro. *Nucleic Acids Res.* **25:** 917–918.

Lindahl P., Hellstrom M., Kalen M., Karlsson L., Pekny M., Pekna M., Soriano P., and Betsholtz C. 1998. Paracrine PDGF-B/PDGF-Rβ signaling controls mesangial cell development in kidney glomeruli. *Development* **125:** 3313–3322.

Mintz B. 1962. Formation of genetically mosaic mouse embryos. *Am. Zool.* **2:** 432.

Nagy A. and Rossant J. 1999. Production and analysis of ES-cell aggregation chimeras. In *Gene targeting: A practical approach* (ed. A. Joyner), pp. 177–206. Oxford University Press, New York.

———. 2001. Chimaeras and mosaics for dissecting complex mutant phenotypes. *Int. J. Dev. Biol.* **45:** 577–582.

Nagy A., Rossant J., Nagy R., Abramow-Newerly W., and Roder J.C. 1993. Derivation of completely cell culture-derived mice from early-passage embryonic stem cells. *Proc. Natl. Acad. Sci.* **90:** 8424–8428.

Nagy A., Gocza E., Diaz E.M., Prideaux V.R., Ivanyi E., Markkula M., and Rossant J. 1990. Embryonic stem cells alone are able to support fetal development in the mouse. *Development* **110:** 815–821.

Nicolson G.L., Yanagimachi R., and Yanagimachi H. 1975. Ultrastructural localization of lectin-binding sites on the zonae pellucidae and plasma membranes of mammalian eggs. *J. Cell. Biol.* **66:** 263–274.

Papaioannou V. and Johson R. 1993. Production of chimeras and genetically defined offspring from targeted ES cells. In *Gene targeting: A practical approach*, (ed. A. Joyner), pp. 107–146: IRL Press at Oxford University Press, United Kingdom.

Rivera-Perez J.A., Mallo M., Gendron-Maguire M., Gridley T., and Behringer R.R. 1995. Goosecoid is not an essential component of the mouse gastrula organizer but is required for craniofacial and rib development. *Development* **121:** 3005–3012.

Robertson E., Bradley A., Kuehn M., and Evans M. 1986. Germ-line transmission of genes introduced into cultured pluripotential cells by retroviral vector. *Nature* **323:** 445–448.

Rossant J. and Spence A. 1998. Chimeras and mosaics in mouse mutant analysis. *Trends Genet.* **14:** 358–363.

Schuster-Gossler K., Lee A.W., Lerner C.P., Parker H.J., Dyer V.W., Scott V.E., Gossler A., and Conover J.C. 2001. Use of coisogenic host blastocysts for efficient establishment of germline chimeras with C57BL/6J ES cell lines. *Biotechniques* **31:** 1022–1024.

Simpson E.M., Linder C.C., Sargent E.E., Davisson M.T., Mobraaten L.E., and Sharp J.J. 1997. Genetic variation among 129 substrains and its importance for targeted mutagenesis in mice. *Nat. Genet.* **16:** 19–27.

Solter D and Knowles B.B. 1975. Immunosurgery of mouse blastocyst. *Proc. Natl. Acad. Sci.* **72:** 5099–5102.

Tanaka M., Gertsenstein M., Rossant J., and Nagy A. 1997. Mash2 acts cell autonomously in mouse spongiotrophoblast development. *Dev. Biol.* **190:** 55–65.

Tanaka S., Kunath T., Hadjantonakis A.K., Nagy A., and Rossant J. 1998. Promotion of trophoblast stem cell proliferation by FGF4. *Science* **282:** 2072–2075.

Tarkowski A.K. 1961. Mouse chimeras developed from fused eggs. *Nature* **184:** 1286–1287.

Tarkowski A.K., Witkowska A., and Opas J. 1977. Development of cytochalasin in B-induced tetraploid and diploid/tetraploid mosaic mouse embryos. *J. Embryol. Exp. Morphol.* **41:** 47–64.

Threadgill D.W., Dlugosz A.A., Hansen L.A., Tennenbaum T., Lichti U., Yee D., LaMantia C., Mourton T., Herrup K., Harris R.C. et al. 1995. Targeted disruption of mouse EGF receptor: Effect of genetic background on mutant phenotype. *Science* **269:** 230–234.

Varlet I., Collignon J., and Robertson E.J. 1997. Nodal expression in the primitive endoderm is required for specification of the anterior axis during mouse gastrulation. *Development* **124:** 1033–1044.

Wagner E.F., Keller G., Gilboa E., Ruther U., and Stewart C. 1985. Gene transfer into murine stem cells and mice using retroviral vectors. *Cold Spring Harbor Symp. Quant. Biol.* **50:** 691–700.

Wang Z.Q., Kiefer F., Urbanek P., and Wagner E.F. 1997. Generation of completely embryonic stem cell-derived mutant mice using tetraploid blastocyst injection. *Mech. Dev.* **62:** 137–145.

Wiley L.M., Spindle A.I., and Pederson R.A. 1978. Morphology of isolated mouse inner cell masses developing in vitro. *Dev. Biol.* **6:** 1–10.

Wood S.A., Allen N.D., Rossant J., Auerbach A., and Nagy A. 1993. Non-injection methods for the production of embryonic stem cell-embryo chimaeras. *Nature* **365:** 87–89.

Yamamoto H., Flannery M.L., Kupriyanov S., Pearce J., McKercher S.R., Henkel G.W., Maki R.A., Werb Z. and Oshima R.G. 1998. Defective trophoblast function in mice with a targeted mutation of Ets2. *Genes Dev.* **12:** 1315–1326.

Ziomek C.A. and Johnson M.H. 1980. Cell surface interaction induces polarization of mouse 8-cell blastomeres at compaction. *Cell* **2:** 935–942.

Detection and Analysis of Mouse Genome Alterations and Specific Sequences

T HIS CHAPTER DESCRIBES methods for identifying mice and for analyzing transgenic or targeted mutation carrier embryos or animals. Included are the current methodologies that apply to easily readable tagging of animals. The analysis of DNA samples isolated from tissues by polymerase chain reaction (PCR) and Southern blot analysis or by detecting reporter gene expression is described. Some experiments may involve collection of mouse embryos before they have come to term, and thus smaller-scale methods for screening DNA from tissue fragments or extraembryonic membranes are also included. Methods are given for measuring transgene copy number, determining the number of integration sites, and localizing transgenes to specific chromosomes. Additional proto-

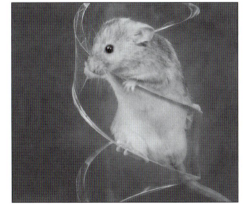

cols for analyzing transgenes, including staining transgenic embryos and tissues for reporter expression such as LacZ and human placental alkaline phosphatase, and for visualizing transgene expression (in situ hybridization, immunohisto-chemistry, fluorescent proteins) can be found in Chapter 16.

CONTENTS

IDENTIFICATION OF ANIMALS

Permanent individual tagging is essential for almost all the aspects of animal management involving transgenic mice. Current methods have different advantages and disadvantages and allow identification tagging at different developmental stages. The application of one method over another depends on the special needs of a particular experiment as well as on local systems and institutional regulations.

One of the most common mouse identification methods is ear punching. Figure 12.1 shows two commonly used numerical codes. Both schemes allow mice to be numbered from 1 to 99 and are sufficient for most purposes when combined with cage card identification and records of sex, age, and coat color. Small ear punches are available from Fisher Scientific (01-337B). A stainless steel ear punch with a handle can be purchased from Roboz Surgical Instruments (http://www.roboz.com), or AgnTho's AB, Sweden (http://www.agnthos.se). Animals as young as 2 weeks old can be ear-punched. Combination of early identification with genotyping of animals from the resulting ear tissue saves space because only positively identified animals are weaned. However, identification and tail clipping for DNA sample are often done at weaning (3–4 weeks of age).

The ear-punch method does not allow unique marking of every mouse in a colony. Toe clipping can generate a larger number series and allows identifying animals as young as newborns. However, this procedure is not permitted in

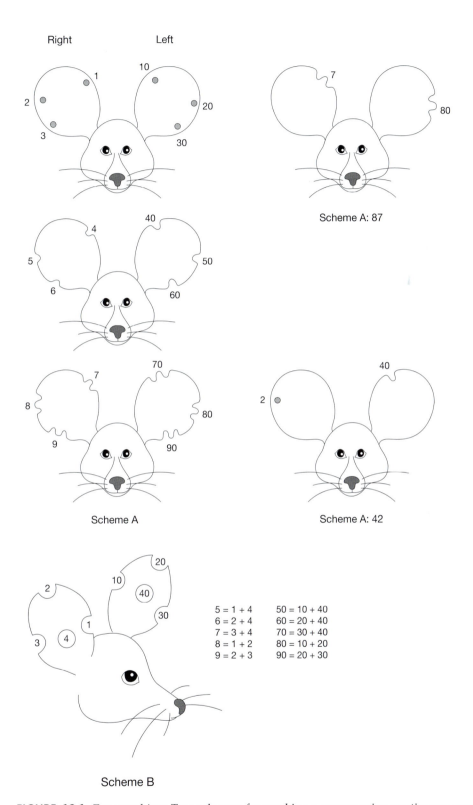

FIGURE 12.1. Ear punching: Two schemes for marking a mouse using continuous number series from 1 to 99.

some animal facilities because toe clipping of the forepaw can potentially reduce the mouse's ability to grip surfaces and groom itself. Nevertheless, because of the limited alternatives for identifying neonatal mice, toe clipping may be an option if your institution's regulations allow it. Toe clipping should be performed only on neonates or juveniles not older than 10–12 days and can be done without anesthesia. Only the third phalanx of the toe is excised to preclude the nail from regenerating.

Other alternatives for unique number identifications include ear tags, tattooing, and subcutaneous implants of microchips. Small animal ear tags contain number–letter combinations, require an applicator, and are suitable for post-weanling mice. Ear tags are available from National Band & Tag Company (1005-1; http://www.nationalband.com). It is important to disinfect the applicator and ear tag prior to insertion to avoid bacterial infection and inflammation of the ear. Care should be taken to place a tag vertically, slightly below the center of the ear, leaving a small amount of space between the margin of the ear and inside edge of the tag (see Fig 12.2).

Tattoo machines are available from AIMS, Animal Identification and Marking Systems (tel: 908-884-9105), or Ancare Corporation. Subcutaneous microchip implants are available from various companies (e.g., Bio Medic Data Systems: http://www.bmds.com; Destron Fearing: http://www.destron-fearing.com; Electronic ID: http://www.electronicidinc.com; Plexx: http://www.plexx.nl). Despite its relatively high cost, certain risk of tissue reaction, incidence of tumor development, microchip failure, and migration problems, this technology can be useful for large-scale, long-term studies.

TISSUE SAMPLES AND DNA PREPARATION FOR GENOTYPING OR CHARACTERIZATION OF TRANSGENES

A small sample of dispensable tissue is collected for transgene detection, characterization, or genotyping of potential transgenic animals. These samples are the source for preparing DNA, which is used either in Southern blotting, dot blotting, or PCR. The two blotting assays require high-quality DNA preparation (see Protocol 12.1 and 12.2), but this is not necessarily required for PCR. A simple cell

FIGURE 12.2. Correct placement of ear tag.

or tissue lysate can provide sufficient quality and amount of template DNA for PCR (see Protocol 12.3).

The tissue from an ear punch or a clipped toe yields sufficient DNA for PCR analysis. Southern blotting usually requires a larger amount of DNA, which can be obtained from tail tips. A small piece from the tip of the tail is removed and processed as described in Protocol 12.1. Marking of animals with ear punch or toe clipping ideally can be combined with obtaining tissues for DNA, since most marking methods automatically provide a small piece of tissue. In this case, the tissues are placed into labeled microcentrifuge tubes on ice for immediate processing (see Protocol 12.3B) followed by PCR (Protocol 12.4). Great care should be taken when cleaning tools between samples to avoid risk of carryover contamination.

There is no consensus regarding the administration or nonadministration of anesthesia for tail tip excision. The procedure should be performed according to local regulations and your animal care committee's guidelines. Generally, tail excision should be performed on mice between 2 and 4 weeks of age because younger mice heal faster and bleed less. If tail clipping is done on animals older than 4 weeks, a fast-acting inhalant or local anesthetic (e.g., ethyl chloride) may be recommended. Sterilized sharp surgical scissors, scalpels, or razor blades are usually used for tail clipping. Instruments are sterilized between use for each animal (e.g., in a glass bead sterilizer) or disinfected with 70% alcohol. The use of a glass bead sterilizer (see Chapter 17) has the additional advantage of eliminating bleeding due to cauterization.

Often, genotyping of prenatal embryos is required, and a small part, such as yolk sac, allantois, or limb bud, can be used for this purpose. The yolk sac of midgestation or later-stage embryos provides a sufficient amount of DNA for Southern analysis (see Protocol 12.2). Small tissues of a few hundred cells are used for genotyping early postimplantation- and preimplantation- (Latham et al. 2000) stage embryos by PCR. It is worth noting here that PCR could work on templates prepared from fixed and/or processed embryos as well; for example, early postimplantation embryos that have gone through the long process of in situ RNA hybridization (see Comment after Protocol 12.3).

DETECTION AND ANALYSIS OF TRANSGENE PRODUCED BY PRONUCLEAR INJECTION

Potential founder transgenic mice produced by pronuclear injection may be initially identified by Southern blot analysis (Southern 1975). Using PCR as a primary transgene detection method is recommended only if its reliability has been well established in former experiments (Saiki et al. 1985). If this is not the case, then the development of the PCR screen should follow the successful Southern detection of the transgene or targeted allele by using the Southern blot for both positive and negative controls. For PCR analysis, it is necessary to choose and synthesize two primers that will amplify a specific band of ~200–1000 base pairs from the transgene. The selection of primers is one of the most important parts of establishing a reliable PCR. The important parameters to look for are the correct length (usually 18–22 base pairs), the primer GC content (somewhere between 40% and 60%), and the correct annealing temperature. There are several publications and manuals (Innis and Gelfand 1990; Innis et al. 1990) that can help in the design of the conditions and primers. In addition, several on-line Web sites offer assistance for PCR primer design, for example, the Primer3 program and Web site (http://www-genome.wi.mit.edu/cgi-bin/primer/primer3_www.cgi).

Once a mouse transgenic line has been bred for 2–3 generations and it is clear that the line carries only a single transgene integration site (see below), PCR analysis is often sufficient for screening the progeny for inheritance of the transgene, as long as PCR results correlate with previous Southern data. However, always keep in mind that this assay is more susceptible to false-positive results due to DNA contamination. On the other hand, the presence of bands of the predicted sizes on a Southern blot is a fairly good indication of a positive transgenic mouse.

The choice of restriction enzymes for Southern blot analysis of the integrated DNA is important, as is the choice of the hybridization probe. Normally, the injected transgene, or a plasmid containing the transgene, can simply be labeled by nick translation or by random priming (Sambrook et al. 2001) and used as the hybridization probe. If the transgene contains a repetitive sequence that would hybridize to related sequences in the host genome, then a subfragment, free of repetitive sequences, may have to be used as a probe instead.

Southern blotting also provides information about the structure, integrity, and copy number of the inserted DNA sequences. When choosing a restriction enzyme, keep in mind that in most cases, the injected transgene will integrate in a head-to-tail array. If a restriction enzyme that has no sites in the transgene DNA is selected, the fragment produced will be larger than the entire head-to-tail array, assuming that the transgene has integrated as a simple array and has not undergone any other rearrangements. Initially, it is preferable to use a restriction enzyme that will yield bands of predictable sizes. A restriction enzyme that cuts once in each injected molecule will yield a band of the same length as the injected fragment if a head-to-tail array has been formed. The enzyme should also yield a "junction fragment(s)" containing a genomic sequence of novel lengths from one or both ends of the array. If a single copy has been integrated, such an enzyme should yield only the junction fragment(s) of novel length. Regardless of whether one copy or a head-to-tail array has integrated, an enzyme that cuts twice in the transgene will yield a predictable band on a Southern blot, derived from an internal fragment in the transgene. In the case of a head-to-tail array, a second predictable fragment representing the junction between adjacent members of the array will be generated. Occasionally, however, head-to-head or tail-to-tail joins can form, and their formation further complicates the band patterns by predictable sizes (i.e., of the head-to-head or tail-to-tail joins).

To determine the number of copies of a transgene that are integrated in the genome of a transgenic mouse, it is first necessary to determine accurately the concentration of a genomic DNA preparation from that mouse. Many DNA preparations, such as those purified from the tail (described below), may be heavily contaminated with RNA. Therefore, the sample UV absorbance at 260 nm will not accurately reflect the DNA concentration, and a fluorometric DNA determination should be performed instead. A very convenient assay to determine DNA concentration uses Hoechst 33258 fluorescent dye (Sigma B2883), which binds specifically to DNA (see, e.g., Brunk et al. 1979; Labarca and Paigen 1980). An estimate of transgene copy number can be obtained by including standard amounts of the injected transgene, each mixed with 5–10 µg of wild-type mouse DNA, in parallel lanes on a Southern blot of transgenic mouse DNAs. A dot-blot assay should be used to provide more accurate DNA measurements. Varying amounts of genomic DNA (typically 0.5, 1, 2, and 4 µg) are spotted onto a nitrocellulose filter for the dot-blot assay (Kafatos et al. 1979); in addition, accurately known amounts of the transgene DNA fragment are mixed with normal mouse DNA and spotted onto the same nitrocellulose filter to produce a standard curve. One copy of a 5- to 10-kb sequence per diploid mouse genome (6×10^9 bp) is roughly one part in 1 million; therefore, the amount of the pure DNA fragment used for the standard curve should cover a range of ~1–100 pg.

The dot blot is hybridized using the transgene (or the plasmid vector, if it was also injected) as a probe, and the hybridization to each dot is quantitated by scintillation counting or by autoradiography and densitometry. If the amount of hybridization is linear with the amount of DNA spotted, for both the transgenic mouse DNA curve and the standard curve, the copy number can be calculated from the slopes of the two lines. If the transgene copy number is very high, the curve generated with transgenic mouse DNA may not be linear because the amount of probe added to the hybridization may not be sufficient to saturate complementary sequences in the mouse DNA. In this case, the procedure must be repeated using either lower amounts of transgenic mouse DNA or a higher concentration of probe, or both.

Note that the apparent copy number measured by either Southern or dot-blot hybridization reflects both the actual copy number per diploid genome and the fraction of cells containing the transgene. If a founder mouse is mosaic and some cells lack the injected gene, the copy number will be underestimated. For this reason, it is preferable to determine gene copy number using transgenic progeny rather than the founder mouse.

Occasionally, a founder transgenic mouse will contain copies of the transgene integrated at two different loci (Lacy et al. 1983; Wagner et al. 1983). This will usually become apparent when the F_1 progeny are analyzed because the two loci will segregate. The different integration loci usually contain different numbers of copies of the transgene (which will be seen by Southern blot analysis, but not by PCR), thus the inheritance of different numbers of copies by progeny provides evidence for multiple integration sites in the founder transgenic mouse. In addition, each integration event will often generate distinct junction fragments, and these may be seen to segregate when progeny DNAs are analyzed by Southern blot analysis.

DETECTION AND ANALYSIS OF ES CELL-MEDIATED GENE/GENOME ALTERATIONS

Characterization of ES cell-mediated genome alterations, including random insertional transgenesis, gene trapping, gene targeting, and site-specific recombinase-mediated changes, is performed mostly at the ES cell level, prior to the introduction of these alterations into the mouse. A detailed characterization requires a larger amount of DNA than is required for the initial detection of the candidates for the desired alteration. Protocol 12.5 describes the preparation of DNA from a 10-cm tissue culture plate, which results in an order of magnitude more DNA than from the methods described in Chapter 10.

Random integration transgenesis performed in ES cells offers selection among a few hundred different integration sites for the desired one, for example, a single copy–single site transgene (Novak et al. 2000; Ding et al. 2001) or expression specificity and level, before the transgene is introduced in vivo. The principle of single copy–single site detection is the same as that described above for pronuclear injection transgenesis. The genomic DNA has to be digested with a restriction enzyme which cuts once in the construct that has been electroporated into the cells. The Southern probe derived from the construct is expected to generate the unique "junction fragment(s)" only and not the fragment that is characteristic of head-to-tail or other types of multiple-copy fusions. The existence of more than the expected one or two junction fragments (depending on the position of the probe relative to the enzyme cutting site in the transgene) indicates multiple-site integrations. Single copy–single site integration is desired, e.g., when the construct contains *loxP* sites, and Cre recombinase-mediated postinsertional modification is designed in the strategy.

The high-throughput screen to identify a homologous recombination-based gene targeting event is based mostly on Southern blot analysis performed on 96-well plate-derived DNA (see Protocol 10.4 in Chapter 10). The amount of DNA per clone allows only one Southern blot to be produced, which most of the time is indicative for homologous recombination through one of the homology arms of the target vector. Additional Southern blotting is needed to confirm the result and to check whether the other chromosomal arm also recombined by homologous recombination. In the latter case, the same recommendations exist: Use the outside probe, if it is available, which will create a unique band besides the wild-type band only if the gene was correctly targeted. In many knockouts, however, an outside probe is not available for diagnosing homologous recombination at the second arm. In this case, an internal probe can be used, which should give an expected size band if homologous recombination has occurred. An internal probe is useful from another perspective. Because such a probe gives a band other than the wild type from all the integration events, including random integration, the presence of an extra band or bands indicates that additional random integration occurred in addition to homologous recombination. These lines should be excluded from further studies if a sufficient number of correctly targeted lines free of random integration are available. If this is not the case, one might "go germ line" with a cell line that also contains a random insertion and select against this transgene after germ-line transmission.

Checking multiple insertions is also important in gene trap approaches. If the DNA is digested with a noncutter or a unique cutter in the trap vector, an internal probe is similarly diagnostic for single copy–single site integration events as described above. The preferred scenario is obviously the simple integration, but frequently more complex integrations have also been characterized if the features of the trapped gene have justified the extra effort (Korn et al. 1992).

Site-specific recombinases can also mediate large genomic changes belonging to the chromosomal aberration category, such as large deletions, duplications, inversions, and reciprocal translocations. The detection of these changes at the molecular level is possible by PCR and Southern blot through the newly formed junction fragment. The most convincing visualization of such aberrations, however, is karyotyping (see Protocol 12.6) or fluorescence in situ hybridization (FISH)-based chromosome painting. Medium-sized deletions (from a few hundred base pairs to a few megabases) can be detected by metaphase or interphase FISH as well.

MOUSE CHROMOSOMES AND KARYOTYPE

The normal mouse mitotic karyotype consists of 40 acrocentric chromosomes that, in an average preparation, range in length from 2 to 5 μm. Because the chromosomes show few distinguishing morphological features after conventional staining (see, e.g., Evans 1981), complete analysis requires chromosome banding. Of the numerous available methods (see, e.g., Miller and Miller 1981), Q and G banding have proved to be the most popular. The latter gives greater resolution and therefore has provided the foundation for the standard idiogram of banding patterns. A list of 312 distinct regions defined by G banding within the karyotype has been published, with the rules for the nomenclature of chromosome anomalies, by Green (1981). As yet, apart from the consistent strain variations observed in the centromeric regions of some homologs, no other variations have been detected in the standard, normal chromosomes within and between strains of the laboratory mouse (see, e.g., Evans 1981), and any strain variations observed should be regarded as either an abnormality or a sign of karyotypic evolution (Fig. 12.3).

Provided there is mitotic activity, chromosome preparations can be made from all tissues of the mouse, whether they are derived from in vivo or in vitro sources. In general, the greater the mitotic activity, the more likely the success of the preparation. Unfortunately, mitotic metaphase arrestants such as Colcemid, colchicine, and vinblastine sulfate, which can greatly enhance the frequency of mitotic spreads, also have the concomitant and deleterious effect of contracting the chromosomes, thus reducing subsequent G-band resolution. If possible, avoid the use of mitotic arrestants; however, if necessary, their use should be limited to low concentrations and brief time exposure (e.g., 0.02 μg/ml final concentration for 1 hour).

A quick and spectacular way to identify chromosomal anomalies, such as translocations, large deletions, and trisomies, in human and mouse was developed in the early 1990s at Lawrence Livermore National Laboratory. Probes with fluorescent labels, or "paints," that are specific to individual chromosomes stain them different colors (Breneman et al. 1993; Antonacci et al. 1995). The procedure is instrument-intensive; therefore, most institutions would outsource this technique.

ANALYSIS OF TRANSGENE EXPRESSION

The crucial point of transgenic production is the expression characteristics of the transgene made. The applications of the transgenic approaches are quite broad, therefore the requirements for expression behavior of a transgene vary between extremes. Thus, there are strategies that require ubiquitous, widespread, lineage-specific, cell-type-specific, or inducible transgene expressions, as well as combinations of these strategies. Characterization of transgene expression is an essential component of most of these strategies.

For transgenes expressing a protein with available specific antibody or with a common epitope tag, the most reliable gene expression assay is western blot-

X 1,300

X 2,600

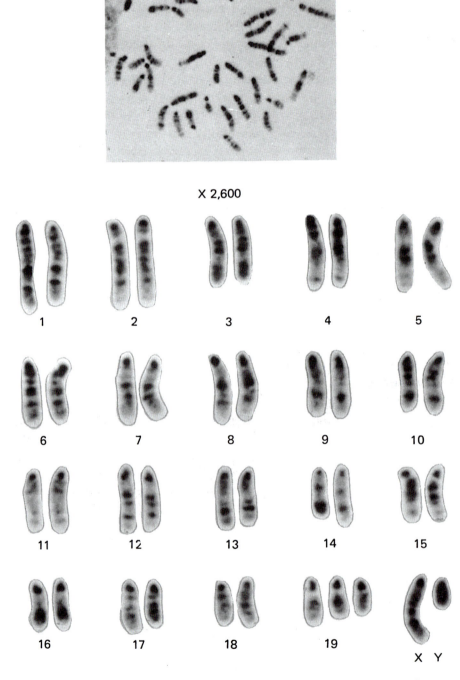

FIGURE 12.3. G-banded cell from an embryonal carcinoma cell line that, when first derived, had 40 normal chromosomes and was XY. After continued culture, the karyotype, although remaining normal as far as the G bands were concerned, evolved by acquiring an extra copy (trisomy) of chromosome 19.

ting, enzyme-linked immunosorbent assay (ELISA), or immunohistochemistry. The most convenient indicator of gene expression, however, is a visible reporter tag, for example, GFP fusion with the protein of interest (Liu et al. 2000; Oh and Eaves 2002; Treloar et al. 2002) or joining the coding region of the gene of interest with a reporter gene (*lacZ* or GFP) by a viral fragment for internal ribosomal entry site (IRES). Such a transgene transcribes a bicistronic mRNA, which allows independent translation of both coding regions, creating two independent proteins. Be aware, however, that the reporter and the transgene of interest are only linked transcriptionally. The translation could be biased between the two coding regions (see Chapter 9 for more details).

The next assay in the order of reliability is checking the transcription from the transgene either by northern blotting or by in situ RNA hybridization. Because transcription does not necessarily mean expression, the pattern of expression determined at the transcript level should be interpreted with caution.

Besides these general comments, this chapter does not give more details about the assays for testing gene expression. See Chapter 16 for protocols for gene expression visualization.

CLONING TRANSGENE/HOST DNA JUNCTIONS

In cases where an integrated transgene has caused an insertional mutation or an ES cell-mediated gene-trap insertion has occurred, it is desirable to determine the flanking DNA or a transcribed region of the trap gene, respectively, as a first step in identifying the disrupted gene responsible for the mutant phenotype. Cloning the flanking DNA is rather straightforward in cases where a retrovirus or a single copy of a transgene has integrated without any rearrangement of the host DNA (see, e.g., Jaenisch et al. 1983); however, cloning of transgene–host DNA junctions is more involved when there are multiple copies of a transgene integrated in a tandem array. The "traditional" method for this scenario is screening of bacteriophage λ libraries with the transgene, identifying λ clones containing the transgene, and obtaining sequence information for the flanking genomic region. This is a tedious approach, but there are strong reasons for using this method (see, e.g., Radice et al. 1991). Additional complications associated with this approach arise from the frequent occurrence of genome rearrangement at the site of integration of pronucleus-injected transgenes. This can make the characterization of the integration site enormously confusing (Brown et al. 1994).

Plasmid rescue could provide a less labor-intensive alternative to cloning flanking genomic regions of a transgene. However, the requirements for plasmid rescue contradict the general suggestion to remove all of the bacterial elements from the transgene (see Chapter 7). The transgene has to contain a functional *Escherichia coli* replicon (i.e., where plasmid or cosmid vector sequences have not been removed from the transgene). Genomic DNA is digested with an enzyme that cuts only once in the transgene, but does not disturb the replicon function. Then the genomic fragments are ligated and introduced into *E. coli*. The recovered plasmid might contain genomic flanking sequences at a probability that negatively correlates with copy number (see, e.g., Woychik et al. 1985; Grant et al. 1990; Singh et al. 1991). Therefore, plasmid clones have to be screened for the presence of mouse DNA. The success of plasmid rescue is reported to be much greater when using *E. coli* strains carrying multiple mutations in the methylation-dependent restriction system (Grant et al. 1990). In the case of single-copy integration, e.g., most of the gene-trap insertions, the plasmid rescue could work at a very high efficiency and be suitable for high-throughput assays (Hicks et al. 1997). If it is known (or assumed) that the terminal copy of the transgene array is intact, or nearly so, the strategy of "inverse PCR" may be used to amplify junction sequences as well as internal transgene sequences, which can then be cloned and analyzed as discussed above (see, e.g., Ochman et al. 1988; von Melchner et al. 1990).

Gene-trap insertion, which creates chimeric transcripts between some of the trapped gene exons and the trapping vector sequence, offers an RT-PCR-based method, called rapid amplification of cDNA ends (RACE), to determine the flanking exon sequences. Depending on the trap vector design, either 5′ or 3′ RACE can be used, based on which flanking side is to be determined (Frohman 1994; Townley et al. 1997). Since the mouse genome sequencing and annotation are close to completion, the flanking sequence provides accurate mapping information as well.

IDENTIFYING HOMOZYGOUS TRANSGENIC MICE OR EMBRYOS

It is occasionally necessary to distinguish mice that are homozygous for a transgene insertion from mice that are heterozygous. These situations include establishing a homozygous transgenic line, eliminating the need to screen each generation; screening for transgenic lines that carry recessive insertional mutations; and analyzing embryos from a transgenic line carrying an insertional mutation. Several available methods are listed.

- *Quantitation of transgene dosage:* Can be used for any transgenic line.

- *Quantitation of transgene product or phenotype:* Requires that the transgene product or phenotype be easily quantitated.

- *In situ hybridization to interphase nuclei:* Most easily applied to high-copy-number transgenic lines. Requires very few cells and is therefore very useful for identifying homozygous transgenic embryos at early postimplantation stages.

- *Test breeding:* Can be used for any transgenic line.

- *Southern blot analysis using a flanking probe:* Can be used only if the mouse DNA flanking the insertion site has been cloned.

- *PCR analysis with a flanking primer:* Can be used only if the mouse DNA flanking the insertion site has been cloned and sequenced. Requires very few cells and is therefore very useful for identifying homozygous transgenic embryos at early postimplantation stages.

Quantitation of Transgene Dosage

Quantitation of transgene dosage can be performed using either Southern or dot-blot analysis. Southern blot analysis can display twofold differences in transgene copy number if the amount of DNA loaded on the gel is very carefully quantitated by fluorometric methods. For this purpose, it is best to use DNA isolated from the mouse tail (which, compared to other tissues, has relatively little contaminating RNA that can interfere with the fluorometric measurement; R.

Woychik, pers. comm.). It is also important to take duplicate samples of each tail DNA preparation for fluorometric analysis and to assay the samples *after* digesting them with the appropriate restriction enzyme; otherwise, sampling errors due to the high viscosity of the undigested DNA can introduce large errors. If the duplicate measurements agree within 10%, adjust the amount of each DNA sample so that equal quantities are loaded in each lane of the agarose gel. When a Southern blot is hybridized, DNA samples from homozygous mice should yield bands of twice the intensity, compared with heterozygous mice.

A variation of this method involves hybridizing the Southern blot with a probe (or mixture of two probes) that will yield distinct bands for the transgene and an endogenous gene (Wagner et al. 1983) and then evaluating the relative intensities of the bands representing the transgene and the endogenous gene by densitometry of the autoradiogram.

The dot-blot method also measures the relative quantity of the transgene in each mouse, using an endogenous gene as an internal standard. An internal standard controls for any errors in measuring DNA concentration and reduces the need to determine DNA concentrations accurately. For this purpose, duplicate dot blots are prepared, each containing a series of dilutions of DNA from the progeny to be tested, as well as from one or two known heterozygous mice. One blot is hybridized with a probe for an endogenous gene. The endogenous gene used as an internal standard should have a reiteration frequency similar to that of the transgene, so that similar amounts of hybridizations (and linear curves; see above) will be obtained with each probe. For mice carrying 1–10 copies of a transgene, any single-copy mouse gene can be used as an internal standard. For mice carrying 10–100 copies of a transgene, a convenient internal standard is a probe for the mouse major urinary protein (MUP) gene family (20–30 copies per haploid genome; Derman 1981; Derman et al. 1981). For 50 to several hundred copies, a probe for a ribosomal RNA gene (Arnheim et al. 1982; Bourbon et al. 1988) works well. For each mouse, the slope of the curve (cpm hybridized/mg DNA) obtained with the transgene probe is divided by the slope obtained with the endogenous probe. The ratio for a homozygote should be twice the ratio for a heterozygote.

PCR can also be applied to detect homozygous transgenic animals (Chatelain et al. 1995), which have limitations when there is a large difference between the copy number of the transgene and that of the reference gene. Real-time PCR seems to compensate for the difference better, and rapid, precise, nonambiguous, and high-throughput identification of homozygosity versus heterozygosity in transgenic animals (Tesson et al. 2002) is possible by this method.

Quantitation of Transgene Product or Phenotype

In some cases where the transgene product is easily quantitated, it may be simpler to detect a twofold increase in the amount of gene product in homozygotes than to detect the increase in the transgene copy number (e.g., human β-globin chains expressed in red blood cells; Costantini et al. 1989). In addition, in some transgenic lines with visible phenotypes (e.g., rescue of albinism in tyrosinase transgenics; Beerman et al. 1990; Tanaka et al. 1990), homozygotes may have a more extreme phenotype than heterozygotes (e.g., darker pigmentation; Mintz and Bradl 1991). Likewise, where a reporter gene, such as *lacZ* or GFP (Hadjantonakis et al. 1998) is employed, the speed and intensity of staining or fluorescence may serve as indicators of homozygosity but require confirmation by test breeding.

In Situ Hybridization to Interphase Nuclei

Transgene arrays can be visualized by in situ hybridization to interphase nuclei with biotinylated probes, using either enzymatic or fluorescent detection systems. Because cells from homozygous transgenic mice contain two transgene arrays, most nuclei display two "signals" upon in situ hybridization, whereas nuclei from heterozygous mice display only one signal (Varmuza et al. 1988; Costantini et al. 1989). Transgene arrays of 50–100 kb (e.g., 5–10 copies of a 10-kb transgene) or longer can be readily detected using streptavidin–horseradish peroxidase and silver enhancement of the 3,3´5´-diaminobenzidine tetrahydrochloride (DAB) signal, and shorter arrays can be detected using fluorescein-conjugated streptavidin. This technique can be used on white blood cells or spleen cells from adult mice or on trophoblast giant cells or amnion cells from postimplantation embryos. When the assay works properly, this method gives clear-cut results; however, it takes some effort to establish the technique so that it works reproducibly. Two or three samples (slides) from each mouse or embryo should be hybridized in the event that some slides do not work.

Test Breeding

Although transmission of the transgene to 100% of the offspring is the definitive test of homozygosity (and similarly, failure to transmit to 100% of offspring indicates heterozygosity), it is too expensive and time-consuming a test for screening large numbers of mice. However, it is important in some instances to confirm a diagnosis of homozygosity by test breeding. If very few "homozygotes" are observed, test breeding will establish whether the mice are actually homozygous. In addition, if two homozygous transgenic mice are to be mated to generate a permanent homozygous line, it is advisable first to confirm the homozygosity of each animal by test breeding, because quantitative methods for identifying homozygotes can sometimes produce erroneous results.

Southern Blot Analysis Using a Flanking Probe

If the host DNA flanking the transgene insertion or a targeted mutation has been cloned, a single-copy mouse DNA probe from the flanking region can be used to identify homozygotes by Southern blot analysis. If used with an appropriate restriction digest, such a probe will detect two bands in heterozygous DNA, one containing the transgene junction and the other containing the preintegration site on the wild-type chromosome; mice homozygous for the transgene insertion will lack the band representing the wild-type locus (see, e.g., Jaenisch et al. 1983). In case of gene targeting, the diagnostic "outside genomic" probe that was used to detect a homologous recombination event in ES cells can be used for determining homozygosity after germ-line transmission as well.

PCR Analysis Using a Flanking Primer

If one or more of the transgene/host DNA junctions have been sequenced, it is convenient to design a PCR to distinguish homozygotes from heterozygotes (see, e.g., Radice et al. 1991). The simplest PCR uses three primers, as shown in Figure 12.4. Amplification of heterozygous transgenic DNA results in two bands, wild-type DNA yielding only band I and homozygous transgenic DNA yielding only band II. This analysis can be performed using very few cells (e.g., fragments of ectoplacental cone from 6.5-day embryos; Radice et al. 1991).

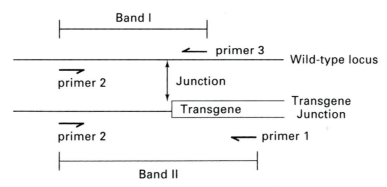

FIGURE 12.4. PCR analysis with a flanking primer. (*Bottom line*) Junction between transgene and flanking mouse DNA; (*top line*) corresponding wild-type mouse locus. Primers 2 and 1 will amplify band II only in samples from heterozygous or homozygous transgenic mice, whereas primers 2 and 3 will amplify the shorter band I in wild-type or heterozygous transgenic samples. Thus, homozygotes will produce only band II, and heterozygotes will produce both bands.

Isolation of High-molecular-weight DNA from Mouse Tail Tips

MATERIALS

EQUIPMENT
Eppendorf Thermomixer R (*optional*)
Incubator/oven, 55°C general use
Microcentrifuge
Microcentrifuge tubes, 1.5-ml
Rotator or shaker (e.g., Fisher Scientific Hematology/Chemistry Mixer 14-059-346)
Surgical scissors or razor blades, sterilized and sharp

REAGENTS
Digestion buffer
    100 mM NaCl
    50 mM Tris-HCl<!>, pH 8
    100 mM EDTA, pH 8
    1% sodium dodecyl sulfate (SDS)<!>
Ethanol<!>, 70% (–20°C)
Isopropanol<!>
Phase Lock Gel Light (PLG Light) (Eppendorf)
Phenol<!>/chloroform<!>/isoamyl alcohol<!>, 25:24:1 (e.g., from Invitrogen Life
    Technologies 1559-3031)
Proteinase K<!>, 25 mg/ml solution (stored at –20°C)
TE buffer (pH 8)
    10 mM Tris<!> (pH 8)
    1 mM EDTA (pH 8)

TISSUES
Mouse tails

CAUTION: See Appendix 2 for appropriate handling of materials marked with <!>.

PROCEDURE

1. Cut 0.5 to 1 cm of tails from 2- to 3-week-old mice. Place the tissue into 1.5-ml microcentrifuge tubes on ice.

   If the samples will not be immediately used, they can be stored at –20°C.

2. Add 620 μl of digestion buffer and 30 μl of 25 mg/ml Proteinase K solution to each tube.

3. Incubate the samples overnight at 55°C, inverting occasionally to mix.

4. Add 650 μl of phenol/chloroform/isoamyl alcohol and 200 μl of PLG Light to the tubes.

5. Rotate the tubes for 60 minutes.

> It is important to close the microcentrifuge tube tightly to prevent spilling of phenol and chloroform during rotation. Phenol is highly corrosive and can cause severe burns. All work with phenol should be done under the hood.

6. Pour the aqueous phase into the fresh microcentrifuge tube.

7. Add an equal volume of isopropanol to each tube and invert the tube until a stringy precipitate forms.

8. Centrifuge the tubes at 10,000 rpm for 10 minutes and then discard the supernatant.

9. Rinse the DNA pellet with 70% ethanol and discard the ethanol.

10. Air-dry the pellet for 10 minutes.

11. Resuspend the DNA pellet in 100–200 µl of TE buffer. Incubate the tubes at 55°C while shaking them gently to make sure DNA is completely dissolved.

12. Measure the DNA concentration. Use 10 µg of the DNA samples for Southern blot analysis and 100 ng to 1 µg of DNA for PCR.

COMMENTS

- For the phenol/chloroform-free method, digest the 0.5-cm tail overnight as in step 1 using the same digestion buffer containing 500 µg per ml Proteinase K (10 µl of the 25 mg/ml stock per 500 µl of tail buffer). The following day, mix the digested tail for 5 minutes (i.e., in a rotator or shaker or Eppendorf Thermomixer). Add 167 µl of 5–6 M NaCl, mix again for 5 minutes, and centrifuge for 10 minutes at full speed. Carefully avoiding the pellet, transfer 500 µl of supernatant into a new microfuge tube (the supernatant may be somewhat viscous, but should be fairly clear, and free of debris). Continue with the isopropanol precipitation in step 7.

- As an alternative to steps 8–10, scoop out the stringy genomic DNA precipitate with a 200-µl pipette tip and rinse by dipping (or dropping) into a microfuge tube containing 500 µl of 70% ethanol. Then transfer the DNA to a microfuge tube and allow the residual ethanol to evaporate (i.e., until the DNA is gel-like but not desiccated). Continue with step 11.

- The QIAGEN DNeasy Tissue Kit or QIAamp DNA Mini Kit (http://www.qiagen.com) can be used for DNA preparation from mouse tails.

Isolation of High-molecular-weight DNA from Embryonic Tissues, Yolk Sac, Umbilical Cord, etc.

MATERIALS

ANIMALS

Mouse embryos (9.5 dpc or later)

EQUIPMENT

Eppendorf Thermomixer R (*optional*)
Incubator/oven, 55°C general use
Microcentrifuge
Microcentrifuge tubes, 1.5-ml
Rotator or shaker (e.g., Fisher Scientific Hematology/Chemistry Mixer (14-059-346)

REAGENTS

Carrier RNA
EDTA, 100 mM
Ethanol<!>, 70%(–20°C)
Isopropanol<!>
Phase Lock Gel Light (PLG Light) (Eppendorf)
Phenol<!>/chloroform<!>/isoamyl alcohol<!>, 25:24:1 (e.g., from Invitrogen Life Technologies 1559-3031)
Proteinase K <!> (10 mg/ml stock in water stored in aliquots at –20°C)
SDS<!>, 0.5%
TE buffer (pH 8): 10 mM Tris (pH 8), 1 mM EDTA (pH 8)
Tris<!>, 50 mM (pH 8)

CAUTION: See Appendix 2 for appropriate handling of materials marked with <!>.

PROCEDURE

1. Dissect embryos from mouse uterus (see Chapter 5) and remove the yolk sacs.

   Embryos or dissected tissues can be fixed for histology or frozen for later RNA extraction, pending the results of the analysis.

2. Place each yolk sac in a minimum volume (~25 µl for 9.5-dpc embryos or 50–100 µl for later-stage embryos) of 50 mM Tris (pH 8), 100 mM EDTA, 0.5% SDS in a microfuge tube.

3. Add Proteinase K to a final concentration of 500 µg/ml, using a 10 mg/ml stock. Incubate the sample at 50°C for several hours or overnight. Agitation is not required.

4. For yolk sacs (9.5 dpc), add 5–10 µg of carrier RNA (e.g., transfer RNA) to aid in later recovery.

5. Process the sample exactly as for tail DNA samples starting at step 4 in Protocol 12.1 (see Comments), scaling down the volumes appropriately. The yield is ~50–100 µg from a 13.5-day yolk sac, 25 µg from a 11.5-day yolk sac, and <5 µg from a 9.5-day yolk sac.

Preparation of Tissue Lysates for PCR Template

Polymerase chain reaction (PCR) can be performed on DNA purified from embryonic or extraembryonic tissues using the preceding protocol. However, for PCR analysis, it is usually not necessary to purify the DNA. The following protocols describe alternative methods to produce tissue lysates suitable for PCR analysis.

PROTOCOL 3A

This protocol can be used to prepare embryonic tissues or yolk sac samples for PCR analysis, but it is also useful for adult tissues (such as tail) that do not lyse completely by boiling in SDS/NaOH. Nonionic detergents are substituted for SDS because they interfere less with the polymerase reaction. This protocol is from Dr. W. Perry (Frederick Cancer Research and Development Center, Maryland), and is based on Higuchi (1988).

MATERIALS

EMBRYOS AND ANIMALS
Mouse embryos (entire or part) or adult mouse tissues

EQUIPMENT
Microtubes with screw-cap, 1.5-ml, sterile (e.g., Sarstedt 72.692.005)
PCR machine
Water bath (boiling)

REAGENTS
PCR tissue homogenization buffer, 10x (The pH is adjusted to 8.3 before addition of gelatin and detergents; this buffer can be stored at room temperature in small aliquots.)
500 mM KCl<!>
100 mM Tris-HCl<!>
0.1 mg/ml gelatin (Sigma G2500)
0.45% Nonidet P-40 (Sigma N6507)
0.45% Tween 20 (Sigma P1379)
500 μg/ml Proteinase K<!>, added just before use (20 mg/ml stock)
Sterile water

CAUTION: See Appendix 2 for appropriate handling of materials marked with <!>.

PROCEDURE

1. Place the tissue in a 1.5-ml sterile screw-capped micro tube. Tissue can be stored frozen in this state until use.

2. To homogenize samples, thaw and add ~20–50 µl of PCR tissue homogenization buffer for embryonic tissues or 200 µl for a 0.5-cm piece of tail. The tissue should be suspended in the buffer and not floating on the surface. Incubate the sample at 55°C for 3–4 hours or until most of the tissue is dissolved (overnight). Samples can be stored frozen after this step if necessary.

3. Boil the tissue homogenate in a boiling water bath for 10 minutes, cool on ice, and spin in a microfuge for a few seconds. Homogenates can be stored at –20°C.

4. Use a 2- to 10-µl sample of boiled tissue homogenate for PCR, bringing the volume to 32–40 µl (use ~2 µl of yolk sac or 5 µl of tail homogenate).

PROTOCOL 3B

This protocol can be used to prepare embryonic tissues, yolk sac, ear punch, toe, or 1- to 2-mm (maximum 5 mm) tail samples for PCR analysis. It is extensively used by the labs of Janet Rossant and Andras Nagy.

MATERIALS

EQUIPMENT
Ear punch (e.g., Fisher Scientific 01-337B), scissors
Microcentrifuge tube, 1.5-ml
Oven/incubator, 55°C, general-use
PCR machine

REAGENTS
Lysis buffer (autoclaved)
    50 mM KCl<!>
    10 mM Tris-HCl<!> (pH 8.3)
    2 mM $MgCl_2$<!>
    0.1 mg/ml gelatin
    0.45% Nonidet P40
    0.45% Tween 20
    Aliquots (e.g., 1 ml) are stored at –20°C.
Proteinase K<!>, 500 µg/ml , added just before use (stock 20 mg/ml)

CAUTION: See Appendix 2 for appropriate handling of materials marked with <!>.

PROCEDURE

1. Immediately before use, add 25 µl of Proteinase K solution (20 mg/ml) to 1 ml of lysis buffer.

2. Place the yolk sac in 50 µl, ear punch or toe tissue in 70–100 µl, *small* piece of tail in 200–300 µl of lysis buffer containing Proteinase K. Samples can be stored frozen until use.

3. Incubate the samples at 55°C for 3–4 hours to overnight. The DNA lysates now can be stored at 4°C if desired.

4. *Optional:* Centrifuge for 2–3 minutes to pellet hair and debris.

5. Incubate the samples at 95°C for 10 minutes before starting PCR amplification to inactivate Proteinase K.

6. Use 1–2 µl of lysate for each PCR.

PROTOCOL 3C

This protocol is suitable for very small fragments of postimplantation mouse embryos, typically containing between 50 and a few hundred cells (e.g., a small fragment of the yolk sac of an embryo or a fragment of the ectoplacental cone or small numbers of cultured cells). It can be used to identify transgenic embryos, using two primers specific for the transgene or to distinguish heterozygous from homozygous transgenic or mutant embryos (see Identifying Homozygous Transgenic Mice, above).

MATERIALS

EMBRYOS

Embryo fragment from postimplantation mouse embryo

EQUIPMENT

Dissecting microscope
Microfuge tube, siliconized
Pasteur pipette, siliconized, drawn-out, mouth-controlled
Water bath (boiling)

REAGENTS

0.05% SDS<!>, 0.035 N NaOH<!>
Tris <!> (pH 8.3) (This is required in addition to the 10 mM Tris in the buffer to neutralize the NaOH in the lysis buffer adequately.)
Water, sterile distilled

CAUTION: See Appendix 2 for appropriate handling of materials marked with <!>.

PROCEDURE

1. Use a siliconized, drawn-out mouth-controlled Pasteur pipette and a dissecting microscope to place the embryo fragment directly into 1 µl of distilled water in a siliconized microfuge tube. For cells, spin down a sufficient number of cells to give a pellet of <0.5 µl in volume, remove all the supernatant, and add 1 µl of water. The embryo fragment or cell pellet should be much less than 1 µl in volume; otherwise, the volumes of water and lysis buffer should be increased. Store at –70°C, if necessary.

2. Lyse the cells by adding 2 µl of 0.05% SDS, 0.035 N NaOH for each 1 µl of water and mix.

   Higher concentrations of SDS or NaOH will inhibit the PCR.

Place the tube in a boiling water bath for 3 minutes (with cap closed and a hole poked in top to prevent popping open). Then cool on ice and spin in a microfuge for a few seconds.

4. Use 3 µl of the boiled lysate for PCR template (see Protocol 12.4 and the Comment).

COMMENT

PCR genotyping of an early postimplantation-stage embryo that has gone through in situ hybridization is possible. The lysate preparation includes rinsing the embryo through several PBT (PBS + 0.1% Tween 20) washes, e.g., in a 24-well plate. Then the embryos are transferred into microcentrifuge tubes along with 10–20 µl, 20-40 µl, and 40–80 µl of PBT for a 6.5-, 7.5-, and 8.5-dpc embryo, respectively. Add 1 µl of Proteinase K (40 mg/ml) into each tube and incubate overnight at 55°C, then heat-inactivate at 95°C for 10 minutes. After the lysate has been centrifuged to separate cellular debris, 1–2 µl of the supernatant is used for PCR. Care should be taken to remove possible contaminating maternal tissue, such as Reichert's membrane or ectoplacental cone.

PROTOCOL 3D

This very popular protocol using hot sodium hydroxide and Tris (HotSHOT) was developed by Truett et al. (2000).

MATERIALS

EQUIPMENT
Heating block, 95°C
Microcentrifuge
Microcentrifuge tubes, 1.5-ml
Vortex

REAGENTS
Sodium hydroxide (NaOH)<!>, 50 mM
Tris <!>, 1 M (pH 8)

TISSUES
Tail tissue, 1–2 mm

CAUTION: See Appendix 2 for appropriate handling of materials marked with <!>.

PROCEDURE

1. Cut a 1- to 2-mm piece of tail tissue and place it into a microcentrifuge tube.
2. Add 200–300 µl of 50 mM NaOH.
3. Heat to 95°C for 10 minutes (if the piece is more than 2 mm, incubate longer).
4. Vortex.

5. Add 50 μl of 1 M Tris (pH 8) to neutralize the NaOH. The pH can be checked at this step; it should be 7.

6. Centrifuge the tube at high speed (e.g., 12,000 rpm) for 6 minutes.

7. Transfer the supernatant to a new tube.

8. Use 1 μl of the supernatant for PCR, or dilute as needed.

Polymerase Chain Reaction

After the DNA template is prepared, a standard PCR is performed. This protocol is designed for ~40-µl reactions.

MATERIALS

EQUIPMENT
PCR machine

REAGENTS
Deoxynucleotide triphosphates (dNTPs)
Paraffin oil, light
10x PCR buffer
    500 mM KCl<!>
    100 mM Tris<!> (pH 8.3)
    15 mM $MgCl_2$<!>
    0.1% (w/v) gelatin (Sigma G2500)
PCR primers
Taq polymerase (e.g., Perkin Elmer AmpliTaq N801-0060)
Water, sterile

CAUTION: See Appendix 2 for appropriate handling of materials marked with <!>.

PROCEDURE

1. Prepare the following mixture in an appropriate volume for the number of reactions planned in the experiment:

   PCR primers at a concentration of 1 µM each
   200 µM of each dNTP
   1x PCR buffer
   Taq polymerase, 2.5 U/30 µl of mixture
   sterile water (free of any possible contamination) to final volume
   For the lysate prepared according to Protocol 3C, final concentrations are:
     1x PCR buffer
     0.05 M Tris (pH 8.3)
     200 µM of each dNTP
     oligonucleotide primers (100-200 ng each)
     Taq polymerase, 1 U/37 µl mixture

2. Add 30 µl of this mixture to each 2- to 10-µl sample of DNA or lysate preparation, bringing the volume to 32–40 µl. Mix well, microfuge for a few seconds, cover with light paraffin oil, and run the PCR.

Preparation of DNA for Southern or PCR from ES or Other Cultured Cells

MATERIALS

EQUIPMENT
Eppendorf Thermomixer R (*optional*)
Incubator/oven, 55°C general use
Microcentrifuge
Microcentrifuge tubes, 1.5-ml
Rotator or shaker (e.g., Fisher Scientific Hematology/Chemistry Mixer 14-059-346)

REAGENTS
Ethanol<!>, 70% (–20°C)
Ethanol<!>, 100% (–20°C)
Lysis buffer:
    100 mM NaCl
    10 mM Tris-HCl <!> (pH 8)
    25 mM EDTA (pH 8)
    0.5% sodium dodecyl sulfate (SDS)<!>
Phase Lock Gel Light (PLG Light) (Eppendorf)
25:24:1 Phenol<!>/chloroform<!>/isoamyl alcohol<!> (e.g., Invitrogen Life
    Technologies 15593-031)
Phosphate-buffered saline (PBS; see Appendix 1), ice-cold
Proteinase K<!> added fresh from the stock (25 mg/ml) stored at –20°C to 500 µg/ml
Sodium acetate<!>, 3 M (pH 5.2)
TE buffer <!> (pH 8) (see Appendix 1)

CAUTION: See Appendix 2 for appropriate handling of materials marked with <!>.

PROCEDURE

1. Culture the cells in a 10-cm tissue culture plate to reach 80–90% confluence. Trypsinize adherent cells and harvest cells from culture plate as described in Chapter 8 and place them in a 1.5-ml microcentrifuge tube. Centrifuge the cells at 1000 rpm for 5 minutes and discard the supernatant.

2. Wash the cell pellet by resuspending it with 5 ml of ice-cold PBS. Centrifuge at 1000 rpm for 5 minutes and discard the supernatant.

3. Resuspend the cell pellet in 300 µl of lysis buffer. (For larger numbers of cells use 1 ml of lysis buffer/10^8 cells.)

4. Incubate the samples at 55°C for 12–16 hours.

5. Add 100 µl of PLG Light and 300 µl of phenol/choroform/ isoamyl alcohol to the samples. Rotate for 60 minutes.

Close the microcentrifuge tubes tightly to prevent spilling of phenol and chloroform. Phenol is highly corrosive and can cause severe burns. All work with phenol should be done under the hood.

6. Centrifuge the tubes at 10,000 rpm for 5 minutes and transfer the top aqueous layer from each sample to a fresh tube.

7. Add 1/10 volume of 3 M sodium acetate (pH 5.2) to each sample, and then add 3 volumes of 100% ethanol. The DNA should immediately form a stringy precipitate.

8. Centrifuge the tubes at 10,000 rpm for 10 minutes and discard the supernatant.

9. Wash the DNA pellet with 1 ml of 70% ethanol and discard the supernatant.

10. Air-dry the samples for about 10 minutes.

11. Resuspend the DNA in TE buffer. To facilitate solubilization, the DNA samples can be shaken gently at 55°C for 4 hours. Store DNA samples at 4°C.

COMMENT

It is important to dissolve DNA completely before using it for Southern analysis or PCR. Storage of DNA samples at 4°C overnight helps their solubilization.

Karyotyping Mouse Cells

The majority of mouse chromosome preparations for banding are now made by air-drying and, in essence, require the production of a cell suspension as a starting point. Some samples such as blood cultures, ascitic fluids, or cells growing in suspension will already be in suspension; others, such as bone marrow, solid tumors, or cells growing as attached layers in culture must be converted to suspensions. This can be achieved by using a variety of disaggregation methods, some of which are described by Cronmiller and Mintz (1978). The basic steps in karyotyping and banding embryonal carcinoma cells are outlined below. For additional variations of these essential technical steps, see McBurney and Rogers (1982) and Robertson et al. (1983). Information was provided by E.P. Evans, MRC Radiobiology Unit, Harwell, Didcot, Oxfordshire, United Kingdom.

MATERIALS

CELLS

Embryonic stem cells (in tissue culture dishes in growth medium)

Exponentially growing cultures will give best results.

EQUIPMENT

Coplin staining jars with lids (Lipshaw 107)
Desk lamp
Forceps
Glass slides, acid-washed, grease-free (for preparation, see below)
Glass tube, 2-ml conical
Green filter for microscope
Kimwipes or any coarse, grease-free tissues
Kodak technical pan film (2415)
Microscope with 100x oil immersion objective and 40x phase-contrast objective
Oil immersion lens (100x)
Pasteur pipettes, drawn (~0.2-ml capacity, drawn to deliver 10-μl drops)
Water bath at 60–65°C
Whatman 3MM filter paper

REAGENTS

Absolute alcohol:concentrated HCl<!> (1:1)
Absolute alcohol:diethyl ether <!>(1:1)
Deionized water
Demecolcine <!> (100x stock is 2 μg/ml in distilled water; Sigma D7385)
Fresh absolute methanol:glacial acetic acid<!> (3:1) at room temperature (fixative)
Giemsa stain<!> (dilute 1:20 with deionized water) (Sigma GS-500)
Kodak HC110 developer

NaCl, 0.85% (w/v)
Phosphate buffer (pH 6.8)
Potassium chloride (KCl) <!>, 0.56% (w/v) (hypotonic solution for swelling cells)
Saline
2x SSC (see Appendix 1)
0.025% Trypsin (Difco 1:250) in 0.85% NaCl. The age of the trypsin powder is not
   important, but the solution should be made ~30 minutes before use.

CAUTION: See Appendix 2 for appropriate handling of materials marked with <!>.

PROCEDURE

Preparation of Cells

1. Trypsinize cells that have been growing in tissue culture dishes with or without pretreatment with a mitoic arrestant. Pipette to give a single cell suspension. Add medium and serum and centrifuge at 1000 rpm in a 2-ml conical glass tube for 5 minutes.

2. Resuspend the cell pellet gently in 1 ml of aqueous 0.56% (w/v) KCl by adding drops and flicking the tube. Do not pipette up and down. Add more KCl to a final volume of 4 ml. Leave at room temperature (21–23°C) for 6 minutes to swell the cells.

3. Centrifuge the cells again at 1000 rpm for 5 minutes. Carefully remove the supernatant, and fix the cell pellet in 1 ml of fresh absolute methanol:glacial acetic acid (3:1) fixative at room temperature. Resuspend by flicking the tube. Make up to 4 ml. Then centrifuge to pellet the cells. Repeat three times.

4. Centrifuge the cells at 1000 rpm for 5 minutes and then resuspend them in a small volume (0.5 ml) of fixative.

Slide Making

1. Clean commercially purchased precleaned slides further by leaving them overnight in a mixture of absolute alcohol and concentrated HCl (1:1). The next day, wash the slides in running tap water, rinse in deionized water, and store in absolute alcohol and diethyl ether (1:1).

2. Shortly before using the slides, use forceps to remove slides from the mixture of absolute alcohol and diethyl either. Wipe the slides dry on Kimwipes or some other coarse, grease-free tissues, avoiding finger contact with the slide to prevent making fingerprints on the slide surfaces.

3. To make chromosome spreads on these slides, add a row of three drops of cell suspension from a prepared Pasteur pipette (~0.2-ml capacity, drawn to deliver 10-μl drops), allowing them to spread to their maximum size. Let the drops dry until interference rings are visible at their periphery. Drying can be assisted by blowing on the slide while holding it up to the heat generated by a desk lamp bulb.

4. Once dry, inspect the slides under low-power phase-contrast microscopy (final magnification 160x). If the density of cells and chromosome spreads is insufficient, pellet the cells, resuspend in a smaller volume of fixative, and repeat with fresh drops.

5. For counting chromosomes without banding, stain in Giemsa for about 15 minutes and photograph with green filter.

G Banding

1. Place suitably aged slides (see Comments) in 2x SSC in a Coplin staining jar with a lid in a water bath at 60–65°C for 1.5 hours. Then cool the slides to room temperature by running tap water over the closed jar. Transfer the slides to 0.85% (w/v) NaCl at room temperature for 5 minutes.

2. Drain the slides by touching them onto filter paper. Place them on a flat surface and flood the chamber with 0.025% trypsin in 0.85% NaCl for 15–20 seconds.

   The trypsin exposure time is critical: Underexposure preserves chromosome morphology but gives poorly differentiated bands, and overexposure distorts morphology and eliminates most of the bands. Optimum trypsin times are known to vary among laboratories. A test slide should be treated for the minimal suggested time of 15 seconds to establish the best treatment time for the rest of the slides.

   In the laboratory of E.P. Evans, the optimal trypsin exposure time has been established as between 15 and 20 seconds for mouse chromosomes and has remained constant, irrespective of the source of the mitotic cells, for the last decade.

3. Stop tryptic activity by placing the slides back into normal saline. Then rinse slides in phosphate buffer (pH 6.8) and stain in fresh Giemsa stain in 5 mM phosphate buffer (pH 6.8). After 10 minutes in the stain, monitor wet slides under low-power, bright-field microscopy (160x) for staining intensity. Because, upon drying, wet slides gain contrast, care should be taken not to overstain the cells as this will reduce G-band differentiation. If necessary, repeat staining until adequate results are achieved and then quickly rinse slides in phosphate buffer (pH 6.8) and blow-dry with a current of cool air.

4. Examine unmounted slides with a 100x oil immersion lens.

   The majority of modern, readily available immersion oils that are declared "PCB-free" also have the unfortunate property of removing Giemsa stain after a few hours of exposure. Although direct viewing of slides under an oil immersion lens gives a higher optical resolution, it is wise to mount slides if they are to be kept.

COMMENTS

- Slide preparations are best made within 0–3 hours of the final fixation, but they can be made up to 7 days afterward if the material is kept at 4°C and the cell pellet is resuspended in fresh fixative before attempting to spread the cells. The longer the material is kept, the poorer the spreading quality of the mitotic cells. If possible, at least five slides should be made from each sample to guarantee success.

- Before G banding, slides should be "aged" for between 3 and 21 days by leaving them in a closed box at room temperature. Fresh slides give poor G-band resolution. Maximum G-band resolution is achieved at ~10 days after slide preparation. Beyond this time, resolution slowly decreases until, after several weeks in storage, the chromosomes either fail to band and stain uniformly or show "pseudo-bands" that are not significant to the standard idiogram.

REFERENCES

Antonacci R., Marzella R., Finelli P., Lonoce A., Forabosco A., Archidiacono N., and Rocchi M. 1995. A panel of subchromosomal painting libraries representing over 300 regions of the human genome. *Cytogenet Cell Genet.* **68:** 25–32.

Arnheim N., Treco D., Taylor B., and E.M. Eicher. 1982. Distribution of ribosomal gene length variants among mouse chromosomes. *Proc. Natl. Acad. Sci.* **79:** 4677–4680.

Beermann F., Ruppert S., Hummler E., Bosch F.X., Muller G., Ruther U., and Schutz G. 1990. Rescue of the albino phenotype by introduction of a functional tyrosinase gene into mice. *EMBO J.* **9:** 2819–2826.

Bourbon H., Michot B., Hassouna N., Feliu J., and Bachellerie J.P. 1988. Sequence and secondary structure of the 5′ external transcribed spacer of mouse pre-rRNA. *DNA* **7:** 181–191.

Breneman J.W., Ramsey M.J., Lee D.A., Eveleth G.G., Minkler J.L., and Tucker J.D. 1993. The development of chromosome-specific composite DNA probes for the mouse and their application to chromosome painting. *Chromosoma* **102:** 591–598.

Brown A., Copeland N.G., Gilbert D.J., Jenkins N.A., Rossant J., and Kothary R. 1994. The genomic structure of an insertional mutation in the dystonia musculorum locus. *Genomics* **20:** 371–376.

Brunk C.F., Jones K.C., and James T.W. 1979. Assay for nanogram quantities of DNA in cellular homogenates. *Anal. Biochem.* **92:** 497–500.

Chatelain G., Brun G., and Michel D. 1995. Screening of homozygous transgenic mice by comparative PCR. *BioTechniques* **18:** 958–960, 962.

Costantini F., Radice G., Lee J.L., Chada K.K., Perry W., and Son H.J. 1989. Insertional mutations in transgenic mice. *Prog. Nucleic Acid Res. Mol. Biol.* **36:** 159–169.

Cronmiller C. and Mintz B. 1978. Karyotypic normalcy and quasi-normalcy of developmentally totipotent mouse teratocarcinoma cells. *Dev. Biol.* **67:** 465–477.

Derman E. 1981. Isolation of a cDNA clone for mouse urinary proteins: Age and sex-related expression of mouse urinary protein genes is transcriptionally controlled. *Proc. Natl. Acad. Sci.* **78:** 5425–5429.

Derman E., Krauter K., Walling L., Weinberger C., Ray M., and Darnell J.E. 1981. Transcriptional control in the production of liver-specific mRNAs. *Cell* **23:** 731–739.

Ding H., Roncari L., Shannon P., Wu X., Lau N., Karaskova J., Gutmann D.H., Squire J.A., Nagy A., and Guha A. 2001. Astrocyte-specific expression of activated p21-ras results in malignant astrocytoma formation in a transgenic mouse model of human gliomas. *Cancer Res.* **61:** 3826–3836.

Evans E.P. 1981. Karyotype of the house mouse. *Symp. Zool. Soc. Lond.* **47:** 127–139.

Frohman M. A. 1994. On beyond classic RACE (rapid amplification of cDNA ends). *PCR Methods Appl.* **4:** S40–58.

Grant S.G., Jessee J., Bloom F.R., and Hanahan D. 1990. Differential plasmid rescue from transgenic mouse DNAs into *Escherichia coli* methylation-restriction mutants. *Proc. Natl. Acad. Sci.* **87:** 4645–4649.

Green M.C., ed. 1981. *Genetic variants and strains of the laboratory mouse.* Gustav Fischer Verlag, Stuttgart.

Hadjantonakis A.K., Gertsenstein M., Ikawa M., Okabe M., and Nagy A. 1998. Generating green fluorescent mice by germline transmission of green fluorescent ES cells. *Mech. Dev.* **76:** 79–90.

Hicks G.G., Shi E.G., Li X.M., Li C.H., Pawlak M., and Ruley H.E. 1997. Functional genomics in mice by tagged sequence mutagenesis. *Nat. Genet.* **16:** 338–344.

Higuchi R. 1988. Rapid, efficient DNA extraction for PCR from cells or blood. *Perkin Elmer Cetus Amplifications* **2:** 1–3.

Innis M.A. and Gelfand D.H. 1990. Optimization of PCRs. In *PCR protocols* (ed. M.A. Innis, D.H. Gelfand, and J.J. Sninsky), pp. 3–12. Academic Press, New York.

Innis M.A., Gelfand D.H., Sninsky J.J., and White T., eds. 1990. *PCR protocols: A guide to methods and applications.* Academic Press, San Diego.

Jaenisch R., Harbers K., Schnieke A., Lohler J., Chumakov I., Jahner D., Grotkopp D., and Hoffman E. 1983. Germline integration of Moloney murine leukemia virus at the Mov 13 locus leads to recessive lethal mutation and early embryonic death. *Cell* **32:** 209–216.

Kafatos F.C., Jones W.C., and Efstratiadis A. 1979. Determination of nucleic acid sequence homologies and relative concentrations by a dot hybridization procedure. *Nucleic Acids Res.* **7:** 1541–1552.

Korn R., Schoor M., Neuhaus H., Henseling U., Soininen R., Zachgo J., and Gossler A. 1992. Enhancer trap integrations in mouse embryonic stem cells give rise to staining patterns in chimaeric embryos with a high frequency and detect endogenous genes. *Mech. Dev.* **39:** 95–109.

Labarca C. and Paigen K. 1980. A simple, rapid and sensitive DNA assay procedure. *Anal. Biochem.* **102:** 344–352.

Lacy E., Roberts S., Evans E.P., Burtenshaw M.D., and Costantini F. 1983. A foreign beta-globin gene in transgenic mice: Integration at abnormal chromosomal positions and expression in inappropriate tissues. *Cell* **34:** 343–358.

Latham K.E., Patel B., Bautista F.D., and Hawes S.M. 2000. Effects of X chromosome number and parental origin on X-linked gene expression in preimplantation mouse embryos. *Biol. Reprod.* **63:** 64–73.

Liu X., Constantinescu S.N., Sun Y., Bogan J.S., Hirsch D., Weinberg R.A., and Lodish H.F. 2000. Generation of mammalian cells stably expressing multiple genes at predetermined levels. *Anal. Biochem.* **280:** 20–28.

McBurney M.W. and Rogers B. 1982. Isolation of male embryonal carcinoma cells lines and their chromosome replication patterns. *Dev. Biol.* **89:** 503–508.

Miller D.A. and Miller O.J. 1981. Cytogenetics. In *The mouse in biomedical research* (ed. H.L. Foster et al.), vol. 1, pp. 241–261. Academic Press, New York.

Mintz B. and Bradl M. 1991. Mosaic expression of a tyrosinase fusion gene in albino mice yields a heritable striped coat color pattern in transgenic homozygotes. *Proc. Natl. Acad. Sci.* **88:** 9643–9647.

Novak A., Guo C., Yang W., Nagy A., and Lobe C.G. 2000. Z/EG, a double reporter mouse line that expresses enhanced green fluorescent protein upon Cre-mediated excision. *Genesis* **28:** 147–155.

Ochman H., Gerber A.S., and Hartl D.L. 1988. Genetic applications of an inverse polymerase chain reaction. *Genetics* **120:** 621–623.

Oh I.H. and Eaves C.J. 2002. Overexpression of a dominant negative form of STAT3 selectively impairs hematopoietic stem cell activity. *Oncogene* **21:** 4778–4787.

Radice G., Lee J., and Costantini F. 1991. Hβ58, an insertional mutation affecting early post-implantation development of the mouse embryo. *Development* **111:** 801–811.

Robertson E.J., Kaufman M.H., Bradley A., and Evans M.J. 1983. Isolation, properties, and karyotype analysis of pluripotential (EK) cell lines from normal and parthenogenetic embryos. *Cold Spring Harbor Conf. Cell Prolif.* **10:** 647–663.

Saiki R.K., Scharf S., Faloona F., Mullis K.M., Horn G.T., Erlich H.A., and Arnheim N. 1985. Enzymatic amplification of β-globin genomic sequences and restriction site analysis for diagnosis of sickle cell anemia. *Science* **230:** 1350–1354.

Sambrook J. and Russell D. 2001. *Molecular cloning: A laboratory manual*, 3rd edition. Cold Spring Harbor Laboratory Press, Cold Spring Harbor, New York.

Singh G., Supp D.M., Schreiner C., McNeish J., Merker H.J., Copeland N.G., Jenkins N.A., Potter S.S., and Scott W. 1991. legless insertional mutation: Morphological, molecular, and genetic characterization. *Genes Dev.* **5:** 2245–2255.

Southern E. 1975. Detection of specific sequences among DNA fragments separated by gel electrophoresis. *J. Mol. Biol.* **98:** 503–517.

Tanaka S., Yamamoto H., Takeuchi S., and Takeuchi T. 1990. Melanization in albino mice transformed by introducing cloned mouse tyrosinase gene. *Development* **108:** 223–227.

Tesson L., Heslan J.M., Menoret S., and Anegon I. 2002. Rapid and accurate determination of zygosity in transgenic animals by real-time quantitative PCR. *Transgenic Res.* **11:** 43–48.

Townley D.J., Avery B.J., Rosen B., and Skarnes W.C. 1997. Rapid sequence analysis of gene trap integrations to generate a resource of insertional mutations in mice. *Genome Res.* **7:** 293–298.

Treloar H.B., Feinstein P., Mombaerts P., and Greer C.A. 2002. Specificity of glomerular targeting by olfactory sensory axons. *J. Neurosci* **22:** 2469–2477.

Truett G.E., Heeger P., Mynatt R.L., Truett A.A., Walker J.A., and Warman M.L. 2000. Preparation of PCR-quality mouse genomic DNA with hot sodium hydroxide and tris (HotSHOT). *BioTechniques* **29:** 52, 54.

Varmuza S., Prideaux V., Kothary R., and Rossant J. 1988. Polytene chromosomes in mouse trophoblast giant cells. *Development* **102:** 127–134.

von Melchner H., Reddy S., and Ruley H.E. 1990. Isolation of cellular promoters by using a retrovirus promoter trap. *Proc. Natl. Acad. Sci.* **87:** 3733–3737.

Wagner E.F., Covarrubias L., Stewart T.A., and Mintz B. 1983. Prenatal lethalities in mice homozygous for human growth hormone gene sequences integrated in the germ line. *Cell* **35:** 647–655.

Woychik R.P., Stewart J.A., Davis L.G., D'Eustachio P., and Leder P. 1985. An inherited limb deformity created by insertional mutagenesis in a transgenic mouse. *Nature* **318:** 36–40.

Parthenogenesis, Pronuclear Transfer, and Mouse Cloning

THIS CHAPTER DESCRIBES METHODS TO CREATE MOUSE embryos exclusively from maternally or paternally derived genomes—parthenogenotes and gynogenotes or androgenotes, respectively. Parthenogenotes are easily generated by briefly exposing oocytes to ethanol. Gynogenotes and androgenotes are generated by physically transferring pronuclei between zygotes. Although diploid parthenogenotes, gynogenotes, and androgenotes can be generated using these methods, each type of embryo eventually fails because of imbalances in genomic imprinting (see Chapter 2). Remarkably, viable and fertile mice can be generated by somatic cell cloning in enucleated oocytes, presumably because the transplanted

Courtesy of Ryuzo Yanagimachi

somatic cell genome becomes reprogrammed in the oocyte. Both direct injection of donor nuclei and cell fusion have been used successfully to generate cloned mice. Mouse cloning is now relatively routine, albeit still inefficient. The cloning of mice from ES cell nuclei brings together the powerful genetic manipulations of ES cells with animal cloning technologies.

CONTENTS

GENERATING PARTHENOGENOTES

Numerous studies have been initiated to investigate the influence of the maternal and paternal genomes on early mammalian development. For this type of study, parthenogenetic embryos provide a unique source of preimplantation and early postimplantation embryos that (by definition) develop in the absence of any contribution from a male gamete (see Chapter 2, Imprinting).

Embryonic stem (ES) cells have been derived from either haploid or diploid (cytochalasin-D-treated) parthenogenetically activated oocytes (Kaufman 1983; Robertson et al. 1983; Mann 1992). Because the ES cells derived from haploid oocytes were found to be diploid, the genome presumably underwent duplication without cell division at some stage prior to, or shortly after, ES cell establishment.

The technique for activating oocytes with ethanol is described in several references (e.g., Kaufman 1978a,b) and in Protocol 13.1.

PERFORMING PRONUCLEAR TRANSFER

In this method (McGrath and Solter 1983), pronuclei are removed without penetrating the plasma membrane of the zygote. Instead, they are withdrawn individually or together into a membrane-bound karyoplast that can then be fused with a recipient enucleated zygote using inactivated Sendai virus or electrofusion. Preincubation of the embryos in the presence of the cytoskeletal inhibitors cytochalasin B and Colcemid is critical for the survival of the embryos during this procedure. Cell fusion induced by Sendai virus injected within the zona pellucida was first described by Lin et al. (1973). This procedure is described in Protocol 13.2.

CLONING MICE

The first somatic clone offspring of mice were obtained by directly injecting donor nuclei into recipient enucleated oocytes (Wakayama et al. 1998). When this method is used (the so-called "Honolulu method"), the donor nuclei readily and completely condense within the enucleated metaphase-II-arrested oocytes, which contain high levels of M-phase-promoting factor (MPF). It is believed that the complete condensation of the donor chromosomes promotes complete reprogramming of the donor genome within the mouse oocytes. Another key to the success of mouse cloning is the use of micropipettes attached to a piezo impact driving micromanipulation device (Prime Tech, Japan; EXFO Burleigh, New York; Maerzheuser, Germany). This system saves a significant amount of time during the micromanipulation of oocytes and, thus, minimizes the loss of oocyte viability in vitro. For example, a group of 20 oocytes can be enucleated within 10 minutes by an experienced operator.

So far, the nuclei from cumulus cells, ES cells, immature Sertoli cells, fetal and adult fibroblast cells, neuronal cells, fetal gonadal cells, and lymphocytes have been demonstrated to support full-term development of mouse embryos (Ogura et al. 2001; Hochedlinger and Jaenisch 2002). Among them, cumulus cells have been most conventionally used because of their easy availability. The F_1 hybrid genotypes are apparently superior to the inbred genotypes in terms of developmental efficiency and normality of newborns (Ogura et al. 2001; Wakayama and Yanagimachi 2001a).

More recently, the Honolulu method was modified to employ a conventional electrofusion technique instead of microinjection for mouse cloning (Ogura et al. 2000a). This method provides a technical advantage over the original intracytoplasmic method when donor cells are too large to safely inject into oocytes. Either method results in the birth of normal-looking offspring, which soon recover respiratory ability and active movement shortly after Caesarean section, as long as freshly recovered cells or short-term cultured cells are used as donors (Inoue et al. 2002). The technical developments and practical applications of mouse nuclear transfer including blastomere cloning have been described elsewhere (Ogura et al. 2001). The cloning technique is described in Protocol 13.3.

Many of the factors responsible for successful mouse cloning remain unknown. However, what has recently become evident is that the already-known factors that may affect embryo development, e.g., the quality of embryo culture medium (water, reagents, etc.), the time of handling in vitro, and choice of recipient females, may more critically influence cloning success than those in other conventional mouse embryo techniques. Therefore, in addition to becoming adept at the micromanipulation techniques, mouse cloners are also strongly encouraged to maximize the efficiency of all of the aspects involved in embryo development experiments.

Ethanol-induced Parthenogenetic Activation of Oocytes

Information for this protocol was supplied by M.H. Kaufman, Department of Anatomy, University Medical School, Edinburgh EH8 9AG, United Kingdom. See comment on activation rate of oocytes on page 545.

MATERIALS

ANIMALS
Female mice (8–12 weeks of age)

EQUIPMENT
Hypodermic needle, 26-gauge, 1/2-inch
Incubator, humidified 37°C, 5% CO_2, 95% air
Microscope with phase-contrast, Nomarski, or Hoffman optics
Syringe, 1- or 3-ml
Tissue culture dishes, 3-cm sterile plastic

REAGENTS
Ethanol, fresh 7% (analytical reagent grade) <!>, in Dulbecco's phosphate-buffered saline (PBS) (pH 7.2; see Appendix 1)
Human chorionic gonadotropin (hCG) (Sigma C8554) (see Chapter 3)
Hyaluronidase (e.g., Sigma H3884) solution in M2 medium (see Appendix 1)
Light paraffin oil (e.g., Fisher O121-1; BDH 29436)
M2 and M16 media (see Chapter 4)
Pregnant mare serum gonadotropin (PMSG) (Sigma G4527; Calbiochem 367222) (see Chapter 3)

CAUTION: See Appendix 2 for appropriate handling of materials marked with <!>.

PROCEDURE

1. Inject female mice with PMSG and hCG as described previously (see Inducing Superovulation, Chapter 3). Sacrifice the mice 17 hours after the administration of hCG and recover the oocytes surrounded by cumulus cells (see Recovering Preimplantation Embryos, Chapter 4).

2. Release the oocytes with their cumulus cells attached into freshly prepared 7% ethanol in Dulbecco's PBS in a 3-cm sterile tissue culture dish and let stand for 5 minutes at room temperature. Wash the oocytes through three changes of Dulbecco's PBS and two changes of M2 medium in 3-cm tissue culture dishes. Then transfer the cumulus masses individually into drops of M16 medium under light paraffin oil and place in the incubator at 37°C.

3. After 5 hours, remove the cumulus cells by hyaluronidase treatment (see Collecting Zygotes and Removing Cumulus Cells with Hyaluronidase, Chapter 4). Classify the activated oocytes under phase-contrast, Nomarski, or Hoffman optics. For classification of oocytes, see Figure 13.1. Place the different classes of oocytes into drops of M16 medium under light paraffin oil.

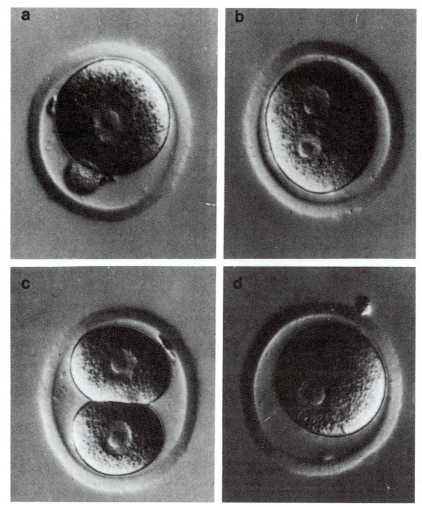

FIGURE 13.1. The four classes of parthenogenetic oocytes that can be distinguished at 4–6 hours after ethanol-induced activation. (*a*) Single pronuclear haploid oocyte with extruded second polar body (uniform haploid). (*b*) Double pronuclear presumptive diploid oocyte (heterozygous diploid). (*c*) Immediate cleavage embryo with two approximately equal-size blastomeres (mosaic haploid). (*d*) Single pronuclear diploid oocyte (heterozygous diploid).

COMMENTS

- Parthenogenetically activated oocytes derived from certain F_1 hybrid females (e.g., C57BL/6 x CBA) can be cultured from the one-cell stage. Oocytes from other sources, for example, inbred strains, suffer a two-cell block and will not proceed beyond the first mitotic division when retained in culture using M16 medium (see Chapter 4). Further development of these oocytes can be obtained by transferring them to the oviducts of 0.5-dpc pseudopregnant recipient female mice.

- The activation rate of oocytes is ~80–85% when exposed to ethanol. Normally, the highest proportion of the activated population consists of haploid parthenogenotes that develop a single pronucleus following extrusion of the second polar body. For experiments requiring a large amount of diploid parthenogenetic embryos, diploidization of this class of oocyte can be achieved by suppression of the formation of the second polar body by incubation in medium containing cytochalasin D (Kaufman 1978a,b) or cytochalasin B (Barton et al. 1987).

Pronuclear Transplantation in the Mouse Embryo

Information for this method was provided by James McGrath (Department of Comparative Medicine, Yale University, New Haven, Connecticut), Davor Solter (Max Planck Institute for Immunobiology, Freiberg, Germany), and Jeff Mann (Beckman Research Institute, City of Hope, Duarte, California). The protocol is presented as the following steps:

1. Isolating embryos

2. Making a holding pipette

3. Making an enucleation/injection pipette

4. Enucleating a zygote

5. Preparing inactivated Sendai virus

6. Introducing pronuclei into enucleated zygotes

STEP 1: ISOLATING EMBRYOS

MATERIALS

EMBRYOS
One-cell embryos

EQUIPMENT
Humidified incubator at 37°C, 5% CO_2, 95% air
Microdrop cultures (see Chapter 4, Setting Up Microdrop Cultures)

REAGENTS
Cytoskeletal inhibitor stocks: Cytochalasin B <!> (5 mg/ml in dimethylsulfoxide <!> [DMSO] [1000x solution]) and either Colcemid (0.1 µg/ml) or nocodazole (3 mg/ml, 10,000x solution). Store cytochalasin B and nocodazole solutions in DMSO at –70°C.
M16 medium (see Chapter 4) plus 5 µg of cytochalasin B/ml and 0.1 µg of Colcemid/ml (Sigma D7385) (nocodazole [0.3 µg/ml; Sigma M1404] can be substituted for Colcemid)
M16 medium (culture medium) (see Chapter 4)

CAUTION: See Appendix 2 for appropriate handling of materials marked with <!>.

PROCEDURE

1. Isolate one-cell embryos and remove their cumulus cells as described in Chapter 4.

2. Place the embryos in microdrop cultures (see Chapter 4).

3. Prior to microsurgery, incubate the embryos in M16 medium plus 5 μg of cytochalasin B/ml and 0.1 μg of Colcemid/ml in a humidified incubator with 5% CO_2, 95% air at 37°C for 15–45 minutes.

STEP 2: MAKING A HOLDING PIPETTE

A holding pipette is used to fix the zygote in place throughout the microsurgery. Make the holding pipette as described in Chapter 7 (Protocol 7.8).

STEP 3: MAKING AN ENUCLEATION/INJECTION PIPETTE

An enucleation/injection pipette is used to remove or introduce the pronuclei.

MATERIALS

EQUIPMENT

Borosilicate glass capillary tubing, 1.0 mm outside diameter, 0.6 mm inside diameter, and 1.0 mm outside diameter, 0.8 mm inside diameter (Drummond Scientific 9-000-2141 and 9-000-2171)

Leica instrument holder

Microforge (De Fonbrune model: Alcatel CIT or Technical Products International or Bachofer)

Modeling clay

Pipette grinder (e.g., Effenberger; available from Bachofer)

Pipette puller (see Chapter 7)

Platinum/iridium wire 90%/10% mixture, outside diameter ~0.10, to replace microforge filaments (e.g., Clark Electromedical Instruments 101R-5T; The Wildinson Co., to specifications)

Polyethylene tubing

Syringe

Tissues

REAGENTS

Distilled water

Nonidet P-40 (NP-40) detergent (Sigma N6507)

Tween 80 (Sigma P1754)

PROCEDURE

1. Use a pipette puller to pull a piece of capillary tubing (1.0 mm outside diameter, 0.6 mm inside diameter) (see Chapter 7). Using a microforge, break the tip of the pipette on a glass anvil to give a 10–15-μm outside diameter at the tip of the pipette (Fig. 13.2).

2. Bevel the pipette tip with an Effenberger grinder using water as a lubricant.

Heating filament   Glass bead

FIGURE 13.2. Breaking the tip of a micropipette using a De Fonbrune microforge. A "glass anvil" is prepared on the tip of a thick filament (0.3 mm) by fusing a small fragment of glass to the filament and allowing it to melt into a rounded ball. The glass anvil will remain in place as long as the filament is not heated to too high a temperature. The micropipette is positioned above the glass anvil and the filament is heated to a temperature at which the pipette will just fuse to the glass anvil, but will not be distorted (this temperature must be determined empirically). The pipette is fused to the glass anvil and then the heat to the filament is turned off. As the filament cools and contracts slightly, it will break the pipette at the point of contact, leaving a flush end.

3. Sharpen the pipette tip so that it will penetrate the zona pellucida.

   a. Position the beveled pipette tip vertically above the thin filament.

   b. Heat the filament to a temperature at which the tip will just fuse to the filament (to be determined empirically).

   c. Touch the pipette tip to the thin filament to sharpen it. The air jets of the microforge must be blowing on the filament so that it heats the pipette only by conduction and not by convection.

   d. When the pipette fuses to the filament, raise it away from the filament, pulling out a short spike of glass at the tip (Fig. 13.3).

4. Fit the pipette into a Leica instrument-holder attached to polyethylene tubing and a syringe.

   a. Draw in and expel out of the pipette tip a few times a solution of 1.25% (v/v) Tween 80 in 0.2 μl of filtered distilled water.

   b. Remove the pipette from the solution and continue to expel air for ~20 seconds. Use a tissue to absorb any excess liquid on the outside of the pipette.

STEP 4: ENUCLEATING A ZYGOTE

In this step, the two pronuclei are removed without penetrating the plasma membrane.

FIGURE 13.3. Sharpening the enucleation/injection pipette. The beveled pipette is positioned vertically above the thin filament. The filament is heated to a temperature at which the pipette will just fuse to the filament (to be determined empirically). The air jets of the microforge must be blowing on the filament so that it heats only by conduction and not by convection. When the pipette fuses to the filament, it is raised away from the filament, pulling out a short spike of glass at the tip.

MATERIALS

EMBRYOS

One-cell embryos (incubated in the presence of cytoskeletal inhibitors, see p. 547)

EQUIPMENT

Beaudouin or comparable syringe (Alcatel CIT)
Enucleation/injection pipette
Holding pipettes (see Chapter 7)
Manipulation chamber (The microscope/micromanipulator setup described in Chapter 7 for pronuclear injection can be adapted for pronuclear transfer, where a Beaudouin syringe controls the amount of suction on the enucleation pipette.)
Silicone oil (200 fluid, 20 cs viscosity; Dow Corning)
Transfer pipettes

REAGENTS

M2 medium (see Chapter 4) plus 5 μg of cytochalasin B<!>/ml and 0.1 μg of Colcemid/ml (Sigma D7385) (nocodazole [0.3 μg/ml: Sigma M1404] can be substituted for Colcemid)

CAUTION: See Appendix 2 for appropriate handling of materials marked with <!>.

PROCEDURE

1. Use a transfer pipette to place one-cell embryos incubated in M2 medium plus 5 μg of cytochalasin B/ml and 0.1 μg of Colcemid/ml in the manipulation chamber. All manipulations are performed in M2 medium plus 5 μg of cytochalasin B/ml and 0.1 μg of Colcemid/ml.

2. Using a holding pipette (see Chapter 7) to keep the embryo in position, penetrate the zona pellucida of the embryo with an enucleation/injection pipette containing low-viscosity silicone oil. This may require considerable deformation of the zona pellucida (Fig. 13.4A–C). Avoid penetrating the embryo plasma membrane (see Comment below).

3. Once the enucleation/injection pipette tip is in the perivitelline space (PVS), advance the tip further to a point adjacent to one of the two pronuclei. Using a Beaudouin syringe to control suction, sequentially draw the plasma membrane overlying the pronucleus, a small volume of cytoplasm, and then the pronucleus into the pipette (Fig. 13.4D). Then move the pipette tip to a point adjacent to the remaining pronucleus and repeat this step (Fig. 13.4E).

4. Withdraw the enucleation/injection pipette from the embryo (Fig. 13.4F). As the pipette is withdrawn, a cytoplasmic bridge will extend from the karyoplast within the pipette to the enucleated embryo within the zona pellucida. This bridge will stretch to a fine thread, pinch off, and reseal the plasma membrane (see Fig. 13.5A).

COMMENT

In the presence of cytoskeletal inhibitors, the embryo offers little resistance to the advancing micropipette. These inhibitors thus help to prevent penetration of the plasma membrane.

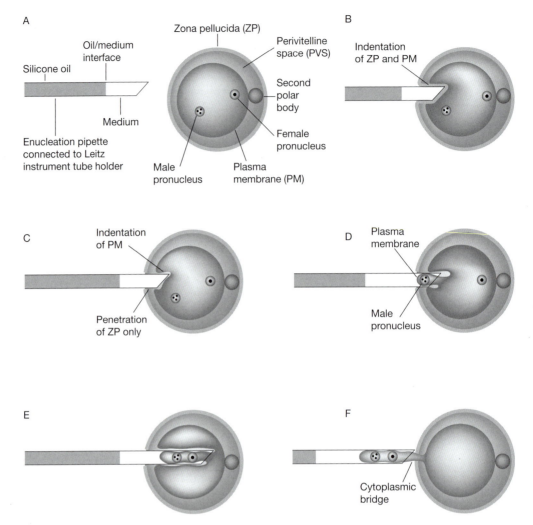

FIGURE 13.4. Technique for enucleating a zygote by withdrawing pronuclei into a karyoplast. (ZP) Zona pellucida; (PVS) perivitelline space; (PM) plasma membrane.

STEP 5: PREPARING INACTIVATED SENDAI VIRUS

Sendai virus is inactivated for use in step 6, below. This procedure is as described by Neff and Enders (1968).

MATERIALS

> **EQUIPMENT**
> Pipette, 100-ml
> Water bath at 37°C
>
> **REAGENTS**
> Acetone<!>–dry ice bath
> 1% β-Propiolactone (BPL) <!> (Sigma P5773)
> Phosphate-buffered saline (PBS), cold (pH 7.2) (see Appendix 1)
> Sendai virus

CAUTION: See Appendix 2 for appropriate handling of materials marked with <!>.

PROCEDURE

1. Prepare Sendai virus as described by Graham (1971) and Giles and Ruddle (1973).

2. Place the Sendai virus suspension on ice. Add 1% BPL (diluted in cold PBS) to the suspension to achieve a final concentration of 0.025% BPL.

3. Store the virus–BPL mixture at 4°C for 24 hours. Then place the mixture in a water bath at 37°C for 20 minutes.

4. Aliquot (100 ml) inactivated Sendai virus, fast-freeze in an acetone–dry ice bath, and store at –70°C.

STEP 6: INTRODUCING PRONUCLEI INTO ENUCLEATED ZYGOTES

Figure 13.5A–F illustrates how the pronuclear karyoplast obtained above can be introduced into a second enucleated one-cell-stage embryo. However, similar manipulations with nuclei obtained from other sources are also possible.

MATERIALS

EMBRYOS
Enucleated one-cell-stage embryos
Pronuclear karyoplast

EQUIPMENT
Enucleation/injection pipette
Flame-drawn Pasteur pipette
Holding pipette (see Chapter 7)
Humidified incubator at 37°C, 5% CO_2, 95% air
Manipulation chamber
Tissue culture dishes

REAGENTS
Inactivated Sendai virus
M2 medium (see Chapter 4) plus 5 µg of cytochalasin B/ml <!> and 0.1 µg of Colcemid/ml (Sigma D7385) (nocodazole [0.3 µg/ml; Sigma M1404] can be substituted for Colcemid)
M2 and M16 media (see Chapter 4)

PROCEDURE

1. Place an enucleation/injection pipette containing a pronuclear karyoplast into a microdrop of inactivated Sendai virus (~3000 HAU/ml) as prepared on p. 550. Draw a small volume of virus (approximately equal to the volume of the karyoplast) into the pipette (Fig. 13.5B).

2. Place the virus/karyoplast-loaded enucleation/injection pipette into a microdrop containing a second enucleated embryo. Use a holding pipette to position the enucleated embryo so that the previous site of penetration (Fig.

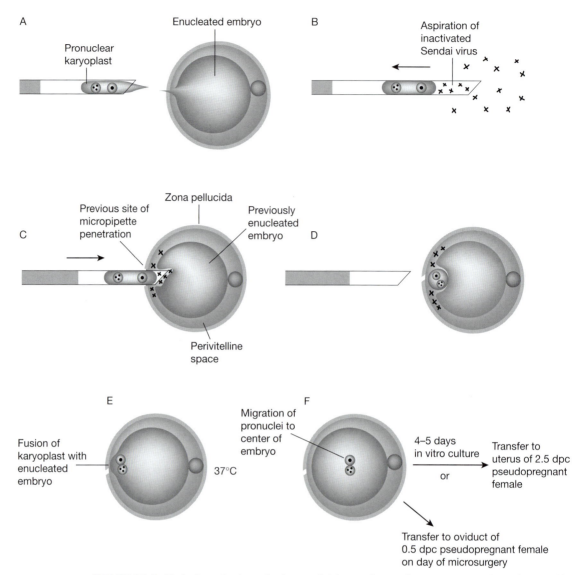

A Pronuclear karyoplast Enucleated embryo

B Aspiration of inactivated Sendai virus

C Previous site of micropipette penetration Zona pellucida Previously enucleated embryo Perivitelline space

D

E Fusion of karyoplast with enucleated embryo 37°C

F Migration of pronuclei to center of embryo 4–5 days in vitro culture Transfer to uterus of 2.5 dpc pseudopregnant female or Transfer to oviduct of 0.5 dpc pseudopregnant female on day of microsurgery

FIGURE 13.5. Technique for introducing nuclei from a karyoplast into an enucleated one-cell-stage embryo by inactivated Sendai-virus-mediated fusion.

13.5C) is accessible to the enucleation/injection pipette containing the karyoplast and virus. Advance the pipette through the hole in the zona pellucida, and sequentially inject the virus and karyoplast into the perivitelline space (PVS) (Fig. 13.5C,D).

3. Continue to manipulate additional enucleated embryos in the manipulation chamber, exposing them to the cytoskeletal inhibitors for a maximum of 1.5 hours. Then transfer all successfully manipulated embryos into 2 ml of M2 medium and let stand for 5 minutes to wash out the inhibitors.

4. Transfer the manipulated embryos with a flame-drawn Pasteur pipette to M16 medium and incubate them in a humidified incubator with 5% CO_2 at 37°C for ~1 hour. Fusion of the karyoplast should occur within 1 hour (Fig. 13.5E,F).

COMMENT

Embryos that undergo fusion should develop normally to the blastocyst stage. Abnormal cleavage of embryos indicates that the cytoskeletal inhibitors may have been toxic to the embryo.

Cloning Mice

This protocol was provided by Atsuo Ogura, RIKEN Bioresource Center, Japan, with some modifications of the protocols originally developed by Wakayama et al. (1998). It is presented as the following steps:

1. Preparing micropipettes

2. Setting up the enucleation and injection micropipettes

3. Collecting and enucleating oocytes

4. Preparing nucleus donor cells

5. Injecting donor nuclei

6. Activating embryos and culturing

7. Transferring cloned embryos

MATERIALS

Materials needed throughout the protocol are listed here. Additional materials for specific steps are listed with each step.

EQUIPMENT

Falcon 1008 and 1007 plastic dishes

Injectors (Narishige or Eppendorf or Leica)

> To increase the survival rate of oocytes after injection, the injection micropipette must move precisely along the x axis without any wavering, especially when the pipette is being withdrawn from the ooplasm. For this reason, three-dimensional micromanipulators with separate x- and y-axis handles are recommended (for example, Narishige MO-202U joystick micromanipulators, the large-handled type, Fig. 13.6). Microinjectors for injection and enucleation must be the oil-filled type rather than the air-filled type; Narishige IM-6 and IM-9 and Eppendorf CellTram Oil work best. Injectors that are connected to the holding pipette can be the air-filled type.

Inverted microscope (Nikon, Olympus, or Leica) with Nomarski or Hoffman optics

Microforge (Narishige or De Fonbrune)

Micromanipulators (Narishige or Leica)

Piezo impact drive system (Prime Tech)

Pipette puller (Sutter Instrument)

REAGENTS

CZB (embryo culture)

CZB-HEPES (embryo manipulation)

CZB-Sr (embryo activation)

Polyvinylpyrrolidone (PVP) <!>

Silicon oil (Sigma-Aldrich 14615-3) or mineral oil (Nakalai Tesque 26132-35)

The embryo-tested oil is used to overlay the drops of culture medium. Oocytes and embryos are cultured in 5–6.5% CO_2 in air at 37–37.5°C.

Chatot-Ziomek-Bavister (CZB)-based media (see Chapter 4) are used with slight modifications for culturing cloned embryos and micromanipulating oocytes in vitro. The two media described are made every 2 weeks (Table 13.1). They are sterilized with Millipore filters (0.45 μm), dispensed into 5-ml plastic tubes (Falcon 2057), and stored at 4°C until they are used. Glutamine and $CaCl_2$ (or $SrCl_2$) are added on the day of the experiment.

CAUTION: See Appendix 2 for appropriate handling of materials marked with <!>.

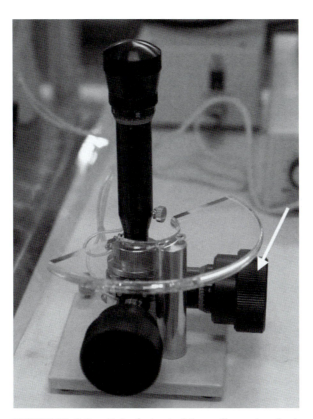

FIGURE 13.6. Joystick of a typical three-dimensional micromanipulator (Narishige MO-202U). The *x*-axis handle (*arrow*) for fine movement is necessary for efficient operation of the piezo-driven micromanipulator.

TABLE 13.1. Formulations of CZB media for mouse cloning

	CZB Embryo culture		CZB-HEPES Embryo handling		CZB-Sr Embryo activation	
	mM	mg/100 ml	mM	mg/100 ml	mM	mg/100 ml
NaCl	82.0	478.9	82.0	478.9	82.0	478.9
KCl<!>	4.9	36.3	4.9	36.3	4.9	36.3
KH$_2$PO$_4$<!>	1.2	15.9	1.2	15.9	1.2	15.9
MgSO$_4$•7H$_2$O	1.2	29.1	1.2	29.1	1.2	29.1
NaHCO$_3$<!>	25.0	210.0	15.0	126.0	25.0	210.0
Glucose	5.6	100.0	5.6	100.0	5.6	100.0
Sodium pyruvate	0.3	2.9	0.3	2.9	0.3	2.9
CaCl$_2$•2H$_2$O[a]	1.7	25.1	1.7	25.1		
SrCl$_2$•6H$_2$O[a]					2.5–10.0	66.7–266.6
HEPES			10	238.0		
Glutamine[a]	1.0	14.6	1.0	14.6	1.0	14.6
Sodium lactate (60% syrup)	20	370 µl	20	370 µl	20	370 µl
EDTA	0.10	3.8	0.10	3.8	0.10	3.8
Polyvinyl alcohol[b]		10.0		10.0		10.0
BSA		5 mg/ml				5 mg/ml

[a]CaCl$_2$, SrCl$_2$, and glutamine are added from 100x, 10x, and 100x stock solutions, respectively.
[b]Polyvinyl alcohol (cold water soluble) is first dissolved in ultrapure water at 80°C for 1 hour.

CAUTION: See Appendix 2 for appropriate handling of materials marked with <!>.

COMMENTS

- KSOM (see Chapter 4) medium can also be used successfully to culture cloned mouse embryos.

- The CZB-HEPES micromanipulation medium does not contain BSA, which can decrease the survival rate of oocytes after injection. Bovine calf serum (10–20%) may be used instead of BSA. This often increases the survival rate (Rybouchkin and Dhont 2000), but this seems to depend on the operator.

- The 10–12% polyvinylpyrrolidone (PVP) (360 kD) solution is prepared in CZB-HEPES medium. PVP must be dissolved completely by shaking overnight, because unevenly dissolved PVP solutions may damage the plasma membranes of the oocytes and donor cells. The solution can be stored for at least 6 months.

STEP 1: PREPARING MICROPIPETTES

MATERIALS

EQUIPMENT
Glass capillary tubing (Sutter Instrument B100-75-10 or Drummond Microcaps 50 µl, 1-000-0500)
Microforge

PROCEDURE

1. For efficient enucleation and injection with a piezo drive micromanipulation device (Fig. 13.7), the wall of the enucleation and injection pipettes must be thin (less than one-fourth of the radius). The tips of the pipettes must be par-

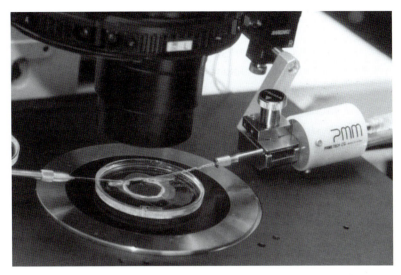

FIGURE 13.7. The impact unit of the piezo-driven micromanipulator system (*right*) attached to an injection pipette holder on an inverted microscope.

allel. Using tapered tips often results in low oocyte-survival rates after enucleation or injection.

2. Break the tips of the pipettes vertically using a microforge to create a blunt end (see Chapter 7 and Fig. 13.2).

> Blunt-ended pipettes are able to break the zona pellucida and oolemma while causing minimal damage to the oocyte. The inner diameter of the enucleation pipettes is between 7 and 8 μm.

> It is possible to thin the wall of the capillary with hydrofluoric acid<!>, which may slightly improve the oocyte survival rate, but it is not necessary.

3. The inner diameter of the injection micropipettes should be adjusted according to the size and hardness of the donor cells. Inner diameters of 4–5 μm for cumulus cells and primordial germ cells, 3–4 μm for immature Sertoli cells, and 7–9 μm for fibroblast cells are recommended.

4. Holding pipettes are prepared the same way as for other mouse embryo manipulations (see Chapter 7).

STEP 2: SETTING UP THE ENUCLEATION AND INJECTION MICROPIPETTES

MATERIALS

EQUIPMENT
Falcon 35-1006 petri dish
Plastic capillary tube

REAGENTS
Mercury <!>

CAUTION: See Appendix 2 for appropriate handling of materials marked with <!>.

PROCEDURE

1. Load a small quantity of mercury (1–2 mm long when in the micropipette) into the enucleation/injection micropipette from the proximal end using a small, flexible plastic capillary tube.

2. Attach the enucleation/injection micropipette that is connected to the oil-filled injector to the instrument holder of the piezo impact drive. After expelling the air from the tip of the enucleation/injection micropipette, wash the inside of the tip several times with oil while it is in the micromanipulation chamber (Fig. 13.8A).

3. Wash the enucleation/injection micropipette in PVP medium until the mercury inside the pipette can be moved smoothly and without any time lag (Fig. 13.8B).

   Take care to completely remove oil adhering to the inside of the pipette.

4. Place a small volume of oil between the PVP and the CZB-HEPES medium to prevent the mercury from contaminating the micromanipulation medium (Fig. 13.8C).

COMMENTS

- This step has a significant effect on the efficiency of nuclear transfer. If the setup is done correctly and carefully, 100–150 oocytes can be injected before the injection pipette needs to be changed.

- The top of a Falcon 35-1006 dish is used as the micromanipulation chamber. If Nomarski (DIC) optics are used, replace the center of the chamber with a glass coverslip. Place drops (2–4 μl each) of the following media on the dish and cover with silicon or mineral oil:

  (1) PVP drops for pipette cleaning, (2) drops of CZB-HEPES containing cytochalasin B<!> for enucleation of oocytes, and (3) CZB-HEPES drops for donor nucleus injection.

STEP 3: COLLECTING AND ENUCLEATING OOCYTES

MATERIALS

ANIMALS
Mature females (8–12 weeks of age)

EQUIPMENT
Incubator, 37°C 5% CO_2 humidified
Microdrop culture dish
Micromanipulation chamber (see above)

REAGENTS
CZB
CZB-HEPES
Pregnant mare serum gonadotropin (PMSG) (see Chapter 3)
Human chorionic gonadotropin (hCG) (see Chapter 3)
Hyaluronidase

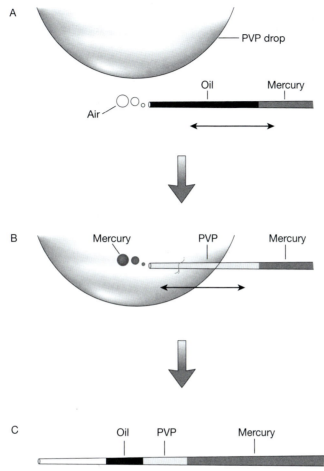

FIGURE 13.8. Setting up an enucleation pipette. The procedure is the same for injection pipettes. See the text for details. Note that PVP drop and pipette are under oil.

PROCEDURE

1. Inject mature females intraperitoneally with 7.5 I.U. of PMSG and 7.5 I.U. of hCG, with a 48-hour period between injections (see Chapter 3).

2. Collect mature oocytes that are surrounded by cumulus cells from the ampullae of the oviducts 15–17 hours after the hCG injection (see Chapter 3).

3. Place the collected oocytes in CZB medium containing 0.1% hyaluronidase and place them in a 37°C 5% CO_2 humidified incubator until the cumulus cells disperse (2–3 minutes). Then wash the oocytes and place them in a microdrop culture dish.

4. Place a group of 10–20 oocytes into a drop of CZB-HEPES medium containing cytochalasin B in a micromanipulation chamber (see above).

5. Hold the oocytes with the metaphase plate between the 2 o'clock and 4 o'clock positions.

6. Place the tip of the enucleation pipette onto the surface of the zona pellucida.

7. Use a few piezo pulses (controller settings: speed 3–6, intensity 1–4) to advance the pipette while applying a very slight negative pressure. To avoid

damaging the oolemma, apply a slight positive pressure when the peri-vitelline space narrows under the impacted area just before the entire zona is pierced.

8. Remove the chromosomes by suction, along with a small volume of cytoplasm.

9. After a group of oocytes has been enucleated, wash the oocytes several times in drops of fresh CZB medium. Leave the oocytes in the incubator for 30 minutes to 2 hours before donor nucleus injection (or electrofusion).

COMMENTS

- The strain of mouse from which recipient oocytes are collected is selected on the basis of developmental ability, tolerance to in vitro handling (in particular, intracytoplasmic injection), and visibility of metaphase II (M II) chromosomes of the oocytes. B6D2F1 (C57BL/6 x DBA/2) oocytes are the best recipients and fulfill all of these requirements. Because oocytes from other strains may also offer these advantages, researchers can select the strain according to their purposes.

  The following is a brief guide to the relevant strain characteristics:

  1. High developmental ability: B6D2F1, B6C3HF1, B6CBAF1, and probably other F_1 hybrids.
  2. Tolerance to intracytoplasmic injection: B6D2F1, DBA/2 (best).
  3. Visibility of M II chromosomes: B6D2F1, DBA/2, ICR, and 129.

- Oocytes retrieved from the oviducts sooner than recommended (e.g., 13 hours) after hCG treatment generally show better developmental ability, but are less tolerant to intracytoplasmic injection.

- Metaphase II chromosomes are visible without the use of fluorescent dye under Nomarski or Hoffman optics. Oocytes with a clear cytoplasm (such as B6D2F1 oocytes; see above) can be more easily enucleated than those with granulated masses (such as B6 and B6CBAF1 oocytes).

- Warming the microscope stage is recommended, because the spindle microtubules will remain well polymerized and form a small clear area, which can be easily distinguished from the surrounding opaque cytoplasm.

- Oocytes are very sensitive and can easily lyse just after enucleation. However, their viability is usually restored after they have been incubated for 30 minutes to 2 hours.

STEP 4: PREPARING NUCLEUS DONOR CELLS

Collect cumulus cells from freshly ovulated oocytes by treatment with 0.1% bovine testicular hyaluronidase (Sigma H3884) in CZB medium.

STEP 5: INJECTING DONOR NUCLEI

PROCEDURE

1. Gently mix the donor cumulus cells into the PVP drops in the micromanipulation chamber.

2. Place a group of 10–20 enucleated oocytes in a drop of CZB-HEPES in a micromanipulation chamber.

3. Remove the nuclei from the donor cells by gently aspirating them in and out of the injection pipette. Then draw up the donor nuclei in a line within the injection pipette (Fig. 13.9a,b).

4. After advancing the injection pipette through the zona pellucida with a few piezo pulses, as described above, insert the pipette deep into the ooplasm while pushing forward a donor nucleus to the tip of the pipette (Fig. 13.9c,d).

5. After leaving the pipette inside the oocyte for a few seconds, apply a single piezo pulse of minimal intensity that is determined empirically. The plasma membrane will be punctured at the pipette tip, as evidenced by a rapid relaxation of the membrane. Expel the nucleus into the ooplasm with a minimal amount of medium (Fig. 13.9d–f).

6. Gently withdraw the pipette (relatively rapidly at first, and then slowly).

7. Leave the injected oocytes at room temperature for 5–10 minutes. Then wash and culture the injected oocytes in CZB medium until they are activated with strontium (see step 6).

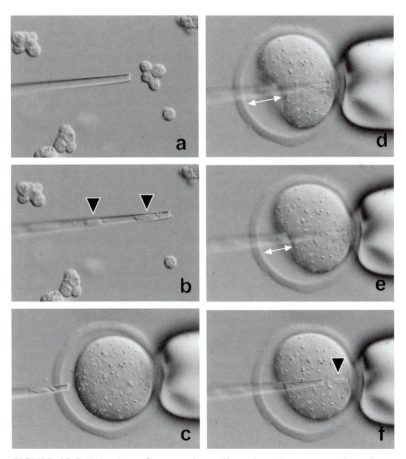

FIGURE 13.9. Injection of a cumulus cell nucleus into an enucleated oocyte. (*a*) Dispersed cumulus cells; (*b*) cumulus cell nuclei (*arrowheads*) in injection pipette; (*c*) piercing the zona pellucida with the injection pipette; (*d*) pushing the injection pipette deep into the enucleated oocyte but not breaking the oolema; (*e*) application of the piezo pulse breaks the oolema, causing a visual relaxation of the membrane (compare the length of double arrows in *d* and *e*); (*f*) injection of the donor cumulus cell nucleus (*arrowhead*) into the ooplasm.

COMMENTS

- It is recommended that cumulus cells from F_1 hybrid mice be used for initial mouse cloning attempts.
- In the original trials of intracytoplasmic injection of mouse oocytes, the micromanipulation temperature was maintained at 17–18°C to enhance the survival rate. However, this is not necessary, and the operation can be performed at room temperature without any problems. A heated stage may cause adverse effects such as oocyte lysis.
- For successful transfer of nuclei from more solid cells, such as fibroblasts, it is necessary to remove the entire plasma membrane before injection. If the rate of nuclear formation is poor (<50%) among surviving oocytes, nuclear transfer by electrofusion is recommended (Ogura et al. 2000a). For transfer of nuclei from soft cells, such as cumulus cells (Wakayama et al. 1998), embryonic stem cells (Wakayama et al. 1999), immature Sertoli cells (Ogura et al. 2000b), or primordial germ cells (Lee et al. 2002), partial damage to the plasma membrane is usually sufficient for the nuclei to intermingle with the ooplasm.
- When the operator becomes sufficiently skilled, several (3–10) nuclei can be aspirated into the injection pipette and injected at one time. This operation saves time handling the oocytes in vitro.

STEP 6: ACTIVATING EMBRYOS AND CULTURING

Reconstructed embryos require a signal to activate development. This is induced by culturing the reconstructed embryos in medium containing strontium.

MATERIALS

REAGENTS

Strontium chloride (Sigma S0390)
Cytochalasin B<!> (Sigma C6762)

CAUTION: See Appendix 2 for appropriate handling of materials marked with <!>.

PROCEDURE

1. Approximately 1 hour after nuclear transfer, activate the oocytes in Ca^{++}-free CZB medium containing 2.5–10 mM strontium chloride (CZB-Sr) and 5 µg/ml cytochalasin B at 37°C in 5% CO_2.

2. About 1 hour later, place the activated oocytes in CZB medium containing 5 µg/ml cytochalasin B, and culture for 5 hours.

3. Wash the reconstructed embryos several times in fresh CZB medium, and then culture until the embryos are transferred.

COMMENTS

- The optimal concentration of strontium chloride and time of exposure to the activation medium may vary between laboratories. Contamination of the CZB-Sr medium for activation by Ca^{++} carryover from the CZB medium decreases the effect of Sr^{++}. Therefore, the reconstructed embryos should be washed free of CZB before being activated in CZB-Sr.

- Cytochalasin B is added to the activation medium to prevent extrusion of the donor chromosomes as a polar body.

- The effects of cytochalasin B and cytochalasin D on cloning experiments do not appear to be different.

- It is important that the cytochalasin be washed away thoroughly and quickly, because concentrations of cytochalasin below 5 µg/ml often cause deformation of the oocyte surface.

- The presence of 1% DMSO<!> in the CZB-Sr activation medium can enhance the reconstructed embryos to develop into blastocysts (Wakayama and Yanagimachi 2001b).

- The success of nuclear transfer can be roughly assessed by examining the pseudopronuclei of the reconstructed embryos. The embryos should have two or three well-developed pseudopronuclei when they are retrieved from the cytochalasin-containing medium. If no or only one nucleus has formed, the oocytes were activated upon nuclear transfer because of inappropriate in vitro handling. Such embryos rarely develop to term (Wakayama and Yanagimachi 2001b).

STEP 7: TRANSFERRING CLONED EMBRYOS

Transfer the cloned embryos into the oviducts or uteri as in other experiments (see Chapter 6).

Special care must be exercised, as described below in the Comments, because cloned embryos are much more sensitive to changes in physical and physiological conditions than are embryos produced by conventional in vitro techniques, such as in vitro fertilization.

COMMENTS

- ICR females have proven to be among the best recipients for cloned embryo transfer. They should be 2.5–4 months of age and weigh 30–35 grams. The use of suboptimal females can result in the loss of entire batches of transferred cloned embryos prior to implantation (no implantation sites). It is recommended that ICR females from different vendors be tested.

- The lower body temperature induced in recipient females by anesthesia may cause poor embryonic development. Avertin anesthesia works well. The use of appropriate heaters also helps to maintain body temperature (see Chapter 6).

- Two combinations of embryonic stage and recipient transfer site are recommended for embryo transfer: (1) cloned four-cell embryos (3 days) are transferred into the oviducts of 0.5-dpc pseudopregnant recipient females and (2) cloned morulae/blastocysts (4 days) are transferred into uteri of 2.5-dpc pseudopregnant recipient females. Low implantation rates have been obtained by transferring cloned two-cell embryos (Day 2) into the oviducts of 0.5-dpc pseudopregnant recipient females.

REFERENCES

Barton S.C., Norris M.L., and Surani M.A. 1987. Nuclear transplantation in fertilised and parthenogenetically activated eggs. In Monk M. (ed.) *Mammalian development: A practical approach*, pp. 235–253. IRL Press, Oxford.

Giles R.E. and Ruddle F.H. 1973. Production of Sendai virus for cell fusion. *In Vitro* **9:** 103–107.

Graham C.F. 1971. Virus assisted fusion of embryonic cells. *Acta Endocrinol.* (suppl.) **153:** 154–167.

Hochedlinger K. and Jaenisch R. 2002. Monoclonal mice generated by nuclear transfer from mature B and T donor cells. *Nature* **415:** 967–969.

Inoue K., Kohda T., Lee J., Ogonuki N., Mochida K., Noguchi Y., Tanemura K., Kaneko-Ishino T., Ishino F., and Ogura A. 2002. Faithful expression of imprinted genes in cloned mice. *Science* **295:** 297.

Kaufman M.H. 1978a. The experimental production of mammalian parthogenetic embryos. In *Methods in mammalian reproduction* (ed. J. C. Daniel), p. 21–47. Academic Press, New York.

———. 1978b. The chromosome complement of single pronuclear haploid mouse embryos following activation by ethanol treatment. *J. Embryol. Exp. Morphol.* **71:** 139–154.

———. 1983. *Early mammalian development: Parthenogenetic studies.* Cambridge University Press, Cambridge.

Lee J., Inoue K., Ono R., Ogonuk R., Kohda K., Kaneko-Ishino T., Ogura A., and Ishino F. 2002. Erasing genomic imprinting memory in mouse clone embryos produced from day 11.5 primordial germ cells. *Development* **129:** 1807–1817.

Lin T.P., Florence J., and Jo O. 1973. Cell fusion induced by a virus within the zona pellucida of mouse eggs. *Nature* **242:** 47–49.

Mann J.R. 1992. Properties of androgenetic and parthenogenetic mouse embryonic stem cell lines—Are genetic imprints conserved? *Semin. Dev. Biol.* **3:** 77–85.

McGrath J. and Solter D. 1983. Nuclear transplantation in the mouse embryo by microsurgery and cell fusion. *Science* **220:** 1300–1302.

Neff J.M. and Enders J.F. 1968. Poliovirus replication and cytogenicity in monolayer hamster cell cultures fused with beta propiolactone-inactivated Sendai virus. *Proc. Soc. Exp. Biol. Med.* **127:** 260–267.

Ogura A., Ogonuki N., Takano K., and Inoue K. 2001. Microinsemination, nuclear transfer, and cytoplasmic transfer: The application of new reproductive engineering techniques to mouse genetics. *Mamm. Genome* **12:** 803–812.

Ogura A., Inoue K., Takano K., Wakayama T., and Yanagimachi R. 2000a. Birth of mice after nuclear transfer by electrofusion using tail tip cells. *Mol. Reprod. Dev.* **57:** 55–59.

Ogura A., Inoue K., Ogonuki N., Noguchi A., Takano K., Nagano R., Suzuki O., Lee J., Ishino F., and Matsuda J. 2000b. Production of male clone mice from fresh, cultured, and cryopreserved immature Sertoli cells. *Biol. Reprod.* **62:** 1579–1584.

Robertson E.J., Evans M.J., and Kaufman M.H. 1983. X chromosome instability in pluripotential stem cell lines derived from parthogenic embryos. *J. Embryol. Exp. Morphol.* **74:** 297–309.

Rybouchkin A. and Dhont M. 2000. Nuclear transfer into mouse oocytes by a conventional method of injection. *Theriogenology* **53:** 241.

Wakayama T. and Yanagimachi R. 2001a. Mouse cloning with nucleus donor cells of different age and type. *Mol. Reprod. Dev.* **58:** 376–383.

———. 2001b. Effect of cytokinesis inhibitors, DMSO and the timing of oocyte activation on mouse cloning using cumulus cell nuclei. *Reproduction* **122:** 49–60.

Wakayama T., Mombaerts P., Rodriguez I., Perry A.C.F., and Yanagimachi R. 1999. Mice cloned from embryonic stem cells. *Proc. Natl. Acad. Sci.* **96:** 14984–14989.

Wakayama T., Perry A.C.F., Zuccotti M., Johnson K.R., and Yanagimachi R. 1998. Full-term development of mice from enucleated oocytes injected with cumulus cell nuclei. *Nature* **394:** 369–374.

14

Assisted Reproduction

Ovary Transplantation, In Vitro Fertilization, Artificial Insemination, and Intracytoplasmic Sperm Injection

TRANSGENE EXPRESSION OR MUTATIONS can compromise male and female mouse viability and/or fertility, often making their maintenance by standard breeding schemes very difficult. In this chapter, assisted reproduction methods are described that can be used when male or female mice are unable to reproduce on their own. Basically, these methods rescue the germ line of mice that cannot or will not breed, or that breed poorly. Even the germ cells of newborn mice can be rescued using these assisted reproduction protocols. In addition, these methods can be coupled with frozen sperm or frozen ovaries to retrieve strains that have been archived by cryopreservation (see Chapter 15). Remarkably, "dead" sperm (freeze-dried) can be used to regenerate mice. Furthermore,

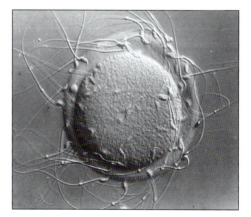

Courtesy of Paul Wassarman

in vitro fertilization and artificial insemination can be used to quickly expand the numbers of individuals in a particular mouse strain. If these various types of assisted reproduction methods fail, one could potentially rescue a very valuable mouse using animal cloning (see Chapter 13).

CONTENTS

INFERTILITY

The following discussion focuses on situations in which a very important male or female mouse is not breeding. A progressive series of methods is proposed from simple to more complex to rescue the germ line of the animal. The normal adult male and female reproductive urogenital systems are shown in Figure 14.1.

Female Infertility

The following course should be followed if female infertility is suspected:

1. If an adult female mouse has not become pregnant after being paired with a male for an extended period of time, the male should be replaced and the female subsequently checked for a copulation plug.

   • If a plug is obtained, wait for the female to give birth.

   • If there is no pregnancy, another copulation plug should be obtained and the female observed for a subsequent pregnancy.

2. If there is still no pregnancy, then the female should be checked to see whether she has ovaries and a reproductive tract (oviducts and uterus). This can be accomplished surgically without sacrificing the female by exposing the reproductive tract as if performing an embryo transfer (see Chapter 6).

   • If there are no ovaries, the only recourse is to save somatic tissues for cloning (see Chapter 13).

   • If there are ovaries but no (or abnormally developed) uterus or oviducts, ovary transplantation should be performed (see Ovary Transplantation, below).

   • If the female has ovaries, a uterus, and oviducts, the reproductive organs can be returned to the body cavity, the skin wound clipped, and the female allowed to recover.

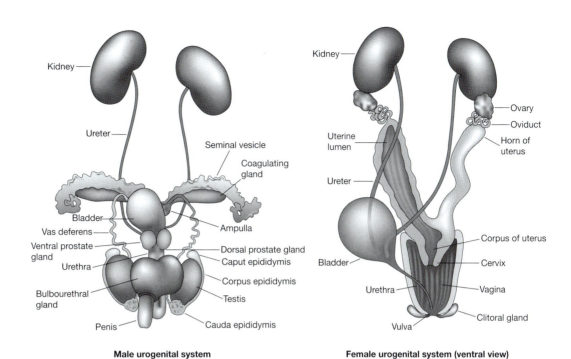

FIGURE 14.1. Adult male and female mouse urogenital systems.

3. As a next step, low doses of pregnant mare serum (PMS) and human chorionic gonadotropin (hCG) can be injected to induce estrus (see Artificial Insemination, below) and the female mated with a proven stud male.

   • If a plug is obtained, the female should be left to give birth.

   • If there is no birth, more terminal strategies must be employed as follows:

   —Superovulating doses of hormones can be injected and hopefully a plug will be obtained after pairing with a proven stud male.

   —One or both oviducts can be surgically recovered to collect zygotes (see Chapter 4). At the same time, the ovaries can be transplanted into a wild-type recipient or cryopreserved. If zygotes are recovered, they can then be transferred into the oviducts of 0.5-dpc pseudopregnant recipient females (see Chapter 6).

   • If no plug is obtained, a new trial can be performed 2 weeks later. If still no plug is obtained, the ovaries can be transplanted to a wild-type recipient female (see Ovary Transplantation, below).

4. If all else fails, the somatic cells of the female mouse can be saved for cloning.

   If an important female mouse becomes sick prior to sexual maturity, the best way to rescue her germ line is by ovary transplantation (see Protocol 14.1). Even neonatal ovaries can be successfully transplanted. If there is no histocompatible recipient available, you can save the ovary by cryopreservation (see Chapter 15).

Male Infertility

If an adult male mouse is not successfully impregnating multiple females, the following course can be pursued:

1. Replace the females with new 6- to 8-week-old females that are in estrus; the females should be subsequently checked for copulation plugs.

   • If plugs are obtained, allow the females to give birth.

   • If there are no births, superovulate 3-week-old hybrid females and place them with the male (see Chapter 3).

     —If there is a plug, collect the embryos and, if fertilized (two pronuclei), transfer them to the oviducts of 0.5-dpc pseudopregnant recipient females (see Chapter 6). This can be repeated to obtain a sufficient number of transfers and progeny from the male.

     —If plugs are not obtained repeatedly using the superovulated weanling females, or if the plugged females do not yield embryos, it is possible that sperm cannot exit the male or that there is a sperm deficiency.

2. The male should be checked to see whether he has normally developed testes and other reproductive organs (epididymides, vasa deferentia, and seminal vesicles). This can be accomplished surgically without sacrificing the male by exposing the reproductive organs as if performing a vasectomy (see Chapter 6).

   • If there are no testes, but externally the animal is phenotypically male, the testes may have degenerated. Regardless of the reason for the absence of testes, the only recourse is to save somatic tissues for cloning (see Chapter 13).

   • It is possible that the testes are of normal size but the cauda epididymides are small and translucent. (In a normal male, the cauda epididymides are large and opaque because of the presence of sperm.) This condition should be noted. The reproductive organs should be returned to the body cavity, the body wall muscle stitched up, and the skin clipped; the male should be allowed to recover.

3. If the surgical examination of the reproductive organs indicated that there was plenty of sperm in the cauda epididymides, either an artificial insemination or in vitro fertilization procedure (see Artificial Insemination and In Vitro Fertilization, below) should be performed. The epididymides can be surgically recovered without having to sacrifice the male.

4. If the surgical examination of the reproductive organs indicated that there was little sperm in the cauda epididymides, consider the intracytoplasmic sperm injection (ICSI) method, which requires few sperm or can even be performed with testicular germ cells (see Intracytoplasmic Sperm Injection, below).

5. Finally, if all else fails, animal cloning should be considered (see Chapter 13).

STRATEGIES TO RESCUE THE GERM LINE OF MICE

Ovary Transplantation

The technique for ovary transplantation described in Protocol 14.1 can be used to rescue the germ line of mutant or transgenic female mice that are unable to breed, or breed poorly (Jones and Krohn 1960). The same technique is used to recover cryopreserved mouse strains from frozen ovaries (Sztein et al. 1998, 1999). In this procedure, the ovaries of the recipient female (see Fig. 14.2) are removed and the donor ovaries are inserted into the ovarian bursal membrane, which physically holds the donor ovary in place. With time, a blood supply to the transplanted ovary will be reestablished and the donor ovary will respond to pituitary hormones to ovulate mature oocytes. After a period of recovery, the transplanted female is bred with males to produce progeny. The age of the donor ovary is not critical, and ovaries from newborn pups have been successfully transplanted (Rivera-Pérez et al. 1995). The recipient females can be prepubertal (3–4 weeks old) or mature mice of a strain that is histocompatible with the donor ovary. For example, if the donor is a genetic mixture of two inbred strains, the F_1 hybrid of

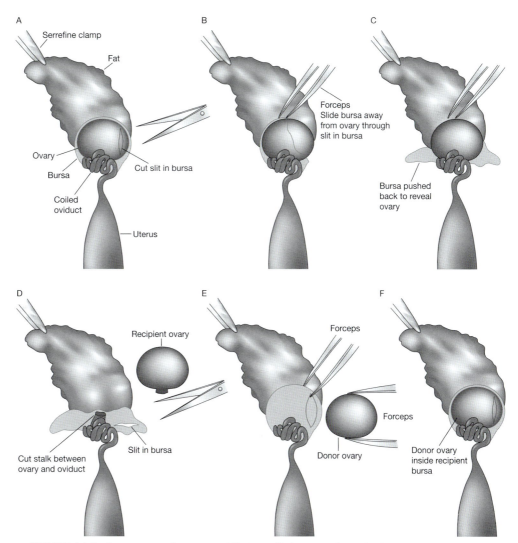

FIGURE 14.2. Ovary transplantation. (*A*) A small slit is made in the bursal membrane on the side opposite to the infundibulum using watchmaker's forceps or fine spring scissors. (*B*) Forceps are used to push away the bursal membrane to completely expose the ovary (*C*). (*D*) Fine scissors are used to excise the recipient ovary by cutting the connecting stalk between the ovary and the oviduct, being careful not to cut the bursa. (*E*) Forceps are used to hold the opening of the bursal membrane, and the donor ovary is inserted into the bursa through the opening. (*F*) The donor ovary is held in place by the bursal membrane.

the two strains can be used as a histocompatible recipient. Alternatively, immunodeficient recipients (e.g., nude mice) can be used.

In Vitro Fertilization

In vitro fertilization (IVF) involves fertilization of mature oocytes with capacitated sperm in a tissue culture dish (see Protocol 14.2). This technique can generate large numbers of cleavage-stage embryos without using a significant number of single-caged stud males for mating. In addition, sperm penetration is synchronous during incubation with mature oocytes, leading to synchronous development, unlike fertilization in vivo, when natural ovulation is usually spread over time. Another use of IVF is to generate offspring from cryopreserved sperm and from mice that, for one reason or another, will not mate or carry litters to term. Young are derived from the in vitro-fertilized oocytes by transfer to pseudopregnant recipient females as described in Embryo Transfer (see Protocol 6.3).

Recently, with the development of mouse sperm cryopreservation (see Chapter 15), IVF has become an important way to recover cryopreserved mouse strains. The success of mouse IVF is critically dependent on the medium and the manner in which both oocytes and the sperm are handled. Media traditionally used for mouse IVF are simple balanced salt solutions supplemented with energy sources, including modified Whitten's medium (Hoppe and Pitts 1973) or modified Whittingham Tyrode's solution (Whittingham 1971; Fraser and Drury 1975). These early media used different pyruvate and lactate concentrations for insemination and embryo culture. Later modifications (mT6) included omission of pyruvate and lactate (Fraser 1984, 1993) (see Table 14.1). Complex media initially developed for somatic cells and then adapted for IVF are Ham's F10 and Eagle's modified essential medium (MEM). MEM with Earle's balanced salt solution, supplemented with essential and nonessential amino acids, 0.23 mM pyruvate, and 0.01 mM EDTA, has also been successfully used for mouse IVF (Ho et al. 1995; Thornton et al. 1999).

The first specific medium for human IVF, based on the chemical composition of human tubal fluid (HTF), was formulated by Quinn et al. (1985) (Table 14.2). Mouse oocytes have been successfully fertilized in HTF (Nakagata 1996). HTF medium became extensively used for human IVF, and at the present time it seems to work better for a wider variety of mouse strains than any other medium, according to data from The Jackson Laboratory. Recently, the original formulation of HTF medium has been further modified (Quinn 2000). It now contains EDTA and glutamine and has a lower phosphate concentration (Enhance HTF, Conception Technologies). Modifications to the original HTF in Quinn's Advantage Fertilization Medium (SAGE BioPharma) include the addition of citrate, EDTA, nonessential amino acids, taurine, alanyl-glutamine, lower phosphate, increased magnesium, and lactate in the form of calcium lactate. Mouse embryos provide an effective and sensitive quality control bioassay for media used and procedures performed in human artificial reproduction (ART) clinics (see Chapter 4). The use of recently modified human IVF medium with frozen/thawed mouse sperm for IVF has not yet been evaluated.

Because the majority of mouse strains have embryos that exhibit the "two-cell-stage block" in media that support IVF (Biggers 1998), early two-cell-stage or pronuclear one-cell-stage embryos must be removed from the fertilization medium to a culture medium. Embryos can be cultured in sequential media, CZB, or KSOM supplemented with amino acids (see Chapter 4). Summers et al. (2000)

TABLE 14.1. Composition of modified Tyrode's medium (mT6)

Stock A (10× solution lasts up to 3 months in refrigerator)

Component	mM	g/100 ml
NaCl	124.57	7.280
KCl	2.64	0.200
$MgCl_2 \cdot 6H_2O$	0.49	0.100
$NaH_2PO_4 \cdot 2H_2O$	0.36	0.056
Glucose	5.56	1.000
Penicillin	100 U/ml	0.06 (from 10^5 IU)

Stock B (10× solution lasts up to 2 weeks in refrigerator)

Component	mM	g/100ml
$NaHCO_3$	25	2.106
Phenol red		0.010

Stock C (100× solution lasts up to 3 months in refrigerator)

Component	mM	g/10 ml
$CaCl_2 \cdot 2H_2O$	1.71	0.252

mT6 (lasts up to 1 week in refrigerator)

Stock	ml
A	1.0
B	1.0
C	0.1
Water	7.9
BSA	40 mg

Data from Whittingham (1971) and Fraser (1984, 1993).
Molarity values represent the final concentration of the component in the medium.

recently demonstrated that mouse oocytes could be fertilized and cultured to the blastocyst stage in modified KSOM, containing 5.56 mM glucose instead of 0.2 mM, 4 mg/ml of BSA instead of 1 mg/ml, and supplemented with amino acids.

TABLE 14.2. Composition of HTF medium

Component	Sigma catalog number	Molecular weight	mM	Grams/liter
NaCl	S5886	58.450	101.6	5.9375
KCl	P5405	74.557	4.69	0.3496
KH_2PO_4	P5655	136.091	0.37	0.0504
$MgSO_4 \cdot 7H_2O$	M1880	246.500	0.2	0.0492
Na lactate 60% syrup (ml/liter)	L1375	112.100	21.4	2.3989 or 3.998 g of 60% syrup (~3.42 ml)
Na pyruvate	P4562	110.000	0.33	0.0365
Glucose	G6152	179.860	2.78	0.5
$NaHCO_3$	S5761	84.020	25.0	2.1
$CaCl_2 \cdot 2H_2O$	C7902	147.200	2.04	0.3
Penicillin-G	P7794			0.075
Streptomycin sulfate	S9137			0.05
Phenol red (1%) (ml/liter)	P0290			0.2 ml
BSA				4.0

Data from Quinn et al. (1985).

COMMENTS

- Fertilization medium is prepared the same way as other embryo culture media (see Chapter 4). The quality of the chemicals, water, protein, and shelf-life in the refrigerator and in the incubator must all be considered.

- BSA quality is of utmost importance for successful IVF, and different batches of BSA may need to be tested for optimal performance (e.g., Sigma A3311, AlbuMAX from Gibco/Invitrogen). BSA from Equitech-Bio (BAC62-0050) is currently used by The Jackson Laboratory.

- A 60% syrup of sodium lactate has a large osmotic effect and must be measured very accurately.

- Medium should be gassed with 5% CO_2 in air, filtered through a 0.22-μm filter, stored at 4°C for no longer than 2 weeks, and regassed if opened. HTF medium is also available commercially from Specialty Media (MR-070), Irvine Scientific, etc.

- If desired, HTF medium may be substituted with modified MEM for insemination (Ho et al. 1995; Thornton et al. 1999) as follows: 100 ml of MEM (e.g., Sigma M4655) is supplemented with 2.5 mg of pyruvate, 0.38 mg of EDTA, 300 mg of BSA, and antibiotics. After fertilization, embryos are transferred to KSOM for overnight culture (Thornton et al. 1999).

- mT6 medium (see Table 14.1) is preferred for IVF of cumulus-denuded oocytes using frozen/thawed sperm (see Protocol 15.3, Method 2, and Comments to Protocol 14.2).

- The low in vitro fertilization ability of frozen/thawed sperm can in some cases be overcome by the relatively simple procedure of partial zona dissection (PZD) (Nakagata et al. 1997). Alternatively, individual frozen/thawed spermatozoa can be microinjected under the zona pellucida.

IMPORTANT CONSIDERATIONS

- The number of motile sperm is reduced almost 50% after cryopreservation, and there is a great variation in fertilization rate between mouse strains when frozen sperm is used (Sztein et al. 2000). The frozen sperm from C57BL/6 and 129S3/SvImJ (129S1) produced the lowest IVF rates from all tested strains (Sztein et al. 2000). The best results are obtained from F_1 hybrids. For more details, see Sztein et al. (2000).

- IVF is strongly influenced by genetic background, handling, and culture conditions; in addition, minor modifications in a protocol, such as the time of oocyte collection and sperm capacitation, may have drastic effects on results. Potential variability in fertilization rate between individual animals should also be kept in mind.

- Optimally, males should be older than 12 weeks but younger than 6 months and of proven fertility. They should be individually caged for at least 2 weeks, mated 5–7 days prior to the experiment, but not mated within the last 3 days.

- Oocyte donors are superovulated (see Chapter 3). The dose of hormones (2.5–10 IU) is established experimentally depending on the strain and age of the females. It is important to inject hormones 48 hours apart, although the timing between hormones is strain-dependent. It is critical that the hCG injection occur 13–14 hours before the planned oocyte collection; i.e., shortly after ovulation. Trying to recover oocytes too early after hCG injection may result in low yield because ovulation might not yet have been completed. If collection takes place too late, the zona pellucida becomes harder and not penetrable to sperm. The exact time varies by strain and must be determined empirically.

- As discussed in Chapter 4, temperature fluctuations during in vitro manipulations of oocytes may disrupt meiotic spindles and contribute to abnormal chromosome distribution and failed or abnormal fertilization (Pickering et al. 1990; Almeida and Bolton 1995). Minimize the time that dishes spend outside the incubator to maintain pH equilibration and temperature of the medium. Keep everything at 37°C as much as possible by using a microscope heating stage or small gas incubator immediately beside the microscope. The time between sacrificing the females and placing the cumulus masses into the fertilization dish should be limited (ideally to 5 minutes). Billups-Rothenberg modular incubator chambers (MIC) that are gassed with a 5% CO_2, 5% O_2, 90% N_2 mixture, sealed, and incubated at 37°C are used for IVF and embryo culture at The Jackson Laboratory.

- It is very important to use wide bore (e.g. Rainin HR-250W, 1000W) instead of regular 200-µl and 1000-µl pipette tips for sperm transfer and to pipet gently, thus minimizing potential damage.

- Although only sperm from cauda epididymides is traditionally used for IVF, the collection of sperm from both the vasa deferentia and cauda epididymides seems to improve the overall motility rate (Sztein et al. 2000).

- The fact that alcohol can parthenogenetically activate oocytes (see Chapter 13) and lead to misinterpretation of IVF results should be kept in mind.

Artifical Insemination

Artificial insemination is used to impregnate female mice by means of a solution containing motile sperm (Snell et al. 1944; Wolfe 1967). This procedure is different from IVF because fertilization occurs in vivo directly within the female reproductive tract. Either freshly collected or frozen/thawed sperm can be used. In addition, both nonsurgical (Protocol 14.3) and surgical (Protocol 14.4) methods can be used to artificially inseminate female mice. For nonsurgical artificial insemination, a blunt needle is inserted through the cervical canal, usually with the aid of a speculum to deliver sperm into the uterus. This method requires practice but can be quickly mastered. The inseminated female is then immediately paired with a vasectomized male for mating to induce pseudopregnancy. We also describe a simple surgical method of artificial insemination that requires skills familiar to individuals who routinely generate transgenic mice, including mating females with vasectomized males to induce pseudopregnancy and oviduct transfer (see Chapter 6). This surgical method requires less sperm and may be useful when frozen sperm is required (Nakagata 1992, 1995; De Repentigny and Kothary 1996).

Intracytoplasmic Sperm Injection

Intracytoplasmic sperm injection (ICSI) is the injection of a single spermatozoon directly into the cytoplasm of an oocyte using an injection pipette (see Protocol 14.5). This technique was originally developed in the hamster (Uehara and Yanagimachi 1976) and has been successfully applied to the mouse (Kimura and Yanagimachi 1995a,b) and many other species, including humans. ICSI requires only one spermatozoon per oocyte for fertilization. Spermatozoa that are used for ICSI do not need to be motile or alive. Indeed, for mouse ICSI, the sperm tail is typically broken off prior to injection into the oocyte, resulting essentially in a nuclear transfer. ICSI works as long as the sperm nuclei are "intact" in terms of genetic integrity (Kuretake et al. 1996; Wakayama and Yanagimachi 1998; Wakayama et al. 1998). For example, freeze-dried sperm are dead in the conventional sense, but have been successfully used for ICSI to generate live mice (Wakayama and Yanagimachi 1998). The technique is applicable for both hybrid and inbred mice (Kawase et al. 2001; Kusakabe et al. 2001; Szczygiel et al. 2002). For the history and biology of ICSI, see Yanagimachi (1998, 2001).

Ovary Transplantation

Information for this protocol was supplied by Richard Behringer, Department of Molecular Genetics, M.D. Anderson Cancer Center, Houston, Texas 77030.

MATERIALS

ANIMALS
Donor female mouse or frozen/thawed ovaries (see Protocol 15.4)
Recipient female mice (histocompatible with donor)

EQUIPMENT
Forceps, watchmaker's #5
Hypodermic needle, 26-gauge 1/2-inch
Needle, curved surgical (size 10, triangular, pointed)
Scissors
Scissors, fine spring
Suture, surgical silk (size 5-0)
Syringe, 1-ml
Weight scale
Wound clips and applier

REAGENTS
Anesthetic (see Appendix 1)
M2 medium or PBS at room temperature
Phosphate-buffered saline (PBS)

PROCEDURE

1. Sacrifice or anesthetize the donor mouse (see Chapter 4) and remove one or both ovaries. The ovaries can immediately be transplanted into a recipient female or held for a short time in culture medium or PBS.

   Alternatively, thaw frozen ovaries and allow them to rehydrate in M2 medium for at least 10 minutes before the surgery (see Protocol 15.4).

2. Anesthetize the recipient mouse.

3. Surgically expose the ovary and uterus as in embryo transfer (see Embryo Transfer, Chapter 6).

4. Use fine spring scissors to make a small incision in the bursa on the side opposite the opening of the oviduct (infundibulum) (Fig. 14.2A).

It is important to make the incision in the bursa no larger than the minimum needed to remove the recipient's ovary so that the transplanted donor's ovary will be held in place by the bursa. Use watchmaker's forceps to slip the ovary through the incision (Fig. 14.2B,C) and remove it by cutting the supporting stalk (Fig. 14.2D). This will cause bleeding that will fill the bursal sac.

5. Use watchmaker's forceps to insert the donor's ovary into the bursal sac (Fig. 14.2E,F).

6. Gently return the ovary and the uterus to the body cavity. Sew up the body wall with one or two stitches and close the skin with wound clips.

7. Surgically expose the ovary and the uterus on the other side of the animal (see step 3). Ligate, cut, or cauterize the oviduct. Gently return the ovary and uterus to the body cavity. Sew up the body wall with one or two stitches and close the skin with the wound clips.

The aim of this part of the procedure is to have a normal, hormonally functioning ovary in place but whose eggs cannot be fertilized and reach the uterus. It is also possible to replace both of the ovaries of the recipient with donor ovaries.

8. Because each donor has two ovaries, the final result will be two recipient female mice. It is also possible to cut ovaries into smaller pieces for transplantation into additional recipients.

COMMENTS

- It is possible to stitch the bursa containing the transplanted ovary using 10-0 Vicryl suture to hold the ovary in place.
- Usually, we wait 3 weeks after the surgery before breeding the ovary-transplanted females.
- Newborn ovaries are transferred into 3-week-old recipients because the recipient bursa is smaller than in older females. A glass pipette can be used to place the newborn ovaries into the bursa.
- Transplanted ovaries should last for the reproductive life of the transplanted recipient female. If the recipient stops bearing litters, it is possible to collect the transplanted ovaries and then transplant them again into another recipient female.

In Vitro Fertilization

The IVF protocol described here was provided by Carlisle Landel, The Jackson Laboratory, Bar Harbor, Maine 04609, and is based on methods described by Fraser and Drury (1975) and Sztein et al. (1997, 2000); http://www.jax.org/resources/documents/cryo/ivf.html. A different method used by the Monash Institute is described in Comments, below.

MATERIALS

ANIMALS
Female mice
Male mice for fresh sperm collection or frozen sperm

EQUIPMENT
Centrifuge (*optional*)
Clean cage
Forceps, watchmaker's #5
Heating stage or block at 37°C
Hypodermic needles, 25- or 30-gauge, 1/2 inch
Incubator, humidified 37°C, with 5% CO_2, 95% air or 5% CO_2, 5% O_2, 90% N_2
Organ culture dishes
Scissors, fine dissection
Syringe, 1-ml
Tissue culture dishes, 35-, 60-mm plastic center-well organ culture (e.g., BD-Falcon 35-3001, 35-3004, 35-3037)
Transfer pipettes consisting of mouth or hand-held pipette assembly and pulled capillary (see Chapter 4)
Warming pad
Water bath
Wide-bore pipette tips (e.g., Rainin HR-250W, 1000W)

REAGENTS
Human chorionic gonadotropin (hCG)
Human tubal fluid (HTF) medium (as described above)
KSOM-AA medium (see Chapter 4)
Paraffin oil, pre-gassed embryo-tested light (e.g., Sigma M8410)
Pregnant mare serum gonadotropin (PMSG)

PROCEDURE

Day 1

Inject oocyte donors with PMSG 48 hours before hCG regardless of the light cycle in the animal room. Injections are done according to the planned oocyte collection, and insemination should be as close as possible to 13–14 hours post-hCG injection (e.g., hCG injection at 8 p.m., insemination at 9 a.m. the following morning).

Day 3

1. Inject oocyte donors with hCG 48 hours after PMS injection.

2. Prepare the fertilization and culture dishes, as listed below, on the afternoon of the day prior to the experiment. The medium should be covered with pre-gassed embryo-tested light mineral oil; 35-mm (e.g., BD-Falcon 35-3001) or center-well organ culture dishes (BD-Falcon 35-3037) are convenient for this step.

   - *Fresh sperm dish:* 1 ml of HTF.

   - *Egg collection dish:* 1 ml of HTF. HTF-HEPES can also be used for this purpose.

   - *Fertilization dish* (1 per three females): 250 or 500 µl of HTF.

   - *Wash dish* (1 per each fertilization dish): 5 drops × 250 µl of HTF in a 60-mm dish. It is also convenient to use the same 60-mm dish for both fertilization (central drop) and washes (surrounding drops).

   - *Culture dish* (1 per each fertilization dish): 5 drops × 250 µl of KSOM-AA in a 60-mm dish. It is also possible to use smaller drops in a 35-mm dish.

Day 4

If fresh sperm is used:

Fresh sperm must be capacitated by incubating the concentrated suspension at 37°C for at least 1 hour (the time may vary between strains) before insemination.

1. Sacrifice the male about 12–13 hours after the females were injected with hCG.

2. Immediately dissect out the cauda epididymides and vasa deferentia, removing as much fat and blood vessels as possible (see Fig. 6.2A in Chapter 6). Place them in the fresh sperm dish. Mince the cauda making 5–7 slashes with a 30-gauge needle on a syringe. Using forceps, gently squeeze out the sperm from the vasa deferentia. Minimizing the trauma by making a single slit instead of mincing may improve the results.

3. Gently shake the dish with the sperm suspension for 30 seconds (*optional*) and place the dish in the incubator for at least 1 hour for capacitation. Using an aliquot of diluted sperm suspension, assess the motility and determine the sperm concentration with a hemocytometer. It is necessary to obtain a final motile sperm concentration of 1×10^6 to 2.5×10^6 sperm/ml for fertilization.

If frozen/thawed sperm is used:

Freezing/thawing of mouse sperm (Chapter 15) results in capacitation-like changes, and therefore preincubation of frozen/thawed sperm is unnecessary (Fuller and Whittingham 1996).

1. Remove the cryovial with frozen sperm suspension from liquid nitrogen and place it into a water bath at 37°C until the ice crystals are melted (~2 minutes).

2. *Optional:* Centrifuge the sperm sample at 735*g* for 4 minutes. Carefully add 50 µl of HTF medium and very carefully flick the tube to resuspend the sperm pellet (do not pipette).

Although it is beneficial for embryo development to remove any cryoprotectant, centrifugation may reduce sperm viability in some cases. This step is no longer used by The Jackson Laboratory.

3. Evaluate the morphology and motility of the sperm and immediately use for IVF.

In vitro fertilization:

1. Add an aliquot of fresh or thawed sperm to each fertilization dish using wide-bore pipette tips (usually 10 μl, or more if the concentration is low).

2. At 13 hours post-hCG administration, sacrifice three females. Quickly dissect out oviducts and place them into a drop of HTF medium. Tear the ampullae to release cumulus masses as described earlier (Collecting Zygotes, Protocol 4.9). Transfer cumulus masses from all three females to a single fertilization dish using a wide-bore pipette tip. Repeat until oocytes from all donors are collected. Then distribute the oocytes among all fertilization dishes.

3. Incubate the fertilization dishes at 37°C, 5% CO_2, 95% air or 5% CO_2, 5% O_2, 90% N_2 for 4–6 hours.

4. Remove the dishes from the incubator, and wash the oocytes through several drops of HTF medium to remove the excess of sperm and debris. At this time, it should be possible to observe the presence of two pronuclei and the extruded second polar body in fertilized oocytes.

5. Although washed fertilized oocytes from nonblocking strains may be left to culture overnight in fresh HTF microdrops, it is recommended to use KSOM supplemented with amino acids for culture of embryos from all strains, especially from those exhibiting two-cell block in HTF medium. It is important to wash embryos through several drops of equilibrated KSOM-AA medium before transferring them into the final microdrop for overnight culture.

Day 5

1. Count the two-cell-stage embryos in the morning.

Apparent two-cell-stage embryos can develop overnight due to parthenogenetic activation or fragmentation of unfertilized oocytes (see Fig. 4.8, p. 200). Therefore, fertility rates based on the number of two-cell-stage embryos must be interpreted with caution. Transfer two-cell-stage embryos to KSOM-AA microdrop dishes (if this was not already done the previous day) for culture to later stages. Alternatively, transfer two-cell-stage embryos to the oviduct of 0.5-dpc pseudopregnant recipient females as described in Protocol 6.3.

COMMENTS

- The use of oocytes without cumulus cells may facilitate fertilization using frozen/thawed sperm (Luis Gabriel Sanchez-Partida and Alan Trounson, Monash Institute of Reproduction and Development, Clayton, Australia, pers. comm.).
  —Add 2 μl of a frozen/thawed sperm suspension to 20 μl of equilibrated drops of fertilization medium (HTF or mT6; see Tables 14.1 and 14.2) under oil.

—Transfer 10 cumulus-denuded oocytes (see Chapter 4) into each drop and incubate as described above for 3–4 hours.

—Wash the oocytes into culture medium (e.g., KSOM) and return them to the incubator for subsequent development.

Nonsurgical Artificial Insemination

This protocol is based on the method of Dziuk and Runner (1960) described in Rafferty (1970).

MATERIALS

ANIMALS
Adult female mice
Male mice (do not use males that have mated within the previous 3 days or have not mated for more than 1 week) or frozen sperm (see Chapter 15)

EQUIPMENT
Forceps, watchmaker's #5
Hypodermic needle, 22-gauge (blunted and bent at 90° angle).
Incubator, humidified 37°C, 5% CO_2, 95% air
Scissors, dissection
Speculum (see Fig. 14.3)
Syringe, 1-ml
Tissue culture dish, sterile plastic

REAGENTS
Human chorionic gonadotropin (hCG) (e.g., Sigma C8554)
Human tubal fluid (HTF) medium (see In Vitro Fertilization section, above)
Light paraffin oil (Fisher O121-1; BDH 29436)
Pregnant mare serum gonadotropin (PMSG) (e.g., Sigma G4527; Calbiochem 367222)

PROCEDURE

1. Inject female mice with PMSG and hCG as described (see Chapter 3, Inducing Superovulation, p. 148), except use only 0.5–1.0 IU of each hormone. This causes the females to ovulate normal numbers of oocytes.

2. Sacrifice male mice 12 hours after the females have been injected with hCG. Dissect the vas deferens or cauda epididymis (Fig. 14.3), removing as much excess fat as possible. Place the vas deferens or cauda epididymis into a 500-µl drop of pregassed HTF medium under light paraffin oil at 37°C. Gently squeeze out the sperm using watchmaker's forceps. It is essential that the sperm then be incubated for 1.5 hours at 37°C to capacitate them. This is not necessary if thawed frozen sperm are used.

3. Mount the speculum in a clamp on a ring stand with overhead illumination (Fig. 14.3).

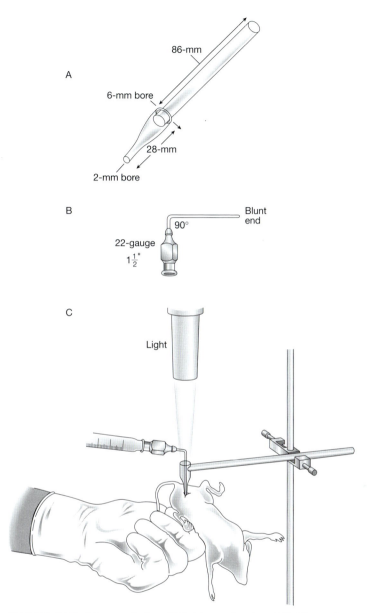

FIGURE 14.3. Nonsurgical artificial insemination. (*A*) Dimensions of the glass speculum to visualize the cervix through the vagina. (*B*) Blunted and bent needle used for artificial insemination. (*C*) Insemination method. The mouse is held and the mounted speculum is inserted into the vagina with the opening of the speculum against the cervix. The insemination needle is inserted through the speculum through the cervix into the uterus, and 50 μl of sperm suspension is injected. ([*A, B*] Redrawn, with permission, from Wolfe 1967; [*C*] redrawn, with permission, from Rafferty 1970.)

4. Firmly grasp a female and hold upside down such that the speculum is inserted into the mouse's vagina with the opening against the cervix.

5. Gently insert the blunt needle through the cervical canal into the uterus. The operator should feel a lower resistance as the needle moves through the cervical canal into the uterus.

6. Inject 0.05 ml of sperm suspension. This amount is sufficient to transfer sperm into both uterine horns because the volume is too great to be contained in a single horn (Leckie et al. 1973).

7. Immediately after insemination, pair the females with vasectomized males to obtain a plug. Even if no plug is obtained, litters can still be produced, but at a reduced rate (~10%).

COMMENTS

- The day of the artificial insemination can be considered day 1 of the pregnancy.
- Artificial insemination can be performed 13–15 hours after estimated ovulation. The midpoint of ovulation is about 12 hours post-hCG injection.
- Insemination with 3×10^6 to 10×10^6 sperm is recommended. Fertilization is significantly less efficient using 10^6 sperm and rare when less than 10^5 sperm are transferred.

Surgical Artificial Insemination

This protocol is based on the method described by De Repentigny and Kothary (1996).

MATERIALS

ANIMALS

Adult female mice
Male mice (do not use males that have mated within the previous 3 days or have not mated for more than 1 week) or frozen sperm (see Chapter 15)
Vasectomized male mice

EQUIPMENT

Forceps, watchmaker's #5
Fiber optic illuminator (very useful)
Hypodermic needle, 26-gauge, 1/2 inch
Incubator, humidified 37°C, 5% CO_2, 95% air
Needle, curved surgical (size 10, triangular, pointed)
Pipette, finely drawn, prewarmed
Scissors, dissection
Serrefine clamp or baby Dieffenbach clip (1.5 inch or smaller) (e.g., Roboz Surgical Instrument RS7440; Weiss B950B, or Fine Science Tools 18050-35)
Stereomicroscopes (ideally one for the surgery and one for loading sperm into the transfer pipette) with transmitted and reflected light
Suture, surgical silk (size 5-0)
Syringe, 1-ml
Tissue culture dish, sterile plastic
Weight scale
Wound clips and applier

REAGENTS

Anesthetic (see Appendix 1)
Human tubal fluid (HTF) medium (see In Vitro Fertilization)
Light paraffin oil (Fisher O121-1; BDH 29436)

PROCEDURE

1. The day before the insemination, breed females with vasectomized males in the late afternoon (~4 p.m.). The next morning, identify the females that have been plugged.

2. Sacrifice the fertile male.

3. Collect sperm from the cauda epididymis (as described for IVF above) into a prewarmed and pregassed 100-μl drop of HTF medium under mineral oil in a sterile culture dish. Capacitate the sperm in the incubator for 2.5–4.5 hours. The average sperm concentration should be ~10^7/ml.

4. Anesthetize the female that was plugged by the vasectomized male.

5. Surgically expose the reproductive tract as if performing an oviduct transfer (see Protocol 6.3).

6. Tear or cut the ovarian bursa to expose the infundibulum.

7. Load 2–5 μl of capacitated sperm suspension (2×10^4 to 5×10^4 sperm) into a prewarmed finely drawn glass transfer pipette (see Fig. 14.4) and transfer the sperm into the ampulla of the oviduct.

8. If desired, repeat steps 5–7 for the other oviduct.

9. Replace the reproductive organs in the body cavity.

10. Stitch the body wall and clip the skin with wound clips.

11. Place the mouse on a warm pad in a clean cage to recover. Follow animal care committee regulations on surgery and postsurgical care.

COMMENT

Sperm transfers into the oviducts of females were performed between 10 a.m. and 12 noon. This technique yielded a ~70% pregnancy rate when fresh sperm was used. Utilizing frozen sperm yielded a ~20% pregnancy rate, although current improvements in sperm freezing protocols may lead to an increased impregnation rate.

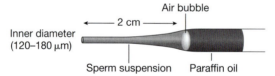

FIGURE 14.4. Dimensions of the pipette used for transferring sperm into the oviduct for surgical artificial insemination.

| # Intracytoplasmic Sperm Injection

The ICSI procedure described here was provided by Monika A. Szczygiel and Ryuzo Yanagimachi, Institute for Biogenesis Research, John A. Burns Medical School, University of Hawaii, Honolulu, Hawaii 96822.

MATERIALS

MICROMANIPULATION SETUP

Dissecting microscope

> Any dissecting microscope with proper magnification and transmission and incident illumination is sufficient.

Inverted microscope with Hoffmann interference contrast or Nomarski optics

> Olympus (Model IX70, Olympus, Japan), Nikon (Diaphot 300, Nikon, Japan), or Zeiss (Axiovert 200, Zeiss, Germany), or any microscope commonly used for pronuclear DNA injection (see Chapter 7) should be sufficient for ICSI.

Micromanipulators (left and right) with suction/injection system

> Manual (Narishige, Tokyo, Japan) or electronic (Eppendorf, Germany) micromanipulators both work well. Narishige and Eppendorf pump systems are both reliable. The choice depends on personal preference.

Piezo impact-driving micromanipulation device

> Some researchers use standard micromanipulation devices for ICSI, but we find the piezo impact-driving micromanipulation device to be much more reliable. The piezo device can drive the injection pipette a minute distance very quickly. The advantages of piezo ICSI are (1) the zona pellucida is drilled without deformation, (2) the sperm tail can be easily removed prior to injection, (3) injection is quick and egg survival is high. The piezo impact drive unit model PMAS-CT150 (Prime Tech, Ibasaki, Japan) works well.

STEP 1: MAKING HOLDING AND INJECTION PIPETTES

A holding pipette is used to secure the oocyte in one position during micromanipulation. Making holding pipettes for ICSI is exactly as described for making holding pipettes for pronuclear microinjection of fertilized oocytes (see Chapter 7). The injection pipette is used to separate the sperm head from the tail, to penetrate the zona pellucida and the oocyte plasma membrane (oolemma), and to deposit the sperm head into the oocyte.

MATERIALS

EQUIPMENT

Borosilicate glass capillary tube (Sutter Instruments, O.D. 1.0 mm, I.D. 0.75, 10-cm length; B100-75-10)
Microforge (Model MF-79, Narishige, Japan)
Micropipette puller (Model P-97, Sutter Instruments)
Needle, 16-gauge
Syringe, 10 ml, with attached rubber tubing for washing
Syringe, 1 ml

REAGENTS

Distilled water
Fluoric acid<!> (6%) in water
Mercury<!> (Fisher Scientific M-140)

CAUTION: See Appendix 2 for appropriate handling of materials marked with <!>.

PROCEDURE

1. Pull a capillary with a micropipette puller. Suggested program settings are: Heat 752, Pull 65, Velocity 130, Time 20. See Protocol 7.9 for instructions on the use of the puller and running a ramp test.

2. Break the pulled capillary using a glass anvil on the heating filament of the microforge (see Fig. 12.2) at the point where the outer diameter of the capillary is ~8 μm.

3. Bend the pipette using the microforge. The angle depends on the type of micromanipulator to be used and personal preference.

4. Rinse the injection pipettes briefly (a few seconds) with a fluoric acid solution to remove glass dust and to make the wall of the pipette tip thinner. Then rinse repeatedly (30 seconds) with distilled water. Use a syringe with attached rubber tubing for pipette cleaning and washing.

5. Fill the wider end of the pipette with mercury using a 16-gauge needle attached to a 1-ml syringe so that it fills ~5–10 mm the length of the pipette. Mercury is moved to the tip after the pipette is mounted on the micromanipulator.

6. Injecting pipettes can be stored in a clean airtight container until use, but prolonged storage is not recommended.

COMMENTS

- Injection pipettes prepared as described above are used for ICSI in combination with a piezo impact drive unit and are blunt. If a piezo impact drive is not used, the tip of the injection pipette must be cut, beveled with a microgrinder, and sharpened (see Chapter 13). High serum concentration in the operational medium may increase oocyte survival rate (Suzuki and Yanagimachi 1997).

- Mercury must be handled and disposed of with maximum care according to local regulations. When the surface of mercury in the storage bottle is oxidized (film formation), remove the film using stationary adhesive tape.

STEP 2: PREPARATION OF OOCYTES

MATERIALS

ANIMALS
Female mice of a chosen strain, 2 months old or older (preferably 2–4 months old)

EQUIPMENT
Dissecting microscope with transmission illuminator
Incubator, humidified CO_2 (5% CO_2, balance air)
Mouth pipettes (made from Drummond "Microcaps," 100 μl, 116 mm, 1-000-1000)

REAGENTS
Bovine testicular hyaluronidase (359 USP units/mg, Sigma H3506)

> A working solution is: 1 mg/ml in CZB-HEPES; stock solution, x 100 in water. The solution can be stored at –20°C in small tubes for months. After thawing, it can be kept at 4°C for about 1 week.

CZB medium (with glucose) for oocyte culture (Table 14.3)
CZB-HEPES medium for oocyte handling (Table 14.3)
Dulbecco's phosphate-buffered saline (PBS) (Gibco 14080-055)
Human chorionic gonadotropin (hCG) (e.g., Calbiochem 230734)
Pregnant mare serum gonadotropin (PMSG) (e.g., Calbiochem 367222)

PROCEDURE

1. Induce female mice to superovulate by intraperitoneal (IP) injection of 5 IU of PMSG followed by an IP injection of 5 IU of hCG 48 hours later.

2. Sacrifice the females 14–15 hours post-hCG administration. Isolate their oviducts (see Protocol 4.9) and place them in 10 ml of warm PBS.

3. Transfer the oviducts to a drop of CZB-HEPES containing hyaluronidase (prewarmed to 37°C). Release the cumulus–oocyte complexes into the medium by rupturing the ampulla of each oviduct with a 25-gauge needle. Discard the oviduct and tissue debris. Cumulus cells disperse within ~5 minutes of incubation in the hyaluronidase solution.

4. Transfer the oocytes from the hyaluronidase solution into a CZB-HEPES drop. Wash the oocytes by passing them through a series of CZB-HEPES drops. Discard immature, fragmented, or abnormal-looking oocytes. Use only those with clearly visible polar bodies.

5. Store the oocytes in CZB medium in a humidified 5% CO_2 incubator at 37°C for a period not extending beyond 1 hour. Immediate use is recommended.

COMMENT

The B6D2F1 (C57BL/6 x DBA/2) hybrid strain is used routinely in our laboratory, but other hybrids and inbred strains (e.g., C57BL/6J, FVB/N, 129Sv/J, BALB/c) can be used as well.

TABLE 14.3. CZB/CZB-HEPES for ICSI

	CZB Medium		
	mM	Molecular weight	g/liter
NaCl	81.62	58.450	4.76
KCl	4.83	74.557	0.36
KH_2PO_4	1.18	136.091	0.16
$MgSO_4•7H_2O$	1.18	246.470	0.29
$NaHCO_3$	25.00	84.020	2.11
$CaCl_2•2H_2O$	1.70	147.200	0.25
Na_2-EDTA•$2H_2O$	0.11	372.20	0.04
L-Glutamine	1.00	146.10	0.15
Na lactate	28.00	112.100	3.14
(60% syrup, d=1.32 g/l, 5.229 g or 3.96 ml)			
Na pyruvate	0.27	110.00	0.03
Glucose	5.55	179.86	1.00
Penicillin	—	—	0.05
Streptomycin	—	—	0.07
BSA	—	—	4.00

CZB-HEPES Medium

HEPES is added; $NaHCO_3$ is reduced. BSA is substituted by PVA.

HEPES-Na (basic)	20.00	260.300	0.52
$NaHCO_3$	5.00	84.020	0.42
PVA	—	—	0.10

pH 7.4; Osmolarity: 266–276
Working solution: Store in refrigerator up to 2 weeks
Frozen stocks: Store 3 months or more

10x Stocks for CZB and CZB-HEPES Media for ICSI

Stock A (10x):	*g/100 ml*
NaCl	4.76
KCl	0.36
KH_2PO_4	0.16
$MgSO_4•7H_2O$	0.29
Na lactate	3.139 (5.229 g or 3.96 ml of 60% syrup)
Glucose	1.00
Penicillin/streptomycin (1000x)	100 μl
Stock B (10x)	*g/100 ml*
$NaHCO_3$	2.101
Phenol red	0.001
Stock C (100x)	*g/10 ml*
Na Pyruvate	0.0300
Stock D (100x)	*g/10 ml*
$CaCl_2•2H_2O$	0.2517
Stock E (10x, pH 7.4)	*g/100 ml*
HEPES-Na	5.2060
Phenol red	0.001
Stock F (10,000x)	*g/10 ml*
100 mM Na_2EDTA•$2H_2O$	0.3720
Stock G (200x)	*g/10 ml*
200 mM L-glutamine	0.2920
PVA	
5% PVA in distilled water	0.5g/10 ml

(continued on facing page)

TABLE 14.3. CZB/CZB-HEPES for ICSI (*continued*)

	CZB		CZB-HEPES	
Stock	*ml/50ml*	*ml/100ml*	*ml/50ml*	*ml/100 ml*
A (salts)	5.0	10.0	5.0	10.0
B (NaHCO$_3$)	5.0	10.0	1.0	2.0
C (Na Pyr)	0.5	1.0	0.5	1.0
D (CaCl$_2$)	0.5	1.0	0.5	1.0
E (HEPES)	—	—	5.0	10.0
F (EDTA)	0.055	0.11	0.055	0.11
G (L-Glu)	0.25	0.5	0.25	0.5
PVA	—	—	0.005 g or 100 µl 5%	0.01 g or 200 µl 5%
BSA	0.25	0.5	—	—
H$_2$O	38.595	77.19	37.595	75.19
Chemicals	**Sources**			
NaCl	Sigma	S5886		
KCl	Sigma	P5405		
KH$_2$PO$_4$	Sigma	P5655		
MgSO$_4$•7H$_2$O	Sigma	M1880		
NaHCO$_3$	Sigma	S5761		
CaCl$_2$•2H$_2$O	Sigma	C7902		
HEPES-Na (basic)	Sigma	H3784		
Na$_2$-EDTA•2H$_2$O	Sigma	E5134		
L-Glutamine	Sigma	G8540		
Na lactate	Sigma	L7900		
Na pyruvate	Sigma	P4562		
Glucose	Sigma	G6152		
Penicillin	Sigma	P4687		
Streptomycin	Sigma	S6501		
Phenol red	Sigma	P3532		
PVA	Sigma	P8136		

STEP 3: PREPARATION OF SPERMATOZOA

MATERIALS

ANIMALS
Male mice of a chosen strain, 2 months old or older

EQUIPMENT
Forceps
Microcentrifuge tubes, 1.5-ml
Scissors, fine dissection

PROCEDURE

1. Sacrifice a male mouse and remove the caudae epididymides. While holding the proximal (narrow) part of the epididymis with a pair of forceps, make a vertical incision in the distal (larger) part of the epididymis using a pair of fine scissors. Squeeze the dense sperm mass out of the epididymis.

2. Place sperm masses at the bottom of a 1.5-ml microcentrifuge tube containing 0.5 ml of CZB-HEPES medium, allowing spermatozoa to swim into the

medium for ~10 minutes at 37°C. Collect the upper fraction with actively motile spermatozoa and transfer it to a new tube.

3. Use spermatozoa immediately for ICSI.

STEP 4: ICSI PROCEDURE

MATERIALS

EQUIPMENT
Inverted microscope
Micromanipulation system
Operational dish (cover of the 100 x 15 mm petri dish, Falcon OPTILUX 351001)
Piezo impact drive unit
Pumps

REAGENTS
CZB medium
CZB-HEPES medium
CZB-HEPES with 12% (w/v) polyvinylpyrrolidone <!> (PVP; ICN, K-90, m.w. 360,000; 102787)

> Ready-to-use solution can be stored in 1.5-ml tubes at –20°C. After thawing, it can be kept in 4°C for ~1 week.

Mineral oil (Squibb or Sigma M8410)

CAUTION: See Appendix 2 for appropriate handling of materials marked with <!>.

PROCEDURE

1. Micromanipulation setup:

   a. Mount the holding pipette in the metal pipette holder of the micromanipulator.

   b. Mount the injection pipette in the metal pipette holder of the micromanipulator. The metal holder is equipped with a piezo-driven unit (Fig. 14.5).

2. Preparation of the ICSI dish: The arrangement of the drops in the ICSI dish depends on personal preferences. An example is shown in Figure 14.6.

3. Separation of the sperm head from the tail:

   a. Add a small volume (~1 µl) of fresh sperm suspension to the first PVP drop and mix thoroughly.

   b. Transfer 10 or more motile spermatozoa from the PVP drop to the next PVP drop.

   c. Aspirate a single, motile spermatozoon (tail first) into the injection pipette (Fig. 14.7A). Position the spermatozoon so that its neck (junction between head and tail) is at the opening of the pipette (Fig. 14.7B).

FIGURE 14.5. ICSI micromanipulation set up with a piezo impact driving unit (*arrow*).

   d. Apply a few piezo pulses to the neck to separate the head from the tail (Fig. 14.7C). Push the tail out of the pipette (Fig. 14.7D) and repeat the procedure for the next spermatozoon.

   e. Suck all of the sperm heads into the pipette (Fig. 14.7E). Up to 10 sperm heads can be assembled in the pipette prior to ICSI.

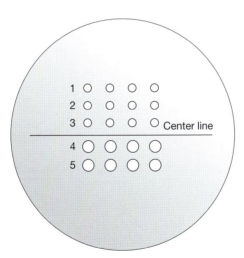

FIGURE 14.6. Diagram of the dish for ICSI. A center line is drawn on the back of the dish. Three upper rows (1, 2, and 3) of drops are 5 μl of PVP medium for sperm storage and handling. Sperm motility is reduced in this high-viscosity medium. The two lower rows (4 and 5) of drops are 20 μl of CZB-HEPES. The row (3) of drops just above the center line is for bulk sperm storage. Ten or more motile spermatozoa at a time are transferred from the bulk sperm storage drop into another PVP drop (row 2) where the sperm heads are separated from the tails. The PVP drops in the top row (1) are for washing the injection pipette from time to time, by pushing mercury out of the pipette while applying piezo pulses. CZB-HEPES drops in the row (4) just below the center line are for oocyte storage and sperm injection. Sperm-injected oocytes are temporarily stored in these drops before further culture. The drops in the bottom row (5) are for positioning the holding and injecting pipettes (when electronic programmable micromanipulators are used) and for storing the pipettes mounted on the micromanipulators prior to injection.

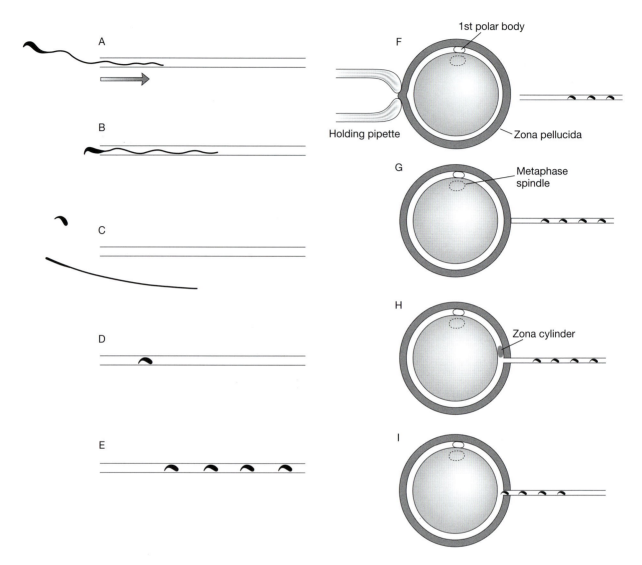

FIGURE 14.7. ICSI procedure. (*A, B*) Aspiration of a spermatozoon, tail first, into the injection pipette. (*C*) Piezo pulses separate the sperm head and tail. (*D, E*) A series of sperm heads are aspirated into the injection pipette. (*F*) Orientation of holding and injection pipettes relative to the oocyte to be injected. (*G, H*) "Coring" the zona pellucida with the injection pipette. The zona cylinder fragment is expelled into the perivitelline space. (*I, J*) Pushing the injection pipette into the oocyte. (*K*) Piezo pulses to the injection pipette puncture the oocyte plasma membrane. (*L*) A sperm head is expelled into the oocyte cytoplasm and the pipette is subsequently removed (*M*). (*Figure continues on facing page.*)

    f. After all of the sperm heads in the pipette are used up for ICSI, repeat the procedure.

    Kimura and Yanagimachi (1995a) originally injected the entire spermatozoon into an oocyte after "scoring" the sperm tail and damaging the sperm tail plasma membrane. Because the mouse sperm tail is very long and injecting the entire length of the tail is difficult, injection of the sperm head only is recommended. Unlike other mammals (e.g., cattle or human), the mouse sperm centrosome is not essential for embryo development.

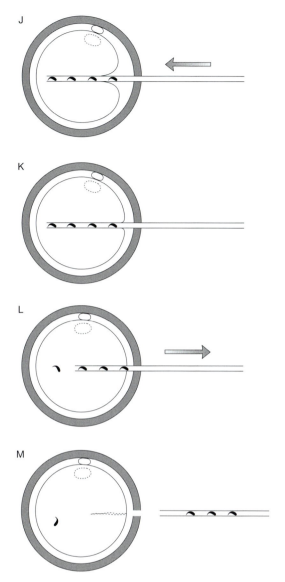

FIGURE 14.7. (*See facing page for legend.*)

4. Sperm injection into oocytes:

   a. Hold an oocyte gently using the holding pipette in such a way that the metaphase II spindle, seen as a semitransparent zone within the oocyte, is at the 6 or 12 o'clock position. The first polar body is usually (but not always) near the metaphase spindle and can be a landmark of the latter. The oocyte should remain in contact with the bottom of the dish for additional support during injection. Keep the tip of the injection pipette as well as the zona pellucida and the oocyte's plasma membrane in sharp focus during the procedure (we routinely use a 20x objective and 10x ocular for injection). The orientation of the holding and injecting pipettes for the injection procedure is shown in Figure 14.7F.

   b. After the tip of the injection pipette is brought into contact with the zona pellucida (Fig. 14.7G), advance the pipette while applying several piezo pulses and a light negative suction at the same time.

c. When the tip of the pipette passes through the zona, expel a cylindrical piece of the zona from the injection pipette into the perivitelline space (Fig. 14.7H).

d. Push one sperm head forward until it is near the tip of the pipette (Fig. 14.7I), and quickly advance the pipette through the oocyte until the pipette tip almost reaches the opposite side of the oocyte's cortex (Fig. 14.7J).

e. Apply one or two piezo pulses to puncture the oolemma at the pipette tip, which is indicated by a rapid relaxation of the oolemma surrounding the pipette (Fig. 14.7K).

f. Expel a sperm head into the ooplasm with a minimum amount of accompanying sperm suspension medium (Fig. 14.7L). Retrieve excess suspension medium that was injected. Withdraw the pipette gently, but swiftly (Fig. 14.7M).

g. Repeat the procedure for all oocytes available.

h. After injection, transfer the oocytes into CZB medium under oil and store in a humidified 5% CO_2 incubator at 37°C.

COMMENTS

- It is very important to minimize the time between sperm head and tail separation and sperm head injection into oocytes.

- Testicular spermatozoa (sperm heads) can also be injected into oocytes with the method described above. Testicular spermatozoa are either immotile or weakly motile. Make sure that their plasma membranes are intact (spermatozoa are alive) before sperm head separation.

- Correct mounting of the injection pipette to the manipulator via the piezo impact drive unit is critical. If the piezo impact drive unit is tightened to the manipulator too strongly, electric pulses are not transmitted properly.

STEP 5: CULTURE OF ICSI OOCYTES AND EMBRYO TRANSFER

MATERIALS

EMBRYOS
Oocytes from ICSI procedure

EQUIPMENT
Dissecting microscope with transillumination

REAGENTS
CZB medium
Mineral oil

PROCEDURE

1. Approximately 1 hour after ICSI, examine oocytes for their survival under the dissecting microscope with transillumination. Discard oocytes that have lysed.

2. Transfer the surviving oocytes into 50-µl drops of equilibrated CZB medium under oil (10 oocytes per drop) to continue culture in a humidified incubator.

3. Between 5 and 6 hours after ICSI, examine the oocytes and determine the number of activated eggs—those with two pronuclei and the second polar body. The second polar body should be compact and distinct, whereas the first polar body would be degenerative. Both proper magnification and angle of illumination are important to visualize polar bodies and pronuclei clearly.

4. Examine embryo development daily. It should follow the schedule: 2-cell (24 hours post-ICSI), 4-cell (48 hours), morula (72 hours), and blastocyst (96 hours).

5. Perform embryo transfer either at the two-cell stage or at the morula/blastocyst stage (see Chapter 6).

COMMENTS

ICSI-produced embryos should develop in vitro with an efficiency comparable to normally fertilized, unmanipulated oocytes. It should be noted that embryos of some strains (e.g., C57BL/6J) do not develop well in commonly used mouse culture media. For such strains, embryo transfer at the two-cell stage is recommended.

REFERENCES

Almeida P.A. and Bolton V.N. 1995. The effect of temperature fluctuations on the cytoskeletal organisation and chromosomal constitution of the human oocyte. *Zygote* **3:** 357–365.

Biggers J.D. 1998. Reflections on the culture of the preimplantation embryo. *Int. J. Dev. Biol.* **42:** 879–884.

De Repentigny Y. and Kothary R. 1996. An improved method for the artificial insemination of mice—Oviduct transfer of spermatozoa. *Trends Genet.* **12:** 44–45.

Dziuk P.J. and Runner M.N. 1960. Recovery of blastocysts and induction of implantation following artificial insemination of immature mice. *J. Reprod. Fertil.* **1:** 321–331.

Fraser L.R. 1984. Mouse sperm capacitation in vitro involves loss of a surface-associated inhibitory component. *J. Reprod. Fertil.* **72:** 373–384.

———. 1993. In vitro capacitation and fertilization. *Methods Enzymol.* **225:** 239–253.

Fraser L.R. and Drury L.M. 1975. The relationship between sperm concentration and fertilization in vitro of mouse eggs. *Biol. Reprod.* **13:** 513–518.

Fuller S.J. and Whittingham D.G. 1996. Effect of cooling mouse spermatozoa to 4 degrees C on fertilization and embryonic development. *J. Reprod. Fertil.* **108:** 139–145.

Ho Y., Wigglesworth K., Eppig J.J., and Schultz R.M. 1995. Preimplantation development of mouse embryos in KSOM: Augmentation by amino acids and analysis of gene expression. *Mol. Reprod. Dev.* **41:** 232–238.

Hoppe P.C. and Pitts S. 1973. Fertilization in vitro and development of mouse ova. *Biol. Reprod.* **8:** 420–426.

Jones E.C. and Krohn P.L. 1960. Orthotopic ovarian transplantation in mice. *J. Endocrinol.* **20:** 135–146.

Kawase Y., Iwata T., Toyoda Y., Wakayama T., Yanagimachi R., and Suzuki H. 2001. Comparison of intracytoplasmic sperm injection for inbred and hybrid mice. *Mol. Reprod. Dev.* **60:** 74–78.

Kimura Y. and Yanagimachi R. 1995a. Intracytoplasmic sperm injection in the mouse. *Biol. Reprod.* **52:** 709–720.

———. 1995b. Mouse oocytes injected with testicular spermatozoa or round spermatids can develop into normal offspring. *Development* **121:** 2397–2405.

Kuretake S., Kimura Y., Hoshi K., and Yanagimachi R. 1996. Fertilization and development of mouse oocytes injected with isolated sperm heads. *Biol. Reprod.* **55:** 789–795.

Kusakabe H., Szczygiel M.A., Whittingham D.G., and Yanagimachi R. 2001. Maintenance of genetic integrity in frozen and freeze-dried mouse spermatozoa. *Proc. Natl. Acad. Sci.* **98:** 13501–13506.

Leckie P.A., Watson J.G., and Chaykin S. 1973. An improved method for the artificial insemination of the mouse (*Mus musculus*). *Biol. Reprod.* **9:** 420–425.

Nakagata N. 1992. Production of normal young following insemination of frozen-thawed mouse spermatozoa into fallopian tubes of pseudopregnant females. *Jikken Dobutsu* **41:** 519–522.

———. 1995. Studies on cryopreservation of embryos and gametes in mice. *Exp. Anim.* **44:** 1–8.

———. 1996. Use of cryopreservation techniques of embryos and spermatozoa for production of transgenic (Tg) mice and for maintenance of Tg mouse lines. *Lab. Anim. Sci.* **46:** 236–238.

Nakagata N., Okamoto M., Ueda O., and Suzuki H. 1997. Positive effect of partial zona-pellucida dissection on the in vitro fertilizing capacity of cryopreserved C57BL/6J transgenic mouse spermatozoa of low motility. *Biol. Reprod.* **57:** 1050–1055.

Pickering S.J., Braude P.R., Johnson M.H., Cant A., and Currie J. 1990. Transient cooling to room temperature can cause irreversible disruption of the meiotic spindle in the human oocyte. *Fertil. Steril.* **54:** 102–108.

Quinn P. 2000. Review of media used in ART laboratories. *J. Androl.* **21:** 610–615.

Quinn P., Kerin J.F., and Warnes G.M. 1985. Improved pregnancy rate in human in vitro fertilization with the use of a medium based on the composition of human tubal fluid. *Fertil. Steril.* **44:** 493–498.

Rafferty Jr. K.A. 1970. *Methods in experimental embryology of the mouse.* John Hopkins Press, Baltimore.

Rivera-Perez J.A., Mallo M., Gendron-Maguire M., Gridley T., and Behringer R.R. 1995. Goosecoid is not an essential component of the mouse gastrula organizer but is required for craniofacial and rib development. *Development* **121:** 3004–3012.

Snell G.D., Hummel K.P., and Abelmann W.H. 1944. A technique for the artificial insemination of mice. *Anat. Rec.* **90:** 243–253.

Summers M.C., McGinnis L.K., Lawitts J.A., Raffin M., and Biggers J.D. 2000. IVF of mouse ova in a simplex optimized medium supplemented with amino acids. *Hum. Reprod.* **15:** 1791–1801.

Suzuki K. and Yanagimachi R. 1997. Beneficial effect of medium with high concentration serum for direct sperm injection into mouse oocytes using a conventional pipette. *Zygote* **5:** 111–116.

Szczygiel M.A., Kusakabe H., Yanagimachi R, Whittingham D.G. 2002. Intracytoplasmic sperm injection is more efficient than in vitro fertilization for generating mouse embryos from cryopreserved spermatozoa. *Biol. Reprod.* **67:** 1278–1284.

Sztein J.M., Farley J.S., and Mobraaten L.E. 2000. In vitro fertilization with cryopreserved inbred mouse sperm. *Biol. Reprod.* **63:** 1774–1780.

Sztein J.M., Farley J.S., Young A.F., and Mobraaten L.E. 1997. Motility of cryopreserved mouse spermatozoa affected by temperature of collection and rate of thawing. *Cryobiology* **35:** 46–52.

Sztein J.M., McGregor T.E., Bedigian H.J., and Mobraaten L.E. 1999. Transgenic mouse strain rescue by frozen ovaries. *Lab. Anim. Sci.* **49:** 99–100.

Sztein J.M., Sweet H., Farley J., and Mobraaten L. 1998. Cryopreservation and orthotopic transplantation of mouse ovaries: new approach in gamete banking. *Biol. Reprod.* **58:** 1071–1074.

Thornton C.E., Brown S.D., and Glenister P.H. 1999. Large numbers of mice established by in vitro fertilization with cryopreserved spermatozoa: Implications and applications for genetic resource banks, mutagenesis screens, and mouse backcrosses. *Mamm. Genome* **10:** 987–992.

Uehara T. and Yanagimachi R. 1976. Microsurgical injection of spermatozoa into hamster eggs with subsequent transformation of sperm nuclei into male pronuclei. *Biol. Reprod.* **15:** 467–447.

Wakayama T., and Yanagimachi R. 1998. Development of normal mice from oocytes injected with freeze-dried spermatozoa. *Nat. Biotechnol.* **16:** 639–641.

Wakayama T., Whittingham D.G., and Yanagimachi R. 1998. Production of normal offspring from mouse oocytes injected with spermatozoa cryopreserved with or without cryoprotection. *J. Reprod. Fertil.* **112:** 11–17.

Whittingham D.G. 1971. Culture of mouse ova. *J. Reprod. Fertil. Suppl.* **14:** 7–21.

Wolfe H.G. 1967. Artificial insemination of the laboratory mouse (*Mus musculus*). *Lab. Animal Care* **17:** 426–432.

Yanagimachi R. 1998. Intracytoplasmic sperm injection experiments using the mouse as a model. *Hum. Reprod.* **13:** 87–98.

———. 2001. Gamete manipulation for development: New method for conception. *Reprod. Fertil. Dev.* **13:** 3–14.

Cryopreservation, Rederivation, and Transport of Mice

T HE TREMENDOUS INCREASE IN THE NUMBER and variety of new mouse strains with induced mutations being generated by different laboratories places an enormous strain on animal facilities. Cryopreservation reduces maintenance costs and safeguards valuable mouse lines against loss through infection, disease, breeding failure, catastrophe, accidents such as cage flooding, and genetic drift. This chapter briefly discusses different methods of cryopreservation and provides slow freezing and vitrification protocols used for mouse embryos. It also contains protocols for the cryopreservation of sperm and ovaries that can supplement embryo cryopreserva-

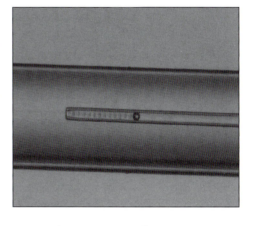

tion. Information about in vitro fertilization and ovary transplantation necessary for successful recovery of cryopreserved germ plasm can be found in Chapter 14 describing assisted reproduction methods. Rederivation and transport of mouse strains are discussed at the end of Chapter 15.

CONTENTS

EMBRYO CRYOPRESERVATION

Mouse embryos were first frozen successfully to –196°C in 1972 (Whittingham et al. 1972; Wilmut 1972a,b). The method involved slow cooling (0.2–2.0°C/minute) in the presence of dimethylsulfoxide (DMSO) and slow warming (4–25°C/minute) and was soon adapted for freezing oocytes from mice (Whittingham 1977) and human embryos (Trounson and Mohr 1983). Since then, mouse embryo cryopreservation has become routine in many laboratories (Pomeroy 1991), and it is now possible to freeze all preimplantation stages of embryos. Multiple protocols using a variety of cryoprotectants, such as glycerol, propylene glycol, ethylene glycol, sucrose, and raffinose, exist at the present time (Shaw et al. 2000a). Methods of cryopreservation can be divided into two general categories: equilibrium and nonequilibrium (Mazur 1990).

Equilibrium Methods of Cryopreservation

In equilibrium methods, embryos are cooled slowly (0.2–2.0°C/minute) to low subzero temperatures (–80°C) in moderate concentrations of cryoprotectant; embryo dehydration occurs during this process. The slow cooling procedure is relatively time-consuming but very reliable and can be used for different embryo stages. One of the slow cooling procedures is detailed in Protocol 15.1.

Several steps are required for successful cryopreservation by equilibrium methods:

• Exposure to a cryoprotectant solution.

• Cooling to very low subzero temperature in a way that permits dehydration. Dehydration may occur prior to cooling or during the cooling process itself. "Seeding" by touching the sample's container with cold forceps or a pipette is done to induce ice crystal formation early in the cooling and to initiate dehydration.

- Storage in liquid nitrogen (LN_2).

- Warming to physiological temperatures and gradual removal of the cryoprotectant to prevent damage by its toxicity or by osmotic shock. The formation
  of large ice crystals is thought to be the major cause of cell death after thawing,
  either by mechanical damage or osmotic stress upon melting of the ice (Jondet
  et al. 1984).

For comprehensive theoretical aspects of embryo cryopreservation, see Leibo
and Mazur (1971), Leibo et al. (1974, 1978), Mazur et al. (1984), Wilmut (1985),
Leibo (1986), and Pomeroy (1991).

Nonequilibrium Methods of Cryopreservation

In nonequilibrium cooling, embryos are exposed to high concentrations of
cryoprotectant and cooled rapidly (>200°C/minute); e.g., by transferring them
into LN_2 (–196°C) or the vapor phase of LN_2 (–150°C) (Rall 1987; Rall et al. 1987;
and Trounson et al. 1987). The rapid cooling methods for embryo cryopreservation are fast and simple because they eliminate the need for controlled-
rate freezers and seeding procedures, which are necessary for slow cooling methods. However, they are much more sensitive to minor variations in a protocol,
demand greater expertise, and leave less room for error. Detailed instructions for
one of the rapid cooling methods are given in Protocol 15.2.

The rapid cooling methods take advantage of concentrated solutions of cryoprotectants to reduce the water content of the cell before cooling commences,
thus preventing the formation of ice crystals. Embryos in plastic insemination
straws are briefly equilibrated in a mixture of highly concentrated cryoprotectants, and, after sufficient dehydration of the cytoplasm, they are plunged directly into the LN_2. When the cryoprotectant concentration is higher than 40% (v/v),
the solution supercools and becomes so viscous that it solidifies into a glass-like
state without forming ice crystals (Leibo et al. 1978; Rall and Fahy 1985; Kasai et
al. 1990). This process is called vitrification, which literally means glass formation. Because no ice formation occurs during vitrification and the cryoprotectant
solution does not change from a liquid to a crystalline state and back, it is more
appropriate to use the term "cooling" instead of "freezing," and "warming"
instead of "thawing," when discussing the vitrification methods. Vitrified
embryos must be stored below the glass transition temperature and should never
be warmed above –140°C to prevent injury from devitrification.

Other rapid cooling methods use lower concentrations of cryoprotectant,
allowing ice crystal formation during cooling or warming. These methods are
also successfully used for mouse embryo cryopreservation (Trounson et al. 1987;
Shaw et al. 1991a,b; Nakagata 1996; Nakao et al. 1997; Van den Abbeel et al. 1997).
The rapid cooling of mouse embryos in 4.5 M DMSO, in which ice crystals are
formed only upon warming, was shown to be safe and efficient; whereas cooling
at lower DMSO concentrations, in which ice is formed during the cooling step,
was associated with severe chromosome damage and reduced viability in vitro
and in vivo (Shaw et al. 1991b).

For theoretical aspects of nonequilibrium methods, see Rall (1987) and
Mazur (1990).

Factors to Consider in Embryo Cryopreservation

Protocols for mouse embryo cryopreservation vary in terms of cryoprotectant solution, cooling and warming rates, methods to add and remove cryoprotectant, use of cryovials or insemination straws, and the stage of preimplantation development at which the embryos are frozen. For a review of these issues, see Shaw et al. (2000a). Eight-cell- and morula-stage embryos collected at 2.5 dpc are often the preferred stages for embryo storage, mainly for the following practical reasons:

• They are robust and less sensitive than one- and two-cell-stage embryos.

• They tolerate handling and cryopreservation well.

• Embryos with one or two damaged blastomeres still have the potential to form a viable fetus (Liu et al. 1993b).

Usually cleavage-stage embryos are obtained from superovulated (see Chapter 3) donors 67–77 hours after human chorionic gonadotropin (hCG) administration, as described in Chapter 4. Natural mating may be used if the strain does not superovulate well (see Chapter 3).

The ultimate test of embryo viability is the recovery of live mice. Embryos can be transferred immediately after the thawing to the oviduct of a 0.5-dpc pseudopregnant recipient (Chapter 6) (Renard and Babinet 1984). Culture for an additional 24 hours provides a further test of the embryos' viability as judged by continuous development to the blastocyst stage. An in vitro culture step has an additional benefit of minimizing the transfer of potential pathogens (Hill and Stalley 1991) and should be recommended when working with embryos from contaminated donors. However, one needs to be cautious to keep the zona pellucida intact during the embryo culture (Carthew et al. 1985). (See more details about the rederivation procedure below, p. 604.) Each mouse strain that is to be cryopreserved and discontinued as a live stock should be first tested in actual laboratory conditions, including the recovery of live pups from cryopreserved embryos.

The number of embryos that needs to be frozen to ensure the reliable re-establishment of a breeding colony depends on such factors as the viability of thawed embryos, which may vary among protocols and strains, and the genotype, which may make it necessary to freeze more embryos from heterozygous donors because some of the frozen embryos would be wild type.

The expected viability of thawed embryos should be 80–90%, with at least 25% of thawed embryos yielding live-born pups (Mobraaten 1999). Optimally, 500–600 embryos need to be frozen, with at least 250–300 embryos being carriers; i.e., carrying the genetic alteration. This is a very general estimate, and that number may vary greatly between facilities. On the basis of these numbers, often the mouse colony to be discontinued will have to be expanded prior to embryo cryopreservation. For a review of factors influencing the efficiency of cryopreservation, see Rall et al. (2000).

SPERM CRYOPRESERVATION

Sperm cryopreservation requires fewer donor animals than embryo cryopreservation and is rather straightforward and fast. However, re-establishment of the mouse line by in vitro fertilization (IVF) and embryo transfer is as labor- and resource-intensive as embryo cryopreservation. Although sperm cryopreservation is a more cost-efficient alternative, the choice should be based on such factors as:

- Necessity to preserve a homozygous mutation and/or a particular background
- Comparison of the survival rate of sperm and embryos after freezing
- Ability to obtain sufficient numbers of embryos for freezing, and fertility
- Susceptibility of sperm to cryopreservation depending on the mouse strain (Sztein et al. 2000a)
- Variation in embryo viability between genotypes (Schmidt et al. 1985, 1987)

Several protocols of mouse sperm cryopreservation were published during the 1990s (Penfold and Moore 1993; Fuller and Whittingham 1996). However, the successful and reproducible cryopreservation of mouse sperm became a reality with the use of novel cryoprotectant consisting of 3% skim milk and 18% raffinose in water (Tada et al. 1990; Takeshima et al. 1991). This method was further improved by N. Nakagata (Nakagata 1992, 1996; Nakagata et al. 1992); normal offspring were derived from cryopreserved oocytes fertilized in vitro by cryopreserved sperm (Nakagata 1993). This method has become widely used by laboratories around the world, specifically in large mutagenesis programs (Marschall and Hrabe de Angelis 1999; Marschall et al. 1999; Thornton et al. 1999). Another successful method of sperm cryopreservation, using a mixture of raffinose and glycerol, supplemented with egg yolk, was reported by Songsasen et al. (1997) and Songsasen and Leibo (1997a,b).

The sperm cryopreservation technique still requires some optimization because its efficacy seems to be restricted to mice with hybrid and mixed backgrounds. It is less reliable for inbred strains, particularly C57BL/6 and 129S3/SvImJ (129S1, new nomenclature), both of which are extensively used in transgenesis (Sztein et al. 2000a). The low fertilizing ability of cryopreserved sperm may be overcome using partial zona dissection of oocytes (Nakagata et al. 1997); or the labor-intensive intracytoplasmic sperm injection (ICSI) method (Wakayama et al. 1998; Wakayama and Yanagimachi 1998) (see Chapter 14).

Protocol 15.3 (Method 1) is based on the original "Nakagata" method (Takeshima et al. 1991) with modifications by Sztein et al. (1997) involving the use of larger volumes (1 ml) and cryovials instead of straws, as well as the removal of cryoprotectant by centrifugation prior to IVF. This method has been successfully used by The Jackson Laboratory (Sztein et al. 2000a) and in a mutagenesis program (Thornton et al. 1999). However, recently (November 2001), the centrifugation step was removed from the protocol used by The Jackson Laboratory because of its potential to reduce sperm viability.

On the other hand, the original protocol (Takeshima et al. 1991; Nakagata and Takeshima 1993; Nakagata 1995; Nakagata et al. 1997) involves the initial sperm collection in a much smaller volume (~100 µl) and the use of 1–2 µl for IVF without any centrifugation. With some modifications, this original protocol has been successfully used in another mutagenesis program (Marschall et al. 1999) and was recently described again in detail by Nakagata (2000). This procedure, with some modifications currently used by the Monash Institute of Reproduction and Development, is described in Protocol 15.3 (Method 2).

As discussed above, the successful recovery of mouse strains from cryopreserved spermatozoa requires IVF, which contributes to a great variation in fertilization rates depending on mouse strains, thus restricting the use of the technique to mice of hybrid and mixed backgrounds.

OVARY CRYOPRESERVATION

It is possible to cryopreserve mouse oocytes (Johnson 1989; Nakagata 1989; Shaw et al. 1991, 2000b). However, oocytes are easily damaged by temperatures below 37°C and by cryoprotectants (Shaw and Trounson 1989; Carroll et al. 1990). Whole mouse ovaries can be successfully cryopreserved (Gunasena et al. 1997; Sztein et al. 1998, 1999, 2000b; Kagabu and Umezu 2000; Shaw and Trounson 2002). Ovary cryopreservation may supplement existing options of embryo and sperm cryopreservation (Shaw et al. 2000b; Shaw and Nakagata 2002; Shaw and Trounson 2002). One such method is given in Protocol 15.4.

STORAGE AND RECORDS

Cryopreserved samples are stored in LN_2 at –196°C, ideally in two storage locations for security. Samples should not be warmed above –140°C at any time during the storage. The choice of a storage tank depends on the required storage capacity and insulation properties of the container. If there is a risk that the samples are infected, they should be stored in the vapor phase of LN_2 because pathogens can spread between samples in LN_2 (Tedder et al. 1995). Recent study has shown that contaminated liquid-phase nitrogen may be a potential source of infection for cryopreserved embryos (Bielanski et al. 2000). An accidental crack or leak in a sealed freezing container may lead to a contamination of embryos; therefore, hermetically sealed containers or secondary protective containers to avoid a direct contact of liquid nitrogen with freezing medium are recommended (Kuleshova and Shaw 2000). Electrical freezers with LN_2 backup, which maintain the temperature at –140 to –150°C, also may be used.

It is absolutely essential to keep full records of all cryopreserved samples, including strain information, location, and, most importantly, the methods used for freezing and thawing, for future retrieval.

REDERIVATION OF MOUSE STRAINS

It is generally accepted that research animals should be free of infectious agents that could cause disease during experiments. New strains of conventional or infected mice can be introduced into the specific-pathogen-free (SPF) colony (see Chapter 3) by hysterectomy of pregnant females just before parturition and using clean foster parents to raise the young. However, there is a concern that some viral (Carthew et al. 1985) and bacterial infections (Hill and Stalley 1991) can be transmitted vertically. Whittingham (1979) has used embryo transfer as a means of rederiving contaminated animals into the SPF colony. In contrast to hysterectomy, embryo transfer has the advantage of avoiding postimplantation, vertically transmitted infections.

Pathogens that can be eliminated by the embryo transfer method include mouse hepatitis virus (MHV) (Carthew et al. 1985; Reetz et al. 1988) and Sendai (parainfluenza type 1) virus (Carthew et al. 1983; Okamoto et al. 1990). Contaminated rat (Rouleau et al. 1993) and mouse (Suzuki et al. 1996a) strains have been successfully rederived and introduced into SPF colonies by the embryo transfer method. IVF, cryopreservation, transport of cryopreserved embryos, and their transfer after thawing have been used for rederivation of contaminated wild mice *Mus musculus molossinus* and *Mus musculus castaneus* (Suzuki et al. 1996b).

There are two issues of major importance concerning rederivation by embryo transfer: extensive washing of collected embryos and the presence of intact zona pellucida. Mouse embryos hatched from the zona pellucida in vitro in the presence of MHV were shown to have infected trophoblast cells, whereas zona-intact embryos incubated with MHV for 48 hours were resistant to infection (Carthew et al. 1985). Although flushing medium was contaminated with MHV, zona-intact embryos collected from infected donors were not contaminated and no virus was detected in the medium after embryos were washed through three changes of medium. When embryos were transferred with medium contaminated with MHV, the virus was transmitted to the foster mothers, whereas fetal and decidual tissues were not infected (Carthew et al. 1985).

Mycoplasma pulmonis presents an additional concern because it adheres to the cell surface, can be isolated from some zona-free embryos after washing, and may possibly penetrate the zona (Hill and Stalley 1991). Therefore, direct freezing and transfer of embryos without prior incubation, despite washing, may present a risk of *M. pulmonis* contamination. However, the embryo freezing and transfer can eradicate *M. pulmonis* from infected colonies if the embryos are cultured for at least 24 hours in culture medium containing antibiotics (Hill and Stalley 1991).

The International Embryo Transfer Society (IETS) has developed protocols and recommendations for the safe handling, collection, and transfer of embryos based on the assumptions that most pathogens are unable to penetrate an intact zona pellucida and infect preimplantation-stage embryos. The sequential (10 times) embryo-washing procedure is effective in removing adhered pathogens from the intact zona (Stringfellow and Seidel 1998).

Generally the rederivation procedure by embryo transfer consists of several steps:

1. Collection of preimplantation-stage embryos (1.5–3.5 dpc) from contaminated or unknown status donors.

2. Extensive washes of zona-intact embryos.

3. There are several options for this step:

   a. Culture overnight (*optional*), but 3.5-dpc embryos should not be used for culture as they may hatch from the zona pellucida.

   b. Shipment of freshly collected embryos.

   c. Cryopreservation and storage or shipment of cryopreserved embryos.

4. Embryo transfer of collected embryos into clean recipients.

5. Quarantine of the recipients.

6. Health status test.

A designated person should dissect contaminated donors in the containment (quarantine) area. The embryos should be collected in a laminar flow hood. It is often convenient to avoid personnel exposure to contaminated donors and have preimplantation-stage embryos shipped from the original source for the embryo transfer in a laminar flow hood or in a designated HEPA-filtered operation room. Thus, the combination of cryopreservation with embryo transfer into clean recipients may be used to eliminate pathogens in the mouse colonies. Similarly, when embryo donors' housing and manipulations during transgenic production are done outside the barrier, embryos can be brought inside the barrier via embryo transfers the same way as during the regular rederivation procedure. The isolation and screening of foster mothers after weaning may be required in this case.

TRANSPORT OF MOUSE EMBRYOS

When manipulated embryos or embryos for rederivation are transferred into SPF barrier facilities, special considerations are necessary to make sure that they arrive behind the barrier or in the sterile hood where transfer takes place in a sterile condition and without being damaged by the transport and possible sterilization/cleaning procedures. Protocol 15.5 gives a method to pack embryos in such a manner that both requirements are met.

Freshly collected preimplantation mouse embryos from 1.5 to 3.5 dpc can be safely shipped at ambient temperature within 48 hours. Because the blastocyst stage is the most robust, it is commonly used for longer transportation overseas (embryos do not hatch during this time). Any other preimplantation embryo stage, possibly with the exception of zygotes, can withstand transportation within 24 hours. Collected embryos should be washed vigorously through several (10) drops of medium and placed into a sterile cryovial/cryotube filled to the top (leaving practically no space for air) with equilibrated overnight embryo culture medium (e.g., KSOM-AA). Equilibrated ES cell medium (see Chapter 8) without leukemia inhibitory factor (LIF) can be used for the transport of blastocysts. The vial should be closed tightly and placed into a protected envelope for courier shipment. A small Styrofoam box with an "ice pack" warmed to 37°C is recommended for use during hot summer or cold winter months. On arrival, the embryos should immediately be placed into fresh medium and then transferred into recipients as soon as possible (see Chapter 6). Additional culture overnight is not recommended.

Cryopreserved embryos must be handled extremely carefully and may not be warmed above −140°C at any time during storage, or their viability will be severely compromised. Vapor shipper containers (e.g., MVE Biological Products, http://www.chart-ind.com) for safe transportation at cryogenic temperatures allow the shipment of cryopreserved samples with a "nonhazardous" classification. It is absolutely essential for successful recovery of the shipped cryopreserved embryos to have the original cryopreservation protocol and to use the recommended thawing procedure. There are multiple protocols available, and it is not recommended to mix these protocols.

Embryo Cryopreservation by Slow Freezing

This procedure is the one that has been used in The Jackson Laboratory for many years (http://www.jax.org/resources/documents/cryo/slow.html). It is based on the protocol originally described by Whittingham et al. (1972). See also Mobraaten (1999). Other methods of slow freezing are reported by Schmidt et al. (1987), Liu et al. (1993a), Dinnyes et al. (1995), and Shaw et al. (1995, 2000a).

It is important to note that if straws are substituted for vials, or if a different controlled-rate freezer is used in the procedure described below, adjustments may be needed in timing and freezing and thawing rates due to the changes in heat transfer rate.

MATERIALS

ANIMALS
Pseudopregnant females at 0.5 or 2.5 dpc for oviduct or uterine transfer (see Chapters 3 and 6)
Superovulated donors at 2.5 dpc (see Chapter 3)

CRYOGENIC EQUIPMENT
CryoMed 1010 programmable freezing controller (Forma Scientific) (see Comment)
Cryotubes (Nunc 12-565-170N)
Forceps (large)
Gloves and goggles for handling LN_2
Liquid nitrogen (LN_2)<!>
LN_2 containers (Dewar flasks) for benchtop work
LN_2 storage containers (e.g., MVE, Taylor-Wharton Cryogenics)
Permanent markers or other means of labeling
Storage goblets and canes or boxes
Tongs with long handles to manipulate samples

GENERAL EQUIPMENT FOR EMBRYO HANDLING
Culture dishes, sterile (e.g., Falcon 3001, 3004, 3037)
Embryo-handling pipette (see Protocol 4.6)
Flushing needle (see Protocol 4.10)
Microdrop cultures (see Protocol 4.5)
Microscopes with transmitted and reflected light
Pasteur pipette
Selectapette (Clay-Adams)(optional)
Surgical instruments
Syringes, 1 ml with 26- and 30-gauge needles
Watch glass or depression slide

REAGENTS

D-PBS supplemented with 1000 mg/liter dextrose (e.g., Sigma G6152), 36 mg/ml sodium pyruvate, penicillin G (0.075 mg/ml), streptomycin sulfate (0.05 mg/ml) (see Appendix 1 and Table 4.3)

- containing 3 mg/ml of embryo-tested bovine serum albumin (BSA) (used for embryo handling)
- BSA-free (used for preparation of 2 M DMSO cryoprotectant)

Dimethyl sulfoxide (DMSO) <!> m.w. 78.13 (e.g., Sigma D8779 or D2650)

To make up 10 ml of 2 M DMSO, measure out 1.56 g of DMSO (~1.42 ml) in a tube and add 8.58 ml of BSA-free D-PBS.

Embryo-tested light mineral oil (e.g., Sigma M8410) (see Chapter 4)
Ethanol, 95% <!>
KSOM-AA medium (as described in Chapter 4)

CAUTION: See Appendix 2 for appropriate handling of materials marked with <!>.

PROCEDURE

Freezing

1. Precool the freezing machine to −6°C.

2. Collect eight-cell-stage embryos in the morning of 2.5 dpc as described (see Protocol 4.10) using D-PBS supplemented with 1000 mg/ml dextrose, 36 mg/liter sodium pyruvate, 3 mg/ml BSA, and antibiotics (see Appendix 1 and Table 4.3).

3. Transfer embryos into a 2-ml cryotube containing 0.1 ml of D-PBS and place the tube on ice until all embryos have been collected and are ready for freezing. The number of embryos placed into one cryotube depends on their expected viability and should be sufficient for transfer into at least two recipients (e.g., around 30 embryos).

4. Gently add 0.1 ml of protein-free D-PBS containing 2 M DMSO at 0°C to bring the final concentration of DMSO to 1 M. Equilibrate at 0°C for at least 30 minutes.

5. Transfer the cryotubes to a salt and ice bath maintaining −6°C (or to the freezing machine at −6°C). Equilibrate for 2 minutes.

6. Seed the contents of the cryotube by touching the surface of the medium with an ice crystal from the tip of a Pasteur pipette. (Draw a small amount of PBS into the tip of the Pasteur pipette by capillary action, and put it inside a glass test tube that is partially immersed in a salt and ice bath maintained at −10°C.) Use a separate pipette for each seeding. Seeding starts ice nucleation of the medium with minimal damage to the embryos caused by the formation of ice crystals. Cap the cryotubes and transfer them to a controlled-rate freezer precooled to −6°C.

   Cryotubes (or straws) also can be seeded by touching the outer surface of the cryotube leveled with the meniscus of cryoprotectant, away from embryos, with a metal object (e.g., long-handled forceps) precooled in LN$_2$.

7. Lower the temperature in a controlled-rate freezer at the rate of 0.5°C per minute to –80°C. If the freezing machine does not have a temperature display, monitor the temperature using a thermocouple inserted into a control cryotube containing 0.2 ml of 95% ethanol.

8. When the temperature reaches –80°C, transfer the cryotubes to LN_2. It may be most convenient to place them into a small LN_2 flask first and then transfer the samples to a large LN_2 container for long-term storage. Record strain information, number of embryos, location, etc.

Thawing

1. Remove the cryotube from the LN_2 and allow it to warm at ambient temperature until all ice crystals have thawed (usually 12–15 minutes).

2. When completely thawed, slowly add 0.8 ml of D-PBS in dropwise fashion to dilute the DMSO.

3. Transfer the contents of the cryotube to a dish, embryological watch glass, or depression slide and observe the embryos. A 1-ml Selectapette (Clay-Adams) may be used.

4. Wash the embryos through several drops of pre-equilibrated KSOM-AA medium and culture them for a few hours or overnight in microdrops covered with embryo-tested oil as described in Protocol 4.5. An alternate way of embryo culture in glass tubes is described at

   http://www.jax.org/resources/documents/cryo/slow.html.

5. Transfer the embryos into the oviduct or uterus of pseudopregnant females as described in Protocol 6.3 or 6.4.

COMMENTS

All freezing parameters described above are for the CryoMed 1010 programmable freezing controller (Forma Scientific) used in The Jackson Laboratory. Many varieties of controlled-rate freezers are commercially available (see partial list in Chapter 17) and may be divided into two large groups: alcohol-based and LN_2-based. Alcohol-based freezers are generally less expensive than LN_2-based freezers and can be used reliably and cost-effectively for embryo cryopreservation in insemination straws or sealed glass ampules. However, they may not be the best choice for embryo cryopreservation in cryotubes because the cryotubes do not provide a tight seal to prevent alcohol penetration inside the vial. Liquid nitrogen-based freezers are a better option for embryo cryopreservation in plastic cryotubes.

Embryo Cryopreservation by Rapid Cooling

The vitrification protocol described below was provided by Jillian Shaw (Monash Institute of Reproduction and Development, Clayton VIC, Australia) and is based on a method originally reported by Kasai et al. (1990) and described in detail by Shaw and Kasai (2001). Other efficient vitrification methods can be found in Ali and Shelton (1993), Tada et al. (1993), and Rall and Wood (1994).

MATERIALS

ANIMALS

Pseudopregnant females at 0.5 or 2.5 dpc for oviduct or uterine transfer (see Chapters 3 and 6)

Superovulated donors at 2.5 dpc (see Chapter 3)

CRYOGENIC EQUIPMENT

Forceps, large

Gloves and goggles for handling LN_2

Heat sealer (or forceps and Bunsen/ethanol burner)

Insemination straws, 0.25-cc clear plastic (e.g., IMV, AA 201 L'Aigle, France), storage goblets and canes, Visitubes, daisy goblets, canisters, etc. (available from various suppliers, e.g., CryoBioSystem Division of IMV Technologies, http://www.cryobiosystem-imv.com; Minitube of America, Verona, WT 608-845-1502; Meditech IST Canada, Montreal, QC 514-683-0037; Conception Technologies, CA 1-800-995-8081, http://www.conceptiontechnologies.com)

Liquid nitrogen (LN_2) <!>

LN_2 containers (Dewar flasks) for benchtop work

LN_2 storage containers (e.g., MVE, Taylor-Wharton Cryogenics)

Permanent markers or printer for labels (e.g., Brother P-touch-1800 available from office supply companies)

Styrofoam box (thick-walled) or other insulated container with a lid that can hold LN_2 and accommodate the straws horizontally

Styrofoam or polystyrene boat floating on the surface of LN_2 (see details in Preparation of cooling/warming container)

Tongs with long handles to manipulate samples

(*Optional*) Critoseal (Oxford Labware 8889-215003, e.g., through VWR 15407-103), polyvinyl alcohol<!>, cold water soluble (PVA, e.g., Sigma P8136) or polyvinylpyrrolidone<!> (PVP, e.g., Sigma PVP-4) powder

GENERAL EQUIPMENT

Culture dishes, sterile (e.g., Falcon 3001, 3004, 3037)

Embryo-handling pipette (see Protocol 4.6 and Fig. 6.1)

The inner diameter of the drawn-over-the-flame Pasteur pipette or capillary should be only slightly larger than an embryo to allow the embryos to be as closely packed as possible, with a minimum amount of culture medium between them.

Flushing needle (30-gauge 1/2 inch prepared as described in Chapter 4)
Graduated tube, 10-ml (with lid)
Microscopes with transmitted and reflected light
Scissors
Surgical instruments
Syringe (1-cc) with inserted cut yellow tip or flexible tubing to fit the straw (airtight joint) (used to move the solutions in and out of straws)
Syringes, 1 ml with 26- and 30-gauge needles
Timer
Water bath at 20°C

REAGENTS

Cryopreservation solution, EFS40 at room temperature (see below for preparation)
EFS40 contains 40% v/v ethylene glycol <!>, 18% w/v Ficoll, and 0.3 M sucrose
Dilution solution, 0.5 M sucrose in M2 medium at room temperature

Sucrose (Sigma S9378)	1.71 g
M2 (with 4 mg/ml BSA)	to 10 ml

Filter through a 0.2-μm filter. Prepare fresh as needed or store aliquots at –20°C.
KSOM-AA medium (as described in Chapter 4)
M2 with 4 mg/ml BSA or any other HEPES-buffered medium at room temperature (as described in Chapter 4)
Mineral oil, embryo-tested light (e.g., Sigma M8410).

CAUTION: See Appendix 2 for appropriate handling of materials marked with <!>.

PROCEDURE

Preparation of EFS40 Solution

Originally, EFS40 solution was formulated with modified D-PBS (Kasai et al. 1990). However, unlike M2, PBS-based solutions tend to precipitate when frozen. Any other embryo-holding HEPES-buffered solution can be used in the formulation; 4 mg/ml BSA or 1–10% serum may be included in EFS40, but it is not essential. Protein- and serum-free medium is wholly defined; thus, when it is used, batch variations and potential exportation problems are avoided. M2 or other HEPES-buffered media may be made as described in Chapter 4 or purchased. If BSA is used, it should be embryo-tested.

Ethylene glycol (Sigma E9129)	4.0 ml
Ficoll 70,000 MW (Sigma F2878)	1.8 g
Sucrose (Sigma S9378)	1.026 g
M2 medium	to 10 ml

with or without BSA or serum (see comment above)

1. Add ethylene glycol to Ficoll in a 10-ml graduated tube with a tight lid and let it stand until the Ficoll has dissolved (1 hour or more). Add sucrose. When the sucrose has dissolved, add M2. (It is also possible to add ethylene glycol to pre-weighed Ficoll and sucrose and, once the three ingredients have dissolved, to make up to volume with M2 medium.)

2. Filter through a 0.45-μm filter and aliquot. (Gentle warming makes the solution less viscous.)

3. Store at –20°C. The solution can be stored for at least a month. If it has been made correctly, EFS40 should remain liquid at –20°C.

> The cryoprotectant solution should remain clear and transparent throughout the cooling and warming steps. If it turns opaque, the solution or cooling/warming rates are incorrect and embryo survival will be compromised.

Preparation of Cooling/Warming Container

1. Precool a thick-walled Styrofoam box or other insulated container with a lid with LN$_2$ for at least 30 minutes before use; it should contain at least 5 cm of LN$_2$. The box should be wide enough to accommodate the straws horizontally.

2. Float a flat Styrofoam or polystyrene "boat" on the surface of LN$_2$. The temperature on the surface of the boat should be less than –150°C; therefore, the boat should not be too thick (e.g., 1 cm). The distance from the top of the boat to the lip of the container must be more than 5 cm (e.g., 10 cm). The boat should have supports or dividers that can hold straws with embryos horizontally above the surface (for more uniform cooling) without touching each other. Two empty straws placed alongside the boat and short pieces of straws inserted upright along sides of the two straws should prevent straws with embryos from falling over the edge.

Cooling Procedure

1. Collect eight-cell- to morula-stage embryos on 2.5 dpc as described (Chapter 4) using M2 or any other HEPES-buffered medium supplemented with 4 mg/ml. Wash the embryos through several drops of fresh M2 medium to remove all debris.

2. Select embryos with normal morphology for cryopreservation. Keep them in M2 medium at room temperature or place them in an incubator in embryo culture medium until ready for cryopreservation.

3. Label the straws with all necessary information using a permanent marker or printed labels.

4. Load each straw with 30 µl of EFS40 solution by connecting the straw to a 1-ml syringe with an inserted cut yellow tip or any flexible tubing that fits the straw. A 15-mm column corresponding to 30 µl of solution is followed by a 5-mm air space near the opening.

> It is also possible (but more complicated) to load the straw with both EFS40 and dilution solution (sucrose) to help expel the embryos after warming and to allow the immediate dilution of the cryoprotectant. Great care should be taken to prevent the EFS40 solution from being diluted with sucrose solution during loading. A 0.5 M sucrose solution (65 mm or ~125 µl) is loaded close to the cotton plug of the straw using a syringe with a needle inserted into the straw and then carefully withdrawn without touching the sides of the straw. The sucrose solution will touch the plug and seal the top end of the straw. The EFS40 solution is separated from the sucrose solution by two air bubbles that prevent the two solutions from mixing in the straw. The EFS40 solution should remain clear during cooling and warming, whereas the dilution solution should turn opaque with ice crystal formation.

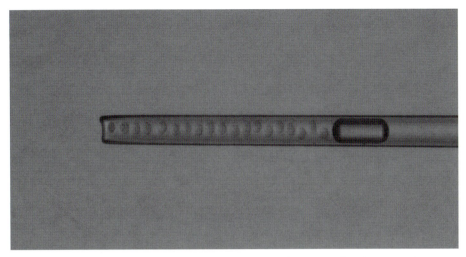

FIGURE 15.1. Finely drawn capillary with aspirated embryos in M2.

5. Aspirate a small group of embryos (e.g., 8–12) in an absolutely *minimal* volume of M2 using a finely drawn capillary (see Fig. 15.1).

   It is very important to have embryos loaded as closely as possible to minimize the dilution of EFS40 with M2, which can compromise the embryos' viability. Air bubbles picked into the capillary help to control the flow by reducing capillary action and to visualize the embryo's transfer into the straw.

6. Place the straw horizontally, and do not touch the region of the straw containing EFS40 (to keep it at room temperature).

7. Gently, but rapidly, insert the tip of the pipette deep into the cryoprotectant, and expel the embryos into the EFS40 while smoothly withdrawing the pipette. The embryos should be spaced uniformly through the cryoprotectant solution. The air bubbles serve as a guide during this procedure.

8. Start the timer (1 minute).

9. Seal the straw at both ends using a heat sealer (the temperature should be ~95°C). Alternatively, clamp the ends of the straw with wide forceps prewarmed to >100°C. The straw can also be sealed using critoseal or PVA or PVP powder. If PVA or PVP powder is used, place an additional drop (~5 μl or 5 mm long) of cryoprotectant at the open end. Then push the straw into the PVA or PVP powder, which swells and forms a plug. However, keep in mind that neither of these powders works as well with cryoprotectant solutions as they do with water.

10. Once the embryos have been equilibrated with the cryoprotectant for 1 minute at room temperature (25°C), place the straw on the Styrofoam boat in the vapor phase of the liquid nitrogen.

   If the room temperature is lower than 25°C (e.g., 20°C), the exposure time in EFS40 can be extended to 1.5–2 minutes. If the room temperature is 30°C, the exposure time must be limited to 30 seconds.

11. Cool the straws in the vapor phase for 3–5 minutes, and then plunge them into the LN_2.

12. Place the straws into LN_2 storage. Precool the goblet, cane, and forceps in LN_2. Handle the straws using precooled forceps. Keep the region of the straw containing embryos as cold as possible at all times.

    Once a straw is brought above the lip of the Dewar, the temperature may rise by as much as 100°C in 5 seconds. To obtain consistent results, avoid such extreme temperature fluctuations.

Warming Procedure

1. Prepare the following items necessary to warm the straws:

    Styrofoam box with floating boat containing LN_2

    water bath at 20°C

    M2 medium

    3 x 35-mm dishes for each straw to be warmed:

    > one dish containing 3 ml of 0.5 M sucrose (1 ml of sucrose is necessary for each 10 μl of cryoprotectant),

    > two dishes containing 3 ml of M2 each

    timers

    scissors to cut straws

    1-ml syringe with inserted cut yellow tip or flexible tubing to fit the straw

    microdrop dish with KSOM-AA or other embryo culture medium equilibrated overnight (see Chapter 4) if embryos are to be cultured

2. Transfer the straws from the storage tank to the Dewar container with LN_2 using precooled forceps. Do not touch the straws with your fingers!

    CAUTION: Handle the straws with care because LN_2 may be trapped inside the straw and cause an explosion. To minimize the risk for injury, keep the straw in the nitrogen vapor just above the level to allow LN_2 to escape.

3. Place the straws on a boat floating on the surface of LN_2 inside the Styrofoam box. Leave for 3–5 minutes.

4. Using forceps, rapidly but gently lift the straw from the boat and hold it steady in air for 5–10 seconds.

5. Fully immerse the straw into a 20°C water bath. The EFS40 solution should remain clear and transparent throughout all steps of cooling, storage, and warming.

6. Remove the straw from the water bath as soon as the solution has liquefied (~5 seconds). Be careful not to warm EFS40 with your fingers.

7. Quickly wipe the straw and cut the seals from both ends with sharp scissors while holding the straw horizontally. Remove the upper seal first, then remove the lower seal over the dilution solution. Do not bend or flex the straw; hold it firmly near the end, because flicking the straw may cause embryos to be lost.

8. Connect a 1-ml syringe with the end of the straw farthest from the EFS40 solution. Slowly tilt the straw and lower the opening until it touches the bottom of the tissue culture dish containing 0.5 M sucrose in M2 medium.

9. Very gently push the syringe plunger to expel the contents of the straw. This is a very crucial step! Set the timer for 3 minutes and start. Leave the dish undisturbed to prevent excessive rapid mixing!

   To remove embryos that may be stuck to the sides of the straw, gently aspirate 0.5 M sucrose solution back into the straw (from the part of the dish without the embryos). Leave the straw lying horizontally until the end of the 3 minutes.

10. After 3 minutes, gently shake the dish to mix the dilution and cryoprotectant solutions. Count and check the embryos. Decant the residual sucrose solution remaining in the straw if some embryos are missing.

11. After a total of 5 minutes in 0.5 M sucrose, transfer all of the embryos (in the minimal volume) to a prepared dish containing 3 ml of M2. Leave the embryos in M2 for 5 minutes.

12. After 5 minutes in M2, wash the embryos in fresh M2 (last dish containing 3 ml). Select embryos with normal morphology for immediate transfer into pseudopregnant females, or wash them through several drops of equilibrated culture medium and culture overnight (e.g., in KSOM-AA).

COMMENTS

- The described procedure should work well for eight-cell- and morula-stage embryos. The survival rate for other stages may be lower (Miyake et al. 1993; Zhu et al. 1993). However, mouse blastocysts may be cryopreserved using this protocol if they are pre-equilibrated in 1.5 M ethylene glycol solution at room temperature (Shaw et al. 2000a).

- Although stepwise cooling and warming used in this protocol are beneficial, it is also possible to immerse the straw slowly into the liquid nitrogen immediately at the end of the equilibration period; similarly, rapid warming may be done by immersing the straw directly into a 20°C water bath (Shaw et al. 2000a).

Sperm Cryopreservation

Two methods of mouse sperm cryopreservation are described here: using cryo-tubes (Method 1) and using insemination straws (Method 2).

MATERIALS

ANIMALS

Sexually mature male mice at least 8 weeks old, optimally 3–5 months old, proven breeders, successfully mated no longer than 7 days and not mated 2–3 days prior to the experiment

EQUIPMENT

Cryocanister, metal, with storage goblet for straws
Cryotubes, 1.8-ml (Nunc 12-565-170N) or 0.25-ml straws (e.g., IMV, AA 201 L'Aigle, France)
Culture dishes (e.g., Falcon 35-3001, 35-3037)
Dewar flask, small, for LN_2
Dissection scissors
Freezing device, deep polystyrene box with lid suitable for holding LN_2
Heating block (*optional*)
Incubator, 5% CO_2 in air at 37°C
Microfuge tubes, 1.5 ml (e.g., eppendorf)
Microscopes with transmitted and reflected light
Needles, 30-gauge
Pipette tips, wide bore (Rainin HR-250W, 1000W)
Pipetter, 200 µl (e.g., Gilson P200)
Polystyrene box for LN_2
Polyvinyl alcohol <!>, cold-water soluble (PVA, e.g., Sigma P8136)
Storage LN_2 containers
Surgical instruments (including watchmaker's forceps)
Thermocouple
Water bath, 37°C

REAGENTS

Cryoprotective agent (CPA) (18% raffinose, 3% skim milk in water)

Raffinose (Sigma R0250, R7630)	1.8 g
Skim milk (Difco (B-D) Betalab 0032-17-3)	0.3 g
Ultrapure water (e.g., Sigma W1503)	to 10 ml
Osmolarity, 400–410 mOsm	

Liquid nitrogen (LN_2) <!>
CPA for Method 1

1. Prepare 6% skim milk solution by dissolving 18 g of skim milk in ultrapure water to a final volume of 300 ml.

2. Centrifuge for 1 hour at 15,000g at 4°C.

3. Carefully decant the supernatant using a pipette rather than pouring off.

4. Take 200 ml of the supernatant, add 125 ml of water and 72 g of raffinose, and top to a final volume of 400 ml with water. Stir at room temperature until the raffinose dissolves.

5. Filter through a 0.22-μm filter using a filter unit (prefilter through a 0.45-μm filter to prevent clogging if necessary).

6. Store the aliquots at 4°C for 10 days or up to 3 months at –20°C/–70°C.

CPA for Method 2

1. Place 8 ml of water in a screw-topped conical 15-ml Falcon tube (2097).

2. Add 1.8 g of raffinose, place the tube into an 80°C water bath for 30 seconds, and dissolve by gentle inversion.

3. Add 0.3 g of skim milk to the solution and mix gently.

4. Make up to a 10-ml final volume.

5. Centrifuge at 18,000g for 15 minutes at 20°C.

6. Remove the supernatant to a new tube.

7. Centrifuge again and/or at higher speed until supernatant becomes clearer (*optional*).

8. Filter the clear supernatant through a 0.22-μm filter. Store the aliquots at 4°C for up to 10 days.

CAUTION: See Appendix 2 for appropriate handling of materials marked with <!>.

COMMENT

The first method makes clearer CPA because the skim milk forms solid pellets at a less dense solution in the absence of raffinose. Discard any unused CPA to prevent contamination.

METHOD 1

This procedure, currently used by The Jackson Laboratory, was provided by Carlisle Landel (http://www.jax.org/resources/documents/cryo/sperm.html).

Sperm Collection and Freezing

1. Thaw an aliquot of cryoprotectant solution and bring it to 37°C in an incubator or on a heating block. Mix by inversion if there is any precipitation. Pipette CPA into a 35-mm dish (Falcon 3001) or organ culture center well dish (Falcon 35-3037) and place it on the heating block at 37°C.

2. Prepare the "freezing device." Place a platform (e.g., the insert from a Gilson 200-µl tips box) into the polystyrene box. This acts as a support for the cryotube rack. Carefully pour LN_2 into the polystyrene box just about to cover the platform. Place a cryotube rack on top of the platform so that it is suspended in liquid nitrogen vapor. Replace the lid on the polystyrene box and allow it to fill with vapor for at least 20–30 minutes before starting. It must contain at least 5 cm of LN_2. Replenish the LN_2 as necessary during the freezing session, but do not allow the level to rise above the platform.

3. Aseptically dissect the vasa deferentia and caudae epididymides from one male mouse and clean off all fat and blood (e.g., on a tissue); examine them under a microscope (see Fig. 14.1).

4. Place the epididimydes and vasa deferentia into 1 ml of prewarmed CPA solution. Using the point of a 30-gauge needle, slice the epididymides 3–5 times. "Walk down" along the vas deferens and squeeze sperm gently by using forceps or a pair of 30-gauge needles.

   > Although only sperm from caudae epididymides is traditionally used for IVF, the collection of sperm from both the vasa deferentia and caudae epididymides might improve the overall motility rate (Sztein et al. 2000a).

5. Incubate the tissue and sperm for 10–15 minutes in a CO_2 incubator at 37°C.

6. Remove the epididymal and vas tissue from the suspension by scraping them to one side of the dish. Gently mix the sperm in the dish to equilibrate distribution of the sperm in CPA. Using wide-bore pipette tips, quickly aliquot 100 µl into prelabeled cryotubes. Replace the screw cap and tighten to seal the cryotube.

7. Place the cryotubes directly into the precooled freezing device in LN_2 vapors (about –120°C) and leave for 10 minutes (cooling rate of –20 to –40°C per minute).

8. Plunge the samples into LN_2 and store there until required.

Thawing

1. Using forceps, remove a frozen sample from LN_2, hold it in air for 30 seconds, then thaw rapidly by placing it in a 37°C water bath until the ice crystals are melted (~2 minutes).

   > **CAUTION:** Take special care that the tube has not filled with LN_2 before plunging it into the water bath or the tube may explode. If LN_2 is present in the tube, keep it in the vapor phase, and wait for the LN_2 to evaporate first.

2. When the sample has thawed (all ice crystals have melted), immediately add sperm aliquots (e.g., 10 µl or more if the concentration is low) to the IVF dishes (see Protocol 14.2).

METHOD 2

The procedure below, based on the original Nakagata protocol (Nakagata 2000) was provided by L. Gabriel Sanchez-Partida (The Monash Institute of Reproduction and Development, Clayton VIC, Australia). CPA is prepared by the second method.

Sperm Collection and Freezing

1. Transfer 130 µl of CPA per male into a 1.5-ml microfuge tube and warm to 37°C.

2. Prepare the freezing device. Place a metal cryo canister containing a storage goblet for straws into a polystyrene box with LN_2 and adjust its height until –150°C is reached.

   Replenish the LN_2 as needed during the freezing session, but do not allow the temperature to go above or below –150°C prior to introduction of the straws. Monitor the temperature with a thermocouple inserted into the straw.

3. Aseptically dissect the vasa deferentia and caudae epididymides from one male and clean off all fat and blood (see Fig. 14.1). Make one cut per cauda using scissors and place epididymal and vas tissues into the microfuge tube containing 130 µl of CPA.

4. Incubate for 10 minutes in 5% CO_2 in air at 37°C.

5. Remove the epididymal and vas tissue from the suspension with a pair of watchmaker's forceps.

6. Check for sperm motility. If there are no motile cells, collect sperm from another mouse.

7. Label the straws with male identification, strain, and date.

8. Once the incubation time has been completed, connect a 200-µl pipettor to a straw by inserting the 200-µl tip into the cotton plug side of the straw.

9. Aspirate successively: 100 µl of preferred fertilization medium, 20 µl of air, 10 µl of sperm suspension, 20 µl of air, and 10 µl of preferred fertilization medium using a micropipette, and a tip inserted into the cotton plug side of the straw.

   100 µl of medium is loaded into the straw to prevent it from floating when plunged into the liquid nitrogen. Loading also can be done using a 1-ml syringe connected to a straw with flexible tubing or cut 200 µl tip as described in Protocol 15.2, p. 610.

10. Seal the end of the straw with PVA by pushing the end containing 10 µl of medium into the powder. Alternatively, use heat sealer or PVP for sealing as described in Protocol 15.2. Both ends of the straws can be sealed if desired.

11. Prepare up to ten straws at a time.

12. Place the straws at –150°C in liquid nitrogen vapor.

13. After 10 minutes, plunge the straws into LN_2 (–196°C).

14. Store until required.

Thawing

1. Insert one end of a pair of forceps into the straw to prevent the cotton plug from popping off if only one end was sealed, remove the straw from LN_2, hold it in the air for 5 seconds, and then rapidly place it in a 37°C water bath.

   CAUTION: Handle the straws with care, because LN_2 may be trapped inside the straw and cause an explosion. To minimize the risk for injury, keep the straw in the nitrogen vapor for a few seconds just above the level to allow LN_2 to escape.

2. When the sample has thawed (30–60 seconds), dry the straw, cut above the PVA seal, and transfer the sperm suspension into a microcentrifuge tube. Avoid contact between thawed sperm and the medium in the straw.

3. Assess sperm motility and morphology. If the sperm show good motility, add 2 µl of frozen-thawed sperm into 20-µl fertilization drop (see Protocol 14.2, Comments). If the motility is low, use 4 µl of sperm suspension per 40-µl fertilization drop.

Ovary Cryopreservation

This procedure was described by Sztein et al. (1998) and provided by The Jackson Laboratory (http://www.jax.org/resources/documents/cryo/ovary.html).

MATERIALS

ANIMALS
Female mice

EQUIPMENT
Culture dishes (e.g., 35 mm)
Cryomed 1010 controlled-rate freezer with freezing chamber model 8024 (Forma Scientific)
Cryoprotectant: 1.5 M dimethyl sulfoxide (DMSO) <!> in M2 medium supplemented with 10% FBS (Sigma) at room temperature (~1.1 ml of DMSO and 8.9 ml of M2)
Cryotubes, 1.8-ml (Nunc 12-565-170N)
LN$_2$ storage containers
Pasteur pipettes
Salt/ice bath at –6°C and –10°C
Surgical instruments

REAGENTS
M2 medium

CAUTION: See Appendix 2 for appropriate handling of materials marked with <!>.

PROCEDURE

Freezing

1. Aseptically remove the ovaries from the bursa and place them in a sterile disposable 35-mm culture dish containing 2 ml of M2 medium.

2. Cut each ovary into halves unless they are unusually small.

3. Transfer each half-ovary into prelabeled 1.8-ml Nunc cryotubes containing 200 µl of cryoprotectant and keep the tubes at room temperature (23°C) for 10 minutes.

4. Place the cryotubes on ice (0–0.5°C) for ~45 minutes.

5. Transfer to –6°C (a freezing machine or an ice/salt bath) and hold for 5 minutes.

6. Seed the surface of the medium inside the cryotube by touching it with a Pasteur pipette containing a small amount of cryoprotectant at the tip and precooled in a glass tube in ice/salt bath at –10°C.

7. Place the cryotubes into a controlled rate freezer at –6°C.

8. Cool at a rate of 0.5°C/per minute to –80°C.

9. Transfer cryotubes to LN_2 storage containers.

Thawing and Ovary Transplantation

1. Thaw the cryotubes at room temperature (23°C) until ice crystals are melted.

2. Remove the cryoprotectant and replace it with 0.2 ml of M2 medium.

3. Allow to rehydrate for 10 minutes before the surgery.

4. In the surgery, one of the host ovaries is ligated, the other one is removed, and the thawed donor half-ovary is transplanted orthotopically into one empty bursa (see Protocol 14.1).

COMMENTS

An alternative protocol for mouse ovary cryopreservation using 1.5 M DMSO (Sigma D9779) and 0.1 M sucrose (Sigma S9378) in protein-free PBS (Sigma D5773) has been described by Shaw et al. (2000c). The ovaries are equilibrated for 30 minutes on ice in pre-cooled (0 to 1°C) cryoprotectant solution, and then transferred to cryovials containing 0.5 ml of cryoprotectant, and placed at –6°C in a controlled-rate freezer. After 5 minutes at –6°C, vials are seeded, cooled at a rate of 0.3°C/minute to –40°C, and then plunged into LN_2. The freezing of mouse ovarian tissue can also be done in a Nalge Nunc (5100-0001) cryo container available from Fisher (15-350-50), VWR (55710-200), or Sigma (Mr. Frosty C1562). It is placed into a –80°C freezer overnight with a cooling rate of 1°C/minute (J. Shaw, pers. comm.). Thawing is performed gradually: first in 0.75 M DMSO/0.25 M sucrose solution for 10 minutes, then in 0.25 M sucrose for 10 minutes before being transferred into PBS.

Packaging and Transport of Embryos through Sterilization Ports

This protocol describes a system of multiple sterile dishes and plastic bags for packaging and transporting embryos.

MATERIALS

EQUIPMENT

Flow hood, sterile, used only for embryo manipulations
Glass embryo-handling capillaries (see Protocol 4.6), sterilized by 3 hours of baking in the oven, NOT autoclaved
Organ culture center well dishes (Falcon 35-3037), new sterile pack
Paper tissues (e.g., Kleenex), autoclaved
Petri dishes, 12-cm glass and 15-cm glass, sterile-packed and autoclaved
Plastic autoclave bags (2), with several stripes of autoclave tape loosely attached to the side, sterile-packed and autoclaved

REAGENTS

M2 medium, sterile-filtered, prewarmed to 37°C
PBS or ddH$_2$O, sterile

PROCEDURE

1. Filter-sterilize the M2 medium an additional time.

2. Move the embryos and all packaging to the sterile hood. Open the packages right at the front shield, only allowing sterile material to enter the hood.

3. Wash the embryos in several drops of M2 medium.

4. Transfer the embryos to the central well of the Falcon 3037 organ culture dish filled with M2 medium.

5. Place the paper tissue in the 12-cm petri dish.

6. Add a few drops of PBS or ddH$_2$O to the paper tissue to moisten it well.

7. Place the dish containing embryos on the moist paper tissue in the 12-cm petri dish.

8. Place the 12-cm petri dish in the 15-cm petri dish.

9. Place the three dishes (now within each other) in two layers of plastic autoclave bags. Seal the bags with the pre-autoclaved autoclave tape (the tape remains sticky also after one round of autoclave treatment).

10. Move the package out of the hood, and seal the outside bag with masking tape.

    This package can be safely passed through a sterilization port using such agents as Clydox to sterilize the outside surface. The embryos can be kept in the package for up to 2 hours, provided that the temperature does not drop below room temperature.

COMMENT

Plastic petri dishes can be used for multiple-layer packaging, but glass dishes are heavier and provide steadier transport. Tightly closed cryovials filled with medium to the top can be used as an alternative to multiple-layer packaging of dishes.

REFERENCES

Ali J. and Shelton J.N. 1993. Vitrification of preimplantation stages of mouse embryos. *J. Reprod. Fertil.* **98:** 459–465.

Beilanski A., Nadin-Davis S., Sapp T., and Lutze-Wallace C. 2000. Viral contamination of embryos cryopreserved in liquid nitrogen. *Cryobiology* **40:** 110–116.

Carroll J., Depypere H., and Matthews C.D. 1990. Freeze-thaw-induced changes of the zona pellucida explains decreased rates of fertilization in frozen-thawed mouse oocytes. *J. Reprod. Fertil.* **90:** 547–553.

Carthew P., Wood M.J., and Kirby C. 1983. Elimination of Sendai (parainfluenza type 1) virus infection from mice by embryo transfer. *J. Reprod. Fertil.* **69:** 253–257.

Carthew P., Wood M.J., and Kirby C. 1985. Pathogenicity of mouse hepatitis virus for preimplantation mouse embryos. *J. Reprod. Fertil.* **73:** 207–213.

Dinnyes A., Wallace G.A., and Rall W.F. 1995. Effect of genotype on the efficiency of mouse embryo cryopreservation by vitrification or slow freezing methods. *Mol. Reprod. Dev.* **40:** 429–435.

Fuller S.J. and Whittingham D.G. 1996. Effect of cooling mouse spermatozoa to 4°C on fertilization and embryonic development. *J. Reprod. Fertil.* **108:** 139–145.

Gunasena K.T., Villines P.M., Critser E.S., and Critser J.K. 1997. Live births after autologous transplant of cryopreserved mouse ovaries. *Hum. Reprod.* **12:** 101–106.

Hill A.C. and Stalley G.P. 1991. *Mycoplasma pulmonis* infection with regard to embryo freezing and hysterectomy derivation. *Lab. Anim. Sci.* **41:** 563–566.

Johnson M.H. 1989. The effect on fertilization of exposure of mouse oocytes to dimethyl sulfoxide: An optimal protocol. *J. In Vitro Fert. Embryo Transf.* **6:** 168–175.

Jondet M., Dominique S., and Scholler R. 1984. Effects of freezing and thawing on mammalian oocyte. *Cryobiology* **21:** 192–199.

Kagabu S. and Umezu M. 2000. Transplantation of cryopreserved mouse, Chinese hamster, rabbit, Japanese monkey and rat ovaries into rat recipients. *Exp. Anim.* **49:** 17–21.

Kasai M., Komi J.H., Takakamo A., Tsudera H., Sakurai T., and Machida T. 1990. A simple method for mouse embryo cryopreservation in a low toxicity vitrification solution, without appreciable loss of viability. *J. Reprod. Fertil.* **89:** 91–97.

Kuleshova L.L. and Shaw J.M. 2000. A strategy for rapid cooling of mouse embryos within a double straw to eliminate the risk of contamination during storage in liquid nitrogen. *Human Reproduction* **15:** 2604–2609.

Leibo S.P. 1986. Cryobiology: Preservation of mammalian embryos. *Basic Life Sci.* **37:** 251–272.

Leibo S.P. and Mazur P. 1971. The role of cooling rates in low-temperature preservation. *Cryobiology* **8:** 447–452.

Leibo S.P., Mazur P., and Jackowski S.C. 1974. Factors affecting survival of mouse embryos during freezing and thawing. *Exp. Cell. Res.* **89:** 79–88.

Leibo S.P., McGrath J.J., and Cravalho E.G. 1978. Microscopic observation of intracellular ice formation in unfertilized mouse ova as a function of cooling rate. *Cryobiology* **15:** 257–271.

Liu J., Van den Abbeel E., and Van Steirteghem A. 1993a. Assessment of ultrarapid and slow freezing procedures for 1-cell and 4-cell mouse embryos. *Hum. Reprod.* **8:** 1115–1119.

———. 1993b. The in-vitro and in-vivo developmental potential of frozen and non-frozen biopsied 8-cell mouse embryos. *Hum. Reprod.* **8:** 1481–1486.

Marschall S. and Hrabe de Angelis M. 1999. Cryopreservation of mouse spermatozoa: Double your mouse space. *Trends Genet.* **15:** 128–131.

Marschall S., Huffstadt U., Balling R., and Hrabe de Angelis M. 1999. Reliable recovery of inbred mouse lines using cryopreserved spermatozoa. *Mamm. Genome* **10:** 773–776.

Mazur P. 1990. Equilibrium, quasi-equilibrium, and nonequilibrium freezing of mammalian embryos. *Cell Biophys.* **17:** 53–92.

Mazur P., Rall W.F., and Leibo S.P. 1984. Kinetics of water loss and the likelihood of intracellular freezing in mouse ova. Influence of the method of calculating the temperature dependence of water permeability. *Cell Biophys.* **6:** 197–213.

Miyake T., Kasai M., Zhu S.E., Sakurai T., and Machida T. 1993. Vitrification of mouse oocytes and embryos at various stages in an ethylene glycol based solution by a simple method. *Theriogenology* **40:** 121–134.

Mobraaten L.E. 1999. Cryopreservation in a transgenic program. *Lab. Animal* Jan: 15–18.

Nakagata N. 1989. High survival rate of unfertilized mouse oocytes after vitrification. *J. Reprod. Fertil.* **87:** 479–483.

———. 1992. Production of normal young following insemination of frozen-thawed mouse spermatozoa into fallopian tubes of pseudopregnant females. *Jikken Dobutsu* **41:** 519–522.

———. 1993. Production of normal young following transfer of mouse embryos obtained by in vitro fertilization between cryopreserved gametes. *J. Reprod. Fertil.* **99:** 77–80.

———. 1995. Studies on cryopreservation of embryos and gametes in mice. *Exp. Anim.* **44:** 1–8.

———. 1996. Use of cryopreservation techniques of embryos and spermatozoa for production of transgenic (Tg) mice and for maintenance of Tg mouse lines. *Lab. Anim. Sci.* **46:** 236–238.

———. 2000. Cryopreservation of mouse spermatozoa. *Mamm. Genome* **11:** 572–576.

Nakagata N. and Takeshima T. 1993. Cryopreservation of mouse spermatozoa from inbred and F1 hybrid strains. *Jikken Dobutsu* **42:** 317–320.

Nakagata N., Okamoto M., Ueda O., and Suzuki H. 1997. Positive effect of partial zona-pellucida dissection on the in vitro fertilizing capacity of cryopreserved C57BL/6J transgenic mouse spermatozoa of low motility. *Biol. Reprod.* **57:** 1050–1055.

Nakagata N., Matsumoto K., Anzai M., Takahashi A., Takahashi Y., Matsuzaki Y., and Miyata K. 1992. Cryopreservation of spermatozoa of a transgenic mouse. *Jikken Dobutsu* **41:** 537–540.

Nakao K., Nakagata N., and Katsuki M. 1997. Simple and efficient vitrification procedure for cryopreservation of mouse embryos. *Exp. Anim.* **46:** 231–234.

Okamoto M., Matsushita S., and Matsumoto T. 1990. Cleaning of Sendai virus-infected mice by embryo transfer technique. *Jikken Dobutsu* **39:** 601–603.

Penfold L.M. and Moore H.D. 1993. A new method for cryopreservation of mouse spermatozoa. *J. Reprod. Fertil.* **99:** 131–134.

Pomeroy K.O. 1991. Cryopreservation of transgenic mice. *Genet. Anal. Tech. Appl.* **8:** 95–101.

Rall W.F. 1987. Factors affecting the survival of mouse embryos cryopreserved by vitrification. *Cryobiology* **24:** 387–402.

Rall W.F. and Fahy G.M. 1985. Ice-free cryopreservation of mouse embryos at –196°C by vitrification. *Nature* **313:** 573–575.

Rall W.F. and Wood M.J. 1994. High in vitro and in vivo survival of day 3 mouse embryos vitrified or frozen in a non-toxic solution of glycerol and albumin. *J. Reprod. Fertil.* **101:** 681–688.

Rall W.F., Wood M.J., Kirby C., and Whittingham D.G. 1987. Development of mouse embryos cryopreserved by vitrification. *J. Reprod. Fertil.* **80:** 499–504.

Rall W.F., Schmidt P.M., Lin X., Brown S.S., Ward A.C., and Hansen C.T. 2000. Factors affecting the efficiency of embryo cryopreservation and rederivation of rat and mouse models. *Ilar J.* **41:** 221–227.

Reetz I.C., Wullenweber-Schmidt M., Kraft V., and Hendrich H.J. 1988. Rederivation of inbred strains of mice by means of embryo transfer. *Lab. Anim. Sci.* **38:** 696–701.

Renard J.P. and Babinet C. 1984. High survival of mouse embryos after rapid freezing and thawing inside plastic straws with 1-2 propanediol as cryoprotectant. *J. Exp. Zool.* **230:** 443–448.

Rouleau A.M., Kovacs P.R., Kunz H.W., and Armstrong D.T. 1993. Decontamination of rat embryos and transfer to specific pathogen-free recipients for the production of a breeding colony. *Lab. Anim. Sci.* **43:** 611–615.

Schmidt P.M., Hansen C.T., and Wildt D.K. 1985. Viability of frozen-thawed mouse embryos is affected by genotype. *Biol. Reprod.* **32:** 507–514.

Schmidt P.M., Schiewe M.C., and Wildt D.E. 1987. The genotypic response of mouse embryos to multiple freezing variables. *Biol. Reprod.* **37:** 1121–1128.

Shaw J.M. and Kasai M. 2001. Embryo cryopreservation for transgenic mouse lines. *Methods Mol. Biol.* **158:** 397–419.

Shaw J.M. and Nakagata N. 2002. Cryopreservation of transgenic mouse lines. *Methods Mol. Biol.* **180:** 207–228.

Shaw J.M. and Trounson A.O. 1989. Parthenogenetic activation of unfertilized mouse oocytes by exposure to 1,2-propanediol is influenced by temperature, oocyte age, and cumulus removal. *Gamete Res.* **24:** 269–279.

———. 2002. Ovarian tissue transplantation and cryopreservation. Application to maintenance and recovery of transgenic and inbred mouse lines. *Methods Mol. Biol.* **180:** 229–251.

Shaw J.M., Diotallevi L., and Trounson A.O. 1991a. A simple rapid 4.5 M dimethyl-sulfoxide freezing technique for the cryopreservation of one-cell to blastocyst stage preimplantation mouse embryos. *Reprod. Fertil. Dev.* **3:** 621–626.

Shaw J.M., Oranratnachai A., and Trounson A.O. 2000a. Cryopreservation of oocytes and embryos. In Handbook of in vitro fertilization, 2nd edition (ed. A. Trounson and D. Gardner), pp. 373–412. CRC Press, New York.

———. 2000b. Fundamental cryobiology of mammalian oocytes and ovarian tissue. *Theriogenology* **53:** 59–72.

Shaw J.M., Ward C., and Trounson A.O. 1995. Evaluation of propanediol, ethylene glycol, sucrose and antifreeze proteins on the survival of slow-cooled mouse pronuclear and 4-cell embryos. *Hum. Reprod.* **10:** 396–402.

Shaw J.M., Wood C., and Trounson A.O. 2000c. Transplantation and cryopreservation of ovarian tissue. In *Handbook of in vitro fertilization,* 2nd edition. (ed. A. Trounson and D. Gardner), pp. 413–430. CRC Press, New York.

Shaw J.M., Kola I., MacFarlane D.R., and Trounson A.O. 1991b. An association between chromosomal abnormalities in rapidly frozen 2-cell mouse embryos and the ice-forming properties of the cryoprotective solution. *J. Reprod. Fertil.* **91:** 9–18.

Shaw P.W., Fuller B.J., Bernard A., and Shaw R.W. 1991. Vitrification of mouse oocytes: improved rates of survival, fertilization, and development to blastocysts. *Mol. Reprod. Dev.* **29:** 373–378.

Songsasen N. and Leibo S.P. 1997a. Cryopreservation of mouse spermatozoa. I. Effect of seeding on fertilizing ability of cryopreserved spermatozoa. *Cryobiology* **35:** 240–254.

———. 1997b. Cryopreservation of mouse spermatozoa. II. Relationship between survival after cryopreservation and osmotic tolerance of spermatozoa from three strains of mice. *Cryobiology* **35:** 255–269.

Songsasen N., Betteridge K.J., and Leibo S.P. 1997. Birth of live mice resulting from oocytes fertilized in vitro with cryopreserved spermatozoa. *Biol. Reprod.* **56:** 143–152.

Stringfellow D. and Seidel S.M. 1998. *Manual of the international embryo transfer society,* 3rd Edition. IETS, Savoy, Illinois.

Suzuki H., Yorozu K., Watanabe T., Nakura M., and Adachi J. 1996a. Rederivation of mice by means of in vitro fertilization and embryo transfer. *Exp. Anim.* **45:** 33–38.

Suzuki H., Nakagata N., Anzai M., Tsuchiya K., Nakura M., Yamaguchi S., and Toyoda Y. 1996b. Transport of wild mice genetic material by in vitro fertilization, cryopreservation, and embryo transfer. *Lab. Anim. Sci.* **46:** 687–688.

Sztein J.M., Farley J.S., and Mobraaten L.E. 2000a. In vitro fertilization with cryopreserved inbred mouse sperm. *Biol. Reprod.* **63:** 1774–1780.

Sztein J.M., Farley J.S., Young A.F., and Mobraaten L.E. 1997. Motility of cryopreserved mouse spermatozoa affected by temperature of collection and rate of thawing. *Cryobiology* **35:** 46–52.

Sztein J.M., McGregor T.E., Bedigian H.J., and Mobraaten L.E. 1999. Transgenic mouse strain rescue by frozen ovaries. *Lab. Anim. Sci.* **49:** 99–100.

Sztein J., Sweet H., Farley J., and Mobraaten L. 1998. Cryopreservation and orthotopic transplantation of mouse ovaries: New approach in gamete banking. *Biol. Reprod.* **58:** 1071–1074.

Sztein J.M., O'Brien M.J., Farley J.S., Mobraaten L.E., and Eppigg J.J. 2000b. Rescue of oocytes from antral follicles of cryopreserved mouse ovaries: Competence to undergo maturation, embryogenesis, and development to term. *Hum. Reprod.* **15:** 567–571.

Tada N., Sato M., Amman E.K., and Ogawa S. 1993. A simple and rapid method for cryopreservation of mouse 2-cell embryos by vitrification: Beneficial effect of sucrose and raffinose on their cryosurvival rate. *Theriogenology* **40:** 333–344.

Tada N., Sato M., Yamanoi J., Mizorogi T., Kasai K., and Ogawa S. 1990. Cryopreservation of mouse spermatozoa in the presence of raffinose and glycerol. *J. Reprod. Fertil.* **89:** 511–516.

Takeshima T., Nakagata N, and Ogawa S. 1991. Cryopreservation of mouse spermatozoa. *Jikken Dobutsu* **40:** 493–497.

Tedder R.S., Zuckerman M.A., Goldstone A.H., Hawkins A.E., Fielding A., Briggs E.M., Irwin D., Blair S., Gorman A.M., Patterson K.G. et al. 1995. Hepatitis B transmission from contaminated cryopreservation tank. *Lancet* **346:** 137–140.

Thornton C.E., Brown S.D., and Glenister P.H. 1999. Large numbers of mice established by in vitro fertilization with cryopreserved spermatozoa: Implications and applications for genetic resource banks, mutagenesis screens, and mouse backcrosses. *Mamm. Genome* **10:** 987–992.

Trounson A. and Mohr L. 1983. Human pregnancy following cryopreservation, thawing and transfer of an eight-cell embryo. *Nature* **305:** 707–709.

Trounson A., Peura A., and Kirby C. 1987. Ultrarapid freezing: A new low-cost and effective method of embryo cryopreservation. *Fertil. Steril.* **48:** 843–850.

van den Abbeel E., van der Elst J., van der Linden M., and van Steirteghem A.C. 1997. High survival rate of one-cell mouse embryos cooled rapidly to –196°C after exposure to a propylene glycol-dimethylsulfoxide-sucrose solution. *Cryobiology* **34:** 1–12.

Wakayama T. and Yanagimachi R. 1998. Development of normal mice from oocytes injected with freeze-dried spermatozoa. *Nat. Biotechnol.* **16:** 639–641.

Wakayama T., Whittingham D.G., and Yanagimachi R. 1998. Production of normal offspring from mouse oocytes injected with spermatozoa cryopreserved with or without cryoprotection. *J. Reprod. Fertil.* **112:** 11–17.

Whittingham, D. G. 1977. Fertilization in vitro and development to term of unfertilized mouse oocytes previously stored at –196°C. *J. Reprod. Fertil.* **49:** 89–94.

———. 1979. In-vitro fertilization, embryo transfer and storage. *Br. Med. Bull.* **35:** 105–111.

Whittingham D.G., Leibo S.P., and Mazur P. 1972. Survival of mouse embryos frozen to –196° and –269°C. *Science* **178:** 411–414.

Wilmut I. 1972a. The effect of cooling rate, warming rate, cryoprotective agent and stage of development on survival of mouse embryos during freezing and thawing. *Life Sci. II* **11:** 1071–1079.

———. 1972b. The low temperature preservation of mammalian embryos. *J. Reprod. Fertil.* **31:** 513–514.

———. 1985. Cryopreservation of mammalian eggs and embryos. *Dev. Biol.* **4:** 217–247.

Zhu S.E., Kasai M., Otoge H., Sakurai T., and Machida T. 1993. Cryopreservation of expanded mouse blastocysts by vitrification in ethylene glycol-based solutions. *J. Reprod. Fertil.* **98:** 139–145.

Techniques for Visualizing Gene Products, Cells, Tissues, and Organ Systems

MUCH VALUABLE INFORMATION ABOUT the specificity of expression of endogenous and foreign genes can be obtained by analyzing RNA and protein extracted from tissues at different stages of embryonic development. However, the amount of material available from mouse embryos is usually very small, and the gene may be active in only a minority of cells. To overcome these problems, it is necessary to use in situ hybridization and immunohistochemistry of whole embryos and embryo sections to localize gene transcripts and proteins, respectively, in specific cells. This chapter provides a simple guide to this technically demanding and specialized field. Protocols are also provided for a number of other useful

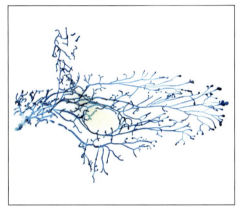

Courtesy of Heiko Schweizer

techniques for visualizing the expression of specific gene products in differentiated tissues, including lacZ (β-galactosidase), alkaline phosphatase, and fluorescent proteins. Additional methods are provided to visualize tissue and organ systems, including the developing skeleton and vasculature.

CONTENTS

VISUALIZING GENE PRODUCTS

The developing mouse embryo has an intrinsic beauty that can be visualized at many levels using a variety of techniques. Whole-mount embryos and embryo sections can be processed to visualize RNA transcripts and protein in specific cells and tissues. In addition, simple histochemical stains can be used to reveal the expression of transgene reporters such as β-galactosidase and alkaline phosphatase (AP). Recently, methods using fluorescent protein reporters have been developed for use in mice to visualize gene expression in living tissues. With practice, the methods described in this chapter will consistently yield high-quality data to allow you to understand the mechanisms of development that result in the formation of a mouse.

Isolating Total RNA from Mouse Embryos or Fetal Tissues

Protocol 16.1 is suitable for almost any size tissue. It is particularly useful for fetal organs or whole embryos, as it can be performed in a volume as small as 0.5 ml. The minimum amount of tissue used should be ten 7.5-dpc embryos, one or two 8.5-dpc embryos, or approximately one-tenth of a 12.5-dpc embryo. If smaller amounts of tissue are homogenized in a 0.5-ml volume, the recovery of RNA may be less efficient than when larger amounts are used.

Analyzing Glucose Phosphate Isomerase Isozymes in Chimeric Tissues by Electrophoresis

Mouse strains carry different alleles at the ubiquitously expressed *Gpi1* (glucose phosphate isomerase) locus (*Gpi1*[a], *Gpi1*[b], *Gpi1*[c]). This is the basis for a widely used method for determining the genotypic composition of different tissues in mouse chimeras, as described in Protocol 16.2. To determine chimerism by this method, it is necessary to separate the differently charged isozymes from tissue homogenates electrophoretically and to visualize them using a color reaction. For the strain distribution of the three variants, see Lyon et al. (1996). Because GPI is a dimer, tissues that normally form by cell fusion (e.g., skeletal muscle) have a heterodimeric form of GPI in a chimera.

General Techniques for Immunohistochemistry and In Situ Hybridization of Mouse Embryo Sections

Described in Protocols 16.3–16.7 are some of the basic equipment, supplies, and protocols needed to establish immunohistochemical and in situ hybridization techniques in a laboratory. Sources for histology equipment, stains, waxes, and chemicals include Raymond Lamb, BDH, Surgipath Medical Industries, Polysciences, and Roboz Surgical Instrument.

Tissue Fixation

For RNA in situ hybridization, the usual fixative is freshly prepared 4% paraformaldehyde in RNase-free phosphate-buffered saline (PBS). The choice of fixative for immunohistochemistry depends on the nature of the antigen and must be determined empirically at the start of the project. The most commonly used fixatives in addition to 4% paraformaldehyde are Bouin's fixative and methanol/dimethyl sulfoxide (DMSO) (4:1) (see Protocol 16.3 for preparation of these fixatives and Protocol 16.4 for handling blastocysts for fixation).

Dehydrating in Alcohol

The general principle is to transfer the fixed tissue through increasing concentrations of alcohol, usually 70%, 80% (*optional*), 90%, 95%, 100% (absolute) two times, or 1:1 ethanol/xylene one time, and then into two changes of xylene. It is important to use fresh xylene for the last step, not the xylene used for dewaxing sections (see below). AmeriClear histology clearing solvent (Scientific Products C4200-1), which is nontoxic, can be used instead of xylene. One hour in each solution is sufficient. Specimens can be stored in 100% ethanol at –20°C. For in situ hybridization, solutions are kept at 4°C, but for immunohistochemistry, room temperature is satisfactory. For small embryos and tissues, the transfers must be done by hand (see Protocol 16.5). However, automated rotary tissue processors are available that are suitable for routine dehydration of larger specimens.

Embedding in Wax

Several types of paraffin waxes are available for embedding samples. The type chosen depends on melting temperature and whether it has been pastillated (molded into small pellets). Good results have been obtained with Paraplast tissue-embedding medium, melting point 56°C (Fisher 12-646-106), and BDH pastillated Fibro-wax, melting temperature 56–58°C (Formulation Raymond Lamb 36142). The wax is melted, filtered through coarse Whatman filter paper, and stored indefinitely in an oven or embedding center (see Protocol 16.5 for details). Do not reuse the wax.

Cutting Sections

The specimen blocks are fixed to the microtome blocks or holding cassettes with molten wax. Care must be taken in positioning the specimen to give the best results. It is important to trim away the excess wax around the specimen with a sharp razor blade so that the maximum number of sections fit on a slide and so that the sections hold together in a ribbon as they come off the microtome (see Protocol 16.6).

Preparing Glass Slides and Coverslips

Precleaned glass slides with a frosted end on one side (e.g., Fisher 12-544-3) are of high enough quality for use in both in situ and immunohistochemical techniques. However, for in situ hybridization, the slides need to be treated with diethyl pyrocarbonate (DEPC), so that any RNase attached to them is destroyed, and with 3-triethoxysilylpropylamine (TESPA, Sigma A3648) or poly-L-lysine (Sigma P1274), so that the sections adhere tightly and do not detach during subsequent extensive washing procedures. There are advantages and disadvantages to each method. In brief, TESPA-treated slides can be stored for a long time, but the sections do not adhere tightly until after drying. Poly-L-lysine-coated slides need to be made fresh, but the sections adhere immediately on contact with the surface. Precoated slides can also be purchased commercially. What is clear, however, is that *some* pretreatment is absolutely necessary. Protocol 16.7 describes these treatments.

Dewaxing and Rehydrating Sections Prior to In Situ Hybridization or Staining

The procedure in Protocol 16.8 describes how to remove wax from embryo or tissue sections that have been affixed to glass slides. Traditionally, xylene has been used for this purpose, but less toxic solutions can also be employed (e.g., AmeriClear, Scientific Products). The embryo or tissue sections are then progressively rehydrated for compatibility with subsequent alcohol stains, aqueous stains, immunohistochemistry, or in situ hybridization.

Immunohistochemistry of Embryo Sections

The technique in Protocol 16.9 is used for localizing antigen in cells and tissues using embryo sections attached to glass slides. General procedures for preparing sections of wax-embedded embryos are described in Protocol 16.6. For the preparation, purification, storage, and use of primary and secondary antibodies; the appropriate controls for the specificity of the staining reaction; and general troubleshooting, see Harlow and Lane (1988) and below. The optimal fixation procedure and the stability of the antigen to dehydration and wax embedding must be determined empirically. The method outlined here uses alkaline-phosphatase-coupled secondary antibody. Horseradish-peroxidase-coupled secondary antibody can be used as an alternative procedure as described below for the whole-mount procedure. The advantages and disadvantages of different staining techniques are given below and in Harlow and Lane (1988).

Immunohistochemistry of Whole-mount Embryos

The technique presented in Protocol 16.10 is ideal for obtaining an overall, three-dimensional picture of the distribution of an antigen in embryos from the pre-streak stage to ~10.5 dpc (Fig. 16.1). It relies on fixing and permeabilizing the embryos so that antibodies can penetrate all of the tissues. However, antibodies will not penetrate the embryo completely if it is too large, and so it is better to dissect specific organs (e.g., brain, lungs, and gut) and process them separately.

It is also extremely important to wash the embryos thoroughly to remove all unbound antibodies. There will always be a problem with antibody trapping in cavities such as the ventricles of the brain. If overall background staining in the tissues is high, even with thorough washing, try reducing the concentration of the primary antibody. For a general discussion of immunohistochemistry protocols, purification, storage, and use of antibodies, appropriate controls for the specificity of the staining reaction, and general troubleshooting, see Harlow and Lane (1988).

In Situ Hybridization of Embryo and Tissue Sections with RNA Probes

This procedure (Protocol 16.11) is used for in situ hybridization of embryo and tissue sections with ^{35}S-labeled single-stranded antisense RNA probes (riboprobes) (Fig. 16.2). It should be used in conjunction with the section on General Techniques for Immunohistochemistry and In Situ Hybridization (see p. 632).

Protocols have also been developed for in situ hybridization to tissue sections using nonradiolabeled RNA probes that can be detected with antibodies coupled to alkaline phosphatase and a chromogenic substrate. This nonradioactive method has the advantage that the results can be obtained relatively quickly, but the sensitivity is probably lower than with radioactive probes. In addition, care must be taken to optimize the amount of each probe used in the hybridization reaction (for details, see Strahle et al. 1994).

General Procedures for Avoiding Contamination with RNase

- Whenever possible, use disposable plastic containers and pipettes. Glass staining jars, slide racks, and any other containers should be filled with distilled water in which 0.1% DEPC (Sigma D5758) has been freshly dissolved. After a minimum of 30 minutes, pour off the water, dry in a clean area, wrap in aluminum foil, and bake at 180°C for 2 hours. Other glassware (e.g., funnels) should be wrapped in foil and baked in the same manner.

- Remove RNase from distilled water and autoclavable solutions by adding DEPC to a final concentration of 0.1%. Store DEPC at 4°C, but allow the container to warm to room temperature before opening it (be careful!). Pipette the required amount of DEPC and mix well by stirring so that it disperses in the aqueous solution. After a minimum of 30 minutes, destroy the DEPC by autoclaving the solution. Do not use DEPC for treating Tris-containing solutions because it will react with the Tris.

- Solutions that cannot be autoclaved (e.g., those that contain protein) should be made in DEPC-treated water that has been autoclaved and cooled. Use a baked glass or disposable plastic container. If possible, sterilize the solution by filtering through a 0.22-μm membrane filter.

Fixing, Wax Embedding, and Sectioning

For in situ hybridization, embryos should be fixed in *freshly prepared* 4% paraformaldehyde in DEPC-treated phosphate-buffered saline (PBS) at 4°C. (For preparation of the fixative, see the General Techniques section, Protocol 16.3.) Best histological results are obtained when the specimens are dehydrated, embedded in wax, sectioned at 5–7 μm, placed on poly-L-lysine- or TESPA-coated slides, dewaxed, and rehydrated with DEPC-treated water and PBS. These procedures are all described in the General Techniques section (see Protocols 16.3–16.7). Note that for in situ hybridization, great care should be taken to keep the wax blocks, microtome knife, and ribbons free of RNase. DEPC-treated water should be used in the flotation bath or spotted onto the slides.

In Situ Hybridization of Whole-mount Embryos with RNA Probes

The method presented in Protocol 16.12 is based on hybridization of fixed and detergent-permeabilized embryos with single-stranded RNA probes (riboprobes) synthesized in vitro using UTP labeled with the plant steroid digoxigenin (DIG-11-UTP). The DIG-labeled RNA is then localized using a conjugate of anti-DIG Fab antibody and calf intestinal alkaline phosphatase. Enzyme activity of the reporter is detected by a color reaction, resulting in the formation of a water-insoluble purple/blue precipitate (Fig. 16.3). Alternatively, biotin-labeled RNA probes are localized using a streptavidin–β-galactosidase complex and X-gal (5-bromo-4-chloro-3-indolyl-β-D-galactopyranoside) as the substrate (Herrmann 1991).

Protocols have also been developed for simultaneously detecting two different transcripts using double-labeled whole-mount in situ hybridization (Hauptmann and Gerster 1994; Jowett and Lettice 1994). One probe is prepared using fluorescein-labeled UTP, and the other is prepared using digoxigenin-labeled UTP. Probes are detected with anti-fluorescein and anti-digoxigenin antibodies, respectively, coupled to alkaline phosphatase but using different chromogenic substrates.

General Considerations

- All solutions and containers *must be free* of contaminating RNase activity (see Protocol 16.11). Cleanliness is also very important throughout the procedure (e.g., wash powder off gloves) because dirt and debris will stick to the embryos. Embryos are particularly sticky after Proteinase K treatment, so filter all solutions and rinse vials carefully.

- Several types of containers can be used to perform the reactions: (1) screw-capped 2-ml plastic tubes with conical bottoms (embryos can be collected in a small volume by sucking off the medium); (2) 10-ml glass Reacti-Vials (Pierce); and (3) 5-ml glass scintillation vials (Fisher 3333B). In some laboratories, glass containers are acid-washed, siliconized, and DEPC-treated before use; however, acid washing and siliconization may not be necessary. *Note:* Blastocysts can be handled by serial transfer to shallow glass dishes containing the appropriate solutions.

- It is important to perform the following controls: (1) a well-characterized probe with a pattern distinctly different from that of the probe being studied; (2) an antibody conjugate without probe; and (3) neither probe nor antibody added (control for endogenous alkaline phosphatase activity). The specificity of a signal can also be demonstrated using probes generated from two different regions of the same cDNA and competition of the specific signal with an excess of an identical but unlabeled transcript.

Imaging Embryos after Whole-mount In Situ Hybridization

Embryos that are being processed for whole-mount in situ hybridization can be photodocumented after the staining procedure depending on the region of the embryo of interest. The embryos can be photodocumented immediately after the color reaction or after clearing in glycerol solutions. A procedure for this is given in Protocol 16.13.

Sectioning Embryos after Whole-mount In Situ Hybridization

Embryos that have been processed for whole-mount in situ hybridization can be sectioned frozen or after embedding in wax to visualize which tissues express the gene transcripts. Procedures are provided in Protocol 16.14.

Staining for β-Galactosidase Activity

A very convenient reporter gene for use in transgenic mice is the bacterial β-galactosidase (*lacZ*) gene (Fig. 16.4). The *lacZ* gene can be introduced into embryos (1) via DNA constructs microinjected into the zygote, (2) via gene targeting into specific loci "knock-in," or (3) via retroviral infection or DNA electroporation into ES cells or postimplantation embryos. The technique described in Protocol 16.15 is used to visualize bacterial β-galactosidase activity in whole-mount embryos or in frozen sections on slides (Fig. 16.5). Intact embryos can also

FIGURE 16.1. Whole-mount immunohistochemical localization of neurofilament protein in a 10.5-dpc embryo. The brown-colored 2H3 monoclonal antibody (Developmental Studies Hybridoma Bank, Iowa) staining reveals the cranial ganglia and spinal nerves. The embryo has been cleared in BABB. (Image courtesy of Deborah L. Guris and Akira Imamoto, Ben May Institute, University of Chicago.)

FIGURE 16.2. In situ hybridization using a radiolabeled *Sonic hedgehog* (*Shh*) antisense riboprobe of a transverse section of a 9.5-dpc embryo. The section shows that *Shh* transcripts (*red*) are detected in the floor plate of the neural tube and the notochord and posterior forelimb bud. (Image courtesy of Andrew McMahon, Harvard University.)

FIGURE 16.3. Whole-mount in situ hybridization using a digoxigenin-labeled *Brachyury* antisense riboprobe of a 7.5-dpc embryo. This lateral view shows that *Brachyury* transcripts (*purple*) are detected in the primitive streak, node, and head process. (Image courtesy of Elizabeth Robertson, Harvard University.)

FIGURE 16.4. Whole-mount β-galactosidase staining of an 11.0-dpc embryo heterozygous for a targeted *Lmx1b* allele in which an *IRES-lacZ* cassette has been introduced into the 3′ untranslated region. The pattern of β-galactosidase activity in the limbs, spinal cord, mid-hindbrain junction, and periocular mesenchyme faithfully mimics the endogenous *Lmx1b* expression pattern. The embryo has been cleared in BABB. (Image courtesy of Randy Johnson, University of Texas M.D. Anderson Cancer Center.)

FIGURE 16.5. β-galactosidase staining of a fixed frozen transverse section from the head of a 15.5-dpc fetus heterozygous for a targeted allele in which *lacZ* has been knocked-in into the *Goosecoid* locus (Rivera-Pérez et al. 1999). Staining fixed frozen embryo sections circumvents reagent penetration problems that occur when using whole-mount embryos later than 13 dpc. (Image courtesy of Jaime Rivera, University of North Carolina.)

FIGURE 16.6. Whole-mount human placental alkaline phosphatase (AP) staining of a 10.5-dpc embryo carrying a *Hoxa1-AP* transgene. The *Hoxa1* regulatory elements direct AP activity to the gut epithelium, floor plate of the neural tube, and notochord. (Image courtesy of Thomas Lufkin, Mount Sinai School of Medicine.)

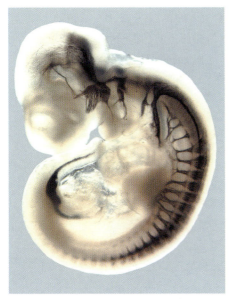

FIGURE 16.1.

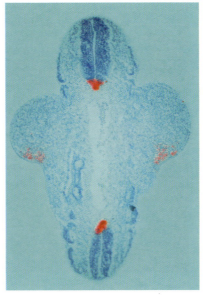

FIGURE 16.2.

FIGURE 16.3.

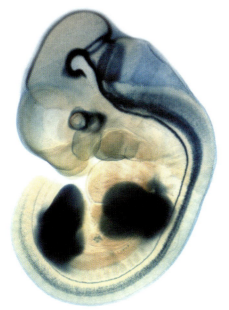

FIGURE 16.4.

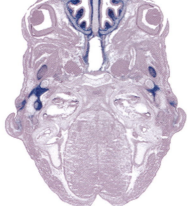

FIGURE 16.5.

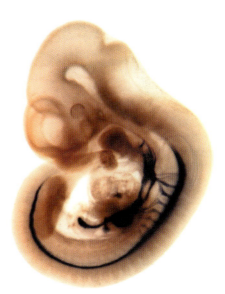

FIGURE 16.6.

(See facing page for legends.)

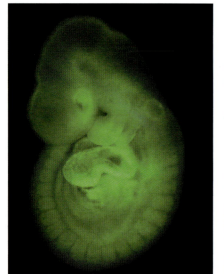

FIGURE 16.7.

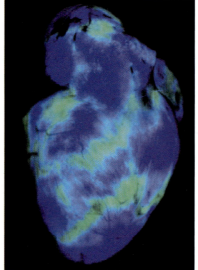

FIGURE 16.8.

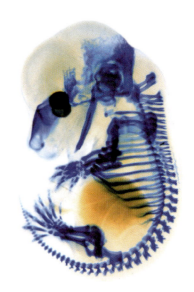

FIGURE 16.9.

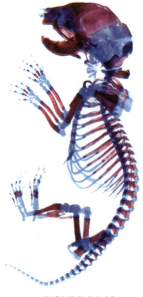

FIGURE 16.10.

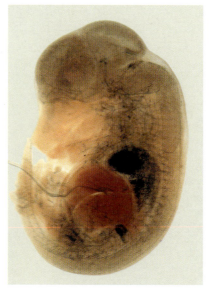

FIGURE 16.11.

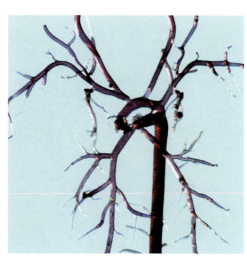

FIGURE 16.12.

(See facing page for legends.)

be sectioned after staining. In addition, cells from *lacZ* transgenic mice can be cultured in vitro and then fixed and stained.

Staining for *lacZ* activity in whole-mount embryos is relatively straightforward up to ~13 dpc. After this stage, penetration of the substrate may become limiting, and endogenous β-galactosidase activity increases, especially in tissues such as bone, kidney, and brain. Background staining as a result of endogenous enzyme activity can be kept to a minimum by carrying out the reaction at pH 7.3, the pH optimum for the bacterial enzyme, and lowering the temperature during staining to 10°C from room temperature. If necessary, add Tris buffer to stabilize the phosphate buffer at pH 7.3.

Staining for Alkaline Phosphatase Activity

The technique described in Protocol 16.16 is used to visualize human placental alkaline phosphatase (AP) activity in transgenic mice (Fig. 16.6). AP activity staining is localized to the cell surface. Endogenous AP activity is eliminated by heat inactivation that spares the heat-stable human placental AP activity.

VISUALIZING CELLS

Visualizing Fluorescent Proteins

Recently, green fluorescent protein (GFP), from the jellyfish *Aequorea victoria*, has joined the set of reporters available for gene expression detection and determination of cellular identities in the mouse. This reporter is unique compared to other reporters because its detection does not require fixation, and it is possible to visualize the protein by shining a light of specific wavelength coupled with an excitation/emission separator filter onto the live specimen (Figs. 16.7–16.8).

FIGURE 16.7. Visualization of enhanced green fluorescent protein (EGFP) expression in a 9.5-dpc whole-mount embryo from the B5/EGFP strain (Hadjantonakis et al. 1998). (Image courtesy of Richard Behringer, University of Texas M.D. Anderson Cancer Center.)

FIGURE 16.8. Visualization of enhanced cyan fluorescent protein (ECFP) and enhanced yellow fluorescent protein (EYFP) expression in the heart of an adult chimera. (Image courtesy of Kat Hadjantonakis and Andras Nagy, Mount Sinai Hospital Research Institute.)

FIGURE 16.9. Alcian blue staining of the cartilaginous skeleton at 14.5 dpc. The blue-stained cartilage is visualized by clearing the fetus in BABB. (Image courtesy of Richard Behringer, University of Texas M.D. Anderson Cancer Center.)

FIGURE 16.10. Alizarin red/alcian blue staining of a newborn skeleton. Ossified tissue (predominantly bone) stains red and cartilage stains blue. The soft tissues are cleared in an alkaline solution and glycerol. (Image courtesy of Kazuhisa Nakashima, University of Texas M.D. Anderson Cancer Center.)

FIGURE 16.11. Visualization of the fetal vasculature at 14.5 dpc by India ink injection. The fine carbon particles of India ink are injected into a blood vessel to mark the developing vasculature. The marked blood vessels are then visualized by clearing the fetus in BABB. (Image courtesy of Andras Nagy, Mount Sinai Hospital Research Institute.)

FIGURE 16.12. Visualization of the fetal great blood vessels by plastic casting. Colored plastic solutions can be injected into the heart to fill the great vessels. Once the plastic hardens, the tissues are removed using a caustic to create a plastic cast of the great vessels. (Image courtesy of Chengyu Liu and James Martin, Texas A&M University Institute of Biosciences and Technology.)

GFP has been used for protein localization in the fields of biochemistry, molecular and cellular biology, high-throughput screening, and gene discovery for several years. It has also opened up new dimensions in studying developmental and disease processes in the mouse. For example, it allows the detection of specific gene expression or chimerism in living embryos and in ex vivo organ cultures in real time. The exploration of the entire potential of GFP is ongoing and is expected to have a significant impact on our understanding of normal and disease processes, which were previously inaccessible. The technical aspects of the use of GFP are still under intensive investigation and development.

Mutagenesis studies of the wild-type GFP have resulted in variants with increased thermostability, such as enhanced GFP (EGFP) (Cormack et al. 1996), which has at least one order of magnitude higher quantum yield and an elevated resistance to photobleaching than wild-type GFP. In addition, novel spectral variants have been generated, such as enhanced cyan (ECFP) and yellow (EYFP) fluorescent proteins (Yang et al. 1996; Tsien 1998). Recent cloning of the red fluorescent protein-encoding gene from an anemone, *Discosoma striata* (dsRED), has provided a further red-shifted fluorescent protein. This protein, however, has slow maturation and aggregation-associated cellular toxicity in certain cell types (Baird et al. 2000) including ES cells and early embryonic cell types (K. Vintersten and A. Nagy, unpubl.).

In general, caution should be taken in planning the use of fluorescent proteins. There are a few areas of applications where expression of certain variants might interfere with the biological functions in question.

The basic principle of visualizing fluorescent proteins (FP) is the same as for the fluorophores used in fluorescent microscopy. The fluorescence elicited by an excitation light source of a specific wavelength and the resulting emitted light is observed through a specific filter set, which cuts off the excitation wavelength. Depending on the FP, the excitation and emission wavelengths are different, therefore necessitating different optimal filter sets for visualization. Microscope manufacturers provide basic sets with their products. A broader solution for special needs can be found at Chroma Technology, http://www.chroma.com/.

Protocols 16.17–16.20 describe typical visualization methods of fluorescent proteins. They cover macroscopic, dissecting, and compound microscopy methods.

Macroscopic Observation of GFP Expression in Transgenic Mice

Many transgenes that use GFP as a reporter express the protein in specific areas, thus allowing GFP detection at certain periods of postnatal life. This situation creates a very convenient way of identifying transgenic animals. The newborn stage is the most favored stage because the major organs are visible through the thin and translucent integument. BLS Ltd. (http://www.bls-ltd.com) manufactures several light source designs ranging from the "miner's lamp" style to desk lamps. The designs are excellent for on-site (i.e., animal facility) "genotyping" (GFP expression detection) in postnatal animals.

GFP Microscopy: Dissecting Microscopes

Depending on the procedure using GFP-expressing transgenes, different light sources are recommended. For routine dissection and screening at mid to low microscopic magnification, the light-emitting diode (LED)-based light source is ideal and economical. It allows free and immediate turning on and off, and has a practically infinite lifetime (see BLS, model MAA-03, universal GFP visualizing light source). Another advantage of this device is that the light beam does not open as the magnification is lowered, in contrast to other light sources. As a result, the excitation energy remains the same, but the smaller magnification causes the GFP-positive object to appear more intense. The disadvantage, however, is that the beam is restricted to a partial area of the field of view.

The more traditional mercury, xenon, and halogen light sources achieve more uniform excitation light coverage of the field of view. At higher magnifications, they give a more intensive excitation than a LED-based light source. The price for this benefit, however, is that these light sources are more expensive, more care is needed for their operation and maintenance, and the lifetime of the light bulbs is only a few hundred hours.

GFP Microscopy: Compound, Confocal, and Deconvolution Microscopes

For general fluorescence microscopy purposes, investigators can use normal fluorescein isothiocyanate (FITC) filter sets for viewing GFP and YFP. These filter sets, however, are not optimal. Chroma Technology specializes in making single- and dual-label filter sets for the various forms of GFP, and the company provides very useful information on their Web site to help make the correct choice for special needs (http://www.chroma.com/).

Observation of GFP-fusion Proteins in Living Cells

Frequently, there is a need for GFP detection in mouse cell cultures, including ES cells or their differentiated derivatives, primary and transformed cells. Protocol 16.18 describes how to observe GFP activity in these types of cultures.

VISUALIZING TISSUES AND ORGAN SYSTEMS

Visualizing the Mouse Skeleton Using Histochemical Stains

Three techniques (Protocols 16.21–16.23) are provided to visualize the developing and postnatal mouse skeleton. The vertebrate skeleton forms by endochondral and intramembranous bone formation (for review, see de Crombrugghe et al. 2002). During endochondral bone formation, mesenchyme condensations give rise to cartilages that are eventually replaced by bone. However, there are some permanent cartilages that do not ossify, such as the cartilage of the trachea and articular cartilage of the joints. Intramembranous bone formation occurs directly without a cartilage template. Because the ossified skeleton is a robust tissue, even the skeletons of dead animals can be processed using these methods.

The first method exquisitely reveals the cartilaginous skeleton of embryos between 12.5 and 16.5 dpc (Fig. 16.9) (Jegalian and De Robertis 1992) The cartilaginous skeleton is stained dark blue by alcian blue while the other tissues of the embryo are made transparent by a clearing solution, benzyl alcohol benzyl benzoate (BABB).

At later stages, when the fetuses become bigger, it is better to examine the skeleton using other staining methods that use alkaline treatment. Thus, the second method is used for later fetal stages to newborn and early postnatal stages to simultaneously visualize cartilage, again using alcian blue, and mineralized tissues (notably bone), using alizarin red (Fig. 16.10). Neonatal skeletons are especially well suited for this technique.

The third method is an abbreviated outline of a technique for staining the bone of newborn and adult mice. Times in the various solutions can be modified according to the age of the mouse and other factors (for details of these and other possible modifications, see Green 1952 and Selby 1987).

Photodocumentation of skeleton preparations is best with bright-field optics and transillumination.

Visualizing the Fetal Vasculature by India Ink Injection

The technique described in Protocol 16.24 was first applied in chicken embryos to visualize the circulatory system. The following method is an adaptation of this "classic" method to the mouse. The basic procedure is to dissect 12.5-dpc or older embryos with an intact yolk sac and placenta while the circulation is still active. The next step is to inject India ink slowly into a branch of the vitelline vein of the yolk sac. The carbon particles in the ink will be distributed into the arterial system by the beating heart. After fixation and clearing, the three-dimensional structure of the fetal arterial system becomes visible (Fig. 16.11).

Visualizing the Fetal Great Blood Vessels by Plastic Casting

The great vessels of the mouse can be visualized by injecting colored plastic solutions directly into the heart, thereby filling the vessels (see Protocol 16.25). The plastic solidifies to create a stable cast that is then revealed by removing the mouse tissues with a caustic (Fig. 16.12). It is also possible to cast the vasculature of other organs by modifying the delivery of the liquid plastic. For example, plastic casts of the maternal and fetal vasculature of the placenta can be generated (L. Adamson, pers. comm.).

Isolating Total RNA from Mouse Embryos or Fetal Tissues

MATERIALS

EMBRYOS
Tissue samples

EQUIPMENT
Centrifuge, high-speed refrigerated
Microfuge
Microfuge tubes, 2-ml screw-capped (e.g., Fisher 05-664-34/53)
Pestles, disposable plastic (Kontes Scientific Glassware/Instruments 749520-0000)
Pipettes
Polytron homogenizer with small probe (Brinkmann PTA7) or sonicator with small probe

REAGENTS
Diethyl pyrocarbonate (DEPC)<!>-treated autoclaved water
Ethanol<!>, 70% and 100%
3 M Lithium chloride (LiCl)<!>, 6 M urea <!>

> Use RNase-free urea (e.g., Fisher Scientific AC 327380010) and a stock of solid LiCl handled only under RNase-free conditions. Make in diethyl pyrocarbonate (DEPC)-treated water and filter through a 0.2-μM filter. Use at 0°C. Store at 4°C for up to 1 month.

3 M NaOAc (pH 6). Treat with DEPC.
Phenol<!>/chloroform<!> solution (see Appendix 1)
TE buffer (see Appendix 1)
TES buffer
   10 mM Tris<!> (pH 7.6)
   1 mM EDTA
   0.5% SDS<!>

> Make from RNase-free stocks of Tris-HCl, EDTA, SDS, and DEPC-treated water.

CAUTION: See Appendix 2 for appropriate handling of materials marked with <!>.

PROCEDURE

Autoclave all equipment (i.e., microfuge tubes, pipette tips, etc.) before use.

1. For large embryos or tissues, place tissue and buffer in a 2-ml screw-cap microfuge tube (use a larger tube for more than 0.5 ml of buffer). Use a minimum of 0.5 ml (for at least 5–10 ml/g of tissue) of cold (0°C) 3 M LiCl and 6 M urea. Homogenize for 1–2 minutes using the small probe of the Polytron homogenizer; set the Polytron on the highest speed possible without causing the homogenate to overflow. It is important to shear the DNA so that it does

not coprecipitate with the RNA. Small embryos (7.5 or 8.5 dpc) can be homogenized manually by using a disposable pestle in a 1.5-ml microfuge tube or by vortexing vigorously. Sonicate the homogenate to shear the DNA.

2. Store the homogenate at 0–4°C overnight to allow RNA to precipitate.

   The homogenate may be stored in this form for several days but not indefinitely.

3. Centrifuge the homogenate in a microfuge at 4°C for 15 minutes (or longer for very small samples). For larger volumes, centrifuge at ~10,000 rpm for 10 minutes in a high-speed refrigerated centrifuge.

4. Remove the supernatant by pipetting and discard. Resuspend the pellet by vortexing in 0.5 ml of LiCl/urea (or one-half original buffer volume for larger sample volumes). Recentrifuge the sample for 10 minutes. Discard the supernatant.

5. To pellet, add 0.25 ml of TES buffer and 0.25 ml of phenol/chloroform. Shake and/or vortex until pellet is completely dissolved.

6. Microfuge at room temperature for 5–10 minutes. Transfer the aqueous phase to a new microfuge tube and re-extract with phenol/chloroform. Transfer the aqueous phase to a new tube.

7. To the aqueous phase from step 6, add one-tenth volume (i.e., 25 μl) of 3 M NaOAc (pH 6) pretreated with DEPC and 2 volumes (i.e., 550 μl) of 100% ethanol. Store at –20°C overnight.

8. Microfuge for 5–15 minutes (depending on amount of RNA precipitate). Wash the resulting pellet with 70% ethanol and recentrifuge. Remove the supernatant and dry the pellet in a vacuum. Resuspend the pellet in DEPC-treated water or TE (~25–100 μl) and store at –70°C. RNA is now ready for oligo(dT) selection of poly(A)$^+$ RNA, RNase protection, or other uses.

9. If it is apparent that some DNA has come through in the final ethanol precipitation step, remove it by bringing the RNA in TE to 2 M LiCl and incubating at 4°C overnight. The RNA is then recovered by centrifugation, redissolved in TE, and precipitated with ethanol.

COMMENTS

- The Micro-FastTrack mRNA isolation kit (Invitrogen) has been used successfully to isolate poly(A)$^+$ RNA from single 6.5-dpc or 7.5-dpc embryos or from dissected embryonic tissues too small for the LiCl/urea method of RNA isolation.
- See Table 16.1 for yields of total RNA.

TABLE 16.1. Expected recovery of total RNA from mouse embryos and extraembryonic tissues and fetal and adult tissues

A. Mouse embryo—Stage and tissue	Yield of total RNA
Day 0.5—oocyte	0.4 ng
Day 2.5 (8–16-cell morula)	0.7 ng
Day 3.5 (64-cell blastocyst)	2 ng
Day 6.5—embryo plus ectoplacental cone	0.2 μg
Day 7.5—embryonic region	0.3 μg
Day 7.5—extraembryonic region	0.2 μg
Day 8.5—embryo minus yolk sac	1 μg
Day 8.5—visceral yolk sac	3 μg
Day 9.5—placenta	30 μg
Day 10.5—embryo minus yolk sac	40 μg
Day 11.5—blood	5 μg
Day 12.5—fetus minus yolk sac	260 μg
Day 13.5—fetus	375 μg
Day 13.5—visceral yolk sac	70 μg
Day 13.5—placenta	250 μg
Day 16.5—fetus	2550 μg
Day 16.5—liver	160 μg
Day 17.5—liver	225 μg
Day 17.5—brain	100 μg
Day 17.5—kidney	15 μg
B. Adult tissue (9 weeks of age)	**Yield of total RNA**
Brain	350 μg
Heart	100 μg
Kidney	425 μg
Spleen	300 μg
Salivary gland	350 μg
Preputial gland (male)	25 μg
Lung	90 μg
Thymus	300 μg
Liver	4500 μg
Ovary	25 μg
Testis	200 μg
Bone marrow (1 tibia and 1 femur)	25 μg
Leg muscle (per hind leg)	200 μg

Information provided by James Lee, Mayo Clinic Scottsdale, S.C. Johnson Medical Research Center, Scottsdale, Arizona 85259.

Analyzing Glucose Phosphate Isomerase Isozymes in Chimeric Tissues by Electrophoresis

MATERIALS

EQUIPMENT
Gel box (custom-made [Helena Laboratories] or standard)
Glass plate
Lids of staining troughs (Raymond Lamb E/106)
Microscope slide marked with 10 equal divisions
Microscope slides
Permanent marker
Pipetman or hand-drawn Pasteur pipettes
Three Titan III cellulose acetate plates (60 x 76 mm; Helena Laboratories 3023)
Tissue culture dish, 96-well
Tissues
Whatman 3MM filter paper

REAGENTS
5% Acetic acid<!> in distilled water
Gel buffer (Tris-glycine [pH 8.1])

Tris<!>	3 g
glycine<!>	14.4 g
distilled water	1 liter

Store at 4°C.
Light paraffin oil (Fisher O121-1; BDH 29436) (see Comment)
Phosphate-buffered saline (PBS) (see Appendix 1)
Stain (prepare fresh)

0.2% $MgCl_2$<!>	9 ml
Tris-citrate buffer (pH 8.0)	
(20.1 g of Tris<!>, 8 g of citric acid<!>, 500 ml of distilled water)	1 ml
glucose-6-phosphate dehydrogenase (Sigma G8878)	10 µl
4-nitro blue tetrazolium chloride<!> (NBT*) (2.7 mg/ml)	1 ml
NADP* (2.7 mg/ml)	1 ml
fructose-6-phosphate* (20 mg/ml)	1 ml

At the last minute, add 30 µl of pregnant mare serum (PMS) (10 mg/ml). Store stock solution at 4°C in the dark.
*Can be stored in aliquots at –20°C.

SAMPLES
Control blood sample (equal volumes of blood from parental strain mice homozygous for *Gpi1*[a], *Gpi1*[b], or *Gpi1*[c] diluted 10x in PBS, stored at –20°C)
Embryo tissue samples

CAUTION: See Appendix 2 for appropriate handling of materials marked with <!>.

PROCEDURE

1. Place embryo tissue samples in PBS or distilled water in a 96-well tissue culture dish. Place small fragments (e.g., the allantois from a 9.0-day embryo) in ~15 µl of PBS and cover the drop with light paraffin oil to prevent evaporation. Intact 9.0-day embryos require at least 200 µl of PBS. Freeze-thaw the tissue culture dish at least twice before running samples.

2. Slowly immerse three Titan III cellulose acetate plates in a beaker of gel buffer to ensure that no air bubbles are trapped and soak for 30 minutes before loading samples.

   > For the custom-made Helena gel box, three Titan III plates can be run simultaneously and, excluding two controls per plate, approximately eight samples can be run on each plate. For the standard horizontal gel boxes, the number of plates run simultaneously is determined by the width of the gel box.

3. Remove the first plate and place the cellulose acetate surface down on a clean glass plate. Wipe the plastic (shiny) surface of the plate dry with a tissue and draw a 60-mm line with a permanent marker pen across the plate, ~1 cm from the bottom. Mark 10 equal divisions, leaving ~3 mm at either side. It is convenient to use as a template a microscope slide that already has the divisions marked on it. Number the divisions 1–10 beneath the line and label the plate at the bottom.

4. Turn the plate cellulose-acetate-surface-up and gently blot the bottom third with a tissue. Place a strip of Whatman 3MM filter paper soaked in gel buffer over the opposite end to prevent drying out. Using a pipetman or a series of fine hand-drawn Pasteur pipettes (use a different tip or pipette for each sample), load ~1 µl of control, containing both isozymes (see Materials), onto divisions 1 and 10 of plate. Load 1 µl of each sample onto each of the remaining eight divisions.

5. Place the plate in the gel box containing gel buffer with the plastic (shiny) surface uppermost and the cellulose acetate surfaces at either end resting on wicks (Whatman 3MM) immersed in the gel buffer. The samples will run from positive to negative. To ensure even electrophoresis, apply weight to each plate with two parallel piles of about six microscope slides. Place ice in the central portion of the gel box, under the plates but not touching them, to reduce heating effects and preserve enzyme activity (especially for Gpi1c, which is heat-labile).

6. Load the samples on the remaining plates and then run gel at 200–250 V (4 mA) for 55–60 minutes. Reduce running time for Gpi1c because the enzyme is less stable (i.e., 30 minutes).

7. It is best to stain the gel plates horizontally; the lids of staining troughs are an ideal size for this. Remove the plates and place cellulose acetate surface down, taking care to avoid air bubbles, in ~3 ml of stain. Place a light-tight cover over the plates while staining, but do not move cover once the plates are placed in the stain because this will cause smearing. Stain for 20–40 minutes, depending on the activity in the samples.

8. Rinse the plates in distilled water and fix for 5 minutes in 5% acetic acid in distilled water. Rinse twice in distilled water. Dry vertically.

COMMENTS

- The genotypic composition of the chimeric tissues is determined by comparing artificial mixtures of the two isozyme variants from tissues from the two different strains used either visually (±5%) or by densitometry (Behringer et al. 1984).

- The staining reaction can also be immobilized using agarose (see Eppig et al. 1977).

- Paraffin oil sold by pharmacists for human medicinal use may also be used. Some investigators sterilize the oil by autoclaving. However, others believe that this greatly increases toxicity and is not necessary, since cultures containing antibiotics rarely become infected. In either case, the potential toxicity of batches of oil should be checked on spare embryos.

| Preparing Tissue Fixation Solutions

FRESH 4% PARAFORMALDEHYDE

MATERIALS

> **EQUIPMENT**
> Flask
> Stirring hot plate
>
> **REAGENTS**
> Paraformaldehyde<!> (Sigma P6148 or Fisher Reagent Grade 04042-500) in
> Ca^{++}/Mg^{++}-free PBS (pH 7.4)

CAUTION: See Appendix 2 for appropriate handling of materials marked with <!>.

PROCEDURE

1. Dissolve paraformaldehyde in Ca^{++}/Mg^{++}-free PBS (pH 7.4; see Appendix 1) with stirring or frequent swirling at ~60°C by placing the container in a flask of water on a stirring hot plate. It may take ~1 hour for the paraformaldehyde to dissolve. Cool to 4°C.

2. Store fixative in aliquots at –20°C, but do not freeze and thaw repeatedly because it will oxidize.

3. Fix embryos or tissues at 4°C for 2–16 hours, depending on the size of the specimen. Do not overfix.

> **COMMENTS**
>
> • For in situ hybridization, the PBS should be RNase-free (see In Situ Hybridization with RNA Probes, Protocols 16.11 and 16.12) and stored in baked glass or disposable plastic containers.
> • Alternatively, it is also possible to purchase ready-made solutions of 16% paraformaldehyde stored under inert gas in sealed glass ampoules (Electron Microscopy Sciences 15710) to prepare 4% paraformaldehyde in PBS.

BOUIN'S FIXATIVE

MATERIALS

REAGENTS

Ethanol<!> series (70%, 80%, 90%, 100%)
40% Formaldehyde<!>   25 ml
Glacial acetic acid<!>   5 ml
Saturated picric acid<!>   75 ml
Wax

Saturated picric acid is made by adding powdered picric acid to water until no
more powder dissolves.

CAUTION: See Appendix 2 for appropriate handling of materials marked with <!>.

PROCEDURE

1. Mix reagents listed above.

2. Perform fixation at room temperature for 2–16 hours, depending on the size
   of the tissue or embryo. Do not overfix.

3. After fixation, wash the tissues repeatedly in 70% ethanol, preferably in a
   tube on a rocking platform, until no yellow color leaches from the tissue.
   Continue dehydrating the specimen through 70%, 80%, 90%, absolute (100%)
   alcohol, and xylene and embed in wax.

   It is extremely important to remove excess Bouin's fixative before embedding
   and sectioning samples because crystals of picric acid will destroy the sections.

COMMENTS

Bouin's fixative is a very good histological preservative that can be prepared or purchased
(Polysciences).

METHANOL/DMSO, 4:1

MATERIALS

REAGENTS

DMSO<!>
Methanol<!>

Prepared fixative is stored at 4°C.

CAUTION: See Appendix 2 for appropriate handling of materials marked with <!>.

COMMENTS

This fixative is very mild, but it does not preserve tissue morphology as well as Bouin's or
paraformaldehyde fixatives.

Handling Blastocysts for Fixation

MATERIALS

ANIMALS AND EMBRYOS
Blastocysts
Female mouse

EQUIPMENT
Embryological glass dishes (Raymond Lamb E/90 or E/91)
Pasteur pipette, hand-drawn, siliconized

REAGENTS
Medium containing serum or bovine serum albumin (BSA; Sigma A4378)

PROCEDURE

1. For whole-mount in situ hybridization, use a hand-drawn siliconized Pasteur pipette to transfer blastocysts manually between embryological glass dishes (Raymond Lamb E/90 or E/91) containing the different solutions.

2. For sectioning, it is more convenient to transfer blastocysts, prior to fixation, into the ampulla of the oviduct. The oviduct then serves as a carrier that is easy to handle during subsequent processing for histology. Blastocysts can be transferred to the oviduct from any female mouse, but it is easier to introduce them into the ampulla of oviducts on the day following ovulation (see Protocol 6.1).

3. Remove the oviducts and place them in medium containing serum or bovine serum albumin. Protein in the medium facilitates stable fixation of the blastocysts in the oviduct lumen. Remove and discard most of the distal portion of the oviduct, leaving the infundibulum, ampulla, and ~1 mm of distal tube. (This step is not essential, but it reduces the number of sections that have to be cut to locate the blastocysts.)

4. Use a finely drawn Pasteur pipette to transfer ~10 blastocysts in a minimum volume of medium into the oviduct via the infundibulum. Check that the blastocysts have not blown out of the distal end of the tube and that they are not still in the pipette.

5. Place the oviduct in fixative for the appropriate length of time. It can now be processed for histology like any other tissue.

COMMENTS

Mintz (1971) describes a clever technique to handle preimplantation mouse embryos for fixation and histological processing. Ant cocoons (*Acanthomyops claviger*) spun of fine silk can be purchased from most pet shops, processed very simply, and then used as convenient "baskets" for preimplantation mouse embryos for fixation and subsequent histology.

Embedding in Wax

MATERIALS

EQUIPMENT

Aluminum foil, plastic wrap, or a plastic (Tupperware-type) container
Embedding center (e.g., "Blockmaster II," Raymond Lamb E/66.2 or Leica Histo-
    embedder), which consists of a warming oven for preheating molds and forceps,
    a wax bath and reservoir, a heated work surface under an illuminator and mag-
    nifying lens, and a cold spot for quickly cooling the wax-filled molds once the
    specimen has been positioned.
Molds

> Several different kinds and sizes of molds are available. Baxter markets disposable
> plastic molds of various sizes (M7275-1), and Raymond Lamb markets a variety of
> suitable plastic molds (E10.6 or E10.4).

Needle
Oven, small
Pasteur pipette, 60°C prewarmed
Slide warming plate

> A set of several plates at different temperatures (48–50°C) is ideal.

Whatman filter paper, coarse

REAGENTS

Paraplast tissue embedding medium, melting point 56°C (Fisher 12-646-106) or BDH
    pastillated Fibro-wax, melting temperature 56–58°C (Formulation Raymond
    Lamb 36142)
50:50 Xylene<!>/wax

CAUTION: See Appendix 2 for appropriate handling of materials marked with <!>.

PROCEDURE

1. Transfer samples from xylene to 50:50 xylene/wax at 60°C for 1 hour. Xylene/wax should be kept in a small oven.

2. Perform two changes of neat paraffin wax at 60°C (use a 60°C prewarmed Pasteur pipette so that the wax does not solidify). Be sure to remove all the air bubbles from large embryos.

3. Transfer the embryos or tissue pieces into a mold in an orientation suitable for cutting good sections. (Once the wax cools it becomes opaque.)

4. To reposition the sample in the wax, put the mold back in the oven after transferring the embryo and, once the wax has melted, reposition the embryo in the wax with a hot needle, resting the mold on a 60°C hot plate.

A very useful piece of equipment when handling large numbers of specimens is an "embedding center" (e.g., "Blockmaster II," Raymond Lamb E/66.2 or Leica Histo-embedder).

5. Store specimens at 4°C enclosed in aluminum foil or plastic wrap or in a plastic (Tupperware-type) container.

HANDLING AND EMBEDDING SMALL SPECIMENS

Unless it is important to visualize gene expression in the deciduum and extraembryonic tissues in relation to the implanted embryo, it is advisable to dissect postimplantation embryos from the uterus before fixation and processing. This is because it is difficult to orient small embryos in the wax accurately if they are still inside the deciduum.

- To minimize loss of small postimplantation embryos during the dehydration stages, use glass tubes with conical bottoms (e.g., Reacti-Vials, Pierce 132254) and remove the supernatant rather than transferring the embryos from one tube to another.

- Color the embryos a light pink during the final stages of dehydration by adding a drop of alcoholic eosin stain to embryos in 1 ml of 90% ethanol and swirling for ~1 minute.

- It is often useful, especially when comparing mutant and normal embryos, to position both mutant and normal samples on the same slide and in the same orientation.

- For embryos up to 9.0 dpc, embedding is best done in glass embryo dishes (Raymond Lamb E/90 or E/91), but for later stages, plastic molds are more suitable.

- Because of the shape of intact postimplantation embryos, it is seldom possible to section them in the same axial plane along their length. This situation can be partially remedied by cutting the embryo into two or more pieces along the rostrocaudal axis before fixation.

EMBEDDING SMALL EMBRYOS

ADDITIONAL MATERIALS

EQUIPMENT
Bunsen or spirit lamp flame
Dissecting microscope
Embryological glass dish
Forceps

REAGENTS
Glycerol

1. Wipe a clean embryological glass dish with a tissue soaked in glycerol (only a thin film is required).

2. Fill the dish with molten wax and use this for the second wax change. Transfer the embryos into this dish (~6 embryos can be embedded simultaneously).

3. Reheat the wax in the oven for ~10 minutes.

4. Quickly transfer the dish to the stage of a dissecting microscope and use hot metal forceps or a metal needle (heated in a Bunsen or spirit lamp flame) to waft the embryos to the center of the dish.

5. Wait for a thin layer of wax to solidify on the bottom of the dish before positioning the embryos.

COMMENTS

- It is often useful to arrange the embryos in a grid formation and to note their positions.
- The wax can be remelted with the hot forceps or needle to align the embryos in the same focal plane, such that they will all be sectioned simultaneously.
- For longitudinal sections, the lateral sides of the embryo should lie on the bottom of the embryological dish. For transverse sections, the rostrocaudal axis should be perpendicular to the bottom of the dish.

MATERIALS

EMBRYOS
Wax-embedded specimens

EQUIPMENT
Compressed air spray (e.g., Fisher brand Phfft)
Fine paintbrush or forceps
Glass slides
Kimwipes
Microtome

REAGENTS
Diethyl pyrocarbonate<!> (DEPC)-treated autoclaved water
Tissue section adhesive (e.g., STA-ON, Surgipath SA-1600)
Xylene<!>

CAUTION: See Appendix 2 for appropriate handling of materials marked with <!>.

PROCEDURE

Initial Advice

- *Obtaining continuous ribbons of sections:* The best ribbons (straight and long) are obtained when the trimming is close to the specimen and leaves a perfect square or rectangle with parallel sides. However, it is sometimes convenient to cut one side at a slight angle, so that the section can be correctly repositioned if it detaches from the ribbon. Note that the side of the ribbon facing the blade will be shiny and the other side will be matte. It is very important, for obtaining a single, continuous ribbon, that the bottom edge of the wax block be parallel to the microtome blade once it is mounted. This can be achieved by last-minute trimming, so do not cut away all of the bottom edge in the initial trimming.

- *Best location for sectioning:* Sectioning should be performed in a cool, draft-free room. Wax blocks should be as cold as possible (if necessary, cool on ice). Hold the ribbon with a fine paintbrush or forceps.

- *Choosing a microtome:* The type of microtome used depends largely on the laboratory's expense account. Reichert-Jung 2030 is a good basic model. Use either disposable blades or blades that can be resharpened.

- *Cutting the sections:* Sections of 5–7 μm are suitable for in situ hybridization or immunohistochemistry. Clean the blade with a compressed air spray and wipe with a Kimwipe between ribboning. If sections curl, try wiping the microtome blade with xylene and allow it to dry before sectioning again.

- *Staining sections:* It is often desirable to use only 1 in 10 sections for staining. Place the ribbon on clean, black, waxed paper and cut off the required section or group of sections.

Methods for Transferring Sections to the Slide

Method 1

1. Float ribbons of sections on a water bath (flotation bath) filled either with water containing tissue section adhesive for immunohistochemistry or with DEPC-treated autoclaved water for in situ hybridization. The temperature should be below the melting point of the wax (e.g., 42–45°C or 50°C).

2. After the sections have spread on the surface of the water, hold a slide at a 45° angle and immerse halfway into the water bath. Use fine forceps to guide the leading edge of the ribbon onto the slide.

3. Gently lift the slide out of the water and stand it upright in a block at room temperature to air-dry for 1 hour to overnight.

COMMENTS

- For immunohistochemistry, the addition of a tissue-section adhesive to the water helps the sections adhere tightly to the glass slides and prevents them from detaching during the washing procedures; 10 ml is added to 1 liter of warm water (40–46°C) with gentle stirring, and this mixture is then added to the flotation bath.
- For in situ hybridization, it is advisable after air-drying to heat slides thoroughly on a slide warmer at 48–50°C overnight. This ensures that the sections are stuck on the slide.

Method 2

1. Lift the ribbon directly onto a drop of water on a slide (the water drop can contain STA-ON).

2. Air-dry the slide. After air-drying, dry thoroughly on the slide warmer as described above.

3. Alternatively, place single sections on small drops of water (with STA-ON) in individual circles marked on slides coated with an inert hydrophobic material (slides are available in a variety of circle diameters from Roboz Surgical Instruments). Air-dry the slides until there is no water under the section, and then heat to 42–45°C overnight. If the slide is heated too quickly, bubbles will destroy the sections.

   This alternative method is very convenient for immunohistochemistry (see Protocol 16.9) because the antibody solutions and washes can be applied in small volumes to each circle.

4. Label slides with a pencil on the frosted surface. Pencil will withstand the solvents used for dehydration.

Preparing Glass Slides and Coverslips for In Situ Hybridization

The following techniques are presented in this protocol:

1. Coating slides with TESPA

2. Coating slides with poly-L-lysine

3. Siliconizing coverslips

TECHNIQUE 1: COATING SLIDES WITH TESPA

MATERIALS

EQUIPMENT
Aluminum foil
Glass slides
Oven
Slide rack

REAGENTS
Acetone, 100% <!>
2% TESPA (3-aminopropyl-triethoxysilane, Sigma A3648) in acetone <!>
Water containing 0.1% diethyl pyrocarbonate(DEPC)<!>

CAUTION: See Appendix 2 for appropriate handling of materials marked with <!>.

PROCEDURE

1. To destroy any RNase attached to the glass slides, load the slides into a metal rack (see below) and place in water containing 0.1% DEPC (see Protocols 16.11 and 16.12) for 15–30 minutes.

2. Wrap the entire rack in aluminum foil and bake at 180°C for at least 2 hours.

3. Cool the slides to room temperature and dip them in a solution of 2% TESPA in acetone for 30 seconds; 250 ml of this solution is sufficient for 100–200 slides.

4. Dip the slides twice in 100% acetone for 30 seconds each. Then, rinse in DEPC-treated, autoclaved water for 30 seconds.

5. Dry at 42°C overnight. Store wrapped in foil at room temperature indefinitely.

TECHNIQUE 2: COATING SLIDES WITH POLY-L-LYSINE

MATERIALS

EQUIPMENT
Dust-free container
Glass slides
Slide racks

REAGENTS
Diethyl pyrocarbonate<!> (DEPC)-treated autoclaved water
NH_4Ac, 0.25 M
Poly-L-lysine (50 µg/ml; Sigma P1274)
Tris<!>,10 mM (pH 8.0) made in DEPC-treated water

CAUTION: See Appendix 2 for appropriate handling of materials marked with <!>.

PROCEDURE

1. Make a stock solution of 10 mg/ml poly-L-lysine in DEPC-treated auto-claved water and store at –20°C. Before use, dilute to 50 µg/ml in 10 mM Tris (pH 8.0) made in DEPC-treated water.

2. Pretreat the slides to destroy RNase as described above (see p. 658).

3. The day before use, dip slides in a rack into 0.25 M NH_4Ac, dry at 60°C in a dust-free container, and then soak for 30 minutes in fresh poly-L-lysine (50 µg/ml) at room temperature.

4. Air-dry the slides overnight at room temperature and store them at 4°C in a dust-free container. Use as soon as possible (although slides have been used after storage at 4°C for 2–3 weeks).

TECHNIQUE 3: SILICONIZING COVERSLIPS

MATERIALS

EQUIPMENT
Aluminum foil
Coverslips, high-quality (e.g., Fisher Finest 22 x 22 mm 12-544-10)

REAGENTS
DEPC<!>-treated, autoclaved water
Metal coverslip racks (e.g., Raymond Lamb E/103)
Siliconizing solution (e.g., dimethyldichlorosilane, Sigma D3879)

CAUTION: See Appendix 2 for appropriate handling of materials marked with <!>.

PROCEDURE

1. Coverslips need not be treated to destroy RNase, but they need to be of high quality (e.g., Fisher Finest 22 x 22 mm 12-544-10).

2. Place the coverslips in metal rack and dip racks in a siliconizing solution for ~2 minutes. Rinse in 100% ethanol for ~2 minutes and then in DEPC-treated, autoclaved water for ~2 minutes.

3. Wrap the rack in foil and bake at 180°C for 2–3 hours.

Dewaxing and Rehydrating Sections Prior to In Situ Hybridization or Staining

MATERIALS

EQUIPMENT
Slide rack, metal or glass
Slides
Staining jars, glass

REAGENTS
Ethanol<!>, 100%, 90%, 70% (proceed to alcohol stains), 50%, 30%, water
Phosphate-buffered saline
Xylene<!>

CAUTION: See Appendix 2 for appropriate handling of materials marked with <!>.

PROCEDURE

1. Place the slides in a metal or glass rack and transfer them from one solution to another in glass staining jars.

   A typical sequence is two washes in xylene for 5–10 minutes; then 2–3 minutes in 100% ethanol, 90%, 70% (proceed to alcohol stains), 50%, 30%, water (5 minutes if proceeding to aqueous stains) or PBS (5 minutes if proceeding to in situ hybridization).

Immunohistochemistry of Embryo Sections

MATERIALS

BIOLOGICAL MOLECULES

Primary antibody (e.g., affinity-purified, polyclonal rabbit antiserum)

Secondary antibody (e.g., alkaline-phosphatase-coupled goat anti-rabbit IgG; Boehringer Mannheim 605 220)

EMBRYO

EQUIPMENT

Coverslips

Filter paper, 3MM

Frosted glass slides

> If necessary, coat slides with 3-triethoxysilylpropylamine (TESPA; Sigma A3648) to increase adhesion of the sections or use a tissue-section adhesive such as STA-ON (Surgipath SA-1600) in the water used for applying sections. Slides with individual circles surrounded by a hydrophobic surface are useful for staining individual sections (see General Techniques section above, p. 633).

Humidity box

> This is designed to prevent the slides from drying out during incubation. It can be made by placing a petri dish bottom, 90-mm plastic, with pieces cut out of the side.

Magnetic stirrer

Microtome

PAP pen (Research Products International 195500)

Plexiglas plate on caps of 50-ml tubes in a plastic (Tupperware-type) container

> Water is then poured under the plate. Small-diameter glass rods can be glued to the Plexiglas plate to provide support for the slides.

Slide-warming plate (at ~47°C)

Staining jars containing slide holders

Stirring bar

Whatman 3MM filter paper (or similar for drying the backs and edges of the slides) cut into small squares

REAGENTS

Alkaline phosphatase (AP) buffer

> 100 mM Tris-HCl<!> (pH 9.5)
> 100 mM NaCl
> 5 mM MgCl$_2$ <!>

Alkaline phosphate enzyme substrate: nitro blue tetrazolium<!> (NBT; Sigma N6876)/5-bromo-4-chloro-3-indolyl phosphate<!> (BCIP; Sigma B6149). Alternatively, both prepared solutions can be obtained from Promega (S3771).

NBT: Dissolve 0.5 g of NBT in 10 ml of 70% methylformamide

BCIP: Dissolve 0.5 g of BCIP (disodium salt) in 10 ml of water

Not more than 1 hour before use, prepare working solution in AP buffer as follows: Add 33 µl of NBT to 5 ml of AP buffer and mix well. Add 16.5 µl of BCIP and mix well. Keep protected from light.

Blocking solution (used for blocking nonspecific protein-binding sites in the tissues)

10 mM Tris-HCl<!> (pH 7.4)

100 mM $MgCl_2$<!>

0.5% Tween 20 (Fisher BP337 or Sigma P1379)

1% bovine serum albumin (BSA) (Sigma A4378)

5% fetal bovine serum (FBS) (do not use calf serum; FBS has lower amounts of IgGs)

Store in aliquots at –20°C. Do not freeze-thaw.

Ethanol<!>, 50%, 70%, 90%, and 100%

100% Glycerol or aqueous mounting solution (e.g., Crystal Mount, Fisher BM-M02)

Levamisole, 1 mM (Sigma L9756; used to inhibit TN-alkaline phosphatase [see Hahnel et al. 1990 and below])

4% Paraformaldehyde <!> in phosphate-buffered saline (PBS) (see p. 649)

Stop buffer

20 mM Tris-HCl<!> (pH 8.0)

5 mM EDTA

TBST

10 mM Tris-HCl<!>, (pH 8.0)

150 mM NaCl

0.05% Tween 20 (Fisher BP337 or Sigma P1379)

TM

10 mM Tris-HCl<!> (pH 7.5)

100 mM $MgCl_2$<!>

Xylene <!>

CAUTION: See Appendix 2 for appropriate handling of materials marked with <!>.

PROCEDURE

1. Fix the embryo. When testing an antibody for the first time, try several different fixation procedures. Some of the most commonly used are 4% paraformaldehyde in PBS (pH 7.4), prepared fresh as described in the General Techniques section; Bouin's fixative; and methanol/dimethyl sulfoxide (DMSO) (4:1) (see Protocol 16.3).

2. Dehydrate and embed the specimen in wax as described in Protocol 16.4.

3. Section the specimen to a thickness of 5–7 µm using a microtome (see Protocol 16.6).

4. Collect the sections on a glass slide either by bringing the slide up under sections of ribbon floating on a water bath (~40°C) or by placing a drop of water on the slide and positioning sections on this drop (see Protocol 16.6). In either case, place the slides on a slide warmer (~47°C) until moisture is no longer visible. Allow the slide to air-dry for 1–2 days before staining. Alternatively, the slides can be stored for several months at 4°C or room temperature (depending on the antigen) in a slide box.

5. Dewax the sections by immersing slides in a holder in staining jars containing the following series:

> xylene (5 minutes)
>
> xylene (5 minutes)
>
> 100% ethanol (3 minutes)
>
> 100% ethanol (3 minutes)
>
> 90% ethanol (3 minutes)
>
> 70% ethanol (3 minutes)
>
> 50% ethanol (3 minutes)
>
> TM (5 minutes)
>
> Discard the xylene (see Appendix 2) after ~10–15 uses (or if contaminated with wax or moisture).

6. Wipe the back and edges of the slide with a square of 3MM filter paper. Place the slide on a flat surface and draw around the desired section (or set of sections) with a PAP pen; this prevents the blocking solution from running off the edges of the slide. Cover the section(s) with blocking solution (usually ~500 μl), place the slide in the humidity box, and incubate at room temperature for 20 minutes.

7. Shake the slide with a flick of the wrist to remove the blocking solution. Cover the sections with diluted primary antibody and incubate in the humidity box at room temperature for 30 minutes. If necessary, the incubation time can be increased up to several hours or overnight.

8. Wash the slides twice in 500 ml of TBST in a 500-ml staining jar at room temperature for 10 minutes each time. Then place the slides in a rack on top of a 90-mm plastic petri dish bottom (with openings in the side) inverted over a stirring bar. This will allow the buffer to circulate when the apparatus is placed on the magnetic stirrer.

9. Wipe the back and edges of the slide with a square of 3MM filter paper. Add secondary antibody (e.g., AP-coupled goat anti-rabbit IgG) diluted in TBST. A dilution of 1:1000 is an appropriate starting dilution when using the Boehringer Mannheim antibody. Incubate in the humidity box at room temperature for 30 minutes.

10. Wash the slides immersed in TBST twice for 10 minutes each as in step 8.

11. Wipe the back and edges of the slide with 3MM filter paper and add the alkaline phosphatase enzyme substrate. Incubate at room temperature in a light-protected humidity box for 30 minutes to 1 hour. To prevent overstaining, check the staining periodically in the microscope.

12. Stop enzyme reaction by immersing the slide in Stop buffer for ~10 minutes.

13. Mount the slides in either 100% glycerol or an aqueous mounting solution.

COMMENT

An important caution when using the AP-coupled secondary antibody staining procedure is that some embryonic cells express this enzyme. Control experiments should therefore be carried out in which the secondary antibody is omitted. The expression of three different AP isozymes in the mouse has been studied in detail by Hahnel et al. (1990). The inhibitor levamisole is specific for tissue-nonspecific (TN)-AP and does not inhibit placental (Pl)-AP or embryonic (E)-AP.

Immunohistochemistry of Whole-mount Embryos

This protocol was adapted from Dent et al. (1989) and LeMotte et al. (1989) by Alex Joyner and Nancy Wall.

MATERIALS

BIOLOGICAL MOLECULES

Primary antibody (e.g., an affinity-purified, polyclonal rabbit antiserum)

Secondary antibody (e.g., horseradish peroxidase [HRP]-coupled goat anti-rabbit IgG [e.g., Boehringer Mannheim 605 220 or Jackson ImmunoResearch Laboratories 111-035-003])

EMBRYOS

Embryos of the desired stage

EQUIPMENT

Microfuge tubes, 1.6-ml
Petri dishes, 35-mm glass
Rocking platform
Tubes, 15-ml screw-capped

REAGENTS

BABB (benzyl alcohol<!>:benzyl benzoate<!>; 1:2) (used to clear the embryos after staining and to provide a mountant to observe and photograph the stained embryos)

DAB<!> (3-3´-diaminobenzidine tetrahydrochloride; Sigma D5637). Store desiccated at −20°C. Warm to room temperature before weighing.

DAB/NiCl$_2$<!>. Mix 0.03 g of DAB and 0.03 g of NiCl$_2$ in 50 ml of PBT. Use within 1 hour of preparing and keep protected from light. The nickel enhances the sensitivity of the color reaction and produces a slate-gray to purple precipitate. If necessary, vary the amount of nickel to alter the intensity of the color. Cobalt can be substituted for nickel.

Hydrogen peroxide (H$_2$O$_2$)<!>(generally supplied as a 30% solution) Store at 4°C; the solution lasts ~1 month at this temperature).

Methanol<!>, 50%, 80%, and 100%

Methanol/dimethyl sulfoxide<!> (DMSO; 4:1) (prepare fresh)

Methanol/DMSO/H$_2$O$_2$(4:1:1) (prepare fresh)

4% Paraformaldehyde <!> in PBS (see p. 649)

PBT

> PBS (see p. 667)
> 0.2% bovine serum albumin (BSA) (Sigma A4378)
> 0.5% Triton X-100
> Make fresh before use. (*Note:* This is different from the PBT used in whole-mount in situ hybridization.)

PBSMT
> PBS (see below)
> 2% nonfat instant skim milk (used to block all nonspecific protein-binding sites in the tissues). The brand is important. Carnation gives consistently good results, whereas other brands (e.g., Kroger) do not.
> 0.5% Triton X-100 (used to facilitate permeability of the tissues)
>> Make fresh before use.

Phosphate-buffered saline (PBS [pH 7.4])

NaCl	8 g
KCl<!>	0.2 g
Na_2HPO_4<!>	1.44 g
KH_2PO_4<!>	0.24 g

Make up to 1 liter in distilled water. Adjust pH to 7.4.

CAUTION: See Appendix 2 for appropriate handling of materials marked with <!>.

PROCEDURE

1. Collect the embryos in PBS or tissue culture medium. If the embryos are at the prestreak to early-somite stages, it is advisable to add some serum (~5%) to reduce stickiness (this will cause a protein precipitate in the fixative, but do not be concerned).

2. Dissect away the extraembryonic membranes to facilitate penetration of the antibodies.

   a. Use the tips of fine forceps to flatten early somite-stage mouse embryos (up to ~8 somites) by pushing the edges of the visceral yolk sac into the surface of a plastic petri dish. By "pinning out" the embryos in this way, they can be fixed flat and will remain extended after being detached from the dish. The extended shape allows better viewing of the somites and neural tube after staining.

   b. For older embryos (9.5–10.5 days), it is sometimes best to cut them in half sagitally before or after fixation, making it easier to visualize and photograph staining of bilaterally symmetrical structures such as the cranial nerves, which otherwise appear as "double images."

3. Fix the embryos in freshly prepared methanol/DMSO (4:1) at 4°C overnight. DMSO permeabilizes the tissues.

4. Transfer the embryos into freshly prepared methanol/DMSO/H_2O_2 (4:1:1) at room temperature for 5–10 hours to block any endogenous peroxidase activity (e.g., in red blood cells). Then store the embryos in 100% methanol individually or in small groups. The embryos may be stored at –20°C for at least several months and probably longer.

5. Rehydrate the embryos at room temperature in microfuge tubes in the following series:

   a. 50% methanol (1 ml), 30 minutes with rocking. Remove solution with a Pasteur pipette or micropipetter.

   b. PBS (1 ml), 30 minutes with rocking (if embryos stick to sides, siliconize the tubes).

   c. PBSMT (1 ml), twice for 1 hour with rocking.

     Rocking is important to facilitate penetration of the antibody, so make sure that the solution is mixing well. However, the embryos are fragile once they are fixed, so do not rock them too hard.

6. Incubate the embryos with rocking in microfuge tubes with 1 ml of primary antibody diluted in PBSMT at 4°C overnight. The correct dilution must be determined empirically, but 1:200 is a typical dilution. This and all subsequent procedures can be performed at room temperature if the antibody used is stable.

7. Wash the embryos five times with rocking in PBSMT:

   a. once with 1 ml for 1 hour at 4°C. Transfer to 15-ml tubes.

   b. once with 10 ml for 1 hour at 4°C.

   c. three times in 10 ml for 1 hour at room temperature.

8. Transfer the embryos to microfuge tubes. Incubate them in 1 ml of secondary antibody diluted 1:500 in PBSMT at 4°C overnight with rocking.

9. Wash the embryos again as in step 7.

10. Rinse in 5 ml of PBT and then transfer the embryos to microfuge tubes. Wash them in 1 ml of PBT with rocking for 20 minutes.

11. Incubate the embryos in microfuge tubes with 1 ml of DAB/NiCl$_2$ at room temperature for 30 minutes (this allows full penetration of the substrate into the embryo).

12. Add H$_2$O$_2$ to a 0.03% final concentration and rock until the color intensity in the embryos looks good (i.e., specific regions of staining are obvious and before background staining comes up; usually 2–10 minutes). If necessary, check color in the dissecting microscope. If the color reaction occurs too quickly, use a lower concentration of H$_2$O$_2$.

13. Postfix the embryos in 4% paraformaldehyde in PBS (see Comments).

14. Rinse embryos in the following series:

   a. PBT (1 ml) quick rinse

   b. PBT (1 ml) 30 minutes at room temperature with rocking

   c. 50% methanol (1 ml) 30 minutes at room temperature with rocking

   d. 80% methanol (1 ml) 30 minutes at room temperature with rocking

   e. 100% methanol (1 ml) 30 minutes at room temperature with rocking

15. Change rinse to BABB (500 µl) for 10 minutes. Make observations of stained embryos in a glass petri dish. Do not use polystyrene dishes.

COMMENTS

- Without postfixation, the color of the stained embryos will fade under strong light, particularly light used for photography. Although embryos can be stored in BABB in a dark place, it is advisable to obtain a photographic record as soon as possible. Use a tungsten color film without filters or a daylight color film with a blue filter. For photography, place embryos in a depression slide with a coverslip on top.

An alternative procedure is to use AP-coupled secondary antibody and a substrate that is insoluble in water and alcohol (e.g., Naphthol-AS-MX-phosphate/fast red TR_f).

- If necessary, whole-mount embryos can be sectioned. Transfer the embryos from 100% methanol through the following series:

  a. twice in 100% ethanol for 1 hour

  b. twice in xylene for 1 hour

  c. once in xylene:wax (1:1) for 1 hour

  d. twice in wax for 1 hour

Embed, section, and mount the embryos. For counterstaining, use a light (30–45 seconds) stain with eosin B.

In Situ Hybridization of Embryo and Tissue Sections with RNA Probes

This protocol is presented in the following steps:

1. Preparing the probe
2. Prehybridization
3. Hybridization
4. Posthybridization washing
5. Autoradiography
6. Autoradiography—developing

MATERIALS

EQUIPMENT

A darkroom equipped with the following:

A 15W bulb behind an Ilford 902S safety filter water bath at 43–45°C

Glass or metal plate wrapped in foil to cool the slides (this should be level to ensure even spreading of the emulsion; some workers cool it on ice)

Dipping jars for emulsion (with marks at 6 ml and 12 ml)

Clean untreated slide without sections

Black exposure box with silica gel in lid (Raymond Lamb E/109 or E/99)

Large lightproof sandwich box

Timer, forceps, foil, tape

10- to 20-ml measuring cylinder

10 ml of 2% glycerol solution in a 50-ml plastic conical tube

Ilford K-5 emulsion (Polysciences 02746-50)

REAGENTS

Acetic acid<!>, 1%

Acetic anhydride <!> (Sigma A6404)

Ammonium acetate, 300 mM and 5 M

ATP/CTP/GTP mixture (each at 2.5 mM)

DEPC<!> H_2O

Developer, Kodak D-19

Dextran sulfate, 50% (Sigma D7037 or Fisher BP1585)

Make in water with heating. Filter through 0.22-μm membrane (this takes a very long time). Store in aliquots at –20°C.

Dithiothreitol<!>, 10 mM , 200 mM , and 1 M (DTT; Boehringer Mannheim 708984 or Sigma D9163)

Filter to sterilize and store at –20°C.

Emulsion (e.g., Ilford K5 [Polysciences 02746-05] or Kodak NTB-2 [1654433])

Use only when fresh because old emulsion gives high background. Store at 4°C away from any penetrating light.

Ethanol <!>

Formamide<!>, deionized

   Store in aliquots at –20°C.

Glycerol, 2%

Glycine<!>, 0.2% in PBS

Hydrolysis buffer (200 mM NaHCO$_3$, 200 mM Na$_2$HCO$_3$)

Mounting medium (e.g., Permount <!>; Fisher SP15-100)

NaCl, 0.5 M, in TE (10 mM Tris<!> [pH 7.6], 1 mM EDTA)

Neutralization buffer (0.2 mM NaOAc<!>, 1% glacial acetic acid<!>, 10 mM DTT<!>)

Paraformaldehyde<!>, 4% in PBS (see p. 649)

Placental RNase inhibitor (20–40 U/ml)

Phenol<!>/chloroform<!>

Phosphate-buffered saline (PBS) (pH 7.4)

   10x stock:

      8 g NaCl

      0.2 g KCl<!>

      1.44 g of Na$_2$HPO$_4$<!>

      0.24 g of KH$_2$PO$_4$<!>

      in 100 ml of water (pH 7.4) with HCl

Proteinase K (Sigma P0390) (10 mg/ml stock in water stored in aliquots at –20°C)

Proteinase K buffer (20x: 0.1 M EDTA, 1 M Tris<!> [pH 7.5])

RNA polymerase

RNase A (Sigma Type 1-A, R4875)

   Make a stock solution at 10 mg/ml in water, boil for 2 minutes to destroy contaminating nucleases, and store at –20°C. Take *great care* to keep any pipettes, tubes,
   and containers that have been exposed to RNase separate from other equipment.

RNase-free DNase I

RNase inhibitor

Salts, 10x

   Final concentration is:

      3 M NaCl

      50 mM EDTA

      100 mM Na$_2$HPO$_4$<!>

      0.1 M Tris-HCl<!> (pH 6.8)

      0.2% Ficoll 400 (Sigma F4375)

      0.2% (w/v) polyvinylpyrrolidone <!> (Sigma PVP-360)

      0.2% BSA (fraction V, Sigma A7906)

      Store in aliquots at –20°C.

Sodium acetate<!>, 3 M

Sodium thiosulfate, 30%

SSC, 20x (see Appendix 1)

Toluidine blue stain (0.02-0.5%; Sigma T3260)

Transcription buffer, 5x

      200 mM Tris<!> (pH 8.0)

      40 mM MgCl$_2$<!>

      10 mM spermidine<!>

      250 mM NaCl

Triethanolamine<!>, 0.1 M (Sigma T1377)

tRNA *E. coli* (Boehringer Mannheim 10 mg/ml or Sigma R8759 or R4251)

   Phenol-extract and ethanol-precipitate before making to correct concentration.

[^{35}S]UTP

Xylene<!>

CAUTION: See Appendix 2 for appropriate handling of materials marked with <!>.

PROCEDURE

STEP 1: PREPARING THE PROBE

Use high-quality reagents for the riboprobe reaction. Boehringer Mannheim and Promega supply enzymes and RNase inhibitors that work well. The DNA should be clean, and all reagents, tubes, and pipette tips should be RNase-free. This riboprobe reaction does not use cold UTP, and thus the probes generated have a high specific activity. The lack of cold UTP does not seem to affect the yield. For more considerations on probe preparation, see Protocol 16.12, step 2.

1. Prepare the riboprobe reaction:

linear template (1 µg/µl)	1 µl
5x transcription buffer	4 µl
200 mM DTT	1 µl
placental RNase inhibitor (20–40 U/µl)	1 µl
ATP/CTP/GTP mixture (each at 2.5 mM)	1 µl
[^{35}S]UTP (800 Ci/mmole)	8 µl
RNA polymerase (20–40 U/µl)	1 µl

Incubate at 37°C for 1 hour.

2. Remove the DNA template by adding the following directly to the riboprobe reaction:

RNase inhibitor	1 µl
tRNA (10 mg/ml)	2 µl
RNase-free DNase I (1–5 U/µl)	1 µl

Incubate at 37°C for 15 minutes.

3. Phenol/chloroform-extract by adding:

1 M DTT	1 µl
DEPC-H$_2$O	63 µl
3 M NaOAc	10 µl
phenol/chloroform	100 µl

Vortex well and centrifuge for 10 minutes.

4. Place the aqueous layer after extraction into a new tube and ethanol-precipitate by adding:

5 M NH$_4$OAc	50 µl
100% ethanol	500 µl

Precipitate on dry ice or at –80°C for 15–20 minutes. Microfuge for 30 minutes, wash pellet in cold 70% ethanol (–20°C), and carefully decant or remove the 70% ethanol using a pipette. Air-dry the pellet for ~15–30 minutes and resuspend in 50 µl of 10 mM DTT.

5. Count 1 µl of the resuspended pellet. If radioactivity is less than 2×10^6 cpm/µl, the efficiency of the reaction was less than 50% and the probe should not be used. Typically, 1 µl should yield 3×10^6 to 4×10^6 cpm/µl.

a. When using a template for the first time, 1 μl of the probe should be analyzed on a sequencing gel to check that the probe is full length.

b. The hybridization signal can be increased by increasing the length of the probe, but the probe is usually hydrolyzed to 100–200 bases.

c. Complete the hydrolysis reactions by adding 50 μl of hydrolysis buffer (200 mM NaHCO$_3$, 200 mM Na$_2$HCO$_3$) and incubating at 60°C.

d. The following formula is used to calculate the appropriate time:

$$x = \frac{L_o - L_f}{k\,(L_o L_f)}$$

where L_o is the original transcript length in kilobases, L_f is the final length of transcript in kilobases (100–200 bases), k equals 0.11, and x equals minutes for hydrolysis.

e. After x minutes, add 100 μl of neutralization buffer (0.2 M NaOAc, 1% glacial acetic acid, 10 mM DTT). Precipitate the probe by adding 20 μl of 3 M NaOAc and 500 μl of 100% ethanol and incubating on dry ice or at –80°C for 15–20 minutes.

f. Pellet the precipitate by centrifuging for 30 minutes followed by a 70% ethanol wash. Remove the ethanol and air-dry. Resuspend the pellet in 50 μl of 10 mM DTT and count 1 μl. Approximately 75% of the counts should be recovered.

g. Again, when establishing the technique for the first time, 1 μl of the hydrolyzed probe should be analyzed on a sequencing gel to check that the hydrolysis was successful.

6. Dilute the probe to 10^6 cpm/μl by adding formamide to 50% of the required volume and making up the difference in volume with 200 mM DTT. The diluted probe can be stored at either –20°C or –80°C.

STEP 2: PREHYBRIDIZATION

1. Transfer the slides (as prepared in Protocol 16.7) from PBS into freshly prepared 4% paraformaldehyde in PBS for 20 minutes to refix. Wash twice in PBS.

2. Remove the PBS and incubate at room temperature for ~10 minutes in 20 μg/ml Proteinase K freshly diluted into 5 mM EDTA, 50 mM Tris (pH 7.5). The purpose of this step is to remove protein associated with mRNA that may otherwise block the hybridization reaction. The timing and Proteinase K concentration in this step may have to be varied with different tissues and enzyme batches to obtain the best results.

3. Add 0.2% glycine in 1x PBS for 30 seconds to inhibit Proteinase K.

4. Wash in PBS twice for 30 seconds each.

5. Refix in 4% paraformaldehyde in PBS for 5 minutes.

6. Rinse in PBS for 5 minutes. Then quickly dip in DEPC-treated water.

7. Place the slide rack in 0.1 M triethanolamine. After a 2-minute incubation, add acetic anhydride to a final dilution of 1/400. Stir and let stand 10 minutes. Add another aliquot of acetic anhydride, stir, and let stand 10 minutes.

8. Wash in PBS for 5 minutes.

9. Dehydrate through 30%, 60%, 80%, 95%, 100% ethanol for 2 minutes each.

10. Air-dry for 1 hour. Proceed directly to hybridization. Do not leave overnight.

STEP 3: HYBRIDIZATION

1. Prepare the hybridization mixture. It is a good idea to make ~1.5x more hybridization buffer than is theoretically needed. Estimate 40 µl per 22 x 22-mm coverslip and 80 µl per 22 x 50-mm coverslip. Hybridization mixture for 1 ml:

10x salts	100 µl
deionized formamide	400 µl
50% dextran sulfate	200 µl
tRNA (10 mg/ml)	20 µl
1 M DTT	8 µl
water	72 µl
probe	200 µl

   Heat mixture at 80–100°C for 2–3 minutes and then vortex **vigorously** for several seconds before placing on ice. Failure to mix vigorously may result in an uneven hybridization signal.

2. Carefully pipette the probe mixture onto the sections to avoid making air bubbles. Make sure that all the sections are covered by spreading the mixture with a small piece of Parafilm. Using a needle or similar device for support, gently lower a siliconized coverslip over the sections (see General Techniques section, Protocol 16.7). Label the slides with a pencil on the frosted surface. Place the slides horizontally in a sealed sandwich box on top of paper towels soaked in 1x salts or 1x PBS. Incubate in a humidified incubator at 50°C overnight. Alternatively, seal the box inside several plastic bags and immerse in a water bath. *Make sure the box is horizontal.*

STEP 4: POSTHYBRIDIZATION WASHING

Digestion with RNase A destroys any single-stranded unhybridized probe while leaving double-stranded hybrids intact. Washing steps are very important to remove background, and variations in time and temperature (stringency) may be necessary to achieve optimal results. Some probes give higher backgrounds than others (for discussion of controls, see Protocol 16.12).

1. Incubate slides in 5x SSC, 10 mM DTT at 50°C for 15–30 minutes to dislodge the coverslips.

2. Wash the slides in 50% formamide, 2x SSC, 10 mM DTT at 50°C (low stringency) or 60–65°C (high stringency) for 20–30 minutes.

3. Wash twice in 0.5 M NaCl in TE at 37°C for 5 minutes.

4. Incubate in 0.5 M NaCl in TE containing 20 µg/ml RNase A at 37°C for 30 minutes.

5. Wash in 0.5 M NaCl in TE at 37°C for 30 minutes.

6. Repeat step 2 (*optional*).

7. Wash in 2x SSC at 50°C (low stringency) or 60–65°C (high stringency) for 30 minutes.

8. Wash in 0.1x SSC at 50°C (low stringency) or 60–65°C (high stringency) for 30 minutes.

9. Dehydrate through 30%, 60%, 80%, and 95% ethanol containing 300 mM NH$_4$Ac for 2 minutes each and then twice through 100% ethanol.

10. Air-dry slides for 1–2 hours before autoradiography.

STEP 5: AUTORADIOGRAPHY

1. Make a 1:1 mixture of emulsion:glycerol (Ilford K5 or Kodak NTB-2). First, add solid emulsion to the 15-ml mark in a 50-ml conical tube. Melt at 45°C for 10–15 minutes. Add 1 volume (~10 ml) of 2% glycerol. Heat at 45°C for 5 minutes. Pour into dipping jar (avoid bubbles).

2. Dip a slide into the mixture, count to three, pull it out, cleanly wipe off the back, and place on the glass slide until dry (1 hour). Keep slides horizontal. Be consistent from slide to slide.

3. Load the slides into the exposure boxes. Wrap in aluminum foil, date, and store overnight at room temperature and then at 4°C until ready to use. As a test, it is a good idea to develop one or two slides after a few days (~5–7 days) before processing the whole batch.

STEP 6: AUTORADIOGRAPHY: DEVELOPING

To avoid swelling, contraction, and distortion of the emulsion, all solutions should kept at *exactly* the same temperature before use. Warm the slides to room temperature. Perform the following steps in the darkroom.

1. Place the slides in Kodak D19 developer for 2 minutes. Use ~4 g/250 ml (a lower concentration than recommended by the manufacturer).

2. Place slides in 1% acetic acid for 2 minutes to stop the reaction.

3. Place slides in 30% sodium thiosulfate for 5 minutes to clear the emulsion. Change the solution after each use.

4. Place slides in distilled water for 10 minutes.

5. Place slides in distilled water for 30 minutes (this can be in the light).

6. Stain slides. Good counterstaining can be obtained with 0.02% toluidine blue in water (filtered through a 0.22-μm filter) for 30–60 seconds. This stains nuclei blue.

7. Dehydrate slides through 30%, 60%, 80%, 95%, and 100% ethanol for 1–2 minutes each and then with fresh xylene twice for 5–10 minutes.

8. Mount slides in mounting fluid (e.g., Permount). If the slides turn white at this point, it is necessary to repeat the dehydration steps with fresh solutions. Remove the coverslips by soaking in xylene (overnight may be necessary),

rehydrate the slide by going through decreasing percentages of ethanol to 70%, and then back up through fresh solutions of increasing ethanol percentages into fresh xylene.

In Situ Hybridization of Whole-mount Embryos with RNA Probes

This procedure was provided by Thomas Lufkin, Brookdale Center for Developmental and Molecular Biology, Mount Sinai School of Medicine, New York, NY 10029.

This protocol is presented in the following steps and sections:

1. Preparing embryos

2. Preparing the probe

3. In situ hybridization procedure

STEP 1: PREPARING EMBRYOS

MATERIALS

EMBRYOS

Embryos of the desired stage

REAGENTS

DMEM + 10% serum

Fixative: 4% paraformaldehyde<!> in PBS (see p. 649)

Methanol<!>, 25%, 50%, 75%, and 100%, in PTW. PTW is phosphate-buffered saline (PBS), Ca^{++}/Mg^{++} (see Appendix 1) or Dulbecco's modified Eagle's medium (DMEM) with 10% serum

PBS containing 0.1% Tween 20

CAUTION: See Appendix 2 for appropriate handling of materials marked with <!>.

PROCEDURE

1. Before fixation, dissect the embryos from the decidua (see Chapter 5) in PBS or DMEM with 10% serum (to reduce embryo stickiness) and remove or retract the yolk sac and amnion (these will trap the probe).

2. Wash the embryos in PBS and fix them in 10 ml of fresh fixative at 4°C. Incubation time depends on the size of the embryo (e.g., 45 minutes for 8–8.5 dpc and younger, 1.5–2 hours for 9.5 dpc, and overnight for older embryos). Do not overfix.

3. Wash the fixed embryos twice with 10 ml of PTW on ice.

4. To store the embryos, dehydrate them through a series of methanols in PTW (25%, 50%, 75%, and twice in 100%, 5 minutes each, on ice). For optimal in situ hybridization, embryos can be stored for up to one month in methanol at –20°C. In situ hybridization is still possible but may not be optimal for embryos stored for up to one year in methanol at –20°C.

COMMENT

Perform experiments on groups of 10 or more embryos to allow for accidental losses and to simplify the interpretation of results. The major complication encountered is background caused by trapping of probe and/or antibody in spaces such as the amniotic cavity, heart, or brain ventricles. High-background problems become particularly acute in older embryos; therefore, it is a good idea to puncture the brain and heart (after fixation) with sharp forceps or a syringe needle to allow solutions to exchange and to increase the volume and time of washes. Embryos of 10.5 days or older can be hemisected with a razor blade after they are in 100% methanol and have become firm.

STEP 2: PREPARING THE PROBE

Antisense digoxigenin-labeled riboprobes are synthesized as run-off transcripts from linearized plasmid templates using bacteriophage RNA polymerases (T3, T7, SP6) under standard conditions. Note that the way the probe is handled after synthesis is altered because of the difference in solubilities of the steroid-modified RNA. Plasmids are prepared, for example, by the Qiagen method (see manufacturer's instructions; Qiagen) or CsCl banding and then extracted with phenol after restriction enzyme digestion. Probes ranging from 250 to 1500 bp have been used successfully. To date, no improvement in signal or signal/noise has been seen using probes reduced in size by controlled alkaline hydrolysis. Reaction mixtures are precipitated with ethanol in the presence of NaCl. They are resuspended in DEPC-treated water and are stable at –20°C for many months. Optimal probe concentration is in the 100 ng to 1 µg/ml range and should be determined empirically. See comment (p. 682) about preparing riboprobes.

MATERIALS

EQUIPMENT
Electrophoresis equipment
Microcentrifuge
Microfuge tubes
Water bath, 37°C

REAGENTS
Agarose gel, 1%, containing 0.5 µg/ml ethidium bromide <!>
DIG RNA labeling mix (Boehringer Mannheim 1277 0732)
Ethanol <!>, 70%
NaCl, 0.3 M in 1X TE
Plasmid DNA, linearized

RNA polymerase (SP6 20 U/μl, New England Biolabs M0207S; T7 50 U/μl, New England Biolabs M0251S; T3 50 U/μl, Stratagene 600111)
RNAsin Ribonuclease Inhibitor, 40 U/μl (Promega N2111)
Transcription buffer, 10x (provided with RNA polymerase)

CAUTION: See Appendix 2 for appropriate handling of materials marked with <!>.

PROCEDURE

1. Add the following to a microfuge tube:

    1 μg of linearized plasmid DNA

    2 μl of 10x DIG RNA labeling mix

    2 μl of 10x transcription buffer

    1 μl of RNAsin

    1 μl of RNA polymerase

    H_2O to a give a final volume of 20 μl

2. Mix, spin down quickly, and then incubate for 2 hours in a 37°C water bath. No DNase treatment of the sample is required after this step.

3. Stop and precipitate the reaction with 0.3 M NaCl in 1x TE. Spin in a microcentrifuge for 20 minutes to pellet the precipitate. Remove the supernatant and wash the pellet with 70% ethanol.

4. Resuspend the pellet in 50 μl of H_2O and store at –20°C.

5. The labeling reaction should yield 5–20 μg of labeled RNA (at a concentration of 0.05–0.2 μg/μl). To check this, remove a 1-μl aliquot and electrophorese on a 1% agarose gel containing 0.5 μg/ml ethidium bromide. An RNA band approximately tenfold more intense than the plasmid band should be seen, indicating that ~10 μg of probe has been synthesized.

STEP 3: IN SITU HYBRIDIZATION PROCEDURE

MATERIALS

EQUIPMENT
Microfuge tubes, 1.5-ml, or 4-ml glass vials with screw caps
Nutator
Oven, 65–70°C
Water bath, 65–70°C

REAGENTS
Alkaline phosphatase buffer (NTMT)
    100 mM NaCl
    100 mM Tris<!> (pH 9.5)
    50 mM $MgCl_2$<!>
    0.1% Tween 20
    Best if made fresh from stocks because precipitate tends to form on standing.

BM blocking reagent (Boehringer Mannheim 1096-176)

Make 10% stock in MAB (no Tween-20) heating to dissolve, then autoclave, aliquot, and freeze. Works better than "embryo powder" and is easier and more reproducible to make. We have used MABT thereafter for consistency, but maleate does not buffer well, and alternative buffers probably substitute as well or better.

DIG labeling mix (Boehringer 1 277 073)

Glutaraldehyde<!>(25% stored at room temperature)

Hybridization mix, can be stored at –20°C

1.3x SSC
5 mM EDTA (pH 8.0)
50% formamide<!>
CHAPS, 0.5%, (Sigma C3023)
Heparin, 100 µg/ml (Sigma H9399)
Tween-20, 0.2% (Sigma P7949)
Yeast RNA, 50 µg/ml

MABT (pH 7.5)

Maleic acid<!>, 100 mM (Sigma M0375)
NaCl, 150 mM
Tween-20, 0.1%

Paraformaldehyde<!>, 4% (prepare fresh) (see General Techniques, Protocol 16.3)

PBS*, 10x (100 ml: 8 g of NaCl, 0.2 g of KCl<!>, 1.44 g of Na_2HPO_4<!>, 0.24 g of KH_2PO_4<!>) (see Appendix 1)

Proteinase K (Sigma P2308)

PTW* is PBS with 0.1% Tween 20

Reagents marked with asterisk should be RNase-free.

Purple AP substrate (Boehringer Mannheim 1 442 074)

This is more sensitive than NBT/BCIP.

Sheep anti-DIG Fab conjugated to calf intestinal alkaline phosphatase (Boehringer Mannheim 1093 274)

Sheep serum for diluting the antiserum and blocking nonspecific sites in the embryo

Because the anti-DIG is from sheep, it is advisable to use serum from the same species for this step. However, other sera may also give good results. Heat the serum to 70°C for 30 minutes before use to destroy endogenous alkaline phosphatase activity. If the serum gels (denatures) during heating, it can still be used but must be mixed well after diluting. Alternatively, heat the denatured serum to 55°C only (good results have been obtained by those who have tried it).

PROCEDURE

Day 1

Pretreatments and Hybridization

Steps 1–9 are carried out in either 1.5-ml microtubes or 4-ml glass vials at room temperature unless otherwise stated. In this recipe, rinses are immediate, and washes are for 5 minutes unless otherwise stated.

1. Rehydrate embryos through 75%, 50%, 25% methanol/PTW (allowing embryos to settle), and wash twice with PTW.

2. Treat embryos with 10 µg/ml Proteinase K in PTW. For mouse embryos: 6.5–7.5 dpc, 5 minutes; thereafter 5 minutes extra/day. 9.5 dpc, 15 minutes; 10.5 dpc, 20 minutes. Decrease Proteinase K to 0.5 µg/ml for only 5 minutes for probes for the apical ectodermal ridge (AER) or superficial tissues such as ectoderm. The thickness of the tissue determines the strength and length of Proteinase K treatment. The thinner the tissue, the more care needs to be taken, and Proteinase K treatment may not be required. For thick tissues or organs, a longer Proteinase K treatment may be necessary.

3. Remove Proteinase K, rinse briefly (take care!) with PTW, and post-fix for 20 minutes in 4% paraformaldehyde + 0.1% glutaraldehyde in PTW. A 2 mg/ml amount of glycine can be included in the PTW washes to guarantee the arrest of Proteinase K activity.

4. Rinse and wash once with PTW. Transfer the embryos to 1.5-ml microtubes or 4-ml glass vials.

5. Rinse once with 1:1 PTW/hybridization mix. Let the embryos settle.

6. Rinse with 1 ml of hybridization mix. Let the embryos settle.

7. Replace with 1 ml of hybridization mix and incubate for 1–24 hours at 65–70°C. The embryos can be stored at –20°C before or after prehybridization.

8. Replace with 1 ml of prewarmed hybridization mix and ~1 µg/ml DIG-labeled RNA probe (as little as 0.1 µg/ml can work). Immediately place at 65–70°C.

9. Incubate 12–48 hours (48 hours is preferred) at 65–70°C. Gentle rocking or shaking can be useful.

Day 2

Post-hybridization Washes

After each 65–70°C wash, let the embryos settle by incubating tube vertically at 65–70°C in a heating block, then change supernatants individually so samples do not cool. Keep wash solutions at 65–70°C in water bath.

1. Rinse twice with prewarmed (65–70°C) hybridization mix.

2. Wash twice for 30 minutes each at 65–70°C with 1.5 ml of prewarmed hybridization mix.

3. Wash 10 minutes at 65–70°C with 1.5 ml of prewarmed 1:1 hybridization mix/MABT.

4. Rinse twice with 1.5 ml of MABT.

5. Wash once for 15 minutes with 1.5 ml of MABT.

6. Incubate for 1 hour with 1.5 ml of MABT + 2% Boehringer Blocking Reagent (BBR).

7. Incubate for at least 1 hour in 1.5 ml of MABT + 2% BBR + 20% heat-treated lamb or sheep serum.

8. Incubate overnight at 4°C (or 4 hours at room temperature) in 1 ml of fresh MABT + 2% BBR + 20% sheep serum + 1/2,000 to 1/10,000 dilution of AP-anti-DIG antibody.

Day 3

Post-antibody Washes

1. Rinse three times with 1.5 ml of MABT. Transfer to glass 4-ml or 20-ml scintillation vial.

2. Wash at least five times for 2–4 hours each and then minimally overnight (what is preferred and will lower the background is to wash up to 4 days at 4°C in the cold room and change the wash 2–3 times per day) with 10–20 ml of MABT, by rolling, rotating, or rocking.

Day 4

Histochemistry

1. Wash three times for 10–60 minutes each with 10–20 ml of NTMT.

2. Incubate with enough Purple AP substrate to cover embryos at 4–10°C for 1–4 days until background starts to come up. Wrap vials in foil or keep in dark drawer during this step.

3. When color has developed to the desired extent, rinse once and wash at least twice with PTW. Refix in 4% paraformaldehyde/0.1% glutaraldehyde/PTW for 2 hours (at room temperature) or overnight (at 4°C). Rinse once and wash two times for 10 minutes with PTW. Store at 4°C in PTW + 0.1% azide.

COMMENT

For preparing riboprobes, it is important that the plasmids are completely linearized. A small amount of the undigested plasmid DNA can give rise to very long transcripts, which will incorporate a substantial fraction of the labeled rNTP. After the restriction digestion, extract the linearized plasmids with phenol:chloroform:isoamyl alcohol, ethanol-precipitate, and suspend in the DEPC-treated water. Since the extraneous transcripts have been reported to produce in addition to the expected transcript when the DNA templates contain 3′ overhangs, it is highly recommended that plasmids should not be linearized with any enzyme that leaves a 3′ overhang, such as *Kpn1*, *Pst1*, *SacI*, and *SacII*. If there is no alternative restriction site, the 3′ overhang should be converted to a blunt end using the DNA Polymerase I Klenow fragment or T4 DNA polymerase.

Imaging Embryos after Whole-mount In Situ Hybridization

This protocol was provided by Maki Wakamiya, Department of Molecular Genetics, M. D. Anderson Cancer Center, Houston, Texas 77030.

MATERIALS

EMBRYOS

In situ hybridization-processed embryos

EQUIPMENT

Camera, film or digital
Dissecting microscope with bright-field transillumination and tilt mirror
Forceps, watchmaker's #5
Petri dish, glass
Slide film, Kodak Elite Chrome Tungsten 160T

REAGENTS

Agarose/PBS, 0.4–0.8% in PBT
Glycerol/PBT series, 1:4, 2:3, 3:2, 4:1

PROCEDURE

1. Clear the embryos by washing in a 1:4, 2:3, 3:2, 4:1 glycerol/PBT series. When the embryos are placed in each solution, they will initially float. Wait until the embryos sink before transferring them to the next solution.

   Embryos should be photodocumented immediately because glycerol causes the in situ stain to fade.

2. Photograph the cleared embryos in a clean glass petri dish containing 4:1 glycerol/PBT. The viscosity of the solution helps to maintain the embryos in a desired position.

3. Embryos that have not been cleared can be physically stabilized by placing them in a dish with a layer of 0.4–0.8% agarose/PBS filled with PBT. A small hole in the agarose can be made to cradle the embryo in the desired position.

4. Different lighting methods should be explored. In general, embryos up to 8.5 dpc are best photodocumented using bright-field and transillumination. Use the tilt mirror of the dissecting microscope to enhance contrast. Embryos at 9.5 dpc and older are best photodocumented using dark-field illumination. Thicker or more opaque samples (e.g., 13.5-dpc fetuses) are visualized using reflected illumination, either from above, from the sides, or both.

5. If a film camera is used, multiple photographs are taken using various exposure times in which various parts of the embryo are focused. This should ensure that a subset of the photographs will be correct.

COMMENT

To genotype prestreak and gastrula-stage embryos after whole-mount in situ hybridization, see Chapter 12, Identification and Genotyping of Genetically Engineered Mice.

Sectioning Embryos after Whole-mount In Situ Hybridization

FROZEN SECTIONS

This procedure was provided by Michael Shen, Center for Advanced Biotechnology and Medicine, Rutgers University, Robert Wood Johnson Medical School, Piscataway, New Jersey 08854.

MATERIALS

EMBRYOS
Post-fixed embryos

EQUIPMENT
Aluminum foil
Cryotome
Dissecting microscope
Glass slides
Tube

REAGENTS
Aqua Poly/Mount (Polysciences 18606)
Dry ice<!>
2-Methylbutane<!>
OCT<!> (Tissue-Tek) and 30% sucrose/PBS in 1:1 ratio
Sucrose, 30% in 1x PBS (pH 7.4)

CAUTION: See Appendix 2 for appropriate handling of materials marked with <!>.

PROCEDURE

1. Wash post-fixed embryos through three changes of PBS to remove glycerol.
2. Transfer the embryos into 30% sucrose in 1x PBS (pH 7.4), and incubate with gentle shaking until the embryos sink to the bottom of the tube; this may take up to 2 days, depending on the size of the embryos.
3. Transfer the embryos to a mixture of OCT and 30% sucrose/PBS in a 1:1 ratio and shake gently at room temperature for 2 hours. Embryos can be stored at 4°C in 30% sucrose or the OCT/30% sucrose mixture for up to 1 week.
4. Embed the embryos in OCT under a dissecting microscope to ensure the desired orientation of the specimens, and then snap-freeze by immersion in a solution of 2-methylbutane cooled on dry ice.

5. Store blocks wrapped in aluminum foil at –80°C.

6. Section the blocks on a cryotome onto glass slides.

7. Mount the sections in Aqua Poly/Mount.

WAX SECTIONS

Embryos or tissues should be dehydrated through an ethanol series, cleared in xylene, and embedded in wax. Sections are prepared as described in the General Techniques section (see Protocol 16.6). Eosin is a good counterstain for contrast with the blue/purple precipitate from the alkaline phosphatase reaction. To counterstain, proceed as described below.

MATERIALS

EMBRYOS
Fixed embryos

EQUIPMENT
Slide warmer

REAGENTS
Eosin stock, 1%
Ethanol<!>, 30%, 70%, 95%, 100%
Glacial acetic acid<!>
Permount<!>
Xylene<!>

CAUTION: See Appendix 2 for appropriate handling of materials marked with <!>.

PROCEDURE

1. Dilute 1% eosin stock 1:50 in water and acidify with ~2 drops of glacial acetic acid per 50 ml until the solution changes from red to orange.

2. Wash the slides twice in xylene to remove wax and pass down an ethanol series (see Protocol 16.8) to 30% ethanol.

3. Quickly dip the slides in distilled water and then submerge them in the diluted eosin stock for 30 seconds to 2 minutes (depending on the desired level of staining) with gentle shaking.

4. Quickly dip the slides in 70% ethanol, bypassing the more dilute ethanol concentrations, which will leach out the eosin stain. Dehydrate in 95% and 100% ethanol (two times each). Clear the slide in xylene (second wash in fresh xylene) and mount in Permount. Allow to dry on a slide warmer overnight.

Staining for β-galactosidase (*lacZ*) Activity

The following techniques are persented in this protocol:

1. Staining whole embryos
2. Imaging X-gal-stained embryos
3. Staining frozen sections

TECHNIQUE 1: STAINING WHOLE EMBRYOS

MATERIALS

EQUIPMENT
Coated slides
Cryotome
Materials for sectioning (see General Techniques section, Protocol 16.6)

REAGENTS
Distilled water
Ethanol<!>, 70%, at 4°C
Detergent rinse
    0.1 M phosphate buffer (pH 7.3)
    2 mM $MgCl_2$<!>
    0.01% sodium deoxycholate<!>
    0.02% Nonidet P-40 (NP-40; Sigma N6507)
Fixative for frozen sections
    0.2% paraformaldehyde<!> in 0.1 M PIPES buffer (pH 6.9) (Sigma P9291)
    2 mM $MgCl_2$<!>
    5 mM EGTA

> For making paraformaldehyde solutions, see General Techniques section, Protocol 16.3.

Fixative for staining intact embryos (prepare fresh)
    0.1 M phosphate buffer (pH 7.3)
    0.2% glutaraldehyde<!> (Sigma G6257)
    5 mM EGTA (made from a stock of 0.1 M at pH 8.0)
    2 mM $MgCl_2$<!>
X-gal (5-bromo-4-chloro-3-indolyl-β-D-galactopyranoside)<!> (e.g., Invitrogen Life Technologies 15520034)

> X-gal is very expensive and should be used carefully. Use only glass containers and glass pipettes. Make a 25 mg/ml stock in dimethylformamide<!> (Sigma D8654). Store at –20°C protected from light.

Histoclear (or xylene)<!>
OCT<!> (see Harlow and Lane 1988)
Orange G (1% w/v in 2% tungstophosphoric acid)
PBS (see Appendix 1) plus 2 mM MgCl$_2$<!> and 30% sucrose
Staining solution
    0.1 M phosphate buffer (pH 7.3)
    2 mM MgCl$_2$<!>
    0.01% sodium deoxycholate<!>
    0.02% NP-40
    5 mM potassium ferricyanide<!>
    5 mM potassium ferrocyanide<!>

Dilute X-gal stock to give a final concentration of 1 mg/ml. If staining is to be performed for more than 1 hour, it is best to add Tris (pH 7.3) to the staining solution to give a final concentration of 20 mM. After use, the stain can be filtered and reused and is good for months. Store at 4°C protected from light.

CAUTION: See Appendix 2 for appropriate handling of materials marked with <!>.

PROCEDURE

1. Fixation:

    a. Dissect embryos free of their extraembryonic membranes before transferring them to fixative. Before fixing, slice embryos 13 days or older in half sagittally with a razor blade to facilitate penetration. Alternatively, with embryos from 15 dpc to birth, it is advisable to perfuse the pregnant mouse.

    b. Fix the embryos in a 0.2% glutaraldehyde solution for best results; however, if *lacZ* histochemistry is to be followed by antibody staining, 0.4% paraformaldehyde fixation can be used instead. The fixation time depends on the size of the sample: Tissue culture cells need only 5 minutes, early postimplantation embryos need 10–15 minutes, and embryos up to ~13 dpc require 15–30 minutes.

    c. When examining many embryos at one time, place them individually in wells of a 24-well tissue culture dish and aspirate the fixative solutions with a Pasteur pipette.

2. Rinse the fixed tissues in detergent rinse three times at room temperature for 15–30 minutes.

3. The staining time depends on the size of the sample and the level of bacterial β-galactosidase activity. Incubate small samples at 37°C for 1–3 hours in the dark and older embryos at 37°C for 4–5 hours or longer in the dark. For longer incubations, it is important to add Tris buffer (pH 7.3) to the staining solution. Store at 4°C in 70% ethanol.

4. After staining, the embryos can be sectioned to observe *lacZ* expression at the cellular level.

    a. Dehydrate the embryos through alcohol (70% ethanol, 90% ethanol, 95% ethanol, 100% ethanol to xylene) and embed them in wax (see General

Techniques section, Protocol 16.5). Do not leave the samples in alcohols and solvents longer than necessary because they may leach out the reaction product. Nuclear fast red can be used as a counterstain.

b. When sections are viewed by dark-field microscopy, the *lacZ* stain is pink and the contrast with surrounding tissues is good; therefore, no counterstaining is necessary.

TECHNIQUE 2: IMAGING X-GAL STAINED EMBRYOS

This procedure was provided by Akio Kobayashi, Department of Molecular Genetics, M. D. Anderson Cancer Center, Houston, Texas 77030.

MATERIALS

EMBRYOS
X-gal-stained embryos

EQUIPMENT
Camera, film or digital
Dissecting microscope
Forceps, watchmaker's #5
Slide film, Kodak Elite Chrome Tungsten 160T

REAGENTS
Agarose, 1%, prepared in PBS
Paraformaldehyde<!>, 4% in PBS
Phosphate-buffered saline (PBS)

CAUTION: See Appendix 2 for appropriate handling of materials marked with <!>.

PROCEDURE

1. Postfix X-gal-stained embryos in 4% paraformaldehyde in PBS overnight at 4°C.

This firms up the embryo, making it less fragile to subsequent manipulations, and also removes the yellow color from the X-gal staining solution. The postfixation treatment is also important for subsequent histological analysis and, if desired, facilitates further dissection of the embryo.

2. Rinse the embryo several times in PBS.

3. If internal embryonic structures are being imaged from later-stage embryos, remove unnecessary tissues using watchmaker's #5 forceps.

4. Place the embryo in a petri dish filled with a layer of 1% agarose (prepared in PBS) covered in PBS.

A small hole can be created in the agarose to help cradle the embryo in the desired orientation.

5. Photodocument the embryos using a dissecting microscope.

Different lighting methods should be explored. In general, embryos up to 9.5 dpc are best photodocumented using bright field and transillumination. Use the tilt mirror of the microscope to enhance contrast. Embryos at 10.5 days and older are best photodocumented using dark-field illumination. Thicker or more opaque samples (e.g., 14.5-dpc fetuses) are visualized using reflected illumination, either from above, from the sides, or both.

TECHNIQUE 3: STAINING FROZEN SECTIONS

MATERIALS

EMBRYOS
Freshly fixed embryos

EQUIPMENT
Cryotome

REAGENTS
Detergent rinse (see above for Staining Whole Embryos)
Distilled water
Histoclear (or xylene)<!>
Methanol<!>, 50%, 70%, 100%
OCT<!>
Orange G (1% [w/v] in 2% tungstophosphoric acid)
Paraformaldehyde<!>, 2% solution
Phosphate-buffered saline (PBS) plus 2 mM MgCl$_2$<!>
Phosphate-buffered saline (PBS) plus 2 mM MgCl$_2$<!> and 30% sucrose
Stain (see above for Staining Whole Embryos)
X-gal

CAUTION: See Appendix 2 for appropriate handling of materials marked with <!>.

PROCEDURE

1. Use freshly fixed material. Fix by immersion in 0.2% paraformaldehyde solution (see p. 687) at 4°C. Overnight fixation is sufficient for embryos up to 10 dpc.

2. Incubate in PBS plus 2 mM MgCl$_2$ and 30% sucrose at 4°C overnight.

3. Embed and freeze the tissue in OCT on dry ice (see Harlow and Lane 1988).

4. Section the sample on a cryotome onto coated slides (see General Techniques section, Protocol 16.6).

5. Postfix the sections in 0.2% paraformaldehyde solution on ice for 10 minutes.

6. Rinse the slides in PBS with 2 mM MgCl$_2$, followed by a 10-minute wash in the same solution. Both washes should be on ice.

7. Place the slides in detergent rinse (see p. 687) on ice for 10 minutes.

8. Stain the sections at 37°C for 2–3 hours in the dark.

9. Wash the slides twice in PBS plus 2 mM $MgCl_2$ at room temperature for 5 minutes each.

10. Rinse the slides in distilled water.

11. Counterstain the sections for 30 seconds in Orange G (1% [w/v] in 2% tungstophosphoric acid).

12. Wash three times in distilled water for 5 minutes each.

13. Dehydrate the sections through methanol (5 minutes each in 50%, 70%, and 100% methanol).

14. Clear sections twice for 5 minutes each in Histoclear (or xylene).

15. Mount (e.g., Permount).

Staining for Alkaline Phosphatase Activity

MATERIALS

REAGENTS

AP buffer
  100 mM Tris-HCl<!> (pH 9.5)
  100 mM NaCl
  10 mM MgCl$_2$<!>
BM Purple AP Substrate (Roche Applied Science 1442074)
Paraformaldehyde<!>, 4%, in PBS, freshly prepared
Ethanol<!>, 70%
PBS
PMT detergent rinse (PBS, MgCl$_2$<!>, 2 mM, 0.1% Tween-20)

CAUTION: See Appendix 2 for appropriate handling of materials marked with <!>.

PROCEDURE

1. Fix embryos in 4% paraformaldehyde in PBS for 20 minutes to overnight at 4°C.

2. Rinse the embryos two or three times in PBS.

3. Heat-inactivate endogenous alkaline phosphatase activity by incubating the embryos in PBS at 70–75°C for 10–30 minutes.

4. Rinse the embryos in PBS at room temperature.

5. Wash the embryos in AP buffer for 10 minutes.

6. Stain the embryos in BM Purple AP Substrate at 4°C for 0.5 to 36 hours.

7. Wash the embryos extensively in PMT.

8. Store at 4°C in 70% ethanol.

COMMENTS

• AP-stained embryos can be photodocumented like X-gal-stained embryos (Protocol 16.15, Technique 2).

• It is possible to double-stain whole-mount embryos for both β-galactosidase and AP activities. Embryos are first stained for β-galactosidase activity, rinsed in PBS, and then stained for AP activity (steps 3–8).

Visualization of GFP Expression in Whole-mount Postimplantation-stage Embryos

MATERIALS

EQUIPMENT
Plastic petri dishes, 60 or 100 mm
Surgical instruments

REAGENTS
Paraformaldehyde<!>, fresh 4%
Phosphate buffer, 0.1 M (pH 7.3)

CAUTION: See Appendix 2 for appropriate handling of materials marked with <!>.

PROCEDURE

1. Dissect the embryos from the uterus and remove maternal tissues.

2. Wash the embryos through cold phosphate buffer to remove cellular debris and blood.

3. *Optional:* If the embryos are to be processed further for histology, fix them in paraformaldehyde at 4°C for the minimum amount of time required. (Do not overfix because this will inhibit the activity of the fluorescent protein.) Change the fix to phosphate buffer.

4. Observe GFP activity under the dissecting scope equipped with a GFP-visualizing light source.

COMMENTS

- If the GFP expression is specific to certain organs, tissues, or cell types, and fine dissection of the embryo is required, it is advantageous to perform the dissection under GFP-visualizing conditions.

- Dissection of very early postimplantation embryos, before obvious decidual reaction (i.e., 5.5 dpc), is difficult due to the extremely small size of the implanted embryo. GFP transgenic embryos can facilitate the dissection of this stage if the recovery is performed under the GFP-visualizing microscope.

Observation of GFP-fusion Proteins in Living Cells

MATERIALS

CELLS

Live cells growing in tissue culture

EQUIPMENT

Microscope (either dissecting or inverted compound) with GFP visualization capabilities

Tissue culture dishes or glass-bottomed tissue culture dishes (MatTek Corporation) or multi-chamber culture plates (Nalgene Labtek 2 chambers)

REAGENTS

Culture medium appropriate for the cells used

Phenol-red-free cell growth medium (Gibco) or PBS with Ca^{++} and Mg^{++}

PROCEDURE

1. Grow the cells to the desired stage, confluence, or time.

2. Replace the culture medium with phenol-red-free medium or PBS with Ca^{++} and Mg^{++}.

3. Place the chamber or tissue culture plate on the microscope stage and use appropriate filters to visualize GFP.

4. Replace the phenol-red-free medium with culture medium and place the chamber or tissue culture plate back into the incubator.

COMMENTS

- If GFP observation takes place outside a sterile hood, care should be taken to retain the sterility of the cell culture.

- It is essential to observe GFP expression in phenol-red-free conditions because phenol red causes significant background fluorescence.

- If a low-power dissecting microscope is used, be aware that dust particles attached to the bottom of the plate or chamber frequently autofluoresce and could be mistaken for a positive cell or colony. The best way to avoid being misled is to check frequently whether the microscope is perfectly focused on the cell layer by turning the normal light on and off.

Intracellular Observation of GFP in Fixed Cells

MATERIALS

EQUIPMENT

Coverslips, prepared sterile (coating might be necessary for cell attachment)
Microscope equipped with GFP visualization
Microscope slides
Whatman filter paper

REAGENTS

Mounting medium (Molecular Probes Inc., Prolong Antifade Kit, P-7481)
Paraformaldehyde<!>, 2%, in PBS (pH 7.4)

CAUTION: See Appendix 2 for appropriate handling of materials marked with <!>.

PROCEDURE

1. Place a coverslip into a tissue culture plate and plate the cells on the coverslip.

2. Grow the cells for desired confluence or length of time.

3. Fix the cells in 2% paraformaldehyde in PBS (pH 7.4) for 15 minutes at room temperature.

4. Rinse the cells in 4 ml of PBS (pH 7.4) and then quickly wash the coverslips in 2 ml of PBS (pH 7.4) twice for 5 minutes at room temperature.

5. Place 10 µl of mounting medium onto the middle of a glass microscopy slide.

6. Blot the edges of the coverslip gently against a piece of filter paper to remove excessive PBS.

7. Place the coverslip with the attached cells facing down onto the mounting medium.

8. Observe the fluorescence.

COMMENTS

- Do not fix in methanol or acetic acid because organic solvents destroy the light emission of GFP.
- This protocol can be used not only for GFP spectral variants but for dsRed as well.

Fixation and Paraffin-embedding for GFP Visualization

This protocol was provided by Hugo Vankelecom, Laboratory of Cell Pharmacology, University of Leuven, Leuven, Belgium

MATERIALS

EMBRYOS

Fresh embryos or tissue

EQUIPMENT

Dark box to keep the material in the dark as much as possible during processing

REAGENTS

Ethanol<!>, 70%, 98%, and 100%

Mounting medium (Molecular Probes, Prolong Antifade Kit P-7481)

Paraffin (Paraplast Plus, Sherwood Medical), at a temperature no higher than 58°C

Paraformaldehyde<!> solution, 4% (pH 7–7.5) in PBS

Phosphate-buffered saline (PBS) (pH 7.4)

Xylene<!> (instead of butanol<!>) gives brighter signals. Alternatively, AmeriClear histology clearing solvent can be used (Scientific Products C4200-1), which is nontoxic.

CAUTION: See Appendix 2 for appropriate handling of materials marked with <!>.

PROCEDURE

Fixation

1. Dissect the embryos or tissue and then transfer to PBS (pH 7.4, at room temperature).

2. Remove the cell debris with a PBS wash.

3. Keep the material in the dark as much as possible.

4. Discard the PBS and add 4% paraformaldehyde.

5. Fixation time depends on the size of the tissue. Two hours is sufficient for 200–400 Mm, 4 hours for 1–2 mm.

6. Discard fix, add PBS, and leave for 10 minutes at room temperature; repeat this step twice.

Paraffin-embedding

1. Remove PBS as much as possible and dehydrate using the following series:

   a. 2x 70% ethanol (leave at room temperature for 15 minutes each time)

   b. 2x 98% ethanol (leave at room temperature for 15 minutes each time)

   c. 2x 100% ethanol (leave at room temperature for 15 minutes each time)

2. Transfer the specimen in 100% ethanol to a metal holder.

3. Remove as much ethanol as possible, add xylene, and leave for 15 minutes at room temperature. Repeat this step once.

4. Remove as much xylene as possible, add paraffin (temperature 56–58°C) and leave for 10–30 minutes at 56–58°C.

5. Remove as much paraffin as possible, add new paraffin (56–58°C), and leave for 10–30 minutes at 56–58°C; repeat this step two or three times.

6. Let the embryos or tissues set at room temperature from 1 hour to overnight.

7. Keep refrigerated until.

8. Proceed with microtome sections.

9. Observe fluorescence.

COMMENTS

- The GFP emission is significantly lower if the pH of paraformaldehyde solution is below 6.5 or above 8.
- Do not expose the specimen to temperatures higher than 58°C. Higher temperatures may weaken the signal.
- Use xylene instead of butanol for embedding. It gives brighter signals.
- Fixation time depends on the size of the tissue.

Alcian Blue Staining of the Fetal Cartilaginous Skeleton

This protocol is as described by Jegalian and De Robertis (1992).

MATERIALS

EMBRYOS
Mouse embryos

REAGENTS
Acetic acid<!>, 5%
0.05% alcian blue 8GX (Fisher), 5% acetic acid (best if made fresh or at least periodically)
Ammonium hydroxide<!> (NH₄OH), 0.1%, 70% ethanol<!>
Benzyl alcohol<!>/benzyl benzoate<!> (BABB), 1:2
Bouin's fixative (Polysciences)
Methanol<!>

CAUTION: See Appendix 2 for appropriate handling of materials marked with <!>.

PROCEDURE

1. Dissect 12.5- to 16.5-dpc embryos in PBS and remove extraembryonic membranes, which can be used for genotyping if necessary.

2. Fix the embryos in Bouin's solution for 2 hours.

3. Wash the embryos in a solution of 70% ethanol plus 0.1% NH₄OH for ~24 hours until the embryos appear white with no remaining yellow color. Six to eight changes may be required.

4. Equilibrate in 5% acetic acid (two changes, each for one hour).

5. Stain the embryos in 0.05% alcian blue in 5% acetic acid for 2 hours.

6. Wash the embryos twice with 5% acetic acid for 1 hour each.

7. Clear embryos by washing them in methanol (two times 1 hour) and then in 1:2 benzyl alcohol/benzylbenzoate (BABB). Use glass containers (e.g., scintillation vials) when using BABB.

COMMENT

Staining has been shown to last at least 2 years in BABB. However, staining has been observed to weaken over the course of months. Therefore, it is advisable to document results immediately.

Alcian Blue/Alizarin Red Staining of Cartilage and Bone

This procedure was provided by Dmitry Ovchinnikov, Institute for Molecular Bioscience, University of Queensland, Brisbane, Australia.

MATERIALS

EMBRYOS

Fetus or neonate

EQUIPMENT

Forceps and scissors (size depends on age of fetus)
Tubes, 50-ml screw-cap, clear plastic

REAGENTS

Acetone<!>
Alcian blue 8GX (Sigma A3157), 15 mg
Alcian blue stain
Alizarin red, 50 mg (Sigma A5533) per liter of 1% potassium hydroxide (KOH)
Alizarin red stain
Ethanol<!>, 95%, 70%
Glacial acetic acid<!>, 20 ml
Glycerol
Potassium hydroxide (KOH)<!>, 1% (w/v), in distilled water

CAUTION: See Appendix 2 for appropriate handling of materials marked with <!>.

PROCEDURE

1. Isolate the fetuses or euthanize neonates according to local regulations. Place the fetuses or neonates in tap water for 1–24 hours (*optional*).

2. Scald the fetuses or pups in hot tap water (65–70°C) for 20–30 seconds for easier maceration of the tissue later. Carefully peel off the skin with forceps.

3. Use forceps to eviscerate the embryos, including the contents of the peritoneal and pleural cavities.

4. Fix the embryos in 95% ethanol overnight. Be sure to remove all of the bubbles from the body cavity.

5. Transfer the embryos to acetone at room temperature overnight to remove fat.

6. Rinse briefly in deionized water. Stain for *cartilage* by placing the embryo in enough alcian blue stain to cover the body completely; 24 hours should be sufficient.

A 50-ml screw-capped plastic tube is convenient, or for newborn mice or younger fetuses, a 6-well tissue culture plate is useful. Be sure to remove all of the bubbles from the body cavity.

7. Wash the embryos in 70% ethanol for 6–8 hours. A few changes of the ethanol should be sufficient.

8. Clear the sample by placing it in 1% KOH overnight or until the tissues are visibly cleared.

9. Counterstain bone in alizarin red stain overnight. Younger fetuses need less time.

10. Clear the samples by placing them in 1% KOH/20% glycerol for 2 days or more. When the samples are appropriately cleared, place them in a 1:1 mixture of glycerol and ethanol for documentation and storage.

COMMENTS

- The limbs can be removed from the skeleton preparation very easily with forceps. This facilitates visualization of the rest of the body or documentation of the limbs in isolation.
- The rib cage can be dissected away from the skeleton preparation by cutting the ribs close to the vertebrae. Place a glass slide on top of the isolated rib cage to create a flat mount for photodocumentation.

Alizarin Red Staining of Postnatal Bone

MATERIALS

ANIMALS
Mouse, postnatal ages

EQUIPMENT
Forceps and scissors (size depends on age of mouse)

REAGENTS
Alizarin red, 50 mg (Sigma A5533) per liter of 1–2% KOH (prepare fresh)
Glycerol
Potassium hydroxide<!>(KOH), 0.5%, 1.0%, and 2.0%, in deionized or distilled water

CAUTION: See Appendix 2 for appropriate handling of materials marked with <!>.

PROCEDURE

1. Sacrifice the mouse humanely (e.g., with CO_2 to avoid damaging the neck bones). Begin the skinning procedure with a crosswise cut in the mid-dorsal region and then tear ventrally without further cutting. Remove the skin from the anterior and head by making a cut from the mid-dorsal region to the tip of the nose, followed by peeling the skin. Remove the viscera, eyes, salivary gland, etc., and as much fat and loose tissue as possible at this stage. There is no need to fix the specimen before proceeding to step 2.

   For convenience, the mouse can be frozen before skinning.

2. Place the mouse in 1% KOH for 4–5 days. For small mice (0–3 weeks), use 0.5% KOH. To speed up the process with older mice, use 2% KOH. The total time depends on the size of the animal, the strength of the KOH, and the ambient temperature. If maceration is prolonged, the skeleton will begin to fall apart; if it is too short, the connective tissue will stain red, giving a high background.

   The integrity of connective tissue on ribs 9–13 is a good indicator of when to stop.

3. Stain the skeleton in alizarin red for 2–5 days, until the bone is red. Excess stain can be removed with 1–2% KOH.

4. Transfer the skeleton to glycerol.

Visualizing the Fetal Vasculature by India Ink Injection

This procedure was provided by Andras Nagy, Mount Sinai Hospital Research Institute, Toronto, Canada.

MATERIALS

ANIMALS
Pregnant female with 12.5-dpc or later-stage embryos

EQUIPMENT
Hypodermic needle, 26-gauge
Mouth pipette assembly
Pasteur pipettes, finely drawn, with 300- to 400-μm tips
Petri dishes, 10-cm
Surgical instruments for embryo recovery
Warm plate set to 37°C

REAGENTS
India ink (such as Eberhard Faber, Black India 4415, ITEM 44001)
Phosphate-buffered saline (PBS), warm (37°C), with Ca++ and Mg++

PROCEDURE

1. Euthanize the pregnant female according to local regulations. Quickly dissect the embryos with an intact yolk sac and placenta attached.

2. Place the embryos immediately into a petri dish with warm PBS on a warm plate.

3. Fill the tip of a finely drawn Pasteur pipette with India ink (about 50 μl).

4. Place one embryo in a separate dish containing warm PBS and position it under the dissecting scope to obtain a clear view of a yolk sac vitelline vein with active blood flow.

5. Cut into one of the vitelline veins, which is still in the yolk sac wall, by gently poking the hypodermic needle through the membrane. The entry point can be located before the final branching-in into the descending main vein. Hold the embryo in position with forceps.

6. Carefully insert the tip of the glass pipette through the cut into the vein (toward the embryo) and push it through the vessel until the tip is just before the merging branch point.

7. Expel ink into the vessel with minute puffs of breath. Wait until the previous amount of ink has been flushed in by the bloodstream coming from the other branch merging at the branch point before adding more ink. The total amount of ink injected depends on your specific purpose and can be visually monitored in the embryo proper.

COMMENTS

- Only India ink can be used for this procedure because it is composed of small carbon particles. The carbon particles in the India ink range in size. Depending on your goal, you can filter or centrifuge the ink to isolate larger or smaller carbon particles. Very fine carbon particles will be able to move more easily through the blood vessels.

- The ink is distributed only if circulation is active. Therefore, working fast and keeping the embryos warm is critical.

- The amount of ink injected into the embryo determines the level of vessel labeling. Small amounts only label the large connecting vessels to the heart, whereas larger amounts of ink mark the branching small vessels. It is up to the investigator to decide what is the optimal amount of labeling needed for their study.

- After injection, it is possible to dissect the organs of interest from the embryos and study their vessel structure in isolation.

Visualizing the Fetal Great Blood Vessels by Plastic Casting

This procedure was provided by Chengyu Liu and Jim Martin, Institute of Biosciences and Technology, Texas A&M University, Houston, Texas 77030.

MATERIALS

ANIMALS
Mouse fetuses, 18.5-dpc

EQUIPMENT
Hypodermic needle, 27-gauge, 1/2-inch
Incubator set at 50°C
Syringe, 1-ml

REAGENTS
Batson's No. 17 Anatomical Corrosion Kit (Polysciences 07349)
Ice-cold water
Phosphate-buffered saline (PBS)

PROCEDURE

1. Isolate 18.5-dpc fetuses and place them in ice-cold water to stop the heart-beat.

2. Remove the sternum to expose the heart.

3. Prepare the casting dyes by mixing, in order, 20 ml of Base Solution A, 1 ml of Red Color, 1.5 ml of Catalyst B, and 1 drop of Promoter C from the kit.

4. Slowly inject PBS into the left ventricle to remove blood and potential blood clots in the proximal regions of the great vessels.

5. Immediately inject the casting dye into the left ventricle. Because the ductus arteriosus is still open at 18.5 dpc, injection into the left ventricle alone is sufficient to fill both the systemic and pulmonary circulatory systems. Continue to inject the casting dye until it is difficult to inject any more solution. Hold the needle inside the ventricle for about 1 minute and then withdraw it slowly.

6. Place the fetus into ice-cold water for half an hour to 1 hour. Then transfer it to distilled water and soak overnight at 4°C.

7. Place the embryos into 6-well cell culture dishes and completely cover with Maceration Solution from the kit. Incubate at 50°C without shaking for 12–72

hours until all of the tissues are gone. After 12 hours, wash the casts with distilled water and change the Maceration Solution if some tissue still remains. Transfer the casts to distilled water.

8. The final cast is then documented. The casts can be stored in distilled water at 4°C for at least 1 year.

REFERENCES

Baird G.S., Zacharias D.A., and Tsien R.Y. 2000. Biochemistry, mutagenesis, and oligomerization of DsRed, a red fluorescent protein from coral. *Proc. Natl. Acad. Sci.* **97:** 11984–11989.

Behringer R.R., Eldridge P.W., and Dewey M.J. 1984. Stable genotypic composition of blood cells in allophenic mice derived from congenic C57BL/6 strains. *Dev. Biol.* **101:** 251–256.

Cormack B.P., Valdivia R., and Falkow S. 1996. FACS-optimized mutants of the green fluorescent protein (GFP). *Gene* **173:** 33–38.

Dent J.A., Polson A.G., and Klymkowsky M.W. 1989. A whole-mount immunocytochemical analysis of the expression of the intermediate filament protein vimentin in *Xenopus. Development* **105:** 61–74.

de Crombrugghe B., Lefebvre V., and Nakashima K. 2002. Deconstructing the molecular biology of cartilage and bone formation. In *Mouse development* (ed. Rossant J. and Tam P.P.L.), pp. 279–295. Academic Press, San Diego.

Green M.C. 1952. A rapid method for clearing and staining specimens for the demonstration of bone. *Ohio J. Sci.* **52:** 31–33.

Hadjantonakis A.K., Gertsenstein M., Ikawa M., Okabe M., and Nagy A. 1998. Generating green fluorescent mice by germline transmission of green fluorescent ES cells. *Mech. Dev.* **76:** 79–90.

Hahnel A.C., Rappolee D.A., Millan J.L., Manes T., Ziomek C.A., Theodosiou N.G., Werb Z., Pedersen R.A., and Schultz G.A. 1990. Two alkaline phosphatase genes are expressed during early development in the mouse embryo. *Development* **110:** 555–564.

Harlow E. and Lane D. 1988. *Antibodies: A laboratory manual.* Cold Spring Harbor Laboratory, Cold Spring Harbor, New York.

Hauptmann G. and Gerster T. 1994. Two-color whole-mount in situ hybridization to vertebrate and *Drosophila* embryos. *Trends Genet.* **10:** 266.

Jegalian B.G. and De Robertis E.M. 1992. Homeotic transformations in the mouse induced by overexpression of a human *Hox3.3* transgene. *Cell* **71:** 901–910.

Jowett T. and Lettice L. 1994. Whole-mount in situ hybridization on zebrafish embryos using a mixture of digoxigenin and fluorescein-labelled probes. *Trends Genet.* **10:** 73–74.

LeMotte P.K., Kuroiwa A., Fessler L.I., and Gehring W.J. 1989. The homeotic gene *Sex Combs Reduced* of *Drosophila*: Gene structure and embryonic expression. *EMBO J.* **8:** 219–227.

Lyon M.F., Rastan S., and Brown S.D.M., eds. 1996. *Genetic variants and strains of the laboratory mouse*, 3rd ed. Oxford University Press, England.

Mintz B. 1971. Allophenic mice of multi-embryo origin. In *Methods in mammalian embryology.* (ed. Daniel J.C., Jr.), pp. 186–214. W.H. Freeman and Company, San Francisco.

Rivera-Pérez J.A., Wakamiya M., and Behringer R.R. 1999. *Goosecoid* acts cell autonomously in mesenchyme-derived tissues during craniofacial development. *Development* **126:** 3811–3821.

Selby P.B. 1987. A rapid method for preparing high quality alizarin stained skeletons of adult mice. *Stain Technol.* **62:** 143-146.

Strahle U., Blader P., Adam J., and Ingham P.W. 1994. A simple and efficient procedure for non-isotopic in situ hybridization to sectioned material. *Trends Genet.* **10:** 75–76.

Tsien R.Y. 1998. The green fluorescent protein. *Annu. Rev. Biochem.* **67:** 509–544.

Yang T.-T., Parisa S., Green G., Kitts P.A., Chen Y.-T., Lybarger L., Chervenak R., Patterson G.H., Piston D.W., and Kain S.R. 1998. Improved fluorescence and dual color detection with enhanced blue and green variants of the green fluorescent protein. *J. Biol. Chem.* **273:** 8212–8216.

Setting Up a Micromanipulation Lab

I N THIS CHAPTER, THE EQUIPMENT NEEDED for performing the various techniques described in the previous chapters of this book is discussed. Some basic concepts and important points for consideration are provided for establishing a micromanipulation lab. The market for supplies in the biomedical field is rapidly changing, new and better products are being released constantly, companies are fusing and changing names, and the locations of Web sites tend to change very rapidly. In addition, each laboratory is located in a unique environment, aiming for a certain range of techniques to be performed. Therefore, the recommendations given in this chapter are 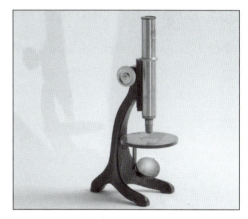 intended as a baseline for making choices in each particular case. The reader is encouraged to consult Appendix 3 for further information regarding suppliers.

CONTENTS

MICE, HUSBANDRY, AND IDENTIFICATION SYSTEMS

Inbred, Outbred, Hybrid, and Genetically Modified Mice

Mice for biomedical research can be purchased from a number of suppliers. Alternatively, a breeding colony can be maintained at an animal facility located at your institution. The decision about which alternative to choose should made in agreement with your local animal facility (see Chapter 3 for discussions on setting up and maintaining mouse colonies).

The choice of which mouse strain to use will depend entirely on the design of a particular experiment. A general discussion can be found in Chapter 3, and more-detailed recommendations can be found in the chapters describing each technique. Laboratory mice are available as either inbred, outbred (or random-bred), or hybrid genetic background strains. Inbred mice are an essential tool for investigating mutants without the influence of genetic diversity. These inbred strains, however, often show reduced breeding performance, and they sometimes display behavioral peculiarities (such as insufficient nursing ability or increased aggressivity). The C57BL/6 strain is probably the most widely used inbred strain, both as a host embryo for embryonic stem (ES) cell injection and for back-crossing of existing mutants. Outbred, or random-bred, mice of, for example, the CD1 or ICR stocks, do not possess a homogeneous genotype; however, they usually show excellent breeding performance, very well-developed mothering instincts, the capability of raising large litters, and a docile behavior. For these reasons, they are often used as foster mothers for transferred embryos. Hybrid mice are produced by crossing two inbred strains. These mice show a typical "hybrid vigor" that results in high reproductive performance and often very high quality and large numbers of embryos upon superovulation. For this reason, pronuclear injection is classically performed using F_1 hybrids as embryo donors. An ever-increasing variety of genetically modified mice is available both commercially from vendors like The Jackson Labs and through academic sources.

Consult Appendix 3 for further information regarding suppliers. Another very useful resource for suppliers of laboratory animals and all related items can be found at the LabAnimal Web site http://guide.labanimal.com/guide.-index.html.

Caging, Husbandry, and Identification

Mice can be housed in a variety of caging systems, ranging from simple open grid cages, over-filter tops, ventilated cabinets, and individually ventilated caging (IVC) systems to isolators of varying size. The more the mice are protected from each other and their environment, the lower the risk that a pathogen may spread throughout the colony. The above-mentioned systems are increasingly more expensive as more protection is offered, and the choice will depend on the local situation (type of animal facility, pathogen load, etc.) and the research to be carried out (for example, a pathogen-free colony may be vitally important for immunology studies). Whichever caging type is chosen, great care should be taken to provide the animals with comfortable living conditions. This is not only important from the animal welfare point of view, but also for optimizing breeding performance, minimizing disease spread and loss of valuable animals, and assuring the relevance of the research findings. Commonly used IVC caging units are manufactured by BioZone, Techniplast, Thoren, Charles River, Lab Products, and Allentown.

Disinfection of animal facilities and items entering the facility (such as embryos packed for transport as described in Chapter 15) is often a matter of debate. A product with which the authors have had good experience is Clidox-S, a tuberculocidal, bactericidal, virucidal, and fungicidal agent (see http://www.-pharmacal.com/prodp2.htm).

Several systems are available for the identification of mice (see Chapter 12 for more details). The ear punch (Fisher 01-337B) method is widely used in many research facilities; it is simple and can be applied quickly, but the numbering system (which is not universal!) may lead to confusion if not recorded very carefully. (A stainless steel ear punch with a handle can be purchased from AgnTho's AB and Roboz Surgical Instrument.) Small animal ear tags are available from International Market Supply and National Band & Tag. These ear tags are relatively cheap and easy to apply, and the risk of misreading the number is minimal. Note, however, that if the tag is applied incorrectly, it may fall off. Tattoo machines, which are available from AIMS and Ancare, can be used to mark very young mice; some training and experience are needed to use the apparatus correctly and efficiently. Subcutaneous microchip implants are available from companies such as Bio Medic Data Systems, Destron Fearing, Electronic ID, and Plexx. Microchips are considerably more expensive, but they may be a very good choice for identification of animals in long-term studies, since a large amount of information can be stored on the chips.

THE MICROMANIPULATION LAB AND ITS ENVIRONMENT

General Considerations

Equipping a micromanipulation laboratory is a challenging task that should be planned with great care because a number of very expensive items need to be considered. Micromanipulation requires high concentration and precise manual skills. Care should be taken to provide as undisturbed an environment as possible (e.g., it is not advisable to place the workplace in a highly frequented laboratory, or near doors that are used frequently).

It is practical to locate the micromanipulation laboratory in convenient proximity to the areas where the following procedures are carried out: (1) embryo transfer, (2) tissue culture areas (in case ES cell injection/aggregation is to be performed), (3) molecular biology equipment for DNA preparation if pronuclear injection will be performed, and (4) media preparation (chemicals, water, storage). These recommendations, of course, apply only in cases where the various procedures are performed on site, and not if, e.g., media are purchased ready made or DNAs and ES cells are delivered already prepared to the laboratory.

If the manipulated embryos are to be transferred into recipient mice in a barrier animal facility, additional precautions should be taken into consideration. The micromanipulation lab may be located behind or outside the barrier (see Chapter 3 for further details). It is difficult to judge whether the former or latter poses a greater risk of introducing pathogens. If the laboratory is located outside a barrier, the following important considerations should be taken into account:

- The lab should be regarded as a clean area, in the same category as a tissue culture lab.

- The incoming air should be High Efficiency Particulate Air (HEPA) filtered.

- Absolutely no tissues from mice that are not specific pathogen-free (SPF) should enter the micromanipulation area. Special care should be taken in cases where the same lab is used for rederivation of pathogen-loaded lines (see Chapter 15). Embryos should be collected in a Class II biological safety cabinent (BSC) to protect the operator and the environment from exposure to pathogens (see the section on laminar flow hoods and biosafety cabinets, below).

- No personnel entering any non-SPF animal facilities should have permission to enter the micromanipulation lab within at least 24 hours, according to local regulations.

- Embryos should always be cultured in media containing antibiotics.

General Equipment for the Microinjection Laboratory

Depending on which tasks will be performed, the list of equipment needed obviously varies, but the major components in most cases consist of the following:

- +4°C and –20°C refrigerator/freezers
- Electrophoresis and electroelution equipment
- Water baths
- Centrifuges
- CO_2 incubators
- Laminar flow hoods or biological safety cabinets

Refrigerators and Freezers

Basic equipment such as +4°C refrigerators and –20°C and –80°C freezers are essential components of the micromanipulation laboratory equipment. It is advisable to use a dedicated +4°C refrigerator for tissue culture/embryo culture media and solutions, and not to store these in locations containing toxic or possibly contaminated compounds. The –20°C freezers should be of the non-self-defrosting type to reduce temperature variations as much as possible. Rising and falling temperatures (also within limited ranges) are major reasons for compound degradation in media and solutions. Freezers should be equipped with automatic alarm systems to detect electrical failure in enough time that stored items can be rescued. For further information about –80°C freezers and LN_2 storage, see the Cryopreservation section, below.

Incubators

Humidified CO_2 incubators are needed for culture of embryos, ES/TS cells, and MEFs. To avoid contamination, the same incubator should not be used for these various applications. This is true not only for embryo versus cell culture, but also for primary cells (such as MEFs) versus established cell lines (ES/TS cells). A separate incubator should also be used for cells that have not yet been screened for the presence of *Mycoplasma* and other pathogens (see Chapter 8 for further discussion on screening of cell lines).

The concentration of CO_2 is generally automatically regulated and monitored by a system in the incubator. However, these gauges should not be relied upon; instead, the CO_2 level should be controlled directly in the incubator to ensure accurate pH levels. A simple apparatus, the FYRITE test kit, is available from Bacharach.

In cases when the incubator door is opened frequently, incubators that have multiple chambers within a single incubator may be useful to avoid rapid reduction of the CO_2 level. Modular incubator chambers (MIC) from, e.g., Billups-Rothenberg, which can be sealed, gassed, and placed at 37°C, may be a convenient alternative for use in embryo culture. Small microincubators can also be obtained from Harvard Apparatus.

Some incubators offer a sterilization program; others are equipped with copper plates or even a completely copper inner surface. These options all serve to minimize or prevent the risk of fungal and/or bacterial contamination. However, one should be aware that such extra features never replace the need for very strict and accurate sterile techniques as well as regular cleaning of the incubator!

For postimplantation embryo culture (see Chapter 5), an automatically regulated CO_2 supply is not needed. However, it will be necessary to rotate the culture vessels at a relatively high speed. This can be achieved either by placing a roller apparatus in a conventional incubator or by using a hybridization incubator adjusted to the high rotation speed of 30 rpm. Miniroller apparatuses are available from BTC Engineering, Precision Scientific, and Wheaton. A series of gas mixtures are required for in vitro development of postimplantation mouse embryos. Specified gas mixtures can be obtained from several suppliers. Generally, the quantities required are very small, which makes the choice of 5-kg containers most practical. Note that later embryonic stages require a higher oxygen content, which poses a safety hazard to the location where the pressurized containers are stored. All gas bottles with 40% O_2 content or higher should be equipped with a specially designed regulator valve that should never be greased when fitted! No open flame, electrostatic sparks, or flammable or combustible compounds should be in the area of these cylinders.

Laminar Flow Hoods and Biosafety Cabinets

Embryo freezing, cell culture, and, in some cases, embryo transfer surgeries, all require a sterile flow hood. The choice of hood should be made taking into account local legislation, which may prescribe which type and class should be used for certain applications.

Laminar flow hoods (also called clean benches) with both horizontal and vertical flow provide protection for the biological specimen by capturing the room air, passing it through a HEPA filter, and directing the filtered air in a laminar flow fashion (in parallel planes at a uniform velocity) over the entire work surface. The HEPA filter removes nearly all of the bacteria from the air. However, such clean benches must not be used when working with any form of biohazard, such as potentially infectious or allergenic agents, because the operator and the environment will be exposed to these agents. Clean benches may be used for aseptic medium preparation and dispensing, as well as for noninfectious animal handling (e.g., surgery) and tissue culture where allowed by local regulations.

The same principle of laminar air flow that aids in the capture and removal of airborne contaminants is incorporated into biocontainment technology. Class II biological safety cabinets (BSC) protect the specimen, the user, and the environment by passing particulate-free air across the work surface and refiltering it before it is exhausted from the building. Airflow is drawn around the operator into the front grille of the cabinet to provide personnel protection. In addition, the downward laminar flow of HEPA-filtered air protects the biological specimen by minimizing the chance of cross-contamination along the work surface of the cabinet. Because cabinet air has passed through the exhaust HEPA filter, it is contaminant-free (environmentally protected), and may be recirculated back into the laboratory (Type A BSC) or ducted out of the building (Type B BSC).

Traditionally, Class II BSCs are most commonly used for tissue culture. Class II BSCs can also be modified for special tasks. For example, the manufacturer can modify the front sash to accommodate the eyepieces of a microscope. Appropriate certification is required to ensure that the basic systems operate properly after modification. Proper aseptic practices and procedures should be strictly followed to achieve maximum containment and sterility.

Electroporation and Electroelution Equipment

Electroporation of postimplantation embryos can be performed by the use of an electroporator from NEPA Gene. For electroporation of ES cells, electroporators are available from a variety of manufacturers such as, e.g., BioRad.

For electroelution, the authors have had good experience with a simple and relatively low-cost device, the BioTrap from Schleicher & Schuell.

Stereomicroscopes and Their Equipment

Dissection Microscopes

Dissection microscopes are available from several companies, such as Leica, Nikon, Olympus, and Zeiss. For an efficient workflow, an appropriate number of dissection microscopes should be available. Generally speaking, dissection microscopes will be needed for the following procedures:

- *Embryo collection, sorting, washing, aggregation, electrofusion, and picking ES colonies.* A working distance of at least 5–6 cm is required. The optics should give a high-resolution image, at a magnification of at least 40x. A lower magnification of 16x or 25x should also be available to give a better overview. A zoom capability is highly recommended, and a light box with adjustable transmitted light from below is absolutely essential. A frosted glass plate is suggested for visualizing the zona pellucida of preimplantation-stage embryos properly.

- *Microsurgery procedures such as embryo and ovary transfer.* A large working distance is the most important aspect for these techniques. A maximum magnification of 10–20x is fully adequate; however, a zoom capability is preferable also for this application. If the transfers are performed in a hood, it may be advisable to look for a model with extra-long eyepieces to allow the mouse to be fully protected behind the airflow during surgery. Cold-light illumination is essential to provide good lighting of the surgery area without drying out the tissues by excessive heat release. Many laboratories find it practical to use two dissection microscopes in parallel for embryo transfer: one for loading the capillaries and a second for the actual surgery. It is important to have both transmitted and outside (ring, fiber optics, or other) sources of light if the same microscope is used for both embryo loading and surgery.

- *Postimplantation embryo dissection.* The working distance should be as large as possible, even if the resolution will be slightly reduced. Light- and dark-field illumination from below through an appropriate light box are required, as well as a light-ring or arms from above. It is important to find an illumination method that provides a good-quality image of thick, three-dimensional structures. A fluorescence device for GFP visualization is often a very useful addition. Microscope models that are already equipped with UV light and filters are available from most manufacturers, although this feature may increase the price considerably. In addition, UV light is damaging to embryos, and is better avoided if working with postimplantation embryo culture. An inexpensive and very good alternative is a LED-based device available from BLS that can be fitted to most microscope models. This model offers the advantage of turning the light on and off without delay; this light source also has a very long lifetime.

A discussion bridge or video camera with screen is helpful if teaching will be performed. Most companies selling dissection microscopes also have models with discussion bridges. Both video cameras and screens can usually be purchased through the major microscope companies (such as Leica, Nikon, Olympus, and Zeiss). With the use of simple adapter rings, the same camera can usually be moved among several microscopes of different models, reducing the investment costs. The screens most commonly sold with these cameras are small, and, although the resolution is high, the screen may be inadequate for teaching larger groups of people. A solution here could be to invest in a larger screen, which may be fitted either in a rotatable arm in the ceiling or on a shelf high up in a corner of the lab where it does not take up valuable space.

Light sources and fiber optic illuminators are available from Sutter Instrument, The Glass Worx, and Schott-Fostec LCC. Replacement bulbs for most microscope models, such as mercury arc lamps and high-intensity xenon halogen bulbs, can be obtained from Osram/Sylvania.

Inverted Microscopes

An inverted microscope of high quality is required for pronuclear injection, blastocyst injection, intracytoplasmic sperm injection (ICSI), and nuclear transfer (NT) (see Chapters 7, 13, and 14). The tools for micromanipulation have to be fixed in relation to the stage, and focusing is done by moving the objective tube.

Micromanipulations require that the stage supporting the specimen remains stationary. Inverted fixed-stage microscopes available from Leica, Nikon, Olympus, and Zeiss are the optimal choice. Upright microscopes with fixed stages generally provide short working distances for micromanipulation at higher magnifications. Standard microscopes that are focused by moving the stage are suitable for routine tissue culture observations.

Some high-power objectives may not reach through the glass slide or the dish in which the micromanipulation chamber is placed. Long-working-distance condensers and objectives, which are available from most manufacturers, provide a good solution to this problem. It is also important that the condenser has a working distance long enough to accommodate the injection chamber, holding capillary, and injection needle conveniently.

Each micromanipulation technique requires slightly different equipment in terms of optics and lenses:

- *ES cell injection into blastocyst-stage embryos.* The magnification required is somewhat lower than for pronuclear injection (most experimenters prefer to use a 20x objective with 10x eyepieces). Phase-contrast optics is preferred, since this technique provides a clear "halo" appearance around three-dimensional shapes. By making use of this feature, it is possible to select for viable ES cells for injection.

- *Pronuclear injection, NT, and ICSI.* The magnification required for these techniques is higher than for blastocyst injection (most experimenters prefer to use a 32x or 40x objective with 10x eyepieces). Optimally, the choice of optics should be differential interference contrast (DIC). Nomarski DIC optics is expensive, but provides the best possible image. Hoffman modulation contrast optics provides a cheaper workable alternative, although a less optimal image is provided. Nomarski optics must be used with glass, whereas Hoffman's can be used with plastic.

- In addition to high-power objectives, a low-power (4–10x) objective is necessary to move embryos around in the injection chamber and can also be used for transferring embryos in and out of the chamber.

Depending on the individual setup, it may be practical to place the control knob of the microscope stage on the left-hand side instead of the usual right-hand side, thus freeing up the right hand for the injection control. With some practice, the left hand can perform the control of the stage movement and suction/release of the injection needle simultaneously. However, the opposite arrangement is very difficult to use.

Some investigators find a blue filter comfortable for the eye when looking through phase-contrast optics for longer periods of time. These filters can be purchased from all microscope companies and are available in several color variants and darkness. Image enhancement optical systems such as polarizing filters, analyzing prisms, and objectives have to be perfectly aligned and adjusted to give the desired image quality. These adjustments should be set professionally when the microscope is used for the first time and checked regularly during further use.

Heating and Cooling Stages

For ES cell injection, ICSI, NT, and on-stage culture of embryos or cells, it may be necessary to control the temperature on the microscope stage. Several options are available for this purpose, some simple, and others very sophisticated and expensive. The major manufacturers of inverted microscopes also offer models with a controlled environment on the stage. These systems offer not only control of the temperature, but also adjust humidity and CO_2 levels. The apparatus on the microscope stage is, on the other hand, large and it may be difficult to use it conveniently in combination with micromanipulators. Smaller and simpler solutions can either be homemade by a well-equipped mechanical workshop, or purchased from, e.g., Biomedical Instruments. This variant consists of a hollow metal frame (fitted to the stage) in which water of a specific temperature can be circulated.

Vibration-free Tables

It is most important to make sure that the microinjection takes place in a vibration-free environment. If the building is relatively free of vibrations, it may be sufficient to use a sturdy table with a heavy stone plate placed on the tabletop, with a shock-absorbing material (for example, insulating mat used for camping) placed between the table and the stone plate. As an alternative, the microscope may be placed on a heavy metal plate with an ~4-cm-diameter hole at each corner. Conventional tennis balls are then placed under the plate in the holes, making the microscope rest on a vibration-reducing surface provided by the balls. Make sure that all sources of occasional vibration (e.g., vibrations from centrifuges or other large-scale equipment) nearby are avoided. Even if all these requirements are met, it might be impossible to ensure a vibration-free environment. In this case, a vibration-free table should be used. Such tables are available from, e.g., Newport, Biotech Products, and Technical Manufacturing. Most of the companies that manufacture inverted microscopes specifically for microinjection have partners that provide vibration reduction tables. If a complete microinjection workstation is purchased from Leica, Nikon, Olympus, or Zeiss, it may be a good idea to discuss this issue directly with those companies.

Micromanipulation Equipment

Equipment necessary for micromanipulation of embryos can be organized in a variety of different ways. Major factors to consider when planning and deciding on individual items are: (1) already-existing equipment, (2) which procedures will be performed and in which amount, (3) location and physical properties of the microinjection laboratory, and (4) personal preference of the operator. The best advice is to establish contact with laboratories equipped with items of interest for gaining personal experience, and/or to make use of manufacturers' offers to provide individual items or complete sets of equipment for a limited trial period in one's own laboratory. Various types of injection chambers are described in Chapter 7.

Micromanipulators

Micromanipulators are essential tools for pronuclear and blastocyst injection, NT, and ICSI. Usually, one micromanipulator is mounted on each side of the microscope stage; one micromanipulator controls the holding pipette and the other controls the injection pipette. Most of the available models consist of a joystick, which allows simultaneous movement in two dimensions, and an additional vertical level adjuster. This control device is attached to the micromanipulator instrument holder (and so to the injection needle or holding capillary) through either of three modes: (1) direct mechanical connection such as, e.g., Leica (Leitz) mechanical manipulator, (2) hydraulic such as, e.g., Narishige, or (3) electronic connection such as, e.g., Eppendorf TransferMan NK. All systems work well, although they each have some advantages and disadvantages. The Leitz (Leica) mechanical micromanipulator was introduced in the early 1950s and since then has been very popular. These mechanical manipulators have a nearly unlimited lifetime; unless they are dropped or mishandled, they usually never fail. However, when mechanical micromanipulators such as Leica are used with modern inverted microscopes, they must be raised on supports or equipped with so-called "swan-neck extensions" to provide access to the stage. The Narishige hydraulic system developed in the early 1980s is classically preferred by most experimenters in the human ART field because it provides very precise remote-controlled operation. Electronic devices have the advantage of programmable positions, which may facilitate techniques like pronuclear injection with a very high efficiency. Manipulators are available from many suppliers, including Eppendorf, Fine Science Tools, Leica/Leitz, Narishige, Sutter Instrument, and The Glass Worx.

Piezo Impact Drives

A piezoelectric mechanism incorporated into the micromanipulator provides a rapid precise motion or continuous vibration regulated by electronic control. Such systems have been used mainly for electrophysiological studies that require the insertion of ultrafine microelectrodes.

Piezoelectric microdrives mounted on standard or piezo-enhanced micromanipulators enhance the penetration of the zona pellucida by large-sized microtools. Piezo impact mechanisms are an essential tool for performing NT and ICSI. These devices consist of a holder for the microinjection needle equipped with a so-called piezo crystal. This crystal enlarges very rapidly when an electric current is applied, and so allows for an extremely precise and fast forward–backward movement of the injection needle. Using piezo-actuated microinjection, it is possible to penetrate the zona pellucida and the cytoplasmic membrane in a very smooth way, resulting in a significant increase in the survival rate of injected embryos. Piezo impact devices are available from Prime Tech Ibasaki, Japan (http://www.primetech-jp.com); BioMedical Instruments (distributor of Maerzhaeuser), and EXFO Burleigh Products Group (http://www.exfo.com/ http://www.myneurolab.com). Several companies and retailers selling micromanipulators can equip them with piezo impact drives.

A word of caution: Some of the piezo models stabilize the injection needle by partly filling it with mercury. Mercury<!> is highly toxic and should be handled with extreme care!

Microinjection Units

Micromanipulators aid in the fine movement of microtools—injection needles and holding capillaries. The control of suction and release for these devices is, however, performed by special control units. Such units often consist of glass syringes under the control of a variety of micrometers and are available from, e.g., Eppendorf, Narishige, Stoelting, and Sutter Instruments. Usually the entire system is filled with an incompressible liquid, such as silicon, paraffin oil, or Fluorinert, which is free of any air bubbles. Some laboratories prefer to use homemade variants, in which a gas-tight Hamilton syringe with Teflon plunger is connected to a micrometer screw, and further to the instrument holder. These simple and cheap solutions may work very well, but do require an experienced operator when set up and used for the first time. As discussed earlier, the mouth pipette may provide very precise control of a holding pipette, but manual hydraulic or pneumatic (such as CellTramAir from Eppendorf) systems are used more often.

A very common system traditionally used for pronuclear injection consists of a 50-cc glass syringe connected with tubing to the pipette holder. The syringe and tubing are filled with air, and the micropipette with internal fiber (filament) is filled with DNA. The plunger of the syringe is compressed by hand to provide pressure for injection. However, electronic compressed gas injectors are used most often, both for pronuclear injection of zygotes and for injection of postimplantation embryos. These injectors are available from several companies, such as Eppendorf (FemtoJet) and Narishige. They can be easily adjusted to provide a constant flow pressure. Manual hydraulic microinjectors also provide adequate control for continuous flow in pronuclear injections. Constant flow in manual hydraulic devices is obtained by applying constant pressure to the syringe. The injection pipette is filled from the front by immersing its tip into DNA solution and applying negative pressure.

A system similar to the one described above for pronuclear injection (consisting of a 10- or 50-cc glass syringe, tubing filled with air, and injection micropipette filled with oil) may also be used for ES cell injection. However, very reliable, albeit expensive, manual hydraulic microinjectors such as CellTram Vario (Eppendorf) are used more often for ES cell injection.

Electrofusion and Aggregation

Electrofusion apparatus for production of tetraploid embryos is available from, e.g., Allentown PA, Biotechnologies and Experimental Research, BLS, Braun Biotech, Brinkmann, Cyto, and Pulse Sciences. Aggregation needles for preparation of depression wells in plastic dishes can be purchased from BLS, Hungary (http://www.bls-ltd.com). These needles seem to be very simple, and indeed it is possible to manufacture them in a well-equipped mechanical workshop. However, it is important to keep in mind that the exact shape of the tip is extremely important for the final result. Wells that are too deep will make the harvesting of aggregated embryos difficult; wells that are too shallow, on the other hand, do not hold the aggregates in place. Finally, the walls of the wells must be absolutely smooth and free of any cracks so as not to damage the embryos.

Production of Microtools from Glass Capillaries

Glass capillaries for holding embryos during microinjection and for their injection can be purchased ready-made from several suppliers such as Biomedical Instruments, Eppendorf, Humagen, and The Glass Worx. These capillaries/needles are usually of high quality, and can be specifically custom-made, but they are rather expensive. If a large-scale operation is planned, where the use of various needles/capillaries will be high, consideration should be given to preparing glass capillaries on site. The advantage of in-lab preparation is both economic savings on the long run, and the possibility of preparing capillaries to custom shapes and sizes. However, the time invested in capillary preparation should also be taken into consideration.

A glass capillary puller is needed for on-site production of capillaries. The final shape of holding capillaries and injection needles may be further adjusted by the use of microforges and micro-grinders. Glass capillary pullers, microforges, bevelers, and grinders are available from Alcatel, Bachofer, De Fonbrune, Narishige, Prior Instruments, Research Precision Instruments, Stoelting, Sutter Instrument, and Technical Products International.

Glass Capillary Pullers

Glass capillary pullers are expensive, but they provide means for customizing needles in an almost unlimited manner to suit personal preference and novel applications. Horizontal, electronically programmable pullers are recommended, as they have proven to produce high-quality, consistent results.

Glass capillaries used for pulling needles can be purchased from Clark Electromedical Instruments, FHC, Leica, Science Bioproducts, Stoelting, Sutter Instrument, The Glass Worx, World Precision Instrument, and others. The most common type of glass is borosilicate. Inexpensive general-use capillary tubing may be used for holding pipettes but should generally not be used for preparation of injection needles. One should keep in mind that glass formulation and exact parameters may vary among companies, causing variation during pipette production. The filling fiber (filament) in the lumen of the capillary tubing provides capillary action and increases the speed and ease of loading very finely tipped micropipettes used for pronuclear injection. Fiber-free capillaries are used for other applications. General guidelines for the use of pullers are discussed in Chapter 7.

Microforges

A microforge essentially consists of a heating filament and very finely adjustable holders for the microcapillary. The temperature of the filament is controlled by an electronic system with an adjustment knob; there is often a foot pedal for hands-free control. A magnification system and micron scale are built into the eyepieces. The glass ball is formed on the filament by allowing a needle tip to melt on hot filament. Four major techniques may be performed on the microforge: (1) heat polishing, (2) bending, (3) breaking, and (4) pulling. By decreasing the distance between filament and needle tip at various angles, the tip and shaft shape can be adjusted in numerous ways. A microforge is also used for creating the narrow tip of a holding pipette and the spike on some injection needles. (For preparation of holding capillaries and injection needles, see Chapters 7 and 10.)

Bevelers and Grinders

Microgrinders and bevelers are used to obtain a very sharp, straight-angled tip by careful grinding. These instruments require practice to operate well, and the choice of model is very much a question of experience and personal preference.

Fine Instruments and Tools

Each micromanipulation application requires a set of specialized tools, all of which are listed in the relevant chapters in this book. Some of these instruments, tools, and consumable items are common for many applications and are mentioned here as a means of quick overview.

Mouth pipetting devices—although prohibited by some local legislation—are still the most widely used technical equipment for collecting, sorting, washing, and transferring embryos. Very convenient mouthpieces can be purchased from HPI, and suitable flexible polyethylene tubing (e.g., 0.76-mm inside diameter, 1.22-mm outside diameter) can be purchased from numerous lab tech suppliers.

Surgical-grade stainless steel instruments to be used for surgeries such as embryo transfer, ovary transplantation, and vasectomy can be purchased from companies such as Bayer, Fine Science Tools, Fischer, Roboz Surgical Instruments, and World Precision Instruments. Although the instruments needed for each procedure vary somewhat (see Chapter 6), the basic set of instruments is broadly the same in many cases:

- *Forceps and scissors for cutting the skin.* These should be replaced or sharpened regularly because cutting through hair will soon make the scissors blades dull. Do not use these instruments for deeper incisions to avoid any possible transmission of bacteria from the skin and hair to inner organs.

- *Fine dissection scissors for cutting the abdominal wall and blunt fine forceps with serrated tips for grasping organs.* These are necessary for the collection of tissues and transfer of embryos to the uterus. If correctly handled, these instruments will remain functional for a long time.

- *Two pairs of watchmaker's #5 ultra-fine-point forceps and one extremely fine pointed scissors.* These are for opening the bursa around the ovary and dissection of early-stage postimplantation embryos. These instruments have to be fine and sharp at their tips to function properly when dissecting delicate tissues of early postimplantation stage embryos. Scissors and forceps of this quality should be handled with utmost care when used, cleaned, and stored. The slightest collision of the tips with a solid surface (such as a microscope stage) will render them useless for this application. Thin tubing or plastic pipette tips should be used to protect the instruments during storage. Experience has shown that these instruments often have to be replaced regularly. An economical alternative is to polish damaged forceps tips with an oilstone and fine sandpaper, thus allowing them to be used further for less delicate procedures (see below).

- *One pair of #5 forceps.* Used for holding the infundibulum during oviduct flushing and transfer, these forceps should be fine enough to grasp one oviduct coil, but preferably somewhat blunt so as not to cause any damage.

- *Tissue clamps.* These are used to hold fat pads during embryo transfer, ovariectomy, and vasectomy. A special very small clamp called a bulldog-type serrated Serrefine clamp or baby Dieffenbach clip (28 mm or 1.5 inch) is preferred. The clamp should be able to be opened by sideward twisting (this feature will make cleaning and sterilization much easier).

Other, more specialized fine tools such as Desmarres chalazion forceps, ophthalmic suture needles, tungsten needles, and tungsten wires can be obtained from Gallenkamp, Goodfellow Metals, and Holborn Surgical and Veterinary Instruments.

Suture material can be obtained from several companies providing medical supplies (e.g., Holborn Surgical and Veterinary Instruments). For sealing abdominal walls and bursa, an absorbable suture (such as Vicryl) is preferred. For skin sutures, however, a nonabsorbable suture is recommended, as these are generally stronger and more durable. Wound clips are a convenient and safe alternative to skin sealing suture and can be purchased from Clay Adams.

Sterilization of surgical instruments is best done by autoclaving. However, this procedure is time-consuming and is not suitable for use between individual mice if a large number of animals are being treated. The classical alternative of ethanol dipping followed by flaming is both potentially dangerous (a fire in a constant air flow hood) and also damaging to the instruments. A safe, effective, and convenient alternative is hot bead sterilization. These devices consist of a heater with a well filled with glass beads. The beads are heated to 250°C, and metal instruments require ~20 seconds' contact with the beads for complete sterilization. Hot bead sterilizers are available from Fine Science Tools and Inotech Biosystems International. Note that instruments should be cleaned before they are sterilized in the hot beads to avoid baking blood on them.

Small, anesthetized animals tend to lose their body temperature very easily. Cleaning with ethanol and the constant airflow in hoods further removes heat from the body surface. It is essential to counteract this phenomenon either by placing the sleeping animals under warming lamps or, better, onto a self-regulating heating pad. Such pads of different sizes are available from Fine Science Tools.

Consumables/Plasticware

Disposable items such as syringes, needles, pipettes, and plasticware can be purchased from a variety of sources, the only criteria being the best price and delivery that can be negotiated in each case. Some investigators prefer to use specific brands of plasticware (such as Corning, Falcon, or Nunc), but we believe that such preferences are mainly grounded on subjective views, and that all major brands of disposables today are of comparably high quality. Having said that, the type and grade of disposable tissue culture dishes are of major importance. Cell culture should generally be performed on tissue culture (TC)-grade, coated dishes to allow the attachment of cells to the plastic surface. Removal of the zona pellucida from preimplantation-stage embryos, however, should be performed either in bacteriological-grade dishes or in the noncoated lid of TC-grade dishes to prevent embryos from sticking to the plastic and becoming damaged.

So-called organ culture dishes (Falcon 35-3037) are very useful for a variety of applications in the micromanipulation lab. These dishes feature a center well with a volume of 1 ml. Embryos can be cultured in these dishes without the need to cover the surface with oil. They are also well suited for sperm preparation. Finally, embryos can conveniently be transported using these dishes as described in Chapter 15.

Cryopreservation

Freezers

Embryo cryopreservation can be carried out according to a number of different protocols (see Chapter 15). So-called slow-freezing techniques all require a controlled-rate freezer, either alcohol or liquid nitrogen (LN_2)-based. Alcohol-based freezers are generally less expensive and are widely used for embryo cryo-preservation in insemination straws or sealed glass ampules. For freezing in cry-ovials, however, LN$_2$-based freezers are preferable. Alcohol-based freezers are available from FTS Systems, whereas LN$_2$-based freezers can be purchased from Biogenics, Cryologic, Thermo Forma, and TS Scientific. It is mandatory that the freezing protocol parameters be adjusted carefully before starting to use new equipment!

Storage

Cryopreserved cells may be stored at –80°C for shorter periods (maximum 2 months), but should thereafter be moved to either LN$_2$ or –150°C freezers. LN$_2$ storage requires the regular monitoring of liquid nitrogen levels either manually or, better, by automatic control systems. Electric –150°C freezers offer a practical solution for long-term storage of 96-well plates, but they are, on the other hand, dependent on electric supply, although they may have LN$_2$ backup.

Cryopreserved embryos should never be warmed above –140°C, including when they are transported. Because LN$_2$ is a dangerous compound, it is often very difficult to find a suitable mode of transport. A solution to this problem is offered by so-called vapor shippers. These are small LN$_2$ storage canisters spe-cially made for road and air transport. A thick inner wall consisting of a foam material absorbs the liquid nitrogen and maintains the frozen embryos in the vapor phase for up to 48 hours (some models even longer). It does take consid-erable time to fill these transportation modules, but they can then safely and reli-ably be used to ship embryos, even overseas.

Small benchtop Dewar flasks as well as a large variety of LN$_2$ containers for long-term storage can be purchased from MVE and Taylor-Wharton Cryogenics. Plastic insemination straws, storage goblets, canisters, and canes are available from various suppliers such as Conception Technology, CryoBioSystem Division of IMV Technologies, Meditech IST Canada, and Minitube of America.

INFORMATION RESOURCES

In the following section, the reader can find a list of books with information relat-ed to the previous chapters. Addresses and links to Web-based resources such as databases, discussion groups, and so forth can be found on the Web site for the manual (see Appendix 3).

Books

Copp A.J. and Cockroft D.L., eds. 1990. *Postimplantation mammalian embryos: A practical approach.* IRL Press.

Falconer D.S. and MacKay T.F.C. 1996. *Introduction to quantitative genetics*, 4th Edition. Addison-Wesley, Boston.

Festing W.F.W. 1992. Origins and characteristics of inbred strains of mice, 14th listing. *Mouse Genome* **90:** 231–352.
Listed are 140 inbred strains of mice, together with notes about their origins and vital statistics. This listing is regularly updated in the journal *Mouse Genome*.

Gilbert S.F. 2000. *Developmental biology*, 6th Edition. Sinauer Associates, Sunderland, Massachusetts.

Jackson I.J. and Abbott C.M., eds. 2000. *Mouse genetics and transgenics: A practical approach*. Oxford University Press, New York.

Joyner A.L., ed. 2000. *Gene targeting: A practical approach*, 2nd Edition. Oxford University Press, New York.

Kaufman M.H. 1992. *The atlas of mouse development*. Academic Press.

Kaufman M.H. and Bard J.B.L. 1999. *The anatomical basis of mouse development*. Academic Press.

Lewin B. 2000. *Genes VII*. Oxford University Press, New York.
An electronic, updated version can be accessed through the molecular biology module at http://www.ergito.com/index.jsp

Lyon M.F. and Searle A.G. 1995. *Genetic variants and strains of the laboratory mouse*, 3rd Edition. Oxford University Press, New York.

Roberts R. 1990. *The mouse: Its reproduction and development*. Oxford Scientific Press.

Robertson E.J., ed. 1992. *Teratocarcinomas and embryonic stem cells: A practical approach*. IRL Press.

Rossant J. and Tam P., eds. 2002. *Mouse development: Patterning, morphogenesis and organogenesis*. Academic Press.

Sambrook J. and Russell D. 2001. *Molecular cloning: A Laboratory manual*, 3rd Edition. Cold Spring Harbor Laboratory Press, Cold Spring Harbor, New York.

Silver L.M. 1995. *Mouse genetics: Concepts and applications*. Oxford University Press, New York.
An electronic version is available at http://www.informatics.jax.org/silver/

Suckow M.A., Danneman P., and Brayton C. 2001. *The laboratory mouse*. CRC Press.

Sundberg J.P., ed. 1994. *Handbook of mouse mutations with skin and hair abnormalities. Animal models and biomedical tools*. CRC Press, Boca Raton, Florida.

Theiler K. 1989. *The house mouse: Atlas of embryonic development*. Springer Verlag.

Torres R.M. and Kühn R. 1997. *Laboratory protocols for conditional gene targeting*. Oxford University Press, New York.

Turksen K., ed. 2002. Embryonic stem cells: Methods and protocols. *Methods. Mol. Biol.* **185.**

Wassarman P.M. and DePamphilis M.L., eds. 1993. Guide to techniques in mouse development. *Methods. Enzymol.* **225.**

DISCLAIMER

The suppliers cited in this chapter and in Appendix 3 constitute a nonexclusive list of suppliers. Much of the equipment listed can be purchased from suppliers other than those cited in this manual, and equivalent equipment may be produced by companies other than those cited. These suggestions provide a guide to what is needed, but investigators should survey the marketplace before making purchases. The cited examples are of models that the authors know work, but these choices do not mean there are not other equally good ones.

Buffers and Solutions

CONTENTS

All solutions should be made in 2x glass-distilled or Millipore Q water and stored in disposable plastic or clean glass containers absolutely free of detergent residue. *Note:* Reagents used for preparation of embryo culture media, and solutions for embryo manipulation (such as Acid Tyrode and Mannitol solution) should be designated for these purposes only. Any trace amounts of contamination may be deleterious for embryonic development. The quality of water used for embryo culture and manipulations is of utmost importance. For details, see Chapter 14.

CAUTION: See Appendix 2 for appropriate handling of materials marked with <!>.

Acidic Tyrode Solution for Removing Zonae

	g/100 ml
NaCl	0.800
KCl<!>	0.020
$CaCl_2 \cdot 2H_2O$	0.024
$MgCl_2 \cdot 6H_2O$<!>	0.010
glucose	0.100
polyvinylpyrrolidone (PVP) <!>	0.400

Prepare at room temperature and adjust to pH 2.5 with Analar HCl (BDH). The PVP is added to increase viscosity and reduce embryo stickiness. Filter-sterilize and store in aliquots at –20°C. Acidic Tyrode's solution is also available from Sigma (T1788) and Specialty Media (MR-004D)

Alkaline Phosphatase Buffers

For immunohistochemistry of embryo sections:
   100 mM Tris-HCl<!> (pH 9.5)
   100 mM NaCl
   5 mM $MgCl_2$<!>
For in situ hybridization of intact embryos with RNA probes (NTMT without Tween 20)
   100 mM Tris<!> (pH 9.5)
   100 mM NaCl
   50 mM $MgCl_2$<!>

This buffer is best if made fresh from stock solutions because precipitate tends to form on standing.

For in situ hybridization of intact embryos with RNA probes (NTMT):
   100 mM NaCl
   100 mM Tris<!> (pH 9.5)
   50 mM $MgCl_2$<!>
   0.1% Tween 20 (Fisher BP337-100 or Sigma P1379)

Best if made fresh from stock solutions because precipitate tends to form on standing.

Alkaline Phosphatase Staining Solution

For culture of primordial germ cells from embryos of different stages:

Solution	Amount	Final concentration
1 M Tris-maleate <!> (pH 9.0) (Sigma T3128)	1.25 ml	25 mM
α-naphthyl phosphate (Sigma N7255)	20 mg	0.4 mg/ml
Fast-Red TR salt<!> (Aldrich 20,155-3)	50 mg	1 mg/ml
1 M $MgCl_2$<!>	400 µl	8 mM
2x glass-distilled or Millipore Q H_2O		to 50 ml

Adjust pH of the Tris-maleate by mixing 1 M Tris-maleate and 1 M Tris base (Fisher BP152-1).

Anesthetics

Tribromoethanol (Avertin)

A stock of 100% avertin is prepared by mixing 10 g of 2,2,2-tribromoethyl alcohol<!> (Aldrich T4,840-2) with 10 ml of *tert*-amyl alcohol<!> (Aldrich 24,048-6). Make sure it is fully dissolved (heat to 50°C; use a microwave oven or magnetic stirrer overnight). (*Note:* If 2,2,2-tribromoethanol is dark in color, it should be recrystallized before use; see below). To use, dilute 100% stock to 1.2–2.5%, v/v, in water or isotonic saline, stirring vigorously until it is dissolved. Filter-sterilize through 0.2-micron filter (filtering also removes remaining undissolved crystals). Both the 100% stock and working solutions are stored in the dark at 4°C to prevent decomposition to the irritant by-products, dibromoacetic aldehyde and hydrobromic acid (Papaioannou and Fox 1993). The working solution is stable for a couple of months, but some labs prefer to prepare it fresh from stock aliquots immediately prior to the experiment.

Ready-to-use 1.25%, v/v, working solution may be prepared by mixing 2.5 g of 2,2,2-tribromoethanol with 5 ml of *tert*-amyl alcohol and dissolving it in 200 ml of water using a magnetic stirrer. The administration of more dilute solutions of the compound may be recommended because higher concentrations seem to be more frequently associated with postanesthetic mortality (Flecknell 1993). The **pH of final working solution should be <5; it is considered toxic if pH is >5.**

The proper dose of tribromoethanol (125–250 mg/kg body weight) may vary with different preparations and should be redetermined each time a new 100% stock is made or a working solution has been stored for more than 2 months. To test, inject several mice with doses ranging from 0.014 to 0.018 ml (2.5%) per gram of body weight. The typical dose for 1.25% solution is 0.02 ml per gram of body weight. The dose should be sufficient to give complete anesthesia, but it is also important to check the health and survival of the mice for 3–4 days afterward. With some batches of 2,2,2-tribromoethanol, the mice may become sick and die several days after being anesthetized. Recrystallizing the 2,2,2-tribromoethanol usually solves this problem (see below).

There are conflicting reports in the literature regarding the use of tribromoethanol. It has been reported to cause acute peritoneal inflammation and fibrinous serositis of the abdominal organs (Zeller 1998). However, with minimal precautions to prevent decomposition (storage in the dark at 4°C), Avertin is effective and simple to use; it provides rapid induction and deep surgical anesthesia in mice followed by fast postoperative recovery and low morbidity and mortality (Papaioannou and Fox 1993; Weiss and Zimmermann 1999). Numerous transgenic facilities have successfully used Avertin for more than 15 years.

To recrystallize 2,2,2-tribromoethanol (procedure from N. Lonberg, GenPharm Corp., Mountain View, California):

1. Dissolve 50 g of 2,2,2-tribromoethanol in 500 ml of boiling petroleum ether<!> (**petroleum ether, NOT ethyl ether–do not use ethyl ether!!!**) or hexane (boiling point 69°C) on a stirring hot plate in a fume hood. *Caution:* These solvents are extremely flammable; exercise extreme caution.

2. Add a full spatula of charcoal.

3. Filter through fluted filter paper in a glass funnel preheated to 65°C into a second beaker or flask. Cool on ice to 30°C.

4. Pour off the supernatant.

5. Break up crystals with a glass rod or metal spatula and dry thoroughly under vacuum overnight. Store at 4°C.

Ketamine/Xylazine

The Ketamine/Xylazine combination is considered to be a very reliable anesthetic for mouse surgery (Erhardt 1984). The recommended dose varies: It is 100–200 mg/kg body weight for Ketamine and 5–16 mg/kg body weight for Xylazine injected intraperitoneally. Although some labs use Ketamine at a dose of 35–50 mg/kg body weight, Ketamine at 50 mg/kg combined with 10 mg/kg Xylazine does not produce a consistent reliable level of immobilization or anesthesia in Syrian hamsters, and a dose of 150 mg/kg Ketamine combined with 10 mg/kg Xylazine has been suggested by Payton et al. (1993).

The most widely used dose of Ketamine/Xylazine for mouse surgery is 100 mg/kg and 10 mg/kg body weight, respectively (Flecknell 1993). Duration of the effect may be extended by increasing the proportion of Xylazine or by an additional dose of Ketamine.

Premedication with anticholinergic drugs such as atropine (0.02–0.05 mg/kg body weight) is often used to prevent bradycardia (slow heart rate) caused by Xylazine, and to control excessive bronchial and salivary secretions caused by Ketamine (Nowrouzian 1981; Magoon et al. 1988).

The recipe below provides an onset of 3–5 minutes with 30–40 minutes duration of surgical anesthesia.

Ketamine (50 mg/ml) (Vetalar, Ketaset, Ketalar)	2 ml (100 mg)
Xylazine (20 mg/ml) (Rompun)	0.8 ml (16 mg)
Water (sterile)	to 10 ml

Store at 4°C for a maximum of 2 weeks. Inject 0.1 ml per 10 grams of body weight (100 mg/kg Ketamine, 16 mg/kg Xylazine).

Bovine Serum Albumin

Bovine serum albumin (BSA) is available in either powdered (Fraction V, Sigma A 9647) or crystalline (Pentex, Miles Laboratories 81-001) forms. Some batches of Fraction V may contain spermine oxidase, which inhibits embryo development beyond the eight-cell stage (D.G. Whittingham, pers. comm.). Some laboratories use Sigma A4378 and AlbuMAX from Invitrogen Life Technologies. Embryo-tested BSA (fraction V) is available from Sigma (A3311). All batches of BSA should be tested for toxicity before use. See Chapter 4 for details.

Ca⁺⁺/Mg⁺⁺-free PBS

See PBS below.

Ca⁺⁺/Mg⁺⁺-free Tyrode Ringer's Saline (pH 7.6–7.7)

Component	g/liter
NaCl	8.0
KCl<!>	0.3
$NaH_2PO_4 \cdot 5H_2O$	0.093
KH_2PO_4<!>	0.025
$NaHCO_3$	1.0
Glucose	2.0

Chicago Sky Blue 6B<!>, also called Pontamine Sky Blue<!> (Sigma C8679)

Dissolve in isotonic saline to provide a 1% solution (1 g/100 ml saline). Filter the solution through a Whatman filter paper (Whatman International, Maidstone, United Kingdom, 1001110). Store in a glass bottle at room temperature.

Hyaluronidase

Use Type IV-S from bovine testes (e.g., Sigma H3884 or embryo-tested H4272). Prepare a stock solution at 10 mg/ml in water, M2, or any other HEPES-buffered embryo culture medium (see Chapter 4). Filter-sterilize, aliquot, and store at –20°C for months. Dilute to ~300 µg/ml in M2 or other HEPES, buffered embryo culture medium with BSA for removing cumulus cells. A 0.5–1 mg/ml final concentration of hyaluronidase also may be used; more concentrated stock solutions (e.g. 100x) are made in this case.

Mannitol, 0.3 M (Sigma M4125)

Dissolve in ultrapure water, add 0.3% BSA (Sigma A3311), and filter through a 0.22-µm Millipore filter. Store in aliquots at –20°C. Use a fresh aliquot for each experiment.

Methylene Blue Solution

Dissolve methylene blue<!> (Basic Blue 9) powder (Sigma M9140) in 0.5 M sodium acetate <!> (pH 5.2), to a final concentration of 0.1% w/v. Make sure to dissolve the powder completely by stirring for 30 minutes. Stain agarose gels for 20 minutes. Destain the gel in distilled water for 30 minutes, changing the water every 10 minutes.

Microinjection Buffers for Pronuclear Injection of Zygotes

See Chapter 7, Protocols 7.6 and 7.7.

Pancreatin/Trypsin Solution for Separating Germ and Tissue Layers

Modified from Levak-Svajger et al. (1969).

	g/20 ml	Final concentration
pancreatin	0.50	2.5%
trypsin	0.10	0.5%
polyvinylpyrrolidone<!> (*optional*)	0.10	0.5%

Make in Ca^{++}/Mg^{++}-free Tyrode Ringer's saline. The suspension will be difficult to filter-sterilize through a 0.45-µm Millipore filter without low-speed centrifugation or prefiltering through a Whatman No. 1 filter. Store sterile in small aliquots at –20°C.

PBSMT for Immunohistochemistry of Whole-mount Embryos

phosphate-buffered saline (see below)

2% nonfat instant skim milk (used to block all nonspecific protein-binding sites in the tissues). The brand is important. Carnation gives consistently good results, whereas other brands (e.g., Kroger) do not.

0.5% Triton X-100 (used to facilitate permeability of the tissues)

Make fresh before use.

PBT

For immunohistochemistry of whole-mount embryos:
phosphate-buffered saline (pH 7.4) (see below)
0.2% bovine serum albumin (BSA) (Sigma A4378)
0.5% Triton X-100

Make fresh before use.

For in situ hybridization of whole-mount embryos:
phosphate-buffered saline (see below)
0.1% Tween 20

Phenol/Chloroform Solution for Isolating Total RNA from Mouse Embryos or Fetal Tissues

Phenol<!> should be equilibrated with 1 M Tris<!> (pH 8) (keep separate stock for RNA work) and stored at –20°C. Before use, mix with an equal volume of chloroform<!> and add 2% isoamyl alcohol<!> (store final mixture also at –20°C).

Phosphate-buffered Saline

The basic formulation for phosphate-buffered saline (PBS) is given in Sambrook and Russell (2001). This is Ca^{++}/Mg^{++}-free phosphate-buffered saline.

NaCl	8 g/liter
KCl<!>	0.2 g/liter
Na_2HPO_4<!>	1.44 g/liter
KH_2PO_4<!>	0.2 µg/liter

Adjust pH with HCl and make up to 1 liter with distilled water. When used for tissue culture, the pH is 7.2. For other uses (e.g., in situ hybridization), the pH is 7.4. Dulbecco's solution A (PBSA) formulation is given below (Dulbecco and Vogt 1954; Spector et al. 1998). Unless otherwise stated, these are the formulations denoted by PBS or D-PBS in this manual.

NaCl	8 g/liter
KCl<!>	0.2 g/liter
Na_2HPO_4<!> (anhydrous)	1.15 g/liter
KH_2PO_4<!> (anhydrous)	0.2 g/liter

Adjust pH and make up to 1 liter with distilled water. Variations in the formulation include the use of $Na_2HPO_4 \cdot 7H_2O$ (2.16 g/liter) or $Na_2HPO_4 \cdot 12H_2O$ (2.88 g/liter) instead of anhydrous dibasic sodium phosphate. Ca^{++}/Mg^{++}-free D-PBS for tissue culture can be purchased already made up as a sterile 1x or 10x solution (e.g., Invitrogen Life Technologies 14190, 14200 or Specialty Media BSS-1006, BSS-2010) .

If Ca^{++} and Mg^{++} are added, the amounts are 0.133 g of CaCl$_2$•2H$_2$O or 0.1 g of CaCl$_2$ (anhydrous) and 0.10 g of MgCl$_2$•6H$_2$O per liter (Dulbecco and Vogt 1954; Spector et al. 1998). This is the same as PBSABC of Dulbecco and can be purchased made up as 1x or 10x solutions (e.g., Invitrogen Life Technologies 14040, 14080 or Specialty Media BSS-1005, BSS-6010).

D-PBS with Ca^{++} and Mg^{++} containing 1 g of glucose and 0.036 g of sodium pyruvate per liter (e.g., Invitrogen Life Technologies 14287) may be used for handling and cryopreservation of preimplantation-stage embryos when supplemented with 3 g/liter BSA, optional phenol red, and antibiotics (e.g., Specialty Media MR-006). This modified D-PBS is also called PB1 medium (see Chapters 4 and 15).

Note that for in situ hybridization, some laboratories may use a different formulation of phosphate-buffered saline, e.g., NaCl (7.6 g/liter), Na$_2$HPO$_4$ (3.8 g/liter), and NaH$_2$PO$_4$ (0.42 g/liter). The pH will be 7.4 based on the ratio of monobasic and dibasic sodium phosphate. This formulation appears to give results in biochemical studies identical to the one given above. **It is not suitable for tissue culture.**

Pronase Solution

Use protease from *Streptomyces griseus* (Calbiochem 537088; Boehringer Mannheim 165 921; Sigma P5147). Prepare a 0.5% solution in M2 medium for removing zonae as an alternative to Acidic Tyrode's solution (see Chapter 11). If necessary, 0.5% polyvinylpyrrolidone<!> can be added to reduce stickiness of the embryos. Because Pronase is a crude enzyme preparation, it probably should be incubated for 30 minutes at room temperature to destroy contaminating nucleases, etc. Centrifuge to remove insoluble material, filter-sterilize, and store in aliquots at –20°C.

Saline/EDTA Buffer plus Glucose for Isolation of Germ Cells and Tissue Culture

	g/100 ml
EDTA (disodium salt)	0.02
NaCl	0.80
KCl<!>	0.02
Na$_2$HPO$_4$<!> (anhydrous)	0.115
KH$_2$PO$_4$<!>	0.02
phenol red<!>	0.001
glucose	0.02

The final EDTA concentration is 0.02%. Check that pH is 7.2. Filter-sterilize or autoclave (121°C, 15 psi for 15 minutes). Store at room temperature. The glucose can be omitted for tissue culture alone.

20× SSC

Dissolve 175.3 g of NaCl and 88.2 g of sodium citrate in 800 ml of H$_2$O. Adjust the pH to 7.0 with a few drops of a 10 N solution of NaOH<!>. Adjust the volume to 1 liter with H$_2$O. Dispense into aliquots. Sterilize by autoclaving.

20× SSPE

Dissolve 175.3 g of NaCl, 27.6 g of $NaH_2PO_4 \cdot H_2O$<!>, and 7.4 g of EDTA in 800 ml of H_2O. Adjust the pH to 7.4 with NaOH<!> (~6.5 ml of a 10 N solution). Adjust the volume to 1 liter with H_2O. Dispense into aliquots. Sterilize by autoclaving.

TAE (Tris-acetate/EDTA) Buffer

1× solution:
40 mM Tris acetate<!>
1 mM EDTA (pH 8.0)

TE (Tris/EDTA) Buffer

10 mM Tris-Cl<!> (pH 8.0)
1 mM EDTA (pH 8.0)

0.25% Trypsin in Tris-Saline for Tissue Culture

Trypsin stock solution for tissue culture 0.25% trypsin 1:250 in Tris-saline (or any other well-buffered isotonic salt solution). Addition of antibiotics is optional.

g/100 ml	
NaCl	8.00
KCl<!>	0.40
Na_2HPO_4	0.10
glucose	1.00
Trizma base (Fisher BP152-1)	3.00
phenol red<!>	0.010
penicillin G	0.060 (final conc. 100 U/ml)
streptomycin<!>	0.100
trypsin (Difco 0152, 1:250)	2.5 g

(dissolve in small volume of H_2O before adding)

Adjust pH to ~7.6. Filter-sterilize and aliquot into sterile containers. Store at –20°C. This stock is diluted 1:4 in saline/EDTA buffer (see above) before use.

Trypsin/EDTA Solution

To detach cells from tissue culture dishes and to dissociate cells from one another in Ca^{++}/Mg^{++}-free buffered saline solution: final concentrations 0.05% trypsin, 0.02% (0.53 mM) EDTA. Dilute 10× stock of trypsin/EDTA from supplier (e.g., Invitrogen Life Technologies, 15400-054)) into 1× Dulbecco's PBS. This diluted solution can be aliquoted (e.g., 5–10 ml) and stored frozen. Ready-to-use 0.05% trypsin, 0.53 mM EDTA is commercially available (e.g., Invitrogen Life Technologies 25300-054). For routine culture of ES cell lines, dissociating ES colonies after selection, and deriving ES cells de novo: final concentrations 0.25% trypsin, 0.04% EDTA also available from Invitrogen Life Technologies (25200-056). Dilute in Hank's or Tris-buffered saline. Some laboratories use 0.5% trypsin.

REFERENCES

Dulbecco R. and Vogt M. 1954. Plaque formation and isolation of pure lines with poliomyelitis viruses. *J. Exp. Med.* **98:** 167.

Erhardt W., Hebestedt A., Aschenbrenner G., Pichotka B., and Blumel G. 1984. A comparative study with various anesthetics in mice (pentobarbitone, ketamine-xylazine, carfentanyl-etomidate). *Res. Exp. Med.* **184:** 159–169.

Flecknell P.A. 1993. Anesthesia and perioperative care. *Methods Enzymol.* **225:** 16–33.

Leval-Svajger B., Svajger A., and Skreb N. 1969. Separation of germ layers in presomite rat embryos. *Experientia* **25:** 1311–1312.

Magoon K.E., Hsu W.H., and Hembrough F.B. 1988. The influence of atropine on the cardiopulmonary effects of a xylazine-ketamine combination in dogs. *Arch. Int. Pharmacodyn. Ther.* **293:** 143–153.

Nowrouzian I., Schels H.F., Ghodsian I., and Karimi H. 1981. Evaluation of the anaesthetic properties of ketamine and a ketamine/xylazine/atropine combination in sheep. *Vet. Rec.* **108:** 354–356.

Papaioannou V.E. and Fox J.G. 1993. Efficacy of tribromoethanol anesthesia in mice. *Lab. Anim. Sci.* **43:** 189–192.

Payton A.J., Forsythe D.B., Dixon D., Myers P.H., and Clark J.A. 1993. Evaluation of ketamine-xylazine in Syrian hamsters. *Cornell Vet.* **83:** 153–161.

Sambrook J. and Russell D. 2001. *Molecular cloning: A laboratory manual,* 3rd edition. Cold Spring Harbor Laboratory Press, Cold Spring Harbor, New York,

Spector D.L., Goldman R., and Leinwand L. 1998. *Cells: A laboratory manual,* Volumes 1–3. Cold Spring Harbor Laboratory Press, Cold Spring Harbor, New York.

Weiss J. and Zimmermann F. 1999. Tribromoethanol (Avertin) as an anaesthetic in mice. *Lab. Anim.* **33:** 192–193.

Zeller W., Meier G., Burki K., and Panoussis B. 1998. Adverse effects of tribromoethanol as used in the production of transgenic mice. *Lab. Anim.* **32:** 407–413.

Cautions

The following general cautions should always be observed.

- **Become completely familiar with the properties of substances used before** beginning the procedure.

- **The absence of a warning** does not necessarily mean that the material is safe, since information may not always be complete or available.

- **If exposed** to toxic substances, contact your local safety office immediately for instructions.

- **Use proper disposal procedures** for all chemical, biological, and radioactive waste.

- **For specific guidelines on appropriate gloves**, consult your local safety office.

- **Handle concentrated acids and bases** with great care. Wear goggles and appropriate gloves. A face shield should be worn when handling large quantities. Do not mix strong acids with organic solvents as they may react. Sulfuric acid and nitric acid especially may react highly exothermically and cause fires and explosions. Do not mix strong bases with halogenated solvent as they may form reactive carbenes which can lead to explosions.

- **Never pipette** solutions using mouth suction. This method is not sterile and can be dangerous. Always use a pipette aid or bulb.

- **Keep halogenated and nonhalogenated** solvents separately (e.g., mixing chloroform and acetone can cause unexpected reactions in the presence of bases). Halogenated solvents are organic solvents such as chloroform, dichloromethane, trichlorotrifluoroethane, and dichloroethane. Some nonhalogenated solvents are pentane, heptane, ethanol, methanol, benzene, toluene, N,N-dimethylformamide (DMF), dimethyl sulfoxide (DMSO), and acetonitrile.

- **Laser radiation**, visible or invisible, can cause severe damage to the eyes and skin. Take proper precautions to prevent exposure to direct and reflected beams. Always follow manufacturers' safety guidelines and consult your local safety office. See caution below for more detailed information.

- **Flash lamps**, due to their light intensity, can be harmful to the eyes. They also may explode on occasion. Wear appropriate eye protection and follow the manufacturers' guidelines.

- **Photographic fixatives and developers** also contain chemicals that can be harmful. Handle them with care and follow manufacturers' directions.

735

- **Power supplies and electrophoresis equipment** pose serious fire hazard and electrical shock hazards if not used properly.

- **Microwave ovens and autoclaves** in the lab require certain precautions. Accidents have occurred involving their use (e.g., to melt agar or Bacto-agar stored in bottles or to sterilize). If the screw top is not completely removed and there is not enough space for the steam to vent, the bottles can explode and cause severe injury when the containers are removed from the microwave or autoclave. Always completely remove bottle caps before microwaving or autoclaving. An alternative method for routine agarose gels that do not require sterile agar is to weigh out the agar and place the solution in a flask.

- **Ultra-sonicators** use high-frequency sound waves (16–100 kHz) for cell disruption and other purposes. This "ultrasound," conducted through air, does not pose a direct hazard to humans, but the associated high volumes of audible sound can cause a variety of effects, including headache, nausea, and tinnitus. Direct contact of the body with high-intensity ultrasound (not medical imaging equipment) should be avoided. Use appropriate ear protection and display signs on the doors of laboratories where the units are used.

- **Use extreme caution when handling cutting devices** such as microtome blades, scalpels, razor blades, or needles. Microtome blades are extremely sharp! Use care when sectioning. If you are unfamiliar with their use, have someone demonstrate proper procedures. For proper disposal, use the "sharps" disposal container in your lab. Discard used needles *unshielded*, with the syringe still attached. This prevents injuries (and possible infections; see Biological Safety) while manipulating used needles, since many accidents occur while trying to replace the needle shield. Injuries may also be caused by broken Pasteur pipettes, coverslips, or slides.

- **Handle and store pressurized gas containers** with caution as they may contain flammable, toxic, or corrosive gases; asphyxiants; or oxidizers. For proper procedures, consult the Material Safety Data Sheet that must be provided by your vendor.

- **Procedures for the humane treatment of animals** must be observed at all times. Consult your local animal facility for guidelines.

GENERAL PROPERTIES OF COMMON CHEMICALS

The hazardous materials list can be summarized in the following categories:

- Inorganic acids, such as hydrochloric, sulfuric, nitric, or phosphoric, are colorless liquids with stinging vapors. Avoid spills on skin or clothing. Spills should be diluted with large amounts of water. The concentrated forms of these acids can destroy paper, textiles, and skin, as well as cause serious injury to the eyes.

- Inorganic bases such as sodium hydroxide are white solids that dissolve in water and under heat development. Concentrated solutions will slowly dissolve skin and even fingernails.

- Salts of heavy metals are usually colored powdered solids that dissolve in water. Many of them are potent enzyme inhibitors and therefore toxic to humans and to the environment (e.g., fish and algae).

- Most organic solvents are flammable volatile liquids. Avoid breathing the vapors, which can cause nausea or dizziness. Also avoid skin contact.

- Other organic compounds, including organosulfur compounds, such as mercaptoethanol or organic amines, can have very unpleasant odors. Others are highly reactive and should be handled with appropriate care.

- If improperly handled, dyes and their solutions can stain not only your sample, but also your skin and clothing. Some of them are also mutagenic (e.g., ethidium bromide), carcinogenic, and toxic.

- All names ending with "ase" (e.g., catalase, β-glucuronidase, or zymolyase) refer to enzymes. There are also other enzymes with nonsystematic names like pepsin. Many of them are provided by manufacturers in preparations containing buffering substances, etc. Be aware of the individual properties of materials contained in these substances.

- Toxic compounds are often used to manipulate cells. They can be dangerous and should be handled appropriately.

- Be aware that several of the compounds listed have not been thoroughly studied with respect to their toxicological properties. Handle each chemical with the appropriate respect. Although the toxic effects of a compound can be quantified (e.g., LD_{50} values), this is not possible for carcinogens or mutagens where one single exposure can have an effect. Also realize that dangers related to a given compound may depend on its physical state (fine powder vs. large crystals/diethylether vs. glycerol/dry ice vs. carbon dioxide under pressure in a gas bomb). Anticipate under which circumstances during an experiment exposure is most likely to occur and how best to protect yourself and your environment.

HAZARDOUS MATERIALS

Note: In general, proprietary materials are not listed here. Follow the manufacturer's safety guidelines that accompany the product.

Acetic acid (glacial) is highly corrosive and must be handled with great care. Liquid and mist cause severe burns to all body tissues. It may be harmful by inhalation, ingestion, or skin absorption. Wear appropriate gloves and goggles and use in a chemical fume hood. Keep away from heat, sparks, and open flame.

Acetic anhydride is extremely destructive to the skin, eyes, mucous membranes, and upper respiratory tract. It may be harmful by inhalation, ingestion, or skin absorption. Wear appropriate gloves and safety glasses, and use in a chemical fume hood.

Acetone causes eye and skin irritation and is irritating to mucous membranes and the upper respiratory tract. Do not breathe the vapors. It is also extremely flammable. Wear appropriate gloves and safety glasses.

3-Aminopropyltriethoxysilane, TESPA, *see* **Silane**

Ammonium hydroxide, NH_4OH, is a solution of ammonia in water. It is caustic and should be handled with great care. As ammonia vapors escape from the solution, they are corrosive, toxic, and can be explosive. Use only with mechanical exhaust. Wear appropriate gloves and use only in a chemical fume hood.

Ammonium sulfate, (NH₄)₂SO₄, may be harmful by inhalation, ingestion, or skin absorption. Wear appropriate gloves and safety glasses.

Amyl alcohol is extremely flammable and may be harmful by inhalation, ingestion, or skin absorption. It may cause irritation to the skin, eyes, and respiratory tract and affects the central nervous system. Use only with adequate ventilation. Wear appropriate gloves and safety glasses. Keep away from heat, sparks, and open flame.

BCIP, *see* **5-Bromo-4-chloro-3-indolyl-phosphate**

BCIG, *see* **5-Bromo-4-chloro-3-indolyl-β-D-galactopyranoside**

Benzyl alcohol is an irritant and may be harmful by inhalation, ingestion, or skin absorption. Wear appropriate gloves and safety glasses. Keep away from heat, sparks, and open flame.

Benzyl benzoate is an irritant and may be harmful by inhalation, ingestion, or skin absorption. Avoid contact with the eyes. Wear appropriate gloves and safety glasses.

5-Bromo-4-chloro-3-indolyl-β-D-galactopyranoside, BCIG, X-gal, is toxic to the eyes and skin and may be harmful by inhalation, ingestion, or skin absorption. Wear appropriate gloves and safety goggles.

5-Bromo-4-chloro-3-indolyl-phosphate, BCIP, is toxic and may be harmful by inhalation, ingestion, or skin absorption. Wear appropriate gloves and safety glasses. Do not breathe the dust.

Butanol is irritating to the mucous membranes, upper respiratory tract, skin, and especially the eyes. Avoid breathing the vapors. Wear appropriate gloves and safety glasses and use in a chemical fume hood. Butanol is also highly flammable. Keep away from heat, sparks, and open flame.

Carbon dioxide, CO₂, in all forms may be fatal by inhalation, ingestion, or skin absorption. In high concentrations, it can paralyze the respiratory center and cause suffocation. Use only in well-ventilated areas. In the form of dry ice, contact with carbon dioxide can also cause frostbite. Do not place large quantities of dry ice in enclosed areas such as cold rooms. Wear appropriate gloves and safety goggles.

CHCl₃, see **Chloroform**

C₆H₅CH₃, *see* **Toluene**

CH₃CH₂OH, *see* **Ethanol**

Chicago Sky Blue is a possible mutagen and may be harmful by inhalation, ingestion, or skin absorption. Wear appropriate gloves and safety glasses and use in a chemical fume hood. Do not breathe the dust.

Chloroform, CHCl₃, is irritating to the skin, eyes, mucous membranes, and respiratory tract. It is a carcinogen and may damage the liver and kidneys. It is also volatile. Avoid breathing the vapors. Wear appropriate gloves and safety glasses and always use in a chemical fume hood.

Citric acid is an irritant and may be harmful by inhalation, ingestion, or skin absorption. It poses a risk of serious damage to the eyes. Wear appropriate gloves and safety goggles. Do not breathe the dust.

CO₂, *see* **Carbon dioxide**

Colchicine is highly toxic, may be fatal, and may cause cancer and heritable genetic damage. It may be harmful by inhalation, ingestion, or skin absorption. Wear appropriate gloves and safety glasses and use only in a chemical fume hood. Do not breathe the dust.

Cytochalasin B may be fatal by inhalation, ingestion, or skin absorption, and is a possible teratogen with possible irreversible effects. Do not breathe the dust. Wear appropriate gloves and safety goggles and use only in a chemical fume hood.

DAB, *see* **3,3′-Diaminobenzidine tetrahydrochloride**

DAPI, *see* **4′,6-Diamidine-2′phenylindole dihydrochloride**

Demecolcine, *see* **Colchicine**

DEPC, *see* **Diethyl pyrocarbonate**

4′,6-Diamidine-2′phenylindole dihydrochloride, DAPI, is a possible carcinogen. It may be harmful by inhalation, ingestion, or skin absorption. It may also cause irritation. Avoid breathing the dust and vapors. Wear appropriate gloves and safety glasses and use in a chemical fume hood.

3,3′-Diaminobenzidine tetrahydrochloride, DAB, is a carcinogen. Handle with extreme care. Avoid breathing vapors. Wear appropriate gloves and safety glasses and use in a chemical fume hood.

Diethyl ether, Et₂O or (C₂H₅)₂O, is extremely volatile and flammable. It is irritating to the eyes, mucous membranes, and skin. It is also a CNS depressant with anesthetic effects. It may be harmful by inhalation, ingestion, or skin absorption. Avoid breathing the vapors. Wear appropriate gloves and safety glasses and always use in a chemical fume hood. Explosive peroxides can form during storage or on exposure to air or direct sunlight. Keep away from heat, sparks, and open flame.

Diethyl pyrocarbonate, DEPC, is a potent protein denaturant and is a suspected carcinogen. Aim bottle away from you when opening it; internal pressure can lead to splattering. Wear appropriate gloves, safety goggles, and lab coat and use in a chemical fume hood.

Dimethyldichlorosilane is extremely flammable and corrosive, therefore causing severe burns. It may be harmful by inhalation, ingestion, or skin absorption. Wear appropriate gloves and safety goggles and use only in a well-ventilated area.

N,N-**Dimethylformamide, DMF or HCON(CH$_3$)$_2$,** is a possible carcinogen and is irritating to the eyes, skin, and mucous membranes. It can exert its toxic effects through inhalation, ingestion, or skin absorption. Chronic inhalation can cause liver and kidney damage. Wear appropriate gloves and safety glasses and use in a chemical fume hood.

Dimethyl sulfoxide, DMSO, may be harmful by inhalation or skin absorption. Wear appropriate gloves and safety glasses and use in a chemical fume hood. DMSO is also combustible. Store in a tightly closed container. Keep away from heat, sparks, and open flame.

Dithiothreitol, DTT, is a strong reducing agent that emits a foul odor. It may be harmful by inhalation, ingestion, or skin absorption. When working with the solid form or highly concentrated stocks, wear appropriate gloves and safety glasses and use in a chemical fume hood.

DMF, *see* **N,N-Dimethylformamide**

DMSO, *see* **Dimethyl sulfoxide**

Dry ice, *see* **Carbon dioxide**

DTT, *see* **Dithiothreitol**

EtBr, *see* **Ethidium bromide**

Ethanol, EtOH or CH$_3$CH$_2$OH, may be harmful by inhalation, ingestion, or skin absorption. Wear appropriate gloves and safety glasses.

Ether, *see* **Diethyl ether**

Ethidium bromide, EtBr, is a powerful mutagen and is toxic. Consult the local institutional safety officer for specific handling and disposal procedures. Avoid breathing the dust. Wear appropriate gloves when working with solutions that contain this dye.

Ethylene glycol may be harmful by inhalation, ingestion, or skin absorption. Wear appropriate gloves and safety glasses and use in a chemical fume hood.

EtOH, *see* **Ethanol**

Et$_2$O or (C$_2$H$_5$)$_2$O, *see* **Diethyl ether**

Fast Red may cause methemoglobinemia through overexposure. It may be harmful by inhalation, ingestion, or skin absorption. Wear appropriate gloves and safety glasses.

Fluoric acid is extremely hazardous in both liquid and vapor forms. It is corrosive and poisonous and may be fatal. It is harmful by inhalation, ingestion, or skin absorption. Wear appropriate gloves and safety goggles and use in a chemical fume hood. Do not breathe the vapor. Keep away from heat, sparks, and open flame.

Formaldehyde, HCHO, is highly toxic and volatile. It is also a possible carcinogen. It is readily absorbed through the skin and is irritating or destructive to the skin, eyes, mucous membranes, and upper respiratory tract. Avoid breathing the vapors. Wear appropriate gloves and safety glasses and always use in a chemical fume hood. Keep away from heat, sparks, and open flame.

G418 (an aminoglycosidic antibiotic) is toxic and may cause harm to the unborn child. It may be harmful by inhalation, ingestion, or skin absorption. Wear appropriate gloves and safety goggles and use in a chemical fume hood. Do not breathe the dust.

Ganciclovir, GCV, is highly toxic, may cause heritable genetic damage, and may impair fertility. It may be harmful by inhalation, ingestion, or skin absorption. Wear appropriate gloves and safety goggles. Do not breathe the dust.

GCV, *see* **Ganciclovir**

Giemsa may be fatal or cause blindness by ingestion and is toxic by inhalation and skin absorption. There is a possible risk of irreversible effects. Wear appropriate gloves and safety goggles and use only in a chemical fume hood. Do not breathe the dust.

Glacial acetic acid, *see* **Acetic acid (glacial)**

Glassware, pressurized, must be used with extreme caution. Autoclave and cool sealed bottles in metal containers, pressurize bottles behind Plexiglas shields, and encase 20-liter bottles in wire mesh. Handle glassware under vacuum, such as desiccators, vacuum traps, drying equipment, or a reactor for working under argon atmosphere, with appropriate caution. Always wear safety glasses.

Glutaraldehyde is toxic. It is readily absorbed through the skin and is irritating or destructive to the skin, eyes, mucous membranes, and upper respiratory tract. Wear appropriate gloves and safety glasses and always use in a chemical fume hood.

Glycine may be harmful by inhalation, ingestion, or skin absorption. Wear gloves and safety glasses. Avoid breathing the dust.

HCHO, *see* **Formaldehyde**

HCl, *see* **Hydrochloric acid**

H$_3$COH, *see* **Methanol**

HCON(CH₃)₂, *see* **Dimethylformamide**

Hg, *see* **Mercury**

H₂O₂, *see* **Hydrogen peroxide**

HOCH₂CH₂SH, *see* **β-Mercaptoethanol**

Hydrochloric acid, HCl, is volatile and may be fatal if inhaled, ingested, or absorbed through the skin. It is extremely destructive to mucous membranes, upper respiratory tract, eyes, and skin. Wear appropriate gloves and safety glasses and use with great care in a chemical fume hood. Wear goggles when handling large quantities.

Hydrofluoric acid is extremely toxic, corrosive, and can cause severe burns. It may be harmful by inhalation, ingestion, and skin absorption. Wear appropriate gloves and safety goggles and use only in a chemical fume hood.

Hydrogen peroxide, H₂O₂, is corrosive, toxic, and extremely damaging to the skin. It may be harmful by inhalation, ingestion, and skin absorption. Wear appropriate gloves and safety glasses and use only in a chemical fume hood.

Hygromycin B is highly toxic and may be fatal if inhaled, ingested, or absorbed through the skin. Wear appropriate gloves and safety goggles and use only in a chemical fume hood. Do not breathe the dust.

Isoamyl alcohol, IAA, may be harmful by inhalation, ingestion, or skin absorption and presents a risk of serious damage to the eyes. Wear appropriate gloves and safety goggles. Keep away from heat, sparks, and open flame.

Isopentane, 2-methylbutane, is extremely flammable. Keep away from heat, sparks, and open flame. It may be harmful by inhalation, ingestion, or skin absorption. Wear appropriate gloves and safety glasses.

Isopropanol is flammable and irritating. It may be harmful by inhalation, ingestion, or skin absorption. Wear appropriate gloves and safety glasses. Do not breathe the vapor. Keep away from heat, sparks, and open flame.

KCl, *see* **Potassium chloride**

K₃Fe(CN)₆, *see* **Potassium ferricyanide**

K₄Fe(CN)₆·3H₂O, *see* **Potassium ferrocyanide**

KH₂PO₄/K₂HPO₄/K₃PO₄, *see* **Potassium phosphate**

KOH, *see* **Potassium hydroxide**

LiCl, *see* **Lithium chloride**

Liquid nitrogen, LN₂, can cause severe damage due to extreme temperature. Handle frozen samples with extreme caution. Do not breathe the vapors. Seepage of liquid nitrogen into frozen vials can result in an exploding tube upon removal from liquid nitrogen. Use vials with O-rings when possible. Wear cryo-mitts and a face mask.

Lithium chloride, LiCl, is an irritant to the eyes, skin, mucous membranes, and upper respiratory tract. It may be harmful by inhalation, ingestion, or skin absorption. Wear appropriate gloves, safety goggles, and use in a chemical fume hood. Do not breathe the dust.

Lithium dodecyl sulfate may be harmful by ingestion, inhalation, or skin absorption. Wear appropriate gloves when weighing, and use in a chemical fume hood.

LN$_2$, *see* **Liquid nitrogen**

Magnesium chloride, MgCl$_2$, may be harmful by inhalation, ingestion, or skin absorption. Wear appropriate gloves and safety glasses and use in a chemical fume hood.

Magnesium sulfate, MgSO$_4$, may be harmful by inhalation, ingestion, or skin absorption. Wear appropriate gloves and safety glasses and use in a chemical fume hood.

Maleic acid is toxic and harmful by inhalation, ingestion, or skin absorption. Reaction with water or moist air can release toxic, corrosive, or flammable gases. Do not breathe the vapors or dust. Wear appropriate gloves and safety glasses.

MeOH or H$_3$COH, *see* **Methanol**

β-Mercaptoethanol, 2-Mercaptoethanol, or **HOCH$_2$CH$_2$SH,** may be fatal if inhaled or absorbed through the skin and is harmful if ingested. High concentrations are extremely destructive to the mucous membranes, upper respiratory tract, skin, and eyes. β-Mercaptoethanol has a very foul odor. Wear appropriate gloves and safety glasses and always use in a chemical fume hood.

Mercury, Hg, may be fatal if inhaled, ingested, or absorbed through the skin. It presents a long-term danger, since mercury accumulates in the liver and interferes with its function. Wear appropriate gloves and safety glasses and use in a chemical fume hood. Because of its high vapor pressure, spills of mercury should be cleaned immediately using mercury-absorbing reagents.

Methanol, MeOH or H$_3$COH, is poisonous and can cause blindness. It may be harmful by inhalation, ingestion, or skin absorption. Adequate ventilation is necessary to limit exposure to vapors. Avoid inhaling these vapors. Wear appropriate gloves and goggles and use only in a chemical fume hood.

2-Methylbutane, *see* **Isopentane**

Methylene blue is irritating to the eyes and skin. It may be harmful by inhalation, ingestion, or skin absorption. Wear appropriate gloves and safety glasses.

Methylformamide, *see* **Dimethylformamide**

MgCl$_2$, *see* **Magnesium chloride**

MgSO$_4$, *see* **Magnesium sulfate**

Mitomycin C is a carcinogen. It may be fatal by inhalation, ingestion, or skin absorption. Do not breathe the dust. Wear appropriate gloves and safety glasses and use only in a chemical fume hood.

Na$_2$HPO$_4$, *see* **Sodium hydrogen phosphate**

NaH$_2$PO$_4$/Na$_2$HPO$_4$/Na$_3$PO$_4$, *see* **Sodium phosphate**

NaOAc, *see* **Sodium acetate**

NaOH, *see* **Sodium hydroxide**

NBT, *see* **4-Nitro blue tetrazolium chloride**

Neomycin may be harmful by inhalation, ingestion, or skin absorption. Wear appropriate gloves and safety glasses.

NH$_4$OH, *see* **Ammonium hydroxide**

(NH$_4$)$_2$SO$_4$, *see* **Ammonium sulfate**

Nickel chloride, NiCl$_2$, is toxic and may be harmful by inhalation, ingestion, or skin absorption. Do not breathe the dust. Wear appropriate gloves and safety glasses.

NiCl$_2$, *see* **Nickel chloride**

4-Nitro blue tetrazolium chloride, NBT, may be harmful by inhalation, ingestion, or skin absorption. Wear appropriate gloves and safety glasses.

OCT is composed of polyvinyl alcohol, polyethylene glycol, and dimethyl benzyl ammonium chloride. Follow the manufacturer's guidelines for handling OCT.

Paraformaldehyde is highly toxic. It is readily absorbed through the skin and is extremely destructive to the skin, eyes, mucous membranes, and upper respiratory tract. Avoid breathing the dust. Wear appropriate gloves and safety glasses and use in a chemical fume hood. Paraformaldehyde is the undissolved form of formaldehyde.

PBL, *see* **β-Propiolactone**

Permount, *see* **Toluene**

Petroleum ether is highly flammable and may cause central nervous system depression. It is a poison and may be harmful by inhalation, ingestion, or skin absorption. Wear appropriate gloves and safety glasses and use with appropriate ventilation. Keep away from heat, sparks, and open flame.

Phenol is extremely toxic, highly corrosive, and can cause severe burns. It may be harmful by inhalation, ingestion, or skin absorption. Wear appropriate gloves, goggles, protective clothing, and always use in a chemical fume hood. Rinse any areas of skin that come in contact with phenol with a large volume of water and wash with soap and water; do not use ethanol!

Phenol red may be harmful by inhalation, ingestion, or skin absorption. Wear appropriate gloves and safety glasses and use in a chemical fume hood.

Picric acid powder, Trinitrophenol, is caustic and potentially explosive if it is dissolved and then allowed to dry out. Care must be taken to ensure that stored solutions do not dry out. Handle all concentrated acids with great care. It is also highly toxic and may be harmful by inhalation, ingestion, or skin absorption. Wear appropriate gloves and goggles.

Polyvinyl alcohol may be harmful by inhalation, ingestion, or skin absorption. Wear appropriate gloves and safety glasses.

Polyvinylpyrrolidone, PVP, may be harmful by inhalation, ingestion, or skin absorption. Wear appropriate gloves and safety glasses and use in a chemical fume hood.

Pontamine Sky Blue, *see* **Chicago Sky Blue**

Potassium chloride, KCl, may be harmful by inhalation, ingestion, or skin absorption. Wear appropriate gloves and safety glasses.

Potassium ferricyanide, $K_3Fe(CN)_6$, may be fatal by inhalation, ingestion, or skin absorption. Wear appropriate gloves and safety glasses and always use with extreme care in a chemical fume hood. Keep away from strong acids.

Potassium ferrocyanide, $K_4Fe(CN)_6 \cdot 3H_2O$, may be fatal by inhalation, ingestion, or skin absorption. Wear appropriate gloves and safety glasses and always use with extreme care in a chemical fume hood. Keep away from strong acids.

Potassium hydroxide, KOH and KOH/methanol, is highly toxic and may be fatal if swallowed. It may be harmful by inhalation, ingestion, or skin absorption. Solutions are corrosive and can cause severe burns. It should be handled with great care. Wear appropriate gloves and safety goggles.

Potassium phosphate, $KH_2PO_4/K_2HPO_4/K_3PO_4$, may be harmful by inhalation, ingestion, or skin absorption. Wear appropriate gloves and safety glasses. Do not breathe the dust. *$K_2HPO_4 \cdot 3H_2O$ is dibasic and KH_2PO_4 is monobasic.*

β-Propiolactone, PBL, is a carcinogen and mutagen and is highly toxic. It may be fatal if inhaled and is also corrosive. It may be harmful by inhalation, ingestion, or skin absorption. Wear appropriate gloves and safety goggles and use only in a chemical fume hood. Keep away from heat, sparks, and open flame. Do not breathe the dust.

Proteinase K is an irritant and may be harmful by inhalation, ingestion, or skin absorption. Wear appropriate gloves and safety glasses.

Puromycin is toxic and may be carcinogenic. It may be harmful by inhalation, ingestion, or skin absorption. Wear appropriate gloves and safety glasses.

PVP, *see* **Polyvinylpyrrolidone**

SDS, *see* **Sodium dodecyl sulfate**

Silane is extremely flammable and corrosive. It may be harmful by inhalation, ingestion, or skin absorption. Keep away from heat, sparks, and open flame. The vapor is irritating to the eyes, skin, mucous membranes, and upper respiratory tract. Wear appropriate gloves and safety goggles and always use in a chemical fume hood.

Sodium acetate, NaOAc, *see* **Acetic acid**

Sodium deoxycholate is irritating to mucous membranes and the respiratory tract and may be harmful by inhalation, ingestion, or skin absorption. Wear appropriate gloves and safety glasses when handling the powder. Do not breathe the dust.

Sodium dodecyl sulfate, SDS, is toxic, an irritant, and poses a risk of severe damage to the eyes. It may be harmful by inhalation, ingestion, or skin absorption. Wear appropriate gloves and safety goggles. Do not breathe the dust.

Sodium hydrogen phosphate, Na$_2$HPO$_4$ or **sodium phosphate, dibasic,** may be harmful by inhalation, ingestion, or skin absorption. Wear appropriate gloves and safety glasses and use in a chemical fume hood.

Sodium hydroxide, NaOH, and solutions containing NaOH are highly toxic and caustic and should be handled with great care. Wear appropriate gloves and a face mask. All other concentrated bases should be handled in a similar manner.

Sodium phosphate, NaH$_2$PO$_4$/Na$_2$HPO$_4$/Na$_3$PO$_4$, is an irritant to the eyes and skin. It may be harmful by inhalation, ingestion, or skin absorption. Wear appropriate gloves and safety goggles. Do not breathe the dust.

Spermidine may be corrosive and harmful by inhalation, ingestion, or skin absorption. Wear appropriate gloves and safety glasses and use in a chemical fume hood.

Streptomycin is toxic and a suspected carcinogen and mutagen. It may cause allergic reactions. It may be harmful by inhalation, ingestion, or skin absorption. Wear appropriate gloves and safety glasses.

TAE buffer contains **Tris-acetatae** and **EDTA**

TE buffer contains **Tris-acetate.**

tert-**Amyl alcohol,** *see* **Amyl alcohol**

TESPA, *see* **Silane**

Tetrahydrochloride may be harmful by inhalation, ingestion, or skin absorption. Wear appropriate gloves and safety glasses.

Toluene, $C_6H_5CH_3$, vapors are irritating to the eyes, skin, mucous membranes, and upper respiratory tract. Toluene can exert harmful effects by inhalation, ingestion, or skin absorption. Do not inhale the vapors. Wear appropriate gloves and safety glasses and use in a chemical fume hood. Toluene is extremely flammable. Keep away from heat, sparks, and open flame.

Triethanolamine may be harmful by inhalation, ingestion, or skin absorption. Wear appropriate gloves and safety glasses and use only in a chemical fume hood.

Trihydrochloride may be harmful by inhalation, ingestion, or skin absorption. Wear appropriate gloves and safety glasses.

Trinitrophenol, *see* **Picric acid**

Tris may be harmful by inhalation, ingestion, or skin absorption. Wear appropriate gloves and safety glasses.

UV light and/or **UV radiation** is dangerous and can damage the retina. Never look at an unshielded UV light source with naked eyes. Examples of UV light sources that are common in the laboratory include hand-held lamps and transilluminators. View only through a filter or safety glasses that absorb harmful wavelengths. UV radiation is also mutagenic and carcinogenic. To minimize exposure, make sure that the UV light source is adequately shielded. Wear protective appropriate gloves when holding materials under the UV light source.

X-gal may be toxic to the eyes and skin. Observe general cautions when handling the powder. Note that stock solutions of X-gal are prepared in DMF, an organic solvent. For details, *see N,N*-**dimethylformamide (DMF).** *See also* **5-Bromo-4-chloro-3-indolyl-β-D-galactopyranoside (BCIG).**

Xylene is flammable and may be narcotic at high concentrations. It may be harmful by inhalation, ingestion, or skin absorption. Wear appropriate gloves and safety glasses and use only in a chemical fume hood. Keep away from heat, sparks, and open flame.

Zymolyase may be harmful by inhalation, ingestion, or skin absorption. Wear appropriate gloves and safety glasses.

Suppliers

Some suppliers are listed in the text with their addresses and are also included on the Cold Spring Harbor Laboratory Press Mouse Manual Web site at: http://www.mousemanual.org. Other suppliers mentioned in this manual can be found in the BioSupplyNet Source Book and on the Web site at:

http://www.biosupplynet.com

If a copy of the BioSupplyNet Source Book was not included with this manual, a free copy can be ordered by any of the following methods:

- Complete the Free Source Book Request Form found at the Web site at:
    http://www.biosupplynet.com
- E-mail a request to info@biosupplynet.com
- Fax a request to 1-919-659-2199.

Index

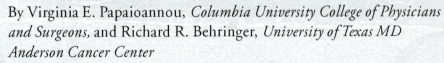